INTEGRATED ELECTRONICS: ANALOG AND DIGITAL CIRCUITS AND SYSTEMS

McGRAW-HILL ELECTRICAL AND ELECTRONIC ENGINEERING SERIES

Frederick Emmons Terman, *Consulting Editor*
W. W. Harman, J. G. Truxal, and R. A. Rohrer, *Associate Consulting Editors*

INTEGRATED ELECTRONICS: ANALOG AND DIGITAL CIRCUITS AND SYSTEMS

Jacob Millman, Ph.D.

Professor of Electrical Engineering
Columbia University

Christos C. Halkias, Ph.D.

Professor of Electrical Engineering
Columbia University

McGRAW-HILL BOOK COMPANY

New York   St. Louis   San Francisco   Düsseldorf   Johannesburg   Kuala Lumpur
London   Mexico   Montreal   New Delhi   Panama   Rio de Janeiro   Singapore
Sydney   Toronto

To our wives
SALLY
and
DEMETRA

INTEGRATED ELECTRONICS: ANALOG AND
DIGITAL CIRCUITS AND SYSTEMS

Library of Congress Catalog Card Number 79-172657

07-042315-6

1617 FGRFGR   832

CONTENTS

chapter **4**   Diode Circuits   **87**

5   Transistor Characteristics   **118**

chapter 9   Transistor Biasing and Thermal Stabilization   282

10   Field-effect Transistors   310

11   The Transistor at High Frequencies   348

chapter **12**   Multistage Amplifiers   **372**

13   Feedback Amplifiers   **408**

14   Stability and Oscillators   **447**

chapter **15**   Operational Amplifiers   501

16   Integrated Circuits as Analog System Building Blocks   537 chapter **17**   Integrated Circuits as Digital System Building Blocks   **593**

18   Power Circuits and Systems   **677**

chapter **19**   Semiconductor-device Physics   **726**

Appendices

PREFACE

This book was written primarily as a text for a first course in electronics for electrical engineering students. It should also be of interest to physics majors and to practicing engineers and scientists who wish to update their knowledge of semiconductor electronics and, particularly, of integrated circuits.

The following basic approach is used: Each device is introduced by presenting a simple physical picture of the internal behavior of the device. This discussion leads to a characterization of the device in terms of appropriate external variables and allows small-signal and large-signal models to be constructed. An analysis of the device as a circuit element in both analog and digital applications is then made. Combinations of devices, particularly in integrated-circuit form, are exploited as basic system building blocks. In summary: the reader is taken step-by-step from semiconductor physics to devices, models, circuits, and systems. Electronic design in the 1970's will depend predominantly upon the use of integrated circuits. Hence, the authors feel that the IC implementation of a device, circuit or system should be introduced into the EE curriculum as early as possible.

Almost every semiconductor device of importance (from dc through the video frequency range) is considered. These include sensistors, the p-n diode, the breakdown diode, the tunnel diode, the Schottky diode, the photodiode, the varactor diode, light-emitting diode, the bipolar junction transistor (BJT), the photo-transistor, the junction field-effect transistor (JFET), the metal-oxide-semiconductor field-effect transistor (MOSFET), the silicon controlled rectifier, and integrated resistors and capacitors. Primary emphasis is on the bipolar junction transistor in both the analog and digital mode.

The principal concern in the book is upon the analysis and design of electronic circuits and subsystems. The approach is to consider each circuit first on a descriptive basis in order to provide a clear understanding and intuitive feeling for its behavior. Only after obtaining such a qualitative insight into the operation of the circuit is mathematics (through simple differential equations) used to express quantitative relationships.

A wide variety of circuits are analyzed. Among these are rectifiers (including capacitor filters), clippers, sampling gates, logic gates (AND, OR, NOT, NAND, DTL, TTL, etc.), the low-frequency BJT CE stage, the emitter follower, the Darlington pair, cascaded

(BJT) stages (properly biased), the low-frequency CS and CD field effect transistor stages, digital MOSFET circuits, the FET at high frequencies, the BJT at high frequencies (both CE and CC configurations), frequency response of multistage RC-coupled BJT and FET stages, feedback amplifiers at low and high frequencies, compensated feedback amplifiers, oscillators, inverting and noninverting operational amplifiers, differential amplifiers, compensated OP AMPS, analog computer circuits, active filters, tuned amplifiers, logarithmic amplifiers, sample and hold circuits, comparators, square-wave generators, triangular generators, flip-flops, shift registers, multiplexers, decoders, binary adders, counters, memories (ROM and RAM), digital-to-analog converters, analog-to-digital converters, power amplifiers, voltage regulators, and SCR power control circuits.

Methods of analysis and features which are common to many different devices and circuits are emphasized. For example, Kirchhoff's, Thevenin's, Norton's, and Miller's theorems are utilized throughout the text. Concepts such as the load line (or load curve), the large-scale piecewise-linear approximations, and the small-scale linear models are used again and again. Pole-zero functions, Bode plots and the dominant-pole concept are exploited in connection with the study of the frequency response and compensation of multistage amplifiers. Computer-aided analysis is resorted to wherever the circuit complexity justifies it. Calculations of input and output resistances, as well as current and voltage gains are made for a wide variety of amplifiers.

For the most part, real (commercially available) device characteristics are employed. In this way the reader may become familiar with the order of magnitude of device parameters, the variability of these parameters within a given type and with a change of temperature, the effect of the inevitable shunt capacitances in circuits, and the effect of input and output resistances and loading on circuit operation. These considerations are of utmost importance to the student or the practicing engineer since the circuits to be designed must function properly and reliably in the physical world, rather than under hypothetical or ideal circumstances.

There are over 720 homework problems, which will test the student's grasp of the fundamental concepts enunciated in the book and will give him experience in the analysis and design of electronic circuits. In almost all numerical problems realistic parameter values and specifications have been chosen. An answer book is available for students, and a solutions manual may be obtained from the publisher by an instructor who has adopted the text. Each chapter is followed by many review (quiz) questions.

Considerable thought was given to the pedagogy of presentation, to the explanation of circuit behavior, to the use of a consistent system of notation, to the care with which diagrams are drawn, to the many illustrative examples worked out in detail in the text, and to the review questions at the end of each chapter. It is hoped that these will facilitate the use of the book in self-study and that the practicing engineer will find the text useful in updating himself in this fast-moving field.

The authors adapted and rewrote much of the material in their book "Electronic Devices and Circuits" (McGraw-Hill Book Company, New York, 1967). The new text differs from the old in a large number of respects as follows: Whereas about 25 percent of the 1967 book was on vacuum tubes, the present text makes no mention of these devices, since tubes are no longer considered important by the electronic designer (except possibly for very high voltage or power applications). The introductory physics has been simplified and shortened, while circuits have been treated in greater breadth and depth. Those who wish to pursue the physics more thoroughly are referred to the last chapter in the book where a comprehensive treatment of semiconductor device physics is presented. Much greater emphasis is placed upon integrated circuits; not only are the fabrication and characteristics of IC's discussed, but many examples are given of subsystems in integrated circuit form (such as logic families, logic functions, and OP AMPS). The chapter on feedback amplifiers at low frequencies has been completely rewritten so that all four topologies are treated in a consistent and simplified manner. The high-frequency response of cascaded stages and of feedback amplifiers (which was slighted in the 1967 text) is given an extensive treatment now which includes a discussion of stability and compensation. There is a chapter on logic circuits and another on IC's as digital system building blocks. A complete chapter is devoted to the operational amplifier and another on IC's as analog system building blocks.

The publishers sent a questionnaire to the (more than 100) professors who had adopted "Electronic Devices and Circuits," asking for desirable deletions, additions, revisions, etc. in this book. We are indebted to the large number of people who answered, and the present book reflects strongly the suggestions made by them. In particular, we wish to acknowledge the detailed constructive criticisms made by Professors R. S. Bennett, W. L. Brown, and D. E. Franklin. The authors are grateful to Drs. M. C. Teich and E. S. Yang who read portions of the manuscript and offered helpful suggestions for improvements. We thank Professor H. Taub also because some of our material parallels that in Millman and Taub's "Pulse, Digital, and Switching Waveforms" (McGraw-Hill Book Company, New York, 1965). We appreciate the considerable assistance given us by J. Derby, G. A. Katopis, and J. J. Werner in carrying out the computer solutions, laboratory verification of the performance of some of the circuits, and the solutions of the homework problems. We express our thanks to Miss S. Silverstein for her skillful service in typing the manuscript.

Jacob Millman
Christos C. Halkias

1 / ENERGY BANDS IN SOLIDS

In this chapter we begin with a review of the basic atomic properties of matter leading to discrete electronic energy levels in atoms. We find that these energy levels are spread into energy bands in a crystal. This band structure allows us to distinguish between an insulator, a semiconductor, and a metal.

1-1   CHARGED PARTICLES

The charge, or quantity, of negative electricity and the mass of the electron have been found to be 1.60×10^{-19} C (coulomb) and 9.11×10^{-31} kg, respectively. The values of many important physical constants are given in Appendix A, and a list of conversion factors and prefixes is given in Appendix B. Some idea of the number of electrons per second that represents current of the usual order of magnitude is readily possible. For example, since the charge per electron is 1.60×10^{-19} C, the number of electrons per coulomb is the reciprocal of this number, or approximately, 6×10^{18} Further, since a current of 1 A (ampere) is the flow of 1 C/s, then a current of only 1 pA (1 pico-ampere, or 10^{-12} A) represents the motion of approximately 6 million electrons per second. Yet a current of 1 pA is so small that considerable difficulty is experienced in attempting to measure it.

The charge of a positive ion is an integral multiple of the charge of the electron, although it is of opposite sign. For the case of singly ionized particles, the charge is equal to that of the electron. For the case of doubly ionized particles, the ionic charge is twice that of the electron.

The mass of an atom is expressed as a number that is based on the choice of the atomic weight of oxygen equal to 16. The mass of a hypothetical atom of atomic weight unity is, by this definition, one-sixteenth that of the mass of monatomic oxygen and has been calculated to be 1.66×10^{-27} kg. Hence, *to calculate the mass in kilograms*

1

of any atom, it is necessary only to multiply the atomic weight of the atom by 1.66 × 10⁻²⁷ kg. A table of atomic weights is given in Table 1-1 on p. 12.

The radius of the electron has been estimated as 10^{-15} m, and that of an atom as 10^{-10} m. These are so small that all charges are considered as mass points in the following sections.

In a semiconductor crystal such as silicon, two electrons are shared by each pair of ionic neighbors. Such a configuration is called a *covalent bond*. Under certain circumstances an electron may be missing from this structure, leaving a "hole" in the bond. These vacancies in the covalent bonds may move from ion to ion in the crystal and constitute a current equivalent to that resulting from the motion of free positive charges. The magnitude of the charge associated with the hole is that of a free electron. This very brief introduction to the concept of a hole as an effective charge carrier is elaborated upon in Chap. 2.

1-2   FIELD INTENSITY, POTENTIAL, ENERGY

By definition, *the force* **f** *(newtons) on a unit positive charge in an electric field is the electric field intensity* **ε** *at that point*. Newton's second law determines the motion of a particle of charge q (coulombs), mass m (kilograms), moving with a velocity **v** (meters per second) in a field **ε** (volts per meter).

$$\mathbf{f} = q\mathbf{\varepsilon} = m\frac{d\mathbf{v}}{dt} \tag{1-1}$$

The mks (meter-kilogram-second) rationalized system of units is found to be most convenient for subsequent studies. Unless otherwise stated, this system of units is employed throughout this book.

Potential By definition, *the potential V (volts) of point B with respect to point A is the work done* against *the field in taking a unit positive charge from A to B*. This definition is valid for a three-dimensional field. For a one-dimensional problem with A at x_o and B at an arbitrary distance x, it follows that†

$$V \equiv -\int_{x_o}^{x} \varepsilon \, dx \tag{1-2}$$

where ε now represents the X component of the field. Differentiating Eq. (1-2) gives

$$\varepsilon = -\frac{dV}{dx} \tag{1-3}$$

The minus sign shows that the electric field is directed from the region of higher potential to the region of lower potential. In three dimensions, the electric field equals the negative gradient of the potential.

---

† The symbol $\equiv$ is used to designate "equal to by definition."

By definition, *the potential energy U (joules) equals the potential multiplied by the charge q under consideration,* or

$$U = qV \tag{1-4}$$

If an electron is being considered, q is replaced by $-q$ (where q is the *magnitude* of the electronic charge) and U has the same shape as V but is inverted.

The law of conservation of energy states that the total energy W, which equals the sum of the potential energy U and the kinetic energy $\frac{1}{2}mv^2$, remains constant. Thus, at any point in space,

$$W = U + \tfrac{1}{2}mv^2 = \text{constant} \tag{1-5}$$

As an illustration of this law, consider two parallel electrodes (A and B of Fig. 1-1a) separated a distance d, with B at a negative potential V_d with respect to A. An electron leaves the surface of A with a velocity v_o in the direction toward B. How much speed v will it have if it reaches B?

From the definition, Eq. (1-2), it is clear that only differences of potential have meaning, and hence let us arbitrarily ground A, that is, consider it to be at zero potential. Then the potential at B is $V = -V_d$, and the potential energy is $U = -qV = qV_d$. Equating the total energy at A to that at B gives

$$W = \tfrac{1}{2}mv_o{}^2 = \tfrac{1}{2}mv^2 + qV_d \tag{1-6}$$

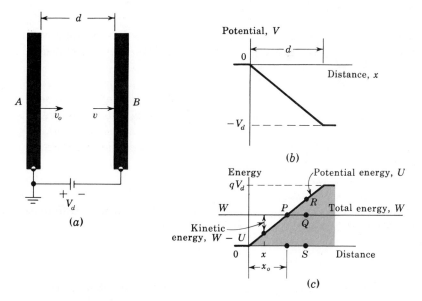

Fig. 1-1   (a) An electron leaves electrode A with an initial speed v_o and moves in a retarding field toward plate B; (b) the potential; (c) the potential-energy barrier between electrodes.

This equation indicates that v must be less than v_o, which is obviously correct since the electron is moving in a repelling field. Note that the final speed v attained by the electron in this conservative system is independent of the form of the variation of the field distribution between the plates and depends only upon the magnitude of the potential difference V_d. Also, if the electron is to reach electrode B, its initial speed must be large enough so that $\frac{1}{2} mv_o{}^2 > qV_d$. Otherwise, Eq. (1-6) leads to the impossible result that v is imaginary. We wish to elaborate on these considerations now.

The Concept of a Potential-energy Barrier For the configuration of Fig. 1-1a with electrodes which are large compared with the separation d, we can draw (Fig. 1-1b) a linear plot of potential V versus distance x (in the interelectrode space). The corresponding potential energy U versus x is indicated in Fig. 1-1c. Since potential is the potential energy per unit charge, curve c is obtained from curve b by multiplying each ordinate by the charge on the electron (a negative number). Since the total energy W of the electron remains constant, it is represented as a horizontal line. The kinetic energy at any distance x equals the difference between the total energy W and the potential energy U at this point. This difference is greatest at O, indicating that the kinetic energy is a maximum when the electron leaves the electrode A. At the point P this difference is zero, which means that no kinetic energy exists, so that the particle is at rest at this point. This distance x_o is the maximum that the electron can travel from A. At point P (where $x = x_o$) it comes momentarily to rest, and then reverses its motion and returns to A.

Consider a point such as S which is at a greater distance than x_o from electrode A. Here the total energy QS is less than the potential energy RS, so that the difference, which represents the kinetic energy, is negative. This is an impossible physical condition, however, since negative kinetic energy ($\frac{1}{2}mv^2 < 0$) implies an imaginary velocity. We must conclude that the particle can never advance a distance greater than x_o from electrode A.

The foregoing analysis leads to the very important conclusion that the shaded portion of Fig. 1-1c can never be penetrated by the electron. Thus, at point P, the particle acts *as if* it had collided with a solid wall, hill, or barrier and the direction of its flight had been altered. *Potential-energy barriers* of this sort play important role in the analyses of semiconductor devices.

It must be emphasized that the words "collides with" or "rebounds from" a potential "hill" are convenient descriptive phrases and that an actual encounter between two material bodies is not implied.

1-3   THE eV UNIT OF ENERGY

The joule (J) is the unit of energy in the mks system. In some engineering power problems this unit is very small, and a factor of 10^3 or 10^6 is introduced to convert from watts (1 W = 1 J/s) to kilowatts or megawatts, respectively.

However, in other problems, the joule is too large a unit, and a factor of 10^{-7} is introduced to convert from joules to ergs. For a discussion of the energies involved in electronic devices, even the erg is much too large a unit. This statement is not to be construed to mean that only minute amounts of energy can be obtained from electron devices. It is true that each electron possesses a tiny amount of energy, but as previously pointed out (Sec. 1-1), an enormous number of electrons are involved even in a small current, so that considerable power may be represented.

A unit of work or energy, called the *electron volt* (eV), is defined as follows:

$$1 \text{ eV} \equiv 1.60 \times 10^{-19} \text{ J}$$

Of course, any type of energy, whether it be electric, mechanical, thermal, etc., may be expressed in electron volts.

The name *electron volt* arises from the fact that, if an electron falls through a potential of one volt, its kinetic energy will increase by the decrease in potential energy, or by

$$qV = (1.60 \times 10^{-19} \text{ C})(1 \text{ V}) = 1.60 \times 10^{-19} \text{ J} = 1 \text{ eV}$$

However, as mentioned above, the electron-volt unit may be used for any type of energy, and is not restricted to problems involving electrons.

A potential-energy barrier of E (electron volts) is equivalent to a potential hill of V (volts) if these quantities are related by

$$qV = 1.60 \times 10^{-19} E \tag{1-7}$$

Note that V and E are *numerically* identical but dimensionally different.

1-4   THE NATURE OF THE ATOM

We wish to develop the band structure of a solid, which will allow us to distinguish between an insulator, a semiconductor, and a metal. We begin with a review of the basic properties of matter leading to discrete electronic energy levels in atoms.

Rutherford, in 1911, found that the atom consists of a nucleus of positive charge that contains nearly all the mass of the atom. Surrounding this central positive core are negatively charged electrons. As a specific illustration of this atomic model, consider the hydrogen atom. This atom consists of a positively charged nucleus (a proton) and a single electron. The charge on the proton is positive and is equal in magnitude to the charge on the electron. Therefore the atom as a whole is electrically neutral. Because the proton carries practically all the mass of the atom, it will remain substantially immobile, whereas the electron will move about it in a closed orbit. The force of attraction between the electron and the proton follows Coulomb's law. It can be shown from classical mechanics that the resultant closed path will be a circle or an ellipse under the action of such a force. This motion is exactly analogous to that of the planets about the sun, because in both cases the force varies inversely as the square of the distance between the particles.

Assume, therefore, that the orbit of the electron in this planetary model of the atom is a circle, the nucleus being supposed fixed in space. It is a simple matter to calculate its radius in terms of the total energy W of the electron. The force of attraction between the nucleus and the electron of the hydrogen atom is $q^2/4\pi\epsilon_o r^2$, where the electronic charge q is in coulombs, the separation r between the two particles is in meters, the force is in newtons, and ϵ_o is the permittivity of free space.† By Newton's second law of motion, this must be set equal to the product of the electronic mass m in kilograms and the acceleration v^2/r toward the nucleus, where v is the speed of the electron in its circular path, in meters per second. Then

$$\frac{q^2}{4\pi\epsilon_o r^2} = \frac{mv^2}{r} \tag{1-8}$$

Furthermore, the potential energy of the electron at a distance r from the nucleus is $-q^2/4\pi\epsilon_o r$, and its kinetic energy is $\frac{1}{2}mv^2$. Then, according to the conservation of energy,

$$W = \tfrac{1}{2}mv^2 - \frac{q^2}{4\pi\epsilon_o r} \tag{1-9}$$

where the energy is in joules. Combining this expression with (1-8) produces

$$W = -\frac{q^2}{8\pi\epsilon_o r} \tag{1-10}$$

which gives the desired relationship between the radius and the energy of the electron. This equation shows that the total energy of the electron is always negative. The negative sign arises because the potential energy has been chosen to be zero when r is infinite. This expression also shows that the energy of the electron becomes smaller (i.e., more negative) as it approaches closer to the nucleus.

The foregoing discussion of the planetary atom has been considered only from the point of view of classical mechanics. However, an accelerated charge must radiate energy, in accordance with the classical laws of electromagnetism. If the charge is performing oscillations of a frequency f, the radiated energy will also be of this frequency. Hence, classically, it must be concluded that the frequency of the emitted radiation equals the frequency with which the electron is rotating in its circular orbit.

There is one feature of this picture that cannot be reconciled with experiment. If the electron is radiating energy, its total energy must decrease by the amount of this emitted energy. As a result the radius r of the orbit must decrease, in accordance with Eq. (1-10). Consequently, as the atom radiates energy, the electron must move in smaller and smaller orbits, eventually falling into the nucleus. Since the frequency of oscillation depends upon the size

† The numerical value of ϵ_o is given in Appendix A.

of the circular orbit, the energy radiated would be of a gradually changing frequency. Such a conclusion, however, is incompatible with the sharply defined frequencies of spectral lines.

The Bohr Atom The difficulty mentioned above was resolved by Bohr in 1913. He postulated the following three fundamental laws:

1. Not all energies as given by classical mechanics are possible, but the atom can possess only certain discrete energies. While in states corresponding to these discrete energies, the electron does *not* emit radiation, and the electron is said to be in a *stationary*, or nonradiating, state.

2. In a transition from one stationary state corresponding to a definite energy W_2 to another stationary state, with an associated energy W_1, radiation will be emitted. The frequency of this radiant energy is given by

$$f = \frac{W_2 - W_1}{h} \tag{1-11}$$

where h is Planck's constant in joule-seconds, the W's are expressed in joules, and f is in cycles per second, or hertz.

3. A stationary state is determined by the condition that the angular momentum of the electron in this state is quantized and must be an integral multiple of $h/2\pi$. Thus

$$mvr = \frac{nh}{2\pi} \tag{1-12}$$

where n is an integer.

Combining Eqs. (1-8) and (1-12), we obtain the radii of the stationary states (Prob. 1-13), and from Eq. (1-10) the energy level in joules of each state is found to be

$$W_n = -\frac{mq^4}{8h^2\epsilon_o^2} \frac{1}{n^2} \tag{1-13}$$

Then, upon making use of Eq. (1-11), the exact frequencies found in the hydrogen spectrum are obtained—a remarkable achievement. The radius of the lowest state is found to be 0.5 Å.

1-5   ATOMIC ENERGY LEVELS

For each integral value of n in Eq. (1-13) a horizontal line is drawn. These lines are arranged vertically in accordance with the numerical values calculated from Eq. (1-13). Such a convenient pictorial representation is called an *energy-level diagram* and is indicated in Fig. 1-2 for hydrogen. The number to the left of each line gives the energy of this level in electron volts. The number immediately to the right of a line is the value of n. Theoretically, an infinite number of levels exist for each atom, but only the first five and the

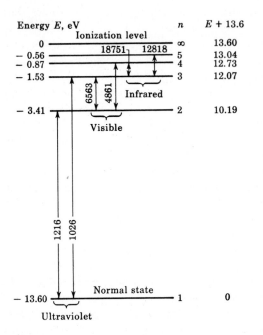

Fig. 1-2 The lowest five energy levels and the ionization level of hydrogen. The spectral lines are in angstrom units.

level for $n = \infty$ are indicated in Fig. 1-2. The horizontal axis has no significance here, but in extending such energy-level diagrams to solids, the X axis will be used to represent the separation of atoms within a crystal (Fig. 1-3) or the distance within a solid. In such cases the energy levels are not constant, but rather are functions of x.

It is customary to express the energy value of the stationary states in electron volts E rather than in joules W. Also, it is more common to specify the emitted radiation by its wavelength λ in angstroms rather than by its frequency f in hertz. In these units, Eq. (1-11) may be rewritten in the form

$$\lambda = \frac{12,400}{E_2 - E_1} \tag{1-14}$$

Since only differences of energy enter into this expression, the zero state may be chosen at will. It is convenient and customary to choose the lowest energy state as the zero level. Such a normalized scale is indicated to the extreme right in Fig. 1-2. The lowest energy state is called the *normal*, or *ground*, level, and the other stationary states of the atom are called *excited, radiating, critical,* or *resonance,* levels.

As the electron is given more and more energy, it moves into stationary states which are farther and farther away from the nucleus. When its energy is large enough to move it completely out of the field of influence of the ion, it becomes "detached" from it. The energy required for this process to occur is called the *ionization potential* and is represented as the highest state in the energy-level diagram; 13.60 eV for hydrogen.

Collisions of Electrons with Atoms   The foregoing discussion shows that energy must be supplied to an atom in order to excite or ionize the atom.   One of the most important ways to supply this energy is by electron impact. Suppose that an electron is accelerated by the potential applied to a discharge tube.   The energy gained from the field may then be transferred to an atom when the electron collides with the atom.   If the bombarding electron has gained more than the requisite energy from the discharge to raise the atom from its normal state to a particular resonance level, the amount of energy in excess of that required for excitation will be retained by the incident electron as kinetic energy after the collision.

If an impinging electron possesses an amount of energy at least equal to the ionization potential of the gas, it may deliver this energy to an electron of the atom and completely remove it from the parent atom.   Three charged particles result from such an ionizing collision: two electrons and a positive ion.

The Photon Nature of Light   Assume that an atom has been raised from the ground state to an excited level by electron bombardment.   The mean life of an excited state ranges from 10^{-7} to 10^{-10} s, the excited electron returning to its previous state after the lapse of this time.   In this transition, the atom must lose an amount of energy equal to the difference in energy between the two states that it has successively occupied, this energy appearing in the form of radiation.   According to the postulates of Bohr, this energy is emitted in the form of a photon of light, the frequency of this radiation being given by Eq. (1-11), or the wavelength by Eq. (1-14).   The term *photon* denotes an amount of radiant energy equal to the constant h times the frequency.   This quantized nature of an electromagnetic wave was first introduced by Planck, in 1901, in order to verify theoretically the blackbody radiation formula obtained experimentally.

The photon concept of radiation may be difficult to comprehend at first. Classically, it was believed that the atoms were systems that emitted radiation *continuously* in all directions.   According to the foregoing theory, however, this is not true, the emission of light by an atom being a discontinuous process. That is, the atom radiates only when it makes a transition from one energy level to a lower energy state.   In this transition, it emits a definite amount of energy of one particular frequency, namely, one photon hf of light.   Of course, when a luminous discharge is observed, this discontinuous nature of radiation is not suspected because of the enormous number of atoms that are radiating energy and, correspondingly, because of the immense number of photons that are emitted in unit time.

Spectral Lines   The arrows in Fig. 1-2 represent six possible transitions between stationary states.   The attached number gives the wavelength of the emitted radiation.   For example, the ultraviolet line 1,216 Å is radiated when the hydrogen atom drops from its first excited state, $n = 2$, to its normal state, $n = 1$.

Another important method, called *photoexcitation*, by which an atom may be elevated into an excited energy state, is to have radiation fall on the gas. An atom may absorb a photon of frequency f and thereby move from the level of energy W_1 to the high energy level W_2, where $W_2 = W_1 + hf$. An extremely important feature of excitation by photon capture is that *the photon will not be absorbed unless its energy corresponds exactly to the energy difference between two stationary levels of the atom with which it collides.* For example, if a normal hydrogen atom is to be raised to its first excited state by means of radiation, the wavelength of this light must be 1,216 Å (which is in the ultraviolet region of the spectrum).

When a photon is absorbed by an atom, the excited atom may return to its normal state in one jump, or it may do so in several steps. If the atom falls into one or more excitation levels before finally reaching the normal state, it will emit several photons. These will correspond to energy differences between the successive excited levels into which the atom falls. None of the emitted photons will have the frequency of the absorbed radiation! This *fluorescence* cannot be explained by classical theory, but is readily understood once Bohr's postulates are accepted.

Photoionization   If the frequency of the impinging photon is sufficiently high, it may have enough energy to ionize the atom. The photon vanishes with the appearance of an electron and a positive ion. Unlike the case of photoexcitation, the photon need not possess an energy corresponding exactly to the ionization energy of the atom. It need merely possess *at least* this much energy. If it possesses more than ionizing energy, the excess will appear as the kinetic energy of the emitted electron and positive ion. It is found by experiment, however, that the maximum probability of photoionization occurs when the energy of the photon is equal to the ionization potential, the probability decreasing rapidly for higher photon energies.

Wave Mechanics   Since a photon is absorbed by only one atom, the photon acts as if it were concentrated in a very small volume of space, in contradiction to the concept of a wave associated with radiation. De Broglie, in 1924, postulated that the dual character of wave and particle is not limited to radiation, but is also exhibited by particles such as electrons, atoms, or macroscopic masses. He postulated that a particle of momentum $p = mv$ has a wavelength λ associated with it given by

$$\lambda = \frac{h}{p} \tag{1-15}$$

We can make use of the wave properties of a moving electron to establish Bohr's postulate that a stationary state is determined by the condition that the angular momentum must be an integral multiple of $h/2\pi$. It seems reasonable to assume that an orbit of radius r will correspond to a stationary state if it contains a standing-wave pattern. In other words, a stable orbit is one whose circumference is exactly equal to the electronic wavelength λ, or to $n\lambda$, where n is an integer (but not zero). Thus

$$2\pi r = n\lambda = \frac{nh}{mv} \tag{1-16}$$

Clearly, Eq. (1-16) is identical with the Bohr condition [Eq. (1-12)].

Schrödinger carried the implication of the wave nature of matter further and developed a wave equation to describe electron behavior in a potential field $U(x, y, z)$. The solution of this differential equation is called the wave function, and it determines the probability density at each point in space of finding the electron with total energy W. If the potential energy, $U = -q^2/4\pi\epsilon_0 r$, for the electron in the hydrogen atom is substituted into the Schrödinger equation, it is found that a meaningful physical solution is possible only if W is given by precisely the energy levels in Eq. (1-13), which were obtained from the simpler Bohr picture of the atom.

1-6   ELECTRONIC STRUCTURE OF THE ELEMENTS

The solution of the Schrödinger equation for hydrogen or any multielectron atom requires three quantum numbers. These are designated by n, l, and m_l and are restricted to the following integral values:

$$n = 1, 2, 3, \ldots$$
$$l = 0, 1, 2, \ldots, (n - 1)$$
$$m_l = 0, \pm 1, \pm 2, \ldots, \pm l$$

To specify a wave function completely it is found necessary to introduce a fourth quantum number. This *spin* quantum number m_s may assume only two values, $+\frac{1}{2}$ or $-\frac{1}{2}$ (corresponding to the same energy).

The Exclusion Principle   The periodic table of the chemical elements (given in Table 1-1) may be explained by invoking a law enunciated by Pauli in 1925. He stated that *no two electrons in an electronic system can have the same set of four quantum numbers, n, l, m_l, and m_s.* This statement that no two electrons may occupy the same quantum state is known as the *Pauli exclusion principle.*

Electronic Shells   All the electrons in an atom which have the same value of n are said to belong to the same *electron shell*. These shells are identified by the letters K, L, M, N, $\ldots$, corresponding to $n = 1, 2, 3, 4, \ldots$, respectively. A shell is divided into *subshells* corresponding to different values of l and identified as s, p, d, f, $\ldots$, corresponding to $l = 0, 1, 2, 3, \ldots$, respectively. Taking account of the exclusion principle, the distribution of

TABLE 1-1 Periodic table of the elements†

Period	Group IA	Group IIA	Group IIIB	Group IVB	Group VB	Group VIB	Group VIIB	Group VIII			Group IB	Group IIB	Group IIIA	Group IVA	Group VA	Group VIA	Group VIIA	Inert gases
1	H 1 1.01																	He 2 4.00
2	Li 3 6.94	Be 4 9.01											B 5 10.81	C 6 12.01	N 7 14.01	O 8 16.00	F 9 19.00	Ne 10 20.18
3	Na 11 22.99	Mg 12 24.31											Al 13 26.98	Si 14 28.09	P 15 30.97	S 16 32.06	Cl 17 35.45	Ar 18 39.95
4	K 19 39.10	Ca 20 40.08	Sc 21 44.96	Ti 22 47.90	V 23 50.94	Cr 24 52.00	Mn 25 54.94	Fe 26 55.85	Co 27 58.93	Ni 28 58.71	Cu 29 63.54	Zn 30 65.37	Ga 31 69.72	Ge 32 72.59	As 33 74.92	Se 34 78.96	Br 35 79.91	Kr 36 83.80
5	Rb 37 85.47	Sr 38 87.62	Y 39 88.90	Zr 40 91.22	Nb 41 92.91	Mo 42 95.94	Tc 43 (99)	Ru 44 101.07	Rh 45 102.90	Pd 46 106.4	Ag 47 107.87	Cd 48 112.40	In 49 114.82	Sn 50 118.69	Sb 51 121.75	Te 52 127.60	I 53 126.90	Xe 54 131.30
6	Cs 55 132.90	Ba 56 137.34	La 57 138.91	Hf 72 178.49	Ta 73 180.95	W 74 183.85	Re 75 186.2	Os 76 190.2	Ir 77 192.2	Pt 78 195.09	Au 79 196.97	Hg 80 200.59	Tl 81 204.37	Pb 82 207.19	Bi 83 208.98	Po 84 (210)	At 85 (210)	Rn 86 (222)
7	Fr 87 (223)	Ra 88 (226)	Ac 89 (227)	Th 90 232.04	Pa 91 (231)	U 92 238.04	Np 93 (237)	Pu 94 (242)	Am 95 (243)	Cm 96 (247)	Bk 97 (247)	Cf 98 (251)	Es 99 (254)	Fm 100 (253)	Md 101 (256)	No 102 (254)	Lw 103 (257)	

The Rare Earths

Ce 58 140.12	Pr 59 140.91	Nd 60 144.24	Pm 61 (147)	Sm 62 150.35	Eu 63 151.96	Gd 64 157.25	Tb 65 158.92	Dy 66 162.50	Ho 67 164.93	Er 68 167.26	Tm 69 168.93	Yb 70 173.04	Lu 71 174.97

† The number to the right of the symbol for the element gives the atomic number. The number below the symbol for the element gives the atomic weight.

TABLE 1-2   Electron shells and subshells

Shell......	K	L		M			N			
n........	1	2		3			4			
l.........	0	0	1	0	1	2	0	1	2	3
Subshell...	s	s	p	s	p	d	s	p	d	f
m_l........	0	0	0, ±1	0	0, ±1	0, ±1, ±2	0	0, ±1	0, ±1, ±2	0, . . . , ±3
Number of electrons	2	2	6	2	6	10	2	6	10	14
	2	8		18			32			

electrons in an atom among the shells and subshells is indicated in Table 1-2. Actually, seven shells are required to account for all the chemical elements, but only the first four are indicated in the table.

There are two states for $n = 1$ corresponding to $l = 0$, $m_l = 0$, and $m_s = \pm\frac{1}{2}$. These are called the 1s states. There are two states corresponding to $n = 2$, $l = 0$, $m_l = 0$, and $m_s = \pm\frac{1}{2}$. These constitute the 2s subshell. There are, in addition, six energy levels corresponding to $n = 2$, $l = 1$, $m_l = -1$, 0, or $+1$, and $m_s = \pm\frac{1}{2}$. These are designated as the 2p subshell. Hence, as indicated in Table 1-2, the total number of electrons in the L shell is $2 + 6 = 8$. In a similar manner we may verify that a d subshell contains a maximum of 10 electrons, an f subshell a maximum of 14 electrons, etc.

The atomic number Z gives the number of electrons orbiting about the nucleus. Let us use superscripts to designate the number of electrons in a particular subshell. Then sodium, Na, for which $Z = 11$, has an electronic configuration designated by $1s^2 2s^2 2p^6 3s^1$. Note that Na has a single electron in the outermost unfilled subshell, and hence is said to be monovalent. This same property is possessed by all the alkali metals (Li, Na, K, Rb, and Cs), which accounts for the fact that these elements in the same group in the periodic table (Table 1-1) have similar chemical properties.

The inner-shell electrons are very strongly bound to an atom, and cannot be easily removed. That is, the electrons closest to the nucleus are the most tightly bound, and so have the lowest energy. Also, atoms for which the electrons exist in closed shells form very stable configurations. For example, the inert gases He, Ne, A, Kr, and Xe, all have either completely filled shells or, at least, completely filled subshells.

Carbon, silicon, germanium, and tin have the electronic configurations indicated in Table 1-3. Note that each of these elements has completely filled subshells except for the outermost p shell, which contains only two of the six possible electrons. Despite this similarity, carbon in crystalline form (diamond) is an insulator, silicon and germanium solids are semiconductors, and tin is a metal. This apparent anomaly is explained in the next section.

TABLE 1-3   Electronic configuration in Group IVA

Element	Atomic number	Configuration
C	6	$1s^2 2s^2 2p^2$
Si	14	$1s^2 2s^2 2p^6 3s^2 3p^2$
Ge	32	$1s^2 2s^2 2p^6 3s^2 3p^6 3d^{10} 4s^2 4p^2$
Sn	50	$1s^2 2s^2 2p^6 3s^2 3p^6 3d^{10} 4s^2 4p^6 4d^{10} 5s^2 5p^2$

1-7   THE ENERGY–BAND THEORY OF CRYSTALS

X-ray and other studies reveal that most metals and semiconductors are crystalline in structure. A crystal consists of a space array of atoms or molecules (strictly speaking, ions) built up by regular repetition in three dimensions of some fundamental structural unit. The electronic energy levels discussed for a single free atom (as in a gas, where the atoms are sufficiently far apart not to exert any influence on one another) do not apply to the same atom in a crystal. This is so because the potential characterizing the crystalline structure is now a periodic function in space whose value at any point is the result of contributions from every atom. When atoms form crystals, it is found that the energy levels of the inner-shell electrons are not affected appreciably by the presence of the neighboring atoms. However, the levels of the outer-shell electrons are changed considerably, since these electrons are shared by more than one atom in the crystal. The new energy levels of the outer electrons can be determined by means of quantum mechanics, and it is found that coupling between the outer-shell electrons of the atoms results in a *band* of closely spaced energy states, instead of the widely separated energy levels of the isolated atom (Fig. 1-3). A qualitative discussion of this energy-band structure follows.

Consider a crystal consisting of N atoms of one of the elements in Table 1-3. Imagine that it is possible to vary the spacing between atoms without altering the type of fundamental crystal structure. If the atoms are so far apart that the interaction between them is negligible, the energy levels will coincide with those of the isolated atom. The outer two subshells for each element in Table 1-3 contain two s electrons and two p electrons. Hence, if we ignore the inner-shell levels, then, as indicated to the extreme right in Fig. 1-3a, there are $2N$ electrons completely filling the $2N$ possible s levels, all at the same energy. Since the p atomic subshell has six possible states, our imaginary crystal of widely spaced atoms has $2N$ electrons, which fill only one-third of the $6N$ possible p states, all at the same level.

If we now decrease the interatomic spacing of our imaginary crystal (moving from right to left in Fig. 1-3a), an atom will exert an electric force on its neighbors. Because of this coupling between atoms, the atomic-wave

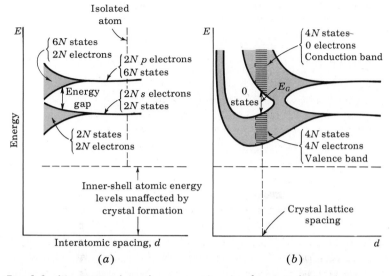

Fig. 1-3 Illustrating how the energy levels of isolated atoms are split into energy bands when these atoms are brought into close proximity to form a crystal.

functions overlap, and the crystal becomes an electronic *system* which must obey the Pauli exclusion principle. Hence the $2N$ degenerate s states must spread out in energy. The separation between levels is small, but since N is very large ($\sim 10^{23}$ cm^{-3}), the total spread between the minimum and maximum energy may be several electron volts if the interatomic distance is decreased sufficiently. This large number of discrete but closely spaced energy levels is called an *energy band,* and is indicated schematically by the lower shaded region in Fig. 1-3a. The $2N$ states in this band are completely filled with $2N$ electrons. Similarly, the upper shaded region in Fig. 1-3a is a band of $6N$ states which has only $2N$ of its levels occupied by electrons.

Note that there is an energy gap (a forbidden band) between the two bands discussed above and that this gap decreases as the atomic spacing decreases. For small enough distances (not indicated in Fig. 1-3a but shown in Fig. 1-3b) these bands will overlap. Under such circumstances the $6N$ upper states merge with the $2N$ lower states, giving a total of $8N$ levels, half of which are occupied by the $2N + 2N = 4N$ available electrons. At this spacing each atom has given up four electrons to the band; these electrons can no longer be said to orbit in s or p subshells of an isolated atom, but rather they belong to the crystal as a whole. In this sense the elements in Table 1-3 are tetravalent, since they contribute four electrons each to the crystal. The band these electrons occupy is called the *valence band.*

If the spacing between atoms is decreased below the distance at which the bands overlap, the interaction between atoms is indeed large. The energy-band structure then depends upon the orientation of the atoms relative to one another in space (the crystal structure) and upon the atomic number, which determines the electrical constitution of each atom. Solutions of Schrödinger's equation are complicated and have been obtained approximately for only relatively few crystals. These solutions lead us to expect an energy-band diagram somewhat as pictured[1] in Fig. 1-3b. At the crystal-lattice spacing (the dashed vertical line), we find the valence band *filled* with $4N$ electrons separated by a forbidden band (no allowed energy states) of extent E_G from an *empty* band consisting of $4N$ additional states. This upper vacant band is called the *conduction band*, for reasons given in the next section.

1-8 INSULATORS, SEMICONDUCTORS, AND METALS

A very poor conductor of electricity is called an *insulator;* an excellent conductor is a *metal;* and a substance whose conductivity lies between these extremes is a *semiconductor.* A material may be placed in one of these three classes, depending upon its energy-band structure.

Insulator The energy-band structure of Fig. 1-3b at the normal lattice spacing is indicated schematically in Fig. 1-4a. For a diamond (carbon) crystal the region containing no quantum states is several electron volts high ($E_G \approx 6$ eV). This large forbidden band separates the filled valence region from the vacant conduction band. The energy which can be supplied to an electron from an applied field is too small to carry the particle from the filled into the vacant band. Since the electron cannot acquire sufficient applied energy, conduction is impossible, and hence diamond is an *insulator.*

Semiconductor A substance for which the width of the forbidden energy region is relatively small (~ 1 eV) is called a *semiconductor.* Graphite, a

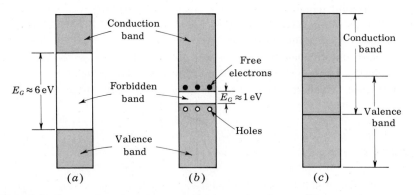

Fig. 1-4 Energy-band structure of (a) an insulator, (b) a semiconductor, and (c) a metal.

crystalline form of carbon but having a crystal symmetry which is different from diamond, has such a small value of E_G, and it is a semiconductor. The most important practical semiconductor materials are germanium and silicon, which have values of E_G of 0.785 and 1.21 eV, respectively, at 0°K. Energies of this magnitude normally cannot be acquired from an applied field. Hence the valence band remains full, the conduction band empty, and these materials are insulators at low temperatures. However, the conductivity increases with temperature, as we explain below. These substances are known as *intrinsic (pure) semiconductors.*

As the temperature is increased, some of these valence electrons acquire *thermal* energy greater than E_G, and hence move into the conduction band. These are now free electrons in the sense that they can move about under the influence of even a small applied field. These free, or conduction, electrons are indicated schematically by dots in Fig. 1-4b. The insulator has now become slightly conducting; it is a *semiconductor.* The absence of an electron in the valence band is represented by a small circle in Fig. 1-4b, and is called a *hole.* The phrase "holes in a semiconductor" therefore refers to the empty energy levels in an otherwise filled valence band.

The importance of the hole is that it may serve as a carrier of electricity, comparable in effectiveness with the free electron. The mechanism by which a hole contributes to conductivity is explained in Sec. 2-2. We also show in Chap. 2 that if certain impurity atoms are introduced into the crystal, these result in allowable energy states which lie in the forbidden energy gap. We find that these impurity levels also contribute to the conduction. A semiconductor material where this conduction mechanism predominates is called an *extrinsic (impurity) semiconductor.*

Since the band-gap energy of a crystal is a function of interatomic spacing (Fig. 1-3), it is not surprising that E_G depends somewhat on temperature. It has been determined experimentally that E_G decreases with temperature, and this dependence is given in Sec. 2-5.

Metal   A solid which contains a partly filled band structure is called a *metal.* Under the influence of an applied electric field the electrons may acquire additional energy and move into higher states. Since these mobile electrons constitute a current, this substance is a conductor and the partly filled region is the conduction band. One example of the band structure of a metal is given in Fig. 1-4c, which shows overlapping valence and conduction bands.

REFERENCES

1. Adler, R. B., A. C. Smith, and R. L. Longini: "Introduction to Semiconductor Physics," vol. 1, p. 78, Semiconductor Electronics Education Committee, John Wiley & Sons, Inc., New York, 1964.

2. Shockley, W.: "Electrons and Holes in Semiconductors," D. Van Nostrand Company, Inc., Princeton, N.J., 1963.

REVIEW QUESTIONS

1-1 Define *potential energy* in words and as an equation.

1-2 Define an *electron volt*.

1-3 State Bohr's three postulates for the atom.

1-4 Define a *photon*.

1-5 Define (*a*) *photoexcitation;* (*b*) *photoionization*.

1-6 State the *Pauli exclusion principle*.

1-7 Give the electronic configuration for an atom of a specified atomic number Z. For example, $Z = 32$ for germanium.

1-8 Explain why the energy levels of an atom become energy bands in a solid.

1-9 What is the difference between the band structure of an insulator and of a semiconductor?

1-10 What is the difference between the band structure of a semiconductor and of a metal?

1-11 Explain why a semiconductor acts as an insulator at $0°K$ and why its conductivity increases with increasing temperature.

1-12 What is the distinction between an intrinsic and an extrinsic semiconductor?

1-13 Define a *hole* in a semiconductor.

2 / TRANSPORT PHENOMENA IN SEMICONDUCTORS

The current in a metal is due to the flow of negative charges (*electrons*), whereas the current in a semiconductor results from the movement of both electrons and positive charges (*holes*). A semiconductor may be doped with impurity atoms so that the current is due predominantly either to electrons or to holes. The transport of the charges in a crystal under the influence of an electric field (a *drift* current), and also as a result of a nonuniform concentration gradient (a *diffusion* current), is investigated.

2-1 MOBILITY AND CONDUCTIVITY

In the preceding chapter we presented an energy-band picture of metals, semiconductors, and insulators. In a metal the outer, or valence, electrons of an atom are as much associated with one ion as with another, so that the electron attachment to any individual atom is almost zero. In terms of our previous discussion this means that the band occupied by the valence electrons may not be completely filled and that there are no forbidden levels at higher energies. Depending upon the metal, at least one, and sometimes two or three, electrons per atom are free to move throughout the interior of the metal under the action of applied fields.

Figure 2-1 is a two-dimensional schematic picture of the charge distribution within a metal. The shaded regions represent the net positive charge of the nucleus and the tightly bound inner electrons. The black dots represent the outer, or valence, electrons in the atom. It is these electrons that cannot be said to belong to any particular atom; instead, they have completely lost their individuality and can wander freely about from atom to atom in the metal. Thus a metal is

19

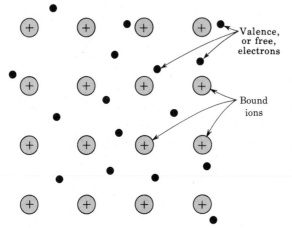

Fig. 2-1   Schematic arrangement of the atoms in one
plane in a metal, drawn for
monovalent atoms.   The
black dots represent the
electron gas, each atom
having contributed one electron to this gas.

visualized as a region containing a periodic three-dimensional array of heavy,
tightly bound ions permeated with a swarm of electrons that may move about
quite freely.   This picture is known as the *electron-gas* description of a metal.

According to the electron-gas theory of a metal, the electrons are in
continuous motion, the direction of flight being changed at each collision
with the heavy (almost stationary) ions.   The average distance between collisions is called the *mean free path*.   Since the motion is random, then, on an
average, there will be as many electrons passing through unit area in the metal ·
in any direction as in the opposite direction in a given time.   Hence the
average current is zero.

Let us now see how the situation is changed if a constant electric field
ε (volts per meter) is applied to the metal.   As a result of this electrostatic
force, the electrons would be accelerated and the velocity would increase
indefinitely with time, were it not for the collisions with the ions.   However,
at each inelastic collision with an ion, an electron loses energy, and a steady-
state condition is reached where a finite value of *drift speed* v is attained.[1]
This drift velocity is in the direction opposite to that of the electric field.
The speed at a time t between collision is at, where $a = q\varepsilon/m$ is the acceleration.
Hence the average speed v is proportional to ε.   Thus

$$v = \mu\varepsilon \tag{2-1}$$

where μ (square meters per volt-second) is called the *mobility* of the electrons.

According to the foregoing theory, a steady-state drift speed has been
superimposed upon the random thermal motion of the electrons.   Such a
directed flow of electrons constitutes a current.   We now calculate the magnitude of the current.

Current Density   If N electrons are contained in a length L of conductor
(Fig. 2-2), and if it takes an electron a time T s to travel a distance of L m in

Fig. 2-2   Pertaining to the calculation of current density.

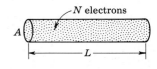

the conductor, the total number of electrons passing through any cross section of wire in unit time is N/T.   Thus the total charge per second passing any area, which, by definition, is the current in amperes, is

$$I = \frac{Nq}{T} = \frac{Nqv}{L} \qquad (2\text{-}2)$$

because L/T is the average, or *drift*, speed v m/s of the electrons.   By definition, the current density, denoted by the symbol J, is the current per unit area of the conducting medium.   That is, assuming a uniform current distribution,

$$J \equiv \frac{I}{A} \qquad (2\text{-}3)$$

where J is in amperes per square meter, and A is the cross-sectional area (in meters) of the conductor.   This becomes, by Eq. (2-2),

$$J = \frac{Nqv}{LA} \qquad (2\text{-}4)$$

From Fig. 2-2 it is evident that LA is simply the volume containing the N electrons, and so N/LA is the electron concentration n (in electrons per cubic meter).   Thus

$$n = \frac{N}{LA} \qquad (2\text{-}5)$$

and Eq. (2-4) reduces to

$$J = nqv = \rho v \qquad (2\text{-}6)$$

where $\rho \equiv nq$ is the charge density, in coulombs per cubic meter, and v is in meters per second.

This derivation is independent of the form of the conducting medium. Consequently, Fig. 2-2 does not necessarily represent a wire conductor.   It may represent equally well a portion of a gaseous-discharge tube or a volume element of a semiconductor.   Furthermore, neither ρ nor v need be constant, but may vary from point to point in space or may vary with time.

Conductivity   From Eqs. (2-6) and (2-1)

$$J = nqv = nq\mu\mathcal{E} = \sigma\mathcal{E} \qquad (2\text{-}7)$$

where

$$\sigma = nq\mu \qquad (2\text{-}8)$$

is the *conductivity* of the metal in (ohm-meter)$^{-1}$.   Equation (2-7) is recognized as Ohm's law, namely, the conduction current is proportional to the applied voltage. As already mentioned, the energy which the electrons acquire from the applied field is, as a result of collisions, given to the lattice ions. Hence power is dissipated within the metal by the electrons, and the power density (Joule heat) is given by $J\mathcal{E} = \sigma\mathcal{E}^2$ (watts per cubic meter).

2-2   ELECTRONS AND HOLES IN AN INTRINSIC SEMICONDUCTOR[1]

From Eq. (2-8) we see that the conductivity is proportional to the concentration n of free electrons. For a good conductor, n is very large ($\sim 10^{28}$ electrons/m³); for an insulator, n is very small ($\sim 10^7$); and for a semiconductor, n lies between these two values. The valence electrons in a semiconductor are not free to wander about as they are in a metal, but rather are trapped in a bond between two adjacent ions, as explained below.

The Covalent Bond   Germanium and silicon are the two most important semiconductors used in electronic devices. The crystal structure of these materials consists of a regular repetition in three dimensions of a unit cell having the form of a tetrahedron with an atom at each vertex. This structure is illustrated symbolically in two dimensions in Fig. 2-3. Germanium has a total of 32 electrons in its atomic structure, arranged in shells as indicated in Table 1-3. As explained in Sec. 1-7, each atom in a germanium crystal contributes four valence electrons, so that the atom is tetravalent. The inert ionic core of the germanium atom carries a positive charge of $+4$ measured in units of the electronic charge. The binding forces between neighboring atoms result from the fact that each of the valence electrons of a germanium atom is shared by one of its four nearest neighbors. This *electron-pair*, or *covalent*, *bond* is represented in Fig. 2-3 by the two dashed lines which join each atom to each of its neighbors. The fact that the valence electrons serve to bind one atom to the next also results in the valence electron being tightly bound to the

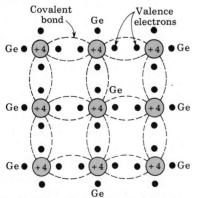

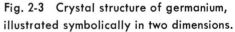

Fig. 2-3 Crystal structure of germanium, illustrated symbolically in two dimensions.

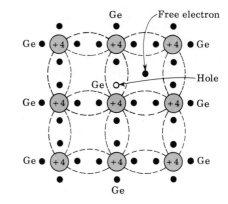

Fig. 2-4 Germanium crystal with a
broken covalent bond.

nucleus.   Hence, in spite of the availability of four valence electrons, the
crystal has a low conductivity.

The Hole   At a very low temperature (say 0°K) the ideal structure of
Fig. 2-3 is approached, and the crystal behaves as an insulator, since no free
carriers of electricity are available.   However, at room temperature, some of
the covalent bonds will be broken because of the thermal energy supplied to the
crystal, and conduction is made possible.   This situation is illustrated in Fig.
2-4.   Here an electron, which for the far greater period of time forms part of a
covalent bond, is pictured as being dislodged, and therefore free to wander in
a random fashion throughout the crystal.   The energy E_G required to break
such a covalent bond is about 0.72 eV for germanium and 1.1 eV for silicon
at room temperature.   The absence of the electron in the covalent bond is
represented by the small circle in Fig. 2-4, and such an incomplete covalent
bond is called a *hole*.   The importance of the hole is that it may serve as a
carrier of electricity comparable in effectiveness with the free electron.

The mechanism by which a hole contributes to the conductivity is qualitatively as follows: When a bond is incomplete so that a hole exists, it is
relatively easy for a valence electron in a neighboring atom to leave its covalent
bond to fill this hole.   An electron moving from a bond to fill a hole leaves a
hole in its initial position.   Hence the hole effectively moves in the direction
opposite to that of the electron.   This hole, in its new position, may now be
filled by an electron from another covalent bond, and the hole will correspondingly move one more step in the direction opposite to the motion of the electron.   Here we have a mechanism for the conduction of electricity which does
not involve *free* electrons.   This phenomenon is illustrated schematically in
Fig. 2-5, where a circle with a dot in it represents a completed bond, and an
empty circle designates a hole.   Figure 2-5a shows a row of 10 ions, with a
broken bond, or hole, at ion 6.   Now imagine that an electron from ion 7
moves into the hole at ion 6, so that the configuration of Fig. 2-5b results.
If we compare this figure with Fig. 2-5a, it looks as if the hole in (a) has

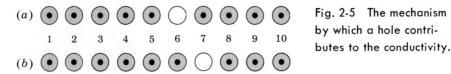

Fig. 2-5 The mechanism by which a hole contributes to the conductivity.

moved toward the right in (*b*) (from ion 6 to ion 7).   This discussion indicates that the motion of the hole in one direction actually means the transport of a negative charge an equal distance in the opposite direction.   So far as the flow of electric current is concerned, the hole behaves like a positive charge equal in magnitude to the electronic charge.   We can consider that the holes are physical entities whose movement constitutes a flow of current.   The heuristic argument that a hole behaves as a *free* positive charge carrier may be justified by quantum mechanics.[1]   An experimental verification of this concept is given in Sec. 2-6.

In a pure (*intrinsic*) semiconductor the number of holes is equal to the number of free electrons.   Thermal agitation continues to produce new hole-electron pairs, whereas other hole-electron pairs disappear as a result of recombination.   The hole concentration p must equal the electron concentration n, so that

$$n = p = n_i \tag{2-9}$$

where n_i is called the *intrinsic concentration*.

Effective Mass[2]   We digress here briefly to discuss the concept of the effective mass of the electron and hole.   It is found that, when quantum mechanics is used to specify the motion within the crystal of a free or conduction electron or hole on which an external field is applied, it is possible to treat the hole and electron as imaginary *classical particles* with effective positive masses m_p and m_n, respectively.   This approximation is valid provided that the externally applied fields are much weaker than the internal *periodic* fields produced by the lattice structure.   In a perfect crystal these imaginary particles respond only to the external fields.

A wave-mechanical analysis[1] shows that a bound or valence electron *cannot* be treated as a classical particle.   This difficulty is bypassed by ignoring the bound electrons and considering only the motion of the holes.   In summary, the effective-mass approximation removes the quantum features of the problem and allows us to use Newton's laws to determine the effect of external forces on the (free) electrons and holes within a crystal.

2-3   DONOR AND ACCEPTOR IMPURITIES

If, to intrinsic silicon or germanium, there is added a small percentage of trivalent or pentavalent atoms, a *doped, impure,* or *extrinsic,* semiconductor is formed.

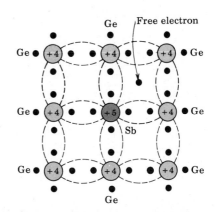

Fig. 2-6 Crystal lattice with a germanium atom displaced by a pentavalent impurity atom.

Donors   If the dopant has five valence electrons, the crystal structure of Fig. 2-6 is obtained.   The impurity atoms will displace some of the germanium atoms in the crystal lattice.   Four of the five valence electrons will occupy covalent bonds, and the fifth will be nominally unbound and will be available as a carrier of current.   The energy required to detach this fifth electron from the atom is of the order of only 0.01 eV for Ge or 0.05 eV for Si.   Suitable pentavalent impurities are antimony, phosphorus, and arsenic.   Such impurities donate excess (negative) electron carriers, and are therefore referred to as *donor*, or *n*-type, impurities.

When donor impurities are added to a semiconductor, allowable energy levels are introduced a very small distance below the conduction band, as is shown in Fig. 2-7.   These new allowable levels are essentially a discrete level because the added impurity atoms are far apart in the crystal structure, and hence their interaction is small.   In the case of germanium, the distance of the new discrete allowable energy level is only 0.01 eV (0.05 eV in silicon) below the conduction band, and therefore at room temperature almost all the "fifth" electrons of the donor material are raised into the conduction band.

If intrinsic semiconductor material is "doped" with *n*-type impurities, not only does the number of electrons increase, but the number of holes decreases below that which would be available in the intrinsic semiconductor. The reason for the decrease in the number of holes is that the larger number of electrons present increases the rate of recombination of electrons with holes.

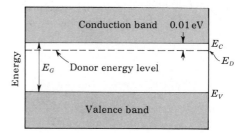

Fig. 2-7 Energy-band diagram of *n*-type semiconductor.

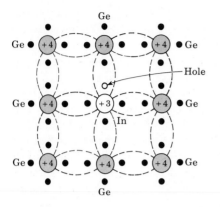

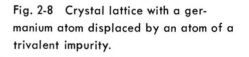

Fig. 2-8 Crystal lattice with a germanium atom displaced by an atom of a trivalent impurity.

Acceptors   If a trivalent impurity (boron, gallium, or indium) is added to an intrinsic semiconductor, only three of the covalent bonds can be filled, and the vacancy that exists in the fourth bond constitutes a hole. This situation is illustrated in Fig. 2-8. Such impurities make available positive carriers because they create holes which can accept electrons. These impurities are consequently known as *acceptor*, or *p*-type, impurities. The amount of impurity which must be added to have an appreciable effect on the conductivity is very small. For example, if a donor-type impurity is added to the extent of 1 part in 10^8, the conductivity of germanium at 30°C is multiplied by a factor of 12.

When acceptor, or *p*-type, impurities are added to the intrinsic semiconductor, they produce an allowable discrete energy level which is just above the valence band, as shown in Fig. 2-9. Since a very small amount of energy is required for an electron to leave the valence band and occupy the acceptor energy level, it follows that the holes generated in the valence band by these electrons constitute the largest number of carriers in the semiconductor material.

The Mass-action Law   We noted above that adding *n*-type impurities decreases the number of holes. Similarly, doping with *p*-type impurities decreases the concentration of free electrons below that in the intrinsic semiconductor. A theoretical analysis (Sec. 2-12) leads to the result that, under

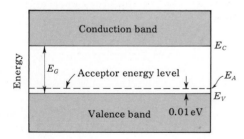

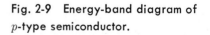

Fig. 2-9   Energy-band diagram of *p*-type semiconductor.

thermal equilibrium, the product of the free negative and positive concentrations is a constant independent of the amount of donor and acceptor impurity doping. This relationship is called the *mass-action law* and is given by

$$np = n_i{}^2 \qquad (2\text{-}10)$$

The intrinsic concentration n_i is a function of temperature (Sec. 2-5).

We have the important result that the doping of an intrinsic semiconductor not only increases the conductivity, but also serves to produce a conductor in which the electric carriers are either predominantly holes or predominantly electrons. In an *n*-type semiconductor, the electrons are called the *majority carriers*, and the holes are called the *minority carriers*. In a *p*-type material, the holes are the majority carriers, and the electrons are the minority carriers.

2-4   CHARGE DENSITIES IN A SEMICONDUCTOR

Equation (2-10), namely, $np = n_i{}^2$, gives one relationship between the electron n and the hole p concentrations. These densities are further interrelated by the law of electrical neutrality, which we shall now state in algebraic form: Let N_D equal the concentration of donor atoms. Since, as mentioned above, these are practically all ionized, N_D positive charges per cubic meter are contributed by the donor ions. Hence the total positive-charge density is $N_D + p$. Similarly, if N_A is the concentration of acceptor ions, these contribute N_A negative charges per cubic meter. The total negative-charge density is $N_A + n$. Since the semiconductor is electrically neutral, the magnitude of the positive-charge density must equal that of the negative concentration, or

$$N_D + p = N_A + n \qquad (2\text{-}11)$$

Consider an *n*-type material having $N_A = 0$. Since the number of electrons is much greater than the number of holes in an *n*-type semiconductor ($n \gg p$), then Eq. (2-11) reduces to

$$n \approx N_D \qquad (2\text{-}12)$$

In an n-type material the free-electron concentration is approximately equal to the density of donor atoms.

In later applications we study the characteristics of *n*- and *p*-type materials connected together. Since some confusion may arise as to which type is under consideration at a given moment, we add the subscript n or p for an *n*-type or a *p*-type substance, respectively. Thus Eq. (2-12) is more clearly written

$$n_n \approx N_D \qquad (2\text{-}13)$$

The concentration p_n of holes in the *n*-type semiconductor is obtained from Eq. (2-10), which is now written $n_n p_n = n_i{}^2$. Thus

$$p_n = \frac{n_i{}^2}{N_D} \qquad (2\text{-}14)$$

Similarly, for a p-type semiconductor,

$$n_p p_p = n_i^2 \qquad p_p \approx N_A \qquad n_p = \frac{n_i^2}{N_A} \tag{2-15}$$

It is possible to add donors to a p-type crystal or, conversely, to add acceptors to n-type material. If equal concentrations of donors and acceptors permeate the semiconductor, it remains intrinsic. The hole of the acceptor combines with the conduction electron of the donor to give no additional free carriers. Thus, from Eq. (2-11) with $N_D = N_A$, we observe that $p = n$, and from Eq. (2-10), $n^2 = n_i^2$, or $n = n_i$ = the intrinsic concentration.

An extension of the above argument indicates that if the concentration of donor atoms added to a p-type semiconductor exceeds the acceptor concentration $(N_D > N_A)$, the specimen is changed from a p-type to an n-type semiconductor. [In Eqs. (2-13) and (2-14) N_D should be replaced by $N_D - N_A$.]

2-5   ELECTRICAL PROPERTIES OF GE AND SI

A fundamental difference between a metal and a semiconductor is that the former is *unipolar* [conducts current by means of charges (electrons) of one sign only], whereas a semiconductor is *bipolar* (contains two charge-carrying "particles" of opposite sign).

Conductivity   One carrier is negative (the free electron), of mobility μ_n, and the other is positive (the hole), of mobility μ_p. These particles move in opposite directions in an electric field $\mathcal{E}$, but since they are of opposite sign, the current of each is in the same direction. Hence the current density J is given by (Sec. 2-1)

$$J = (n\mu_n + p\mu_p)q\mathcal{E} = \sigma\mathcal{E} \tag{2-16}$$

where n = magnitude of free-electron (negative) concentration
p = magnitude of hole (positive) concentration
σ = conductivity

Hence $\sigma = (n\mu_n + p\mu_p)q$ $\qquad\qquad\qquad\qquad\qquad\qquad\qquad$ (2-17)

For the pure semiconductor, $n = p = n_i$, where n_i is the intrinsic concentration.

Intrinsic Concentration   With increasing temperature, the density of hole-electron pairs increases and, correspondingly, the conductivity increases. In Sec. 19-5 it is found that the intrinsic concentration n_i varies with T as

$$n_i^2 = A_o T^3 \epsilon^{-E_{GO}/kT} \tag{2-18}$$

where E_{GO} is the energy gap at $0°K$ in electron volts, k is the Boltzman constant in $eV/°K$ (Appendix A), and A_o is a constant independent of T. The constants E_{GO}, μ_n, μ_p, and many other important physical quantities for germanium and

TABLE 2-1   Properties of germanium and silicon†

Property	Ge	Si
Atomic number.....................	32	14
Atomic weight.....................	72.6	28.1
Density, g/cm³.....................	5.32	2.33
Dielectric constant (relative).........	16	12
Atoms/cm³.........................	4.4×10^{22}	5.0×10^{22}
E_{GO}, eV, at 0°K....................	0.785	1.21
E_G, eV, at 300°K...................	0.72	1.1
n_i at 300°K, cm⁻³.................	2.5×10^{13}	1.5×10^{10}
Intrinsic resistivity at 300°K, Ω-cm....	45	230,000
μ_n, cm²/V-s at 300°K..............	3,800	1,300
μ_p, cm²/V-s at 300°K..............	1,800	500
D_n, cm²/s $= \mu_n V_T$.................	99	34
D_p, cm²/s $= \mu_p V_T$.................	47	13

† G. L. Pearson and W. H. Brattain, History of Semiconductor Research, *Proc. IRE*, vol. 43, pp. 1794–1806, December, 1955.   E. M. Conwell, Properties of Silicon and Germanium, Part II, *Proc. IRE*, vol. 46, no. 6, pp. 1281–1299, June, 1958.

silicon are given in Table 2-1.   Note that germanium has of the order of 10^{22} atoms/cm³, whereas at room temperature (300°K), $n_i \approx 10^{13}$/cm³.   Hence only 1 atom in about 10^9 contributes a free electron (and also a hole) to the crystal because of broken covalent bonds.   For silicon this ratio is even smaller, about 1 atom in 10^{12}

The Energy Gap   The forbidden region E_G in a semiconductor depends upon temperature, as pointed out in Sec. 1-7.   Experimentally it is found that, for silicon,[3]

$$E_G(T) = 1.21 - 3.60 \times 10^{-4}T \qquad (2\text{-}19)$$

and at room temperature (300°K), $E_G = 1.1$ eV.   Similarly, for germanium,[4]

$$E_G(T) = 0.785 - 2.23 \times 10^{-4}T \qquad (2\text{-}20)$$

and at room temperature, $E_G = 0.72$ eV.

The Mobility   This parameter μ varies[3] as T^{-m} over a temperature range of 100 to 400°K.   For silicon, $m = 2.5$ (2.7) for electrons (holes), and for germanium, $m = 1.66$ (2.33) for electrons (holes).   The mobility is also found[4] to be a function of electric field intensity and remains constant only if $\mathcal{E} < 10^3$ V/cm in n-type silicon.   For $10^3 < \mathcal{E} < 10^4$ V/cm, μ_n varies approximately as $\mathcal{E}^{-\frac{1}{2}}$.   For higher fields, μ_n is inversely proportional to $\mathcal{E}$ and the carrier speed approaches the constant value of 10^7 cm/s.

EXAMPLE (*a*) Using Avogadro's number, verify the numerical value given in Table 2-1 for the concentration of atoms in germanium. (*b*) Find the resistivity of intrinsic germanium at 300°K. (*c*) If a donor-type impurity is added to the extent of 1 part in 10^8 germanium atoms, find the resistivity. (*d*) If germanium were a monovalent metal, find the ratio of its conductivity to that of the *n*-type semiconductor in part *c*.

Solution   *a*. A quantity of any substance equal to its molecular weight in grams is a *mole* of that substance. Further, a mole of any substance contains the same number of molecules as a mole of any other material. This number is called *Avogadro's number* and equals 6.02×10^{23} molecules per mole (Appendix A). Thus, for monatomic germanium (using Table 2-1),

$$\text{Concentration} = 6.02 \times 10^{23}\,\frac{\text{atoms}}{\text{mole}} \times \frac{1\ \text{mole}}{72.6\ \text{g}} \times \frac{5.32\ \text{g}}{\text{cm}^3} = 4.41 \times 10^{22}\,\frac{\text{atoms}}{\text{cm}^3}$$

b. From Eq. (2-17), with $n = p = n_i$,

$$\sigma = n_i q(\mu_n + \mu_p) = (2.5 \times 10^{13}\ \text{cm}^{-3})(1.60 \times 10^{-19}\ \text{C})(3{,}800 + 1{,}800)\,\frac{\text{cm}^2}{\text{V-s}}$$

$$= 0.0224\ (\Omega\text{-cm})^{-1}$$

$$\text{Resistivity} = \frac{1}{\sigma} = \frac{1}{0.0224} = 44.6\ \Omega\text{-cm}$$

in agreement with the value in Table 2-1.

   c. If there is 1 donor atom per 10^8 germanium atoms, then $N_D = 4.41 \times 10^{14}$ atoms/cm³. From Eq. (2-12) $n \approx N_D$ and from Eq. (2-14)

$$p = \frac{n_i^2}{N_D} = \frac{(2.5 \times 10^{13})^2}{4.41 \times 10^{14}} = 1.42 \times 10^{12}\ \text{holes/cm}^3$$

Since $n \gg p$, we can neglect p in calculating the conductivity. From Eq. (2-17)

$$\sigma = nq\mu_n = 4.41 \times 10^{14} \times 1.60 \times 10^{-19} \times 3{,}800 = 0.268\ (\Omega\text{-cm})^{-1}$$

The resistivity $= 1/\sigma = 1/0.268 = 3.72\ \Omega\text{-cm}$.
   NOTE: The addition of 1 donor atom in 10^8 germanium atoms has multiplied the conductivity by a factor of $44.6/3.72 = 11.7$.
   d. If each atom contributed one free electron to the "metal," then

$$n = 4.41 \times 10^{22}\ \text{electrons/cm}^3$$

and

$$\sigma = nq\mu_n = 4.41 \times 10^{22} \times 1.60 \times 10^{-19} \times 3{,}800$$
$$= 2.58 \times 10^7\ (\Omega\text{-cm})^{-1}$$

Hence the conductivity of the "metal" is higher than that of the *n*-type semiconductor by a factor of

$$\frac{2.58 \times 10^7}{0.268} \approx 10^8$$

2-6   THE HALL EFFECT

If a specimen (metal or semiconductor) carrying a current I is placed in a transverse magnetic field **B**, an electric field **ε** is induced in the direction perpendicular to both I and **B**.   This phenomenon, known as the *Hall effect*, is used to determine whether a semiconductor is *n*- or *p*-type and to find the carrier concentration.   Also, by simultaneously measuring the conductivity σ, the mobility μ can be calculated.

The physical origin of the Hall effect is not difficult to find.   If in Fig. 2-10 I is in the positive X direction and **B** is in the positive Z direction, a force will be exerted in the negative Y direction on the current carriers.   The current I may be due to holes moving from left to right or to free electrons traveling from right to left in the semiconductor specimen.   Hence, independently of whether the carriers are holes or electrons, they will be forced downward toward side 1 in Fig. 2-10.   If the semiconductor is *n*-type material, so that the current is carried by electrons, these electrons will accumulate on side 1, and this surface becomes negatively charged with respect to side 2. Hence a potential, called the *Hall voltage*, appears between surfaces 1 and 2.

If the polarity of V_H is positive at terminal 2, then, as explained above, the carriers must be electrons.   If, on the other hand, terminal 1 becomes charged positively with respect to terminal 2, the semiconductor must be *p*-type. These results have been verified experimentally, thus justifying the bipolar (two-carrier) nature of the current in a semiconductor.

If I is the current in a *p*-type semiconductor, the carriers might be considered to be the *bound* electrons jumping from right to left.   Then side 1 would become negatively charged.   However, experimentally, side 1 is found to become positive with respect to side 2 for a *p*-type specimen.   This experiment confirms the quantum-mechanical fact noted in Sec. 2-2 that the hole acts like a classical *free* positive-charge carrier.

Experimental Determination of Mobility   In the equilibrium state the electric field intensity ε due to the Hall effect must exert a force on the carrier which just balances the magnetic force, or

$$q\varepsilon = Bqv \qquad\qquad\qquad (2\text{-}21)$$

where q is the magnitude of the charge on the carrier, and v is the drift speed. From Eq. (1-3), $\varepsilon = V_H/d$, where d is the distance between surfaces 1 and 2.

Fig. 2-10 Pertaining to the Hall effect. The carriers (whether electrons or holes) are subjected to a magnetic force in the negative Y direction.

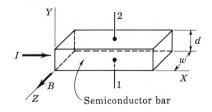

From Eq. (2-6), $J = \rho v = I/wd$, where J is the current density, ρ is the charge density, and w is the width of the specimen in the direction of the magnetic field. Combining these relationships, we find

$$V_H = \mathcal{E}d = Bvd = \frac{BJd}{\rho} = \frac{BI}{\rho w} \qquad (2\text{-}22)$$

If V_H, B, I, and w are measured, the charge density ρ can be determined from Eq. (2-22).

It is customary to introduce the Hall coefficient R_H defined by

$$R_H \equiv \frac{1}{\rho} \qquad (2\text{-}23)$$

Hence $R_H = \dfrac{V_H w}{BI} \qquad (2\text{-}24)$

If conduction is due primarily to charges of one sign, the conductivity σ is related to the mobility μ by Eq. (2-8), or

$$\sigma = \rho\mu \qquad (2\text{-}25)$$

If the conductivity is measured together with the Hall coefficient, the mobility can be determined from

$$\mu = \sigma R_H \qquad (2\text{-}26)$$

We have assumed in the foregoing discussion that all particles travel with the mean drift speed v. Actually, the current carriers have a random thermal distribution in speed. If this distribution is taken into account, it is found that Eq. (2-24) remains valid provided that R_H is defined by $3\pi/8\rho$. Also, Eq. (2-26) must be modified to $\mu = (8\sigma/3\pi)R_H$.

Applications Since V_H is proportional to B (for a given current I), then the Hall effect has been incorporated into a magnetic field meter. Another instrument, called a *Hall-effect multiplier*, is available to give an output proportional to the product of two signals. If I is made proportional to one of the inputs and if B is linearly related to the second signal, then, from Eq. (2-22), V_H is proportional to the product of the two inputs.

2-7   CONDUCTIVITY MODULATION

Since the conductivity σ of a semiconductor is proportional to the concentration of free carriers [Eq. (2-17)], σ may be increased by increasing n or p. The two most important methods for varying n and p are to change the temperature or to illuminate the semiconductor and thereby generate new hole-electron pairs.

Thermistors   The conductivity of germanium (silicon) is found from Eq. (2-18) to increase approximately 6 (8) percent per degree increase in temperature.   Such a large change in conductivity with temperature places a limitation upon the use of semiconductor devices in some circuits.   On the other hand, for some applications it is exactly this property of semiconductors that is used to advantage.   A semiconductor used in this manner is called a *thermistor*. Such a device finds extensive application in thermometry, in the measurement of microwave-frequency power, as a thermal relay, and in control devices actuated by changes in temperature.   Silicon and germanium are not used as thermistors because their properties are too sensitive to impurities.   Commercial thermistors consist of sintered mixtures of such oxides as NiO, Mn_2O_3, and Co_2O_3.

The exponential decrease in resistivity (reciprocal of conductivity) of a semiconductor should be contrasted with the small and almost linear increase in resistivity of a metal.   An increase in the temperature of a metal results in greater thermal motion of the ions, and hence decreases slightly the mean free path of the free electrons.   The result is a decrease in the mobility, and hence in conductivity.   For most metals the resistance increases about 0.4 percent/°C increase in temperature.   It should be noted that a thermistor has a negative coefficient of resistance, whereas that of a metal is positive and of much smaller magnitude.   By including a thermistor in a circuit it is possible to compensate for temperature changes over a range as wide as 100°C.

A heavily doped semiconductor can exhibit a positive temperature coefficient of resistance, for under these circumstances the material acquires metallic properties and the resistance increases because of the decrease in carrier mobility with temperature.   Such a device, called a *sensistor* (manufactured by Texas Instruments), has a temperature coefficient of resistance of +0.7 percent/°C (over the range from −60 to +150°C).

Photoconductors   If radiation falls upon a semiconductor, its conductivity increases.   This photoconductive effect is explained as follows: Radiant energy supplied to the semiconductor ionizes covalent bonds; that is, these bonds are broken, and hole-electron pairs in excess of those generated thermally are created.   These increased current carriers decrease the resistance of the material, and hence such a device is called a *photoresistor*, or *photoconductor*. For a light-intensity change of 100 fc,† the resistance of a commercial photoconductor may change by several kilohms.

In Fig. 2-11 we show the energy diagram of a semiconductor having both acceptor and donor impurities.   If photons of sufficient energies illuminate this specimen, *photogeneration* takes place and the following transitions are possible: An electron-hole pair can be created by a high-energy photon, in what is called *intrinsic excitation;* a photon may excite a donor electron into the conduction band; or a valence electron may go into an acceptor state.   The

† fc is the standard abbreviation for foot-candle.

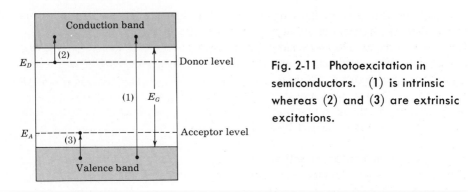

Fig. 2-11 Photoexcitation in semiconductors. (1) is intrinsic whereas (2) and (3) are extrinsic excitations.

last two transitions are known as *impurity excitations*. Since the density of states in the conduction and valence bands greatly exceeds the density of impurity states, photoconductivity is due principally to intrinsic excitation.

Spectral Response The minimum energy of a photon required for intrinsic excitation is the forbidden-gap energy E_G (electron volts) of the semiconductor material. The wavelength λ_c of a photon whose energy corresponds to E_G is given by Eq. (1-14), with $E_1 - E_2 = E_G$. If λ_c is expressed in microns† and E_G in electron volts,

$$\lambda_c = \frac{1.24}{E_G} \tag{2-27}$$

If the wavelength λ of the radiation exceeds λ_c, then the energy of the photon is less than E_G and such a photon cannot cause a valence electron to enter the conduction band. Hence λ_c is called the *critical*, or *cutoff*, *wavelength*, or *long-wavelength threshold*, of the material. For Si, $E_G = 1.1$ eV and $\lambda_c = 1.13$ μm, whereas for Ge, $E_G = 0.72$ eV and $\lambda_c = 1.73$ μm at room temperature (Table 2-1).

The spectral-sensitivity curves for Si and Ge are plotted in Fig. 2-12, indicating that a photoconductor is a frequency-selective device. This means

† 1 micron = 1 micrometer = 1 μm = 10^{-6} m.

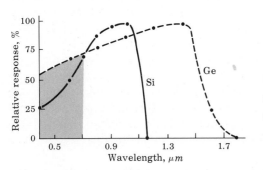

Fig. 2-12 Relative spectral response of Si and Ge. (Courtesy of Texas Instruments, Inc.)

that a given intensity of light of one wavelength will not generate the same number of free carriers as an equal intensity of light of another wavelength. In other words, the *photoelectric yield*, or *spectral response*, depends upon the frequency of the incident radiation.   Note that the long-wavelength limit is slightly greater than the values of λ_c calculated above, because of the impurity excitations.   As the wavelength is decreased ($\lambda < \lambda_c$ or $f > f_c$), the response increases and reaches a maximum.   The range of wavelengths of visible light (0.38 to 0.76 μm) is indicated by the shaded region in Fig. 2-12.

Commercial Photoconductive Cells   There are three important types of applications of such a device: It is used (1) to measure a fixed amount of illumination (as with light meter), (2) to record a modulating light intensity (as on a sound track), and (3) as an ON–OFF light relay (as in a digital or control circuit).

The photoconducting device with the widest application is the cadmium sulfide cell.   The sensitive area of this device consists of a layer of chemically deposited CdS, which may contain a small amount of silver, antimony, or indium impurities.   In absolute darkness the resistance may be as high as 2 M, and when stimulated with strong light, the resistance may be less than 10 Ω.

The primary advantages of CdS photoconductors are their high dissipation capability, their excellent sensitivity in the visible spectrum, and their low resistance when stimulated by light.   These photoconductors are designed to dissipate safely 300 mW, and can be made to handle safely power levels of several watts.   Hence a CdS photoconductor can operate a relay directly, without intermediate amplifier circuits.

Other types of photoconductive devices are available for specific applications.   A lead sulfide, PbS, cell has a peak on the sensitivity curve at 2.9 μm, and hence is used for infrared-detection or infrared-absorption measurements. A selenium cell is sensitive throughout the visible end, and particularly toward the blue end of the spectrum.

2-8   GENERATION AND RECOMBINATION OF CHARGES

In Sec. 2-2 we see that in a pure semiconductor the number of holes is equal to the number of free electrons.   Thermal agitation, however, continues to generate g new hole-electron pairs per unit volume per second, while other hole-electron pairs disappear as a result of recombination; in other words, free electrons fall into empty covalent bonds, resulting in the loss of a pair of mobile carriers.   On an average, a hole (an electron) will exist for $\tau_p(\tau_n)$ s before recombination.   This time is called the *mean lifetime* of the hole and electron, respectively.   These parameters are very important in semiconductor devices because they indicate the time required for electron and hole concentrations which have been caused to change to return to their equilibrium concentrations.

Consider a bar of n-type silicon containing the thermal-equilibrium con-

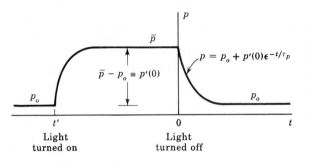

Fig. 2-13 The hole (minority) concentration in an n-type semiconductor bar as a function of time, due to generation and recombination.

centration p_o and n_o. Assume that at $t = t'$ the specimen is illuminated (Fig. 2-13) and that additional hole-electron pairs are generated uniformly throughout the crystal. An equilibrium situation is reached, and the new concentrations are $\bar{p}$ and $\bar{n}$ under the influence of the radiation. The *photo-injected*, or *excess*, concentration is $\bar{p} - p_o$ for holes and $\bar{n} - n_o$ for electrons. Since the radiation causes hole-electron pairs to be created, then clearly,

$$\bar{p} - p_o = \bar{n} - n_o \tag{2-28}$$

Although the increase in hole concentration p equals that for the electron density n, the *percentage* increase for electrons in an n-type semiconductor (where electrons are plentiful) is very small. On the other hand, the percentage increase in holes may be tremendous, because holes are scarce in an n-type crystal. In summary, the radiation affects the majority concentration hardly at all, and therefore we shall limit the discussion to the behavior of the minority carriers.

After a steady state is reached, at $t = 0$ in Fig. 2-13, the radiation is removed. We shall now demonstrate that the excess carrier density returns to zero exponentially with time. To do so we must derive the differential equation which governs the hole concentration as a function of time for $t > 0$ (*when there is no external excitation*).

From the definition of mean lifetime τ_p and *assuming that τ_p is independent of the magnitude of the hole concentration,*

$$\frac{p}{\tau_p} = decrease \text{ in hole concentration per second} \atop due \text{ to recombination} \tag{2-29}$$

From the definition of the generation rate,

$$g = increase \text{ in hole concentration per second} \atop due \text{ to thermal generation} \tag{2-30}$$

Since charge can neither be created nor destroyed, there must be an increase in hole concentration per second of amount dp/dt. This rate must, at every instant of time, equal the algebraic sum of the rates given in Eqs. (2-29) and

(2-30), or

$$\frac{dp}{dt} = g - \frac{p}{\tau_p} \tag{2-31}$$

Under steady-state conditions, $dp/dt = 0$, and with no radiation falling on the sample, the hole concentration p reaches its thermal-equilibrium value p_o. Hence $g = p_o/\tau_p$, and the above equation becomes

$$\frac{dp}{dt} = \frac{p_o - p}{\tau_p} \tag{2-32}$$

The *excess*, or *injected*, carrier density p' is defined as the increase in minority concentration above the equilibrium value. Since p' is a function of time, then

$$p' \equiv p - p_o = p'(t) \tag{2-33}$$

It follows from Eq. (2-32) that the differential equation controlling p' is

$$\frac{dp'}{dt} = -\frac{p'}{\tau_p} \tag{2-34}$$

The rate of change of excess concentration is proportional to this concentration—an intuitively correct result. The minus sign indicates that the change is a decrease in the case of recombination and an increase when the concentration is recovering from a temporary depletion.

Since the radiation results in an initial (at $t \le 0$) excess concentration $p'(0) = \bar{p} - p_o$ and then this excitation is removed, the solution of Eq. (2-34) for $t \ge 0$ is

$$p'(t) = p'(0)\epsilon^{-t/\tau_p} = (\bar{p} - p_o)\epsilon^{-t/\tau_p} = p - p_o \tag{2-35}$$

The excess concentration decreases exponentially to zero ($p' = 0$ or $p = p_o$) with a time constant equal to the mean lifetime τ_p, as indicated in Fig. 2-13. The pulsed-light method indicated in this figure is used to measure τ_p.

Recombination Centers   Recombination is the process where an electron moves from the conduction band into the valence band so that a mobile electron-hole pair disappear. Classical mechanics requires that momentum be conserved in an encounter of two particles. Since the momentum is zero after recombination, this conservation law requires that the "colliding" electron and hole must have equal magnitudes of momentum and they must be traveling in opposite directions. This requirement is very stringent, and hence the probability of recombination by such a direct encounter is very small.

The most important mechanism in silicon or germanium through which holes and electrons recombine is that involving *traps*, or *recombination centers*,[5] which contribute electronic states in the energy gap of the semiconductor. Such a location acts effectively as a third body which can satisfy the conservation-of-momentum requirement. These new states are associated with imperfections in the crystal. Specifically, metallic impurities in the semiconductor are capable of introducing energy states in the forbidden gap. Recombination is affected not only by volume impurities but also by surface imperfections in the crystal.

Gold is extensively used as a recombination agent by semiconductor-device manufacturers. Thus the device designer can obtain desired carrier lifetimes by introducing gold into silicon under controlled conditions.[6] Carrier lifetimes range from nanoseconds (1 ns = 10^{-9} s) to hundreds of microseconds (μs).

2-9 DIFFUSION

In addition to a conduction current, the transport of charges in a semiconductor may be accounted for by a mechanism called *diffusion*, not ordinarily encountered in metals. The essential features of diffusion are now discussed.

It is possible to have a nonuniform concentration of particles in a semiconductor. As indicated in Fig. 2-14, the concentration p of holes varies with distance x in the semiconductor, and there exists a concentration gradient, dp/dx, in the density of carriers. The existence of a gradient implies that if an imaginary surface (shown dashed) is drawn in the semiconductor, the density of holes immediately on one side of the surface is larger than the density on the other side. The holes are in a random motion as a result of their thermal energy. Accordingly, holes will continue to move back and forth across this surface. We may then expect that, in a given time interval, more holes will cross the surface from the side of greater concentration to the side of smaller concentration than in the reverse direction. This net transport of holes across the surface constitutes a current in the positive X direction. It should be noted that this net transport of charge is not the result of mutual repulsion among charges of like sign, but is simply the result of a statistical phenomenon. This diffusion is exactly analogous to that which occurs in a neutral gas if a concentration gradient exists in the gaseous container. The diffusion hole-current density J_p (amperes per square meter) is proportional

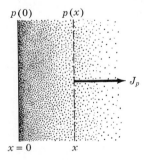

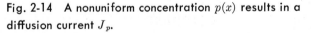

Fig. 2-14 A nonuniform concentration $p(x)$ results in a diffusion current J_p.

to the concentration gradient, and is given by

$$J_p = -qD_p \frac{dp}{dx} \tag{2-36}$$

where D_p (square meters per second) is called the *diffusion constant* for holes. Since p in Fig. 2-14 decreases with increasing x, then dp/dx is negative and the minus sign in Eq. (2-36) is needed, so that J_p will be positive in the positive X direction. A similar equation exists for diffusion electron-current density [p is replaced by n, and the minus sign is replaced by a plus sign in Eq. (2-36)].

Einstein Relationship Since both diffusion and mobility are statistical thermodynamic phenomena, D and μ are not independent. The relationship between them is given by the Einstein equation (Eq. 19-59)

$$\frac{D_p}{\mu_p} = \frac{D_n}{\mu_n} = V_T \tag{2-37}$$

where V_T is the "volt-equivalent of temperature," defined by

$$V_T \equiv \frac{\bar{k}T}{q} = \frac{T}{11,600} \tag{2-38}$$

where $\bar{k}$ is the Boltzmann constant in joules per degree Kelvin. Note the distinction between $\bar{k}$ and k; the latter is the Boltzmann constant in electron volts per degree Kelvin. (Numerical values of $\bar{k}$ and k are given in Appendix A. From Sec. 1-3 it follows that $\bar{k} = 1.60 \times 10^{-19}k$.) At room temperature (300°K), $V_T = 0.026$ V, and $\mu = 39D$. Measured values of μ and computed values of D for silicon and germanium are given in Table 2-1.

Total Current It is possible for both a potential gradient and a concentration gradient to exist simultaneously within a semiconductor. In such a situation the total hole current is the sum of the drift current [Eq. (2-7), with n replaced by p] and the diffusion current [Eq. (2-36)], or

$$J_p = q\mu_p p \mathcal{E} - qD_p \frac{dp}{dx} \tag{2-39}$$

Similarly, the net electron current is

$$J_n = q\mu_n n \mathcal{E} + qD_n \frac{dn}{dx} \tag{2-40}$$

2-10   THE CONTINUITY EQUATION

In Sec. 2-8 it was seen that if we disturb the equilibrium concentrations of carriers in a semiconductor material, the concentration of holes or electrons (which is constant throughout the crystal) will vary with time. In the general

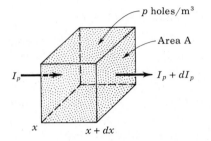

Fig. 2-15 Relating to the conservation of charge.

case, however, the carrier concentration in the body of a semiconductor is a function of both time and distance. We now derive the differential equation which governs this functional relationship. This equation is based on the fact that charge can be neither created nor destroyed, and hence is an extension of Eq. (2-31).

Consider the infinitesimal element of volume of area A and length dx (Fig. 2-15) within which the average hole concentration is p. Assume that the problem is one-dimensional and that the hole current I_p is a function of x. If, as indicated in Fig. 2-15, the current entering the volume at x is I_p at time t and leaving at $x + dx$ is $I_p + dI_p$ at the same time t, there must be dI_p more coulombs per second leaving the volume than entering it (for a positive value of dI_p). Hence the *decrease* in number of coulombs per second within the volume is dI_p. Since the magnitude of the carrier charge is q, then dI_p/q equals the decrease in the number of holes per second within the elemental volume $A\,dx$. Remembering that the current density $J_p = I_p/A$, we have

$$\frac{1}{qA}\frac{dI_p}{dx} = \frac{1}{q}\frac{dJ_p}{dx} = decrease \text{ in hole concentration (holes per unit volume) per second, due to current } I_p \quad (2\text{-}41)$$

From Eq. (2-30) we know that there is an *increase* per second of $g = p_o/\tau_p$ holes per unit volume due to thermal generation, and from Eq. (2-29) a *decrease* per second of p/τ_p holes per unit volume because of recombination. Since charge can neither be created nor destroyed, the *increase* in holes per unit volume per second, dp/dt, must equal the algebraic sum of all the increases listed above, or

$$\frac{\partial p}{\partial t} = \frac{p_o - p}{\tau_p} - \frac{1}{q}\frac{\partial J_p}{\partial x} \quad (2\text{-}42)$$

(Since both p and J_p are functions of both t and x, then partial derivatives are used in this equation.)

The continuity equation is applied to a specific physical problem in the following section and is discussed further in Sec. 19-9. Equation (2-42) is called the *law of conservation of charge*, or the *continuity equation* for charge. This law applies equally well for electrons, and the corresponding equation is obtained by replacing p by n in Eq. (2-42).

2-11   INJECTED MINORITY–CARRIER CHARGE

Consider the physical situation pictured[7] in Fig. 2-16a.   A long semiconductor bar is doped uniformly with donor atoms so that the concentration $n = N_D$ is independent of position.   Radiation falls upon the end of the bar at $x = 0$. Some of the photons are captured by the bound electrons in the covalent bonds near the illuminated surface.   As a result of this energy transfer, these bonds are broken and hole-electron *pairs* are generated.   Let us investigate how the steady-state minority-carrier concentration p varies with the distance x into the specimen.

We shall make the reasonable assumption that the injected minority concentration is very small compared with the doping level; that is, $p' \ll n$. The statement that the minority concentration is much smaller than the majority concentration is called the *low-level injection* condition.   Since the drift current is proportional to the concentration [Eq. (2-16)] and since $p = p' + p_o \ll n$, we shall neglect the *hole* drift current (but not the *electron* drift current) and shall assume that I_p is due entirely to diffusion.   This assumption is justified at the end of this section.   The controlling differential equation for p is

$$\frac{d^2p}{dx^2} = \frac{p - p_o}{D_p \tau_p} \qquad (2\text{-}43)$$

This equation is obtained by substituting Eq. (2-36) for the diffusion current into the equation of continuity [Eq. (2-42)] and setting $dp/dt = 0$ for steady-state operation.   Defining the *diffusion length* for holes L_p by

$$L_p \equiv (D_p \tau_p)^{\frac{1}{2}} \qquad (2\text{-}44)$$

Fig. 2-16   (c) Light falls upon the end of a long semiconductor bar.   This excitation causes hole-electron pairs to be injected at $x = 0$.   (b) The hole (minority) concentration $p(x)$ in the bar as a function of distance x from the end of the specimen.   The injected concentration is $p'(x) = p(x) - p_o$. The radiation injects $p'(0)$ carriers/m³ into the bar at $x = 0$.   [Not drawn to scale since $p'(0) \gg p_o$.]

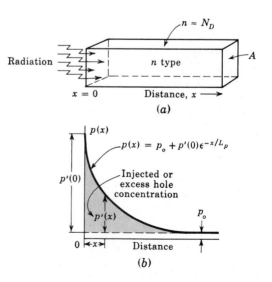

(a)

$p(x) = p_o + p'(0)\epsilon^{-x/L_p}$

Injected or excess hole concentration

(b)

the differential equation for the injected hole concentration $p' = p - p_o$ becomes

$$\frac{d^2p'}{dx^2} = \frac{p'}{L_p{}^2} \qquad (2\text{-}45)$$

The solution of this equation is

$$p'(x) = K_1\epsilon^{-x/L_p} + K_2\epsilon^{+x/L_p} \qquad (2\text{-}46)$$

where K_1 and K_2 are constants of integration. Consider a very long piece of semiconductor extending from $x = 0$ in the positive X direction. Since the concentration cannot become infinite as $x \to \infty$, then K_2 must be zero. We shall assume that at $x = 0$ the injected concentration is $p'(0)$. To satisfy this boundary condition, $K_1 = p'(0)$. Hence

$$p'(x) = p'(0)\epsilon^{-x/L_p} = p(x) - p_o \qquad (2\text{-}47)$$

The hole concentration decreases exponentially with distance, as indicated in Fig. 2-16b. We see that the diffusion length L_p represents the distance into the semiconductor at which the injected concentration falls to $1/\epsilon$ of its value at $x = 0$. In Sec. 19-9 it is demonstrated that L_p also represents the average distance that an injected hole travels before recombining with an electron.

Diffusion Currents The minority (hole) diffusion current is $I_p = AJ_p$, where A is the cross section of the bar. From Eqs. (2-36) and (2-47)

$$I_p(x) = \frac{AqD_p p'(0)}{L_p}\,\epsilon^{-x/L_p} = \frac{AqD_p}{L_p}\,[p(0) - p_o]\epsilon^{-x/L_p} \qquad (2\text{-}48)$$

This current falls exponentially with distance in the same manner that the minority-carrier concentration decreases. This result is used to find the current in a semiconductor diode (Sec. 3-3).

The majority (electron) diffusion current is $AqD_n\, dn/dx$. Assuming that electrical neutrality is preserved under low-level injection, then $n' = p'$, or

$$n - n_o = p - p_o \qquad (2\text{-}49)$$

Since the thermal-equilibrium concentrations n_o and p_o are independent of the position x, then

$$\frac{dn}{dx} = \frac{dp}{dx} \qquad (2\text{-}50)$$

Hence the electron diffusion current is

$$AqD_n \frac{dn}{dx} = AqD_n \frac{dp}{dx} = -\frac{D_n}{D_p} I_p \qquad (2\text{-}51)$$

where $I_p = -AqD_p\, dp/dx$ = the hole diffusion current. The dependence of the diffusion current upon x is given in Eq. (2-48). The magnitude of the ratio of majority to minority diffusion current is $D_n/D_p \sim 2$ for germanium and ~ 3 for silicon (Table 2-1).

Drift Currents Since Fig. 2-16a represents an open-circuited bar, the resultant current (the sum of hole and electron currents) must be zero everywhere. Hence a majority (electron) drift current I_{nd} must exist such that

$$I_p + \left(I_{nd} - \frac{D_n I_p}{D_p} \right) = 0 \quad (2\text{-}52)$$

or

$$I_{nd} = \left(\frac{D_n}{D_p} - 1 \right) I_p \quad (2\text{-}53)$$

From Eq. (2-48) we see that the electron drift current also decreases exponentially with distance.

It is important to point out that *an electric field ε must exist in the bar in order for a drift current to exist.* This field is created internally by the injected carriers. From Eqs. (2-7) and (2-53)

$$\varepsilon = \frac{1}{A q n \mu_n} \left(\frac{D_n}{D_p} - 1 \right) I_p \quad (2\text{-}54)$$

The results obtained in this section are based on the assumption that the hole drift current I_{pd} is zero. Using Eq. (2-7), with n replaced by p, the next approximation for this current is

$$I_{pd} = A q p \mu_p \varepsilon = \frac{p}{n} \frac{\mu_p}{\mu_n} \left(\frac{D_n}{D_p} - 1 \right) I_p \quad (2\text{-}55)$$

Since $p \ll n$, then $I_{pd} \ll I_p$. The hole drift current is negligible compared with the hole diffusion current, thus justifying the assumption that the injected minority-carrier current, under low-level injection, is essentially a diffusion current.

2-12   THE POTENTIAL VARIATION WITHIN A GRADED SEMICONDUCTOR

Consider a semiconductor (Fig. 2-17a) where the hole concentration p is a function of x; that is, the doping is *nonuniform*, or graded.[7] Assume a steady-state situation and zero excitation; that is, no carriers are injected into the specimen from any *external* source. With no excitation there can be no *steady* movement of charge in the bar, although the carriers possess random motion due to thermal agitation. Hence the total hole current must be zero. (Also, the total electron current must be zero.) Since p is not constant, we expect a nonzero hole diffusion current. In order for the total hole current to vanish there must exist a hole drift current which is equal and opposite to the diffusion current. However, a conduction current requires an electric field,

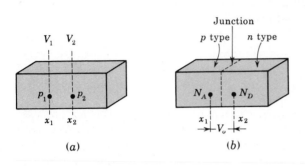

Fig. 2-17 (a) A graded semiconductor: $p(x)$ is not constant. (b) One portion is doped uniformly with acceptor ions and the other section is doped uniformly with donor ions so that a metallurgical junction is formed. A contact potential V_o appears across this step-graded p-n junction.

and hence we conclude that, as a result of the nonuniform doping, an electric field is generated within the semiconductor. We shall now find this field and the corresponding potential variation throughout the bar.

Setting $J_p = 0$ in Eq. (2-39) and using the Einstein relationship $D_p = \mu_p V_T$ [Eq. (2-37)], we obtain

$$\varepsilon = \frac{V_T}{p}\frac{dp}{dx} \tag{2-56}$$

If the doping concentration $p(x)$ is known, this equation allows the built-in field $\varepsilon(x)$ to be calculated. From $\varepsilon = -dV/dx$ we can calculate the potential variation. Thus

$$dV = -V_T\frac{dp}{p} \tag{2-57}$$

If this equation is integrated between x_1, where the concentration is p_1 and the potential is V_1 (Fig. 2-17a), and x_2, where $p = p_2$ and $V = V_2$, the result is

$$V_{21} \equiv V_2 - V_1 = V_T \ln\frac{p_1}{p_2} \tag{2-58}$$

Note that the potential difference between two points depends only upon the concentrations at these two points and is independent of their separation $x_2 - x_1$. Equation (2-58) may be put in the form

$$p_1 = p_2 \epsilon^{V_{21}/V_T} \tag{2-59}$$

This is the Boltzmann relationship of kinetic gas theory.

Mass-action Law Starting with $J_n = 0$ and proceeding as above, the Boltzmann equation for electrons is obtained.

$$n_1 = n_2 \epsilon^{-V_{21}/V_T} \tag{2-60}$$

Multiplying Eqs. (2-59) and (2-60) gives

$$n_1 p_1 = n_2 p_2 \tag{2-61}$$

This equation states that the product np is a constant independent of x, and hence of the amount of doping, under thermal equilibrium. For an intrinsic semiconductor, $n = p = n_i$, and hence

$$np = n_i{}^2 \tag{2-10}$$

which is the law of mass action introduced in Sec. 2-3. An alternative proof is given in Sec. 19-5.

An Open-circuited Step-graded Junction   Consider the special case indicated in Fig. 2-17b.   The left half of the bar is p-type with a constant concentration N_A, whereas the right-half is n-type with a uniform density N_D. The dashed plane is a metallurgical (p-n) junction separating the two sections with different concentration.   This type of doping, where the density changes abruptly from p- to n-type, is called *step grading*.   The step-graded junction is located at the plane where the concentration is zero.   The above theory indicates that there is a built-in potential between these two sections (called the *contact difference of potential V_o*).   Equation (2-58) allows us to calculate V_o.   Thus

$$V_o = V_{21} = V_T \ln \frac{p_{po}}{p_{no}} \tag{2-62}$$

because $p_1 = p_{po} =$ thermal-equilibrium hole concentration in p side and $p_2 = p_{no} =$ thermal-equilibrium hole concentration in n side. From Eq. (2-15) $p_{po} = N_A$, and from Eq. (2-14) $p_{no} = n_i{}^2/N_D$, so that

$$V_o = V_T \ln \frac{N_A N_D}{n_i{}^2} \tag{2-63}$$

The same expression for V_o is obtained from an analysis corresponding to that given above and based upon equating the total electron current I_n to zero (Prob. 2-20).   The p-n junction, both open-circuited and with an applied voltage, is studied in detail in Chaps. 3 and 19.

2-13   RECAPITULATION

The fundamental principles governing the electrical behavior of semi-conductors, discussed in this chapter, are summarized as follows:

1. Two types of mobile charge carriers (positive holes and negative electrons) are available.   This bipolar nature of a semiconductor is to be contrasted with the unipolar property of a metal, which possesses only free electrons.

2. A semiconductor may be fabricated with donor (acceptor) impurities; so it contains mobile charges which are primarily electrons (holes).

3. The intrinsic concentration of carriers is a function of temperature. At room temperature, essentially all donors or acceptors are ionized.

4. Current is due to two distinct phenomena:

a. Carriers drift in an electric field (this conduction current is also available in a metal).

b. Carriers diffuse if a concentration gradient exists (a phenomenon which does not take place in a metal).

5. Carriers are continuously being generated (due to thermal creation of hole-electron pairs) and are simultaneously disappearing (due to recombination).

6. The fundamental law governing the flow of charge is called the *continuity equation*. It is formulated by considering that charge can neither be created nor destroyed if generation, recombination, drift, and diffusion are all taken into account.

7. If minority carriers (say, holes) are injected into a region containing majority carriers (say, an *n*-type bar), then usually the injected minority concentration is very small compared with the density of the majority carriers. For this low-level injection condition the minority current is predominantly due to diffusion; in other words, the *minority* drift current may be neglected.

8. The total majority-carrier flow is the sum of a drift and a diffusion current. The majority conduction current results from a small electric field internally created within the semiconductor because of the injected carriers.

9. The minority-carrier concentration injected into one end of a semiconductor bar decreases exponentially with distance into the specimen (as a result of diffusion and recombination).

10. Across an open-circuited *p-n* junction there exists a contact difference of potential.

These basic concepts are applied in the next chapter to the study of the *p-n* junction diode.

REFERENCES

1. Shockley, W.: "Electrons and Holes in Semiconductors," D. Van Nostrand Company, Inc., Princeton, N.J., reprinted February, 1963.

2. Adler, R. B., A. C. Smith, and R. L. Longini: "Introduction to Semiconductor Physics," vol.1, Semiconductor Electronics Education Committee, John Wiley & Sons, Inc., New York, 1964.

3. Morin, F. J., and J. P. Maita: Conductivity and Hall Effect in the Intrinsic Range of Germanium, *Phys. Rev.*, vol. 94, pp. 1525–1529, June, 1954.

   Morin, F. J., and J. P. Maita: Electrical Properties of Silicon Containing Arsenic and Boron, *Phys. Rev.*, vol. 96, pp. 28–35, October, 1954.

4. Sze, S. M.: "Physics of Semiconductor Devices," Fig. 29, p. 59, John Wiley & Sons, Inc., New York, 1969.

5. Shockley, W., and W. T. Read, Jr.: Statistics of the Recombination of Holes and Electrons, *Phys. Rev.*, vol. 87, pp. 835–842, September, 1952.

Hall, R. N.: Electron-Hole Recombination in Germanium, *Phys. Rev.*, vol. 87, p. 387, July, 1952.

6. Collins, C. B., R. O. Carlson, and C. J. Gallagher: Properties of Gold-doped Silicon, *Phys. Rev.*, vol. 105, pp. 1168–1173, February, 1957.

Bemski, G.: Recombination Properties of Gold in Silicon, *Phys. Rev.*, vol. 111, pp. 1515–1518, September, 1958.

7. Gray, P. E., and C. L. Searle: "Electronic Principles: Physics, Models, and Circuits," John Wiley & Sons, Inc., New York 1969.

REVIEW QUESTIONS

2-1 Give the electron-gas description of a metal.

2-2 (*a*) Define *mobility*. (*b*) Give its dimensions.

2-3 (*a*) Define *conductivity*. (*b*) Give its dimensions.

2-4 Define a *hole* (in a semiconductor).

2-5 Indicate pictorially how a hole contributes to conduction.

2-6 (*a*) Define *intrinsic concentration* of holes. (*b*) What is the relationship between this density and the intrinsic concentration for electrons? (*c*) What do these equal at 0°K?

2-7 Show (in two dimensions) the crystal structure of silicon containing a donor impurity atom.

2-8 Repeat Rev. 2-7 for an acceptor impurity atom.

2-9 Define (*a*) *donor*, (*b*) *acceptor* impurities.

2-10 A semiconductor is doped with both donors and acceptors of concentrations N_D and N_A, respectively. Write the equation or equations from which to determine the electron and hole concentrations (n and p).

2-11 Define the *volt equivalent of temperature*.

2-12 Describe the *Hall effect*.

2-13 What properties of a semiconductor are determined from a Hall effect experiment?

2-14 Given an intrinsic semiconductor specimen, state two physical processes for increasing its conductivity. Explain briefly.

2-15 Is the temperature coefficient of resistance of a semiconductor positive or negative? Explain briefly.

2-16 Answer Rev. 2-15 for a metal.

2-17 (*a*) Sketch the spectral response curve for silicon. (*b*) Explain its shape qualitatively.

2-18 (*a*) Define *long-wavelength, threshold,* or *critical wavelength* for a semiconductor. (*b*) Explain why λ_c exists.

2-19 Define *mean lifetime* of a carrier.

2-20 Explain physically the meaning of the following statement: An electron and a hole recombine and disappear.

2-21   Radiation falls on a semiconductor specimen which is uniformly illuminated, and a steady-state is reached.   At $t = 0$ the light is turned off.   (a) Sketch the minority-carrier concentration as a function of time for $t \geq 0$.   (b) Define all symbols in the equation describing your sketch.

2-22   (a) Define *diffusion constant* for holes.   (b) Give its dimensions.

2-23   Repeat Rev. 2-22 for electrons.

2-24   (a) Write the equation for the net electron current in a semiconductor. What is the physical significance of each term?   (b) How is this equation modified for a metal?

2-25   (a) The *equation of continuity* is a mathematical statement of what physical law?   (b) The left-hand side of this equation for holes is dp/dt.   The right-hand side contains several terms.   State in words (no mathematics) what each of these terms represents physically.

2-26   Light falls upon the end of a long open-circuited semiconductor specimen. (a) Sketch the steady-state minority-carrier concentration as a function of distance. (b) Define all the symbols in the equation describing your sketch.

2-27   Light falls upon the end of a long open-circuited semiconductor bar.   (a) For low-level injection is the minority current due predominantly to drift, diffusion, or both?   (b) Is the majority current due predominantly to drift, diffusion, or both?

2-28   (a) Define a *graded semiconductor*.   (b) Explain why an electric field must exist in a graded semiconductor.

2-29   Consider a step-graded junction under open-circuited conditions.   Upon what four parameters does the contact difference of potential depend?

2-30   State the *mass-action law* as an equation and in words.

2-31   Explain why a contact difference of potential must develop across an open-circuited *p-n* junction.

3 / JUNCTION-DIODE CHARACTERISTICS

In this chapter we demonstrate that if a junction is formed between a sample of p-type and one of n-type semiconductor, this combination possesses the properties of a rectifier. The volt-ampere characteristics of such a two-terminal device (called a *junction diode*) is studied. The capacitance across the junction is calculated.

Although the transistor is a triode (three-terminal) semiconductor, it may be considered as one diode biased by the current from a second diode. Hence most of the theory developed here is utilized in Chap. 5 in connection with the study of the transistor.

3-1   THE OPEN–CIRCUITED p-n JUNCTION

If donor impurities are introduced into one side and acceptors into the other side of a single crystal of a semiconductor, a p-n junction is formed, as in Fig. 2-17b. Such a system is illustrated in more schematic detail in Fig. 3-1a. The donor ion is represented by a plus sign because, after this impurity atom "donates" an electron, it becomes a positive ion. The acceptor ion is indicated by a minus sign because, after this atom "accepts" an electron, it becomes a negative ion. Initially, there are nominally only p-type carriers to the left of the junction and only n-type carriers to the right.

Space-charge Region   Because there is a density gradient across the junction, holes will initially diffuse to the right across the junction, and electrons to the left. We see that the positive holes which neutralized the acceptor ions near the junction in the p-type silicon have disappeared as a result of combination with electrons which have diffused across the junction. Similarly, the neutralizing electrons in

49

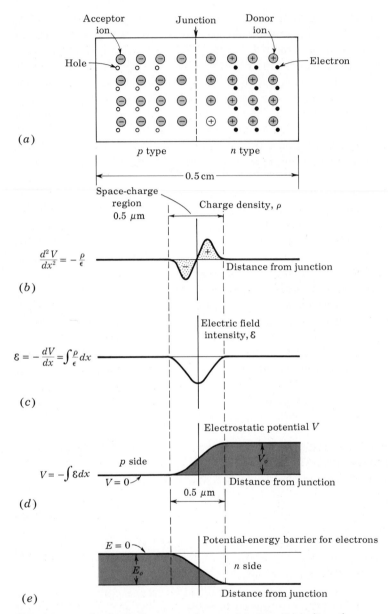

Fig. 3-1 A schematic diagram of a *p-n* junction, including the charge density, electric field intensity, and potential-energy barriers at the junction. Since potential energy = potential × charge, the curve in (d) is proportional to the potential energy for a hole (a positive charge) and the curve in (e) is proportional to the negative of that in (d) (an electron is a negative charge). (Not drawn to scale.)

the n-type silicon have combined with holes which have crossed the junction from the p material. The unneutralized ions in the neighborhood of the junction are referred to as *uncovered charges*. The general shape of the charge density ρ (Fig. 3-1b) depends upon how the diode is doped (a step-graded junction is considered in detail in Sec. 3-7). Since the region of the junction is depleted of mobile charges, it is called the *depletion region*, the *space-charge region*, or the *transition region*. The thickness of this region is of the order of the wavelength of visible light (0.5 micron = 0.5 μm). Within this very narrow space-charge layer there are no mobile carriers. To the left of this region the carrier concentration is $p \approx N_A$, and to its right it is $n \approx N_D$.

Electric Field Intensity The space-charge density ρ is zero at the junction. It is positive to the right and negative to the left of the junction. This distribution constitutes an electrical dipole layer, giving rise to electric lines of flux from right to left, corresponding to negative field intensity $\mathcal{E}$ as depicted in Fig. 3-1c. Equilibrium is established when the field is strong enough to restrain the process of diffusion. Stated alternatively, under steady-state conditions the drift hole (electron) current must be equal and opposite to the diffusion hole (electron) current so that the net hole (electron) current is reduced to zero—as it must be for an open-circuited device. In other words, there is no steady-state movement of charge across the junction.

The field intensity curve is proportional to the integral of the charge density curve. This statement follows from Poisson's equation

$$\frac{d^2V}{dx^2} = -\frac{\rho}{\epsilon} \tag{3-1}$$

where ϵ is the permittivity. If ϵ_r is the (relative) dielectric constant and ϵ_o is the permittivity of free space (Appendix A), then $\epsilon = \epsilon_r \epsilon_o$. Integrating Eq. (3-1) and remembering that $\mathcal{E} = -dV/dx$ gives

$$\mathcal{E} = \int_{x_o}^{x} \frac{\rho}{\epsilon} \, dx \tag{3-2}$$

where $\mathcal{E} = 0$ at $x = x_o$. Therefore the curve plotted in Fig. 3-1c is the integral of the function drawn in Fig. 3-1b (divided by ϵ).

Potential The electrostatic-potential variation in the depletion region is shown in Fig. 3-1d, and is the negative integral of the function $\mathcal{E}$ of Fig. 3-1c. This variation constitutes a potential-energy barrier (Sec. 1-2) against the further diffusion of holes across the barrier. The form of the potential-energy barrier against the flow of electrons from the n side across the junction is shown in Fig. 3-1e. It is similar to that shown in Fig. 3-1d, except that it is inverted, since the charge on an electron is negative. Note the existence, across the depletion layer, of the *contact potential* V_o, discussed in Sec. 2-12.

Summary Under open-circuited conditions the net hole current must be zero. If this statement were not true, the hole density at one end of the semiconductor would continue to increase indefinitely with time, a situation which is obviously physically impossible. Since the concentration of holes in the p side is much greater than that in the n side, a very large hole diffusion current tends to flow across the junction from the p to the n material. Hence an electric field must build up across the junction in such a direction that a hole drift current will tend to flow across the junction from the n to the p side in order to counterbalance the diffusion current. This equilibrium condition of zero resultant hole current allows us to calculate the height of the potential barrier V_o [Eq. (2-63)] in terms of the donor and acceptor concentrations. The numerical value for V_o is of the order of magnitude of a few tenths of a volt.

3-2   THE p-n JUNCTION AS A RECTIFIER[1]

The essential electrical characteristic of a p-n junction is that it constitutes a rectifier which permits the easy flow of charge in one direction but restrains the flow in the opposite direction. We consider now, qualitatively, how this diode rectifier action comes about.

Reverse Bias In Fig. 3-2, a battery is shown connected across the terminals of a p-n junction. The negative terminal of the battery is connected to the p side of the junction, and the positive terminal to the n side. The polarity of connection is such as to cause both the holes in the p type and the electrons in the n type to move away from the junction. Consequently, the region of negative-charge density is spread to the left of the junction (Fig. 3-1b), and the positive-charge-density region is spread to the right. However, this process cannot continue indefinitely, because in order to have a steady flow of holes to the left, these holes must be supplied across the junction from the n-type silicon. And there are very few holes in the n-type side. Hence, nominally, zero current results. Actually, a small current does flow because a small number of hole-electron pairs are generated throughout the crystal as a result of thermal energy. The holes so formed in the n-type silicon will wander over to the junction. A similar remark applies to the electrons

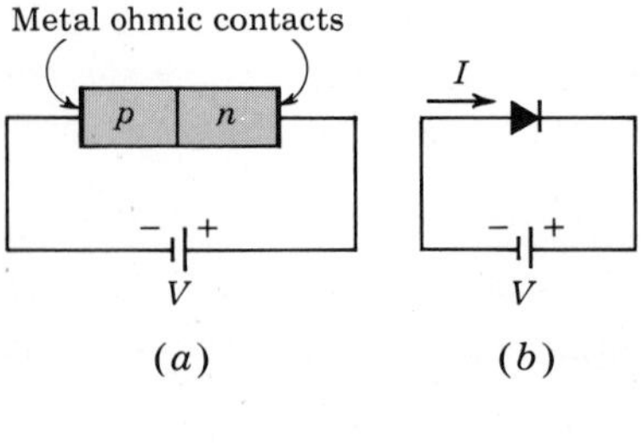

Fig. 3-2 (a) A p-n junction biased in the reverse direction. (b) The rectifier symbol is used for the p-n diode.

Fig. 3-3 (a) A *p-n* junction biased in the forward direction. (b) The rectifier symbol is used for the *p-n* diode.

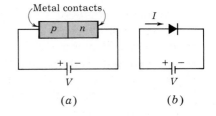

(a)   (b)

thermally generated in the *p*-type silicon. This small current is the diode *reverse saturation current*, and its magnitude is designated by I_o. This reverse current will increase with increasing temperature [Eq. (3-11)], and hence the back resistance of a crystal diode decreases with increasing temperature. From the argument presented here, I_o should be independent of the magnitude of the reverse bias.

The mechanism of conduction in the reverse direction may be described alternatively in the following way: When no voltage is applied to the *p-n* diode, the potential barrier across the junction is as shown in Fig. 3-1*d*. When a voltage V is applied to the diode in the direction shown in Fig. 3-2, the height of the potential-energy barrier is increased by the amount qV. This increase in the barrier height serves to reduce the flow of majority carriers (i.e., holes in *p* type and electrons in *n* type). However, the minority carriers (i.e., electrons in *p* type and holes in *n* type), since they fall down the potential-energy hill, are uninfluenced by the increased height of the barrier. The applied voltage in the direction indicated in Fig. 3-2 is called the *reverse*, or *blocking, bias*.

Forward Bias An external voltage applied with the polarity shown in Fig. 3-3 (opposite to that indicated in Fig. 3-2) is called a *forward* bias. An ideal *p-n* diode has zero ohmic voltage drop across the body of the crystal. For such a diode the height of the potential barrier at the junction will be lowered by the applied forward voltage V. The equilibrium initially established between the forces tending to produce diffusion of majority carriers and the restraining influence of the potential-energy barrier at the junction will be disturbed. Hence, for a forward bias, the holes cross the junction from the *p*-type into the *n*-type region, where they constitute an injected minority current. Similarly, the electrons cross the junction in the reverse direction and become a minority current injected into the *p* side. Holes traveling from left to right constitute a current in the *same* direction as electrons moving from right to left. Hence the resultant current crossing the junction is the *sum* of the hole and electron minority currents. A detailed discussion of the several current components within the diode is given in the next section.

Ohmic Contacts In Fig. 3-2 (3-3) we show an external reverse (forward) bias applied to a *p-n* diode. We have assumed that the external bias voltage appears directly across the junction and has the effect of raising (lowering) the electrostatic potential across the junction. To justify this assumption we must specify how electric contact is made to the semiconductor from the external bias circuit. In Figs. 3-2 and 3-3 we indicate metal contacts with which the homogeneous p-type and n-type materials are provided. We thus see that we have introduced two metal-semiconductor junctions, one at each end of the diode. We naturally expect a contact potential to develop across these additional junctions. However, we shall assume that the metal-semiconductor contacts shown in Figs. 3-2 and 3-3 have been manufactured in such a way that they are nonrectifying. In other words, the contact potential across these junctions is constant, independent of the direction and magnitude of the current. A contact of this type is referred to as an *ohmic contact.*

We are now in a position to justify our assumption that the entire applied voltage appears as a *change* in the height of the potential barrier. Inasmuch as the voltage across the metal-semiconductor ohmic contacts remains constant and the voltage drop across the bulk of the crystal is neglected, approximately the entire applied voltage will indeed appear as a change in the height of the potential barrier at the p-n junction.

The Short-circuited and Open-circuited p-n Junction If the voltage V in Fig. 3-2 or 3-3 were set equal to zero, the p-n junction would be short-circuited. Under these conditions, as we show below, no current can flow ($I = 0$) and the electrostatic potential V_o remains unchanged and equal to the value under open-circuit conditions. If there were a current ($I \neq 0$), the metal would become heated. Since there is no external source of energy available, the energy required to heat the metal wire would have to be supplied by the p-n bar. The semiconductor bar, therefore, would have to cool off. Clearly, under thermal equilibrium the simultaneous heating of the metal and cooling of the bar is impossible, and we conclude that $I = 0$. Since under short-circuit conditions the sum of the voltages around the closed loop must be zero, the junction potential V_o must be exactly compensated by the metal-to-semiconductor contact potentials at the ohmic contacts. Since the current is zero, the wire can be cut without changing the situation, and the voltage drop across the cut must remain zero. If in an attempt to measure V_o we connected a voltmeter across the cut, the voltmeter would read zero voltage. In other words, it is not possible to measure contact difference of potential directly with a voltmeter.

Large Forward Voltages Suppose that the forward voltage V in Fig. 3-3 is increased until V approaches V_o. If V were equal to V_o, the barrier would disappear and the current could be arbitrarily large, exceeding the rating of the diode. As a practical matter we can never reduce the barrier to zero because, as the current increases without limit, the bulk resistance of the crystal, as well as the resistance of the ohmic contacts, will limit the current.   Therefore it is no longer possible to assume that all the voltage V appears as a change across the p-n junction.   We conclude that, as the forward voltage V becomes comparable with V_o, the current through a real p-n diode will be governed by the ohmic-contact resistances and the crystal bulk resistance.   Thus the volt-ampere characteristic becomes approximately a straight line.

3-3   THE CURRENT COMPONENTS IN a p-n DIODE

In the preceding section it was indicated that when a forward bias is applied to a diode, holes are injected into the n side and electrons into the p side. In Sec. 2-11 it was emphasized that under low-level injection conditions such minority currents are due almost entirely to diffusion, so that minority drift currents may be neglected.   From Eq. (2-48) the hole diffusion current in the n-type material† I_{pn} decreases exponentially with distance x into the n-type region and falls to $1/\epsilon$th its peak value in a diffusion length L_p.   This current is plotted in Fig. 3-4, as is also the corresponding electron diffusion current I_{np} in the p-type side.   The doping on the two sides of the junction need not be identical, and it is assumed in this plot that the acceptor concentration is much greater than the donor density, so that the hole current greatly exceeds

---

† Since we must consider both hole and electron currents in each side of the junction, we add the second subscript n to I_p in order to indicate that the hole current in the n-type region is under consideration.   In general, if the letters p and n both appear in a symbol, the first letter refers to the type of carrier and the second to the type of material.

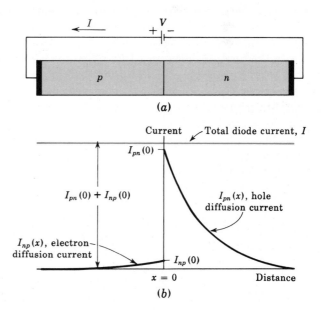

Fig. 3-4 The hole- and electron-current diffusion components vs. distance in a p-n junction diode. The p side is much more heavily doped than the n section. The space-charge region at the junction is assumed to be negligibly small.

the electron current. Also, in Fig. 3-4, the space-charge region has been assumed to be negligibly small, a restriction to be removed soon.

From Eq. (2-48) (with the subscript n added to I_p and to p) the minority (hole) diffusion current at the junction ($x = 0$) is given by

$$I_{pn}(0) = \frac{AqD_p}{L_p} [p_n(0) - p_{no}] \tag{3-3}$$

The Law of the Junction   In the preceding section it was pointed out that a forward bias V lowers the barrier height and allows more carriers to cross the junction. Hence $p_n(0)$ must be a function of V. From the Boltzmann relationship, Eq. (2-59), it seems reasonable that $p_n(0)$ should depend exponentially, upon V. Indeed, in Sec. 19-10, it is found that

$$p_n(0) = p_{no}\epsilon^{V/V_T} \tag{3-4}$$

This relationship, which gives the hole concentration at the edge of the n region (at $x = 0$, just outside of the transition region) in terms of the thermal-equilibrium minority-carrier concentration p_{no} (far away from the junction) and the applied potential V, is called the *law of the junction*. A similar equation with p and n interchanged gives the electron concentration at the edge of the p region in terms of V.

The Total Diode Current   Substituting Eq. (3-4) into Eq. (3-3) yields

$$I_{pn}(0) = \frac{AqD_p p_{no}}{L_p} (\epsilon^{V/V_T} - 1) \tag{3-5}$$

The expression for the electron current $I_{np}(0)$ crossing the junction into the p side is obtained from Eq. (3-5) by interchanging p and n.

Electrons crossing the junction at $x = 0$ from right to left constitute a current in the same direction as holes crossing the junction from left to right. Hence the total diode current I at $x = 0$ is

$$I = I_{pn}(0) + I_{np}(0) \tag{3-6}$$

Since the current is the same throughout a series circuit, I is independent of x and is indicated as a horizontal line in Fig. 3-4b. The expression for the diode current is

$$I = I_o(\epsilon^{V/V_T} - 1) \tag{3-7}$$

where I_o is given in Prob. 3-6 in terms of the physical parameters of the diode.

The Reverse Saturation Current   In the foregoing discussion a positive value of V indicates a forward bias. The derivation of Eq. (3-7) is equally valid if V is negative, signifying an applied reverse-bias voltage. For a reverse bias whose magnitude is large compared with V_T (≈ 26 mV at room temperature), $I \to -I_o$. Hence I_o is called the *reverse saturation current*.

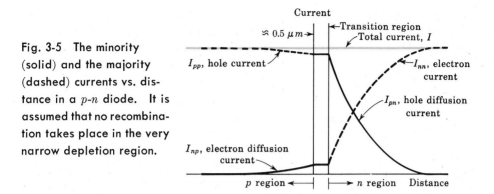

Fig. 3-5 The minority (solid) and the majority (dashed) currents vs. distance in a p-n diode. It is assumed that no recombination takes place in the very narrow depletion region.

Since the thermal-equilibrium concentrations p_{no} and n_{po} depend upon temperature T, then I_o is a function of T. This temperature dependence is derived in Sec. 19-10.

The Majority-carrier Current Components In the n-type region of Fig. 3-4b the total current I is constant and the minority (hole) current I_{pn} varies with x. Clearly, there must exist a majority (electron) current I_{nn} which is a function of x because the diode current I at any position is the sum of the hole and electron currents at this distance. This majority current

$$I_{nn}(x) = I - I_{pn}(x) \tag{3-8}$$

is plotted as the dashed curve in the n region of Fig. 3-5, where the narrow depletion region is also indicated. The majority hole current I_{pp} is similarly shown dashed in the p region of this figure. As discussed in Sec. 2-11, these majority currents are each composed of two current components; one is a drift current and the second is a diffusion current. Recall from Sec. 2-11 that the diffusion electron current is $-(D_n/D_p)I_{pn}$ in the n side.

Note that deep into the p side the current is a drift (conduction) current I_{pp} of holes sustained by the small electric field in the semiconductor (Prob. 3-4). As the holes approach the junction, some of them recombine with the electrons, which are injected into the p side from the n side. The current I_{pp} thus decreases toward the junction (at just the proper gradient to maintain the total current constant, independent of distance). What remains of I_{pp} at the junction enters the n side and becomes the hole diffusion current I_{pn}. Similar remarks can be made with respect to current I_{nn}.

We emphasize that the current in a p-n diode is bipolar in character since it is made up of both positive and negative carriers of electricity. The total current is constant throughout the device, but the proportion due to holes and that due to electrons varies with distance, as indicated in Fig. 3-5.

The Transition Region Since this depletion layer contains very few mobile charges, it has been assumed (Fig. 3-5) that carrier generation and recombination may be neglected within the bulk and on the surface of this region. Such an assumption is valid for a germanium diode, but not for a silicon device. For the latter it is found[2] that Eq. (3-5) must be modified by multiplying V_T by a factor η, where $\eta \approx 2$ for small (rated) currents and $\eta \approx 1$ for large currents.

3-4   THE VOLT–AMPERE CHARACTERISTIC

The discussion of the preceding section indicates that, for a p-n junction, the current I is related to the voltage V by the equation

$$I = I_o(\epsilon^{V/\eta V_T} - 1) \tag{3-9}$$

A positive value of I means that current flows from the p to the n side. The diode is forward-biased if V is positive, indicating that the p side of the junction is positive with respect to the n side. The symbol η is unity for germanium and is approximately 2 for silicon at rated current.

The symbol V_T stands for the volt equivalent of temperature, and is given by Eq. (2-38), repeated here for convenience:

$$V_T \equiv \frac{T}{11,600} \tag{3-10}$$

At room temperature ($T = 300°K$), $V_T = 0.026$ V $= 26$ mV.

The form of the volt-ampere characteristic described by Eq. (3-9) is shown in Fig. 3-6a. When the voltage V is positive and several times V_T,

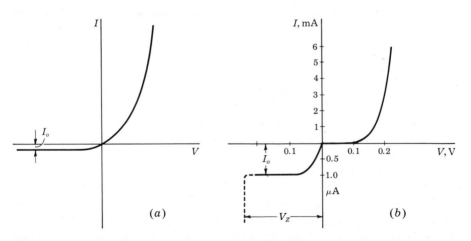

(a) (b)

Fig. 3-6   (a) The volt-ampere characteristic of an ideal p-n diode.   (b) The volt-ampere characteristic for a germanium diode redrawn to show the order of magnitude of currents.   Note the expanded scale for reverse currents.   The dashed portion indicates breakdown at V_Z.

the unity in the parentheses of Eq. (3-9) may be neglected. Accordingly, except for a small range in the neighborhood of the origin, the current increases exponentially with voltage. When the diode is reverse-biased and $|V|$ is several times V_T, $I \approx -I_o$. The reverse current is therefore constant, independent of the applied reverse bias. Consequently, I_o is referred to as the *reverse saturation current.*

For the sake of clarity, the current I_o in Fig. 3-6 has been greatly exaggerated in magnitude. Ordinarily, the range of forward currents over which a diode is operated is many orders of magnitude larger than the reverse saturation current. To display forward and reverse characteristics conveniently, it is necessary, as in Fig. 3-6b, to use two different current scales. The volt-ampere characteristic shown in that figure has a forward-current scale in milliamperes and a reverse scale in microamperes.

The dashed portion of the curve of Fig. 3-6b indicates that, at a reverse-biasing voltage V_Z, the diode characteristic exhibits an abrupt and marked departure from Eq. (3-9). At this critical voltage a large reverse current flows, and the diode is said to be in the *breakdown* region, discussed in Sec. 3-11.

The Cutin Voltage V_γ Both silicon and germanium diodes are commercially available. A number of differences between these two types are relevant in circuit design. The difference in volt-ampere characteristics is brought out in Fig. 3-7. Here are plotted the forward characteristics at room temperature of a general-purpose germanium switching diode and a general-purpose silicon diode, the 1N270 and 1N3605, respectively. The diodes have comparable current ratings. A noteworthy feature in Fig. 3-7 is that there exists a *cutin, offset, break-point,* or *threshold*, voltage V_γ below which the current is very small (say, less than 1 percent of maximum rated value). Beyond

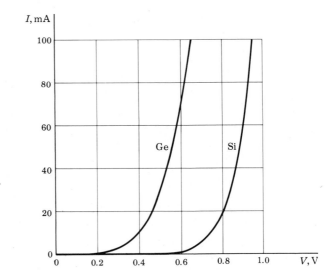

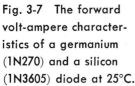

Fig. 3-7 The forward volt-ampere characteristics of a germanium (1N270) and a silicon (1N3605) diode at 25°C.

V_γ the current rises very rapidly.   From Fig. 3-7 we see that V_γ is approximately 0.2 V for germanium and 0.6 V for silicon.

Note that the break in the silicon-diode characteristic is offset about 0.4 V with respect to the break in the germanium-diode characteristic.   The reason for this difference is to be found, in part, in the fact that the reverse saturation current in a germanium diode is normally larger by a factor of about 1,000 than the reverse saturation current in a silicon diode of comparable ratings. I_o is in the range of microamperes for a germanium diode and nanoamperes for a silicon diode at room temperature.

Since $\eta = 2$ for small currents in silicon, the current increases as $\epsilon^{V/2V_T}$ for the first several tenths of a volt and increases as ϵ^{V/V_T} only at higher voltages. This initial smaller dependence of the current on voltage accounts for the further delay in the rise of the silicon characteristic.

Logarithmic Characteristic   It is instructive to examine the family of curves for the silicon diodes shown in Fig. 3-8.   A family for a germanium diode of comparable current rating is quite similar, with the exception that corresponding currents are attained at lower voltage.

From Eq. (3-9), assuming that V is several times V_T, so that we may drop the unity, we have $\log I = \log I_o + 0.434V/\eta V_T$.   We therefore expect in Fig. 3-8, where $\log I$ is plotted against V, that the plots will be straight lines.   We do indeed find that at low currents the plots are linear and correspond to $\eta \approx 2$.   At large currents an increment of voltage does not yield as large an increase of current as at low currents.   The reason for this behavior is to be found in the ohmic resistance of the diode.   At low currents the ohmic drop is negligible and the externally impressed voltage simply decreases the potential barrier at the p-n junction.   At high currents the externally impressed voltage is called upon principally to establish an electric field to

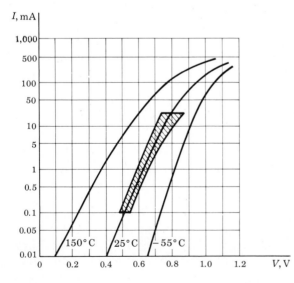

Fig. 3-8 Volt-ampere characteristics at three different temperatures for a silicon diode (planar epitaxial passivated types 1N3605, 1N3606, 1N3608, and 1N3609). The shaded area indicates 25°C limits of controlled conductance. Note that the vertical scale is logarithmic and encompasses a current range of 50,000. (Courtesy of General Electric Company.)

overcome the ohmic resistance of the semiconductor material. Therefore, at high currents, the diode behaves more like a resistor than a diode, and the current increases linearly rather than exponentially with applied voltage.

Reverse Saturation Current   Many commercially available diodes exhibit an essentially constant value of I_o for negative values of V, as indicated in Fig. 3-6. On the other hand, some diodes show a very pronounced increase in reverse current with increasing reverse voltage. This variation in I_o results from leakage across the surface of the diode, and also from the additional fact that new charge carriers may be generated by collision in the transition region at the junction.

3-5   THE TEMPERATURE DEPENDENCE OF THE V/I CHARACTERISTIC

The volt-ampere relationship (3-9) contains the temperature implicitly in the two symbols V_T and I_o. In Sec. 19-11 it is shown that the theoretical variation of I_o with T is 8 percent/°C for silicon and 11 percent/°C for germanium. The performance of commercial diodes is only approximately consistent with these results. The reason for the discrepancy is that, in a physical diode, there is a component of the reverse saturation current due to leakage over the surface that is not taken into account in Sec. 19-11. Since this leakage component is independent of temperature, we may expect to find a smaller rate of change of I_o with temperature than that predicted above. From experimental data we observe that the reverse saturation current increases approximately 7 percent/°C for both silicon and germanium. Since $(1.07)^{10} \approx 2.0$, we conclude that *the reverse saturation current approximately doubles for every 10°C rise in temperature.* If $I_o = I_{o1}$ at $T = T_1$, then at a temperature T, I_o is given by

$$I_o(T) = I_{o1} \times 2^{(T-T_1)/10} \tag{3-11}$$

If the temperature is increased at a fixed voltage, the current increases. However, if we now reduce V, then I may be brought back to its previous value. In Sec. 19-11 it is found that for either silicon or germanium (*at room temperature*)

$$\frac{dV}{dT} \approx -2.5 \text{ mV/°C} \tag{3-12}$$

in order to maintain a constant value of I. It should also be noted that $|dV/dT|$ decreases with increasing T.

3-6   DIODE RESISTANCE

The static resistance R of a diode is defined as the ratio V/I of the voltage to the current. At any point on the volt-ampere characteristic of the diode

(Fig. 3-7), the resistance R is equal to the reciprocal of the slope of a line joining the operating point to the origin. The static resistance varies widely with V and I and is not a useful parameter. The rectification property of a diode is indicated on the manufacturer's specification sheet by giving the maximum forward voltage V_F required to attain a given forward current I_F and also the maximum reverse current I_R at a given reverse voltage V_R. Typical values for a silicon planar epitaxial diode are $V_F = 0.8$ V at $I_F = 10$ mA (corresponding to $R_F = 80\ \Omega$) and $I_R = 0.1\ \mu$A at $V_R = 50$ V (corresponding to $R_R = 500$ M).

For small-signal operation the *dynamic*, or *incremental*, *resistance* r is an important parameter, and is defined as the reciprocal of the slope of the volt-ampere characteristic, $r \equiv dV/dI$. The dynamic resistance is not a constant, but depends upon the operating voltage. For example, for a semiconductor diode, we find from Eq. (3-9) that the dynamic conductance $g \equiv 1/r$ is

$$g \equiv \frac{dI}{dV} = \frac{I_o \epsilon^{V/\eta V_T}}{\eta V_T} = \frac{I + I_o}{\eta V_T} \tag{3-13}$$

For a reverse bias greater than a few tenths of a volt (so that $|V/\eta V_T| \gg 1$), g is extremely small and r is very large. On the other hand, for a forward bias greater than a few tenths of a volt, $I \gg I_o$, and r is given approximately by

$$r \approx \frac{\eta V_T}{I} \tag{3-14}$$

The dynamic resistance varies inversely with current; at room temperature and for $\eta = 1$, $r = 26/I$, where I is in milliamperes and r in ohms. For a forward current of 26 mA, the dynamic resistance is 1 Ω. The ohmic body resistance of the semiconductor may be of the same order of magnitude or even much higher than this value. Although r varies with current, in a small-signal model, it is reasonable to use the parameter r as a constant.

A Piecewise Linear Diode Characteristic   A large-signal approximation which often leads to a sufficiently accurate engineering solution is the *piecewise linear* representation. For example, the piecewise linear approximation for a semiconductor diode characteristic is indicated in Fig. 3-9. The break point is not at the origin, and hence V_γ is also called the *offset*, or *threshold*, *voltage*. The diode behaves like an open circuit if $V < V_\gamma$, and has a constant incremental resistance $r = dV/dI$ if $V > V_\gamma$. Note that the resistance r (also designated as R_f and called the *forward resistance*) takes on added physical significance even for this large-signal model, whereas the static resistance $R_F = V/I$ is not constant and is not useful.

The numerical values V_γ and R_f to be used depend upon the type of diode and the contemplated voltage and current swings. For example, from Fig. 3-7 we find that, for a current swing from cutoff to 10 mA with a germanium diode, reasonable values are $V_\gamma = 0.2$ V and $R_f = 20\ \Omega$, and for a silicon diode, $V_\gamma = 0.6$ V and $R_f = 15\ \Omega$. On the other hand, a better approximation

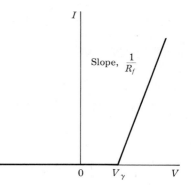

Fig. 3-9 The piecewise linear characterization of a semiconductor diode.

for current swings up to 50 mA leads to the following values; germanium, $V_\gamma = 0.3$ V, $R_f = 6 \ \Omega$; silicon, $V_\gamma = 0.65$ V, $R_f = 5.5 \ \Omega$. For an avalanche diode, discussed in Sec. 3-11, $V_\gamma = V_Z$ and R_f is the dynamic resistance in the breakdown region.

3-7 SPACE–CHARGE, OR TRANSITION, CAPACITANCE C_T

As mentioned in Sec. 3-1, a reverse bias causes majority carriers to move away from the junction, thereby uncovering more immobile charges. Hence the thickness of the space-charge layer at the junction increases with reverse voltage. This increase in uncovered charge with applied voltage may be considered a capacitive effect. We may define an incremental capacitance C_T by

$$C_T = \left| \frac{dQ}{dV} \right| \tag{3-15}$$

where dQ is the increase in charge caused by a change dV in voltage. It follows from this definition that a change in voltage dV in a time dt will result in a current $i = dQ/dt$, given by

$$i = C_T \frac{dV}{dt} \tag{3-16}$$

Therefore a knowledge of C_T is important in considering a diode (or a transistor) as a circuit element. The quantity C_T is referred to as the *transition-region, space-charge, barrier,* or *depletion-region, capacitance.* We now consider C_T quantitatively. As it turns out, this capacitance is not a constant, but depends upon the magnitude of the reverse voltage. It is for this reason that C_T is defined by Eq. (3-16) rather than as the ratio Q/V.

A Step-graded Junction Consider a junction in which there is an abrupt change from acceptor ions on one side to donor ions on the other side. Such a junction is formed experimentally, for example, by placing indium, which is trivalent, against n-type germanium and heating the combination to a high

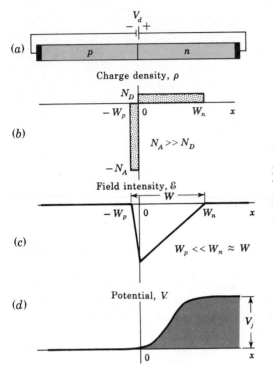

Fig. 3-10 (a) A reverse-biased p-n step-graded junction. (b) The charge density. (c) The field intensity. (d) The potential variation with distance x.

temperature for a short time. Some of the indium dissolves into the germanium to change the germanium from n to p type at the junction. Such a step-graded junction is called an *alloy*, or *fusion*, *junction*. A step-graded junction is also formed between emitter and base of an integrated transistor (Fig. 7-11). It is not necessary that the concentration N_A of acceptor ions equal the concentration N_D of donor impurities. As a matter of fact, it is often advantageous to have an unsymmetrical junction. Figure 3-10 shows the charge density as a function of distance from an alloy junction in which the acceptor impurity density is assumed to be much larger than the donor concentration. Since the net charge must be zero, then

$$N_A W_p = N_D W_n \tag{3-17}$$

If $N_A \gg N_D$, then $W_p \ll W_n \approx W$. The relationship between potential and charge density is given by Eq. (3-1):

$$\frac{d^2 V}{dx^2} = \frac{-q N_D}{\epsilon} \tag{3-18}$$

The electric lines of flux start on the positive donor ions and terminate on the negative acceptor ions. Hence there are no flux lines to the right of the boundary $x = W_n$ in Fig. 3-10, and $\mathcal{E} = -dV/dx = 0$ at $x = W_n \approx W$.

Integrating Eq. (3-18) subject to this boundary condition yields

$$\frac{dV}{dx} = \frac{-qN_D}{\epsilon}(x - W) = -\mathcal{E} \tag{3-19}$$

Neglecting the small potential drop across W_p, we may arbitrarily choose $V = 0$ at $x = 0$. Integrating Eq. (3-19) subject to this condition gives

$$V = \frac{-qN_D}{2\epsilon}(x^2 - 2Wx) \tag{3-20}$$

The linear variation in field intensity and the quadratic dependence of potential upon distance are plotted in Fig. 3-10c and d. These graphs should be compared with the corresponding curves of Fig. 3-1.

At $x = W$, $V = V_j =$ junction, or barrier, potential. Thus

$$V_j = \frac{qN_DW^2}{2\epsilon} \tag{3-21}$$

In this section we have used the symbol V to represent the potential at any distance x from the junction. Hence, let us introduce V_d as the externally applied diode voltage. Since the barrier potential represents a reverse voltage, it is lowered by an applied forward voltage. Thus

$$V_j = V_o - V_d$$

where V_d is a negative number for an applied *reverse* bias and V_o is the contact potential (Fig. 3-1d). This equation confirms our qualitative conclusion that the thickness of the depletion layer increases with applied reverse voltage. We now see that W varies as $V_j^{\frac{1}{2}} = (V_o - V_d)^{\frac{1}{2}}$.

If A is the area of the junction, the charge in the distance W is

$$Q = qN_DWA$$

The transition capacitance C_T, given by Eq. (3-15), is

$$C_T = \left|\frac{dQ}{dV_d}\right| = qN_DA\left|\frac{dW}{dV_j}\right| \tag{3-22}$$

From Eq. (3-21), $|dW/dV_j| = \epsilon/qN_DW$, and hence

$$C_T = \frac{\epsilon A}{W} \tag{3-23}$$

It is interesting to note that this formula is exactly the expression which is obtained for a parallel-plate capacitor of area A (square meters) and plate separation W (meters) containing a material of permittivity ϵ. If the concentration N_A is not neglected, the above results are modified only slightly. In Eq. (3-21) W represents the total space-charge width, and $1/N_D$ is replaced by $1/N_A + 1/N_D$. Equation (3-23) remains valid (Prob. 3-18).

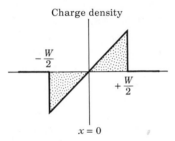

Charge density

$-\dfrac{W}{2}$

$+\dfrac{W}{2}$

$x = 0$

Fig. 3-11   The charge-density variation vs. distance at a linearly graded p-n junction.

A Linearly Graded Junction   A second form of junction is obtained by drawing a single crystal from a melt of germanium whose type is changed during the drawing process by adding first p-type and then n-type impurities. A linearly graded junction is also formed between the collector and base of an integrated transistor (Fig. 7-12). For such a junction the charge density varies gradually (almost linearly), as indicated in Fig. 3-11. If an analysis similar to that given above is carried out for such a junction, Eq. (3-23) is found to be valid where W equals the total width of the space-charge layer. However, it now turns out that W varies as $V_j^{\frac{1}{3}}$ instead of $V_j^{\frac{1}{2}}$.

Varactor Diodes   We observe from the above equations that the barrier capacitance is not a constant but varies with applied voltage. The larger the reverse voltage, the larger is the space-charge width W, and hence the smaller the capacitance C_T. The variation is illustrated for two typical diodes in Fig. 3-12. Similarly, for an increase in forward bias (V_d positive), W decreases and C_T increases.

The voltage-variable capacitance of a p-n junction biased in the reverse direction is useful in a number of circuits. One of these applications is voltage tuning of an LC resonant circuit. Other applications include self-balancing bridge circuits and special types of amplifiers, called *parametric amplifiers*.

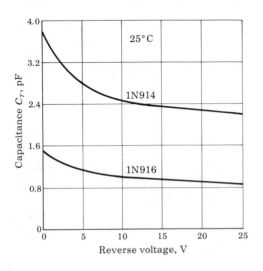

Fig. 3-12   Typical barrier-capacitance variation, with reverse voltage, of silicon diodes 1N914 and 1N916. (Courtesy of Fairchild Semiconductor Corporation.)

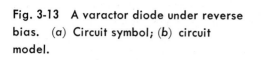

(a)

Fig. 3-13 A varactor diode under reverse
bias. (a) Circuit symbol; (b) circuit
model.

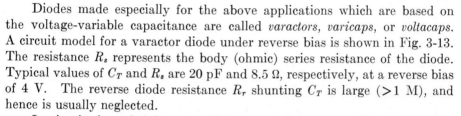

(b)

Diodes made especially for the above applications which are based on
the voltage-variable capacitance are called *varactors, varicaps,* or *voltacaps.*
A circuit model for a varactor diode under reverse bias is shown in Fig. 3-13.
The resistance R_s represents the body (ohmic) series resistance of the diode.
Typical values of C_T and R_s are 20 pF and 8.5 Ω, respectively, at a reverse bias
of 4 V. The reverse diode resistance R_r shunting C_T is large (>1 M), and
hence is usually neglected.

In circuits intended for use with fast waveforms or at high frequencies,
it is required that the transition capacitance be as small as possible, for the
following reason: a diode is driven to the reverse-biased condition when it is
desired to prevent the transmission of a signal. However, if the barrier
capacitance C_T is large enough, the current which is to be restrained by the
low conductance of the reverse-biased diode will flow through the capacitor
(Fig. 3-13b).

3-8   CHARGE–CONTROL DESCRIPTION OF A DIODE

If the bias is in the forward direction, the potential barrier at the junction
is lowered and holes from the p side enter the n side. Similarly, electrons
from the n side move into the p side. This process of *minority-carrier injection*
is discussed in Sec. 2-11. The excess hole density falls off exponentially with
distance, as indicated in Fig. 3-14a. The shaded area under this curve is
proportional to the injected charge.

For simplicity of discussion we assume that one side of the diode, say,
the p material, is so heavily doped in comparison with the n side that the current I is carried across the junction entirely by holes moving from the p to
the n side, or $I = I_{pn}(0)$. The excess minority charge Q will then exist only
on the n side, and is given by the shaded area in the n region of Fig. 3-14a
multiplied by the diode cross section A and the electronic charge q. Hence,
from Eq. (2-47),

$$Q = \int_0^\infty Aqp'(0)\epsilon^{-x/L_p}\,dx = AqL_pp'(0) \qquad (3\text{-}24)$$

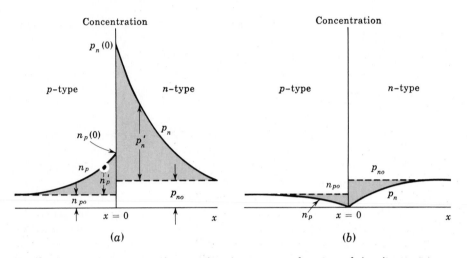

Fig. 3-14 Minority-carrier density distribution as a function of the distance x from a junction. (a) A forward-biased junction; (b) a reverse-biased junction. The excess hole (electron) density $p'_n = p_n - p_{no}$ ($n'_p = n_p - n_{po}$) is positive in (a) and negative in (b). (The transition region is assumed to be so small relative to the diffusion length that it is not indicated in this figure.)

The hole current I is given by $I_p(x)$ in Eq. (2-48), with $x = 0$, or

$$I = \frac{AqD_p p'(0)}{L_p} \tag{3-25}$$

Eliminating $p'(0)$ from Eqs. (3-24) and (3-25) yields

$$I = \frac{Q}{\tau} \tag{3-26}$$

where $\tau \equiv L_p^2/D_p \equiv \tau_p$ = mean lifetime for holes [Eq. (2-44)].

Equation (3-26) is an important relationship, referred to as the *charge-control description of a diode*. It states that the diode current (which consists of holes crossing the junction from the p to the n side) is proportional to the stored charge Q of excess minority carriers. The factor of proportionality is the reciprocal of the decay time constant (the mean lifetime τ) of the minority carriers. Thus, in the steady state, *the current I supplies minority carriers at the rate at which these carriers are disappearing because of the process of recombination.*

Charge Storage under Reverse Bias When an external voltage reverse-biases the junction, the steady-state density of minority carriers is as shown in Fig. 3-14b. Far from the junction the minority carriers are equal to their thermal-equilibrium values p_{no} and n_{po}, as is also the situation in Fig. 3-14a. As the minority carriers approach the junction they are rapidly swept across, and the density of minority carriers diminishes to zero at this junction. This result follows from the law of the junction, Eq. (3-4), since the concentration $p_n(0)$ reduces to zero for a negative junction potential V.

The injected charge under reverse bias is given by the shaded area in Fig. 3-14b. This charge is negative since it represents less charge than is available under conditions of thermal equilibrium with no applied voltage. From Eq. (3-26) with Q negative, the diode current I is negative and, of course, equals the reverse saturation current I_o in magnitude.

The charge-control characterization of a diode describes the device in terms of the current I and the stored charge Q, whereas the equivalent-circuit characterization uses the current I and the junction voltage V. One immediately apparent advantage of this charge-control description is that the exponential relationship between I and V is replaced by the linear dependence of I on Q. The charge Q also makes a simple parameter, the sign of which determines whether the diode is forward- or reverse-biased. The diode is forward-biased if Q is positive and reverse-biased if Q is negative.

3-9   DIFFUSION CAPACITANCE

For a forward bias a capacitance which is much larger than the transition capacitance C_T considered in Sec. 3-8 comes into play. The origin of this larger capacitance lies in the injected charge stored near the junction outside the transition region (Fig. 3-14a). It is convenient to introduce an incremental capacitance, defined as the rate of change of injected charge with voltage, called the *diffusion*, or *storage*, *capacitance* C_D.

Static Derivation of C_D We now make a quantitative study of C_D. From Eqs. (3-26) and (3-13)

$$C_D \equiv \frac{dQ}{dV} = \tau \frac{dI}{dV} = \tau g = \frac{\tau}{r} \tag{3-27}$$

where the diode incremental conductance $g \equiv dI/dV$. Substituting the expression for the diode incremental resistance $r = 1/g$ given in Eq. (3-14) into Eq. (3-27) yields

$$C_D = \frac{\tau I}{\eta V_T} \tag{3-28}$$

We see that *the diffusion capacitance is proportional to the current I*. In the derivation above we have assumed that the diode current I is due to holes only. If this assumption is not satisfied, Eq. (3-27) gives the diffusion capacitance C_{D_p} due to holes only, and a similar expression can be obtained for the diffusion capacitance C_{D_n} due to electrons. The total diffusion capacitance can then be obtained as the sum of C_{D_p} and C_{D_n} (Prob. 3-24).

For a reverse bias, g is very small and C_D may be neglected compared with C_T. For a forward current, on the other hand, C_D is usually much larger than C_T. For example, for germanium ($\eta = 1$) at $I = 26$ mA, $g = 1$ ℧, and $C_D = \tau$. If, say, $\tau = 20$ μs, then $C_D = 20$ μF, a value which is about a million times larger than the transition capacitance.

Despite the large value of C_D, the time constant rC_D (which is of importance in circuit applications) may not be excessive because the dynamic forward resistance $r = 1/g$ is small. From Eq. (3-27),

$$rC_D = \tau \tag{3-29}$$

Hence the diode time constant equals the mean lifetime of minority carriers, which lies in range of nanoseconds to hundreds of microseconds. The importance of τ in circuit applications is considered in the following section.

Diffusion Capacitance for an Arbitrary Input   Consider the buildup of injected charge across a junction as a function of time when the potential is changed. In Fig. 3-15, the curve marked (1) is the steady-state value of p_n' for an applied voltage V. If the voltage at time t is increased by dV in the interval dt, then p_n' changes to that indicated by the curve marked (2) at time $t + dt$. The increase in charge dQ' in the time dt is proportional to the heavily shaded area in Fig. 3-15. Note that the concentration near the junction has increased markedly, whereas p_n far away from the junction has changed very little, because it takes time for the holes to diffuse into the n region. If the applied voltage is maintained constant at $V + dV$, the stored charge will continue to increase. Finally, at $t = \infty$, p_n follows the curve marked (3) [which is given by Eq. (3-4), with V replaced by $V + dV$]. The steady-state injected charge dQ, due to the increase in voltage by dV, is proportional to the total shaded area in Fig. 3-15. Clearly, $dQ - dQ'$ is represented by the lightly shaded area, and hence $dQ > dQ'$.

Since dQ' is the charge injected across the junction in time dt, the current is given by

$$i = \frac{dQ'}{dt} = C_D' \frac{dV}{dt} \tag{3-30}$$

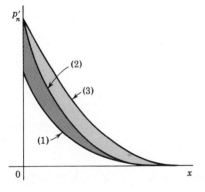

Fig. 3-15 The transient buildup of stored excess charge. The curve marked (1) gives the steady-state value of p_n' at a time t when the voltage is V. If the voltage increases by dV and held at $V + dV$, then p_n' is given by (2) at time $t + dt$ and by (3) at $t = \infty$.

where we define the *small-signal diffusion capacitance* C'_D by $C'_D \equiv dQ'/dV$. Note that the diode current is *not* given by the steady-state charge Q or static capacitance C_D.

$$i \neq \frac{dQ}{dt} \qquad \text{or} \qquad i \neq C_D \frac{dV}{dt} \tag{3-31}$$

Since $dQ' < dQ$, then $C'_D < C_D$.

From the above argument we conclude that the dynamic diffusion capacitance C'_D depends upon how the input voltage varies with time. To find C'_D, the equation of continuity must be solved for the given voltage waveform. This equation controls how p_n varies both as a function of x and t, and from $p_n(x, t)$ we can obtain the current. If the input varies with time in an arbitrary way, it may not be possible to define the diffusion capacitance in a unique manner.

Diffusion Capacitance for a Sinusoidal Input   For the special case where the excitation varies sinusoidally with time, C'_D may be obtained from a solution of the equation of continuity. This analysis is carried out in Sec. 19-12, and C'_D is found to be a function of frequency. At low frequencies

$$C'_D = \tfrac{1}{2}\tau g \qquad \text{if } \omega\tau \ll 1 \tag{3-32}$$

which is half the value found in Eq. (3-27), based upon static considerations. For high frequencies, C'_D decreases with increasing frequency and is given by

$$C'_D = \left(\frac{\tau}{2\omega}\right)^{\frac{1}{2}} g \qquad \text{if } \omega\tau \gg 1 \tag{3-33}$$

3-10   JUNCTION–DIODE SWITCHING TIMES

When a diode is driven from the reversed condition to the forward state or in the opposite direction, the diode response is accompanied by a transient, and an interval of time elapses before the diode recovers to its steady state. The forward recovery time t_{fr} is the time difference between the 10 percent point of the diode voltage and the time when this voltage reaches and remains within 10 percent of its final value. It turns out that t_{fr} does not usually constitute a serious practical problem, and hence we here consider only the more important situation of reverse recovery.

Diode Reverse Recovery Time   When an external voltage forward-biases a p-n junction, the steady-state density of minority carriers is as shown in Fig. 3-14a. The number of minority carriers is very large. These minority carriers have, in each case, been supplied from the other side of the junction, where, being majority carriers, they are in plentiful supply.

If the external voltage is suddenly reversed in a diode circuit which has

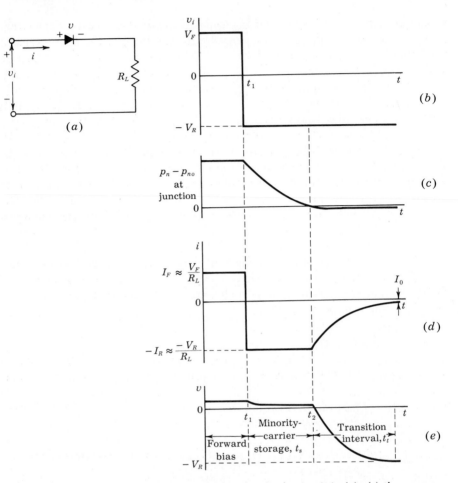

Fig. 3-16   The waveform in (b) is applied to the diode circuit in (a); (c) the excess carrier density at the junction; (d) the diode current; (e) the diode voltage.

been carrying current in the forward direction, the diode current will not immediately fall to its steady-state reverse-voltage value. For the current cannot attain its steady-state value until the minority-carrier distribution, which at the moment of voltage reversal had the form in Fig. 3-14a, reduces to the distribution in Fig. 3-14b. Until such time as the *injected*, or *excess*, *minority-carrier density* $p_n - p_{no}$ (or $n_p - n_{po}$) has dropped nominally to zero, the diode will continue to conduct easily, and the current will be determined by the external resistance in the diode circuit.

Storage and Transition Times   The sequence of events which accompanies the reverse biasing of a conducting diode is indicated in Fig. 3-16. We consider that the voltage in Fig. 3-16b is applied to the diode-resistor circuit in Fig. 3-16a. For a long time, and up to the time t_1, the voltage $v_i = V_F$ has been in the direction to forward-bias the diode. The resistance R_L is assumed large enough so that the drop across R_L is large in comparison with the drop across the diode. Then the current is $i \approx V_F/R_L \equiv I_F$. At the time $t = t_1$ the input voltage reverses abruptly to the value $v = -V_R$. For the reasons described above, the current does not drop to zero, but instead reverses and remains at the value $i \approx -V_R/R_L \equiv -I_R$ until the time $t = t_2$. At $t = t_2$, as is seen in Fig. 3-16c, the minority-carrier density p_n at $x = 0$ has reached its equilibrium state p_{no}. If the diode ohmic resistance is R_d, then at the time t_1 the diode voltage falls slightly [by $(I_F + I_R)R_d$] but does not reverse. At $t = t_2$, when the excess minority carriers in the immediate neighborhood of the junction have been swept back across the junction, the diode voltage begins to reverse and the magnitude of the diode current begins to decrease. The interval t_1 to t_2, for the stored-minority charge to become zero, is called the *storage time* t_s.

The time which elapses between t_2 and the time when the diode has nominally recovered is called the *transition time* t_t. This recovery interval will be completed when the minority carriers which are at some distance from the junction have diffused to the junction and crossed it and when, in addition, the junction transition capacitance across the reverse-biased junction has charged through R_L to the voltage $-V_R$.

Manufacturers normally specify the reverse recovery time of a diode t_{rr} in a typical operating condition in terms of the current waveform of Fig. 3-16d. The time t_{rr} is the interval from the current reversal at $t = t_1$ until the diode has recovered to a specified extent in terms either of the diode current or of the diode resistance. If the specified value of R_L is larger than several hundred ohms, ordinarily the manufacturers will specify the capacitance C_L shunting R_L in the measuring circuit which is used to determine t_{rr}. Thus we find, for the Fairchild 1N3071, that with $I_F = 30$ mA and $I_R = 30$ mA, the time required for the reverse current to fall to 1.0 mA is 50 nsec. Again we find, for the same diode, that with $I_F = 30$ mA, $-V_R = -35$ V, $R_L = 2$ K, and $C_L = 10$ pF $(-I_R = -35/2 = -17.5$ mA), the time required for the diode to recover to the extent that its resistance becomes 400 K is $t_{rr} = 400$ nsec. Commercial switching-type diodes are available with times t_{rr} in the range from less than a nanosecond up to as high as 1 μs in diodes intended for switching large currents.

3-11   BREAKDOWN DIODES[3]

The reverse-voltage characteristic of a semiconductor diode, including the breakdown region, is redrawn in Fig. 3-17a. Diodes which are designed with adequate power-dissipation capabilities to operate in the breakdown region may be employed as voltage-reference or constant-voltage devices. Such diodes are known as *avalanche*, *breakdown*, or *Zener diodes*. They are used

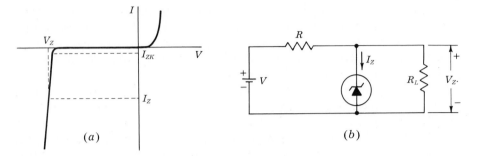

Fig. 3-17   (a) The volt-ampere characteristic of an avalanche, or Zener, diode.
(b) A circuit in which such a diode is used to regulate the voltage across R_L against
changes due to variations in load current and supply voltage.

characteristically in the manner indicated in Fig. 3-17b.   The source V and
resistor R are selected so that, initially, the diode is operating in the breakdown region.   Here the diode voltage, which is also the voltage across the
load R_L, is V_Z, as in Fig. 3-17a, and the diode current is I_Z.   The diode will
now regulate the load voltage against variations in load current and against
variations in supply voltage V because, in the breakdown region, large changes
in diode current produce only small changes in diode voltage.   Moreover, as
load current or supply voltage changes, the diode current will accommodate
itself to these changes to maintain a nearly constant load voltage.   The diode
will continue to regulate until the circuit operation requires the diode current
to fall to I_{ZK}, in the neighborhood of the knee of the diode volt-ampere curve.
The upper limit on diode current is determined by the power-dissipation rating
of the diode.

Avalanche Multiplication   Two mechanisms of diode breakdown for
increasing reverse voltage are recognized.   Consider the following situation:
A thermally generated carrier (part of the reverse saturation current) falls
down the junction barrier and acquires energy from the applied potential.
This carrier collides with a crystal ion and imparts sufficient energy to disrupt
a covalent bond.   In addition to the original carrier, a new electron-hole pair
has now been generated.   These carriers may also pick up sufficient energy
from the applied field, collide with another crystal ion, and create still another
electron-hole pair.   Thus each new carrier may, in turn, produce additional
carriers through collision and the action of disrupting bonds.   This cumulative
process is referred to as *avalanche multiplication*.   It results in large reverse
currents, and the diode is said to be in the region of *avalanche breakdown*.

Zener Breakdown   Even if the initially available carriers do not acquire
sufficient energy to disrupt bonds, it is possible to initiate breakdown through
a direct rupture of the bonds.   Because of the existence of the electric field

at the junction, a sufficiently strong force may be exerted on a bound electron by the field to tear it out of its covalent bond. The new hole-electron pair which is created increases the reverse current. Note that this process, called *Zener breakdown*, does not involve collisions of carriers with the crystal ions (as does avalanche multiplication).

The field intensity ε increases as the impurity concentration increases, for a fixed applied voltage (Prob. 3-25). It is found that Zener breakdown occurs at a field of approximately 2×10^7 V/m. This value is reached at voltages below about 6 V for heavily doped diodes. For lightly doped diodes the breakdown voltage is higher, and avalanche multiplication is the predominant effect. Nevertheless, the term *Zener* is commonly used for the *avalanche*, or *breakdown*, *diode* even at higher voltages. Silicon diodes operated in avalanche breakdown are available with maintaining voltages from several volts to several hundred volts and with power ratings up to 50 W.

Temperature Characteristics A matter of interest in connection with Zener diodes, as with semiconductor devices generally, is their temperature sensitivity. The temperature coefficient is given as the percentage change in reference voltage per centigrade degree change in diode temperature. These data are supplied by the manufacturer. The coefficient may be either positive or negative and will normally be in the range ± 0.1 percent/°C. If the reference voltage is above 6 V, where the physical mechanism involved is avalanche multiplication, the temperature coefficient is positive. However, below 6 V, where true Zener breakdown is involved, the temperature coefficient is negative.

A qualitative explanation of the sign (positive or negative) of the temperature coefficient of V_Z is now given. A junction having a narrow depletion-layer width, and hence high field intensity, will break down by the Zener mechanism. An increase in temperature increases the energies of the valence electrons, and hence makes it easier for these electrons to escape from the covalent bonds. Less applied voltage is therefore required to pull these electrons from their positions in the crystal lattice and convert them into conduction electrons. Thus the Zener breakdown voltage decreases with temperature.

A junction with a broad depletion layer, and therefore a low field intensity, will break down by the avalanche mechanism. In this case we rely on intrinsic carriers to collide with valence electrons and create avalanche multiplication. As the temperature increases, the vibrational displacement of atoms in the crystal grows. This vibration increases the probability of collisions with the lattice atoms of the intrinsic particles as they cross the depletion width. The intrinsic holes and electrons thus have less of an opportunity to gain sufficient energy between collisions to start the avalanche process. Therefore the value of the avalanche voltage must increase with increased temperature.

Dynamic Resistance and Capacitance A matter of importance in connection with Zener diodes is the slope of the diode volt-ampere curve in the operating range. If the reciprocal slope $\Delta V_Z / \Delta I_Z$, called the *dynamic resistance*, is r, then a change ΔI_Z in the operating current of the diode produces a change $\Delta V_Z = r \, \Delta I_Z$ in the operating voltage. Ideally, $r = 0$, corresponding to a volt-ampere curve which, in the breakdown region, is precisely vertical. The variation of r at various currents for a series of avalanche diodes of fixed power-dissipation rating and various voltages show a rather broad minimum in the range 6 to 10 V. This minimum value of r is of the order of magnitude of a few ohms. However, for values of V_Z below 6 V or above 10 V, and particularly for small currents (~ 1 mA), r may be of the order of hundreds of ohms.

Some manufacturers specify the minimum current I_{ZK} (Fig. 3-17a) below which the diode should not be used. Since this current is on the knee of the above curve, where the dynamic resistance is large, then for currents lower than I_{ZK} the regulation will be poor. Some diodes exhibit a very sharp knee even down into the microampere region.

The capacitance across a breakdown diode is the transition capacitance, and hence varies inversely as some power of the voltage. Since C_T is proportional to the cross-sectional area of the diode, high-power avalanche diodes have very large capacitances. Values of C_T from 10 to 10,000 pF are common.

Additional Reference Diodes Zener diodes are available with voltages as low as about 2 V. Below this voltage it is customary, for reference and regulating purposes, to use diodes in the *forward* direction. As appears in Fig. 3-7, the volt-ampere characteristic of a forward-biased diode (sometimes called a *stabistor*) is not unlike the reverse characteristic, with the exception that, in the forward direction, the knee of the characteristic occurs at lower voltage. A number of forward-biased diodes may be operated in series to reach higher voltages. Such series combinations, packaged as single units, are available with voltages up to about 5 V, and may be preferred to reverse-biased Zener diodes, which at low voltages have very large values of dynamic resistance.

When it is important that a Zener diode operate with a low temperature coefficient, it may be feasible to operate an appropriate diode at a current where the temperature coefficient is at or near zero. Quite frequently, such operation is not convenient, particularly at higher voltages and when the diode must operate over a range of currents. Under these circumstances temperature-compensated avalanche diodes find application. Such diodes consist of a reverse-biased Zener diode with a positive temperature coefficient, combined in a single package with a forward-biased diode whose temperature coefficient is negative. As an example, the Transitron SV3176 silicon 8-V reference diode has a temperature coefficient of ± 0.001 percent/°C at 10 mA over the range -55 to $+100$°C. The dynamic resistance is only 1.5 Ω. The temperature coefficient remains below 0.002 percent/°C for currents in the range 8 to 12 mA. The voltage stability with time of some of these reference diodes is comparable with that of conventional standard cells.

When a high-voltage reference is required, it is usually advantageous (except of course with respect to economy) to use two or more diodes in series rather than a single diode. This combination will allow higher voltage, higher dissipation, lower temperature coefficient, and lower dynamic resistance.

3-12   THE TUNNEL DIODE

A *p-n* junction diode of the type discussed in Sec. 3-1 has an impurity concentration of about 1 part in 10^8. With this amount of doping, the width of the depletion layer, which constitutes a potential barrier at the junction, is of the order of a micron. This potential barrier restrains the flow of carriers from the side of the junction where they constitute majority carriers to the side where they constitute minority carriers. If the concentration of impurity atoms is greatly increased, say, to 1 part in 10^3 (corresponding to a density in excess of 10^{19} cm^{-3}), the device characteristics are completely changed. This new diode was announced in 1958 by Esaki,[4] who also gave the correct theoretical explanation for its volt-ampere characteristic.

The Tunneling Phenomenon The width of the junction barrier varies inversely as the square root of impurity concentration [Eq. (3-21)] and therefore is reduced to less than 100 Å (10^{-6} cm). This thickness is only about one-fiftieth the wavelength of visible light. Classically, a particle must have an energy at least equal to the height of a potential-energy barrier if it is to move from one side of the barrier to the other. However, for barriers as thin as those estimated above in the Esaki diode, the Schrödinger equation indicates that there is a large probability that an electron will penetrate *through* the barrier. This quantum-mechanical behavior is referred to as *tunneling*, and hence these high-impurity-density *p-n* junction devices are called tunnel *diodes*. The volt-ampere relationship is explained in Sec. 19-8 and is depicted in Fig. 3-18.

Characteristics of a Tunnel Diode[5] From Fig. 3-18 we see that the tunnel diode is an excellent conductor in the reverse direction (the *p* side of the junction negative with respect to the *n* side). Also, for small forward voltages (up to 50 mV for Ge), the resistance remains small (of the order of 5 Ω). At the *peak current* I_P corresponding to the voltage V_P, the slope dI/dV of the characteristic is zero. If V is increased beyond V_P, the current decreases. As a consequence, the dynamic conductance $g = dI/dV$ is negative. The tunnel diode exhibits a *negative-resistance characteristic* between the peak current I_P and the minimum value I_V, called the *valley current*. At the *valley voltage* V_V at which $I = I_V$, the conductance is again zero, and beyond this point the resistance becomes and remains positive. At the so-called *peak forward voltage* V_F the current again reaches the value I_P. For larger voltages the current increases beyond this value.

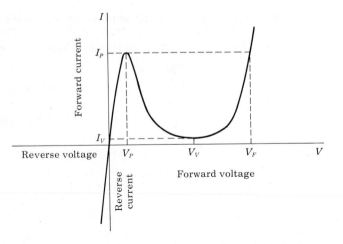

Fig. 3-18   Volt-ampere characteristic of a tunnel diode.

For currents whose values are between I_V and I_P, the curve is triple-valued, because each current can be obtained at three different applied voltages. It is this multivalued feature which makes the tunnel diode useful in pulse and digital circuitry.[6]

The standard circuit symbol for a tunnel diode is given in Fig. 3-19a. The small-signal model for operation in the negative-resistance region is indicated in Fig. 3-19b. The negative resistance $-R_n$ has a minimum at the point of inflection between I_P and I_V. The series resistance R_s is ohmic resistance. The series inductance L_s depends upon the lead length and the geometry of the dipole package. The junction capacitance C depends upon the bias and is usually measured at the valley point. Typical values for these parameters for a tunnel diode of peak current value $I_P = 10$ mA are $-R_n = -30\ \Omega$, $R_s = 1\ \Omega$, $L_s = 5$ nH, and $C = 20$ pF.

One interest in the tunnel diode is its application as a very high-speed switch. Since tunneling takes place at the speed of light, the transient response is limited only by total shunt capacitance (junction plus stray wiring capacitance) and peak driving current. Switching times of the order of a nanosecond are reasonable, and times as low as 50 ps have been obtained. A second application[5] of the tunnel diode is as a high-frequency (microwave) oscillator.

The most common commercially available tunnel diodes are made from germanium or gallium arsenide. It is difficult to manufacture a silicon tunnel

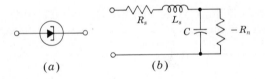

(a)   (b)

Fig. 3-19   (a) Symbol for a tunnel diode; (b) small-signal model in the negative-resistance region.

TABLE 3-1   Typical tunnel-diode parameters

	Ge	GaAs	Si
I_P/I_V..........	8	15	3.5
V_P, V..........	0.055	0.15	0.065
V_V, V..........	0.35	0.50	0.42
V_F, V..........	0.50	1.10	0.70

diode with a high ratio of peak-to-valley current I_P/I_V. Table 3-1 summarizes the important static characteristics of these devices. The voltage values in this table are determined principally by the particular semiconductor used and are almost independent of the current rating. Note that gallium arsenide has the highest ratio I_P/I_V and the largest voltage swing $V_F - V_P \approx 1.0$ V, as against 0.45 V for germanium. The peak current I_P is determined by the impurity concentration (the resistivity) and the junction area. For computer applications, devices with I_P in the range of 1 to 100 mA are most common. The peak point (V_P, I_P), which is in the tunneling region, is not a very sensitive function of temperature. However, the valley point (V_V, I_V), which is affected by the injection current, is quite temperature-sensitive.[5]

The advantages of the tunnel diode are low cost, low noise, simplicity, high speed, environmental immunity, and low power. The disadvantages of the diode are its low output-voltage swing and the fact that it is a two-terminal device. Because of the latter feature, there is no isolation between input and output, and this leads to serious circuit-design difficulties.

3-13   THE SEMICONDUCTOR PHOTODIODE

If a reverse-biased *p-n* junction is illuminated, the current varies almost linearly with the light flux. This effect is exploited in the semiconductor *photodiode*. This device consists of a *p-n* junction embedded in a clear plastic, as indicated in Fig. 3-20. Radiation is allowed to fall upon one surface across the junction. The remaining sides of the plastic are either painted black or enclosed in a metallic case. The entire unit is extremely small and has dimensions of the order of tenths of an inch.

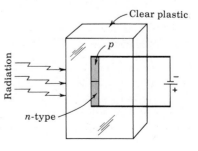

Fig. 3-20 The construction of a semiconductor photodiode.

Volt-Ampere Characteristics If reverse voltages in excess of a few tenths of a volt are applied, an almost constant current (independent of the magnitude of the reverse bias) is obtained. The dark current corresponds to the reverse saturation current due to the thermally generated minority carriers. As explained in Sec. 3-2, these minority carriers "fall down" the potential hill at the junction, whereas this barrier does not allow majority carriers to cross the junction. Now if light falls upon the surface, additional electron-hole pairs are formed. In Sec. 2-8 we note that it is justifiable to consider the radiation solely as a *minority-carrier injector*. These injected minority carriers (for example, electrons in the p side) diffuse to the junction, cross it, and contribute to the current.

The reverse saturation current I_o in a p-n diode is proportional to the concentrations p_{no} and n_{po} of minority carriers in the n and p region, respectively. If we illuminate a reverse-biased p-n junction, the number of new hole-electron pairs is proportional to the number of incident photons. Hence the current under large reverse bias is $I = I_o + I_s$, where I_s, the short-circuit current, is proportional to the light intensity. Hence the volt-ampere characteristic is given by

$$I = I_s + I_o(1 - \epsilon^{V/\eta V_T}) \tag{3-34}$$

where I, I_s, and I_o represent the *magnitude* of the reverse current, and V is positive for a forward voltage and negative for a reverse bias. The parameter η is unity for germanium and 2 for silicon, and V_T is the volt equivalent of temperature defined by Eq. (3-10).

Typical photodiode volt-ampere characteristics are indicated in Fig. 3-21. The curves (with the exception of the dark-current curve) do not pass through the origin. The characteristics in the millivolt range and for positive bias are discussed in the following section, where we find that the photodiode may be used under either short-circuit or open-circuit conditions. It should be

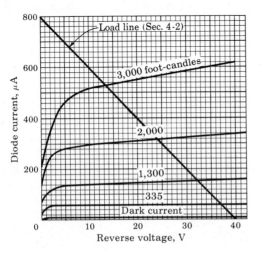

Fig. 3-21 Volt-ampere characteristics for the 1N77 germanium photodiode. (Courtesy of Sylvania Electric Products, Inc.)

Fig. 3-22 Sensitivity of a semiconductor photodiode as a function of the distance of the light spot from the junction.

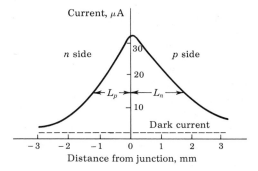

noted that the characteristics drift somewhat with age. The barrier capacitance $C_T \approx 10$ pF, the dynamic resistance $R \approx 50$ M, and the ohmic resistance $r \approx 100$ Ω.

Sensitivity with Position of Illumination   The current in a reverse-biased semiconductor photodiode depends upon the diffusion of minority carriers to the junction.   If the radiation is focused into a small spot far away from the junction, the injected minority carriers can recombine before diffusing to the junction.   Hence a much smaller current will result than if the minority carriers were injected near the junction.   The photocurrent as a function of the distance from the junction at which the light spot is focused is indicated in Fig. 3-22.   The curve is somewhat asymmetrical because of the differences in the diffusion lengths of minority carriers in the p and n sides.   Incidentally, the spectral response of the semiconductor photodiode is the same as that for a photoconductive cell, and is indicated in Fig. 2-12.

The p-n photodiode and, particularly, the improved n-p-n version described in Sec. 5-14 find extensive application in high-speed reading of computer punched cards and tapes, light-detection systems, reading of film sound track, light-operated switches, production-line counting of objects which interrupt a light beam, etc.

3-14   THE PHOTOVOLTAIC EFFECT[8]

In Fig. 3-21 we see that an almost constant reverse current due to injected minority carriers is collected in the p-n photodiode for large reverse voltages. If the applied voltage is reduced in magnitude, the barrier at the junction is reduced.   This decrease in the potential hill does not affect the minority current (since these particles fall down the barrier), but when the hill is reduced sufficiently, some majority carriers can also cross the junction.   These carriers correspond to a forward current, and hence such a flow will reduce the net (reverse) current.   It is this increase in majority-carrier flow which accounts for the drop in the reverse current near the zero-voltage axis in Fig. 3-21.

An expanded view of the origin in this figure is indicated in Fig. 3-23. (Note that the first quadrant of Fig. 3-21 corresponds to the third quadrant of Fig. 3-23.)

The Photovoltaic Potential If a forward bias is applied, the potential barrier is lowered, and the majority current increases rapidly. When this majority current equals the minority current, the total current is reduced to zero. The voltage at which zero resultant current is obtained is called the *photovoltaic* potential. Since, certainly, no current flows under open-circuited conditions, the photovoltaic emf is obtained across the open terminals of a *p-n* junction.

An alternative (but of course equivalent) physical explanation of the photovoltaic effect is the following: In Sec. 3-1 we see that the height of the potential barrier at an open-circuited (nonilluminated) *p-n* junction adjusts itself so that the resultant current is zero, the electric field at the junction being in such a direction as to repel the majority carriers. If light falls on the surface, minority carriers are injected, and since these fall down the barrier, the minority current increases. Since under open-circuited conditions the total current must remain zero, the majority current (for example, the hole current in the *p* side) must increase the same amount as the minority current. This rise in majority current is possible only if the retarding field at the junction is reduced. Hence the barrier height is automatically lowered as a result of the radiation. Across the diode terminals there appears a voltage just equal to the amount by which the barrier potential is decreased. This potential is the photovoltaic emf and is of the order of magnitude of 0.5 V for a silicon and 0.1 V for a germanium cell.

The photovoltaic voltage $V_{\max}$ corresponds to an open-circuited diode. If $I = 0$ is substituted into Eq. (3-34), we obtain

$$V_{\max} = \eta V_T \ln\left(1 + \frac{I_s}{I_o}\right) \qquad (3\text{-}35)$$

Since, except for very small light intensities, $I_s/I_o \gg 1$, then $V_{\max}$ increases logarithmically with I_s, and hence with illumination. Such a logarithmic relationship is obtained experimentally.

Maximum Output Power If a resistor R_L is placed directly across the diode terminals, the resulting current can be found at the intersection of the characteristic in Fig. 3-23 and the load line defined by $V = -IR_L$. If $R_L = 0$, then the output voltage V is zero, and for $R_L = \infty$, the output current I is zero. Hence, for these two extreme values of load, the output power is zero. If for each assumed value of R_L the values of V and I are read from Fig. 3-23 and $P = VI$ is plotted versus R_L, we can obtain the *optimum* load resistance to give maximum output power. For the types LS222 and LS223 photovoltaic light sensors, this optimum load is 3.4 K and $P_{\max} \approx 34$ μW. When the *p-n*

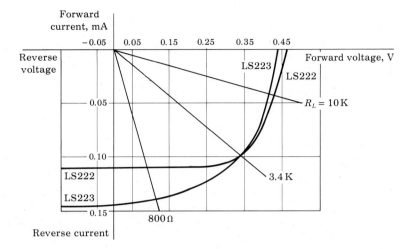

Fig. 3-23 Volt-ampere characteristics for the LS222 and LS223 *p-n* junction photodiodes at a light intensity of 500 fc. (Courtesy of Texas Instruments, Inc.)

photodiode is used as an energy converter (to transform radiant energy into electric energy), the optimum load resistance should be used.

The Short-circuit Current   We see from Fig. 3-23 and Eq. (3-34) that a definite (nonzero) current is obtained for zero applied voltage. Hence a junction photocell can be used under short-circuit conditions. As already emphasized, this current I_s is proportional to the light intensity. Such a linear relationship is obtained experimentally.

Solar-energy Converters   The current drain from a photovoltaic cell may be used to power electronic equipment or, more commonly, to charge auxiliary storage batteries. Such energy converters using sunlight as the primary energy are called *solar batteries* and are used in satellites like the Telstar. A silicon photovoltaic cell of excellent stability and high ($\sim$14 percent) conversion efficiency is made by diffusing a thin *n*-type impurity onto a *p*-type base. In direct noonday sunlight such a cell generates an open-circuit voltage of approximately 0.6 V.

3-15   LIGHT–EMITTING DIODES

Just as it takes energy to generate a hole-electron pair, so energy is released when an electron recombines with a hole. In silicon and germanium this recombination takes place through traps (Sec. 2-8) and the liberated energy goes into the crystal as heat. However, it is found that in other semiconductors, such as gallium arsenide, there is a considerable amount of direct recombination without the aid of traps. Under such circumstances the energy released when an electron falls from the conduction into the valence band appears in the form of radiation. Such a *p-n* diode is called a *light-emitting diode* (*LED*), although the radiation is principally in the infrared. The efficiency of the process of light generation increases with the injected current and with a decrease in temperature. The light is concentrated near the junction because most of the carriers are to be found within a diffusion length of the junction.

Under certain conditions, the emitted light is coherent (essentially monochromatic). Such a diode is called an *injection junction laser*.

REFERENCES

1. Gray, P. E., D. DeWitt, A. R. Boothroyd, and J. F. Gibbons: "Physical Electronics and Circuit Models of Transistors," vol. 2, Semiconductor Electronics Education Committee, John Wiley & Sons, Inc., New York, 1964.

   Shockley, W.: The Theory of *p-n* Junctions in Semiconductor and *p-n* Junction Transistors, *Bell System Tech. J.*, vol. 28, pp. 435–489, July, 1949.

   Middlebrook, R. D.: "An Introduction to Junction Transistor Theory," pp. 115–130, John Wiley & Sons, Inc., New York, 1957.

2. Phillips, A. B.: "Transistor Engineering," pp. 129–133, McGraw-Hill Book Company, New York, 1962.

   Sah, C. T.: Effect of Surface Recombination and Channel on P-N Junction and Transistor Characteristics, *IRE Trans. Electron. Devices*, vol. ED-9, no. 1, pp. 94–108, January, 1962.

3. Corning, J. J.: "Transistor Circuit Analysis and Design," pp. 40–42, Prentice-Hall, Inc., Englewood Cliffs, N.J., 1965.

4. Esaki, L.: New Phenomenon in Narrow Ge *p-n* Junctions, *Phys. Rev.*, vol. 109, p. 603, 1958.

   Nanavati, R. P.: "Introduction to Semiconductor Electronics," chap. 12, McGraw-Hill Book Company, New York, 1963.

5. "Tunnel Diode Manual, TD-30," Radio Corporation of America, Semiconductor and Materials Division, Somerville, N.J., 1963.

   "Tunnel Diode Manual," General Electric Company, Semiconductor Products Dept., Liverpool, N.Y., 1961.

6. Millman, J., and H. Taub: "Pulse, Digital, and Switching Waveforms," chap. 13, McGraw-Hill Book Company, New York, 1965.

7. Shive, J. N.: "Semiconductor Devices," chaps. 8 and 9, D. Van Nostrand Company, Inc., Princeton, N.J., 1959.

8. Rappaport, R.: The Photovoltaic Effect and Its Utilization, *RCA Rev.*, vol. 21, no. 3, pp. 373–397, September, 1959.

Loferski, J. J.: Recent Research on Photovoltaic Solar Energy Converters, *Proc. IEEE*, vol. 51, no. 5, pp. 667–674, May, 1963.

Loferski, J. J., and J. J. Wysocki: Spectral Response of Photovoltaic Cells, *RCA Rev.*, vol. 22, no. 1, pp. 38–56, March, 1961.

REVIEW QUESTIONS

3-1 Consider an open-circuited *p-n* junction. Sketch curves as a function of distance across the junction of space charge, electric field, and potential.

3-2 (*a*) What is the order of magnitude of the space-charge width at a *p-n* junction? (*b*) What does this space charge consist of—electrons, holes, neutral donors, neutral acceptors, ionized donors, ionized acceptors, etc.?

3-3 (*a*) For a reverse-biased diode, does the transition region increase or decrease in width? (*b*) What happens to the junction potential?

3-4 Explain why the *p-n* junction contact potential *cannot* be measured by placing a voltmeter across the diode terminals.

3-5 Explain physically why a *p-n* diode acts as a rectifier.

3-6 (*a*) Write the *law of the junction*. (*b*) Define all terms in this equation. (*c*) What does this equation state for a large forward bias? (*d*) A large reverse bias?

3-7 Plot the minority-carrier current components and the total current in a *p-n* diode as a function of the distance from the junction.

3-8 Plot the hole current, the electron current, and the total current as a function of distance on both sides of a *p-n* junction. Indicate the transition region.

3-9 (*a*) Write the volt-ampere equation for a *p-n* diode. (*b*) Explain the meaning of each symbol.

3-10 Plot the volt-ampere curves for germanium and silicon to the same scale, showing the cutin value for each.

3-11 (*a*) How does the reverse saturation current of a *p-n* diode vary with temperature? (*b*) How does the diode voltage (at constant current) vary with temperature?

3-12 How does the dynamic resistance r of a diode vary with (*a*) current and (*b*) temperature? (*c*) What is the order of magnitude of r for silicon at room temperature and for a dc current of 1 mA?

3-13 (*a*) Sketch the piecewise linear characteristic of a diode. (*b*) What are the approximate cutin voltages for silicon and germanium?

3-14 Consider a step-graded *p-n* junction with equal doping on both sides of the junction ($N_A = N_D$). Sketch the charge density, field intensity, and potential as a function of distance from the junction for a reverse bias.

3-15 (*a*) How does the transition capacitance C_T vary with the depletion-layer width? (*b*) With the applied reverse voltage? (*c*) What is the order of magnitude of C_T?

3-16 What is a *varactor diode?*

3-17 Plot the minority-carrier concentration as a function of distance from a *p-n* junction in the *n* side only for (*a*) a forward-biased junction, (*b*) a negatively biased junction. Indicate the excess concentration and note where it is positive and where negative.

3-18 Under steady-state conditions the diode current is proportional to a charge Q. (a) What is the physical meaning of the factor of proportionality? (b) What charge does Q represent—transition layer charge, injected minority-carrier charge, majority-carrier charge, etc.?

3-19 (a) How does the diffusion capacitance C_D vary with dc diode current? (b) What does the product of C_D and the dynamic resistance of a diode equal?

3-20 What is meant by the *minority-carrier storage time* of a diode?

3-21 A diode in series with a resistor R_L is forward-biased by a voltage V_F. After a steady state is reached, the input changes to $-V_R$. Sketch the current as a function of time. Explain qualitatively the shape of this curve.

3-22 (a) Draw the volt-ampere characteristic of an avalanche diode. (b) What is meant by the *knee* of the curve? (c) By the dynamic resistance? (d) By the temperature coefficient?

3-23 Describe the physical mechanism for avalanche breakdown.

3-24 Describe the physical mechanism for Zener breakdown.

3-25 Draw a circuit which uses a breakdown diode to regulate the voltage across a load.

3-26 Sketch the volt-ampere characteristic of a tunnel diode. Indicate the negative-resistance portion.

3-27 Draw the small-signal model of the tunnel diode operating in the negative-resistance region. Define each circuit element.

3-28 (a) Draw the volt-ampere characteristics of a *p-n* photodiode. (b) Does the current correspond to a forward- or reverse-biased diode? (c) Each curve is drawn for a different value of what physical parameter?

3-29 (a) Write the equation for the volt-ampere characteristic of a photodiode. (b) Define each symbol in the equation.

3-30 (a) Sketch the curve of photodiode current as a function of the position of a narrow light source from the junction. (b) Explain the shape of the curve.

3-31 (a) Define *photovoltaic potential*. (b) What is its order of magnitude? (c) Does it correspond to a forward or a reverse voltage?

3-32 Explain how to obtain maximum power output from a photovoltaic cell.

3-33 What is a *light-emitting diode?*

4 / DIODE CIRCUITS

The p-n junction diode is considered as a circuit element. The concept of "load line" is introduced. The piecewise linear diode model is exploited in the following applications: clippers (single-ended and double-ended), comparators, diode gates, and rectifiers. Half-wave, full-wave, bridge, and voltage-doubling rectification are considered. Capacitor filters are discussed.

Throughout this chapter we shall assume that the input waveforms vary slowly enough so that the diode switching times (Sec. 3-10) may be neglected.

4-1 THE DIODE AS A CIRCUIT ELEMENT

The basic diode circuit, indicated in Fig. 4-1, consists of the device in series with a load resistance R_L and an input-signal source v_i. This circuit is now analyzed to find the instantaneous current i and the instantaneous diode voltage v, when the instantaneous input voltage is v_i.

The Load Line   From Kirchhoff's voltage law (KVL),

$$v = v_i - iR_L \tag{4-1}$$

where R_L is the magnitude of the load resistance. This one equation is not sufficient to determine the two unknowns v and i in this expression. However, a second relation between these two variables is given by the static characteristic of the diode (Fig. 3-7). In Fig. 4-2a is indicated the simultaneous solution of Eq. (4-1) and the diode characteristic. The straight line, which is represented by Eq. (4-1), is called the *load line*. The load line passes through the points $i = 0$,

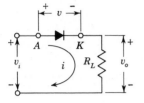

Fig. 4-1 The basic diode circuit. The anode (the p side) of the diode is marked A, and the cathode (the n side) is labeled K.

$v = v_i$, and $i = v_i/R_L$, $v = 0$. That is, the intercept with the voltage axis is v_i, and with the current axis is v_i/R_L. The slope of this line is determined, therefore, by R_L; the negative value of the slope is equal to $1/R_L$. The point of intersection A of the load line and the static curve gives the current i_A that will flow under these conditions. This construction determines the current in the circuit when the instantaneous input potential is v_i.

A slight complication may arise in drawing the load line because $i = v_i/R_L$ is too large to appear on the printed volt-ampere curve supplied by the manufacturer. Under such circumstance choose an arbitrary value of current I' which is on the vertical axis of the printed characteristic. Then the load line is drawn through the point P (Fig. 4-2a), where $i = I'$, $v = v_i - I'R_L$, and through a second point $i = 0$, $v = v_i$.

The Dynamic Characteristic Consider now that the input voltage is allowed to vary. Then the above procedure must be repeated for each voltage value. A plot of current vs. input voltage, called the *dynamic characteristic*, may be obtained as follows: The current i_A is plotted vertically above v_i at point B in Fig. 4-2b. As v_i changes, the slope of the load line does not

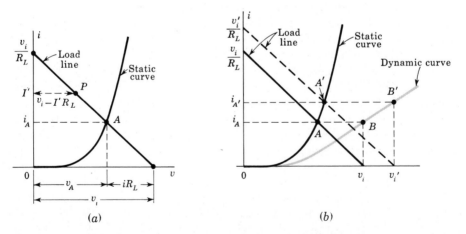

Fig. 4-2 (a) The intersection A of the load line with the diode static characteristic gives the current i_A corresponding to an instantaneous input voltage v_i. (b) The method of constructing the dynamic curve from the static curve and the load line.

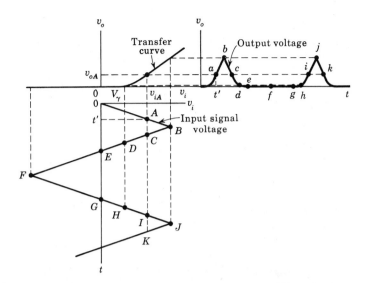

Fig. 4-3 The method of obtaining the output-voltage wave-form from the transfer characteristic for a given input-signal-voltage waveform.

vary since R_L is fixed.   Thus, when the applied potential has the value v_i', the corresponding current is $i_{A'}$.   This current is plotted vertically above v_i' at B'. The resulting curve OBB' that is generated as v_i varies is the dynamic characteristic.

The Transfer Characteristic   The curve which relates the output voltage v_o to the input v_i of any circuit is called the *transfer*, or *transmission, character-istic*.   Since in Fig. 4-1 $v_o = iR_L$, then for this particular circuit the transfer curve has the same shape as the dynamic characteristic.

It must be emphasized that, regardless of the shape of the static volt-ampere characteristic or the waveform of the input signal, the resultant output waveshape can always be found graphically (at low frequencies) from the transfer curve.   This construction is illustrated in Fig. 4-3.   The input-signal waveform (not necessarily triangular) is drawn with its time axis vertically downward, so that the voltage axis is horizontal.   Suppose that the input voltage has the value v_{iA} indicated by the point A at an instant t'.   The corre-sponding output voltage is obtained by drawing a vertical line through· A and noting the voltage v_{oA} where this line intersects the transfer curve.   This value of v_o is then plotted (a) at an instant of time equal to t'.   Similarly, points $b, c, d, \ldots$ of the output waveform correspond to points $B, C, D, \ldots$ of the input-voltage waveform.   Note that $v_o = 0$ for $v_i < V_\gamma$, so that the diode acts as a *clipper* and a portion of the input signal does not appear at the output. Also note the distortion (the deviation from linearity) introduced into the

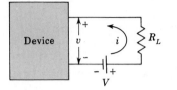

Fig. 4-4 The output circuit of most devices consist of a supply voltage V in series with a load resistance R_L.

output in the neighborhood of $v_i = V_\gamma$ because of the nonlinearity in the transfer curve in this region.

4-2   THE LOAD–LINE CONCEPT

We now show that the use of the load-line construction allows the graphical analysis of many circuits involving devices which are much more complicated than the p-n diode. The external circuit at the output of almost all devices consists of a dc (constant) supply voltage V in series with a load resistance R_L, as indicated in Fig. 4-4. Since KVL applied to this output circuit yields

$$v = V - iR_L \tag{4-2}$$

we once again have a straight-line relationship between output current i and output (device) voltage v. The load line passes through the point $i = 0$, $v = V$ and has a slope equal to $-1/R_L$ *independently of the device characteristics.* A p-n junction diode or an avalanche diode possesses a single volt-ampere characteristic at a given temperature. However, most other devices must be described by a family of curves. For example, refer to Fig. 3-21, which gives the volt-ampere characteristics of a photodiode, where a separate curve is drawn, for each fixed value of light intensity. The load line superimposed upon these characteristics corresponds to a 40-V supply and a load resistance of $40/800$ M $= 50$ K. Note that, from the intersection of the load line with the curve for an intensity $L = 3,000$ fc, we obtain a photodiode current of 530 μA and a device voltage of 13.5 V. For $L = 2,000$ fc, $i = 320$ μA, and $v = 24.0$ V, etc.

The volt-ampere characteristics of a transistor (which is discussed in the following chapter) are similar to those in Fig. 3-21 for the photodiode. However, the independent parameter, which is held constant for each curve, is the input transistor current instead of light intensity. The output circuit is identical with that in Fig. 4-4, and the graphical analysis begins with the construction of the load line.

4-3   THE PIECEWISE LINEAR DIODE MODEL

If the reverse resistance R_r is included in the diode characteristic of Fig. 3-9, the piecewise linear and continuous volt-ampere characteristic of Fig. 4-5a is

Fig. 4-5 (a) The piece-
wise linear volt-ampere
characteristic of a p-n
diode. (b) The large-
signal model in the ON, or
forward, direction (anode
A more positive than V_γ
with respect to the cath-
ode). (c) The model in
the OFF, or reverse, direction $(v < V_\gamma)$.

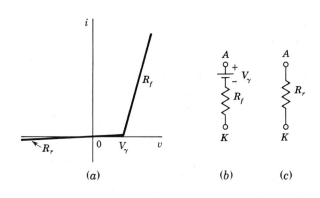

(a)   (b)   (c)

obtained. The diode is a *binary* device, in the sense that it can exist in only one of two possible states; that is, the diode is either ON or OFF at a given time. If the voltage applied across the diode exceeds the cutin potential V_γ with the anode A (the p side) more positive than the cathode K (the n side), the diode is forward-biased and is said to be in the ON state. The large-signal model for the ON state is indicated in Fig. 4-5b as a battery V_γ in series with the low forward resistance R_f (of the order of a few tens of ohms or less). For a reverse bias $(v < V_\gamma)$ the diode is said to be in its OFF state. The large-signal model for the OFF state is indicated in Fig. 4-5c as a large reverse resistance R_r (of the order of several hundred kilohms or more). Usually R_r is so much larger than any other resistance in the diode circuit that this reverse resistance may be considered to be infinite. We shall henceforth assume that $R_r = \infty$, unless otherwise stated.

A Simple Application Consider that in the basic diode circuit of Fig. 4-1 the input is sinusoidal, so that $v_i = V_m \sin \alpha$, where $\alpha = \omega t$, $\omega = 2\pi f$, and f is the frequency of the input excitation. Assume that the piecewise linear model of Fig. 4-5 (with $R_r = \infty$) is valid. The current in the forward direction $(v_i > V_\gamma)$ may then be obtained from the equivalent circuit of Fig. 4-6a. We have

$$i = \frac{V_m \sin \alpha - V_\gamma}{R_L + R_f} \qquad (4\text{-}3)$$

for $v_i = V_m \sin \alpha \geq V_\gamma$ and $i = 0$ for $v_i < V_\gamma$. This waveform is plotted in Fig. 4-6b, where the cutin angle ϕ is given by

$$\phi = \arcsin \frac{V_\gamma}{V_m} \qquad (4\text{-}4)$$

If, for example, $V_m = 2V_\gamma$, then $\phi = 30°$. For silicon (germanium),

$$V_\gamma = 0.6 \text{ V } (0.2 \text{ V}),$$

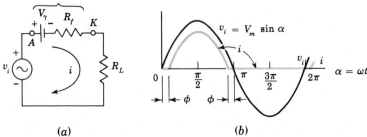

Fig. 4-6 (a) The equivalent circuit of a diode D (in the on state) in series with a load resistance R_L and a sinusoidal voltage v_i. (b) The input waveform v_i and the rectified current i.

and hence a cutin angle of 30° is obtained for very small peak sinusoidal voltages; 1.2 V (0.4 V) for Si (Ge). On the other hand, if $V_m \geq 10$ V, then $\phi \leq 3.5°$ (1.2°) for Si (Ge) and the cutin angle may be neglected; the diode conducts essentially for a full half cycle. Such a rectifier is considered in more detail in Sec. 4-8.

Incidentally, the circuit of Fig. 4-6 may be used to charge a battery from an ac supply line. The battery V_B is placed in series with the diode D, and R_L is adjusted to supply the desired dc (average) charging current. The instantaneous current is given by Eq. (4-3), with V_B added to V_γ.

The Break Region The piecewise linear approximation given in Fig. 4-5a indicates an abrupt discontinuity in slope at V_γ. Actually, the transition of the diode from the off condition to the on state is not abrupt. Therefore the waveform transmitted through a clipper or a rectifier will not show an abrupt change of attenuation at a break point, but instead there will exist a *break* region, that is, a region over which the slope of the diode characteristic changes gradually from a very small to a very large value. We shall now estimate the range of voltage of this break region.

The break point is defined at the voltage V_γ, where the diode resistance changes discontinuously from the very large value R_r to the very small value R_f. Hence, let us arbitrarily define the break region as the voltage change over which the diode resistance is multiplied by some large factor, say 100. The incremental resistance $r \equiv dV/dI = 1/g$ is, from Eq. (3-13),

$$r = \eta \frac{V_T}{I_o} \epsilon^{-V/\eta V_T} \tag{4-5}$$

If $V_1(V_2)$ is the potential at which $r = r_1(r_2)$, then

$$\frac{r_1}{r_2} = \epsilon^{(V_2-V_1)/\eta V_T} \tag{4-6}$$

For $r_1/r_2 = 100$, $\Delta V \equiv V_2 - V_1 = \eta V_T \ln 100 = 0.12$ V for Ge ($\eta = 1$) and 0.24 V for Si ($\eta = 2$) at room temperature. Note that the break region ΔV is only one- or two-tenths of a volt. If the input signal is large compared with this small range, then the piecewise linear volt-ampere approximation and models of Fig. 4-5 are valid.

Analysis of Diode Circuits Using the Piecewise Linear Model Consider a circuit containing several diodes, resistors, supply voltages, and sources of excitation. A general method of analysis of such a circuit consists in assuming (guessing) the state of each diode. For the ON state, replace the diode by a battery V_γ in series with a forward resistance R_f, and for the OFF state replace the diode by the reverse resistance R_r (which can usually be taken as infinite), as indicated in Fig. 4-5b and c. After the diodes have been replaced by these piecewise linear models, the entire circuit is linear and the currents and voltages everywhere can be calculated using Kirchhoff's voltage and current laws. The assumption that a diode is ON can then be verified by observing the sign of the current through it. If the current is in the forward direction (from anode to cathode), the diode is indeed ON and the initial guess is justified. However, if the current is in the reverse direction (from cathode to anode), the assumption that the diode is ON has been proved incorrect. Under this circumstance the analysis must begin again with the diode assumed to be OFF.

Analogous to the above trial-and-error method, we test the assumption that a diode is OFF by finding the voltage across it. If this voltage is either in the reverse direction or in the forward direction but with a voltage less than V_γ, the diode is indeed OFF. However, if the diode voltage is in the forward direction and exceeds V_γ, the diode must be ON and the original assumption is incorrect. In this case the analysis must begin again by assuming the ON state for this diode.

The above method of analysis will be employed in the study of the diode circuits which follows.

4-4   CLIPPING (LIMITING) CIRCUITS

Clipping circuits are used to select for transmission that part of an arbitrary waveform which lies above or below some reference level. Clipping circuits are also referred to as voltage (or current) *limiters*, *amplitude selectors*, or *slicers*.

In the above sense, Fig. 4-1 is a clipping circuit, and input voltages below V_γ are *not* transmitted to the output, as is evident from the waveforms of Figs. 4-3 and 4-6. Some of the more commonly employed clipping circuits are now to be described.

Consider the circuit of Fig. 4-7a. Using the piecewise linear model, the transfer characteristic of Fig. 4-7b is obtained, as may easily be verified. For example, if D is OFF, the diode voltage $v < V_\gamma$ and $v_i < V_\gamma + V_R$. How-

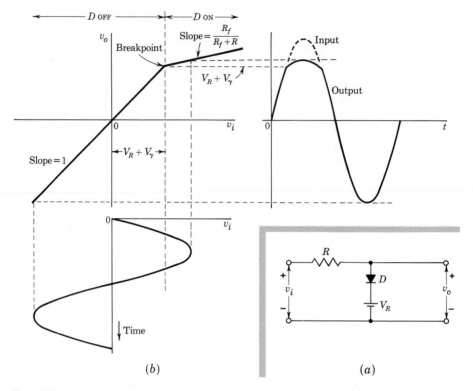

Fig. 4-7 (a) A diode clipping circuit which transmits that part of the waveform more negative than $V_R + V_\gamma$. (b) The piecewise linear transmission characteristic of the circuit. A sinusoidal input and the clipped output are shown.

ever, if D is OFF, there is no current in R and $v_o = v_i$. This argument justifies the linear portion (with slope unity) of the transmission characteristic extending from arbitrary negative values to $v_i = V_R + V_\gamma$. For v_i larger than $V_R + V_\gamma$, the diode conducts, and it behaves as a battery V_γ in series with a resistance R_f, so that increments Δv_i in the input are attenuated and appear at the output as increments $\Delta v_o = \Delta v_i R_f/(R_f + R)$. This verifies the linear portion of slope $R_f/(R_f + R)$ for $v_i > V_R + V_\gamma$ in the transfer curve. Note that the transmission characteristic is piecewise linear and continuous and has a break point at $V_R + V_\gamma$.

Figure 4-7b shows a sinusoidal input signal of amplitude large enough so that the signal makes excursions past the break point. The corresponding output exhibits a suppression of the positive peak of the signal. If $R_f \ll R$, this suppression will be very pronounced, and the positive excursion of the output will be sharply limited at the voltage $V_R + V_\gamma$. The output will appear as though the positive peak had been "clipped off" or "sliced off."

Often it turns out that $V_R \gg V_\gamma$, in which case one may consider that V_R itself is the limiting reference voltage.

In Fig. 4-8a the clipping circuit has been modified in that the diode in Fig. 4-7a has been reversed. The corresponding piecewise linear representation of the transfer characteristic is shown in Fig. 4-8b. In this circuit, the portion of the waveform more positive than $V_R - V_\gamma$ is transmitted without attenuation, but the less positive portion is greatly suppressed.

In Figs. 4-7b and 4-8b we have assumed R_r arbitrarily large in comparison with R. If this condition does not apply, the transmission characteristics must be modified. The portions of these curves which are indicated as having unity slope must instead be considered to have a slope $R_r/(R_r + R)$.

In a transmission region of a diode clipping circuit we require that $R_r \gg R$, for example, that $R_r = kR$, where k is a large number. In the attenuation region, we require that $R \gg R_f$, for example, that $R = kR_f$. From these two equations we deduce that $R = \sqrt{R_f R_r}$ and that $k = \sqrt{R_r/R_f}$. On this

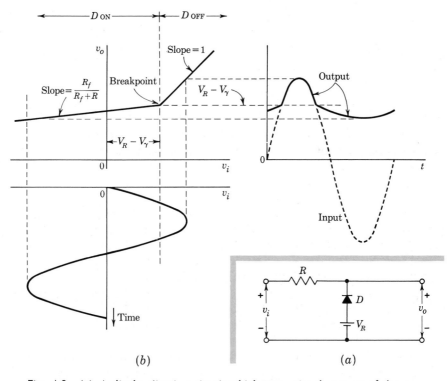

(b)   (a)

Fig. 4-8 (a) A diode clipping circuit which transmits that part of the waveform more positive than $V_R - V_\gamma$. (b) The piecewise linear transmission characteristic of the circuit. A sinusoidal input and the clipped output are shown.

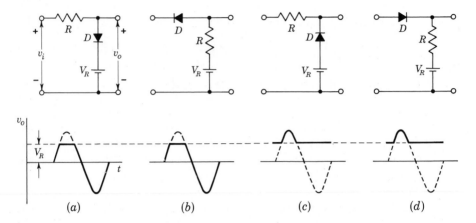

Fig. 4-9 Four diode clipping circuits. In (a) and (c) the diode appears as a shunt element. In (b) and (d) the diode appears as a series element. Under each circuit appears the output waveform (solid) for a sinusoidal input. The clipped portion of the input is shown dashed.

basis we conclude that it is reasonable to select R as the geometrical mean of R_r and R_f. And we note that the ratio R_r/R_f may well serve as a figure of merit for diodes used in the present application.

Additional Clipping Circuits   Figures 4-7 and 4-8 appear again in Fig. 4-9, together with variations in which the diodes appear as series elements. If in each case a sinusoid is applied at the input, the waveforms at the output will appear as shown by the heavy lines. In these output waveforms we have neglected V_γ in comparison with V_R and we have assumed that the break region is negligible in comparison with the amplitude of the waveforms. We have also assumed that $R_r \gg R \gg R_f$. In two of these circuits the portion of the waveform transmitted is that part which lies below V_R; in the other two the portion above V_R is transmitted. In two the diode appears as an element in series with the signal lead; in two it appears as a shunt element. The use of the diode as a series element has the disadvantage that when the diode is OFF and it is intended that there be no transmission, fast signals or high-frequency waveforms may be transmitted to the output through the diode capacitance. The use of the diode as a shunt element has the disadvantage that when the diode is open (back-biased) and it is intended that there be transmission, the diode capacitance, together with all other capacitance in shunt with the output terminals, will round sharp edges of input waveforms and attenuate high-frequency signals. A second disadvantage of the use of the diode as a shunt element is that in such circuits the impedance R_s of the source which supplies V_R must be kept low. This requirement does not arise in circuits where V_R is in series with R, which is normally large compared with R_s.

4-5   CLIPPING AT TWO INDEPENDENT LEVELS

Diode clippers may be used in pairs to perform double-ended limiting at independent levels. A parallel, a series, or a series-parallel arrangement may be used. A parallel arrangement is shown in Fig. 4-10a. Figure 4-10b shows the piecewise linear and continuous input-output voltage curve for the circuit in Fig. 4-10a. The transfer curve has two break points, one at $v_o = v_i = V_{R1}$ and a second at $v_o = v_i = V_{R2}$, and has the following characteristics (assuming $V_{R2} > V_{R1} \gg V_\gamma$ and $R_f \ll R$):

Input v_i	Output v_o	Diode states
$v_i \leq V_{R1}$	$v_o = V_{R1}$	$D1$ ON, $D2$ OFF
$V_{R1} < v_i < V_{R2}$	$v_o = v_i$	$D1$ OFF, $D2$ OFF
$v_i \geq V_{R2}$	$v_o = V_{R2}$	$D1$ OFF, $D2$ ON

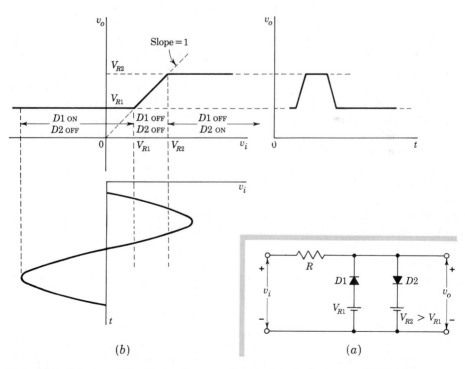

(b)   (a)

Fig. 4-10 (a) A double-diode clipper which limits at two independent levels. (b) The piecewise linear transfer curve for the circuit in (a). The doubly clipped output for a sinusoidal input is shown.

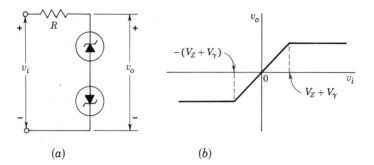

(a)   (b)

Fig. 4-11 (a) A double-ended clipper using avalanche diodes; (b) the transfer characteristic.

The circuit of Fig. 4-10a is referred to as a *slicer* because the output contains a slice of the input between the two reference levels V_{R1} and V_{R2}.

The circuit is used as a means of converting a sinusoidal waveform into a square wave. In this application, to generate a symmetrical square wave, V_{R1} and V_{R2} are adjusted to be numerically equal but of opposite sign. The transfer characteristic passes through the origin under these conditions, and the waveform is clipped symmetrically top and bottom. If the amplitude of the sinusoidal waveform is very large in comparison with the difference in the reference levels, the output waveform will have been *squared*.

Two avalanche diodes in series opposing, as indicated in Fig. 4-11a, constitute another form of double-ended clipper. If the diodes have identical characteristics, a symmetrical limiter is obtained. If the breakdown (Zener) voltage is V_Z and if the diode cutin voltage is V_γ, then the transfer characteristic of Fig. 4-11b is obtained.

Catching or Clamping Diodes Consider that v_i and R in Fig. 4-10a represent Thévenin's equivalent circuit at the output of a device, such as an amplifier. In other words, R is the output resistance and v_i is the open-circuit output signal. In such a situation $D1$ and $D2$ are called *catching diodes*. The reason for this terminology should be clear from Fig. 4-12, where we see

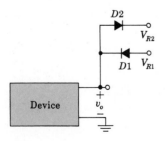

Fig. 4-12 Catching diodes $D1$ and $D2$ limit the output excursion of the device between V_{R1} and V_{R2}.

that $D1$ "catches" the output v_o and does not allow it to fall below $V_{R1} - V_\gamma$, whereas $D2$ "catches" v_o and does not permit it to rise above $V_{R2} + V_\gamma$.

Generally, whenever a node becomes connected through a low resistance (as through a conducting diode) to some reference voltage V_R, we say that the node has been clamped to V_R, since the voltage at that point in the circuit is unable to depart appreciably from V_R. In this sense the diodes in Fig. 4-12 are called *clamping diodes*.

A circuit for clamping the extremity of a periodic waveform to a reference voltage is considered in Sec. 4-11.

4-6   COMPARATORS

The nonlinear circuits which we have used to perform the operation of clipping may also be used to perform the operation of *comparison*. In this case the circuits become elements of a *comparator system* and are usually referred to simply as *comparators*. A comparator circuit is one which may be used to mark the instant when an arbitrary waveform attains some reference level. The distinction between comparator circuits and the clipping circuits considered earlier is that in a comparator there is no interest in reproducing any part of the signal waveform. For example, the comparator output may consist of an abrupt departure from some quiescent level which occurs at the time the signal attains the reference level but is otherwise independent of the signal. Or the comparator output may be a sharp pulse which occurs when signal and reference are equal.

The diode circuit of Fig. 4-13a which we encountered earlier as a clipping circuit is used here in a comparator operation. For the sake of illustration the input signal is taken as a ramp. This input crosses the voltage level $v_i = V_R + V_\gamma$ at time $t = t_1$. The output remains quiescent at $v_o = V_R$ until $t = t_1$, after which it rises with the input signal.

The device to which the comparator output is applied will respond when the comparator voltage has risen to some level V_o above V_R. However, the precise voltage at which this device responds is subject to some variability Δv_o because of gradual changes which result from aging of components, temperature changes, etc. As a consequence (as shown in Fig. 4-13b) there will be a variability Δt in the precise moment at which this device responds and an uncertainty Δv_i in the input voltage corresponding to Δt. Furthermore, if the device responds in the range Δv_o, the device will respond not at $t = t_1$, but at some later time t_2. The situation may be improved by increasing the slope of the rising portion of the output waveform v_o. If the diode were indeed ideal, it would be advantageous to follow the comparator of Fig. 4-13a by an amplifier. However, because of the exponential characteristic of a physical diode, such an anticipated advantage is not realized.[1]

Although an amplifier which follows a diode-resistor comparator does not improve the sharpness of the comparator break, an amplifier *preceding* the

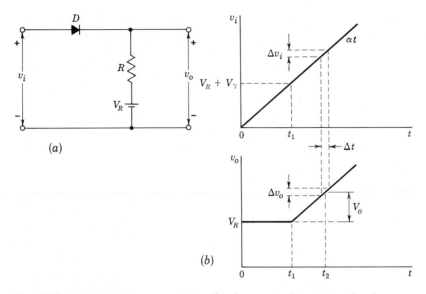

Fig. 4-13 (a) A diode comparator; (b) the comparison operation is illustrated with a ramp input signal v_i, and the corresponding output waveform is indicated.

comparator will do so. Thus, suppose that the input signal to a diode comparator must go through a range Δv_i to carry the comparator through its uncertainty region. Then, if the amplifier has a gain A, the input signal need only go through the range $\Delta v_i/A$ to carry the comparator output through the same voltage range. The amplifier must be direct-coupled and must be extremely stable against drift due to aging of components, temperature change, etc. Such an amplifier is the difference or operational amplifier discussed in Sec. 15-2. Comparators are treated in detail in Sec. 16-11.

4-7 SAMPLING GATE

An ideal sampling gate is a transmission circuit in which the output is an exact reproduction of an input waveform during a selected time interval and is zero otherwise. The time interval for transmission is selected by an externally impressed signal, called the *control*, or *gating*, *signal*, and is usually rectangular in shape. These sampling gates are also referred to as *transmission gates*, or *time-selection circuits*.

A four-diode sampling gate is indicated in Fig. 4-14a. This circuit has the topology of a bridge with the external signal v_s applied at node P_1, the output v_o taken across the load R_L at node P_2, and symmetrical control voltages $+v_c$ and $-v_c$ applied to nodes P_3 and P_4 through the control resistors R_c. The rectangularly shaped v_c, the sinusoidal v_s (it could be of arbitrary waveshape), and the sampled output v_o are drawn in Fig. 4-14b. Note that the period of v_c need not be the same as that of v_s, although in most practical systems the period of v_c would equal or be an integral multiple of that of v_s.

If we assume ideal diodes with $V_\gamma = 0$, $R_f = 0$, $R_r = \infty$, the operation of the circuit is easily understood. During the time interval T_c, when $v_c = V_c$, all four diodes conduct and the voltage across each is zero. Hence nodes P_1 and P_2 are at the same potential and $v_o = v_s$. The output is therefore an exact replica of the input during the selection time T_c. During the time T_n, when $v_c = -V_n$, all four diodes are nonconducting and the current in R_L is zero, so that $v_o = 0$.

We must now justify the statements made in the preceding paragraph that during T_c all diodes conduct and during T_n all diodes are nonconducting. Consider the situation when $v_c = -V_n$, $-v_c = +V_n$, and $v_s = V_s =$ the (positive) peak signal voltage. Let us assume that all diodes are reverse-biased, so that $v_o = 0$, and then justify this assumption (Sec. 4-3). From Fig. 4-14a we see that $D1$ and $D2$ are each reverse-biased by V_n, that $D3$ is reverse-biased by $V_n + V_s$, and that the voltage across $D4$ (in the forward direction) is $V_s - V_n$. Hence diodes $D1$, $D2$, and $D3$ are OFF for any value of V_n or V_s and $D4$ is OFF provided that $V_n \geq V_s$, or the minimum value of V_n is given by

$$(V_n)_{\min} = V_s \qquad\qquad\qquad (4\text{-}7)$$

In other words, there is a restriction on the control-gate amplitude during the nonconducting interval T_n; the minimum value of V_n just equals the peak

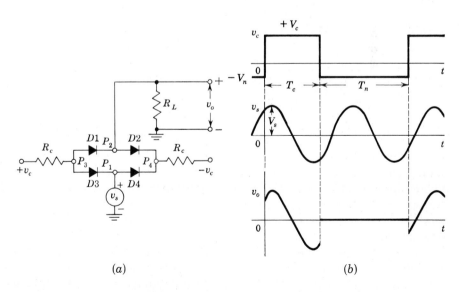

(a)   (b)

Fig. 4-14 (a) A four-diode-bridge sampling gate. (b) The control v_c, the signal v_s, and the output v_o waveforms.

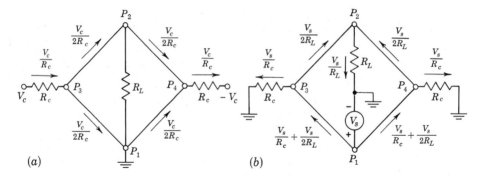

Fig. 4-15 The diodes in Fig. 4-14 are replaced by short circuits. (a) The currents due to V_c; (b) the currents due to V_s.

signal voltage V_s if we require that all four diodes be nonconducting during this interval.

Consider now the situation during T_c, when $v_c = +V_c$, $-v_c = -V_c$, and $v_s = V_s$. We now assume that all four diodes are ON and then determine the restriction required upon V_c so that each diode current is indeed in the forward direction. The current in each diode consists of two components, one due to V_c (as indicated in Fig. 14-15a) and the other due to V_s (as indicated in Fig. 14-15b). The current due to V_c is $V_c/2R_c$ and is in the forward direction in each diode, but the current due to V_s is in the reverse direction in $D3$ (between P_3 and P_1) and in $D2$ (between P_2 and P_4). The larger reverse current is in $D3$ and equals $V_s/R_c + V_s/2R_L$, and hence this quantity must be less than $V_c/2R_c$. The minimum value of V_c is therefore given by

$$(V_c)_{\min} = V_s\left(2 + \frac{R_c}{R_L}\right) \tag{4-8}$$

Balance Conditions Assume that $v_s = 0$ but that the four diodes are not identical in the parameters V_γ and R_f (which are now no longer taken to be zero). Then the bridge will not be balanced and node P_2 is not at the same potential (ground) as P_1. Under these circumstances a portion of the rectangular control waveform appears at the output. In other words, during T_c the output is $v_o = V_c'$, instead of zero. If now the restriction $v_s = 0$ is removed, the sampled portion of the output v_o in Fig. 4-14b will be raised with respect to ground by the voltage V_c', and v_o is said to be "sitting upon a pedestal." Fortunately, all four diodes can be fabricated simultaneously on a tiny chip of silicon by integrated-circuit techniques (Chap. 7), and this ensures matched diodes, so that the pedestal is minimized. It must be emphasized that even with identical diodes a pedestal will exist in the output if the two control waveforms are not balanced (one control signal must be the negative of the other). Other sampling circuits which minimize control-signal imbalance are possible.[2]

4-8   RECTIFIERS

Almost all electronic circuits require a dc source of power.   For portable low-power systems batteries may be used.   More frequently, however, electronic equipment is energized by a *power supply*, a piece of equipment which converts the alternating waveform from the power lines into an essentially direct voltage. The study of ac-to-dc conversion is initiated in this section.

A Half-wave Rectifier   A device, such as the semiconductor diode, which is capable of converting a sinusoidal input waveform (whose average value is zero) into a unidirectional (though not constant) waveform, with a nonzero average component, is called a *rectifier*.   The basic circuit for half-wave rectification is shown in Fig. 4-16.   Since in a rectifier circuit the input $v_i = V_m \sin \omega t$ has a peak value V_m which is very large compared with the cutin voltage V_γ of the diode, we assume in the following discussion that $V_\gamma = 0$.   (The condition $V_\gamma \neq 0$ is treated in Sec. 4-3, and the current waveform is shown in Fig. 4-6b.)   With the diode idealized to be a resistance R_f in the ON state and an open circuit in the OFF state, the current i in the diode or load R_L is given by

$$i = I_m \sin \alpha \quad \text{if } 0 \leq \alpha \leq \pi$$
$$i = 0 \quad \text{if } \pi \leq \alpha \leq 2\pi$$

$$(4\text{-}9)$$

where $\alpha \equiv \omega t$ and

$$I_m \equiv \frac{V_m}{R_f + R_L} \tag{4-10}$$

The transformer secondary voltage v_i is shown in Fig. 4-16b, and the rectified

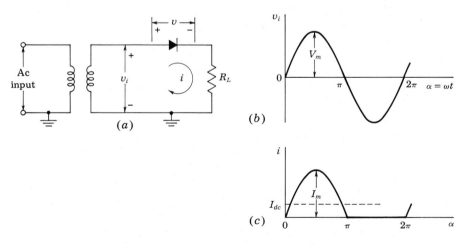

(a)

(b)

(c)

Fig. 4-16 (a) Basic circuit of half-wave rectifier. (b) Transformer sinusoidal secondary voltage v_i. (c) Diode and load current i.

current in Fig. 4-16c. Note that the output current is unidirectional. We now calculate this nonzero value of the average current.

A dc ammeter is constructed so that the needle deflection indicates the average value of the current passing through it. By definition, the average value of a periodic function is given by the area of one cycle of the curve divided by the base. Expressed mathematically,

$$I_{dc} = \frac{1}{2\pi} \int_0^{2\pi} i \, d\alpha \tag{4-11}$$

For the half-wave circuit under consideration, it follows from Eqs. (4-9) that

$$I_{dc} = \frac{1}{2\pi} \int_0^{\pi} I_m \sin \alpha \, d\alpha = \frac{I_m}{\pi} \tag{4-12}$$

Note that the upper limit of the integral has been changed from 2π to π since the instantaneous current in the interval from π to 2π is zero and so contributes nothing to the integral.

The Diode Voltage   The dc output voltage is clearly given as

$$V_{dc} = I_{dc}R_L = \frac{I_m R_L}{\pi} \tag{4-13}$$

However, the reading of a dc voltmeter placed across the diode is *not* given by $I_{dc}R_f$ because the diode cannot be modeled as a constant resistance, but rather it has two values: R_f in the ON state and ∞ in the OFF state.

A dc voltmeter reads the average value of the voltage across its terminals. Hence, to obtain V'_{dc} across the diode, the instantaneous voltage must be plotted as in Fig. 4-17 and the average value obtained by integration. Thus

$$V'_{dc} = \frac{1}{2\pi} \left(\int_0^{\pi} I_m R_f \sin \alpha \, d\alpha + \int_{\pi}^{2\pi} V_m \sin \alpha \, d\alpha \right)$$

$$= \frac{1}{\pi} (I_m R_f - V_m) = \frac{1}{\pi} [I_m R_f - I_m(R_f + R_L)]$$

where use has been made of Eq. (4-10). Hence

$$V'_{dc} = -\frac{I_m R_L}{\pi} \tag{4-14}$$

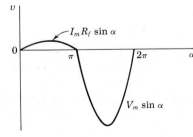

Fig. 4-17   The voltage across the diode in Fig. 4-16.

This result is negative, which means that if the voltmeter is to read upscale, its positive terminal must be connected to the cathode of the diode. From Eq. (4-13) the dc diode voltage is seen to be equal to the negative of the dc voltage across the load resistor. This result is evidently correct because the sum of the dc voltages around the complete circuit must add up to zero.

The AC Current (Voltage) *A root-mean-square ammeter (voltmeter) is constructed so that the needle deflection indicates the effective, or rms, current (voltage).* Such a "square-law" instrument may be of the thermocouple type. By definition, the effective or rms value squared of a periodic function of time is given by the area of one cycle of the curve, which represents the square of the function, divided by the base. Expressed mathematically,

$$I_{\rm rms} = \left(\frac{1}{2\pi}\int_0^{2\pi} i^2\, d\alpha\right)^{\frac{1}{2}} \tag{4-15}$$

By use of Eqs. (4-9), it follows that

$$I_{\rm rms} = \left(\frac{1}{2\pi}\int_0^{\pi} I_m{}^2 \sin^2 \alpha\, d\alpha\right)^{\frac{1}{2}} = \frac{I_m}{2} \tag{4-16}$$

Applying Eq. (4-15) to the *sinusoidal input voltage*, we obtain

$$V_{\rm rms} = \frac{V_m}{\sqrt{2}} \tag{4-17}$$

Most ac meters are of the rectifier type discussed in Sec. 4-9, instead of being true rms reading instruments.

Regulation The variation of dc output voltage as a function of dc load current is called *regulation*. The percentage regulation is defined as

$$\% \text{ regulation} \equiv \frac{V_{\rm no\ load} - V_{\rm load}}{V_{\rm load}} \times 100\% \tag{4-18}$$

where *no load* refers to zero current and *load* indicates the normal load current. For an ideal power supply the output voltage is independent of the load (the output current) and the percentage regulation is zero.

The variation of $V_{\rm dc}$ with $I_{\rm dc}$ for the half-wave rectifier is obtained as follows: From Eqs. (4-12) and (4-10),

$$I_{\rm dc} = \frac{I_m}{\pi} = \frac{V_m/\pi}{R_f + R_L} \tag{4-19}$$

Solving Eq. (4-19) for $V_{\rm dc} = I_{\rm dc}R_L$, we obtain

$$V_{\rm dc} = \frac{V_m}{\pi} - I_{\rm dc}R_f \tag{4-20}$$

This result is consistent with the circuit model given in Fig. 4-18 for the dc voltage and current. Note that the rectifier circuit functions as if it were

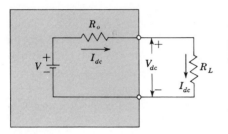

Fig. 4-18 The Thévenin's model which gives the dc voltage and current for a power supply. For the half-wave circuit of Fig. 4-16, $V = V_m/\pi$ and $R_o = R_f$. For the full-wave circuit of Fig. 4-19, $V = 2V_m/\pi$ and $R_o = R_f$. For the full-wave rectifier with a capacitor filter (Sec. 4-10), $V_o = V_m$ and $R_o = 1/4fC$.

a constant (open-circuit) voltage source $V = V_m/\pi$ in series with an effective internal resistance (the *output resistance*) $R_o = R_f$. This model shows that V_{dc} equals V_m/π at no load and that the dc voltage decreases linearly with an increase in dc output current. In practice, the resistance R_s of the transformer secondary is in series with the diode, and in Eq. (4-20) R_s should be added to R_f. The best method of estimating the diode resistance is to obtain a regulation plot of V_{dc} versus I_{dc} in the laboratory. The negative slope of the resulting straight line gives $R_f + R_s$.

Thévenin's Theorem This theorem states that *any two-terminal linear network may be replaced by a generator equal to the open-circuit voltage between the terminals in series with the output impedance seen at this port.* Clearly, Fig. 4-18 represents a Thévenin's model, and hence a rectifier behaves as a linear circuit with respect to average current and voltage.

A Full-wave Rectifier The circuit of a full-wave rectifier is shown in Fig. 4-19a. This circuit is seen to comprise two half-wave circuits so connected that conduction takes place through one diode during one half of the power cycle and through the other diode during the second half of the cycle.

The current to the load, which is the sum of these two currents, has the form shown in Fig. 4-19b. The dc and rms values of the load current and voltage in such a system are readily found to be

$$I_{dc} = \frac{2I_m}{\pi} \qquad I_{rms} = \frac{I_m}{\sqrt{2}} \qquad V_{dc} = \frac{2I_m R_L}{\pi} \qquad (4\text{-}21)$$

where I_m is given by Eq. (4-10) and V_m is the peak transformer secondary voltage from one end to the center tap. Note by comparing Eq. (4-21) with Eq. (4-13) that the dc output voltage for the full-wave connection is twice that for the half-wave circuit.

From Eqs. (4-10) and (4-21) we find that the dc output voltage varies with current in the following manner:

$$V_{dc} = \frac{2V_m}{\pi} - I_{dc}R_f \qquad (4\text{-}22)$$

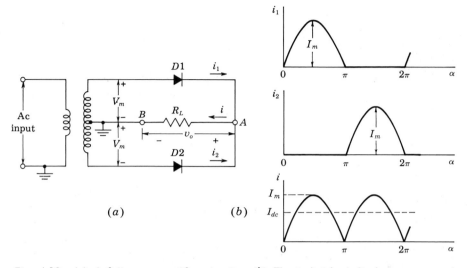

Fig. 4-19 (a) A full-wave rectifier circuit. (b) The individual diode currents and the load current i. The output voltage is $v_o = iR_L$.

This expression leads to Thévenin's dc model of Fig. 4-18, except that the internal (open-circuit) supply is $V = 2V_m/\pi$ instead of V_m/π.

Peak Inverse Voltage   For each rectifier circuit there is a maximum voltage to which the diode can be subjected. This potential is called the *peak inverse voltage* because it occurs during that part of the cycle when the diode is nonconducting. From Fig. 4-16 it is clear that, for the half-wave rectifier, the peak inverse voltage is V_m. We now show that, for a full-wave circuit, twice this value is obtained. At the instant of time when the transformer secondary voltage to midpoint is at its peak value V_m, diode $D1$ is conducting and $D2$ is nonconducting. If we apply KVL around the outside loop and neglect the small voltage drop across $D1$, we obtain $2V_m$ for the peak inverse voltage across $D2$. Note that this result is obtained without reference to the nature of the load, which can be a pure resistance R_L or a combination of R_L and some reactive elements which may be introduced to "filter" the ripple. We conclude that, *in a full-wave circuit, independently of the filter used, the peak inverse voltage across each diode is twice the maximum transformer voltage measured from midpoint to either end.*

4-9   OTHER FULL-WAVE CIRCUITS

A variety of other rectifier circuits find extensive use. Among these are the bridge circuit, several voltage-doubling circuits, and a number of voltage-

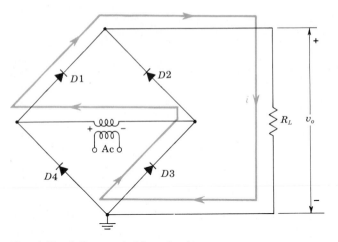

Fig. 4-20   Full-wave bridge circuit.

multiplying circuits.   The bridge circuit finds application not only for power circuits, but also as a rectifying system in rectifier ac meters for use over a fairly wide range of frequencies.

The Bridge Rectifier   The essentials of the bridge circuit are shown in Fig. 4-20.   To understand the action of this circuit, it is necessary only to note that two diodes conduct simultaneously.   For example, during the portion of the cycle when the transformer polarity is that indicated in Fig. 4-20, diodes 1 and 3 are conducting, and current passes from the positive to the negative end of the load.   The conduction path is shown in the figure.   During the next half cycle, the transformer voltage reverses its polarity, and diodes 2 and 4 send current through the load in the same direction as during the previous half cycle.

The principal features of the bridge circuit are the following: The currents drawn in both the primary and the secondary of the supply transformer are sinusoidal, and therefore a smaller transformer may be used than for the full-wave circuit of the same output; a transformer without a center tap is used; and each diode has only transformer voltage across it on the inverse cycle.   The bridge circuit is thus suitable for high-voltage applications.

The Rectifier Meter   This instrument, illustrated in Fig. 4-21, is essentially a bridge-rectifier system, except that no transformer is required.   Instead, the voltage to be measured is applied through a multiplier resistor R to two corners of the bridge, a dc milliammeter being used as an indicating instrument across the other two corners.   Since the dc milliammeter reads average values of current, the meter scale is calibrated to give rms values when a sinusoidal

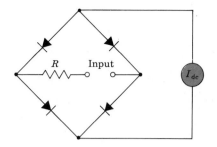

Fig. 4-21   The rectifier voltmeter.

voltage is applied to the input terminals.   As a result, this instrument will not read correctly when used with waveforms which contain appreciable harmonics.

Voltage Multipliers   A common voltage-doubling circuit which delivers a dc voltage approximately equal to twice the transformer maximum voltage at no load is shown in Fig. 4-22.   This circuit is operated by alternately charging each of the two capacitors to the transformer peak voltage V_m, current being continually drained from the capacitors through the load.   The capacitors also act to smooth out the ripple in the output.

This circuit is characterized by poor regulation unless very large capacitors are used.   The inverse voltage across the diodes during the nonconducting cycle is twice the transformer peak voltage.   The action of this circuit will be better understood after the capacitor filter is studied, in the following section.

4-10   CAPACITOR FILTERS

Filtering is frequently effected by shunting the load with a capacitor.   The action of this system depends upon the fact that the capacitor stores energy during the conduction period and delivers this energy to the load during the inverse, or nonconducting, period.   In this way, the time during which the current passes through the load is prolonged, and the ripple is considerably decreased.   The ripple voltage is defined as the deviation of the load voltage from its average or dc value.

Fig. 4-22   The bridge voltage-doubling circuit.   This is the single-phase full-wave bridge circuit of Fig. 4-19 with two capacitors replacing two diodes.

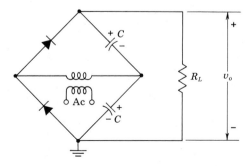

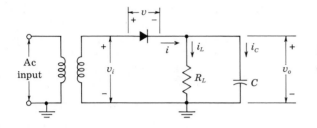

Fig. 4-23 A half-wave capacitor-filtered rectifier.

Consider the half-wave capacitive rectifier of Fig. 4-23. Suppose, first, that the load resistance $R_L = \infty$. The capacitor will charge to the potential V_m, the transformer maximum value. Further, the capacitor will maintain this potential, for no path exists by which this charge is permitted to leak off, since the diode will not pass a negative current. The diode resistance is infinite in the inverse direction, and no charge can flow during this portion of the cycle. Consequently, the filtering action is perfect, and the capacitor voltage v_o remains constant at its peak value, as is seen in Fig. 4-24.

The voltage v_o across the capacitor is, of course, the same as the voltage across the load resistor, since the two elements are in parallel. The diode voltage v is given by

$$v = v_i - v_o \tag{4-23}$$

We see from Fig. 4-24 that the diode voltage is always negative and that the peak inverse voltage is twice the transformer maximum. Hence the presence of the capacitor causes the peak inverse voltage to increase from a value equal to the transformer maximum when no capacitor filter is used to a value equal to twice the transformer maximum value when the filter is used.

Suppose, now, that the load resistor R_L is finite. Without the capacitor input filter, the load current and the load voltage during the conduction period will be sinusoidal functions of time. The inclusion of a capacitor in the circuit results in the capacitor charging in step with the applied voltage. Also, the capacitor must discharge through the load resistor, since the diode will prevent a current in the negative direction. Clearly, the diode acts as a switch which permits charge to flow into the capacitor when the transformer voltage exceeds the capacitor voltage, and then acts to disconnect the power source when the transformer voltage falls below that of the capacitor.

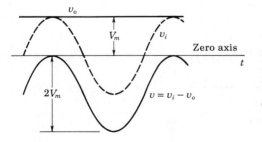

Fig. 4-24 Voltages in a half-wave capacitor-filtered rectifier at no load. The output voltage v_o is a constant, indicating perfect filtering. The diode voltage v is negative for all values of time, and the peak inverse voltage is $2V_m$.

The analysis now proceeds in two steps.   First, the conditions during conduction are considered, and then the situation when the diode is non-conducting is investigated.

Diode Conducting   If the diode drop is neglected, the transformer voltage is impressed directly across the load.   Hence the output voltage is $v_o = V_m \sin \omega t$.   The question immediately arises: Over what interval of time is this equation applicable?   In other words, over what portion of each cycle does the diode remain conducting?   The point at which the diode starts to conduct is called the *cutin point*, and that at which it stops conducting is called the *cutout point*.   The cutout time t_1 and the cutin time t_2 are indicated in Fig. 4-25, where we observe that the output waveform consists of portions of sinusoids (when the diode is ON) joined to exponential segments (when the diode is OFF).

We now calculate the cutout angle by finding the expression for the diode current i and then noting when $i = 0$.   When the diode conducts, $v_o = V_m \sin \omega t$ and i is the sum of the load resistor current i_L and the capacitor current i_C.   Hence

$$i = \frac{v_o}{R_L} + C\frac{dv_o}{dt} = \frac{V_m}{R_L}\sin \omega t + \omega C V_m \cos \omega t \qquad (4\text{-}24)$$

This current is of the form $i = I_m \sin(\omega t + \psi)$, where

$$I_m \equiv V_m\sqrt{\frac{1}{R_L{}^2} + \omega^2 C^2} \qquad \psi \equiv \arctan \omega C R_L \qquad (4\text{-}25)$$

The cutout time t_1 is found by setting $i = 0$ at $t = t_1$, which leads to

$$\omega t_1 = \pi - \psi \qquad (4\text{-}26)$$

for the cutout angle ωt_1 in the first cycle.   The diode current is indicated in Fig. 4-25.

Equation (4-25) shows that the use of a large capacitance to improve the filtering at a given load R_L is accompanied by a high-peak diode current I_m. For a specified average load current, i becomes more peaked and the conduc-

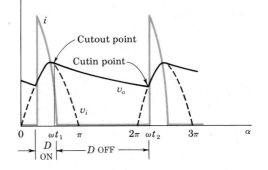

Fig. 4-25   Theoretical sketch of diode current i and output voltage v_o in a half-wave capacitor-filtered rectifier.

tion period decreases as C is made larger. It is to be emphasized that the use of a capacitor filter may impose serious restrictions on the diode, since the average current may be well within the current rating of the diode, and yet the peak current may be excessive.

Diode Nonconducting In the interval between the cutout time t_1 and the cutin time t_2, the diode is effectively out of the circuit, and the capacitor discharges through the load resistor with a time constant CR_L. Thus the capacitor voltage (equal to the load voltage) is

$$v_o = A\epsilon^{-t/CR_L} \tag{4-27}$$

To determine the value of the constant A, note from Fig. 4-25 that at the cutout time $t = t_1$, $v_o = v_i = V_m \sin \omega t_1$. Equation (4-27) thus attains the form

$$v_o = (V_m \sin \omega t_1)\epsilon^{-(t-t_1)/CR_L} \tag{4-28}$$

Since t_1 is known from Eq. (4-26), v_o can be plotted as a function of time. This exponential curve is indicated in Fig. 4-25, and where it intersects the curve $V_m \sin \omega t$ (in the following cycle) is the cutin point t_2. The validity of this statement follows from the fact that at an instant of time greater than t_2, the transformer voltage v_i (the sine curve) is greater than the capacitor voltage v_o (the exponential curve). Since the diode voltage is $v = v_i - v_o$, then v will be positive beyond t_2 and the diode will become conducting. Thus t_2 is the cutin point. No analytic expression can be given for t_2; it must be found graphically by the method outlined above.

Full-wave Circuit Consider a full-wave rectifier with a capacitor filter obtained by placing a capacitor C across R_L in Fig. 4-19. The analysis of this circuit requires a simple extension of that just made for the half-wave circuit. If in Fig. 4-25 a dashed half-sinusoid is added between π and 2π, the result is the dashed full-wave voltage in Fig. 4-26. The cutin point now lies between π and 2π, where the exponential portion of v_o intersects this sinusoid. The cutout point is the same as that found for the half-wave rectifier.

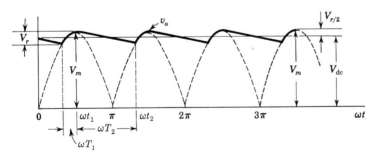

Fig. 4-26 The approximate load-voltage waveform v_o in a full-wave capacitor-filtered rectifier.

Approximate Analysis It is possible to obtain the dc output voltage for given values of the parameters ω, R_L, C, and V_m from the graphical construction indicated in Fig. 4-25. Such an analysis is involved and tedious. Hence we now present an approximate solution which is simple and yet sufficiently accurate for most engineering applications.

We assume that the output-voltage waveform of a full-wave circuit with a capacitor filter may be represented by the approximately piecewise linear curve shown in Fig. 4-26. For large values of C (so that $\omega CR_L \gg 1$) we note from Eqs. (4-25) and (4-26) that $\omega t_1 \to \pi/2$ and $v_o \to V_m$ at $t = t_1$. Also, with C very large, the exponential decay in Eq. (4-28) can be replaced by a linear fall. If the total capacitor discharge voltage (the ripple voltage) is denoted by V_r, then from Fig. 4-26, the average value of the voltage is approximately

$$V_{dc} = V_m - \frac{V_r}{2} \tag{4-29}$$

It is necessary, however, to express V_r as a function of the load current and the capacitance. If T_2 represents the total nonconducting time, the capacitor, when discharging at the constant rate I_{dc}, will lose an amount of charge $I_{dc}T_2$. Hence the change in capacitor voltage is $I_{dc}T_2/C$, or

$$V_r = \frac{I_{dc}T_2}{C} \tag{4-30}$$

The better the filtering action, the smaller will be the conduction time T_1 and the closer T_2 will approach the time of half a cycle. Hence we assume that $T_2 = T/2 = 1/2f$, where f is the fundamental power-line frequency. Then

$$V_r = \frac{I_{dc}}{2fC} \tag{4-31}$$

and from Eq. (4-29),

$$V_{dc} = V_m - \frac{I_{dc}}{4fC} \tag{4-32}$$

This result is consistent with Thévenin's model of Fig. 4-18, with the open-circuit voltage $V = V_m$ and the effective output resistance $R_o = 1/4fC$.

The ripple is seen to vary directly with the load current I_{dc} and also inversely with the capacitance. Hence, to keep the ripple low and to ensure good regulation, very large capacitances (of the order of tens of microfarads) must be used. The most common type of capacitor for this rectifier application is the electrolytic capacitor. These capacitors are polarized, and care must be taken to insert them into the circuit with the terminal marked $+$ to the positive side of the output.

The desirable features of rectifiers employing capacitor input filters are the small ripple and the high voltage at light load. The no-load voltage is equal, theoretically, to the maximum transformer voltage. The disadvantages of this system are the relatively poor regulation, the high ripple at large loads, and the peaked currents that the diodes must pass.

An approximate analysis similar to that given above applied to the half-wave circuit shows that the ripple, and also the drop from no load to a given load, are double the values calculated for the full-wave rectifier.

4-11   ADDITIONAL DIODE CIRCUITS

Many applications depend upon the semiconductor diode besides those already considered in this chapter.   We mention four others below.

Peak Detector   The half-wave capacitor-filtered rectifier circuit of Fig. 4-23 may be used to measure the peak value of an input waveform.   Thus, for $R_L = \infty$, the capacitor charges to the maximum value V_{max} of v_i, the diode becomes nonconducting, and v_o remains at V_{max} (assuming an ideal capacitor with no leakage resistance shunting C).   Refer to Fig. 4-24, where $V_{max} = V_m =$ the peak value of the input sinusoid.   Improved peak detector circuits are given in Sec. 16-13.

In an AM radio the amplitude of the high-frequency wave (called the *carrier*) is varied in accordance with the audio information to be transmitted. This process is called *amplitude modulation*, and such an AM waveform is illustrated in Fig. 4-27.   The audio information is contained in the envelope (the locus, shown dashed, of the peak values) of the modulated waveform. The process of extracting the audio signal is called *detection*, or *demodulation*. If the input to Fig. 4-23 is the AM waveform shown in Fig. 4-27, the output v_o is the heavy-weight curve, provided that the time constant $R_L C$ is chosen properly; that is, $R_L C$ must be small enough so that, when the envelope decreases in magnitude, the voltage across C can fall fast enough to keep in step with the envelope, but $R_L C$ must not be so small as to introduce excessive ripple. The order of magnitude of the frequency of an AM radio carrier is 1,000 kHz, and the audio spectrum extends from about 20 Hz to 20 kHz.   Hence there should be *at least* 50 cycles of the carrier waveform for each audio cycle.   If Fig. 4-27 were drawn more realistically (with a much higher ratio of carrier to audio frequency), then clearly, the ripple amplitude of the demodulated signal

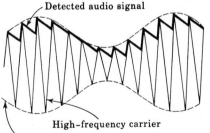

Detected audio signal

High-frequency carrier

Envelope containing audio information

Fig. 4-27 An amplitude-modulated wave and the detected audio signal.   (For ease of drawing, the carrier waveform is indicated triangular instead of sinusoidal and of much lower frequency than it really is, relative to the audio frequency.)

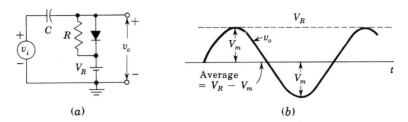

Fig. 4-28 (a) A circuit which clamps to the voltage V_R. (b) The output voltage v_o for a sinusoidal input v_i.

would be very much smaller.   This low-amplitude high-frequency ripple in v_o is easily filtered so that the smoothed detected waveform is an excellent reproduction of the audio signal.   The capacitor-rectifier circuit of Fig. 4-23 therefore also acts as an *envelope demodulator*.

A Clamping Circuit   A function which must be frequently performed with a periodic waveform is the establishment of the recurrent positive or negative extremity at some constant reference level V_R.   Such a clamping circuit is indicated in Fig. 4-28a.   Assuming an ideal diode, the drop across the device is zero in the forward direction.   Hence the output cannot rise above V_R and is said to be *clamped* to this level.   If the input is sinusoidal with a peak value V_m and an average value of zero, then, as indicated in Fig. 4-28b, the output is sinusoidal, with an average value of $V_R - V_m$.   This waveform is obtained subject to the following conditions: the diode parameters are $R_f = 0$, $R_r = \infty$, and $V_\gamma = 0$; the source impedance $R_s = 0$; and the time constant RC is much larger than the period of the signal.   In practice these restrictions are not completely satisfied and the clamping is not perfect; the output voltage rises slightly above V_R, and the waveshape is a somewhat distorted version of the input.[3]

Digital-computer Circuits   Since the diode is a binary device existing in either the ON or OFF state for a given interval of time, it is a very usual component in digital-computer applications.   Such so-called "logic" circuits are discussed in Chap. 6 in conjunction with transistor binary applications.

Avalanche-diode Regulator   The basic circuit used to decrease voltage variations across a load due to variations in output current or supply voltage is given in Fig. 3-17 and discussed in Sec. 3-11.

REFERENCES

1. Millman, J., and H. Taub: "Pulse, Digital, and Switching Waveforms," sec. 7–11, McGraw-Hill Book Company, New York, 1965.

2. Ref. 1, chap. 17.

3. Ref. 1, chap. 3.

REVIEW QUESTIONS

4-1 Explain how to obtain the dynamic characteristic from the static volt-ampere curve of a diode.

4-2 You are given the V-I output characteristic in graphical form of a new device. (a) Sketch the circuit using this device which will require a load-line construction to determine i and v. (b) Is the load line vertical, horizontal, at 135° or 45° for infinite load resistance? (c) For zero load resistance?

4-3 (a) Draw the piecewise linear volt-ampere characteristic of a p-n diode. (b) What is the circuit model for the ON state? (c) The OFF state?

4-4 Consider a circuit consisting of a diode D, a resistance R, and a signal source v_i in series. Define (a) static characteristic; (b) dynamic characteristic; (c) *transfer*, or *transmission*, characteristic. (d) What is the correlation between (b) and (c)?

4-5 Consider the circuit of Rev. 4-4 using a silicon diode and a 100-Ω resistance. (a) Plot approximately the dynamic curve. (b) If the input is $v_i = -1 + A \sin \omega t$, plot the current waveform for $A = 1.5$, 2.0, and 5.0 V.

4-6 For the circuit of Rev. 4-4, $v_i = V_m \sin \omega t$. If the diode is represented by its piecewise linear model, find the current as a function of time and plot.

4-7 What is meant by *break region* of a diode?

4-8 In analyzing a circuit containing several diodes by the piecewise linear method, you assume (guess) that certain of the diodes are ON and others are OFF. Explain carefully how you determine whether or not the assumed state of each diode is correct.

4-9 Consider a series circuit consisting of a diode D, a resistance R, a reference battery V_R, and an input signal v_i. The output is taken across R and V_R in series. Draw the transfer characteristic if the anode of D is connected to the positive terminal of the battery. Use the piecewise linear diode model.

4-10 Repeat Rev. 4-9 with the anode of D connected to the negative terminal of the battery.

4-11 If v_i is sinusoidal and D is ideal (with $R_f = 0$, $V_\gamma = 0$, and $R_r = \infty$), find the output waveforms in (a) Rev. 4-9 and (b) Rev. 4-10.

4-12 Sketch the circuit of a *double-ended clipper* using ideal p-n diodes which limit the output between ± 10 V.

4-13 Repeat Rev. 4-12 using avalanche diodes.

4-14 (a) What is a *comparator circuit*? (b) How does such a circuit differ from a *clipping circuit*?

4-15 (a) What is a *sampling gate*? (b) Sketch the circuit of a four-diode sampling gate.

4-16 Define in words and as an equation (a) dc current I_{dc}; (b) dc voltage V_{dc}; (c) ac current I_{rms}.

4-17 (a) Sketch the circuit for a full-wave rectifier. (b) Derive the expression for (1) the dc current; (2) the dc load voltage; (3) the dc diode voltage; (4) the rms current.

4-18 (a) Define *regulation*. (b) Derive the regulation equation for a full-wave circuit.

4-19 Draw the Thévenin's model for a full-wave rectifier.

4-20 (a) Define *peak inverse voltage*. (b) What is the peak inverse voltage for a full-wave circuit using ideal diodes? (c) Repeat part b for a half-wave rectifier.

4-21 Sketch the circuit of a bridge rectifier and explain its operation.

4-22 Repeat Rev. 4-21 for a rectifier meter circuit.

4-23 Repeat Rev. 4-21 for a voltage multiplier.

4-24 (a) Draw the circuit of a half-wave capacitive rectifier. (b) At no load draw the steady-state voltage across the capacitor and also across the diode.

4-25 (a) Draw the circuit of a full-wave capacitive rectifier. (b) Draw the output voltage under load. Indicate over what period of time the diode conducts. Make no calculations. (c) Indicate the diode current waveform superimposed upon the output waveform.

4-26 For the circuit of Rev. 4-25, derive the expression for (a) the diode current; (b) the cutout angle. (c) How is the cutin angle found?

4-27 (a) Consider a full-wave capacitor rectifier using a large capacitance C. Sketch the approximate output waveform. (b) Derive the expression for the peak ripple voltage. (c) Derive the Thévenin's model for this rectifier.

4-28 For a full-wave capacitor rectifier circuit, list (a) two advantages; (b) three disadvantages.

4-29 Describe (a) *amplitude modulation* and (b) *detection*.

5 / TRANSISTOR CHARACTERISTICS

The physical behavior of a semiconductor triode, called a *bipolar junction transistor* (BJT), is given. The volt-ampere characteristics of this device are studied. It is demonstrated that the transistor is capable of producing amplification. Analytical expressions relating the transistor currents with the junction voltages are indicated. Typical voltage values are given, for the several possible modes of operation.

5-1 THE JUNCTION TRANSISTOR[1]

A junction transistor consists of a silicon (or germanium) crystal in which a layer of n-type silicon is sandwiched between two layers of p-type silicon. Alternatively, a transistor may consist of a layer of p-type between two layers of n-type material. In the former case the transistor is referred to as a p-n-p transistor, and in the latter case, as an n-p-n transistor. The semiconductor sandwich is extremely small, and is hermetically sealed against moisture inside a metal or plastic case. Manufacturing techniques and constructional details for several transistor types are described in Sec. 5-4.

The two types of transistor are represented in Fig. 5-1a. The representations employed when transistors are used as circuit elements are shown in Fig. 5-1b. The three portions of a transistor are known as *emitter, base,* and *collector.* The arrow on the emitter lead specifies the direction of current flow when the emitter-base junction is biased in the forward direction. In *both* cases, however, the emitter, base, and collector currents, I_E, I_B, and I_C, respectively, are assumed positive when the currents flow *into* the transistor. The symbols V_{EB}, V_{CB}, and V_{CE} are the emitter-base, collector-base, and collector-emitter voltages, respectively. (More specifically, V_{EB} represents the voltage *drop* from emitter to base.)

118

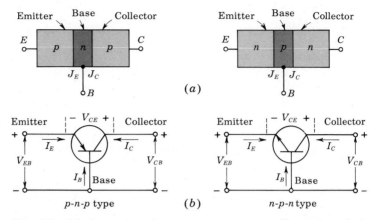

Fig. 5-1   (a) A p-n-p and an n-p-n transistor. The emitter (collector) junction is $J_E(J_C)$.   (b) Circuit representation of the two transistor types.

Open-circuited Transistor   If no external biasing voltages are applied, all transistor currents must be zero. The potential barriers at the junctions adjust to the contact difference of potential V_o—given in Eq. (2-63) (a few tenths of a volt)—required so that no free carriers cross each junction. If we assume a completely symmetrical junction (emitter and collector regions having identical physical dimensions and doping concentrations), the barrier height is identical at the emitter junction J_E and at the collector junction J_C, as indicated in Fig. 5-2a. The narrow space-charge regions at the junctions have been neglected.

Under open-circuited conditions, the minority concentration is constant within each section and is equal to its thermal-equilibrium value, n_{po} in the p-type emitter and collector regions and p_{no} in the n-type base, as shown in Fig. 5-2b. Since the transistor may be looked upon as a p-n diode followed by an n-p diode, much of the theory developed in the preceding chapters for the junction diode will be used to explain the physical behavior of the transistor,

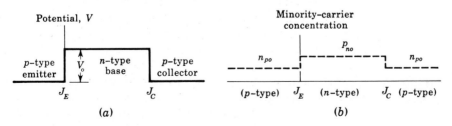

Fig. 5-2   (a) The potential and (b) the minority-carrier density in each section of an open-circuited symmetrical p-n-p transistor.

when voltages are applied so as to disturb it from the equilibrium situation pictured in Fig. 5-2.

The Transistor Biased in the Active Region We may now begin to appreciate the essential features of a transistor as an active circuit element by considering the situation depicted in Fig. 5-3a. Here a p-n-p transistor is shown with voltage sources which serve to bias the emitter-base junction in the forward direction and the collector-base junction in the reverse direction. The potential variation through the biased transistor is indicated in Fig. 5-3b. The dashed curve applies to the case before the application of external biasing voltages (Fig. 5-2a), and the solid curve to the case after the biasing voltages are applied. The externally applied voltages appear, essentially, across the junctions. Hence, as shown in Fig. 5-3b, the forward biasing of the emitter junction lowers the emitter-base potential barrier by $|V_{EB}|$, whereas the reverse biasing of the collector junction increases the collector-base potential barrier by $|V_{CB}|$. The lowering of the emitter-base barrier permits minority-carrier injection; that is, holes are injected into the base, and electrons are injected into the emitter region. The excess holes diffuse across the n-type base,

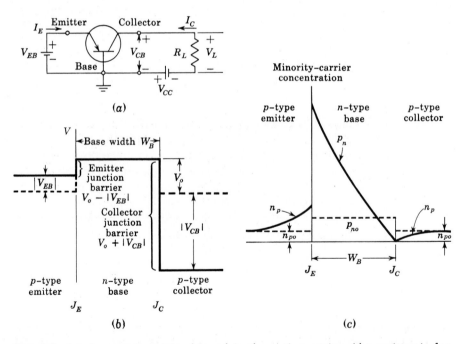

Fig. 5-3 (a) A p-n-p transistor biased in the active region (the emitter is for-ward-biased and the collector is reverse-biased). (b) The potential variation through the transistor. The narrow depletion regions at the junctions are negligi-bly small. (c) The minority-carrier concentration in each section of the tran-sistor. It is assumed that the emitter is much more heavily doped than the base.

where the electric field intensity ε is zero, to the collector junction. At J_C the field is positive and large ($\varepsilon = -dV/dx \gg 0$), and hence holes are accelerated across this junction. In other words, the holes which reach J_C fall down the potential barrier, and are therefore *collected* by the collector. Since the applied potential across J_C is negative, then from the law of the junction, Eq. (3-4), p_n is reduced to zero at the collector as shown in Fig. 5-3c. Similarly, the reverse collector-junction bias reduces the collector electron density n_p to zero at J_C. The minority-carrier-density curves pictured in Fig. 5-3c should be compared with the corresponding concentration plots for the forward- and reverse-biased p-n junction given in Fig. 3-14.

5-2   TRANSISTOR CURRENT COMPONENTS

In Fig. 5-4 we show the various current components which flow across the forward-biased emitter junction and the reverse-biased collector junction. The emitter current I_E consists of hole current I_{pE} (holes crossing from emitter into base) and electron current I_{nE} (electrons crossing from base into the emitter). The ratio of hole to electron currents, I_{pE}/I_{nE}, crossing the emitter junction is proportional to the ratio of the conductivity of the p material to that of the n material. In a commercial transistor the doping of the emitter is made much larger than the doping of the base. This feature ensures (in a p-n-p transistor) that the emitter current consists almost entirely of holes.

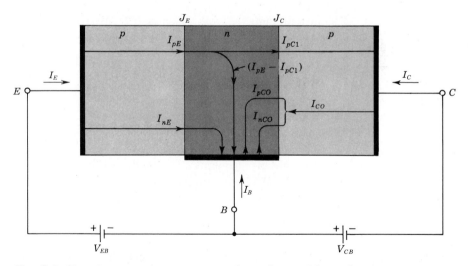

Fig. 5-4 Transistor current components for a forward-biased emitter junction and a reversed-biased collector junction. If a current has a subscript $p(n)$, it consists of holes (electrons) moving in the same (opposite) direction as the arrow indicating the current direction.

Such a situation is desirable since the current which results from electrons crossing the emitter junction from base to emitter does not contribute carriers which can reach the collector.

We assume low-level injection, and hence (Sec. 2-11) the minority current I_{pE} is the hole *diffusion* current into base and its magnitude is proportional to the slope at J_E of the p_n curve [Eq. (2-36)]. Similarly, I_{nE} is the electron *diffusion* current into the emitter, and its magnitude is proportional to the slope at J_E of the n_p curve. Note that I_{pE} and I_{nE} correspond to the minority-carrier diffusion currents $I_{pn}(0)$ and $I_{np}(0)$, respectively, in Fig. 3-4 for the currents crossing a p-n junction. Just as the total current crossing the junction in Fig. 3-4 is $I = I_{pn}(0) + I_{np}(0)$, so the total emitter current in Fig. 5-4 is

$$I_E = I_{pE} + I_{nE} \qquad (5\text{-}1)$$

All currents in this equation are positive for a p-n-p transistor.

Not all the holes crossing the emitter junction J_E reach the collector junction J_C, because some of them combine with the electrons in the n-type base. In Fig. 5-4, let I_{pC1} represent the hole current at J_C as a result of holes crossing the base from the emitter. Hence there must be a bulk recombination hole current $I_{pE} - I_{pC1}$ leaving the base, as indicated in Fig. 5-4 (actually, electrons enter the base region from the external circuit through the base lead to supply those charges which have been lost by recombination with the holes injected into the base across J_E).

Consider, for the moment, an open-circuited emitter, while the collector junction remains reverse-biased. Then I_C must equal the reverse saturation current I_{CO} of the back-biased diode at J_C. This *reverse* current consists of two components, as shown in Fig. 5-4, I_{nCO} consisting of electrons moving from the p to the n region across J_C and a term, I_{pCO}, resulting from holes crossing from n to p across J_C.

$$-I_{CO} = I_{nCO} + I_{pCO} \qquad (5\text{-}2)$$

(The minus sign is chosen arbitrarily so that I_C and I_{CO} will have the same sign.)

Since $I_E = 0$ under open-circuit conditions, no holes are injected across J_E, and hence none can reach J_C from the emitter. Clearly, I_{pCO} results from the small concentration of holes generated thermally within the base.

Now let us return to the situation depicted in Fig. 5-4, where the emitter is forward-biased so that

$$I_C = I_{CO} - I_{pC1} = I_{CO} - \alpha I_E \qquad (5\text{-}3)$$

where α is defined as the fraction of the total emitter current [given in Eq. (5-1)] which represents holes which have traveled from the emitter across the base to the collector. For a p-n-p transistor, I_E is positive and both I_C and I_{CO} are negative, which means that the current in the collector lead is in the direction opposite to that indicated by the arrow of I_C in Fig. 5-4. For an n-p-n transistor these currents are reversed.

The electron current crossing J_C is I_{nCO} and represents electrons diffusing from the collector into the base (and hence a positive current from the base into the collector), and its magnitude is proportional to the slope at J_C of the n_p distribution in Fig. 5-3c. The total diffusion hole current crossing J_C from the base is

$$I_{pC} \equiv I_{pC1} + I_{pCO} \qquad (5\text{-}4)$$

and its magnitude is proportional to the slope at J_C of the p_n distribution in Fig. 5-3c.

Large-signal Current Gain α   From Eq. (5-3) it follows that α may be defined as the ratio of the negative of the collector-current increment from cutoff ($I_C = I_{co}$) to the emitter-current change from cutoff ($I_E = 0$), or

$$\alpha \equiv - \frac{I_c - I_{co}}{I_E - 0} \qquad (5\text{-}5)$$

Alpha is called the *large-signal current gain* of a common-base transistor. Since I_C and I_E have opposite signs (for either a p-n-p or an n-p-n transistor), then α, as defined, is always positive. Typical numerical values of α lie in the range 0.90 to 0.995. It should be pointed out that α is not a constant, but varies with emitter current I_E, collector voltage V_{CB}, and temperature.

A Generalized Transistor Equation   Equation (5-3) is valid only in the *active region*, that is, if the emitter is forward-biased and the collector is reverse-biased. For this mode of operation the collector current is essentially independent of collector voltage and depends only upon the emitter current. Suppose now that we seek to generalize Eq. (5-3) so that it may apply not only when the collector junction is substantially reverse-biased, but also for any voltage across J_C. To achieve this generalization we need only replace I_{CO} by the current in a p-n diode (that consisting of the base and collector regions). This current is given by the volt-ampere relationship of Eq. (3-7), with I_o replaced by $-I_{CO}$ and V by V_C, where the symbol V_C represents the drop across J_C from the p to the n side. The complete expression for I_C for any V_C and I_E is

$$I_C = -\alpha I_E + I_{CO}(1 - \epsilon^{V_C/V_T}) \qquad (5\text{-}6)$$

Note that if V_C is negative and has a magnitude large compared with V_T, Eq. (5-6) reduces to Eq. (5-3). The physical interpretation of Eq. (5-6) is that the p-n junction diode current crossing the collector junction is augmented by the fraction α of the current I_E flowing in the emitter. This relationship is derived in Sec. 19-14 by evaluating quantitatively the various current components introduced above.

The similarity between Eq. (5-6) and Eq. (3-34) for the photodiode should be noted. Both equations represent the volt-ampere characteristics of a reverse-biased diode (with reverse saturation current $I_o = I_{CO}$) subjected to external excitation. In the case of the photodiode, the external stimulus is radiation, which injects minority carriers into the semiconductor, resulting in the current I_s. This component of current, which is proportional to the light intensity, augments the current from the simple p-n junction diode. Similarly, for the transistor, the collector-base diode current is increased by the term $-\alpha I_E$, which is proportional to the "external excitation," namely, the emitter current resulting from the voltage applied between emitter and base.

5-3   THE TRANSISTOR AS AN AMPLIFIER

A load resistor R_L is in series with the collector supply voltage V_{CC} of Fig. 5-3a. A small voltage change ΔV_i between emitter and base causes a relatively large emitter-current change ΔI_E. We define by the symbol α' that fraction of this current change which is collected and passes through R_L, or $\Delta I_C = \alpha' \Delta I_E$. The change in output voltage across the load resistor

$$\Delta V_L = -R_L \Delta I_C = -\alpha' R_L \Delta I_E$$

may be many times the change in input voltage ΔV_i. Under these circumstances, the voltage amplification $A \equiv \Delta V_L / \Delta V_i$ will be greater than unity, and the transistor acts as an amplifier. If the dynamic resistance of the emitter junction is r_e, then $\Delta V_i = r_e \Delta I_E$, and

$$A \equiv - \frac{\alpha' R_L \Delta I_E}{r_e \Delta I_E} = - \frac{\alpha' R_L}{r_e} \tag{5-7}$$

From Eq. (3-14), $r_e = 26/I_E$, where I_E is the quiescent emitter current in milliamperes. For example, if $r_e = 40\ \Omega$, $\alpha' = -1$, and $R_L = 3{,}000\ \Omega$, $A = +75$. This calculation is oversimplified, but in essence it is correct and gives a physical explanation of why the transistor acts as an amplifier. The transistor provides power gain as well as voltage or current amplification. From the foregoing explanation it is clear that current in the low-resistance input circuit is transferred to the high-resistance output circuit. The word "transistor," which originated as a contraction of "transfer resistor," is based upon the above physical picture of the device.

The Parameter α'   The parameter α' introduced above is defined as the ratio of the change in the collector current to the change in the emitter current at constant collector-to-base voltage and is called the *negative of the small-signal short-circuit current transfer ratio, or gain.* More specifically,

$$\alpha' \equiv \frac{\Delta I_C}{\Delta I_E}\bigg|_{V_{CB}} \tag{5-8}$$

On the assumption that α is independent of I_E, then from Eq. (5-3) it follows that $\alpha' = -\alpha$.

5-4   TRANSISTOR CONSTRUCTION

Four basic techniques have been developed for the manufacture of diodes, transistors, and other semiconductor devices. Consequently, such devices may be classified[2,3] into one of the following types: grown, alloy, diffusion, or epitaxial.

Grown Type   The n-p-n grown-junction transistor is illustrated in Fig. 5-5a. It is made by drawing a single crystal from a melt of silicon or germanium whose impurity concentration is changed during the crystal-drawing operation by adding n- or p-type atoms as required.

Alloy Type   This technique, also called the *fused* construction, is illustrated in Fig. 5-5 for a p-n-p transistor. The center (base) section is a thin wafer of n-type material. Two small dots of indium are attached to opposite sides of the wafer, and the whole structure is raised for a short time to a high temperature, above the melting point of indium but below that of germanium. The indium dissolves the germanium beneath it and forms a saturation solution. On cooling, the germanium in contact with the base material recrystallizes, with enough indium concentration to change it from n to p type. The collector is made larger than the emitter, so that the collector subtends a large angle as viewed from the emitter. Because of this geometrical arrangement, very little emitter current follows a diffusion path which carries it to the base rather than to the collector.

Diffusion Type   This technique consists of subjecting a semiconductor wafer to gaseous diffusions of both n- and p-type impurities to form both the emitter and the collector junctions. A *planar* silicon transistor of the diffusion

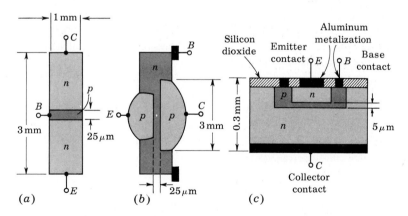

Fig. 5-5 Construction of transistors. (a) Grown, (b) alloy, and (c) diffused planar types. (The dimensions are approximate, and the figures are not drawn to scale.)

type is illustrated in Fig. 5-5c. In this process (described in greater detail in Chap. 7 on integrated-circuit fabrication), the base-collector junction area is determined by a diffusion mask. The emitter is then diffused on the base through a different mask. A thin layer of silicon dioxide is grown over the entire surface and photoetched, so that aluminum contacts can be made for the emitter and base leads (Fig. 5-5c). Because of the passivating action of this oxide layer, most surface problems are avoided and very low leakage currents result. There is also an improvement in the current gain at low currents and in the noise figure.

Epitaxial Type The epitaxial technique (Sec. 7-2) consists of growing a very thin, high-purity, single-crystal layer of silicon or germanium on a heavily doped substrate of the same material. This augmented crystal forms the collector on which the base and emitter may be diffused (Fig. 7-6).

The foregoing techniques may be combined to form a large number of methods for constructing transistors. For example, there are *diffused-alloy* types, *grown-diffused* devices, *alloy-emitter–epitaxial-base* transistors, etc. The special features of transistors of importance at high frequencies are discussed in Chap. 11. The volt-ampere characteristics at low frequencies of all types of junction transistors are essentially the same, and the discussion to follow applies to them all.

Finally, because of its historical significance, let us mention the first type of transistor to be invented. This device consists of two sharply pointed tungsten wires pressed against a semiconductor wafer. However, the reliability and reproducibility of such point-contact transistors are very poor, and as a result these transistors are no longer of practical importance.

5-5   THE COMMON–BASE CONFIGURATION

In Fig. 5-3a, a p-n-p transistor is shown in a *grounded-base* configuration. This circuit is also referred to as a *common-base*, or CB, configuration, since the base is common to the input and output circuits. For a p-n-p transistor the largest current components are due to holes. Since holes flow from the emitter to the collector and down toward ground out of the base terminal, then, referring to the polarity conventions of Fig. 5-1, we see that I_E is positive, I_C is negative, and I_B is negative. For a forward-biased emitter junction, V_{EB} is positive, and for a reverse-biased collector junction, V_{CB} is negative. For an n-p-n transistor all current and voltage polarities are the negative of those for a p-n-p transistor.

From Eq. (5-6) we see that the output (collector) current I_C is completely determined by the input (emitter) current I_E and the output (collector-to-base) voltage $V_{CB} = V_C$. This output relationship may be written in implicit form as

$$I_C = \phi_2(V_{CB}, I_E) \tag{5-9}$$

(This equation is read "I_C is some function ϕ_2 of V_{CB} and I_E.")

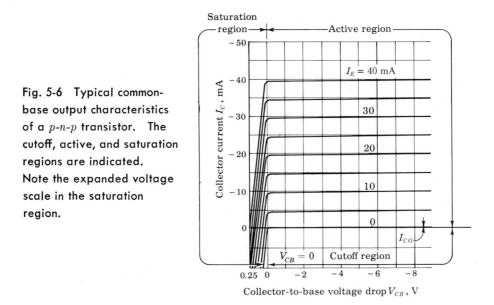

Fig. 5-6 Typical common-base output characteristics of a *p-n-p* transistor. The cutoff, active, and saturation regions are indicated. Note the expanded voltage scale in the saturation region.

Similarly, if we continue to choose V_{CB} and I_E as independent variables, the input (emitter-to-base) voltage V_{EB} is completely determined from these two variables. In implicit form the input characteristic is given by

$$V_{EB} = \phi_1(V_{CB}, I_E) \tag{5-10}$$

The relation of Eq. (5-9) is given in Fig. 5-6 for a typical *p-n-p* germanium transistor and is a plot of collector current I_C versus collector-to-base voltage drop V_{CB}, with emitter current I_E as a parameter. The curves of Fig. 5-6 are known as the *output*, or *collector, static characteristics*. The relation of Eq. (5-10) is given in Fig. 5-7 for the same transistor, and is a plot of emitter-to-base voltage V_{EB} versus emitter current I_E, with collector-to-base voltage V_{CB} as a parameter. This set of curves is referred to as the *input*, or *emitter, static characteristics*. We digress now to discuss a phenomenon which is used to account for the shapes of the transistor characteristics.

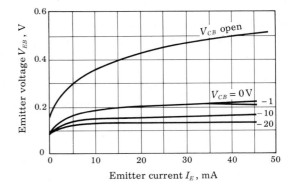

Fig. 5-7 Common-base input characteristics of a typical *p-n-p* germanium junction transistor.

The Early Effect, or Base-width Modulation[4] In Fig. 5-3 the narrow space-charge regions in the neighborhood of the junctions are neglected. This restriction is now to be removed. From Eq. (3-21) we note that the width W of the depletion region of a diode increases with the magnitude of the reverse voltage. Since the emitter junction is forward-biased but the collector junction is reverse-biased in the active region, then in Fig. 5-8 the barrier width at J_E is negligible compared with the space-charge width W at J_C.

The transition region at a junction is the region of uncovered charges on both sides of the junction at the positions occupied by the impurity atoms. As the voltage applied across the junction increases, the transition region penetrates deeper into the collector and base. Because neutrality of charge must be maintained, the number of uncovered charges on each side remains equal. Since the doping in the base is ordinarily substantially smaller than that of the collector, the penetration of the transition region into the base is much larger than into the collector. Hence the collector depletion region is neglected in Fig. 5-8, and all the immobile charge is indicated in the base region.

If the metallurgical base width is W_B, then the effective electrical base width is $W'_B = W_B - W$. This modulation of the effective base width by the collector voltage is known as the *Early effect*. The decrease in W'_B with increasing reverse collector voltage has three consequences: First, there is less chance for recombination within the base region. Hence α increases with increasing $|V_{CB}|$. Second, the concentration gradient of minority carriers p_n is increased within the base, as indicated in Fig. 5-8b. Note that p_n becomes

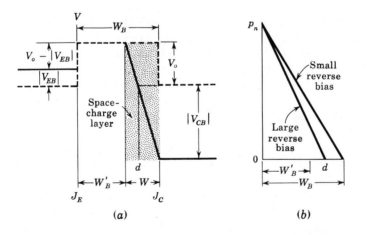

(a)   (b)

Fig. 5-8 (a) The potential variation through a p-n-p transistor. The space-charge width W at the collector junction increases, and hence the effective base width W'_B decreases with increasing $|V_{CB}|$. (Compare with Fig. 5-3.) (b) The injected minority-carrier charge density within the base.

zero at the distance d (between W'_B and W_B), where the potential with respect to the base falls below V_o. At this distance the effective applied potential becomes negative, and the law of the junction, Eq. (3-4), yields $p_n = 0$. Since the hole current injected across the emitter is proportional to the gradient of p_n at J_E, then I_E increases with increasing reverse collector voltage. Third, for extremely large voltages, W'_B may be reduced to zero, as in Fig. 5-24, causing voltage breakdown in the transistor. This phenomenon of *punch-through* is discussed further in Sec. 5-13.

The Input Characteristics   A qualitative understanding of the form of the input and output characteristics is not difficult if we consider the fact that the transistor consists of two diodes placed in series "back to back" (with the two cathodes connected together). In the active region the input diode (emitter-to-base) is biased in the forward direction. The input characteristics of Fig. 5-7 represent simply the forward characteristic of the emitter-to-base diode for various collector voltages. A noteworthy feature of the input characteristics is that there exists a *cutin, offset,* or *threshold,* voltage V_γ below which the emitter current is very small. In general, V_γ is approximately 0.1 V for germanium transistors (Fig. 5-7) and 0.5 V for silicon.

The shape of the input characteristics can be understood if we consider the fact that an increase in magnitude of collector voltage will, by the Early effect, cause the emitter current to increase, with V_{EB} held constant. Thus the curves shift downward as $|V_{CB}|$ increases, as noted in Fig. 5-7. The curve with the collector open represents the characteristic of the forward-biased emitter diode.

The Output Characteristics   Note, as in Fig. 5-6, that it is customary to plot along the abscissa and to the right that polarity of V_{CB} which reverse-biases the collector junction even if this polarity is negative. If $I_E = 0$, the collector current is $I_C = I_{CO}$. For other values of I_E, the output-diode reverse current is augmented by the fraction of the input-diode forward current which reaches the collector. Note also that I_C and I_{CO} are negative for a *p-n-p* transistor and positive for an *n-p-n* transistor.

Active Region   In this region *the collector junction is biased in the reverse direction and the emitter junction in the forward direction.* Consider first that the emitter current is zero. Then the collector current is small and equals the reverse saturation current I_{CO} (microamperes for germanium and nanoamperes for silicon) of the collector junction considered as a diode. Suppose now that a forward emitter current I_E is caused to flow in the emitter circuit. Then a fraction $-\alpha I_E$ of this current will reach the collector, and I_C is therefore given by Eq. (5-3). In the active region, the collector current is essentially independent of collector voltage and depends only upon the emitter current. However, because of the Early effect, we note in Fig. 5-6 that there actually is a small (perhaps 0.5 percent) increase in $|I_C|$ with $|V_{CB}|$. Because α is less than, but almost equal to, unity, the magnitude of the collector current is (slightly) less than that of the emitter current.

Saturation Region   The region to the left of the ordinate, $V_{CB} = 0$, and above the $I_E = 0$ characteristics, in which *both emitter and collector junctions are forward-biased*, is called the *saturation* region.   We say that "bottoming" has taken place because the voltage has fallen near the bottom of the characteristic where $V_{CB} \approx 0$.   Actually, V_{CB} is slightly positive (for a *p-n-p* transistor) in this region, and this forward biasing of the collector accounts for the large change in collector current with small changes in collector voltage.   For a forward bias, I_C increases exponentially with voltage according to the diode relationship [Eq. (3-9)].   A forward bias means that the collector *p* material is made positive with respect to the base *n* side, and hence that hole current flows from the *p* side across the collector junction to the *n* material.   This hole flow corresponds to a positive change in collector current.   Hence the collector current increases rapidly, and as indicated in Fig. 5-6, I_C may even become positive if the forward bias is sufficiently large.

Cutoff Region   The characteristic for $I_E = 0$ passes through the origin, but is otherwise similar to the other characteristics.   This characteristic is not coincident with the voltage axis, though the separation is difficult to show because I_{CO} is only a few nanoamperes or microamperes.   The region below the $I_E = 0$ characteristic, for which the *emitter and collector junctions are both reverse-biased*, is referred to as the *cutoff* region.   The temperature characteristics of I_{CO} are discussed in Sec. 5-7.

5-6   THE COMMON–EMITTER CONFIGURATION

Most transistor circuits have the emitter, rather than the base, as the terminal common to both input and output.   Such a *common-emitter* (CE), or *grounded-emitter*, configuration is indicated in Fig. 5-9.   In the common-emitter (as in the common-base) configuration, the input current and the output voltage are taken as the independent variables, whereas the input voltage and output

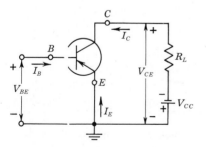

Fig. 5-9   A transistor common-emitter configuration.   The symbol V_{CC} is a positive number representing the magnitude of the supply voltage.

current are the dependent variables. We may write

$$V_{BE} = f_1(V_{CE}, I_B) \qquad\qquad (5\text{-}11)$$

$$I_C = f_2(V_{CE}, I_B) \qquad\qquad (5\text{-}12)$$

Equation (5-11) describes the family of input characteristic curves, and Eq. (5-12) describes the family of output characteristic curves. Typical output and input characteristic curves for a p-n-p junction germanium transistor are given in Figs. 5-10 and 5-11, respectively. In Fig. 5-10 the abscissa is the collector-to-emitter voltage V_{CE}, the ordinate is the collector current I_C, and the curves are given for various values of base current I_B. For a fixed value of I_B, the collector current is not a very sensitive value of V_{CE}. However, the slopes of the curves of Fig. 5-10 are larger than in the common-base characteristics of Fig. 5-6. Observe also that the base current is much smaller than the emitter current.

The locus of all points at which the collector dissipation is 150 mW is indicated in Fig. 5-10 by a solid line $P_C = 150$ mW. This curve is the hyperbola $P_C = V_{CB}I_C \approx V_{CE}I_C =$ constant. To the right of this curve the rated collector dissipation is exceeded. In Fig. 5-10 we have selected $R_L = 500\ \Omega$ and a supply $V_{CC} = 10$ V and have superimposed the corresponding load line on the output characteristics. The method of constructing a load line is identical with that explained in Sec. 4-2 in connection with a diode.

The Input Characteristics   In Fig. 5-11 the abscissa is the base current I_B, the ordinate is the base-to-emitter voltage V_{BE}, and the curves are given for various values of collector-to-emitter voltage V_{CE}. We observe that, with the collector shorted to the emitter and the emitter forward-biased, the input characteristic is essentially that of a forward-biased diode. If V_{BE} becomes zero, then I_B will be zero, since under these conditions both emitter and collector junctions will be short-circuited. In general, increasing $|V_{CE}|$ with constant

Fig. 5-10 Typical common-emitter output characteristics of a p-n-p germanium junction transistor. A load line corresponding to $V_{CC} = 10$ V and $R_L = 500\ \Omega$ is superimposed. (Courtesy of Texas Instruments, Inc.)

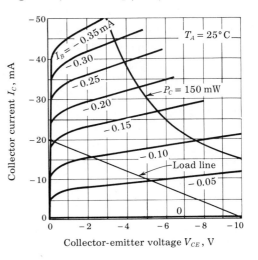

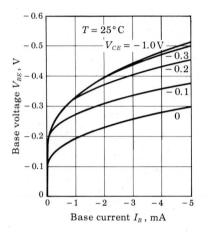

Fig. 5-11 Typical common-emitter input characteristics of the *p-n-p* germanium junction transistor of Fig. 5-10.

V_{BE} causes a decrease in base width W'_B (Fig. 5-8) and results in a decreasing recombination base current. These considerations account for the shape of input characteristics shown in Fig. 5-11.

The input characteristics for silicon transistors are similar in form to those in Fig. 5-11. The only notable difference in the case of silicon is that the curves break away from zero current in the range 0.5 to 0.6 V, rather than in the range 0.1 to 0.2 V as for germanium.

The Output Characteristics This family of curves may be divided into three regions, just as was done for the CB configuration. The first of these, the *active region*, is discussed here, and the *cutoff* and *saturation regions* are considered in the next two sections.

In the active region *the collector junction is reverse-biased and the emitter junction is forward-biased.* In Fig. 5-10 the active region is the area to the right of the ordinate V_{CE} = a few tenths of a volt and above $I_B = 0$. In this region the transistor output current responds most sensitively to an input signal. If the transistor is to be used as an amplifying device without appreciable distortion, it must be restricted to operate in this region.

The common-emitter characteristics in the active region are readily understood qualitatively on the basis of our earlier discussion of the common-base configuration. From Kirchhoff's current law (KCL) applied to Fig. 5-9, the base current is

$$I_B = -(I_C + I_E) \tag{5-13}$$

Combining this equation with Eq. (5-3), we find

$$I_C = \frac{I_{CO}}{1 - \alpha} + \frac{\alpha I_B}{1 - \alpha} \tag{5-14}$$

If we define β by

$$\beta \equiv \frac{\alpha}{1 - \alpha} \tag{5-15}$$

then Eq. (5-14) becomes

$$I_C = (1 + \beta)I_{CO} + \beta I_B \tag{5-16}$$

Note that usually $I_B \gg I_{CO}$, and hence $I_C \approx \beta I_B$ in the active region.

If α were truly constant, then, according to Eq. (5-14), I_C would be independent of V_{CE} and the curves of Fig. 5-10 would be horizontal. Assume that, because of the Early effect, α increases by only one-half of 1 percent, from 0.98 to 0.985, as $|V_{CE}|$ increases from a few volts to 10 V. Then the value of β increases from $0.98/(1 - 0.98) = 49$ to $0.985/(1 - 0.985) = 66$, or about 34 percent. This numerical example illustrates that a very small change (0.5 percent) in α is reflected in a very large change (34 percent) in the value of β. It should also be clear that a slight change in α has a large effect on β, and hence upon the common-emitter curves. Therefore the common-emitter characteristics are normally subject to a wide variation even among transistors of a given type. This variability is caused by the fact that I_B is the difference between large and nearly equal currents, I_E and I_C.

EXAMPLE (a) Find the transistor currents in the circuit of Fig. 5-12a. A silicon transistor with $\beta = 100$ and $I_{CO} = 20\ nA = 2 \times 10^{-5}$ mA is under consideration. (b) Repeat part a if a 2-K emitter resistor is added to the circuit, as in Fig. 5-12b.

Solution a. Since the base is forward-biased, the transistor is not cut off. Hence it must be either in its active region or in saturation. Assume that the transistor operates in the active region. From KVL applied to the base circuit of Fig.

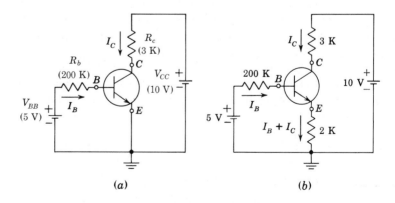

(a)   (b)

Fig. 5-12 An example illustrating how to determine whether or or not a transistor is operating in the active region.

5-12a (with I_B expressed in milliamperes), we have

$$-5 + 200\, I_B + V_{BE} = 0$$

As noted in Table 5-1, a reasonable value for V_{BE} is 0.7 V in the active region. Hence

$$I_B = \frac{5 - 0.7}{200} = 0.0215 \text{ mA}$$

Since $I_{CO} \ll I_B$, then $I_C \approx \beta I_B = 2.15$ mA.

We must now justify our assumption that the transistor is in the active region, by verifying that the collector junction is reverse-biased. From KVL applied to the collector circuit we obtain

$$-10 + 3\, I_C + V_{CB} + V_{BE} = 0$$

or

$$V_{CB} = -(3)(2.15) + 10 - 0.7 = +2.85 \text{ V}$$

For an n-p-n device a positive value of V_{CB} represents ε reverse-biased collector junction, and hence the transistor is indeed in its active region.

Note that I_B and I_C in the active region are independent of the collector circuit resistance R_c. Hence, if R_c is increased sufficiently above 3 K, then V_{CB} changes from a positive to a negative value, indicating that the transistor is no longer in its active region. The method of calculating I_B and I_C when the transistor is in saturation is given in Sec. 5-9.

b. The current in the emitter resistor of Fig. 5-12b is

$$I_B + I_C \approx I_B + \beta I_B = 101\, I_B,$$

assuming $I_{CO} \ll I_B$. Applying KVL to the base circuit yields

$$-5 + 200 I_B + 0.7 + (2)(101\, I_B) = 0$$

or

$$I_B = 0.0107 \text{ mA} \qquad I_C = 100\, I_B = 1.07 \text{ mA}$$

Note that $I_{CO} = 2 \times 10^{-5}$ mA $\ll I_B$, as assumed.

To check for active circuit operation, we calculate V_{CB}. Thus

$$V_{CB} = -3 I_C + 10 - (2)(101\, I_B) - 0.7$$

$$= -(3)(1.07) + 10 - (2)(101)(0.0107) - 0.7 = +3.93 \text{ V}$$

Since V_{CB} is positive, this (n-p-n) transistor is in its active region.

5-7   THE CE CUTOFF REGION

We might be inclined to think that cutoff in Fig. 5-10 occurs at the intersection of the load line with the current $I_B = 0$; however, we now find that appreciable collector current may exist under these conditions. From Eqs.

(5-13) and (5-14), if $I_B = 0$, then $I_E = -I_C$ and

$$I_C = -I_E = \frac{I_{CO}}{1 - \alpha} \equiv I_{CEO} \qquad (5\text{-}17)$$

The actual collector current with collector junction reverse-biased and base open-circuited is designated by the symbol I_{CEO}. Since, even in the neighborhood of cutoff, α may be as large as 0.9 for germanium, then $I_C \approx 10 I_{CO}$ at zero base current. Accordingly, in order to cut off the transistor, it is not enough to reduce I_B to zero. Instead, it is necessary to reverse-bias the emitter junction slightly. We shall define cutoff as the condition where the collector current is equal to the reverse saturation current I_{CO} and the emitter current is zero. It is found (Sec. 19-15) that a reverse-biasing voltage of the order of 0.1 V established across the emitter junction will ordinarily be adequate to cut off a germanium transistor. In silicon, at collector currents of the order of I_{CO}, α is very nearly zero because of recombination[5,6] in the emitter-junction transition region. Hence, even with $I_B = 0$, we find, from Eq. (5-17), that $I_C = I_{CO} = -I_E$, so that the transistor is still very close to cutoff. We verify in Sec. 19-15 that, in silicon, cutoff occurs at $V_{BE} \approx 0$ V corresponding to a base short-circuited to the emitter. *In summary, cutoff means that* $I_E = 0$, $I_C = I_{CO}$, $I_B = -I_C = -I_{CO}$, *and* V_{BE} *is a reverse voltage whose magnitude is of the order of* 0.1 V *for germanium and* 0 V *for a silicon transistor.*

The Reverse Collector Saturation Current I_{CBO}   The collector current in a physical transistor (a real, nonidealized, or commercial device) when the emitter current is zero is designated by the symbol I_{CBO}. Two factors cooperate to make $|I_{CBO}|$ larger than $|I_{CO}|$. First, there exists a leakage current which flows, not through the junction, but around it and across the surfaces. The leakage current is proportional to the voltage across the junction. The second reason why $|I_{CBO}|$ exceeds $|I_{CO}|$ is that new carriers may be generated by collision in the collector-junction transition region, leading to avalanche multiplication of current and eventual breakdown. But even before breakdown is approached, this *multiplication* component of current may attain considerable proportions.

At 25°C, I_{CBO} for a germanium transistor whose power dissipation is in the range of some hundreds of milliwatts is of the order of microamperes. Under similar conditions a silicon transistor has an I_{CBO} in the range of nanoamperes. The temperature sensitivity of I_{CBO} is the same as that of the reverse saturation current I_O of a p-n diode (Sec. 3-5). Specifically, it is found[8] that I_{CBO} approximately doubles for every 10°C increase in temperature for both Ge and Si. However, because of the lower absolute value of I_{CBO} in silicon, these transistors may be used up to about 200°C, whereas germanium transistors are limited to about 100°C.

In addition to the variability of reverse saturation current with temperature, there is also a wide variability of reverse current among samples of a

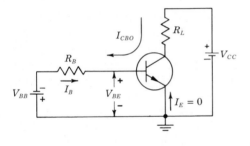

Fig. 5-13 Reverse biasing of the emitter junction to maintain the transistor in cutoff in the presence of the reverse saturation current I_{CBO} through R_B.

given transistor type. For example, the specification sheet for a Texas Instrument type 2N337 grown-diffused silicon switching transistor indicates that this type number includes units with values of I_{CBO} extending over the tremendous range from 0.2 nA to 0.3 μA. Accordingly, any particular transistor may have an I_{CBO} which differs very considerably from the average characteristic for the type.

Circuit Considerations at Cutoff Because of temperature effects, avalanche multiplication, and the wide variability encountered from sample to sample of a particular transistor type, even silicon may have values of I_{CBO} of the order of many tens of microamperes. Consider the circuit configuration of Fig. 5-13, where V_{BB} represents a biasing voltage intended to keep the transistor cutoff. Assume that the transistor is just at the point of cutoff, with $I_E = 0$, so that $I_B = -I_{CBO}$. If we require that at cutoff $V_{BE} \approx -0.1$ V, then the condition of cutoff requires that

$$V_{BE} = -V_{BB} + R_B I_{CBO} \leq -0.1 \text{ V} \tag{5-18}$$

As an extreme example consider that R_B is, say, as large as 100 K and that we want to allow for the contingency that I_{CBO} may become as large as 100 μA. Then V_{BB} must be at least 10.1 V. When I_{CBO} is small, the magnitude of the voltage across the base-emitter junction will be 10.1 V. Hence we must use a transistor whose maximum allowable reverse base-to-emitter junction voltage before breakdown exceeds 10 V. It is with this contingency in mind that a manufacturer supplies a rating for the reverse *breakdown voltage* between emitter and base, represented by the symbol BV_{EBO}. The subscript O indicates that BV_{EBO} is measured under the condition that the collector current is zero. Breakdown voltages BV_{EBO} may be as high as some tens of volts or as low as 0.5 V. If $BV_{EBO} = 1$ V, then V_{BB} must be chosen to have a maximum value of 1 V.

5-8   THE CE SATURATION REGION

In the saturation region *the collector junction (as well as the emitter junction) is forward-biased by at least the cutin voltage.* Since the voltage V_{BE} (or V_{BC})

across a forward-biased junction has a magnitude of only a few tenths of a volt, the $V_{CE} = V_{BE} - V_{BC}$ is also only a few tenths of a volt at saturation. Hence, in Fig. 5-10, the saturation region is very close to the zero-voltage axis, where all the curves merge and fall rapidly toward the origin. A load line has been superimposed on the characteristics of Fig. 5-10 corresponding to a resistance $R_L = 500\ \Omega$ and a supply voltage of 10 V. We note that in the saturation region the collector current is approximately independent of base current, for given values of V_{CC} and R_L. Hence we may consider that the onset of saturation takes place at the knee of the transistor curves in Fig. 5-10. Saturation occurs for the given load line at a base current of -0.17 mA, and at this point the collector voltage is too small to be read in Fig. 5-10. In saturation, the collector current is nominally V_{CC}/R_L, and since R_L is small, it may well be necessary to keep V_{CC} correspondingly small in order to stay within the limitations imposed by the transistor on maximum current and dissipation.

We are not able to read the collector-to-emitter saturation voltage, $V_{CE,\text{sat}}$, with any precision from the plots of Fig. 5-10. We refer instead to the characteristics shown in Fig. 5-14. In these characteristics the 0- to -0.5-V region of Fig. 5-10 has been expanded, and we have superimposed the same load line as before, corresponding to $R_L = 500\ \Omega$. We observe from Figs. 5-10 and 5-14 that V_{CE} and I_C no longer respond appreciably to base current I_B, after the base current has attained the value -0.15 mA. At this current the transistor enters saturation. For $I_B = -0.15$ mA, $|V_{CE}| \approx$ 175 mV. At $I_B = -0.35$ mA, $|V_{CE}|$ has dropped to $|V_{CE}| \approx 100$ mV. Larger magnitudes of I_B will, of course, decrease $|V_{CE}|$ slightly further.

Saturation Resistance For a transistor operating in the saturation region, a quantity of interest is the ratio $V_{CE,\text{sat}}/I_C$. This parameter is called the *common-emitter saturation resistance*, variously abbreviated R_{CS}, R_{CES}, or $R_{CE,\text{sat}}$. To specify R_{CS} properly, we must indicate the operating

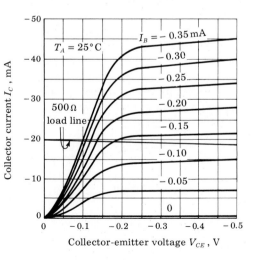

Fig. 5-14 Saturation-region common-emitter characteristics of the type 2N404 germanium transistor. A load line corresponding to V_{CC} = 10 V and R_L = 500 Ω is superimposed. (Courtesy of Texas Instruments, Inc.)

point at which it was determined.   For example, from Fig. 5-14, we find that, at $I_C = -20$ mA and $I_B = -0.35$ mA, $R_{CS} \approx -0.1/(-20 \times 10^{-3}) = 5$ Ω. The usefulness of R_{CS} stems from the fact, as appears in Fig. 5-14, that to the left of the knee each of the plots, for fixed I_B, may be approximated, at least roughly, by a straight line.

The Base-spreading Resistance $r_{bb'}$ Recalling that the base region is very thin (Fig. 5-5), we see that the current which enters the base region across the emitter junction must flow through a long narrow path to reach the base terminal.   The cross-sectional area for current flow in the collector (or emitter) is very much larger than in the base.   Hence, usually the ohmic resistance of the base is very much larger than that of the collector or emitter.   The dc ohmic base resistance, designated by $r_{bb'}$, is called the *base-spreading resistance* and is of the order of magnitude of 100 Ω.

The Temperature Coefficient of the Saturation Voltages   Since both junctions are forward-biased, a reasonable value for the temperature coefficient of either $V_{BE,\text{sat}}$ or $V_{BC,\text{sat}}$ is -2.5 mV/°C.   In saturation the transistor consists of two forward-biased diodes back to back in series opposing.   Hence it is to be anticipated that the temperature-induced voltage change in one junction will be canceled by the change in the other junction.   We do indeed find such to be the case for $V_{CE,\text{sat}}$, whose temperature coefficient is about one-tenth that of $V_{BE,\text{sat}}$.

The DC Current Gain h_{FE} A transistor parameter of interest is the ratio I_C/I_B, where I_C is the collector current and I_B is the base current.   This quantity is designated by β_{dc} or h_{FE}, and is known as the (negative of the) *dc beta*, the *dc forward current transfer ratio*, or the *dc current gain*.

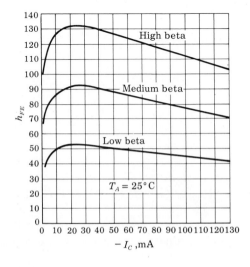

Fig. 5-15   Plots of dc current gain h_{FE} (at $V_{CE} = -0.25$ V) versus collector current for three samples of the type 2N404 germanium transistor. (Courtesy of General Electric Company.)

In the saturation region, the parameter h_{FE} is a useful number and one which is usually supplied by the manufacturer when a switching transistor is involved. We know $|I_C|$, which is given approximately by V_{CC}/R_L, and a knowledge of h_{FE} tells us the minimum base current (I_C/h_{FE}) which will be needed to saturate the transistor. For the type 2N404, the variation of h_{FE} with collector current at a low value of V_{CE} is as given in Fig. 5-15. Note the wide spread (a ratio of 3:1) in the value which may be obtained for h_{FE} even for a transistor of a particular type. Commercially available transistors have values of h_{FE} that cover the range from 10 to 150 at collector currents as small as 5 mA and as large as 30 A.

5-9   TYPICAL TRANSISTOR—JUNCTION VOLTAGE VALUES

The characteristics plotted in Fig. 5-16 of output current I_C as a function of input voltage V_{BE} for *n-p-n* germanium and silicon transistors are quite instructive and indicate the several regions of operation for a CE transistor circuit. The numerical values indicated are typical values obtained experimentally or from the theoretical equations of the following section. (The calculations are made in Sec. 19-15.) Let us examine the various portions of the transfer curves of Fig. 5-16.

The Cutoff Region *Cutoff* is defined, as in Sec. 5-5, to mean $I_E = 0$ and $I_C = I_{CO}$, and it is found that a *reverse* bias $V_{BE,\text{cutoff}} = 0.1$ V (0 V) will cut off a germanium (silicon) transistor.

What happens if a larger reverse voltage than $V_{BE,\text{cutoff}}$ is applied? It turns out that if V_{BE} is reverse biased and much larger than V_T, that the collector current falls slightly below I_{CO} and that the emitter current *reverses* but remains small in magnitude (less than I_{CO}).

Short-circuited Base Suppose that, instead of reverse-biasing the emitter junction, we connect the base to the emitter so that $V_E = V_{BE} = 0$. As indicated in Fig. 5-16, $I_C \equiv I_{CES}$ does not increase greatly over its cutoff value I_{CO}.

Open-circuited Base If instead of a shorted base we allow the base to "float" so that $I_B = 0$, we obtain the $I_C \equiv I_{CEO}$ given in Eq. (5-17). At low currents $\alpha \approx 0.9$ (0) for Ge (Si), and hence $I_C \approx 10I_{CO}(I_{CO})$ for Ge (Si). The values of V_{BE} calculated for this open-base condition $(I_C = -I_E)$ are a few tens of millivolts of *forward* bias, as indicated in Fig. 5-16.

The Cutin Voltage The volt-ampere characteristic between base and emitter at constant collector-to-emitter voltage (Fig. 5-11) is not unlike the volt-ampere characteristic of a simple junction diode. When the emitter junction is reverse-biased, the base current is very small, being of the order of nanoamperes or microamperes, for silicon and germanium, respectively.

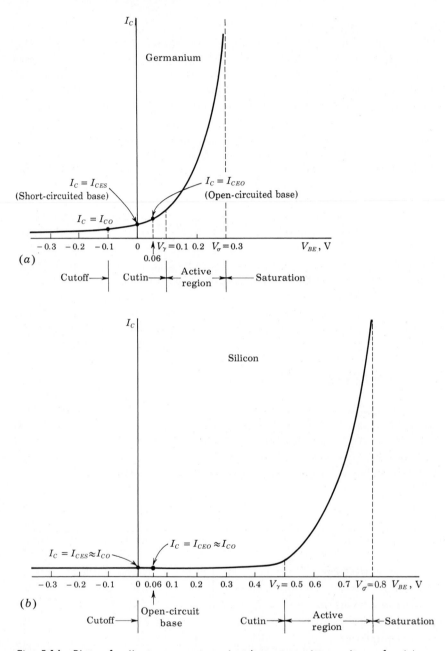

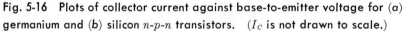

Fig. 5-16   Plots of collector current against base-to-emitter voltage for (a) germanium and (b) silicon n-p-n transistors.   (I_C is not drawn to scale.)

When the emitter junction is forward-biased, again, as in the simple diode, no appreciable base current flows until the emitter junction has been forward-biased to the extent where $|V_{BE}| \geq |V_\gamma|$, where V_γ is called the *cutin voltage*. Since the collector current is nominally proportional to the base current, no appreciable collector current will flow until an appreciable base current flows. Therefore a plot of collector current against base-to-emitter voltage will exhibit a cutin voltage, just as does the simple diode.

In principle, a transistor is in its active region whenever the base-to-emitter voltage is on the forward-biasing side of the cutoff voltage, which occurs at a reverse voltage of 0.1 V for germanium and 0 V for silicon. In effect, however, a transistor enters its active region when $V_{BE} > V_\gamma$.

We may estimate the cutin voltage V_γ by assuming that $V_{BE} = V_\gamma$ when the collector current reaches, say, 1 percent of the maximum (saturation) current in the CE circuit of Fig. 5-9. Typical values of V_γ are 0.1 V for germanium and 0.5 V for silicon.

Figure 5-17 shows plots, for several temperatures, of the collector current as a function of the base-to-emitter voltage at constant collector-to-emitter voltage for a typical silicon transistor. We see that a value for V_γ of the order of 0.5 V at room temperature is entirely reasonable. The temperature dependence results from the temperature coefficient of the emitter-junction diode. Therefore the lateral shift of the plots with change in temperature and the change with temperature of the cutin voltage V_γ are approximately -2.5 mV/°C [Eq. (3-12)].

Saturation Voltages Manufacturers specify saturation values of input and output voltages in a number of different ways, in addition to supplying characteristic curves such as Figs. 5-11 and 5-14. For example, they may specify R_{CS} for several values of I_B or they may supply curves of $V_{CE,\text{sat}}$ and $V_{BE,\text{sat}}$ as functions of I_B and I_C.[9] The saturation voltages depend not only

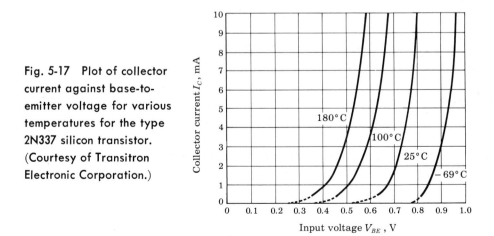

Fig. 5-17 Plot of collector current against base-to-emitter voltage for various temperatures for the type 2N337 silicon transistor. (Courtesy of Transitron Electronic Corporation.)

on the operating point, but also on the semiconductor material (germanium or silicon) and on the type of transistor construction. Typical values of saturation voltages are indicated in Table 5-1.

TABLE 5-1   Typical n-p-n transistor-junction voltages at 25°C

	$V_{CE,\text{sat}}$	$V_{BE,\text{sat}} \equiv V_\sigma$	$V_{BE,\text{active}}$	$V_{BE}\dagger,\text{cutin}} \equiv V_\gamma$	$V_{BE,\text{cutoff}}$
Si	0.2	0.8	0.7	0.5	0.0
Ge	0.1	0.3	0.2	0.1	−0.1

† The temperature variation of these voltages is discussed in Secs. 5-8 and 5-9.

Summary The voltages referred to above and indicated in Fig. 5-16 are summarized in Table 5-1. The entries in the table are appropriate for an n-p-n transistor. For a p-n-p transistor the signs of all entries should be reversed. Observe that the total range of V_{BE} between cutin and saturation is rather small, being only 0.3 V. The voltage $V_{BE,\text{active}}$ has been located somewhat arbitrarily, but nonetheless reasonably, near the midpoint of the active region in Fig. 5-16.

Of course, particular cases will depart from the estimates of Table 5-1. But it is unlikely that the numbers will be found in error by more than 0.1 V.

EXAMPLE (a) The circuits of Fig. 5-12a and b are modified by changing the base-circuit resistance from 200 to 50 K (as indicated in Fig. 5-18). If $h_{FE} = 100$, determine whether or not the silicon transistor is in saturation and find I_B and I_C. (b) Repeat with the 2K emitter resistance added.

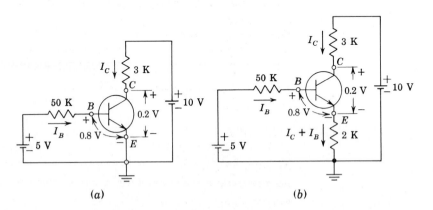

(a)   (b)

Fig. 5-18 An example illustrating how to determine whether or not a transistor is operating in the saturation region.

Solution   Assume that the transistor is in saturation. Using the values $V_{BE,\text{sat}}$ and $V_{CE,\text{sat}}$ in Table 5-1, the circuit of Fig. 5-18a is obtained. Applying KVL to the base circuit gives

$$-5 + 50\, I_B + 0.8 = 0$$

or

$$I_B = \frac{4.2}{50} = 0.084 \text{ mA}$$

Applying KVL to the collector circuit yields

$$-10 + 3\, I_C + 0.2 = 0$$

or

$$I_C = \frac{9.8}{3} = 3.27 \text{ mA}$$

The minimum value of base current required for saturation is

$$(I_B)_{\min} = \frac{I_C}{h_{FE}} = \frac{3.27}{100} = 0.033 \text{ mA}$$

Since $I_B = 0.084 > I_{B,\min} = 0.033$ mA, we have verified that the transistor is in saturation.

b. If the 2-K emitter resistance is added, the circuit becomes that in Fig. 5-18b. Assume that the transistor is in saturation. Applying KVL to the base and collector circuits, we obtain

$$-5 + 50\, I_B + 0.8 + 2\,(I_C + I_B) = 0$$

$$-10 + 3\, I_C + 0.2 + 2\,(I_C + I_B) = 0$$

If these simultaneous equations are solved for I_C and I_B, we obtain

$$I_C = 1.95 \text{ mA} \qquad I_B = 0.0055 \text{ mA}$$

Since $(I_B)_{\min} = I_C/h_{FE} = 0.0195$ mA $> I_B = 0.0055$, the transistor is *not* in saturation. Hence the device must be operating in the active region. Proceeding exactly as we did for the circuit of Fig. 5-12b (but with the 200 K replaced by 50 K), we obtain

$$I_C = 1.71 \text{ mA} \qquad I_B = 0.0171 \text{ mA} = 17 \text{ }\mu\text{A} \qquad V_{CB} = 0.72 \text{ V}$$

5-10   COMMON–EMITTER CURRENT GAIN

Three different definitions of current gain appear in the literature. The interrelationships between these are now to be found.

Large-signal Current Gain β   We define β in terms of α by Eq. (5-15). From Eq. (5-16), with I_{CO} replaced by I_{CBO}, we find

$$\beta = \frac{I_C - I_{CBO}}{I_B - (-I_{CBO})} \tag{5-19}$$

In Sec. 5-7 we define *cutoff* to mean that $I_E = 0$, $I_C = I_{CBO}$, and $I_B = -I_{CBO}$. Consequently, Eq. (5-19) gives the ratio of the collector-current increment to the base-current change from cutoff to I_B, and hence β *represents the* (negative of the) *large-signal current gain of a common-emitter transistor.* This parameter is of primary importance in connection with the biasing and stability of transistor circuits, as discussed in Chap. 9.

DC Current Gain h_{FE} In Sec. 5-8 we define the dc current gain by

$$\beta_{dc} \equiv \frac{I_C}{I_B} \equiv h_{FE} \tag{5-20}$$

In that section it is noted that h_{FE} is most useful in connection with determining whether or not a transistor is in saturation. In general, the base current (and hence the collector current) is large compared with I_{CBO}. Under these conditions the large-signal and the dc betas are approximately equal; then $h_{FE} \approx \beta$.

Small-signal Current Gain h_{fe} We define β' as the ratio of a collector-current increment ΔI_C for a small base-current change ΔI_B (at a given quiescent operating point, at a fixed collector-to-emitter voltage V_{CE}), or

$$\beta' \equiv \frac{\partial I_C}{\partial I_B}\bigg|_{V_{CE}} = h_{fe} \tag{5-21}$$

Clearly, β' is (the negative of) the *small-signal* current gain. If β were independent of current, we see from Eq. (5-20) that $\beta' = \beta \approx h_{FE}$. However, Fig. 5-15 indicates that β is a function of current, and differentiating Eq. (5-16) with respect to I_C gives (with $I_{CO} = I_{CBO}$)

$$1 = (I_{CBO} + I_B)\frac{\partial \beta}{\partial I_C} + \beta \frac{\partial I_B}{\partial I_C} \tag{5-22}$$

The small-signal CE gain β' is used in the analysis of small-signal amplifier circuits and is designated by h_{fe} in Chap. 8. Using Eq. (5-21), and with $\beta' = h_{fe}$ and $\beta = h_{FE}$, Eq. (5-22) becomes

$$h_{fe} = \frac{h_{FE}}{1 - (I_{CBO} + I_B)(\partial h_{FE}/\partial I_C)} \tag{5-23}$$

Since h_{FE} versus I_C given in Fig. 5-15 shows a maximum, h_{fe} is larger than h_{FE} for small currents (to the left of the maximum) and h_{fe} is smaller than h_{FE} for currents larger than that corresponding to the maximum. Over most of the wide current range in Fig. 5-14, h_{fe} differs from h_{FE} by less than 20 percent.

It should be emphasized that Eq. (5-23) is valid in the active region only. From Fig. 5-14 we see that $h_{fe} \to 0$ in the saturation region because $\Delta I_C \to 0$ for a small increment ΔI_B.

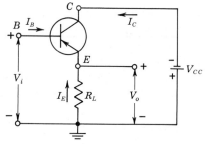

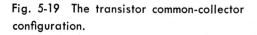

Fig. 5-19 The transistor common-collector
configuration.

5-11   THE COMMON–COLLECTOR CONFIGURATION

Another transistor-circuit configuration, shown in Fig. 5-19, is known as the
common-collector configuration. The circuit is basically the same as the circuit of Fig. 5-9, with the exception that the load resistor is in the emitter
lead rather than in the collector circuit. If we continue to specify the operation of the circuit in terms of the currents which flow, the operation for the
common-collector is much the same as for the common-emitter configuration.
When the base current is I_{CO}, the emitter current will be zero, and no current
will flow in the load. As the transistor is brought out of this back-biased
condition by increasing the magnitude of the base current, the transistor will
pass through the active region and eventually reach saturation. In this condition all the supply voltage, except for a very small drop across the transistor,
will appear across the load.

5-12   ANALYTICAL EXPRESSIONS FOR
TRANSISTOR CHARACTERISTICS

The dependence of the currents in a transistor upon the junction voltages, or
vice versa, may be obtained by starting with Eq. (5-6), repeated here for
convenience:

$$I_C = -\alpha_N I_E - I_{CO}(\epsilon^{V_C/V_T} - 1) \qquad (5\text{-}24)$$

We have added the subscript N to α to indicate that we are using the transistor in the *normal* manner. We must recognize, however, that there is no
essential reason which constrains us from using a transistor in an *inverted*
fashion, that is, interchanging the roles of the emitter junction and the collector junction. From a practical point of view, such an arrangement might
not be as effective as the *normal* mode of operation, but this matter does not
concern us now. With this inverted mode of operation in mind, we may now
write, in correspondence with Eq. (5-24),

$$I_E = -\alpha_I I_C - I_{EO}(\epsilon^{V_E/V_T} - 1) \qquad (5\text{-}25)$$

Here α_I is the *inverted* common-base current gain, just as α_N in Eq. (5-24) is the current gain in *normal* operation. I_{EO} is the emitter-junction reverse saturation current, and V_E is the voltage drop from p side to n side at the emitter junction and is positive for a forward-biased emitter.

Equations (5-24) and (5-25) were derived in a heuristic manner. A physical analysis of the transistor currents by Ebers and Moll[7] verifies these equations (Sec. 19-13). This quantitative study reveals that the parameters α_N, α_I, I_{CO}, and I_{EO} are not independent, but are related by the condition

$$\alpha_I I_{CO} = \alpha_N I_{EO} \tag{5-26}$$

Manufacturer's data sheets often provide information about α_N, I_{CO}, and I_{EO}, so that α_I may be determined. For many transistors I_{EO} lies in the range $0.5I_{CO}$ to I_{CO}.

Since the sum of the three currents must be zero, the base current is given by

$$I_B = -(I_E + I_C) \tag{5-27}$$

If three of the four parameters α_N, α_I, I_{CO}, and I_{EO} are known, the equations in this section allow calculations of the three currents for given values of junction voltages V_C and V_E. Explicit expressions for I_C and I_E in terms of V_C and V_E are found in Sec. 19-13.

In the literature, α_R (*reversed* alpha) and α_F (*forward* alpha) are sometimes used in place of α_I and α_N, respectively.

Reference Polarities The symbol $V_C(V_E)$ represents the drop across the collector (emitter) junction and is positive if the junction is forward-biased. The reference directions for currents and voltages are indicated in Fig. 5-20. Since V_{CB} represents the voltage drop from collector-to-base terminals, then V_{CB} differs from V_C by the ohmic drop in the base-spreading resistance $r_{bb'}$, or

$$V_{CB} = V_C - I_B r_{bb'} \tag{5-28}$$

The Ebers-Moll Model Equations (5-24) and (5-25) have a simple interpretation in terms of a circuit known as the *Ebers-Moll model*.[7] This model is shown in Fig. 5-21 for a *p-n-p* transistor. We see that it involves two ideal diodes placed back to back with reverse saturation currents $-I_{EO}$ and $-I_{CO}$ and two dependent current-controlled current sources shunting the ideal diodes. For a *p-n-p* transistor, both I_{CO} and I_{EO} are negative, so that $-I_{CO}$ and $-I_{EO}$ are positive values, giving the magnitudes of the reverse saturation currents of the diodes. The current sources account for the minority-carrier transport across the base. An application of KCL to the collector node of Fig. 5-21 gives

$$I_C = -\alpha_N I_E + I = -\alpha_N I_E + I_o(\epsilon^{V_C/V_T} - 1)$$

where the diode current I is given by Eq. (3-9). Since I_o is the magnitude

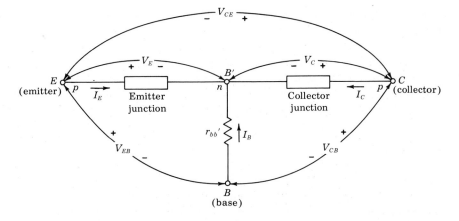

Fig. 5-20 Defining the voltages and currents used in the Ebers-Moll equations. For either a *p-n-p* or an *n-p-n* transistor, a positive value of current means that positive charge flows into the junction and a positive $V_E(V_C)$ means that the emitter (collector) junction is forward-biased (the *p* side positive with respect to the *n* side).

of the reverse saturation, then $I_o = -I_{CO}$. Substituting this value of I_o into the preceding equation for I_C yields Eq. (5-24).

This model is valid for both forward and reverse static voltages applied across the transistor junctions. It should be noted that we have omitted the base-spreading resistance from Fig. 5-20 and have neglected the difference between I_{CBO} and I_{CO}.

Observe from Fig. 5-21 that the dependent current sources can be eliminated from this figure provided $\alpha_N = \alpha_I = 0$. For example, by making the base width much larger than the diffusion length of minority carriers in the base, all minority carriers will recombine in the base and none will survive to reach the collector. For this case the current gain α will be zero. Under these conditions, transistor action ceases, and we simply have two diodes placed back to back. This discussion shows why it is impossible to construct a transistor by simply connecting two separate (isolated) diodes in series opposing. A cascade of two *p-n* diodes exhibits transistor properties (for

Fig. 5-21 The Ebers-Moll model for a *p-n-p* transistor.

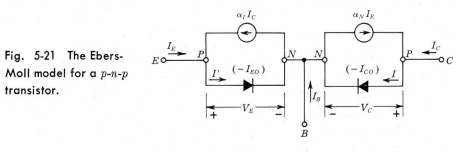

example, it is capable of amplification) only if carriers injected across one junction *diffuse* across the second junction.

Voltages as Functions of Currents We may solve explicitly for the junction voltages in terms of the currents from Eqs. (5-24) and (5-25), with the result that

$$V_E = V_T \ln \left(1 - \frac{I_E + \alpha_I I_C}{I_{EO}} \right) \qquad (5\text{-}29)$$

$$V_C = V_T \ln \left(1 - \frac{I_C + \alpha_N I_E}{I_{CO}} \right) \qquad (5\text{-}30)$$

We now derive the analytic expression for the common-emitter characteristics of Fig. 5-10. The abscissa in this figure is the collector-to-emitter voltage $V_{CE} = V_E - V_C$ for an *n-p-n* transistor and is $V_{CE} = V_C - V_E$ for a *p-n-p* transistor (remember that V_C and V_E are positive at the *p* side of the junction). Hence the common-emitter characteristics are found by subtracting Eqs. (5-29) and (5-30) and by eliminating I_E by the use of Eq. (5-27). The resulting equation can be simplified provided that the following inequalities are valid: $I_B \gg I_{EO}$ and $I_B \gg I_{CO}/\alpha_N$. After some manipulations and by the use of Eqs. (5-15) and (5-26), we obtain (except for very small values of I_B)

$$V_{CE} = \pm V_T \ln \frac{\dfrac{1}{\alpha_I} + \dfrac{1}{\beta_I} \dfrac{I_C}{I_B}}{1 - \dfrac{1}{\beta} \dfrac{I_C}{I_B}} \qquad (5\text{-}31)$$

where

$$\beta_I \equiv \frac{\alpha_I}{1 - \alpha_I} \quad \text{and} \quad \beta_N \equiv \beta \equiv \frac{\alpha}{1 - \alpha} \qquad (5\text{-}32)$$

Note that the $+$ sign in Eq. (5-31) is used for an *n-p-n* transistor, and the $-$ sign for a *p-n-p* device. For a *p-n-p* germanium-type transistor, at $I_C = 0$, $V_{CE} = -V_T \ln (1/\alpha_I)$, so that the *common-emitter characteristics do not pass through the origin*. For $\alpha_I = 0.78$ and $V_T = 0.026$ V, we have $V_{CE} \approx -6$ mV at room temperature. This voltage is so small that the curves of Fig. 5-10 look as if they pass through the origin, but they are actually displaced to the right by a few millivolts.

If I_C is increased, then V_{CE} rises only slightly until I_C/I_B approaches β. For example, even for $I_C/I_B = 0.9\beta = 90$ (for $\beta = 100$),

$$V_{CE} = -0.026 \ln \frac{1/0.78 + 90/3.5}{1 - 0.9} \approx -0.15 \text{ V}$$

This voltage can barely be detected at the scale to which Fig. 5-10 is drawn, and hence near the origin it appears as if the curves rise vertically. However, note that Fig. 5-14 confirms that a voltage of the order of 0.2 V is required for I_C to reach 0.9 of its maximum value.

The maximum value of I_C/I_B is β, and as this value of I_C/I_B is approached,

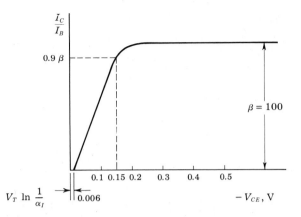

Fig. 5-22 The common-
emitter output characteristic for a p-n-p transistor as
obtained analytically.

$V_{CE} \to -\infty$. Hence, as I_C/I_B increases from 0.9β to β, $|V_{CE}|$ increases from
0.15 V to infinity. A plot of the theoretical common-emitter characteristic is
indicated in Fig. 5-22. We see that, at a fixed value of V_{CE}, the ratio I_C/I_B is
a constant. Hence, for equal increments in I_B, we should obtain equal increments in I_C at a given V_{CE}. This conclusion is fairly well satisfied by the
curves in Fig. 5-10. However, the $I_B = 0$ curve seems to be inconsistent since,
for a constant I_C/I_B, this curve should coincide with the $I_C = 0$ axis. This
discrepancy is due to the approximation made in deriving Eq. (5-31), which is
not valid for $I_B = 0$.

The theoretical curve of Fig. 5-22 is much flatter than the curves of Fig.
5-10 because we have implicitly assumed that α_N is truly constant. As already
pointed out, a very slight increase of α_N with V_{CE} can account for the slopes
of the common-emitter characteristic.

5-13   MAXIMUM VOLTAGE RATING[9]

Even if the rated dissipation of a transistor is not exceeded, there is an upper
limit to the maximum allowable collector-junction voltage since, at high
voltages, there is the possibility of voltage breakdown in the transistor. Two
types of breakdown are possible, *avalanche breakdown*, discussed in Sec. 3-11,
and *reach-through*, discussed below.

Avalanche Multiplication The maximum reverse-biasing voltage which
may be applied before breakdown between the collector and base terminals
of the transistor, under the condition that the emitter lead be open-circuited, is
represented by the symbol BV_{CBO}. This breakdown voltage is a characteristic
of the transistor alone. Breakdown may occur because of avalanche multiplication of the current I_{CO} that crosses the collector junction. As a result
of this multiplication, the current becomes MI_{CO}, in which M is the factor
by which the original current I_{CO} is multiplied by the avalance effect. (We

neglect leakage current, which does not flow through the junction and is therefore not subject to avalanche multiplication.) At a high enough voltage,
namely, BV_{CBO}, the multiplication factor M becomes nominally infinite, and
the region of breakdown is then attained. Here the current rises abruptly,
and large changes in current accompany small changes in applied voltage.

The avalanche multiplication factor depends on the voltage V_{CB} between
collector and base. We shall consider that

$$M \equiv \frac{1}{1 - (V_{CB}/BV_{CBO})^n} \tag{5-33}$$

Equation (5-33) is employed because it is a simple expression which gives
a good empirical fit to the breakdown characteristics of many transistor types.
The parameter n is found to be in the range of about 2 to 10, and controls
the sharpness of the onset of breakdown.

If a current I_E is caused to flow across the emitter junction, then, neglecting the avalanche effect, a fraction αI_E, where α is the common-base current
gain, reaches the collector junction. Taking multiplication into account, I_C
has the magnitude $M\alpha I_E$. Consequently, it appears that, in the presence
of avalanche multiplication, the transistor behaves as though its common-base
current gain were $M\alpha$.

An analysis[9] of avalanche breakdown for the CE configuration indicates
that the collector-to-emitter breakdown voltage *with open-circuited base*, designated BV_{CEO}, is

$$BV_{CEO} = BV_{CBO} \sqrt[n]{\frac{1}{h_{FE}}} \tag{5-34}$$

For an n-p-n germanium transistor, a reasonable value for n, determined
experimentally, is $n = 6$. If we now take $h_{FE} = 50$, we find that

$$BV_{CEO} = 0.52 BV_{CBO}$$

so that if $BV_{CBO} = 40$ V, BV_{CEO} is about half as much, or about 20 V. Ideal-

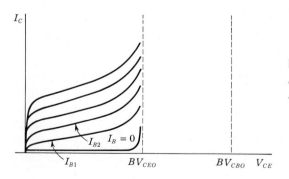

Fig. 5-23 Idealized common-
emitter characteristics extended into the breakdown
region.

er characteristics extended into the breakdown region are
If the base is not open-circuited, these breakdown char-
ified, the shapes of the curves being determined by the
tions. In other words, the maximum allowable collector-
depends not only upon the transistor, but also upon the
s used.

The second mechanism by which a transistor's usefulness
as the collector voltage is increased is called *punch-through,*
d results from the increased width of the collector-junction
ith increased collector-junction voltage (the Early effect).
region at a junction is the region of uncovered charges
junction at the positions occupied by the impurity atoms.
plied across the junction increases, the transition region
nto the base (Fig. 5-8a). Since the base is very thin, it is
noderate voltages, the transition region will have spread
the base to reach the emitter junction, as indicated in
hould be compared with Fig. 5-8. The emitter barrier is
maller than the normal value $V_o - |V_{EB}|$ because the col-
"reached through" the base region. This lowering of the
oltage may result in an excessively large emitter current,
per limit on the magnitude of the collector voltage.

h differs from avalanche breakdown in that it takes place
given by V_j in Eq. (3-21), with $W = W_B$] between collector
and base, and is not dependent on circuit configuration. In a particular
transistor, the voltage limit is determined by punch-through or breakdown,
whichever occurs at the lower voltage.

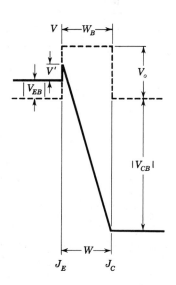

Fig. 5-24 The potential variation through a *p-n-p* transistor after "reach-through" when the effective base width (Fig. 5-8) $W_B' = W_B - W$ has been reduced to zero. The effective emitter barrier is V'.

5-14 THE PHOTOTRANSISTOR

The phototransistor (also called *photoduodiode*) is a much more sensitive semi-conductor photodevice than the *p-n* photodiode. The phototransistor is usually connected in a common-emitter configuration with the base open, and radiation is concentrated on the region near the collector junction J_C, as in Fig. 5-25a. The operation of this device can be understood if we recognize that the junction J_E is slightly forward-biased (Fig. 5-16, open-circuited base), and the junction J_C is reverse-biased (that is, the transistor is biased in the active region). Assume, first, that there is no radiant excitation. Under these circumstances minority carriers are generated thermally, and the electrons crossing from the base to the collector, as well as the holes crossing from the collector to the base, constitute the reverse saturation collector current I_{CO} (Sec. 5-2). The collector current is given by Eq. (5-16), with $I_B = 0$; namely,

$$I_C = (\beta + 1)I_{CO} \tag{5-35}$$

If the light is now turned on, additional minority carriers are photogenerated, and these contribute to the reverse saturation current in exactly the same manner as do the thermally generated minority charges. If the component of the reverse saturation current due to the light is designated I_L, the total collector current is

$$I_C = (\beta + 1)(I_{CO} + I_L) \tag{5-36}$$

We note that, due to transistor action, the current caused by the radiation is multiplied by the large factor $\beta + 1$.

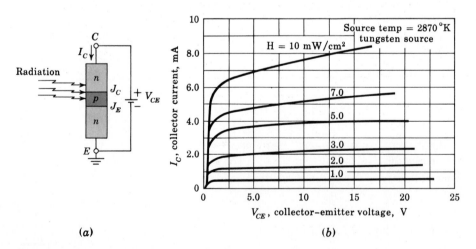

(a)

(b)

Fig. 5-25 (a) A phototransistor. (b) The output characteristics of the MRD 450 *n-p-n* silicon phototransistor. (Courtesy of Motorola Semiconductor Products, Inc.)

Typical volt-ampere characteristics are shown in Fig. 5-23*b* for an *n-p-n* planar phototransistor for different values of illumination intensities. Note the similarity between this family of curves and those in Fig. 5-10 for the CE transistor output characteristics with base current (instead of illumination) as a parameter. It is also possible to bring out the base lead and to inject a base current I_B. The current I_C in Eq. (5-36) is then increased by the term βI_B.

REFERENCES

1. Shockley, W.: The Theory of *p-n* Junctions in Semiconductors and *p-n* Junction Transistors, *Bell System Tech. J.*, vol. 28, pp. 435–489, July, 1949.

   Middlebrook, R. D.: "An Introduction to Junction Transistor Theory," pp. 115–130, John Wiley & Sons, Inc., New York, 1957.

   Terman, F. E.: "Electronic and Radio Engineering," 4th ed., pp. 747–760, McGraw-Hill Book Company, New York, 1955.

   Moll, J. L.: Junction Transistor Electronics, *Proc. IRE*, vol. 43, pp. 1807–1819, December, 1955.

2. Phillips, A. B.: "Transistor Engineering," chap. 1, McGraw-Hill Book Company, New York, 1962.

3. Texas Instruments, Inc.: J. Miller (ed.), "Transistor Circuit Design," chap. 1, McGraw-Hill Book Company, New York, 1963.

4. Early, J. M.: Effects of Space-charge Layer Widening in Junction Transistors, *Proc. IRE*, vol. 40, pp. 1401–1406, November, 1952.

5. Sah, C. T., R. N. Noyce, and W. Shockley: Carrier-generation and Recombination in *p-n* Junctions and *p-n* Junction Characteristics, *Proc. IRE*, vol. 45, pp. 1228–1243, September, 1957.

   Pritchard, R. L.: Advances in the Understanding of the P-N Junction Triode, *Proc. IRE*, vol. 46, pp. 1130–1141, June, 1958.

6. Ref. 2, pp. 236–237.

7. Ebers, J. J., and J. L. Moll: Large-signal Behavior of Junction Transistors, *Proc. IRE*, vol. 42, pp. 1761–1772, December, 1954.

8. Millman, J., and H. Taub: "Pulse, Digital, and Switching Waveforms," p. 196, McGraw-Hill Book Company, New York, 1965.

9. Ref. 8, chap. 6.

REVIEW QUESTIONS

5-1 Draw the circuit symbol for a *p-n-p* transistor and indicate the reference directions for the three currents and the reference polarities for the three voltages.

5-2 Repeat Rev. 5-1 for an *n-p-n* transistor.

5-3 For a *p-n-p* transistor biased in the active region, plot (in each region E, B, and C) (*a*) the potential variation; (*b*) the minority-carrier concentration. (*c*) Explain the shapes of the plots in (*a*) and (*b*).

5-4 (*a*) For a *p-n-p* transistor biased in the active region, indicate the various electron and hole current components crossing each junction and entering (or leaving) the base terminal. (*b*) Which of the currents is proportional to the gradient of p_n at J_E and J_C, respectively? (*c*) Repeat part *b* with p_n replaced by n_p. (*d*) What is the physical origin of the several current components crossing the base terminal?

5-5 (*a*) From the currents indicated in Rev. 5-4 obtain an expression for the collector current I_C. Define each symbol in this equation. (*b*) Generalize the equation for I_C in part *a* so that it is valid even if the transistor is not operating in its active region.

5-6 (*a*) Define the *current gain* α in words and as an equation. (*b*) Repeat part *a* for the parameter α'.

5-7 Describe the fabrication of an alloy transistor.

5-8 For a *p-n-p* transistor in the active region, what is the sign (positive or negative) of I_E, I_C, I_B, V_{CB}, and V_{EB}?

5-9 Repeat Rev. 5-8 for an *n-p-n* transistor.

5-10 (*a*) Sketch a family of CB output characteristics for a transistor. (*b*) Indicate the active, cutoff, and saturation regions. (*c*) Explain the shapes of the curves qualitatively.

5-11 (*a*) Sketch a family of CB input characteristics for a transistor. (*b*) Explain the shapes of the curves qualitatively.

5-12 Explain *base-width modulation* (the Early effect) with the aid of plots of potential and minority concentration throughout the base region.

5-13 Explain qualitatively the three consequences of base-width modulation.

5-14 Define the following regions in a transistor: (*a*) active; (*b*) saturation; (*c*) cutoff.

5-15 (*a*) Draw the circuit of transistor in the CE configuration. (*b*) Sketch the output characteristics. (*c*) Indicate the active, saturation, and cutoff regions.

5-16 (*a*) Sketch a family of CE input characteristics. (*b*) Explain the shape of these curves qualitatively.

5-17 (*a*) Derive the expression for I_C versus I_B for a CE transistor configuration in the active region. (*b*) For $I_B = 0$, what is I_C?

5-18 (*a*) What is the order of magnitude of the reverse collector saturation current I_{CBO} for a silicon transistor? (*b*) How does I_{CBO} vary with temperature?

5-19 Repeat Rev. 5-18 for a germanium transistor.

5-20 Why does I_{CBO} differ from I_{CO}?

5-21 (*a*) Define *saturation resistance* for a CE transistor. (*b*) Give its order of magnitude.

5-22 (*a*) Define *base-spreading resistance* for a transistor. (*b*) Give its order of magnitude.

5-23 What is the order of magnitude of the temperature coefficients of $V_{BE,\text{sat}}$, $V_{BC,\text{sat}}$ and $V_{CE,\text{sat}}$?

5-24· (*a*) Define h_{FE}. (*b*) Plot h_{FE} versus I_C.

5-25 (*a*) Give the order of magnitude of V_{BE} at cutoff for a silicon transistor. (*b*) Repeat part *a* for a germanium transistor. (*c*) Repeat parts *a* and *b* for the cutin voltage.

5-26 Is $|V_{BE,\text{sat}}|$ greater or less than $|V_{CE,\text{sat}}|$? Explain.

5-27 (a) What is the range in volts for V_{BE} between cutin and saturation for a silicon transistor? (b) Repeat part a for a germanium transistor.

5-28 What is the collector current relative to I_{CO} in a silicon transistor (a) if the base is shorted to the emitter? (b) If the base floats? (c) Repeat parts a and b for a germanium transistor.

5-29 Consider a transistor circuit with resistors R_b, R_c, and R_e in the base, collector, and emitter legs, respectively. The biasing voltages are V_{BB} and V_{CC} in base and collector circuits, respectively. (a) Outline the method for finding the quiescent currents, assuming that the transistor operates in the active region. (b) How do you test to see if your assumption is correct?

5-30 Repeat Rev. 5-29, assuming that the transistor is in saturation.

5-31 For a CE transistor define (in words and symbols) (a) β; (b) $\beta_{dc} = h_{FE}$; (c) $\beta' = h_{fe}$.

5-32 Derive the relationship between h_{FE} and h_{fe}.

5-33 (a) For what condition is $\beta \approx h_{FE}$? (b) For what condition is $h_{FE} \approx h_{fe}$?

5-34 (a) Draw the circuit of a CC transistor configuration. (b) Indicate the input and output terminals.

5-35 What is meant by the *inverted mode* of operation of a transistor?

5-36 (a) Write the Ebers and Moll equations. (b) Sketch the circuit model which satisfies these equations.

5-37 Discuss the two possible sources of breakdown in a transistor as the collector-to-emitter voltage is increased.

5-38 (a) Sketch the circuit of a phototransistor. (b) The radiation is concentrated near which junction? Explain why. (c) Describe the physical action of this device.

5-39 (a) For a phototransistor in the active region, write the expression for the collector current. Define all terms. (b) Sketch the family of output characteristics.

6 / DIGITAL CIRCUITS

Even in a large-scale digital system, such as a computer, or a data-processing, control, or digital-communication system, there are only a few basic operations which must be performed. These operations, to be sure, may be repeated very many times. The four circuits most commonly employed in such systems are known as the OR, AND, NOT, and FLIP-FLOP. These are called *logic* gates, or circuits, because they are used to implement Boolean algebraic equations (as we shall soon demonstrate). This algebra was invented by G. Boole in the middle of the nineteenth century as a system for the mathematical analysis of logic.

This chapter discusses in detail the first three basic logic circuits mentioned above. These basic gates are combined into FLIP-FLOPS and other digital-system building blocks in Chap. 17.

6-1 DIGITAL (BINARY) OPERATION OF A SYSTEM

A digital system functions in a binary manner. It employs devices which exist only in two possible states. A transistor is allowed to operate at cutoff or in saturation, but not in its active region. A node may be at a high voltage of, say, 4 ± 1 V or at a low voltage of, say, 0.2 ± 0.2 V, but no other values are allowed. Various designations are used for these two quantized states, and the most common are listed in Table 6-1. In logic, a statement is characterized as *true* or *false*, and this is the first binary classification listed in the table. A switch may be *closed* or *open*, which is the notation under 9, etc. Binary arithmetic and mathematical manipulation of switching or logic functions are best carried out with classification 3, which involves two symbols, 0 (zero) and 1 (one).

156

TABLE 6-1   Binary-state terminology

	1	2	3	4	5	6	7	8	9	10	11
One of the states.....	True	High	1	Up	Pulse	Excited	Off	Hot	Closed	North	Yes
The other state.....	False	Low	0	Down	No pulse	Non-excited	On	Cold	Open	South	No

The binary system of representing numbers will now be explained by making reference to the familiar *decimal system*. In the latter the base is 10 (ten), and ten numerals, 0, 1, 2, 3, . . . , 9, are required to express an arbitrary number. To write numbers larger than 9, we assign a meaning to the *position* of a numeral in an array of numerals. For example, the number 1,264 (one thousand two hundred sixty four) has the meaning

$$1{,}264 \equiv 1 \times 10^3 + 2 \times 10^2 + 6 \times 10^1 + 4 \times 10^0$$

Thus the individual digits in a number represent the coefficients in an expansion of the number in powers of 10. The digit which is farthest to the right is the coefficient of the zeroth power, the next is the coefficient of the first power, and so on.

In the *binary system* of representation the base is 2, and only the two numerals 0 and 1 are required to represent a number. The numerals 0 and 1 have the same meaning as in the decimal system, but a different interpretation is placed on the position occupied by a digit. In the binary system the individual digits represent the coefficients of powers of *two* rather than *ten* as in the decimal system. For example, the decimal number 19 is written in the binary representation as 10011 since

$$10011 \equiv 1 \times 2^4 + 0 \times 2^3 + 0 \times 2^2 + 1 \times 2^1 + 1 \times 2^0$$
$$= \quad 16 \quad + \quad 0 \quad + \quad 0 \quad + \quad 2 \quad + \quad 1 \quad = 19$$

A short list of equivalent numbers in decimal and binary notation is given in Table 6-2.

A general method for converting from a decimal to a binary number is indicated in Table 6-3. The procedure is the following. Place the decimal number (in this illustration, 19) on the extreme right. Next divide by 2 and place the quotient (9) to the left and indicate the remainder (1) directly below it. Repeat this process (for the next column $9 \div 2 = 4$ and a remainder of 1) until a quotient of 0 is obtained. The array of 1's and 0's in the second row is the binary representation of the original decimal number. In this example, decimal 19 = 10011 binary.

A binary digit (a 1 or a 0) is called a *bit*. A group of bits having a significance is a *bite*, *word*, or *code*. For example, to represent the 10 numerals

TABLE 6-2   Equivalent numbers in decimal
and binary notation

Decimal notation	Binary notation	Decimal notation	Binary notation
0	00000	11	01011
1	00001	12	01100
2	00010	13	01101
3	00011	14	01110
4	00100	15	01111
5	00101	16	10000
6	00110	17	10001
7	00111	18	10010
8	01000	19	10011
9	01001	20	10100
10	01010	21	10101

(0, 1, 2, . . . , 9) and the 26 letters of the English alphabet would require 36 different combinations of 1's and 0's. Since $2^5 < 36 < 2^6$, then a minimum of 6 bits per bite are required in order to accommodate all the alphanumeric characters. In this sense a bite is sometimes referred to as a *character* and a group of one or more characters as a *word*.

Logic Systems In a *dc*, or *level-logic*, system a bit is implemented as one of two voltage levels. If, as in Fig. 6-1*a*, the more positive voltage is the 1 level and the other is the 0 level, the system is said to employ dc *positive* logic. On the other hand, a dc *negative*-logic system, as in Fig. 6-1*b*, is one which designates the more negative voltage state of the bit as the 1 level and the more positive as the 0 level. It should be emphasized that the absolute values of the two voltages are of no significance in these definitions. In particular, the 0 state need not represent a zero voltage level (although in some systems it might).

The parameters of a physical device (for example, $V_{CE,\text{sat}}$ of a transistor) are not identical from sample to sample, and they also vary with temperature. Furthermore, ripple or voltage spikes may exist in the power supply or ground leads, and other sources of unwanted signals, called *noise*, may be present in the circuit. For these reasons the digital levels are not specified precisely, but as indicated by the shaded regions in Fig. 6-1, each state is defined by a voltage range about a designated level, such as 4 ± 1 V and 0.2 ± 0.2 V.

TABLE 6-3   Decimal-to-binary conversion

Divide by 2..........	0	1	2	4	9	19 decimal
Remainder..........	1	0	0	1	1	Binary

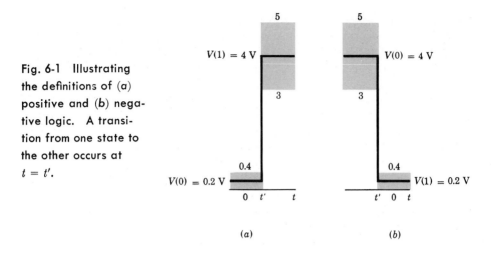

Fig. 6-1 Illustrating the definitions of (a) positive and (b) negative logic. A transition from one state to the other occurs at $t = t'$.

(a)

(b)

In a *dynamic*, or *pulse-logic*, system a bit is recognized by the presence or absence of a pulse. A 1 signifies the existence of a positive pulse in a dynamic positive-logic system; a negative pulse denotes a 1 in a dynamic negative-logic system. In either system a 0 at a particular input (or output) at a given instant of time designates that no pulse is present at that particular moment.

6-2   THE OR GATE[1]

An OR gate has two or more inputs and a single output, and it operates in accordance with the following definition: *The output of an* OR *assumes the 1 state if one or more inputs assume the 1 state.* The n inputs to a logic circuit will be designated by $A, B, \ldots, N$ and the output by Y. It is to be understood that each of these symbols may assume one of two possible values, either 0 or 1. A standard symbol for the OR circuit is given in Fig. 6-2a, together with the Boolean expression for this gate. The equation is to be read "Y equals A or B or $\cdots$ or N." Instead of defining a logical operation in words, an alternative method is to give a *truth table* which contains a tabulation of all possible input values and their corresponding outputs. It

Fig. 6-2 (a) The standard symbol for an OR gate and its Boolean expression; (b) the truth table for a two-input OR gate.

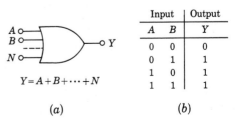

$$Y = A + B + \cdots + N$$

(a)

Input		Output
A	B	Y
0	0	0
0	1	1
1	0	1
1	1	1

(b)

should be clear that the two-input truth table of Fig. 6-2b is equivalent to the above definition of the OR operation.

In a *diode-logic* (DL) system the logical gates are implemented by using diodes. A diode OR for negative logic is shown in Fig. 6-3. The generator source resistance is designated by R_s. We consider first the case where the supply voltage V_R has a value equal to the voltage $V(0)$ of the 0 state for dc logic.

If all inputs are in the 0 state, the voltage across each diode is $V(0) - V(0) = 0$. Since, in order for a diode to conduct, it must be forward-biased by at least the cutin voltage V_γ (Fig. 4-5), none of the diodes conducts. Hence the output voltage is $v_o = V(0)$, and Y is in the 0 state.

If now input A is changed to the 1 state, which for negative logic is at the potential $V(1)$, less positive than the 0 state, then $D1$ will conduct. The output becomes

$$v_o = V(0) - [V(0) - V(1) - V_\gamma] \frac{R}{R + R_s + R_f} \tag{6-1}$$

where R_f is the diode forward resistance. Usually R is chosen much larger than $R_s + R_f$. Under this restriction

$$v_o \approx V(1) + V_\gamma \tag{6-2}$$

Hence the output voltage exceeds the more negative level $V(1)$ by V_γ (approximately 0.2 V for germanium or 0.6 V for silicon). Furthermore, the step in output voltage is *smaller* by V_γ than the change in input voltage.

From now on, unless explicitly stated otherwise, we shall assume $R \gg R_s$ and ideal diodes with $R_f = 0$ and $V_\gamma = 0$. The output, for input A excited, is then $v_o = V(1)$, and the circuit has performed the following logic: if $A = 1$, $B = 0, \ldots, N = 0$, then $Y = 1$, which is consistent with the OR operation.

For the above excitation, the output is at $V(1)$, and each diode, except $D1$, is back-biased. Hence the presence of signal sources at $B, C, \ldots, N$ does not result in an additional load on generator A. Since the OR configuration minimizes the interaction of the sources on one another, this gate is sometimes referred to as a *buffer* circuit. Since it allows several independent sources to be applied at a given node, it is also called a (nonlinear) *mixing* gate.

If two or more inputs are in the 1 state, the diodes connected to these inputs conduct and all other diodes remain reverse-biased. The output is $V(1)$, and again the OR function is satisfied. If for any reason the level $V(1)$ is not identical for all inputs, *the most negative value of $V(1)$ (for negative logic) appears at the output*, and all diodes except one are nonconducting.

A positive-logic OR gate uses the same configuration as that in Fig. 6-3, except that all diodes must be reversed. *The output now is equal to the most positive level $V(1)$* [or more precisely is smaller than the most positive value of $V(1)$ by V_γ]. If a dynamic logic system is under consideration, *the output-*

Fig. 6-3 A diode OR circuit for negative logic. [It is also possible to
choose the supply voltage such that
$V_R > V(0)$, but that arrangement has
the disadvantage of drawing standby
current when all inputs are in the 0
state.]

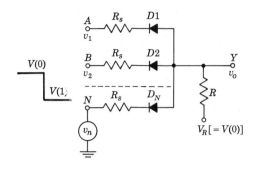

pulse magnitude is (approximately) *equal to the largest input pulse* (regardless
of whether the system uses positive or negative logic).

A second mode of operation of the OR circuit of Fig. 6-3 is possible if V_R
is set equal to a voltage more positive than $V(0)$ by at least V_γ. For this
condition *all diodes conduct in the 0 state,* and $v_o \approx V(0)$ if $R \gg R_s + R_f$. If
one or more inputs are excited, then the diode connected to the most negative
$V(1)$ conducts, the output equals this value of $V(1)$, and all other diodes are
back-biased. Clearly, the OR function has been satisfied.

A third mode of operation of the circuit of Fig. 6-3 results if we select
$V_R < V(0)$. This arrangement has the disadvantage that the output will not
respond until the input falls enough to overcome the initial reverse bias of the
diodes.

Boolean Identities If it is remembered that A, B, and C can take on
only the value 0 or 1, the following equations from Boolean algebra pertaining
to the OR $(+)$ operation are easily verified:

$$A + B + C = (A + B) + C = A + (B + C) \tag{6-3}$$

$$A + B = B + A \tag{6-4}$$

$$A + A = A \tag{6-5}$$

$$A + 1 = 1 \tag{6-6}$$

$$A + 0 = A \tag{6-7}$$

These equations may be justified by referring to the definition of the OR
operation, to a truth table, or to the action of the OR circuits discussed above.

6-3   THE AND GATE[1]

An AND gate has two or more inputs and a single output, and it operates in
accordance with the following definition: *The output of an AND assumes the 1
state if and only if all the inputs assume the 1 state.* A symbol for the AND

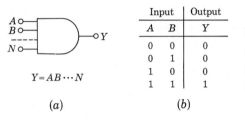

	Input		Output
	A	B	Y
	0	0	0
	0	1	0
	1	0	0
	1	1	1

$$Y = AB \cdots N$$

(a)   *(b)*

Fig. 6-4 (a) The standard symbol for an AND gate and its Boolean expression; the truth table for a two-input AND gate.

circuit is given in Fig. 6-4a, together with the Boolean expression for this gate. The equation is to be read "Y equals A *and* B *and* . . . *and* N." [Sometimes a dot (·) or a cross (×) is placed between symbols to indicate the AND operation.] It may be verified that the two-input truth table of Fig. 6-4b is consistent with the above definition of the AND operation.

A diode-logic (DL) configuration for a negative AND gate is given in Fig. 6-5a. To understand the operation of the circuit, assume initially that all source resistances R_s are zero and that the diodes are ideal. If *any* input is at the 0 level $V(0)$, the diode connected to this input conducts and the output is clamped at the voltage $V(0)$, or $Y = 0$. However, if *all* inputs are at the 1 level $V(1)$, then all diodes are reverse-biased and $v_o = V(1)$, or $Y = 1$. Clearly, the AND operation has been implemented. The AND gate is also called a *coincidence circuit*.

A positive-logic AND gate uses the same configuration as that in Fig. 6-5a, except that all diodes are reversed. This circuit is indicated in Fig. 6-5b and should be compared with Fig. 6-3. It is to be noted that the symbol $V(0)$ in Fig. 6-3 designates the same voltage as $V(1)$ in Fig. 6-5b because each represents the upper binary level. Similarly, $V(1)$ in Fig. 6-3 equals $V(0)$ in Fig. 6-5b, since both represent the lower binary level. Hence these two circuits are identical, and we conclude that *a negative* OR *gate is the same circuit as a positive* AND *gate*. This result is not restricted to diode logic, and by using Boolean algebra, we show in Sec. 6-8 that it is valid independently of the hardware used to implement the circuit.

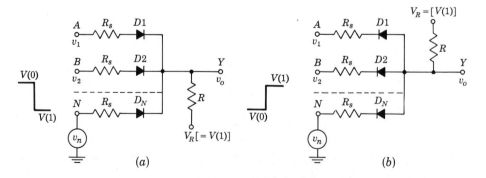

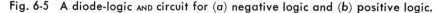

Fig. 6-5 A diode-logic AND circuit for (a) negative logic and (b) positive logic.

In Fig. 6-5b it is possible to choose V_R to be more positive than $V(1)$. If this condition is met, all diodes will conduct upon a coincidence (all inputs in the 1 state) and the output will be clamped to $V(1)$. The output impedance is low in this mode of operation, being equal to $(R_s + R_f)/n$ in parallel with R. On the other hand, if $V_R = V(1)$, then all diodes are cut off at a coincidence, and the output impedance is high (equal to R). If for any reason not all inputs have the same upper level $V(1)$, then the output of the positive AND gate of Fig. 6-5b will equal $V(1)_{min}$, the *least* positive value of $V(1)$. Note that the diode connected to $V(1)_{min}$ conducts, clamping the output to this minimum value of $V(1)$ and maintaining all other diodes in the reverse-biased condition. If, on the other hand, V_R is smaller than all inputs $V(1)$, then all diodes will be cut off upon coincidence and the output will rise to the voltage V_R. Similarly, if the inputs are pulses, then *the output pulse will have an amplitude equal to the smallest input amplitude* [provided that V_R is greater than $V(1)_{min}$].

Boolean Identities   Since A, B, and C can have only the value 0 or 1, the following expressions involving the AND operation may be verified:

$$ABC = (AB)C = A(BC) \tag{6-8}$$

$$AB = BA \tag{6-9}$$

$$AA = A \tag{6-10}$$

$$A1 = A \tag{6-11}$$

$$A0 = 0 \tag{6-12}$$

$$A(B + C) = AB + AC \tag{6-13}$$

These equations may be proved by reference to the definition of the AND operation, to a truth table, or to the behavior of the AND circuits discussed above. Also, by using Eqs. (6-11), (6-13), and (6-6), it can be shown that

$$A + AB = A \tag{6-14}$$

Similarly, if follows from Eqs. (6-13), (6-10), and (6-6) that

$$A + BC = (A + B)(A + C) \tag{6-15}$$

We shall have occasion to refer to the last two equations later.

6-4   THE NOT, OR INVERTER, CIRCUIT

The NOT circuit has a single input and a single output and performs the operation of *logic negation* in accordance with the following definition: *The output of a NOT circuit takes on the 1 state if and only if the input does not take on the 1 state.* The standard to indicate a *logic negation* is a small circle drawn at

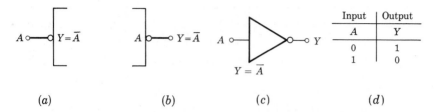

Input	Output
A	Y
0	1
1	0

$Y = \bar{A}$

(a)   (b)   (c)   (d)

Fig. 6-6 Logic negation at (a) the input and (b) the output of a logic block; (c) a symbol often used for a NOT gate and the Boolean equation; (d) the truth table.

the point where a signal line joins a logic symbol. Negation at the input of a logic block is indicated in Fig. 6-6a and at the output in Fig. 6-6b. The symbol for a NOT gate and the Boolean expression for negation are given in Fig. 6-6c. The equation is to be read "Y equals NOT A" or "Y is the complement of A." [Sometimes a prime (') is used instead of the bar (⁻) to indicate the NOT operation.] The truth table is given in Fig. 6-6d.

A circuit which accomplishes a logic negation is called a NOT circuit, or, since it inverts the sense of the output with respect to the input, it is also known as an *inverter*. The output of an INVERTER is relatively more positive if and only if the input is relatively less positive. In a truly binary system only two levels $V(0)$ and $V(1)$ are recognized, and the output, as well as the input, of an inverter must operate between these two voltages. When the input is at $V(0)$, the output must be at $V(1)$, and vice versa. Ideally, then, a NOT circuit inverts a signal while preserving its shape and the binary levels between which the signal operates.

The transistor circuit of Fig. 6-7 implements an inverter for positive logic having a 0 state of $V(0) = V_{EE}$ and a 1 state of $V(1) = V_{CC}$. If the input is low, $v_i = V(0)$, then the parameters are chosen so that the Q is OFF, and hence $v_o = V_{CC} = V(1)$. On the other hand, if the input is high, $v_i = V(1)$, then the circuit parameters are picked so that Q is in saturation and then $v_o = V_{EE} = V(0)$, if we neglect the collector-to-emitter saturation voltage $V_{CE,\text{sat}}$. A detailed calculation of quiescent conditions is made in the following example.

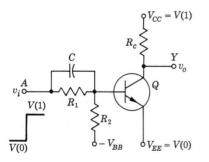

Fig. 6-7 An INVERTER for positive logic. A similar circuit using a p-n-p transistor is used for a negative-logic NOT circuit.

EXAMPLE   If the silicon transistor in Fig. 6-8 has a minimum value of h_{FE} of 30, find the output levels for input levels of 0 and 12 V.

Solution   For $v_i = V(0) = 0$ the open-circuited base voltage V_B is

$$V_B = -12 \times \frac{15}{100 + 15} = -1.56 \text{ V}$$

Since a bias of about 0 V is adequate to cut off a silicon emitter junction (Table 5-1, page 142), then Q is indeed cut off. Hence $v_o = 12$ V for $v_i = 0$.

For $v_i = V(1) = 12$ V let us verify the assumption that Q is in saturation. The minimum base current required for saturation is

$$(I_B)_{\min} = \frac{I_C}{h_{FE}}$$

It is usually sufficiently accurate to use the approximate values for the saturation junction voltages given in Table 5-1, which for silicon are $V_{BE,\text{sat}} = 0.8$ V and $V_{CE,\text{sat}} = 0.2$ V. With these values

$$I_C = \frac{12 - 0.2}{2.2} = 5.36 \text{ mA} \qquad (I_B)_{\min} = \frac{5.36}{30} = 0.18 \text{ mA}$$

$$I_1 = \frac{12 - 0.8}{15} = 0.75 \text{ mA} \qquad I_2 = \frac{0.8 - (-12)}{100} = 0.13 \text{ mA}$$

and

$$I_B = I_1 - I_2 = 0.75 - 0.13 = 0.62 \text{ mA}$$

Since this value exceeds $(I_B)_{\min}$, Q is indeed in saturation and the drop across the transistor is $V_{CE,\text{sat}}$. Hence $v_o = 0.2$ V for $v_i = 12$ V, and the circuit has performed the NOT operation.

If the input to the inverter is obtained from the output of a similar gate, the input levels are $V(0) = V_{CE,\text{sat}} = 0.2$ V and $V(1) = 12$ V. The corresponding output levels are 12 and 0.2 V, respectively.

The capacitor C across R_1 in Fig. 6-7 is added to improve the transient response of the inverter. This capacitor aids in the removal of the minority-carrier charge stored in the base when the signal changes abruptly between

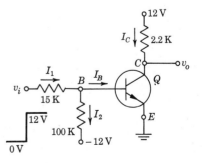

Fig. 6-8   An inverter calculation.

logic states. A discussion of this phenomenon, including rise time, fall time, and storage time, is given in the following section. The order of magnitude of C is 100 pF, but its exact value depends upon the transistor.

Transistor Limitations   There are certain transistor characteristics as well as certain circuit features which must particularly be taken into account in designing transistor inverters.

1. *The back-bias emitter-junction voltage V_{EB}.* This voltage must not exceed the emitter-to-base breakdown voltage BV_{EBO} specified by the manufacturer. For the type 2N914, $BV_{EBO} = 5$ V, and for the 2N1304, $BV_{EBO} = 25$ V. However, for some (diffused-base) transistors, BV_{EBO} may be quite small (less than 1 V).

2. *The dc current gain h_{FE}.* Since h_{FE} decreases with decreasing temperature, the circuit must be designed so that at the lowest expected temperature the transistor will remain in saturation. The maximum value of R_1 is determined principally by this condition.

3. *The reverse collector saturation current I_{CBO}.* Since $|I_{CBO}|$ increases about 7 percent/°C (doubles every 10°C for either germanium or silicon), we cannot continue to neglect the effect of I_{CBO} at high temperatures. At cutoff the emitter current is zero and the base current is I_{CBO} (in a direction opposite to that indicated as I_B in Fig. 6-8). Let us calculate the value of I_{CBO} which just brings the transistor to the point of cutoff. If we assume, as in Table 5-1, that at cutoff, $V_{BE} = 0$ V, then $I_1 = 0$ and the drop across the 100-K resistor is

$$100 \, I_{CBO} = 12 \text{ V} \quad \text{or} \quad I_{CBO} = 0.12 \text{ mA}$$

The ambient temperature at which $I_{CBO} = 0.12$ mA $= 120$ μA is the maximum temperature at which the inverter will operate satisfactorily. A silicon transistor can be operated at temperatures in excess of 185°C.

Boolean Identities   From the basic definition of the NOT, AND, and OR connectives we can verify the following Boolean identities:

$$\bar{\bar{A}} = A \tag{6-16}$$

$$\bar{A} + A = 1 \tag{6-17}$$

$$\bar{A}A = 0 \tag{6-18}$$

$$A + \bar{A}B = A + B \tag{6-19}$$

EXAMPLE   Verify Eq. (6-19).

Solution   Since $B + 1 = 1$ and $A1 = A$, then

$$A + \bar{A}B = A(B + 1) + \bar{A}B = AB + A + \bar{A}B = (A + \bar{A})B + A = B + A$$

where use is made of Eq. (6-17).

6-5   TRANSISTOR SWITCHING TIMES[2]

If a pulse is applied to an inverter, the transistor acts as a switch, since
it operates from cutoff to saturation and then returns to cutoff. We now
turn our attention to the behavior of the transistor inverter as it makes a
transition from one state to the other. We consider the transistor circuit
shown in Fig. 6-9a, driven by the pulse waveform shown in Fig. 6-9b. This
waveform makes transitions between the voltage levels V_2 and V_1. At V_2
the transistor is at cutoff, and at V_1 the transistor is in saturation. The
input waveform v_i is applied between base and emitter through a resistor
R_s, which represents $R_1 \| R_2$ of Fig. 6-7 (assume that C is not present).

The response of the collector current i_C to the input waveform, together
with its time relationship to that waveform, is shown in Fig. 6-9c. The current does not immediately respond to the input signal. Instead, there is a
delay, and the time that elapses during this delay, together with the time
required for the current to rise to 10 percent of its maximum (saturation)
value $I_{CS} \approx V_{CC}/R_L$, is called the *delay time* t_d. The current waveform has
a nonzero *rise time* t_r, which is the time required for the current to rise through
the active region from 10 to 90 percent of I_{CS}. The total *turn-on* time t_{ON} is
the sum of the delay and rise time, $t_{ON} \equiv t_d + t_r$. When the input signal
returns to its initial state at $t = T$, the current again fails to respond immediately. The interval which elapses between the transition of the input
waveform and the time when i_C has dropped to 90 percent of I_{CS} is called the
storage time t_s. The storage interval is followed by the *fall time* t_f, which is
the time required for i_C to fall from 90 to 10 percent of I_{CS}. The *turnoff*

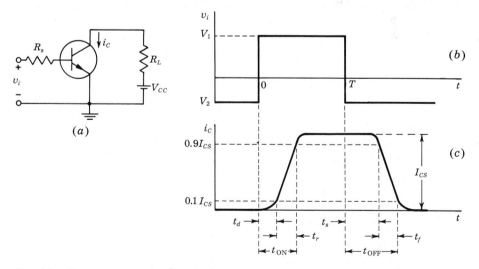

Fig. 6-9   The pulse waveform in (b) drives the transistor in (a) from cutoff to saturation and back again.   (c) The collector-current response to the driving input pulse.

time t_{OFF} is defined as the sum of the storage and fall times, $t_{OFF} \equiv t_s + t_f$. We shall consider now the physical reasons for the existence of each of these times. The actual calculation of the time intervals (t_d, t_r, t_s, and t_f) is complex, and the reader is referred to Ref. 2. Numerical values of delay time, rise time, storage time, and fall time for the Texas Instruments *n-p-n* epitaxial planar silicon transistor 2N3830 under specified conditions can be as low as $t_d = 10$ ns, $t_r = 50$ ns, $t_s = 40$ ns, and $t_f = 30$ ns.

The Delay Time Three factors contribute to the delay time. First, when the driving signal is applied to the transistor input, a nonzero time is required to charge up the emitter-junction transition capacitance so that the transistor may be brought from cutoff to the active region. Second, even when the transistor has been brought to the point where minority carriers have begun to cross the emitter junction into the base, a time interval is required before these carriers can cross the base region to the collector junction and be recorded as collector current. Finally, some time is required for the collector current to rise to 10 percent of its maximum.

Rise Time and Fall Time The rise time and the fall time are due to the fact that, if a base-current step is used to saturate the transistor or return it from saturation to cutoff, the transistor collector current must traverse the active region. The collector current increases or decreases along an exponential curve whose time constant τ_r can be shown to be given by $\tau_r = h_{FE}(C_c R_c + 1/\omega_T)$, where C_c is the collector transition capacitance and ω_T is the radian frequency at which the current gain is unity (Sec. 11-3).

Storage Time The failure of the transistor to respond to the trailing edge of the driving pulse for the time interval t_s (indicated in Fig. 6-9c) results from the fact that a transistor in saturation has a saturation charge of excess minority carriers stored in the base. The transistor cannot respond until this

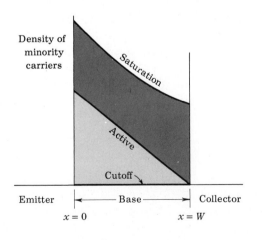

Fig. 6-10 Minority-carrier concentration in the base for cutoff, active, and saturation conditions of operation.

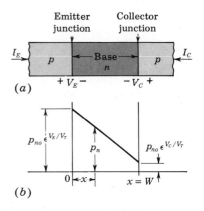

Fig. 6-11 The minority-carrier density in the base region. The concentrations at $x = 0$ and $x = W$ are governed by the law of the junction [(Eq. 3-4)].

saturation excess charge has been removed. The stored charge density in the base is indicated in Fig. 6-10 under various operating conditions.

The concentration of minority carriers in the base region decreases linearly from $p_{no}\epsilon^{V_E/V_T}$ at $x = 0$ to $p_{no}\epsilon^{V_C/V_T}$ at $x = W$, as indicated in Fig. 6-11b. In the cutoff region, both V_E and V_C are negative, and p_n is almost zero everywhere. In the active region, V_E is positive and V_C negative, so that p_n is large at $x = 0$ and almost zero at $x = W$. Finally, in the saturation region, where V_E and V_C are both positive, p_n is large everywhere, and hence a large amount of minority-carrier charge is stored in the base. These densities are pictured in Fig. 6-10.

Consider that the transistor is in its saturation region and that at $t = T$ an input step is used to turn the transistor off, as in Fig. 6-9. Since the turnoff process cannot begin until the abnormal carrier density (the heavily shaded area of Fig. 6-10) has been removed, a relatively long storage delay time t_s may elapse before the transistor responds to the turnoff signal at the input. In an extreme case this storage-time delay may be two or three times the rise or fall time through the active region. It is clear that, when transistor switches are to be used in an application where speed is at a premium, it is advantageous to reduce the storage time. By adding the capacitor C across the base resistor (Fig. 6-7), an impulsive current will flow out of the base at the time T at the end of the pulse. If C is properly chosen,[3] this impulsive current will instantaneously reduce t_s to zero. A method for preventing a transistor from saturating, and thus eliminating storage time, is given in Sec. 7-13.

6-6   THE INHIBIT (ENABLE) OPERATION

A NOT circuit preceding one terminal (S) of an AND gate acts as an *inhibitor*. This modified AND circuit implements the logical statement. *If $A = 1$, $B = 1$, . . . , $M = 1$, then $Y = 1$ provided that $S = 0$. However, if $S = 1$, then the*

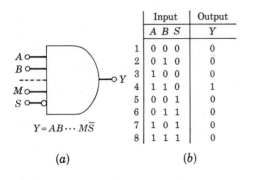

	Input	Output
	A B S	Y
1	0 0 0	0
2	0 1 0	0
3	1 0 0	0
4	1 1 0	1
5	0 0 1	0
6	0 1 1	0
7	1 0 1	0
8	1 1 1	0

$Y = AB \cdots M\bar{S}$

(a)

(b)

Fig. 6-12 (a) The logic block and Boolean expression for an AND with an enable terminal S. (b) The truth table for $Y = AB\bar{S}$. The column on the left numbers the eight possible input combinations.

coincidence of A, B, . . . , M is inhibited, and $Y = 0$. Such a configuration is also called an *anticoincidence* circuit. The logical block symbol is drawn in Fig. 6-12a, together with its Boolean equation. The equation is to be read "Y equals A *and B and* . . . *and M and not S.*" The truth table for a three-input AND gate with one inhibitor terminal (S) is given in Fig. 6-12b.

The terminal S is also called a *strobe* or an *enable input*. The enabling bit $S = 0$ allows the gate to perform its AND logic, whereas the inhibiting bit $S = 1$ causes the output to remain at $Y = 0$, independently of the values of the input bits.

A combination of the AND circuit of Fig. 6-5b and the INVERTER of Fig. 6-8 satisfying the logic given in the truth table (Fig. 6-12b) is indicated in Fig. 6-13. If either input A or B or both are in the 0 state, $V(0) = 0$ V, then at least one of the diodes $D1$ or $D2$ conducts and clamps the output to 0 V, or $Y = 0$. This argument verifies all items in the truth table except lines 4 and 8. Consider now the situation where a coincidence occurs at A and B. If S is in the 0 state, then Q is cut off, and the output of the NOT circuit is $\bar{S} = 1$ (12 V). Hence all

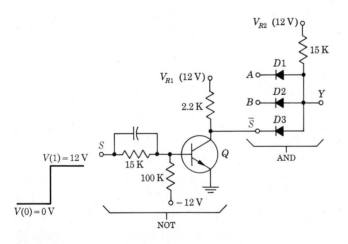

Fig. 6-13 A positive-logic AND circuit with a negation input terminal.

three diodes are reverse-biased and the output rises to 12 V, or $Y = 1$, which verifies line 4 of the truth table.   (If A, B, V_{R1}, and V_{R2} are not all equal to the same voltage, the output will rise to the smallest of these values.)   Finally, consider the condition in line 8 of Fig. 6-12b.   If S is in the 1 state, then Q is driven into saturation, and the output of the transistor drops to 0 V (ideally). Hence $\bar{S} = 0$, D3 conducts, and $Y = 0$, which indeed is the logic in the last row of the truth table.

It is possible to have a two-input AND, one terminal of which is inhibiting. This circuit satisfies the logic: "The output is true (1) if input A is true (1) provided that B is not true (0) [or equivalently, provided that B is false (0)]." Another possible configuration is an AND with more than one inhibit terminal.

In a dynamic system, if an inhibit pulse is to allow none of the signal to be transmitted through the gate, it is necessary that the inhibit pulse begin earlier and last longer than the signal pulses.

6-7   THE EXCLUSIVE OR CIRCUIT

An EXCLUSIVE OR gate obeys the definition: *The output of a two-input EXCLU-SIVE OR assumes the 1 state if one and only one input assumes the 1 state.*   The standard symbol for an EXCLUSIVE OR is given in Fig. 6-14a and the truth table in Fig. 6-14b.   The circuit of Sec. 6-2 is referred to as an INCLUSIVE OR if it is desired to distinguish it from the EXCLUSIVE OR.

The above definition is equivalent to the statement: "If $A = 1$ or $B = 1$ but not simultaneously, then $Y = 1$."   In Boolean notation,

$$Y = (A + B)(\overline{AB}) \tag{6-20}$$

This function is implemented in logic diagram form in Fig. 6-15a.

A second logic statement equivalent to the definition of the EXCLUSIVE OR is the following: "If $A = 1$ and $B = 0$, or if $B = 1$ and $A = 0$, then $Y = 1$." The Boolean expression is

$$Y = A\bar{B} + B\bar{A} \tag{6-21}$$

The block diagram which satisfies this logic is indicated in Fig. 6-15b.

An EXCLUSIVE OR is employed within the arithmetic section of a computer. Another application is as an *inequality comparator, matching circuit,* or *detector* because, as can be seen from the truth table, $Y = 1$ only if $A \neq B$.   This

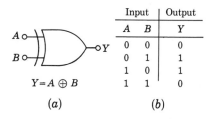

Fig. 6-14 (a) The standard symbol for an EXCLUSIVE OR gate and its Boolean expression; (b) the truth table.

$Y = A \oplus B$

(a)

Input		Output
A	B	Y
0	0	0
0	1	1
1	0	1
1	1	0

(b)

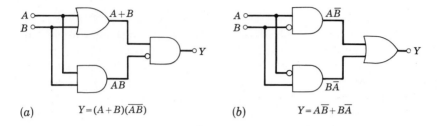

(a) $Y = (A+B)(\overline{AB})$ (b) $Y = A\overline{B} + B\overline{A}$

Fig. 6-15 Two logic block diagrams for the EXCLUSIVE OR gate.

property is used to check for the inequality of two bits. If bit A is not identical with bit B, then an output is obtained. Equivalently, "If A and B are both 1 or if A and B are both 0, then no output is obtained, and $Y = 0$." This latter statement may be put into Boolean form as

$$Y = \overline{AB + \bar{A}\bar{B}} \qquad (6\text{-}22)$$

This equation leads to a third implementation for the EXCLUSIVE OR block, which is indicated by the logic diagram of Fig. 6-16a. An *equality detector* gives an output $Z = 1$ if A and B are both 1 or if A and B are both 0, and hence

$$Z = \bar{Y} = AB + \bar{A}\bar{B} \qquad (6\text{-}23)$$

where use was made of Eq. (6-16). If the output Z is desired, the negation in Fig. 6-16a may be omitted or an additional inverter may be cascaded with the output of the EXCLUSIVE OR.

A fourth possibility for this gate is

$$Y = (A + B)(\bar{A} + \bar{B}) \qquad (6\text{-}24)$$

which may be verified from the definition or from the truth table. This logic is depicted in Fig. 6-16b.

We have demonstrated that there often are several ways to implement a logical circuit. In practice one of these may be realized more advantageously than the others. Boolean algebra is sometimes employed for manipulating a logic equation so as to transform it into a form which is better from the point

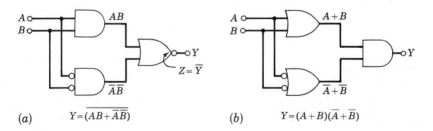

(a) $Y = \overline{(AB + \bar{A}\bar{B})}$ (b) $Y = (A+B)(\bar{A}+\bar{B})$

Fig. 6-16 Two additional logic block diagrams for the EXCLUSIVE OR gate.

of view of implementation in hardware. In the next section we shall verify through the use of Boolean algebra that the four expressions given above for the EXCLUSIVE OR are equivalent.

Two-level Logic Digital design often calls for several gates (AND, OR, or combinations of those) feeding into an OR (or AND) gate. Such a combination is known as *two-level (or two-wide) logic*. The EXCLUSIVE OR circuits of Figs. 6-15 and 6-16 are examples of two-level logic. In the discussion of digital systems in Chap. 17 it is found that the most useful logic array consists of several ANDS which feed an OR which is followed by a NOT gate. This cascade of gates is called an AND-OR-INVERT (AOI) configuration. The detailed circuit topology for an AOI is given in Fig. 17-2.

6-8   DE MORGAN'S LAWS

The statement "If and only if all inputs are true (1), then the output is true (1)" is logically equivalent to the statement "If at least one input is false (0), then the output is false (0)." In Boolean notation this equivalence is written

$$ABC \cdots = \overline{\bar{A} + \bar{B} + \bar{C} + \cdots} \tag{6-25}$$

If we take the complement of both sides of this equation and use Eq. (6-16), we obtain

$$\overline{ABC \cdots} = \bar{A} + \bar{B} + \bar{C} + \cdots \tag{6-26}$$

This equation and its dual,

$$\overline{A + B + C + \cdots} = \bar{A}\bar{B}\bar{C} \cdots \tag{6-27}$$

(which may be proved in a similar manner), are known as De Morgan's laws. These complete the list of basic Boolean identities. For easy future reference, all these relationships are summarized in Table 6-4.

With the aid of Boolean algebra we shall now demonstrate the equivalence of the four EXCLUSIVE OR circuits of the preceding section. Using Eq. (6-26), it is immediately clear that Eq. (6-20) is equivalent to Eq. (6-24). Now the latter equation can be expanded with the aid of Table 6-4 as follows:

$$(A + B)(\bar{A} + \bar{B}) = A\bar{A} + B\bar{A} + A\bar{B} + B\bar{B} = B\bar{A} + A\bar{B} \tag{6-28}$$

This result shows that the EXCLUSIVE OR of Eq. (6-21) is equivalent to that of Eq. (6-24). Finally, applying Eq. (6-27) to Eq. (6-22) gives

$$\overline{AB + \bar{A}\bar{B}} = (\overline{AB})(\overline{\bar{A}\bar{B}}) \tag{6-29}$$

From Eq. (6-26), we have

$$(\overline{AB})(\overline{\bar{A}\bar{B}}) = (\bar{A} + \bar{B})(\bar{\bar{A}} + \bar{\bar{B}}) = (\bar{A} + \bar{B})(A + B) \tag{6-30}$$

TABLE 6-4   Summary of basic Boolean identities

Fundamental laws

OR	AND	NOT
$A + 0 = A$	$A0 = 0$	$A + \bar{A} = 1$
$A + 1 = 1$	$A1 = A$	$A\bar{A} = 0$
$A + A = A$	$AA = A$	$\bar{\bar{A}} = A$
$A + \bar{A} = 1$	$A\bar{A} = 0$	

Associative laws

$$(A + B) + C = A + (B + C) \qquad (AB)C = A(BC)$$

Commutative laws

$$A + B = B + A \qquad AB = BA$$

Distributive law

$$A(B + C) = AB + AC$$

De Morgan's laws

$$\overline{AB \cdots} = \bar{A} + \bar{B} + \cdots$$
$$\overline{A + B + \cdots} = \bar{A}\bar{B} \cdots$$

Auxiliary identities

$$A + AB = A \qquad A + \bar{A}B = A + B$$
$$(A + B)(A + C) = A + BC$$

where use is made of the identity $\bar{\bar{A}} = A$. Comparing Eqs. (6-29) and (6-30) shows that the EXCLUSIVE OR of Eq. (6-22) is equivalent to that of Eq. (6-24).

With the aid of De Morgan's law we can show that *an* AND *circuit for positive logic also functions as an* OR *gate for negative logic.* Let Y be the output and $A, B, \ldots, N$ be the inputs to a positive AND so that

$$Y = AB \cdots N \tag{6-31}$$

Then, by Eq. (6-26),

$$\bar{Y} = \bar{A} + \bar{B} + \cdots + \bar{N} \tag{6-32}$$

If the output and all inputs of a circuit are complemented so that a 1 becomes a 0 and vice versa, then positive logic is changed to negative logic (refer to Fig. 6-1). Since Y and $\bar{Y}$ represent the *same* output terminal, A and $\bar{A}$ the *same* input terminal, etc., the circuit which performs the positive AND logic in Eq. (6-31) also operates as the negative OR gate of Eq. (6-32). Similar reasoning is used to verify that the same circuit is either a negative AND or a positive OR, depending upon how the binary levels are defined. We verified this result for diode logic in Sec. 6-3, but the present proof is independent of how the circuit is implemented.

It should now be clear that it is really not necessary to use all three connectives OR, AND, and NOT. The OR and the NOT are sufficient because, from the De Morgan law of Eq. (6-25), the AND can be obtained from the OR and the NOT, as is indicated in Fig. 6-17a. Similarly, the AND and the NOT may be chosen as the basic logic circuits, and from the De Morgan law of Eq. (6-27), the OR may be constructed as shown in Fig. 6-17b. This figure makes clear

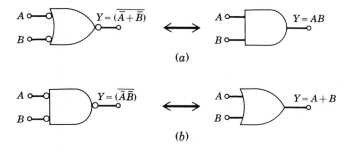

Fig. 6-17 (a) An OR is converted into an AND by inverting all inputs and also the output. (b) An AND becomes an OR if all inputs and the output are complemented.

once again that an OR (AND) circuit negated at input and output performs the AND (OR) logic.

6-9   THE NAND AND NOR DIODE-TRANSISTOR LOGIC (DTL) GATES

In Fig. 6-15a the negation before the second AND could equally well be put at the output of the first AND without changing the logic. Such an AND-NOT sequence is also present in Fig. 6-17b and in many other logic operations. This negated AND is called a NOT-AND, or a NAND, gate. The logic symbol, Boolean expression, and truth table for the NAND are given in Fig. 6-18. The NAND may be implemented by placing a transistor NOT circuit *after* the diode AND in Fig. 6-13. Such a transposition is shown in Fig. 6-19. Circuits involving diodes and transistors as in Fig. 6-19 are called *diode-transistor logic* (DTL) *gates.*

EXAMPLE (a) Verify that the circuit of Fig. 6-19 is a positive NAND for the binary levels 0 and 12 V. Neglect source impedance and junction saturation voltages and diode voltages in the forward direction. Find the minimum h_{FE}. (b) If the drop across a conducting diode is 0.7 V and if the sum of source and diode resistances is 1 K, is the NAND logic satisfied? (c) Will the circuit operate

Fig. 6-18 (a) The logic symbol and Boolean expression for a two-input NAND gate; (b) the truth table.

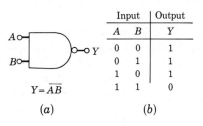

Input		Output
A	B	Y
0	0	1
0	1	1
1	0	1
1	1	0

$Y = \overline{AB}$

(a)   (b)

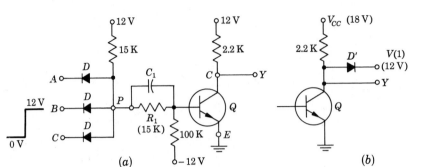

Fig. 6-19 (a) A three-input positive NAND (or negative NOR) gate; (b) a collector catching or clamping diode may be used to improve the characteristics of the output circuit.

properly if the inputs are obtained from the outputs of similar NAND gates? Assume silicon transistors and diodes and neglect collector saturation resistance and diode forward resistance.

Solution   *a.* If any input is at 0 V, then the junction point P of the diodes is at 0 V because a diode conducts and clamps this point to $V(0) = 0$. The base voltage of the transistor is then

$$V_B = -(12)\left(\frac{15}{115}\right) = -1.56 \text{ V}$$

Hence Q is cut off and Y is at 12 V, or $Y = 1$. This result confirms the first three rows of the truth table of Fig. 6-18b.

If all inputs are at $V(1) = 12$ V, assume that all diodes are reverse-biased and that the transistor is in saturation. We shall now verify that these assumptions are indeed correct. If Q is in saturation, then with $V_{BE} = 0$, the voltage at P is $(12)(\frac{15}{30}) = 6$ V. Hence, with 12 V at each input, all diodes are reverse-biased by 6 V. Since the diodes are nonconducting, the two 15-K resistors are in series and the base current of Q is

$$\frac{12}{30} - \frac{12}{100} = 0.40 - 0.12 = 0.28 \text{ mA}$$

Since the collector current is

$$I_C = \frac{12}{2.2} = 5.45 \text{ mA} \quad \text{and} \quad (h_{FE})_{\min} = \frac{5.45}{0.28} = 19$$

then Q will indeed be in saturation if $h_{FE} \geq 19$. Under these circumstances the output is at ground, or $Y = 0$. This result confirms the last row of the truth table.

b. The transistor must be OFF if at least one input is at 0. The worst case occurs when all diodes except one are reverse-biased, because then the voltage at P is a maximum. For this case the Thévenin's equivalent from P to ground is,

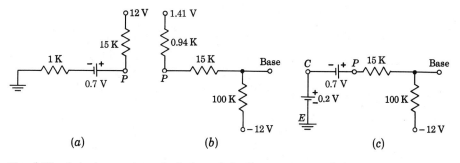

Fig. 6-20 Relating to the calculation of the base voltage of the transistor in the circuit of Fig. 6-19.

from Fig. 6-20a, a voltage of

$$(12)\left(\frac{1}{16}\right) + (0.7)\left(\frac{15}{16}\right) = 0.75 + 0.66 = 1.41 \text{ V}$$

in series with a resistance of $(1)(\frac{15}{16}) = 0.94$ K. The open-circuit voltage at the base of the transistor is, from Fig. 6-20b,

$$V_B = -(12)\left(\frac{15.9}{115.9}\right) + (1.41)\left(\frac{100}{115.9}\right) = -1.65 + 1.21 = -0.44 \text{ V}$$

This voltage is more than adequate to reverse-bias Q, and hence NAND/logic is satisfied.

c. If the inputs are high, the situation is exactly as in part a. With respect to keeping the base node at a low voltage when there is no coincidence, the worst situation occurs when all but one input are high. The low input now comes from a transistor in saturation, and $V_{CE,\text{sat}} \approx 0.2$ V. The open-circuit voltage at the base of Q is, from Fig. 6-20c,

$$V_B = -(12)\left(\frac{15}{115}\right) + (0.9)\left(\frac{100}{115}\right) = -0.78 \text{ V}$$

which cuts off Q and $Y = 1$, as it should.

If we neglect the inherent speed limitations of the transistor the rise time of the output, when Q is cut off, depends upon the shunt capacitance C_s and the collector resistance R_c. If V_{CC} is increased, then for fixed values of C_s and R_c, less time is required to reach the particular voltage at which the next stage switches (is driven into saturation). If such an increased value of V_{CC} [$> V(1)$] is used, a collector clamping diode is often added, as indicated in Fig. 6-19b. This diode limits the collector voltage of Q to $V(1)$ and also prevents the following stage from being driven too heavily into saturation. Secondarily, the diode helps to reduce the time to discharge C_s when Q is driven into saturation by providing a lower collector starting voltage.

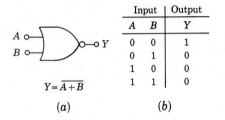

Input		Output
A	B	Y
0	0	1
0	1	0
1	0	0
1	1	0

$Y = \overline{A+B}$

(a)   (b)

Fig. 6-21   (a) The logic symbol and Boolean expression for a two-input NOR gate; (b) the truth table.

A NOR Gate   A negation following an OR is called a NOT-OR, or a NOR gate.   The logic symbol, Boolean expression, and truth table for the NOR are given in Fig. 6-21.   A positive NOR circuit is implemented by a cascade of a diode OR and a transistor INVERTER.

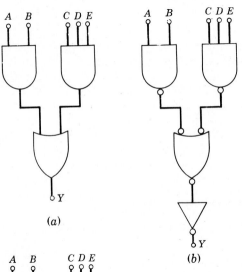

(a)

(b)

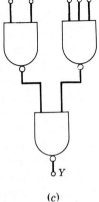

(c)

Fig. 6-22   A two-level AND-OR is equivalent to a NAND-NAND configuration.

The circuits of Figs. 6-13 and 6-19 employ *diode-transistor logic* (DTL). The NAND and NOR may also be implemented in other configurations, as is indicated in Secs. 6-12 and 6-14. With the aid of De Morgan's laws, it can be shown that, regardless of the hardware involved, a positive NAND is also a negative NOR, whereas a negative NAND may equally well be considered a positive NOR.

It is clear that a single input NAND is a NOT. Also, a NAND followed by a NOT is an AND. In Sec. 6-8 it is pointed out that all logic can be performed by using only the two connectives AND and NOT. Therefore we now conclude that, by repeated use of the NAND circuit alone, any logical function can be carried out. A similar argument leads equally well to the result that all logic can be performed by using only the NOR circuit.

EXAMPLE Verify that two-level AND-OR topology is equivalent to a NAND-NAND system.

Solution The AND-OR logic is indicated in Fig. 6-22a. Since $X = \bar{\bar{X}}$, then inverting the output of an AND and simultaneously negating the input to the following OR does not change the logic. These modifications are made in Fig. 6-22b. We have also negated the output of the OR gate and, at the same time, have added an INVERTER to Fig. 6-22b, so that once again the logic is unaffected. An OR gate negated at each terminal is an AND circuit (Fig. 6-17a). Since an AND followed by an INVERTER is a NAND then Fig. 6-22c is equivalent to Fig. 6-22b. Hence, the NAND-NAND of Fig. 6-22c is equivalent to the AND-OR of Fig. 6-22a.

If any of the inputs in Fig. 6-22 are obtained from the output of another gate then the resultant topology is referred to as *three-level logic*.

6-10   MODIFIED (INTEGRATED-CIRCUIT) DTL GATES[5]

Most logic gates are fabricated as an *integrated circuit* (IC). This process is described in the next chapter, where it is demonstrated that all transistors, diodes, resistors, and capacitors in a fairly complicated circuit may be shaped within a tiny chip of single-crystal silicon (approximately 50 mil by 50 mil in surface area and 1 mil thick). It turns out that large values of resistance (above 30 K) and of capacitance (above 100 pF) cannot be fabricated economically. On the other hand, transistors and diodes may be constructed very inexpensively. In view of these facts, the NAND gate of Fig. 6-19 is modified for integrated-circuit implementation by eliminating the capacitor C_1, reducing the resistance values drastically, and using diodes or transistors to replace resistors wherever possible. At the same time the power-supply requirements are simplified so that only a single 5-V supply is used (instead of the three separate voltages of Fig. 6-19b). The resulting circuit is indicated in Fig. 6-23.

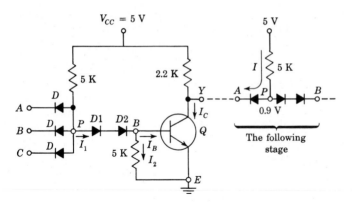

Fig. 6-23 An integrated positive DTL NAND gate.

The operation of this positive NAND gate is easily understood qualitatively. If at least one of the inputs is low (the 0 state), the diode D connected to this input conducts and the voltage V_P at point P is low. Hence diodes $D1$ and $D2$ are nonconducting, $I_B = 0$, and the transistor is OFF. Therefore the output of Q is high and Y is in the 1 state. This logic satisfies the first three rows of the truth table in Fig. 6-18. Consider now the case where all inputs are high (1) so that all input diodes D are cut off. Then V_P tries to rise toward V_{CC}, and a base current I_B results. If I_B is sufficiently large, Q is driven into saturation and the output Y falls to its low (0) state, thus satisfying the fourth row of the truth table.

This NAND gate is considered quantitatively in the following illustrative example. The necessity for using two diodes $D1$ and $D2$ in series is explained. False logic can be caused by switching transients, power-supply noise spikes, coupling between leads, etc. The noise margins that this circuit can tolerate are calculated below.

EXAMPLE (a) For the transistor in Fig. 6-23 assume (Table 5-1) that $V_{BE,\text{sat}} = 0.8$ V, $V_\gamma = 0.5$ V, and $V_{CE,\text{sat}} = 0.2$ V. The drop across a conducting diode is 0.7 V and V_γ (diode) $= 0.6$ V. The inputs of this switch are obtained from the outputs of similar gates. Verify that the circuit functions as a positive NAND and calculate $(h_{FE})_{\min}$. (b) Will the circuit operate properly if $D2$ is not used? (c) If all inputs are high, what is the magnitude of the noise voltage at the input which will cause the gate to malfunction? (d) Repeat part c if at least one input is low. Assume, for the moment, that Q is not loaded by a following stage.

Solution a. The logic levels are $V_{CE,\text{sat}} = 0.2$ V for the 0 state and $V_{CC} = 5$ V for the 1 state. If at least one input is in the 0 state, its diode conducts and $V_P = 0.2 + 0.7 = 0.9$ V. Since, in order for $D1$ and $D2$ to be conducting, a voltage of $(2)(0.7) = 1.4$ V is required, these diodes are cut off, and $V_{BE} = 0$. Since the cutin voltage of Q is $V_\gamma = 0.5$ V, then Q is OFF, the output rises to 5 V, and $Y = 1$. This confirms the first three rows of the NAND truth table.

If all inputs are at $V(1) = 5$ V, then we shall assume that all input diodes are OFF, that $D1$ and $D2$ conduct, and that Q is in saturation. If these conditions are true, the voltage at P is the sum of two diode drops plus $V_{BE,\text{sat}}$, or $V_P = 0.7 + 0.7 + 0.8 = 2.2$ V. The voltage across each input diode is $5 - 2.2 = 2.8$ V in the reverse direction, thus justifying the assumption that D is OFF. We now find $(h_{FE})_{\text{min}}$ to put Q into saturation.

$$I_1 = \frac{V_{CC} - V_P}{5} = \frac{5 - 2.2}{5} = 0.56 \text{ mA}$$

$$I_2 = \frac{V_{BE,\text{sat}}}{5} = \frac{0.8}{5} = 0.16 \text{ mA}$$

$$I_B = I_1 - I_2 = 0.56 - 0.16 = 0.40 \text{ mA}$$

$$I_C = \frac{V_{CC} - V_{CE,\text{sat}}}{2.2} = \frac{5 - 0.2}{2.2} = 2.18 \text{ mA}$$

and

$$(h_{FE})_{\text{min}} = \frac{I_C}{I_B} = \frac{2.18}{0.40} = 5.5$$

If $h_{FE} > (h_{FE})_{\text{min}}$, then $Y = V(0)$ for all inputs at $V(1)$, thus verifying the last line in the truth table in Fig. 6-18.

b. If at least one input is at $V(0)$, then $V_P = 0.2 + 0.7 = 0.9$ V. Hence, if only one diode $D1$ is used between P and B, then $V_{BE} = 0.9 - 0.6 = 0.3$ V, where 0.6 V represents the diode cutin voltage. Since the cutin base voltage is $V_\gamma = 0.5$ V, then theoretically Q is cut off. However, this is not a very conservative design because a small (>0.2 V) spike of noise will turn Q ON. An even more conservative design uses three diodes in series, instead of the two indicated in Fig. 6-23.

c. If all inputs are high, then from part *a,* $V_P = 2.2$ V and each input diode is reverse-biased by 2.8 V. A diode starts to conduct when it is forward-biased by 0.6 V. Hence a negative noise spike in excess of $2.8 + 0.6 = 3.4$ V must be present at the input before the circuit malfunctions. Such a large noise voltage is improbable.

d. If at least one input is low, then from part *a,* $V_P = 0.9$ V and Q is OFF. If a noise spike takes Q just into its active region, $V_{BE} = V_\gamma = 0.5$ and V_P must increase to $0.5 + 0.6 + 0.6 = 1.7$ V. Hence the noise margin is $1.7 - 0.9 = 0.8$ V. If only one diode $D1$ were used, the noise voltage would be reduced by 0.6 V (the drop across $D2$ at cutin) to $0.8 - 0.6 = 0.2$ V. This confirms the value obtained in part *b.*

Fan-out In the foregoing discussion we have unrealistically assumed that the NAND gate is unloaded. If it drives N similar gates, we say that the *fan-out* is N. The output transistor now acts as a *sink* for the current in the input to the gates it drives. In other words, when Q is in saturation ($Y = 0$), the input current I in Fig. 6-23 of a following stage adds to the collector

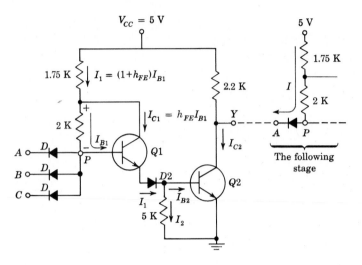

Fig. 6-24 A modified integrated positive DTL NAND gate with increased fan-out.

current of Q. Assume that all the input diodes to a following stage (which is now considered to be a *current source*) are high except the one driven by Q. Then the current in this diode is $I = (5 - 0.9)/5 = 0.82$ mA. This current is called a *standard load*. The total collector current of Q is now $I_C = 0.82N + 2.18$ mA, where 2.18 mA is the unloaded collector current found in part a of the preceding example. Since the base current is almost independent of loading, I_B remains at its previous value of 0.4 mA. If we assume a reasonable value for $(h_{FE})_{min}$ of 30, the fan-out is given by $I_C = h_{FE}I_B$, or

$$I_C = 0.82N + 2.18 = (30)(0.40) = 12.0 \text{ mA} \qquad (6\text{-}33)$$

and $N = 12$. Of course, the current rating of Q must exceed 12.0 mA if we are to drive 12 gates.

The fan-out may be increased considerably by replacing $D1$ by a transistor $Q1$, as indicated in Fig. 6-24. When $Q1$ is conducting, it is in its active region and *not* in saturation. This statement follows from the fact that the current in the 2-K resistance is in the direction to reverse-bias the collector junction of the n-p-n transistor $Q1$. Since the emitter current of $Q1$ supplies the base current of $Q2$, then $Q2$ is driven by a much higher base current than is Q in Fig. 6-23. For the same $(h_{FE})_{min}$ of the output transistors in Figs. 6-23 and 6-24, it is clear that the latter circuit has the larger collector current, and hence the larger fan-out.

EXAMPLE If $(h_{FE})_{min} = 30$, calculate the fan-out N for the NAND gate of Fig. 6-24. From Table 5-1, $V_{BE,active} = 0.7$ V.

Solution   As with the circuit of Fig. 6-23, if any input is low, then $V_P = 0.9$ V, and both $Q1$ and $D2$ are OFF. Hence $V_{BE2} = 0$, $Q2$ is OFF, and $Y = 1$. If, however, all inputs are high, the input diodes are OFF, $Q2$ goes into saturation, and

$$V_P = V_{BE1,active} + V_{D2} + V_{BE2,sat} = 0.7 + 0.7 + 0.8 = 2.2 \text{ V}$$

Since $Q1$ is in its active region, $I_{C1} = h_{FE}I_{B1}$. As indicated in Fig. 6-24, the current in the 2-K resistor is I_{B1} (remember that each D is cut off), and the current in the 1.75-K resistor is $I_1 = I_{B1} + I_{C1} = (1 + h_{FE})I_{B1}$. Applying KVL between V_{CC} and V_P, we have, for $h_{FE} = 30$,

$$5 - 2.2 = (1.75)(31)I_{B1} + 2I_{B1} \tag{6-34}$$

or

$$I_{B1} = 0.050 \text{ mA} \qquad\qquad I_1 = (31)(0.050) = 1.55 \text{ mA}$$

$$I_2 = 0.8/5 = 0.16 \text{ mA} \qquad I_{B2} = 1.55 - 0.16 = 1.39 \text{ mA}$$

The unloaded collector current of $Q2$ is $(5 - 0.2)/2.2 = 2.18$ mA. For *each* gate which it drives, $Q2$ must sink a standard load of

$$I = \frac{5 - 0.7 - 0.2}{1.75 + 2} = 1.10 \text{ mA}$$

Since the maximum collector current is $h_{FE}I_{B2}$, then

$$I_{C2} = (30)(1.39) = 1.10N + 2.18 = 41.7 \text{ mA} \tag{6-35}$$

and $N = 36$. The fan-out has been increased to 36 for the same h_{FE} which resulted in only 12 for the circuit of Fig. 6-22.

The above calculation assumes that the current rating of $Q2$ is at least 41.7 mA. On the other hand, if $(I_{C2})_{max}$ is limited to, say, 15 mA, then $1.10N + 2.18 = 15$, or $N \approx 11$. To drive these 11 gates requires that $h_{FE,min} = I_{C2}/I_{B2} = 15/1.39 = 10.8$, which is a very small number.

The *fan-in M* of a logic gate gives the number of inputs to the switch. For example, in Fig. 6-24, $M = 3$.

Wired Logic   It is possible to connect the outputs of several DTL gates together, as in Fig. 6-25, to perform additional logic without additional hardware. If positive NAND logic is under consideration, this connection is called a *wired*-AND, *phantom*-AND, *dotted*-AND, or *implied*-AND. Thus, if both $Y_1 = 1$ and $Y_2 = 1$, then $Y = 1$, whereas if $Y_1 = 0$ and/or $Y_2 = 0$, then $Y = 0$.

The circuit of Fig. 6-24 also represents a negative NOR gate, and connecting two outputs together as in Fig. 6-25 now represents negative wired-OR logic. If Y_1 and/or Y_2 is in the low state (which is now the 1 state), then Y is also in the low state ($Y = 1$), whereas if Y_1 and Y_2 are both high (the 0 state), then Y is also high ($Y = 0$).

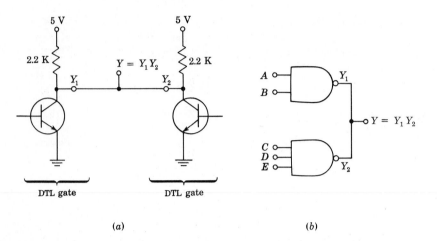

Fig. 6-25 (a) Wired-AND logic is obtained by connecting the outputs of positive NAND gates together; (b) a two-input and a three-input NAND gate wired-AND together to perform the logic in Eq. (6-36).

Consider two positive NAND gates wired-AND together as in Fig. 6-25b. Then $Y_1 = \overline{AB}$ and $Y_2 = \overline{CDE}$. Hence

$$Y = Y_1 Y_2 = (\overline{AB})(\overline{CDE}) = \overline{AB + CDE} \qquad (6\text{-}36)$$

where use is made of De Morgan's law. Note that the wired-AND has led to an implementation of the AOI two-level logic (Sec. 6-7). Because of the + sign in Eq. (6-36), the connection in Fig. 6-25 is often incorrectly referred to as a positive wired-OR.

Note that the wired-AND connection places the collector resistors in Fig. 6-25a in parallel. This reduction in resistance increases the power dissipation in the ON state. In order to avoid this condition *open-collector* gates are available specifically for wired-AND applications.

6-11 HIGH–THRESHOLD–LOGIC (HTL) GATE[5]

In an industrial environment the noise level is quite high because of the presence of motors, high-voltage switches, on-off control circuits, etc. By using a higher supply voltage (15 V instead of 5 V) and a 6.9-V Zener diode in place of $D2$ in the DTL gate of Fig. 6-24, this circuit is converted into the high-noise-immunity gate of Fig. 6-26. The resistances are increased in Fig. 6-26 with respect to those in Fig. 6-24, so that approximately the same currents are obtained in both circuits. The noise margin obtained with this circuit is typically 7 V (Prob. 6-44).

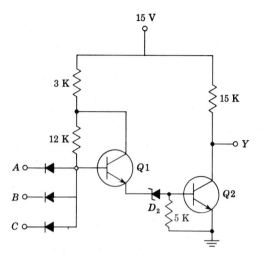

Fig. 6-26 A high-threshold-logic NAND gate.

6-12 TRANSISTOR–TRANSISTOR–LOGIC (TTL) GATE[6]

The fastest-saturating logic circuit is the transistor-transistor-logic gate (TTL, or T²L), shown in Fig. 6-27. This switch uses a multiple-emitter transistor which is easily and economically fabricated using integrated-circuit techniques (Sec. 7-7). The TTL circuit has the topology of the DTL circuit of Fig. 6-23, with the emitter junctions of $Q1$ acting as the input diodes D of the DTL gate and the collector junction of $Q1$ replacing the diode $D1$ of Fig. 6-23. The base-to-emitter diode of $Q2$ is used in place of the diode $D2$ of the DTL gate, and both circuits have an output transistor ($Q3$ or Q).

The explanation of the operation of the TTL gate parallels that of the DTL switch. Thus, if at least one input is at $V(0) = 0.2$ V, then

$$V_P = 0.2 + 0.7 = 0.9V$$

For the collector junction of $Q1$ to be forward-biased and for $Q2$ and $Q3$ to be ON requires about $0.7 + 0.7 + 0.7 = 2.1$ V. Hence these are OFF; the output rises to $V_{CC} = 5$ V, and $Y = V(1)$. On the other hand, if all inputs are high (at 5 V), the input diodes (the emitter junctions) are reverse-biased and V_P rises toward V_{CC} and drives $Q2$ and $Q3$ into saturation. Then the output is $V_{CE,\mathrm{sat}} = 0.2$ V, and $Y = V(0)$ (and V_P is clamped at about 2.3 V).

The above explanation indicates that $Q1$ acts like isolated back-to-back diodes, and not as a transistor. However, we shall now show that, during turnoff, $Q1$ does exhibit transistor action, thereby reducing storage time (Sec. 6-5) considerably. Note that the base voltage of $Q2$, which equals the collector voltage of $Q1$, is at $0.8 + 0.8 = 1.6$ V during saturation of $Q2$ and $Q3$. If now any input drops to 0.2 V, then $V_P = 0.9$ V, and hence the base of $Q1$ is at 0.9 V. At this time the collector junction is reverse-biased by

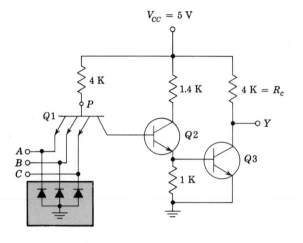

Fig. 6-27 An IC positive TTL NAND gate. (Neglect the diodes in the shaded block.)

$1.6 - 0.9 = 0.7$ V, the emitter junction is forward-biased, and $Q1$ is in its active region. The large collector current of $Q1$ now quickly removes the stored charge in $Q2$ and $Q3$. It is this transistor action which gives TTL the highest speed of any saturated logic.

Clamping diodes (shown in the shaded block in Fig. 6-27) are often included from each input to ground, with the anode grounded. These diodes are effectively out of the circuit for positive input signals, but they limit negative voltage excursions at the input to a safe value. These negative signals may arise from ringing caused by lead inductance resonating with shunt capacitance.

6-13   OUTPUT STAGES

At the output terminal of the DTL or TTL gate there is a capacitive load C_L, consisting of the capacitances of the reverse-biased diodes of the fan-out gates and any stray wiring capacitance. If the collector-circuit resistor of the inverter is R_c (called a *passive pull-up*), then, when the output changes from the low to the high state, the output transistor is cut off and the capacitance charges exponentially from $V_{CE,\text{sat}}$ to V_{CC}. The time constant R_cC_L of this waveform may introduce a prohibitively long delay time into the operation of these gates.

The output delay may be reduced by decreasing R_c, but this will increase the power dissipation when the output is in its low state and the voltage across R_c is $V_{CC} - V_{CE,\text{sat}}$. A better solution to this problem is indicated in Fig. 6-28, where a transistor acts as an *active pull-up* circuit, replacing the passive pull-up resistance R_c. This output configuration is called a *totem-pole* amplifier because the transistor $Q4$ "sits" upon $Q3$. It is also referred to as a power-driver, or power-buffer, output stage.

The transistor $Q2$ acts as a *phase splitter*, since the emitter voltage is out of phase with the collector voltage (for an increase in base current, the emitter voltage increases and the collector voltage decreases). We now explain the operation of this driver circuit in detail, with reference to the TTL gate of Fig. 6-28a.

The output is in the low-voltage state when $Q2$ and $Q3$ are driven into saturation. For this state we should like $Q4$ to be OFF. Is it? Note that the collector voltage V_{CN2} of $Q2$ with respect to ground N is given by

$$V_{CN2} = V_{CE2,sat} + V_{BE3,sat} = 0.2 + 0.8 = 1.0 \text{ V} \qquad (6\text{-}37)$$

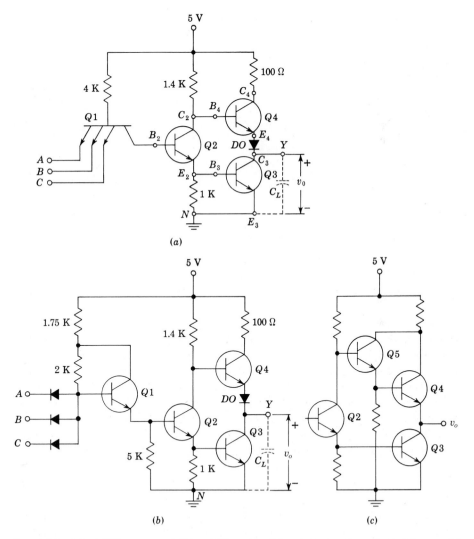

Fig. 6-28 (a) A TTL gate and (b) a DTL gate with a totem-pole output driver. (c) A modified configuration of the output circuit. [The circuit to the left of $Q2$ is identical with that in (a) for TTL or in (b) for DTL.]

Since the base of $Q4$ is tied to the collector of $Q2$, then $V_{BN4} = V_{CN2} = 1.0$ V. *If the output diode DO were missing,* the base-to-emitter voltage of $Q4$ would be

$$V_{BE4} = V_{BN4} - V_{CE3,sat} = 1.0 - 0.2 = 0.8 \text{ V}$$

which would put $Q4$ into saturation. Under these circumstances the steady current through $Q4$ would be

$$\frac{V_{CC} - V_{CE4,sat} - V_{CE3,sat}}{100} = \frac{5 - 0.2 - 0.2}{100} \text{ A} = 46 \text{ mA} \qquad (6\text{-}38)$$

which is excessive and wasted current. The necessity for adding DO is now clear. With it in place, the sum of V_{BE4} and V_{DO} is 0.8 V. Hence both $Q4$ and DO are at cutoff. In summary, if C_L is at the high voltage $V(1)$ and the gate is excited, $Q4$ and DO go off, and $Q3$ conducts. Because of its large active-region current, $Q3$ quickly discharges C_L, and as v_o approaches $V(0)$, $Q3$ enters saturation. The bottom transistor $Q3$ of the totem pole is referred to as a *current sink,* which discharges C_L.

Assume now that with the output at $V(0)$, there is a change of state, because one of the inputs drops to its low state. Then $Q2$ is turned off, which causes $Q3$ to go to cutoff because V_{BE3} drops to zero. The output v_o remains momentarily at 0.2 V because the voltage across C_L cannot change instantaneously. Now $Q4$ goes into saturation and DO conducts, as we can verify:

$$V_{BN4} = V_{BE4,sat} + V_{DO} + v_o = 0.8 + 0.7 + 0.2 = 1.7 \text{ V}$$

and the base and collector currents of $Q4$ are

$$I_{B4} = \frac{V_{CC} - V_{BN4}}{1.4} = \frac{5 - 1.7}{1.4} = 2.36 \text{ mA}$$

$$I_{C4} = \frac{V_{CC} - V_{CE4,sat} - V_{DO} - v_o}{0.1} = \frac{5 - 0.2 - 0.7 - 0.2}{0.1} = 39.0 \text{ mA}$$

Hence, if h_{FE} exceeds $(h_{FE})_{min} = I_{C4}/I_{B4} = 39.0/2.36 = 16.5$, then $Q4$ is in saturation. The transistor $Q4$ is referred to as a *source,* supplying current to C_L. As long as $Q4$ remains in saturation, the output voltage rises exponentially toward V_{CC} with the very small time constant $(100 + R_{CS4} + R_f)C_L$, where R_{CS4} is the saturation resistance (Sec. 5-8) of $Q4$, and where R_f (a few ohms) is the diode forward resistance. As v_o increases, the currents in $Q4$ decrease, and $Q4$ comes out of saturation and finally v_o reaches a steady state when $Q4$ is at the cutin condition. Hence the final value of the output voltage is

$$v_o = V_{CC} - V_{BE4,cutin} - V_{DO,cutin} \approx 5 - 0.5 - 0.6 = 3.9 \text{ V} = V(1) \qquad (6\text{-}39)$$

If the 100-Ω resistor were omitted, there would result a faster change in output from $V(0)$ to $V(1)$. However, the 100-Ω resistor is needed to limit the current spikes during the turn-on and turn-off transients. In particular, $Q3$ does not turn off (because of storage time) as quickly as $Q4$ turns on. With both totem-pole transistors conducting at the same time, the supply voltage would be short-circuited if the 100-Ω resistor were missing. The peak current drawn from the supply during the transient is limited to $I_{C4} + I_{B4} = 39 + 2.4 \approx 41$ mA if the 100-Ω resistor is used. These current spikes generate noise in the power-supply distribution system, and also result in increased power consumption at high frequencies.

Wired Logic It should be emphasized that the wired-AND connection must *not* be used with the totem-pole driver circuit. If the output from one gate is high while that from a second gate is low, and if these two outputs are tied together, we have exactly the situation just discussed in connection with transient current spikes. Hence, if the wired-AND were used, the power supply would deliver a *steady* current of 41 mA under these circumstances.

Alternative Output Stages From the foregoing discussion it should be clear that the diode DO can be moved from the emitter into the base lead of $Q4$. This configuration is used by some manufacturers.

The diode in the base of $Q4$ referred to in the preceding paragraph may be the base-to-emitter diode of an additional transistor, such as $Q5$ in Fig. 6-28c. The combination of $Q5$ and $Q4$, called a *Darlington pair*, is discussed in Sec. 8-6. It has a very high current gain and extremely low output resistance and results in a very high speed gate. Another form of power buffer is given in Prob. 6-45.

Storage time is eliminated by using *Schottky transistors* (which are clamped to prevent saturation, Sec. 7-13) for the transistors in Fig. 6-28.

6-14   RESISTOR–TRANSISTOR LOGIC (RTL) AND DIRECT–COUPLED TRANSISTOR LOGIC (DCTL)

In 1972 TTL was the most popular family, and it had captured about 45 percent of the integrated-circuit digital market. Approximately another 30 percent of the market was taken by DTL, and the remaining 25 percent was divided between several additional families, called *resistor-transistor logic* (RTL), *direct-coupled transistor logic* (DCTL), *emitter-coupled logic* (ECL), and *metal-oxide-semiconductor* (MOS) *logic*. Since MOS logic uses the field-effect transistor (FET), the discussion of these gates is postponed until Chap. 10, where the MOSFET is introduced. Since ECL requires an understanding of the *differential amplifier*, this type of (nonsaturating) logic is considered in Sec. 16-17, after the differential amplifier is studied. A discussion of RTL and DCTL now follows.

Resistor-Transistor Logic (RTL)[5] This configuration is indicated in Fig. 6-29a, which represents a three-input positive NOR gate with a fan-out of 5. If any input is high, the corresponding transistor is driven into saturation and the output is low, $v_o = V_{CE,\text{sat}} \approx 0.2$ V $= V(0)$. If all inputs are low, then all input transistors are cut off by $V_\gamma - V(0) = 0.5 - 0.2 = 0.3$ V and the

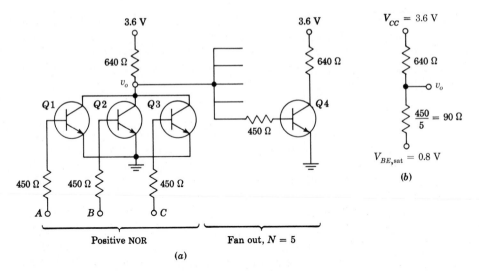

Fig. 6-29 (a) An RTL positive NOR gate with a fan-in of 3 and a fan-out of 5. (b) The equivalent circuit from which to calculate v_o in the high state.

output v_o is high. (Note the low noise margin.) The preceding two statements confirm that the gate performs positive NOR (or negative NAND) logic.

The value of v_o depends upon the fan-out. For example, if $N = 5$, then the output of the NOR gate is loaded by five 450-Ω resistors in parallel (or 90 Ω), which is tied to $V_{BE,sat} \approx 0.8$ V, as shown in Fig. 6-29b. Under these circumstances (using superposition),

$$v_o = \frac{3.6 \times 90}{90 + 640} + \frac{0.8 \times 640}{90 + 640} = 1.14 \text{ V} = V(1) \tag{6-40}$$

This voltage must be large enough so that the base current can drive each of the five transistors into saturation. Since

$$I_B = \frac{1.14 - 0.8}{0.45} = 0.755 \text{ mA} \qquad I_C = \frac{3.6 - 0.2}{0.64} = 5.31 \text{ mA}$$

then the circuit will operate properly if $h_{FE} > (h_{FE})_{min} = 5.31/0.76 = 7.0$.

Resistor-transistor logic uses the minimum space (for a standard digital function) on a silicon wafer, and hence is very economical.

Direct-Coupled Transistor Logic (DCTL)[7]   This configuration is the same as RTL, except that the base resistors are omitted. In Fig. 6-30 the fan-in is 3 and the fan-out is 2.

To verify that the circuit implements positive NOR logic, consider first that all inputs are in the 0 state. Because this low voltage to an input (say, to $Q1$) comes from a saturated transistor (Q') of a preceding state,

$$v_1 = V_{CE,sat} = V(0)$$

Since this voltage is 0.2 V for a saturated silicon transistor, and since the cutin voltage $V_\gamma \approx 0.5$ V, $Q1$ will conduct very little (although the noise margin is only $0.5 - 0.2 = 0.3$ V as it is with RTL logic). Since the current in $Q1$ is almost zero, the output Y tries to raise V_{CC} and $Q4$ and $Q5$ go into saturation. Hence the output Y is clamped at

$$V_{BE,\text{sat}} = V(1) \approx 0.8 \text{ V}$$

for silicon. Thus, with all inputs in the low state, the output is in the high state. Note that the high state is only 0.8 V, independent of V_{CC}.

Consider now that at least one input v_1 is in the high state. Since $Q1$ is fed from Q', Q' is cut off and $Q1$ is driven into saturation. Under these circumstances the output Y is $V_{CE,\text{sat}} = V(0)$. If more than one input is excited, the output will certainly be low. Hence we have confirmed that the NOR function is satisfied.

There are a number of difficulties with DCTL: (1) The reverse saturation current for all fan-in transistors adds in the common collector-circuit resistor R_c. At high enough temperatures the total $I_{CBO}R_c$ drop may be large enough so that the output Y is too low to drive the fan-out transistors into saturation. (2) Because of the direct connection, the base current is almost equal to the collector current (for $V_{CC} \gg V_{CE,\text{sat}}$ and $V_{CC} \gg V_{BE,\text{sat}}$). With a transistor so heavily driven into saturation, very large stored base charge will result, with a corresponding detrimental effect on the switching speed. (3) Since the voltage levels are so low—the total output-voltage step is only of the order of 0.6 V for silicon—then spurious (noise) spikes can be troublesome. (4) The bases of the fan-out transistors are connected together. Since the input characteristics can never be identical, let us assume that $Q4$ has a much lower V_{BE} for a given I_B than does $Q5$. Under these circumstances, $Q4$ will "hog" most of the base current, and it is possible that $Q5$ may not even be driven into saturation. Hence transistors suitable for DCTL must have very

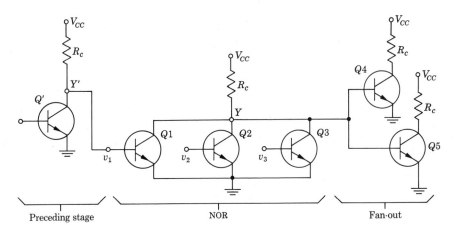

Fig. 6-30 A positive NOR DCTL gate.

close control on uniformity of input characteristics, very low values of I_{CBO}, as large a differential as possible between $V_{BE,\text{sat}}$ and $V_{CE,\text{sat}}$, a large h_{FE}, and a small storage time.

The advantages of DCTL are: (1) Only one low-voltage supply (operation with 1.5 V is possible) is needed; (2) transistors with low breakdown voltages may be used; and (3) the power dissipation is low.

In DCTL a NOR or NAND circuit is possible in which the transistors are in series (totem-pole fashion) rather than in parallel as in Fig. 6-30. Because of the difficulties mentioned above, DCTL is not often used with bipolar transistors, but it is the standard logic with field-effect transistors (Sec. 10-6).

6-15   COMPARISON OF LOGIC FAMILIES

In the discussion of each logic configuration some of its advantages and disadvantages have been listed. An exhaustive comparison is extremely difficult because we must take into account all the following characteristics: (1) speed (propagation time delay), (2) noise immunity, (3) fan-in and fan-out capabilities, (4) power-supply requirements, (5) power dissipation per gate, (6) suitability for integrated fabrication, (7) operating temperature range, (8) number of functions available, and (9) cost. Also to be considered is the personal prejudice of the engineer, who is always strongly influenced by past experience. Items 1 and 8 require some explanation; all others have already been defined or are self-evident.

Propagation Delay   In Fig. 6-9 the ON time t_{ON} of an (n-p-n) inverter is defined as the interval between the application of an ideal pulse input and the time when the output current reaches 90 percent (or the output voltage falls to 10 percent) of its final value. This definition is *not* useful for the propagation-delay time of a logic gate, for two reasons. First, the input to a gate is not a square pulse, but it is a waveform with a nonzero rise time. Second, the input does not have to reach its 90 percent value before the gate changes state. Hence we define *propagation-delay* ON *time* T_{ON} (to distinguish it from t_{ON}) as indicated below.

As the input voltage to a positive NAND gate rises from $V(0)$ toward $V(1)$, then at some *switching threshold voltage* $V'(0)$ (Fig. 6-31), conditions within the gate are modified, so that a change of state of the output from $V(1)$ to $V(0)$ is initiated. Similarly, as the input falls from $V(1)$ toward $V(0)$, then at some other *switching threshold voltage* $V'(1)$, the initiation of the change of state from $V(0)$ to $V(1)$ takes place. We now define (as in Fig. 6-31) the propagation delay ON time T_{ON} as the interval between the time when the input v_i reaches $V'(0)$ and the output falls to $V'(1)$. Also, the propagation delay OFF time T_{OFF} is defined as the interval between the time when the input equals $V'(1)$ and the output rises to $V'(0)$. Because of minority-

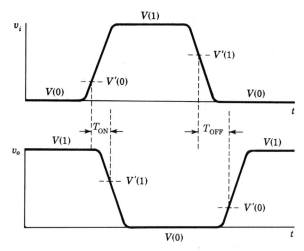

Fig. 6-31 Pertaining to the definitions of the propaga-tion-delay times.

carrier storage time, $T_{\text{ON}} \neq T_{\text{OFF}}$. Hence the *propagation-delay time* T_{PD} is usually defined as the average of these two times, or

$$T_{PD} = \tfrac{1}{2}(T_{\text{ON}} + T_{\text{OFF}}) \tag{6-41}$$

In passing, we note that some authors arbitrarily assume the two threshold voltages to be equal: $V'(0) = V'(1) = \tfrac{1}{2}[V(0) + V(1)]$.

Functions   The basic AND, OR, NAND, and NOR gates are combined in one integrated chip in various combinations to perform specific circuit func-

TABLE 6-5   Comparison of the major IC digital logic families[8]

Logic Parameter	RTL	DTL	HTL	TTL	ECL	MOS	CMOS
Basic gate[a].........	NOR	NAND	NAND	NAND	OR-NOR	NAND	NOR or NAND
Fan-out[b]...........	5	8	10	10	25	20	>50
Power dissipated[c] per gate, mW	12	8–12	55	12–22	40–55	0.2–10	0.01 static 1 at 1 MHz
Noise immunity.......	Nominal	Good	Excellent	Very good	Good	Nominal	Very good
Propagation delay[d] per gate, ns........	12	30	90	12–6	4–1	300	70
Clock rate,[e] MHz.....	8	12–30	4	15–60	60–400	2	5
Number of functions[f].	High	Fairly high	Nominal	Very high	High	Low	Low
Reference...........	Sec. 6-14	Sec. 6-10	Sec. 6-11	Sec. 6-12	Sec. 16-17	Sec. 10-7	Sec. 10-7

[a] Positive logic.

[b] Worst-case number of inputs that the gate can drive.

[c] Typical; affected by temperature and frequency of operation.

[d] Typical for a nominal fan-out.

[e] Maximum frequency at which flip-flops operate.   The actual clock rate is from one-half to one-tenth the frequency listed.

[f] New functions are being added continuously by the manufacturers, except to DTL.   However, TTL functions are compatible with DTL.

tions. These *system building blocks* are discussed in Chap. 17 and include flip-flops, counters, arithmetic functions, decoders, shift registers, etc.

It should be clear that no single logic configuration can be best suited for all applications. A comparison of the major integrated-circuit digital logic families is given in Table 6-5.

REFERENCES

1. Chen, T. C.: Diode Coincidence and Mixing Circuits in Digital Computers, *Proc. IRE*, vol. 38, pp. 511–514, May, 1950.

   Hussey, L. W.: Semiconductor Diode Gates, *Bell System Tech. J.*, vol. 32, no. 5, pp. 1137–1154, September, 1953.

2. Millman, J., and H. Taub: "Pulse, Digital, and Switching Waveforms," chap. 20, McGraw-Hill Book Company, New York, 1965.

3. Ref. 2, sec. 20-19.

4. Masher, D. P.: The Design of Diode-Transistor NOR Circuits, *IRE Trans. Electron. Computers*, vol. EC-9, no. 1, pp. 15–24, March, 1960.

   Todd, C. R.: An Annotated Bibliography on NOR and NAND Logic, *IEEE Trans. Electron. Computers*, vol. EC-12, no. 5, pp. 462–464, October, 1963.

5. Garrett, L. S.: Integrated-circuit Digital Logic Families, Part I, RTL, DTL, and HTL Devices, *IEEE Spectrum*, vol. 7, pp. 46–56, October, 1970.

6. Garret, L. S.: Integrated-circuit Digital Logic Families, Part II, TTL Devices, *IEEE Spectrum*, vol. 7, pp. 63–72, November, 1970.

7. Beter, R. H., W. E. Bradley, R. B. Brown, and M. Rubinoff: Directly Coupled Transistor Circuits, *Electronics*, vol. 28, no. 6, pp. 132–136, June, 1955.

8. Garrett, L. S.: Integrated-circuit Digital Logic Families, Part III, ECL and MOS Devices, *IEEE Spectrum*, vol. 7, p. 41, December, 1970.

REVIEW QUESTIONS

6-1 Express the following decimal numbers in binary form: (a) 28; (b) 100; (c) 5,127.

6-2 Define (a) *positive logic;* (b) *negative logic.*

6-3 Define an OR gate and give its truth table.

6-4 Draw a positive diode OR gate and explain its operation.

6-5 Evaluate the following expressions: (a) $A + 1$; (b) $A + A$; (c) $A + 0$.

6-6 Define an AND gate and give its truth table.

6-7 Draw a positive-diode AND gate and explain its operation.

6-8 Evaluate the following expressions: (a) $A1$; (b) AA; (c) $A0$; (d) $A + AB$.

6-9   Define a NOT gate and give its truth table.

6-10   Draw a positive-logic NOT gate and explain its operation.

6-11   Evaluate the following expressions: (a) $\bar{\bar{A}}$; (b) $\bar{A}A$; (c) $\bar{A} + A$.

6-12   A pulse waveform drives an n-p-n transistor from cutoff into saturation and then back to cutoff. (a) Draw the output current waveshape, lined up in time with the input voltage. (b) Indicate the following times on your sketch: *delay, rise,* ON, *storage, fall,* and OFF.

6-13   (a) What is the physical origin of *storage time?* (b) Is it important in turning a transistor ON or OFF? Explain. (c) Draw the minority-carrier concentration in the base; in the active region and in saturation.

6-14   Define an INHIBITOR and give the truth table for $AB\bar{S}$.

6-15   Define an EXCLUSIVE OR and give its truth table.

6-16   Show two logic block diagrams for an EXCLUSIVE OR.

6-17   Verify that the following Boolean expressions represent an EXCLUSIVE OR: (a) $\overline{AB + \bar{A}\bar{B}}$; (b) $(A + B)(\bar{A} + \bar{B})$.

6-18   State the two forms of De Morgan's laws.

6-19   Show how to implement an AND with OR and NOT gates.

6-20   Show how to implement an OR with AND and NOT gates.

6-21   Define a NAND gate and give its truth table.

6-22   Draw a positive NAND gate with diodes and a transistor (DTL) and explain its operation.

6-23   Define a NOR gate and give its truth table.

6-24   Repeat Rev. 6-22 for a NOR gate.

6-25   Draw the circuit of an IC DTL gate and explain its operation.

6-26   Define (a) *fan-out;* (b) *fan-in;* (c) *standard load;* (d) *current sink;* (e) *current source.*

6-27   What logic is performed if the outputs of two DTL gates are connected together? Explain.

6-28   How does high-threshold logic (HTL) differ from DTL?

6-29   Draw the circuit of a TTL gate and explain its operation.

6-30   Draw a totem-pole output buffer with a TTL gate. Explain its operation.

6-31   Repeat Rev. 6-30 for a DTL gate.

6-32   Draw the circuit of an RTL gate and explain its operation for positive logic.

6-33   Repeat Rev. 6-32 for negative logic.

6-34   Draw a DCTL circuit and explain its operation.

6-35   List three advantages and three disadvantages of DCTL gates.

6-36   Define (a) two *threshold voltages;* (b) *propagation-delay* ON *time;* (c) *propagation-delay* OFF *time;* (d) *propagation-delay time.*

6-37   Give an order of magnitude which is applicable to any of the logic families studied in this chapter for (a) fan-out; (b) power dissipation per gate; (c) propagation delay per gate; (d) clock rate.

7 / INTEGRATED CIRCUITS: FABRICATION AND CHARACTERISTICS

An integrated circuit consists of a single-crystal chip of silicon, typically 50 by 50 mils in cross section, containing both active and passive elements and their interconnections. Such circuits are produced by the same processes used to fabricate individual transistors and diodes. These processes include epitaxial growth, masked impurity diffusion, oxide growth, and oxide etching, using photolithography for pattern definition. A method of batch processing is employed which offers excellent repeatability and is adaptable to the production of large numbers of integrated circuits at low cost. In this chapter we describe the basic processes involved in fabricating an integrated circuit.

7-1 INTEGRATED–CIRCUIT TECHNOLOGY

The fabrication of integrated circuits is based on materials, processes, and design principles which constitute a highly developed semiconductor (planar-diffusion) technology. The basic structure of an integrated circuit is shown in Fig. 7-1b, and consists of four distinct layers of material. The bottom layer ① (6 mils thick) is p-type silicon and serves as a *substrate* upon which the integrated circuit is to be built. The second layer ② is thin (typically 25 μm = 1 mil) n-type material which is grown as a single-crystal extension of the substrate. All active and passive components are built within the thin n-type layer using a series of diffusion steps. These components are transistors, diodes, capacitors, and resistors, and they are made by diffusing p-type and n-type impurities. The most complicated component fabricated is the transistor, and all other elements are constructed with one or more of the processes required to make a transistor. In the fabrication of all the above elements it is necessary to distribute impurities in certain precisely defined regions within the second (n-type) layer. The selective diffusion of impurities is accomplished by using SiO_2 as a barrier which protects portions of the wafer against impurity penetration. Thus the third layer of material ③ is silicon dioxide, and it also provides protection of the semiconductor surface against contamination. In the regions where diffusion is to take place, the SiO_2 layer is etched away, leaving the rest of the wafer protected against diffusion. To permit selective etching, the SiO_2 layer must be subjected to a photolithographic process, described in Sec. 7-4. Finally, a fourth metallic (aluminum) layer ④ is added to supply the necessary interconnections between components.

The p-type substrate which is required as a foundation for the integrated circuit is obtained by growing an ingot (1 to 2 in. in diameter and about 10 in. long) from a silicon melt with a predetermined number of impurities. The crystal ingot is subsequently sliced into round wafers approximately 6 mils thick, and one side of each wafer is lapped and polished to eliminate surface imperfections.

We are now in a position to appreciate some of the significant advantages of the integrated-circuit technology. Let us consider a 1-in.-square wafer divided into 400 chips of surface area 50 mil by 50 mil. We demonstrate in this chapter that a reasonable area under which a component (say, a transistor) is fabricated is 50 mils². Hence each chip (each integrated circuit) contains 50 separate components, and there are $50 \times 400 = 20,000$ components/in.² on each wafer.

If we process 10 wafers in a batch, we can manufacture 4,000 integrated circuits simultaneously, and these contain 200,000 components. Some of the chips will contain faults due to imperfections in the manufacturing process, but the *yield* (the percentage of fault-free chips per wafer) is extremely large.

The following advantages are offered by integrated-circuit technology as compared with discrete components interconnected by conventional techniques:

1. Low cost (due to the large quantities processed).
2. Small size.
3. High reliability. (All components are fabricated simultaneously, and there are no soldered joints.)
4. Improved performance. (Because of the low cost, more complex circuitry may be used to obtain better functional characteristics.)

In the next sections we examine the processes required to fabricate an integrated circuit.

7-2   BASIC MONOLITHIC INTEGRATED CIRCUITS[1,2]

We now examine in some detail the various techniques and processes required to obtain the circuit of Fig. 7-1a in an integrated form, as shown in Fig. 7-1b. This configuration is called a monolithic integrated circuit because it is formed on a single silicon chip. The word "monolithic" is derived from the Greek *monos*, meaning "single," and *lithos*, meaning "stone." Thus a monolithic circuit is built into a single stone, or single crystal.

In this section we describe qualitatively a complete epitaxial-diffused fabrication process for integrated circuits. In subsequent sections we examine in more detail the epitaxial, photographic, and diffusion processes involved. The logic circuit of Fig. 7-1a is chosen for discussion because it contains typical components: a resistor, diodes, and a transistor. These elements (and also capacitors with small values of capacitances) are the components encountered in integrated circuits. The monolithic circuit is formed by the steps indicated in Fig. 7-2 and described below.

Step 1.   Epitaxial Growth   An n-type epitaxial layer, typically 25 μm thick, is grown onto a p-type substrate which has a resistivity of typically

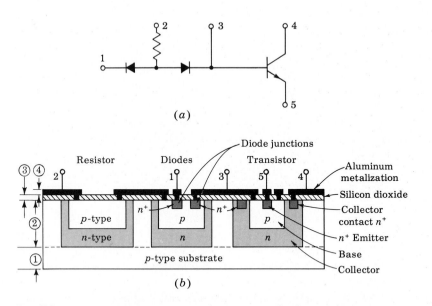

Fig. 7-1   (a) A circuit containing a resistor, two diodes, and a transistor. (b) Cross-sectional view of the circuit in (a) when transformed into a monolithic form (not drawn to scale). The four layers are (1) substrate, (2) n-type crystal containing the integrated circuit, (3) silicon dioxide, and (4) aluminum metalization. (After Phillips.[2] Not drawn to scale.)

10 Ω-cm, corresponding to $N_A = 1.4 \times 10^{15}$ atoms/cm^3. The epitaxial process described in Sec. 7-3 indicates that the resistivity of the n-type epitaxial layer can be chosen independently of that of the substrate. Values of 0.1 to 0.5 Ω-cm are chosen for the n-type layer. After polishing and cleaning, a thin layer (0.5 μm = 5,000 Å) of oxide, SiO$_2$, is formed over the entire wafer, as shown in Fig. 7-2a. The SiO$_2$ is grown by exposing the epitaxial layer to an oxygen atmosphere while being heated to about 1000°C. Silicon dioxide has the fundamental property of preventing the diffusion of impurities through it. Use of this property is made in the following steps.

Step 2. Isolation Diffusion In Fig. 7-2b the wafer is shown with the oxide removed in four different places on the surface. This removal is accomplished by means of a photolithographic etching process described in Sec. 7-4. The remaining SiO$_2$ serves as a mask for the diffusion of acceptor impurities (in this case, boron). The wafer is now subjected to the so-called *isolation diffusion*, which takes place at the temperature and for the time interval required for the p-type impurities to penetrate the n-type epitaxial layer and reach the p-type substrate. We thus leave the shaded n-type regions in Fig. 7-2b. These sections are called *isolation islands*, or *isolated regions*, because they are separated by two back-to-back p-n junctions. Their purpose is to allow electrical isolation between different circuit components. For example, it will become apparent later in this section that a different isolation region must be used for the collector of each separate transistor. The p-type substrate must always be held at a negative potential with respect to the isolation islands in order that the p-n junctions be reverse-biased. If these diodes were to become forward-biased in an operating circuit, then, of course, the isolation would be lost.

It should be noted that the concentration of acceptor atoms ($N_A \approx 5 \times 10^{20}$ cm^{-3}) in the region between isolation islands will generally be much higher (and hence indicated as p^+) then in the p-type substrate. The reason for this higher density is to prevent the depletion region of the reverse-biased isolation-to-substrate junction from extending into p^+-type material (Sec. 3-7) and possibly connecting two isolation islands.

Parasitic Capacitance It is now important to consider that these isolation regions, or junctions, are connected by a significant barrier, or transition capacitance C_{Ts}, to the p-type substrate, which capacitance can affect the operation of the circuit. Since C_{Ts} is an undesirable by-product of the isolation process, it is called the *parasitic capacitance*.

The parasitic capacitance is the sum of two components, the capacitance C_1 from the bottom of the n-type region to the substrate (Fig. 7-2b) and C_2 from the sidewalls of the isolation islands to the p^+ region. The bottom component, C_1, results from an essentially step junction due to the epitaxial growth (Sec. 7-3), and hence varies inversely as the square root of the voltage V between the isolation region and the substrate (Sec. 3-7). The sidewall capacitance C_2 is associated with a diffused graded junction, and it varies as $V^{-\frac{1}{3}}$.

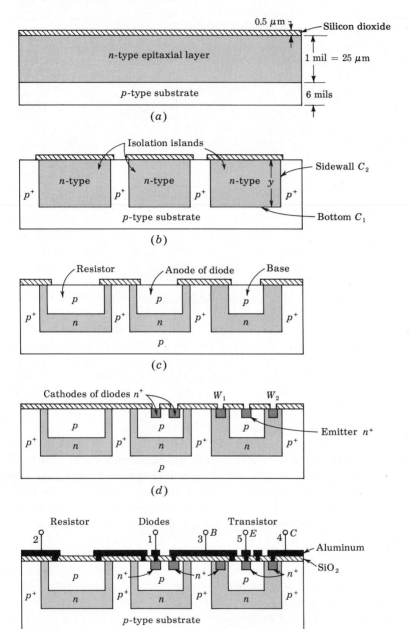

Fig. 7-2 The steps involved in fabricating a monolithic circuit (not drawn to scale). (a) Epitaxial growth; (b) isolation diffusion; (c) base diffusion; (d) emitter diffusion; (e) aluminum metalization.

For this component the junction area is equal to the perimeter of the isolation region times the thickness y of the epitaxial n-type layer. The total capacitance is of the order of a few picofarads.

Step 3. Base Diffusion During this process a new layer of oxide is formed over the wafer, and the photolithographic process is used again to create the pattern of openings shown in Fig. 7-2c. The p-type impurities (boron) are diffused through these openings. In this way are formed the transistor base regions as well as resistors, the anode of diodes, and junction capacitors (if any). It is important to control the depth of this diffusion so that it is shallow and does not penetrate to the substrate. The resistivity of the base layer will generally be much higher than that of the isolation regions.

Step 4. Emitter Diffusion A layer of oxide is again formed over the entire surface, and the masking and etching processes are used again to open windows in the p-type regions, as shown in Fig. 7-2d. Through these openings are diffused n-type impurities (phosphorus) for the formation of transistor emitters, the cathode regions for diodes, and junction capacitors.

Additional windows (such as W_1 and W_2 in Fig. 7-2d) are often made into the n regions to which a lead is to be connected, using aluminum as the ohmic contact, or interconnecting metal. During the diffusion of phosphorus a heavy concentration (called n^+) is formed at the points where contact with aluminum is to be made. Aluminum is a p-type impurity in silicon, and a large concentration of phosphorus prevents the formation of a p-n junction when the aluminum is alloyed to form an ohmic contact.[4]

Step 5. Aluminum Metalization All p-n junctions and resistors for the circuit of Fig. 7-1a have been formed in the preceding steps. It is now necessary to interconnect the various components of the integrated circuit as dictated by the desired circuit. To make these connections, a fourth set of windows is opened into a newly formed SiO_2 layer, as shown in Fig. 7-2e, at the points where contact is to be made. The interconnections are made first, using vacuum deposition of a thin even coating of aluminum over the entire wafer. The photoresist technique is now applied to etch away all undesired aluminum areas, leaving the desired pattern of interconnections shown in Fig. 7-2e between resistors, diodes, and transistors.

In production a large number (several hundred) of identical circuits such as that of Fig. 7-1a are manufactured simultaneously on a single wafer. After the metalization process has been completed, the wafer is scribed with a diamond-tipped tool and separated into individual chips. Each chip is then mounted on a ceramic wafer and is attached to a suitable header. The package leads are connected to the integrated circuit by stitch bonding[1] of a 1-mil aluminum or gold wire from the terminal pad on the circuit to the package lead (Fig. 7-27).

Summary In this section the epitaxial-diffused method of fabricating integrated circuits is described. We have encountered the following processes:

1. Crystal growth of a substrate
2. Epitaxy
3. Silicon dioxide growth
4. Photoetching
5. Diffusion
6. Vacuum evaporation of aluminum

Using these techniques, it is possible to produce the following elements on the same chip: transistors, diodes, resistors, capacitors, and aluminum interconnections.

7-3   EPITAXIAL GROWTH[1]

The epitaxial process produces a thin film of single-crystal silicon from the gas phase upon an existing crystal wafer of the same material. The epitaxial layer may be either p-type or n-type. The growth of an epitaxial film with impurity atoms of boron being trapped in the growing film is shown in Fig. 7-3.

The basic chemical reaction used to describe the epitaxial growth of pure silicon is the hydrogen reduction of silicon tetrachloride:

$$SiCl_4 + 2H_2 \underset{\longleftarrow}{\overset{1200°C}{\longrightarrow}} Si + 4HCl \tag{7-1}$$

Since it is required to produce epitaxial films of specific impurity concentrations, it is necessary to introduce impurities such as phosphine for n-type doping or biborane for p-type doping into the silicon tetrachloride–hydrogen

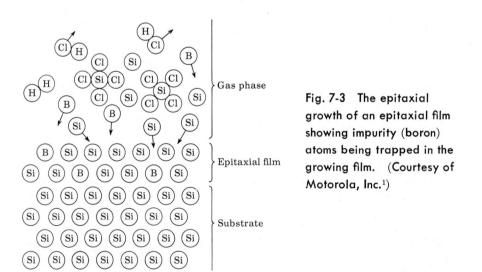

Fig. 7-3 The epitaxial growth of an epitaxial film showing impurity (boron) atoms being trapped in the growing film. (Courtesy of Motorola, Inc.[1])

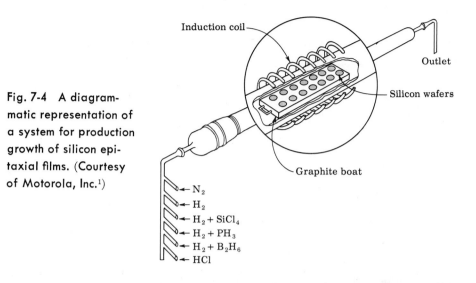

Fig. 7-4 A diagrammatic representation of a system for production growth of silicon epitaxial films. (Courtesy of Motorola, Inc.[1])

Induction coil

Outlet

Silicon wafers

Graphite boat

N_2

H_2

$H_2 + SiCl_4$

$H_2 + PH_3$

$H_2 + B_2H_6$

HCl

gas stream. An apparatus for the production of an epitaxial layer is shown in Fig. 7-4. In this system a long cylindrical quartz tube is encircled by a radio-frequency induction coil. The silicon wafers are placed on a rectangular graphite rod called a *boat*. The boat is inserted in the reaction chamber, and the graphite is heated inductively to about 1200°C. At the input of the reaction chamber a control console permits the introduction of various gases required for the growth of appropriate epitaxial layers. Thus it is possible to form an almost abrupt step *p-n* junction similar to the junction shown in Fig. 3-10.

7-4 MASKING AND ETCHING[1]

The monolithic technique described in Sec. 7-2 requires the selective removal of the SiO_2 to form openings through which impurities may be diffused. The photoetching method used for this removal is illustrated in Fig. 7-5. During the photolithographic process the wafer is coated with a uniform film of a photosensitive emulsion (such as the Kodak *photoresist* KPR). A large black-and-white layout of the desired pattern of openings is made and then reduced photographically. This negative, or stencil, of the required dimensions is placed as a mask over the photoresist, as shown in Fig. 7-5a. By exposing the KPR to ultraviolet light through the mask, the photoresist becomes polymerized under the transparent regions of the stencil. The mask is now removed, and the wafer is "developed" by using a chemical (such as trichloroethylene) which dissolves the unexposed (unpolymerized) portions of the photoresist film and leaves the surface pattern as shown in Fig. 7-5b. The emulsion which was not removed in development is now *fixed*, or *cured*, so that it becomes resistant to the corrosive etches used next. The chip is

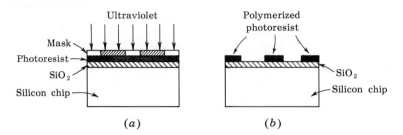

Fig. 7-5 Photoetching technique. (a) Masking and exposure to ultraviolet radiation. (b) The photoresist after development.

immersed in an etching solution of hydrofluoric acid, which removes the oxide from the areas through which dopants are to be diffused. Those portions of the SiO$_2$ which are protected by the photoresist are unaffected by the acid. After etching and diffusion of impurities, the resist mask is removed (stripped) with a chemical solvent (hot H$_2$SO$_4$) and by means of a mechanical abrasion process.

7-5   DIFFUSION OF IMPURITIES[5]

The most important process in the fabrication of integrated circuits is the diffusion of impurities into the silicon chip. We now examine the basic theory connected with this process. The solution to the diffusion equation will give the effect of temperature and time on the diffusion distribution.

The Diffusion Law   The equation governing the diffusion of neutral atoms is

$$\frac{\partial N}{\partial t} = D \frac{\partial^2 N}{\partial x^2}$$

(7-2)

where N is the particle concentration in atoms per unit volume as a function of distance x from the surface and time t, and D is the diffusion constant in area per unit time.

The Complementary Error Function   If an intrinsic silicon wafer is exposed to a volume of gas having a uniform concentration N_o atoms per unit volume of n-type impurities, such as phosphorus, these atoms will diffuse into the silicon crystal, and their distribution will be as shown in Fig. 7-6a. If the diffusion is allowed to proceed for extremely long times, the silicon will become uniformly doped with N_o phosphorus atoms per unit volume. The basic assumptions made here are that the surface concentration of impurity atoms remains at N_o for all diffusion times and that $N(x) = 0$ at $t = 0$ for $x > 0$.

If Eq. (7-2) is solved and the above boundary conditions are applied,

$$N(x, t) = N_o \left(1 - \text{erf} \frac{x}{2\sqrt{Dt}}\right) = N_o \text{ erfc} \frac{x}{2\sqrt{Dt}} \qquad (7\text{-}3)$$

where erfc y means the error-function complement of y, and the *error function* of y is defined by

$$\text{erf } y \equiv \frac{2}{\sqrt{\pi}} \int_0^y \epsilon^{-\lambda^2} d\lambda \qquad (7\text{-}4)$$

and is tabulated in Ref. 3. The function erfc $y = 1 - \text{erf } y$ is plotted in Fig. 7-7.

The Gaussian Distribution   If a specific number Q of impurity atoms per unit area are deposited on one face of the wafer, and then if the material is heated, the impurity atoms will again diffuse into the silicon. When the boundary conditions $\int_0^\infty N(x) \, dx = Q$ for all times and $N(x) = 0$ at $t = 0$ for $x > 0$ are applied to Eq. (7-2), we find

$$N(x, t) = \frac{Q}{\sqrt{\pi Dt}} \epsilon^{-x^2/4Dt} \qquad (7\text{-}5)$$

Equation (7-5) is known as the Gaussian distribution, and is plotted in Fig. 7-6b for two times. It is noted from the figure that as time increases, the surface concentration decreases. The area under each curve is the same, however, since this area represents the total amount of impurity being diffused, and

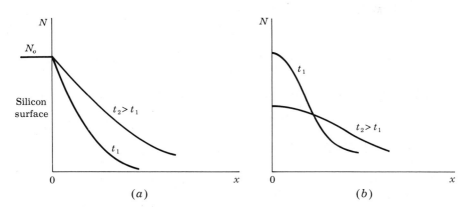

Fig. 7-6   The concentration N as a function of distance x into a silicon chip for two values t_1 and t_2 of the diffusion time.   (a) The surface concentration is held constant at N_o per unit volume.   (b) The total number of atoms on the surface is held constant at Q per unit area.

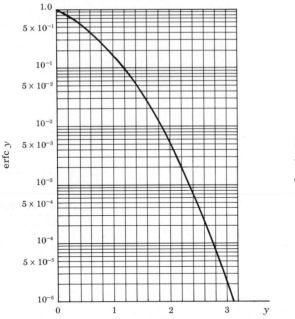

erfc y

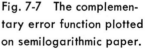

Fig. 7-7 The complementary error function plotted on semilogarithmic paper.

this is a constant amount Q. Note that in Eqs. (7-3) and (7-5) time t and the diffusion constant D appear only as a product Dt.

Solid Solubility[1,6] The designer of integrated circuits may wish to produce a specific diffusion profile (say, the complementary error function of an n-type impurity). In deciding which of the available impurities (such as phosphorus, arsenic, antimony) can be used, it is necessary to know if the number of atoms per unit volume required by the specific profile of Eq. (7-3) is less than the diffusant's *solid solubility*. The solid solubility is defined as the maximum concentration N_o of the element which can be dissolved in the solid silicon at a given temperature. Figure 7-8 shows solid solubilities of some impurity

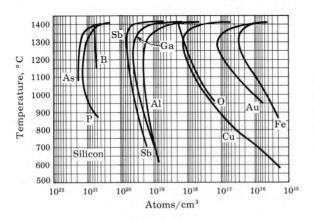

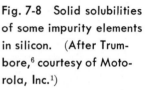

Fig. 7-8 Solid solubilities of some impurity elements in silicon. (After Trumbore,[6] courtesy of Motorola, Inc.[1])

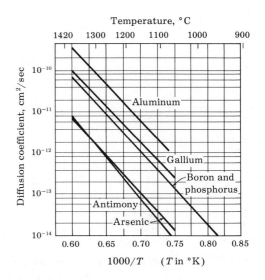

Fig. 7-9 Diffusion coefficients as a function of temperature for some impurity elements in silicon. (After Fuller and Ditzenberger,[5] courtesy of Motorola, Inc.[1])

elements. It can be seen that, since for phosphorus the solid solubility is approximately 10^{21} atoms/cm^3 and for pure silicon we have 5×10^{22} atoms/cm^3, the maximum concentration of phosphorus in silicon is 2 percent. For most of the other impurity elements the solubility is a small fraction of 1 percent.

Diffusion Coefficients   Temperature affects the diffusion process because higher temperatures give more energy, and thus higher velocities, to the diffusant atoms. It is clear that the diffusion coefficient is a function of temperature, as shown in Fig. 7-9. From this figure it can be deduced that the diffusion coefficient could be doubled for a few degrees increase in temperature. This critical dependence of D on temperature has forced the development of accurately controlled diffusion furnaces, where temperatures in the range of 1000 to 1300°C can be held to a tolerance of ± 0.5°C or better. Since time t in Eqs. (7-3) and (7-5) appears in the product Dt, an increase in either diffusion constant or diffusion time has the same effect on diffusant density.

Note from Fig. 7-9 that the diffusion coefficients, for the same temperature, of the n-type impurities (antimony and arsenic) are lower than the coefficients for the p-type impurities (gallium and aluminum), but that phosphorus (n-type) and boron (p-type) have the same diffusion coefficients.

Typical Diffusion Apparatus   Reasonable diffusion times require high diffusion temperatures ($\sim$1000°C). Therefore a high-temperature diffusion furnace, having a closely controlled temperature over the length (20 in.) of the hot zone of the furnace, is standard equipment in a facility for the fabrication of integrated circuits. Impurity sources used in connection with diffu-

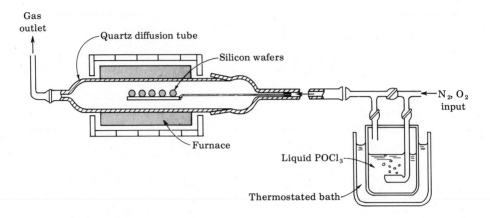

Fig. 7-10   Schematic representation of typical apparatus for POCl₃ diffusion. (Courtesy of Motorola, Inc.[1])

sion furnaces can be gases, liquids, or solids. For example, $POCl_3$, which is a liquid, is often used as a source of phosphorus. Figure 7-10 shows the apparatus used for $POCl_3$ diffusion. In this apparatus a carrier gas (mixture of nitrogen and oxygen) bubbles through the liquid-diffusant source and carries the diffusant atoms to the silicon wafers. Using this process, we obtain the complementary-error-function distribution of Eq. (7-3). A two-step procedure is used to obtain the Gaussian distribution. The first step involves *predeposition*, carried out at about 900°C, followed by *drive-in* at about 1100°C.

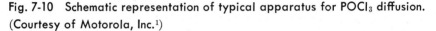

EXAMPLE   A uniformly doped n-type silicon epitaxial layer of 0.5 Ω-cm resistivity is subjected to a boron diffusion with constant surface concentration of 5×10^{18} cm⁻³. It is desired to form a p-n junction at a depth of 2.7 μm. At what temperature should this diffusion be carried out if it is to be completed in 2 hr?

Solution   The concentration N of boron is high at the surface and falls off with distance into the silicon, as indicated in Fig. 7-6a. At that distance $x = x_j$ at which N equals the concentration n of the doped silicon wafer, the net impurity density is zero. For $x < x_j$, the net impurity density is positive, and for $x > x_j$, it is negative. Hence x_j represents the distance from the surface at which a junction is formed. We first find n from Eq. (2-8):

$$n = \frac{\sigma}{\mu_n q} = \frac{1}{(0.5)(1,300)(1.60 \times 10^{-19})} = 0.96 \times 10^{16} \text{ cm}^{-3}$$

where all distances are expressed in centimeters and the mobility μ_n for silicon is taken from Table 2-1, on page 29. The junction is formed when $N = n$. For

$$\text{erfc } y = \frac{N}{N_o} = \frac{n}{N_o} = \frac{0.96 \times 10^{16}}{5 \times 10^{18}} = 1.98 \times 10^{-3}$$

we find from Fig. 7-7 that $y = 2.2$.   Hence

$$2.2 = \frac{x_j}{2\sqrt{Dt}} = \frac{2.7 \times 10^{-4}}{2\sqrt{D \times 2 \times 3,600}}$$

Solving for D, we obtain $D = 5.2 \times 10^{-13}$ cm²/sec.   This value of diffusion constant for boron is obtained from Fig. 7-9 at $T = 1130°C$.

7-6   TRANSISTORS FOR MONOLITHIC CIRCUITS[1,7]

A planar transistor made for monolithic integrated circuits, using epitaxy and diffusion, is shown in Fig. 7-11a.   Here the collector is electrically separated from the substrate by the reverse-biased isolation diodes.   Since the anode of the isolation diode covers the back of the entire wafer, it is necessary to make the collector contact on the top, as shown in Fig. 7-11a.   It is now clear that the isolation diode of the integrated transistor has two undesirable effects: it adds a parasitic shunt capacitance to the collector and a leakage current path.   In addition, the necessity for a top connection for the collector increases the collector-current path and thus increases the collector resistance and $V_{CE,\text{sat}}$.   All these undesirable effects are absent from the discrete epitaxial transistor shown in Fig. 7-11b.   What is then the advantage of the monolithic transistor?   A significant improvement in performance arises from the fact that integrated transistors are located physically close together and their electrical characteristics are closely matched.   For example, integrated transistors spaced within 30 mils (0.03 in.) have V_{BE} matching of better than

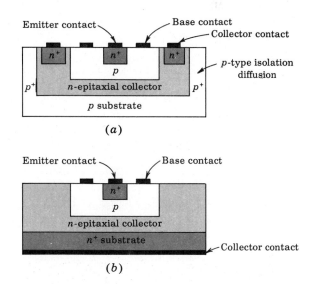

Fig. 7-11   Comparison of cross sections of (a) a monolithic integrated circuit transistor with (b) a discrete planar epitaxial transistor.   [For a top view of the transistor in (a) see Fig. 7-13.]

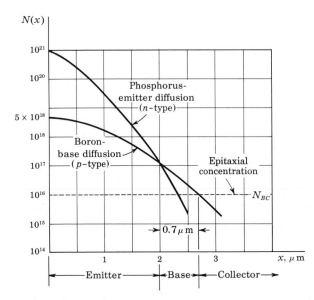

Fig. 7-12 A typical impurity profile in a monolithic integrated transistor. [Note that $N(x)$ is plotted on a logarithmic scale.]

5 mV with less than 10 μV/°C drift and an h_{FE} match of ± 10 percent. These matched transistors make excellent difference amplifiers (Sec. 15-3).

The electrical characteristics of a transistor depend on the size and geometry of the transistor, doping levels, diffusion schedules, and the basic silicon material. Of all these factors the size and geometry offer the greatest flexibility for design. The doping levels and diffusion schedules are determined by the standard processing schedule used for the desired transistors in the integrated circuit.

Impurity Profiles for Integrated Transistors[1] Figure 7-12 shows a typical impurity profile for a monolithic integrated circuit transistor. The background, or epitaxial-collector, concentration N_{BC} is shown as a dashed line in Fig. 7-12. The base diffusion of p-type impurities (boron) starts with a surface concentration of 5×10^{18} atoms/cm³, and is diffused to a depth of 2.7 μm, where the collector junction is formed. The emitter diffusion (phosphorus) starts from a much higher surface concentration (close to the solid solubility) of about 10^{21} atoms/cm³, and is diffused to a depth of 2 μm, where the emitter junction is formed. This junction corresponds to the intersection of the base and emitter distribution of impurities. We now see that the base thickness for this monolithic transistor is 0.7 μm. The emitter-to-base junction is usually treated as a step-graded junction, whereas the base-to-collector junction is considered a linearly graded junction.

EXAMPLE (*a*) Obtain the equations for the impurity profiles in Fig. 7-12. (*b*) If the phosphorus diffusion is conducted at 1100°C, how long should be allowed for this diffusion?

Solution   *a.* The base diffusion specifications are exactly those given in the example on page 208, where we find (with x expressed in micrometers) that

$$y = 2.2 = \frac{2.7}{2\sqrt{Dt}}$$

or

$$2\sqrt{Dt} = \frac{2.7}{2.2} = 1.23 \ \mu m$$

Hence the boron profile, given by Eq. (7-3), is

$$N_B = 5 \times 10^{18} \ \text{erfc} \ \frac{x}{1.23}$$

The emitter junction is formed at $x = 2 \ \mu m$, and the boron concentration here is

$$N_B = 5 \times 10^{18} \ \text{erfc} \ \frac{2}{1.23} = 5 \times 10^{18} \times 2 \times 10^{-2}$$
$$= 1.0 \times 10^{17} \ \text{cm}^{-3}$$

The phosphorus concentration N_P is given by

$$N_P = 10^{21} \ \text{erfc} \ \frac{x}{2\sqrt{Dt}}$$

At $x = 2$, $N_P = N_B = 1.0 \times 10^{17}$, so that

$$\text{erfc} \ \frac{2}{2\sqrt{Dt}} = \frac{1.0 \times 10^{17}}{10^{21}} = 1.0 \times 10^{-4}$$

From Fig. 7-7, $2/(2\sqrt{Dt}) = 2.7$ and $2\sqrt{Dt} = 0.75 \ \mu m$. Hence the phosphorus profile is given by

$$N_P = 10^{21} \ \text{erfc} \ \frac{x}{0.74}$$

b. From Fig. 7-9, at $T = 1100°C$, $D = 3.8 \times 10^{-13} \ \text{cm}^2/\text{sec}$. Solving for t from $2\sqrt{Dt} = 0.74 \ \mu m$, we obtain

$$t = \frac{(0.37 \times 10^{-4})^2}{3.8 \times 10^{-13}} = 3,600 \ s = 60 \ \text{min}$$

Monolithic Transistor Layout[1,2]   The physical size of a transistor determines the parasitic isolation capacitance as well as the junction capacitance. It is therefore necessary to use small-geometry transistors if the integrated circuit is designed to operate at high frequencies or high switching speeds. The geometry of a typical monolithic transistor is shown in Fig. 7-13. The emitter rectangle measures 1 by 1.5 mils, and is diffused into a 2.5- by 4.0-mil base region. Contact to the base is made through two metalized stripes on either side of the emitter. The rectangular metalized area forms the ohmic

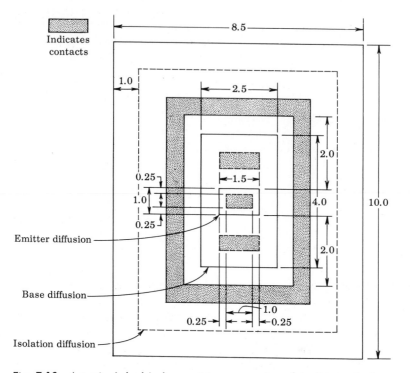

Indicates
contacts

Emitter diffusion

Base diffusion

Isolation diffusion

Fig. 7-13 A typical double-base stripe geometry of an integrated-circuit transistor. Dimensions are in mils. (For a side view of the transistor see Fig. 7-11.) (Courtesy of Motorola Monitor.)

contact to the collector region. The rectangular collector contact of this transistor reduces the saturation resistance. The substrate in this structure is located about 1 mil below the surface. Since diffusion proceeds in three dimensions, it is clear that the *lateral-diffusion* distance will also be 1 mil. The dashed rectangle in Fig. 7-13 represents the substrate area and is 6.5 by 8 mils. A summary of the electrical properties[2] of this transistor for both the 0.5- and the 0.1-Ω-cm collectors is given in Table 7-1.

Buried Layer[1] We noted above that the integrated transistor, because of the top collector contact, has a higher collector series resistance than a similar discrete-type transistor. One common method of reducing the collector series resistance is by means of a heavily doped n^+ "buried" layer sandwiched between the p-type substrate and the n-type epitaxial collector, as shown in Fig. 7-14. The buried-layer structure can be obtained by diffusing the n^+ layer into the substrate before the n-type epitaxial collector is grown or by selectively growing the n^+-type layer, using masked epitaxial techniques.

We are now in a position to appreciate one of the reasons why the integrated transistor is usually of the n-p-n type. Since the collector region is

TABLE 7-1   Characteristics for 1- by 1.5-mil double-base stripe monolithic transistors[2]

Transistor parameter	0.5 Ω-cm	0.1 Ω-cm†
BV_{CBO}, V	55	25
BV_{EBO}, V	7	5.5
BV_{CEO}, V	23	14
$C_{Te,\text{forward bias}}$, pF	6	10
C_{Te} at 0.5 V, pF	1.5	2.5
C_{Te} at 5 V, pF	0.7	1.5
h_{FE} at 10 mA	50	50
R_{CS}, Ω	75	15
$V_{CE,\text{sat}}$ at 5 mA, V	0.5	0.26
V_{BE} at 10 mA, V	0.85	0.85
f_T at 5 V, 5 mA, MHz	440	520

† Gold-doped.

subjected to heating during the base and emitter diffusions, it is necessary that the diffusion coefficient of the collector impurities be as small as possible, to avoid movement of the collector junction. Since Fig. 7-9 shows that n-type impurities have smaller values of diffusion constant D than p-type impurities, the collector is usually n-type. In addition, the solid solubility of some n-type impurities is higher than that of any p-type impurity, thus allowing heavier doping of the n^+-type emitter and other n^+ regions.

Lateral p-n-p Transistor[9]   The standard integrated-circuit transistor is an n-p-n type, as we have already emphasized. In some applications it is required to have both n-p-n and p-n-p transistors on the same chip. The lateral p-n-p structure shown in Fig. 7-15 is the most common form of the integrated p-n-p transistor. This p-n-p uses the standard diffusion techniques as the n-p-n, but the last n diffusion (used for the n-p-n transistor) is eliminated. While the p base for the n-p-n transistor is made, the two adjacent p regions are diffused for the emitter and collector of the p-n-p transistor shown in Fig. 7-15. Note that the current flows *laterally* from emitter to collector. Because of inaccuracies in masking, and because, also, of lateral diffusion, the base width between emitter and collector is large (about 1 mil compared with 1 μm

Fig. 7-14 Utilization of "buried" n^+ layer to reduce collector series resistance.

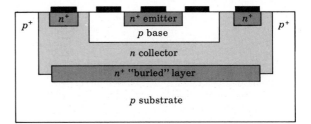

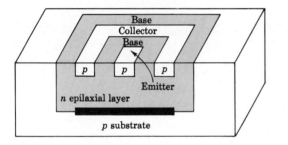

Fig. 7-15 A p-n-p lateral transistor.

for an n-p-n base). Hence the current gain of the p-n-p transistor is very low (0.5 to 5) instead of 50 to 300 for the n-p-n device. Since the base-p resistivity of the n-p-n transistor is relatively high, the collector and emitter resistances of the p-n-p device are high.

Vertical p-n-p Transistor[9] This transistor uses the substrate for the p collector; the n epitaxial layer for the base; and the p base of the standard n-p-n transistor as the emitter of this p-n-p device. We have already emphasized that the substrate must be connected to the most negative potential in the circuit. Hence a vertical p-n-p transistor can be used only if its collector is at a fixed negative voltage. Such a configuration is called an *emitter follower*, and is discussed in Sec. 8-8.

Supergain n-p-n Transistor[9] If the emitter is diffused into the base region so as to reduce the effective base width almost to the point of *punch-through* (Sec. 5-13), the current gain may be increased drastically (typically, 5,000). However, the breakdown voltage is reduced to a very low value (say, 5 V). If such a transistor in the CE configuration is operated in series with a standard integrated CB transistor (such a combination is called a *cascode* arrangement), the superhigh gain can be obtained at very low currents and with breakdown voltages in excess of 50 V.

7-7   MONOLITHIC DIODES[1]

The diodes utilized in integrated circuits are made by using transistor structures in one of five possible connections (Prob. 7-9). The three most popular diode structures are shown in Fig. 7-16. They are obtained from a transistor structure by using the emitter-base diode, with the collector short-circuited to the base (a); the emitter-base diode, with the collector open (b); and the collector-base diode, with the emitter open-circuited (or not fabricated at all) (c). The choice of the diode type used depends upon the application and circuit performance desired. Collector-base diodes have the higher collector-base voltage-breakdown rating of the collector junction ($\sim$12 V minimum), and they are suitable for common-cathode diode arrays diffused within a single isolation

Fig. 7-16 Cross section of various diode structures. (*a*) Emitter-base diode with collector shorted to base; (*b*) emitter-base diode with collector open; (*c*) collector-base diode (no emitter diffusion).

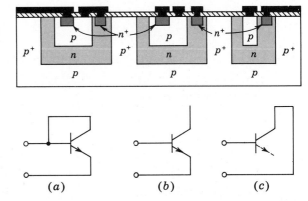

(*a*)   (*b*)   (*c*)

island, as shown in Fig. 7-17*a*. Common-anode arrays can also be made with the collector-base diffusion, as shown in Fig. 7-17*b*. A separate isolation is required for each diode, and the anodes are connected by metalization.

The emitter-base diffusion is very popular for the fabrication of diodes provided that the reverse-voltage requirement of the circuit does not exceed the lower base-emitter breakdown voltage ($\sim$7 V). Common-anode arrays can easily be made with the emitter-base diffusion by using a multiple-emitter transistor within a single isolation area, as shown in Fig. 7-18. The collector may be either open or shorted to the base. The diode pair in Fig. 7-1 is constructed in this manner, with the collector floating (open).

Diode Characteristics The forward volt-ampere characteristics of the three diode types discussed above are shown in Fig. 7-19. It will be observed that the diode-connected transistor (emitter-base diode with collector shorted

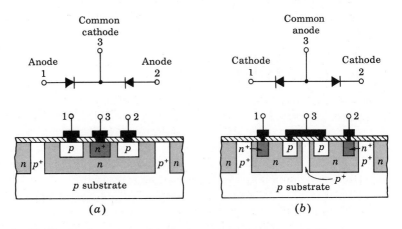

Fig. 7-17 Diode pairs. (*a*) Common-cathode pair and (*b*) common-anode pair, using collector-base diodes.

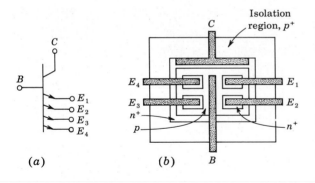

Fig. 7-18 A multiple-emitter n-p-n transistor. (a) Schematic, (b) monolithic surface pattern. If the base is connected to the collector, the result is a multiple-cathode diode structure with a common anode.

to the base) provides the highest conduction for a given forward voltage. The reverse recovery time for this diode is also smaller, one-third to one-fourth that of the collector-base diode.

7-8 INTEGRATED RESISTORS[1]

A resistor in a monolithic integrated circuit is very often obtained by utilizing the bulk resistivity of one of the diffused areas. The p-type base diffusion is most commonly used, although the n-type emitter diffusion is also employed. Since these diffusion layers are very thin, it is convenient to define a quantity known as the *sheet resistance* R_S.

Sheet Resistance If, in Fig. 7-20, the width w equals the length l, we have a square l by l of material with resistivity ρ, thickness y, and cross-sectional area $A = ly$. The resistance of this conductor (in ohms per square) is

$$R_S = \frac{\rho l}{ly} = \frac{\rho}{y} \tag{7-6}$$

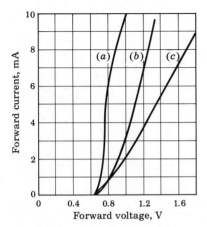

Fig. 7-19 Typical diode volt-ampere characteristics for the three diode types of Fig. 7-16. (a) Base-emitter (collector shorted to base); (b) base-emitter (collector open); (c) collector-base (emitter open). (Courtesy of Fairchild Semiconductor.)

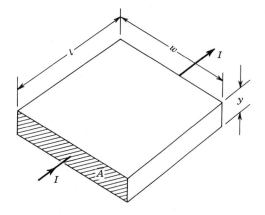

Fig. 7-20 Pertaining to sheet resistance, ohms per square.

Note that R_S is independent of the size of the square. Typically, the sheet
resistance of the base and emitter diffusions whose profiles are given in Fig.
7-12 is 200 Ω/square and 2.2 Ω/square, respectively.

The construction of a base-diffused resistor is shown in Fig. 7-1 and is
repeated in Fig. 7-21a. A top view of this resistor is shown in Fig. 7-21b.
The resistance value may be computed from

$$R = \frac{\rho l}{y w} = R_S \frac{l}{w} \tag{7-7}$$

where l and w are the length and width of the diffused area, as shown in the
top view. For example, a base-diffused-resistor stripe 1 mil wide and 10
mils long contains 10 (1 by 1 mil) squares, and its value is $10 \times 200 = 2,000\,\Omega$.
Empirical[1,2] corrections for the end contacts are usually included in calculations
of R.

Resistance Values Since the sheet resistance of the base and emitter
diffusions is fixed, the only variables available for diffused-resistor design are

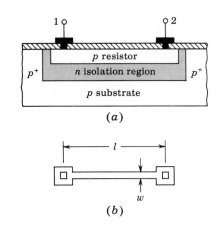

Fig. 7-21 A monolithic resistor. (a) Cross-
sectional view; (b) top view.

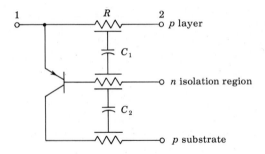

Fig. 7-22 The equivalent circuit of a diffused resistor.

stripe length and stripe width. Stripe widths of less than 1 mil (0.001 in.) are not normally used because a line-width variation of 0.0001 in. due to mask drawing error or mask misalignment or photographic-resolution error can result in 10 percent resistor-tolerance error.

The range of values obtainable with diffused resistors is limited by the size of the area required by the resistor. Practical range of resistance is 20 Ω to 30 K for a base-diffused resistor and 10 Ω to 1 K for an emitter-diffused resistor. The tolerance which results from profile variations and surface geometry errors[1] is as high as ± 10 percent of the nominal value at 25°C, with ratio tolerance of ± 1 percent. For this reason the design of integrated circuits should, if possible, emphasize resistance ratios rather than absolute values. The temperature coefficient for these heavily doped resistors is positive (for the same reason that gives a positive coefficient to the silicon sensistor, discussed in Sec. 2-7) and is $+0.06$ percent/°C from -55 to 0°C and $+0.20$ percent/°C from 0 to 125°C.

Equivalent Circuit A model of the diffused resistor is shown in Fig. 7-22, where the parasitic capacitances of the base-isolation (C_1) and isolation-substrate (C_2) junctions are included. In addition, it can be seen that a parasitic p-n-p transistor exists, with the substrate as collector, the isolation n-type region as base, and the resistor p-type material as the emitter. The collector is reverse-biased because the p-type substrate is at the most negative potential. It is also necessary that the emitter be reverse-biased to keep the parasitic transistor at cutoff. This condition is maintained by placing all resistors in the same isolation region and connecting the n-type isolation region surrounding the resistors to the *most positive* voltage present in the circuit. Typical values of h_{fe} for this parasitic transistor range from 0.5 to 5.

Thin-film Resistors[1] A technique of vapor thin-film deposition can also be used to fabricate resistors for integrated circuits. The metal (usually nichrome NiCr) film is deposited (to a thickness of less than 1 μm) on the silicon dioxide layer, and masked etching is used to produce the desired geometry. The metal resistor is then covered by an insulating layer, and apertures for the ohmic contacts are opened through this insulating layer. Typical sheet-resistance values for nichrome thin-film resistors are 40 to 400 Ω/square, resulting in resistance values from about 20 Ω to 50 K.

7-9   INTEGRATED CAPACITORS AND INDUCTORS[1,2]

Capacitors in integrated circuits may be obtained by utilizing the transition capacitance of a reverse-biased p-n junction or by a thin-film technique.

Junction Capacitors   A cross-sectional view of a junction capacitor is shown in Fig. 7-23a.   The capacitor is formed by the reverse-biased junction J_2, which separates the epitaxial n-type layer from the upper p-type diffusion area.   An additional junction J_1 appears between the n-type epitaxial plane and the substrate, and a parasitic capacitance C_1 is associated with this reverse-biased junction.   The equivalent circuit of the junction capacitor is shown in Fig. 7-23b, where the desired capacitance C_2 should be as large as possible relative to C_1.   The value of C_2 depends on the junction area and impurity concentration.   Since this junction is essentially linearly graded, C_2 is given by Eq. (3-23).   The series resistance R (10 to 50 Ω) represents the resistance of the n-type layer.

It is clear that the substrate must be at the most negative voltage so as to minimize C_1 and isolate the capacitor from other elements by keeping junction J_1 reverse-biased.   It should also be pointed out that the junction capacitor C_2 is polarized since the p-n junction J_2 must always be reverse-biased.

Thin-film Capacitors   A metal-oxide-semiconductor (MOS) nonpolarized capacitor is indicated in Fig. 7-24a.   This structure is a parallel-plate capac-

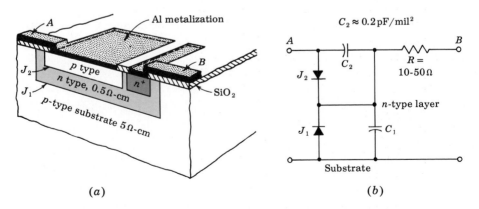

(a)   (b)

Fig. 7-23   **(a) Junction monolithic capacitor.   (b) Equivalent circuit.   (Courtesy of Motorola, Inc.)**

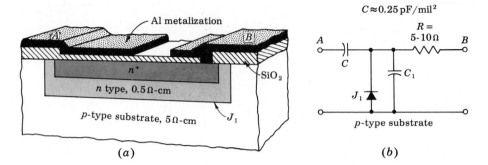

Fig. 7-24   An MOS capacitor.   (a) The structure; (b) the equivalent circuit.

itor with SiO_2 as the dielectric.   A surface thin film of metal (aluminum) is the top plate.   The bottom plate consists of the heavily doped n^+ region that is formed during the emitter diffusion.   A typical value for capacitance[3] is 0.4 pF/mil² for an oxide thickness of 500 Å, and the capacitance varies inversely with the thickness.

The equivalent circuit of the MOS capacitor is shown in Fig. 7-24b, where C_1 denotes the parasitic capacitance J_1 of the collector-substrate junction, and R is the small series resistance of the n^+ region.   Table 7-2 lists the range of possible values for the parameters of junction and MOS capacitors.

TABLE 7-2   Integrated capacitor parameters

Characteristic	Diffused-junction capacitor	Thin-film MOS
Capacitance, pF/mil²..........	0.2	0.25–0.4
Maximum area, mil²..........	2×10^3	2×10^3
Maximum value, pF..........	400	800
Breakdown voltage, V.........	5–20	50–200
Voltage dependence...........	$kV^{-\frac{1}{2}}$	0
Tolerance, percent............	±20	±20

Inductors   No practical inductance values have been obtained at the present time (1972) on silicon substrates using semiconductor or thin-film techniques.   Therefore their use is avoided in circuit design wherever possible. If an inductor is required, a discrete component is connected externally to the integrated circuit.

Characteristics of Integrated Components   Based upon our discussion of integrated-circuit technology, we can summarize the significant characteristics of integrated circuits (in addition to the advantages listed in Sec. 7-1).

1. A restricted range of values exists for resistors and capacitors. Typically, $10\ \Omega \le R \le 30$ K and $C \le 200$ pF.

2. Poor tolerances are obtained in fabricating resistors and capacitors of specific magnitudes. For example, ± 20 percent of absolute values is typical. Resistance ratio tolerance can be specified to ± 1 percent because all resistors are made at the same time using the same techniques.

3. Components have high-temperature coefficients and may also be voltage-sensitive.

4. High-frequency response is limited by parasitic capacitances.

5. The technology is very costly for small-quantity production.

6. No practical inductors or transformers can be integrated.

In the next section we examine some of the design rules for the layout of monolithic circuits.

7-10   MONOLITHIC–CIRCUIT LAYOUT[1,10]

In this section we describe how to transform the discrete logic circuit of Fig. 7-25*a* into the layout of the monolithic circuit shown in Fig. 7-26.

Design Rules for Monolithic Layout   The following 10 reasonable design rules are stated by Phillips[10]:

1. Redraw the schematic to satisfy the required pin connection with the minimum number of crossovers.

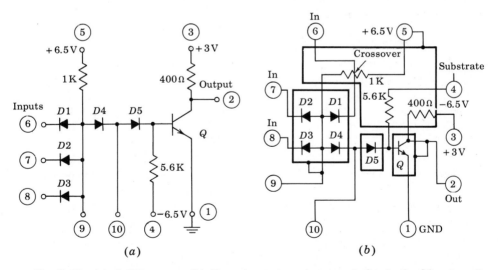

Fig. 7-25 (*a*) A DTL gate. (*b*) The schematic redrawn to indicate the 10 external connections arranged in the sequence in which they will be brought out to the header pins. The isolation regions are shown in heavy outline.

2. Determine the number of isolation islands from collector-potential considerations, and reduce the areas as much as possible.

3. Place all resistors having fixed potentials at one end in the same isolation island, and return that isolation island to the most positive potential in the circuit.

4. Connect the substrate to the most negative potential of the circuit.

5. In layout, allow an isolation border equal to twice the epitaxial thickness to allow for underdiffusion.

6. Use 1-mil widths for diffused emitter regions and $\frac{1}{2}$-mil widths for base contacts and spacings, and for collector contacts and spacings.

7. For resistors, use widest possible designs consistent with die-size limitations. Resistances which must have a close ratio must have the same width and be placed close to one another.

8. Always optimize the layout arrangement to maintain the smallest possible die size, and if necessary, compromise pin connections to achieve this.

9. Determine component geometries from the performance requirements of the circuit.

10. Keep all metalizing runs as short and as wide as possible, particularly at the emitter and collector output connections of the saturating transistor.

Pin Connections   The circuit of Fig. 7-25a is redrawn in Fig. 7-25b, with the external leads labeled 1, 2, 3, . . . , 10 and arranged in the order in which they are connected to the header pins. The diagram reveals that the power-supply pins are grouped together, and also that the inputs are on adjacent pins. In general, the external connections are determined by the system in which the circuits are used.

Crossovers   Very often the layout of a monolithic circuit requires two conducting paths (such as leads 5 and 6 in Fig. 7-25b) to cross over each other. This crossover cannot be made directly because it will result in electric contact between two parts of the circuit. Since all resistors are protected by the SiO_2 layer, any resistor may be used as a crossover region. In other words, if aluminum metalization is run over a resistor, no electric contact will take place between the resistor and the aluminum.

Sometimes the layout is so complex that additional crossover points may be required. A diffused structure which allows a crossover is also possible.[1]

Isolation Islands   The number of isolation islands is determined next. Since the transistor collector requires one isolation region, the heavy rectangle has been drawn in Fig. 7-25b around the transistor. It is shown connected to the output pin 2 because this isolation island also forms the transistor collector. Next, all resistors are placed in the same isolation island, and the island is then connected to the most positive voltage in the circuit, for reasons discussed in Sec. 7-8.

To determine the number of isolation regions required for the diodes, it is necessary first to establish which kind of diode will be fabricated. In this case, because of the low forward drop shown in Fig. 7-19, it was decided to make the common-anode diodes of the emitter-base type with the collector shorted to the base. Since the "collector" is at the "base" potential, it is required to have a single isolation island for the four common-anode diodes. Finally, the remaining diode is fabricated as an emitter-base diode, with the collector open-circuited, and thus it requires a separate isolation island.

The Fabrication Sequence   The final monolithic layout is determined by a trial-and-error process, having as its objective the smallest possible die size. This layout is shown in Fig. 7-26. The reader should identify the four isolation islands, the three resistors, the five diodes, and the transistor. It is interesting to note that the 5.6-K resistor has been achieved with a 2-mil-wide 1.8-K resistor in series with a 1-mil-wide 3.8-K resistor. To conserve space,

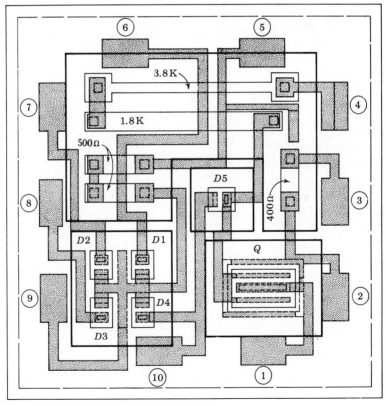

——— Indicates isolation region   ▨ Indicates metalization

Fig. 7-26   Monolithic design layout for the circuit of Fig. 7-25. (Courtesy of Motorola Monitor, Phoenix, Ariz.)

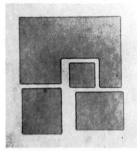

Isolation diffusion

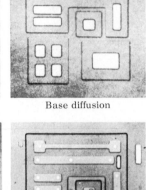

Base diffusion

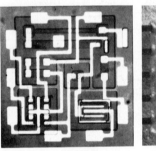

Emitter diffusion

Preohmic etch

Fig. 7-27 Monolithic fabrication sequence for the circuit of Fig. 7-25. (Courtesy of Motorola Monitor, Phoenix, Ariz.)

Metalization

Flat package assembly

the resistor was folded back on itself. In addition, two metalizing crossovers ran over this resistor.

From a layout such as shown in Fig. 7-26, the manufacturer produces the masks required for the fabrication of the monolithic integrated circuit. The production sequence which involves isolation, base, and emitter diffusions, preohmic etch, aluminum metalization, and the flat package assembly is shown in Fig. 7-27.

7-11   ADDITIONAL ISOLATION METHODS

Electrical isolation between the different elements of a monolithic integrated circuit is accomplished by means of a diffusion which yields back-to-back

p-n junctions, as indicated in Sec. 7-2. With the application of bias voltage to the substrate, these junctions represent reverse-biased diodes with a very high back resistance, thus providing adequate dc isolation. But since each p-n junction is also a capacitance, there remains that inevitable capacitive coupling between components and the substrate. These parasitic distributed capacitances thus limit monolithic integrated circuits to frequencies somewhat below those at which corresponding discrete circuits can operate.

Additional methods for achieving better isolation, and therefore improved frequency response, have been developed, and are discussed in this section.

Dielectric Isolation In this process[11] the diode-isolation concept is discarded completely. Instead, isolation, both electrical and physical, is achieved by means of a layer of solid dielectric which completely surrounds and separates the components from each other and from the common substrate. This passive layer can be silicon dioxide, silicon monoxide, ruby, or possibly a glazed ceramic substrate which is made thick enough so that its associated capacitance is negligible.

In a dielectric isolated integrated circuit it is possible to fabricate readily p-n-p and n-p-n transistors within the same silicon substrate. It is also simple to have both fast and charge-storage diodes and also both high- and low-frequency transistors in the same chip through selective gold diffusion—a process prohibited by conventional techniques because of the rapid rate at which gold diffuses through silicon unless impeded by a physical barrier such as a dielectric layer.

An isolation method pioneered by RCA[11] is referred to as SOS (silicon-on-sapphire). On a single-crystal sapphire substrate an n-type silicon layer is grown heteroepitaxially. By etching away selected portions of the silicon, isolated islands are formed (interconnected only by the high-resistance sapphire substrate).

One isolation method employing silicon dioxide as the isolating material is the EPIC process,[12] developed by Motorola, Inc. This EPIC isolation method reduces parasitic capacitance by a factor of 10 or more. In addition, the insulating oxide precludes the need for a reverse bias between substrate and circuit elements. Breakdown voltage between circuit elements and substrate is in excess of 1,000 V, in contrast to the 20 V across an isolation junction.

Beam Leads The beam-lead concept[13] of Bell Telephone Laboratories was primarily developed to batch-fabricate semiconductor devices and integrated circuits. This technique consists in depositing an array of thick (of the order of 1 mil) contacts on the surface of a slice of standard monolithic circuit, and then removing the excess semiconductor from under the contacts, thereby separating the individual devices and leaving them with semirigid beam leads cantilevered beyond the semiconductor. The contacts serve not only as electrical leads, but also as the structural support for the devices;

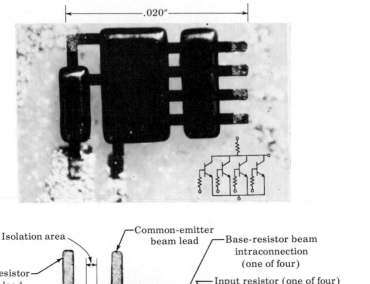

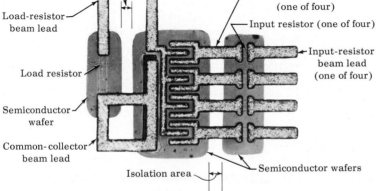

Fig. 7-28 The beam-lead isolation technique. (a) Photomicrograph
of logic circuit connected in a header. (b) The underside of the
same circuit, with the various elements identified. (Courtesy of
Bell Telephone Laboratories.)

hence the name beam leads. Chips of beam-lead circuits are mounted directly
by leads, without 1-mil aluminum or gold wires.

Isolation within integrated circuits may be accomplished by the beam-
lead structure. By etching away the unwanted silicon from under the beam
leads which connect the devices on an integrated chip, isolated pads of silicon
may be attained, interconnected by the beam leads. The only capacitive
coupling between elements is then through the small metal-over-oxide overlay.
This is much lower than the junction capacitance incurred with p-n junction-
isolated monolithic circuits.

It should be pointed out that the dielectric and beam-lead isolation techniques involve additional process steps, and thus higher costs and possible reduction in yield of the manufacturing process.

Figure 7-28 shows photomicrographs of two different views of a logic circuit made using the beam-lead technique. The top photo shows the logic circuit connected in a header. The bottom photo shows the underside of the same circuit with the various elements identified. This device is made using conventional planar techniques to form the transistor and resistor regions. Electrical isolation is accomplished by removing all unwanted material between components. The beam leads then remain to support and intraconnect the isolated components.

Hybrid Circuits[1]   The hybrid circuit as opposed to the monolithic circuit consists of several component parts (transistors, diodes, resistors, capacitors, or complete monolithic circuits), all attached to the same ceramic substrate and employing wire bonding to achieve the interconnections. In these circuits electrical isolation is provided by the physical separation of the component parts, and in this respect hybrid circuits resemble beam-lead circuits.

7-12   LARGE–SCALE AND MEDIUM–SCALE INTEGRATION (LSI AND MSI)

Large-scale integration[11] represents the process of fabricating chips with a large number of components which are interconnected to form complete subsystems or systems. In 1972 commercially available LSI circuits contained, typically, more than 100 gates, or 1,000 individual circuit components. A triple-diffused bipolar transistor requires approximately 50 mil^2 of chip area, whereas a typical MOS transistor (Chap. 10) requires only 3 mil^2. Much higher element densities are possibile with MOS LSI than bipolar LSI circuits

Since LSI is an extension of integrated-circuit techniques, the fabrication is identical with that described in Sec. 7-2. Only the methods of testing and interconnection are different with LSI. There are two principal techniques, called *discretionary wiring* and *fixed interconnection pattern*. The former consists in manufacturing on a single large chip many identical units, called *unit cells*, such as logic gates which are to be interconnected into a system. The cells are then tested by an automatic LSI tester which remembers the location of the "good" cells. The tester is coupled to a digital computer which calculates instructions for a pattern of metalization runs which interconnects the good cells so as to yield the desired system function. This process must be repeated for each LSI wafer, since the patterns of good and bad circuits will differ from wafer to wafer.

A *fixed interconnection pattern* starts with a more complex cell, called a *polycell*, and then interconnects several of these to form a larger system through a fixed interconnection which is less complex than the pattern required for an equal array composed of simpler circuits.

The most common LSI products are read-write memories (R/W), read-only memories (ROM), and shift registers (discussed in Chap. 17).

Medium-scale Integration   MSI devices have a component density less than that of LSI, but in excess of about 100 per chip. These commercially available units include shift registers, counters, decoders, adders, etc. (Chap. 17).

7-13   THE METAL–SEMICONDUCTOR CONTACT[14]

Two types of metal-semiconductor junctions are possible, *ohmic* and *rectifying*. The former is the type of contact desired when a lead is to be attached to a semiconductor. On the other hand, the rectifying contact results in a metal-semiconductor diode (called a *Schottky barrier*), with volt-ampere characteristics very similar to those of a *p-n* diode. The metal-semiconductor diode was investigated many years ago, but until the late 1960s commercial Schottky diodes were not available because of problems encountered in their manufacture. It has turned out that most of the fabrication difficulties are due to surface effects; by employing the surface-passivated integrated-circuit techniques described in this chapter, it is possible to construct almost ideal metal-semiconductor diodes very economically.

As mentioned in Sec. 7-2 (step 4), aluminum acts as a *p*-type impurity when in contact with silicon. If Al is to be attached as a lead to *n*-type Si, an ohmic contact is desired and the formation of a *p-n* junction must be prevented. It is for this reason that n^+ diffusions are made in the *n* regions near the surface where the Al is deposited (Fig. 7-2*d*). On the other hand, if the n^+ diffusion is omitted and the Al is deposited directly upon the *n*-type Si, an equivalent *p-n* structure is formed, resulting in an excellent metal-semiconductor diode. In Fig. 7-29 contact 1 is a Schottky barrier, whereas contact 2 is an ohmic (nonrectifying) contact, and a metal-semiconductor diode exists between these two terminals, with the anode at contact 1. Note that

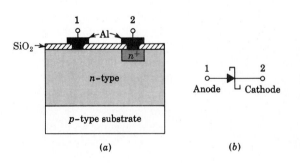

(a)   (b)

Fig. 7-29 (*a*) A Schottky diode formed by IC techniques. The aluminum and the lightly doped *n* region form a rectifying contact 1, whereas the metal and the heavily doped n^+ region form an ohmic contact 2. (*b*) The symbol for this metal-semiconductor diode.

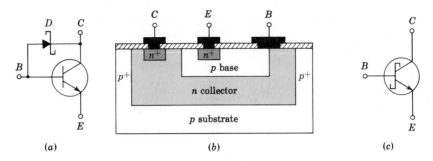

Fig. 7-30   (a) A transistor with a Schottky-diode clamp between base and collector to prevent saturation.   (b) The cross section of a monolithic IC equivalent to the diode-transistor combination in (a).   (c) The Schottky transistor symbol, which is an abbreviation for that shown in (a).

the fabrication of a Schottky diode is actually simpler than that of a p-n diode, which requires an extra (p-type) diffusion.

The external volt-ampere characteristic of a metal-semiconductor diode is essentially the same as that of a p-n junction, but the physical mechanisms involved are more complicated.   Note that in the forward direction electrons from the n-type Si cross the junction into the metal, where electrons are plentiful.   In this sense, this is a majority-carrier device, whereas minority carriers account for a p-n diode characteristic.   As explained in Sec. 3-10, there is a delay in switching a p-n diode from ON to OFF because the minority carriers stored at the junction must first be removed.   Schottky diodes have a negligible storage time t_s because the current is carried predominantly by majority carriers.   (Electrons from the n side enter the aluminum and become indistinguishable from the electrons in the metal, and hence are not "stored" near the junction.)

It should be mentioned that the voltage drop across a Schottky diode is much less than that of a p-n diode for the same forward current.   Thus, a cutin voltage of about 0.3 V is reasonable for a metal-semiconductor diode as against 0.6 V for a p-n barrier.   Hence the former is closer to the ideal diode clamp than the latter.

The Schottky Transistor   To reduce the propogation-delay time in a logic gate, it is desirable to eliminate storage time in all transistors.   In other words, a transistor must be prevented from entering saturation.   This condition can be achieved, as indicated in Fig. 7-30a, by using a Schottky diode as a clamp between the base and collector.   If an attempt is made to saturate this transistor by increasing the base current, the collector voltage drops, D conducts, and the base-to-collector voltage is limited to about 0.4 V.   Since the collector junction is forward-biased by less than the cutin voltage (≈ 0.5 V), the transistor does *not* enter saturation (Sec. 5-8).

With no additional processing steps, the Schottky clamping diode can be fabricated at the same time that the transistor is constructed. As indicated in Fig. 7-30b, the aluminum metalization for the base lead is allowed to make contact also with the n-type collector region (but without an intervening n^+ section). This simple procedure forms a metal-semiconductor diode between base and collector. The device in Fig. 7-30b is equivalent to the circuit of Fig. 7-30a. This is referred to as a *Schottky transistor,* and is represented by the symbol in Fig. 7-30c.

REFERENCES

1. Motorola, Inc. (R. M. Warner, Jr., and J. N. Fordemwalt, eds.): "Integrated Circuits," McGraw-Hill Book Company, New York, 1965.

2. Phillips, A. B.: Monolithic Integrated Circuits, *IEEE Spectrum,* vol. 1, no. 6, pp. 83–101, June, 1964.

3. Jahnke, E., and F. Emde: "Tables of Functions," Dover Publications, New York, 1945.

4. Hunter, L. P.: "Handbook of Semiconductor Electronics," 2d ed., sec. 8, McGraw-Hill Book Company, New York, 1962.

5. Fuller, C. S., and J. A. Ditzenberger: Diffusion of Donor and Acceptor Elements in Silicon, *J. Appl. Phys.,* vol. 27, pp. 544–553, May, 1956.
   Barrer, P. M.: "Diffusion in and through Solids," Cambridge University Press, London, 1951.

6. Trumbore, F. A.: Solid Solubilities of Impurity Elements in Germanium and Silicon, *Bell System Tech. J.,* vol. 39, pp. 205–234, January, 1960.

7. King, D., and L. Stern: Designing Monolithic Integrated Circuits, *Semicond. Prod. Solid State Technol.,* March, 1965.

8. "Custom Microcircuit Design Handbook," Fairchild Semiconductor, Mountain View, Calif., 1963.

9. Hunter, L. P., Ref. 4, sec. 10.1.

10. Phillips, A. B.: Designing Digital Monolithic Integrated Circuits, *Motorola Monitor,* vol. 2, no. 2, pp. 18–27, 1964.

11. Khambata, A. J.: "Introduction to Large-scale Integration," John Wiley & Sons, Inc., New York, 1969.

12. Epic Process Isolates Integrated Circuit Elements with Silicon Dioxide, *Electrotechnol. (New York),* July, 1964, p. 136.

13. Lepselter, M. P., et al.: Beam Leads and Integrated Circuits, *Proc. IEEE,* vol. 53, p. 405, April, 1965.

Lepselter, M. P.: Beam-lead Technology, *Bell System Tech. J.*, February, 1966, pp. 233–253.

14. Yu, A. Y. C.: The Metal-semiconductor Contact: An Old Device with a New Future, *IEEE Spectrum*, vol. 7, no. 3, pp. 83–89, March, 1970.

REVIEW QUESTIONS

7-1   What are the four advantages of integrated circuits?

7-2   List the five steps involved in fabricating a monolithic integrated circuit (IC), assuming you already have a substrate.

7-3   List the five basic processes involved in the fabrication of an IC, assuming you already have a substrate.

7-4   Describe *epitaxial growth.*

7-5   (a) Describe the *photoetching process.* (b) How many masks are required to complete an IC?   List the function performed by each mask.

7-6   (a) Describe the *diffusion process.* (b) What is meant by an *impurity profile?*

7-7   (a) How is the surface layer of SiO_2 formed? (b) How thick is this layer? (c) What are the reasons for forming the SiO_2 layers?

7-8   Explain how isolation between components is obtained in an IC.

7-9   How are the components interconnected in an IC?

7-10   Explain what is meant by *parasitic capacitance* in an IC.

7-11   Give the order of magnitude of (a) the substrate thickness; (b) the epitaxial thickness; (c) the base width; (d) the diffusion time; (e) the diffusion temperature; (f) the surface area of a transistor; (g) the chip size.

7-12   Sketch the cross section of an IC transistor.

7-13   Sketch the cross section of a discrete planar epitaxial transistor.

7-14   List the advantages and disadvantages of an IC vs. a discrete transistor.

7-15   (a) Define *buried layer.* (b) Why is it used?

7-16   Describe a *lateral p-n-p* transistor.   Why is its current gain low?

7-17   Describe a *vertical p-n-p* transistor.   Why is it of limited use?

7-18   Describe a *supergain* transistor.

7-19   (a) How are IC diodes fabricated? (b) Sketch the cross sections of two types of emitter-base diodes.

7-20   Sketch the cross section of a diode pair using collector-base regions if (a) the cathode is common and (b) the anode is common.

7-21   Sketch the top view of a multiple-emitter transistor.   Show the isolation, collector, base, and emitter regions.

7-22   (a) Define *sheet resistance* R_S. (b) What is the order of magnitude of R_S for the base region and also for the emitter region? (c) Sketch the cross section of an IC resistor. (d) What are the order of magnitudes of the smallest and the largest values of an IC resistance?

7-23   (a) Sketch the equivalent circuit of a base-diffused resistor, showing all parasitic elements. (b) What must be done (externally) to minimize the effect of the parasitic elements?

7-24   Describe a *thin-film resistor.*

7-25 (*a*) Sketch the cross section of a junction capacitor.   (*b*) Draw the equivalent circuit, showing all parasitic elements.

7-26 Repeat Rev. 7-25 for an MOS capacitor.

7-27 (*a*) What are the two basic distinctions between a junction and an MOS capacitor?   (*b*) What is the order of magnitude of the capacitance per square mil? (*c*) What is the order of magnitude of the maximum value of C?

7-28 (*a*) To what voltage is the substrate connected? Why? (*b*) To what voltage is the isolation island containing the resistors connected? Why? (*c*) Can several transistors be placed in the same isolation island?   Explain.

7-29 Describe briefly (*a*) *dielectric isolation;* (*b*) *beam-lead isolation.*

7-30 What is meant by a *hybrid circuit?*

7-31 How is an aluminum contact made with *n*-type silicon so that it is (*a*) ohmic; (*b*) rectifying?

7-32 Why is storage time eliminated in a metal-semiconductor diode?

7-33 What is a *Schottky transistor?* Why is storage time eliminated in such a transistor?

7-34 Sketch the cross section of an IC Schottky transistor.

8 / THE TRANSISTOR AT LOW FREQUENCIES

The large-signal response of a transistor is obtained graphically. For small signals the transistor operates with reasonable linearity, and we inquire into small-signal linear models which represent the operation of the transistor in the active region. The parameters introduced in the models presented here are interpreted in terms of the external volt-ampere characteristics of the transistor. A detailed study of the transistor amplifier in its various configurations is made.

Very often, in practice, a number of stages are used in cascade to amplify a signal from a source, such as a phonograph pickup, to a level which is suitable for the operation of another transducer, such as a loudspeaker. In this chapter we consider the problem of cascading a number of transistor amplifier stages. In addition, various special transistor circuits of practical importance are examined in detail. Also, simplified approximate methods of solution are presented. All transistor circuits in this chapter are examined at low frequencies, where the transistor internal capacitances may be neglected.

8-1   GRAPHICAL ANALYSIS OF THE CE CONFIGURATION

It is our purpose in this section to analyze graphically the operation of the circuit of Fig. 8-1. In Fig. 8-2a the output characteristics of a p-n-p germanium transistor and in Fig. 8-2b the corresponding input characteristics are indicated. We have selected the CE configuration because, as we see in Sec. 8-9, it is the most generally useful configuration.

In Fig. 8-2a we have drawn a load line for a 250-Ω load with $V_{CC} = 15$ V. If the input base-current signal is symmetric, the qui-

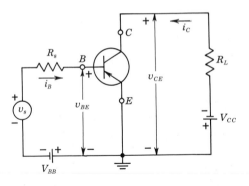

Fig. 8-1   The CE transistor configuration.

escent point Q is usually selected at about the center of the load line, as shown in Fig. 8-2a.   We postpone until Chap. 9 our discussion on biasing of transistors.

Notation   At this point it is important to make a few remarks on transistor symbols.   Specifically, instantaneous values of quantities which vary with time are represented by lowercase letters (i for current, v for voltage, and p for power).   Maximum, average (dc), and effective, or root-mean-square (rms), values are represented by the uppercase letter of the proper symbol (I, V, or P).   Average (dc) values and instantaneous total values are indicated by the uppercase subscript of the proper electrode symbol (B for base, C for collector, E for emitter).   Varying components from some

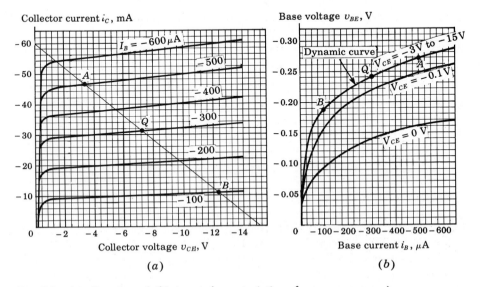

Fig. 8-2   (a) Output and (b) input characteristics of a p-n-p germanium transistor.

quiescent value are indicated by the lowercase subscript of the proper electrode symbol. A single subscript is used if the reference electrode is clearly understood. If there is any possibility of ambiguity, the conventional double-subscript notation should be used. For example, in Figs. 8-3a to d and 8-1, we show collector and base currents and voltages in the common-emitter transistor configuration, employing the notation just described. The collector and emitter current and voltage component variations from the corresponding quiescent values are

$$i_c = i_C - I_C = \Delta i_C \qquad v_c = v_C - V_C = \Delta v_C$$

$$i_b = i_B - I_B = \Delta i_B \qquad v_b = v_B - V_B = \Delta v_B$$

$$(8\text{-}1)$$

The *magnitude* of the supply voltage is indicated by repeating the electrode subscript. This notation is summarized in Table 8-1.

TABLE 8-1   Notation summarized

	Base (collector) voltage with respect to emitter	Base (collector) current toward electrode from external circuit
Instantaneous total value....................	v_B (v_C)	i_B (i_C)
Quiescent value.............................	V_B (V_C)	I_B (I_C)
Instantaneous value of varying component........	v_b (v_c)	i_b (i_c)
Effective value of varying component (phasor, if a sinusoid).................................	V_b (V_c)	I_b (I_c)
Supply voltage (magnitude)....................	V_{BB} (V_{CC})	

The Waveforms   Assume a 200-μA peak sinusoidally varying base current around the quiescent point Q, where $I_B = -300$ μA. Then the extreme points of the base waveform are A and B, where $i_B = -500$ μA and -100 μA, respectively. These points are located on the load line in Fig. 8-2a. We find i_C and v_{CE}, corresponding to any given value of i_B, at the intersection of the load line and the collector characteristics corresponding to this value of i_B. For example, at point A, $i_B = -500$ μA, $i_C = -46.5$ mA, and $v_{CE} = -3.4$ V. The waveforms i_C and v_{CE} are plotted in Fig. 8-3a and b, respectively. We observe that the collector-current and collector-voltage waveforms are not the same as the base-current waveform (the sinusoid of Fig. 8-3c) because the collector characteristics in the neighborhood of the load line in Fig. 8-2a are not parallel lines equally spaced for equal increments in base current. This change in waveform is known as *output nonlinear distortion*.

The base-to-emitter voltage v_{BE} for any combination of base current and collector-to-be-emitter voltage can be obtained from the input characteristic curves. In Fig. 8-2b we show the *dynamic operating curve* drawn for the combinations of base current and collector voltage found along A-Q-B of the

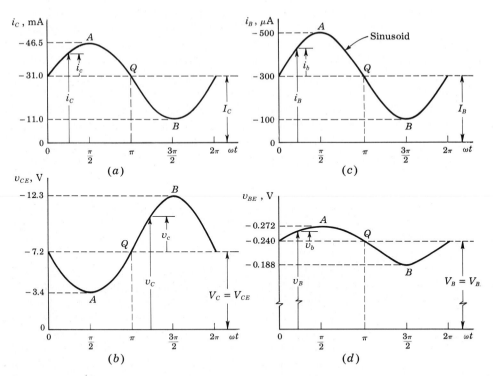

Fig. 8-3 (a,b) Collector and (c,d) base current and voltage waveforms.

load line of Fig. 8-2a. The waveform v_{BE} can be obtained from the dynamic operating curve of Fig. 8-2b by reading the voltage v_{BE} corresponding to a given base current i_B. We now observe that, since the dynamic curve is not a straight line, the waveform of v_b (Fig. 8-3d) will not, in general, be the same as the waveform of i_b. This change in waveform is known as *input nonlinear distortion*. In some cases it is more reasonable to assume that v_b in Fig. 8-3d is sinusoidal, and then i_b will be distorted. The above condition will be true if the sinusoidal voltage source v_s driving the transistor has a small output resistance R_s in comparison with the input resistance R_i of the transistor, so that the transistor input-voltage waveform is essentially the same as the source waveform. However, if $R_s \gg R_i$, the variation in i_B is given by $i_b \approx v_s/R_s$, and hence the base-current waveform is also sinusoidal.

From Fig. 8-2b we see that *for a large sinusoidal base voltage v_b around the point Q, the base-current swing $|i_b|$ is smaller to the left of Q than to the right of Q.* This input distortion tends to cancel the output distortion because, in Fig. 8-2a, the collector-current swing $|i_c|$ for a given base-current swing is larger over the section BQ than over QA. Hence, if the amplifier is biased so that Q is near the center of the i_C-v_{CE} plane, there will be less distortion if the excitation is a sinusoidal base voltage than if it is a sinusoidal base current.

It should be noted here that the dynamic load curve can be approximated by a straight line over a sufficiently small line segment, and hence, if the input signal is small, there will be negligible input distortion under any conditions of operation (current-source or voltage-source driver).

Since the small-signal low-frequency response of a transistor is linear, it can be obtained analytically rather than graphically. As a matter of fact, for very small signals the graphical technique used in connection with Fig. 8-2 would require interpolation between the printed characteristics and would result in very poor accuracy. In the remainder of the chapter we assume small-signal (also called *incremental*) operation, obtain a linear circuit model for the transistor, and then analyze the network analytically.

8-2   TWO–PORT DEVICES AND THE HYBRID MODEL

The terminal behavior of a large class of two-port devices is specified by two voltages and two currents. The box in Fig. 8-4 represents such a two-port network. We may select two of the four quantities as the independent variables and express the remaining two in terms of the chosen independent variables. It should be noted that, in general, we are not free to select the independent variables arbitrarily. For example, if the two-port device is an ideal transformer, we cannot pick the two voltages v_1 and v_2 as the independent variables because their ratio is a constant equal to the transformer turns ratio. If the current i_1 and the voltage v_2 are independent and if the two-port is linear, we may write

$$v_1 = h_{11}i_1 + h_{12}v_2 \qquad\qquad (8\text{-}2)$$

$$i_2 = h_{21}i_1 + h_{22}v_2 \qquad\qquad (8\text{-}3)$$

The quantities h_{11}, h_{12}, h_{21}, and h_{22} are called the h, or *hybrid, parameters* because they are not all alike dimensionally. Let us assume that there are no reactive elements within the two-port network. Then, from Eqs. (8-2) and (8-3), the h parameters are defined as follows:

$$h_{11} \equiv \left.\frac{v_1}{i_1}\right|_{v_2=0} = \text{input resistance with output short-circuit (ohms).}$$

$$h_{12} \equiv \left.\frac{v_1}{v_2}\right|_{i_1=0} = \text{fraction of output voltage at input with input open-circuited, or more simply, reverse-open-circuit voltage amplification (dimensionless).}$$

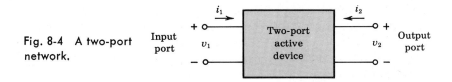

Fig. 8-4   A two-port network.

$$h_{21} \equiv \frac{i_2}{i_1}\Big|_{v_2=0}$$ = negative of current transfer ratio (or current gain) with output short-circuited. (Note that the current into a load across the output port would be the negative of i_2.) This parameter is usually referred to, simply, as the *short-circuit current gain* (dimensionless).

$$h_{22} \equiv \frac{i_2}{v_2}\Big|_{i_1=0}$$ = output conductance with input open-circuited (mhos).

Notation The following convenient alternative subscript notation is recommended by the IEEE Standards[1]:

$i = 11 =$ input $\qquad\qquad\qquad o = 22 =$ output

$f = 21 =$ forward transfer $\qquad r = 12 =$ reverse transfer

In the case of transistors, another subscript (b, e, or c) is added to designate the type of configuration. For example,

$h_{ib} = h_{11b} =$ input resistance in common-base configuration

$h_{fe} = h_{21e} =$ short-circuit forward current gain in common-emitter circuit

Since the device described by Eqs. (8-2) and (8-3) is assumed to include no reactive elements, the four parameters h_{11}, h_{12}, h_{21}, and h_{22} are real numbers and the voltages and currents v_1, v_2, and i_1, i_2 are functions of time. However, if reactive elements had been included in the device, the excitation would be considered to be sinusoidal, the h parameters would in general be functions of frequency, and the voltages and currents would be represented by phasors V_1, V_2, and I_1, I_2.

The Model We may now use the four h parameters to construct a mathematical model of the device of Fig. 8-4. The hybrid circuit for any device characterized by Eqs. (8-2) and (8-3) is indicated in Fig. 8-5. We can verify that the model of Fig. 8-5 satisfies Eqs. (8-2) and (8-3) by writing Kirchhoff's voltage and current laws for the input and output ports, respectively. Note that the input circuit contains a dependent voltage generator whereas the output circuit has a dependent current source.

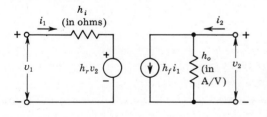

Fig. 8-5 The hybrid model for the two-port network of Fig. 8-4. The parameters h_r and h_f are dimensionless.

8-3   TRANSISTOR HYBRID MODEL

The basic assumption in arriving at a transistor linear model or equivalent circuit is that the variations about the quiescent point are assumed small, so that the transistor parameters can be considered constant over the signal excursion. The operating (quiescent) values of current and voltages are determined by the method employed to bias the transistor (Chap. 9). These values do not enter into an incremental model, which is used only to find small-signal variations about the Q point.

The transistor model presented in this chapter is given in terms of the h parameters, which are *real numbers* at audio frequencies, are easy to measure, can also be obtained from the transistor static characteristic curves, and are particularly convenient to use in circuit analysis and design. Furthermore, a set of h parameters is specified for many transistors by the manufacturers.

To see how we can derive a hybrid model for a transistor, let us consider the common-emitter connection shown in Fig. 8-1. The variables i_B, i_C, $v_{BE} = v_B$, and $v_{CE} = v_C$ represent total instantaneous currents and voltages. From our discussion in Chap. 5 of transistor voltages and currents, we see that we may select the current i_B and voltage v_C as independent variables. Since v_B is some function f_1 of i_B and v_C and since i_C is another function f_2 of i_B and v_C, we may write

$$v_B = f_1(i_B, v_C) \tag{8-4}$$

$$i_C = f_2(i_B, v_C) \tag{8-5}$$

Making a Taylor's series expansion of Eqs. (8-4) and (8-5) around the quiescent point I_B, V_C, and neglecting higher-order terms, we obtain

$$\Delta v_B = \left.\frac{\partial f_1}{\partial i_B}\right|_{V_C} \Delta i_B + \left.\frac{\partial f_1}{\partial v_C}\right|_{I_B} \Delta v_C \tag{8-6}$$

$$\Delta i_C = \left.\frac{\partial f_2}{\partial i_B}\right|_{V_C} \Delta i_B + \left.\frac{\partial f_2}{\partial v_C}\right|_{I_B} \Delta v_C \tag{8-7}$$

The partial derivatives are taken, keeping the collector voltage or the base current constant, as indicated by the subscript attached to the derivative.

The quantities Δv_B, Δv_C, Δi_B, and Δi_C represent the small-signal (incremental) base and collector voltages and currents. According to the notation in Eq. (8-1), we represent them with the symbols v_b, v_c, i_b, and i_c. We may now write Eqs. (8-6) and (8-7) in the following form:

$$v_b = h_{ie}i_b + h_{re}v_c \tag{8-8}$$

$$i_c = h_{fe}i_b + h_{oe}v_c \tag{8-9}$$

where

$$h_{ie} \equiv \frac{\partial f_1}{\partial i_B} = \left.\frac{\partial v_B}{\partial i_B}\right|_{V_C} \qquad h_{re} \equiv \frac{\partial f_1}{\partial v_C} = \left.\frac{\partial v_B}{\partial v_C}\right|_{I_B} \tag{8-10}$$

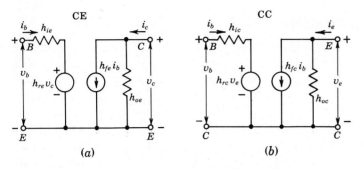

Fig. 8-6 The hybrid small-signal models for (a) the common-emitter and (b) the common-collector configuration.

and

$$h_{fe} \equiv \frac{\partial f_2}{\partial i_B} = \frac{\partial i_C}{\partial i_B}\bigg|_{V_C} \qquad h_{oe} \equiv \frac{\partial f_2}{\partial v_C} = \frac{\partial i_C}{\partial v_C}\bigg|_{I_B} \tag{8-11}$$

The partial derivatives of Eqs. (8-10) and (8-11) define the h parameters for the common-emitter connection. In the next section we show that the above partial derivatives can be obtained from the transistor characteristic curves and that they are real numbers. We now observe that Eqs. (8-8) and (8-9) are exactly the same form as Eqs. (8-2) and (8-3). Hence the model of Fig. 8-5 can be used to represent a transistor.

Note that the partial derivatives in Eqs. (8-10) and (8-11) are taken keeping V_C = constant or I_B = constant. If a parameter is constant, its incremental change is zero. Hence V_C = constant is equivalent to $v_c = 0$ and I_B = constant corresponds to $i_b = 0$. With this notation,

$$h_{re} = \frac{\partial v_B}{\partial v_C}\bigg|_{I_B} = \frac{v_b}{v_c}\bigg|_{i_b=0} \qquad \text{or} \qquad h_{re} = \frac{V_b}{V_c}\bigg|_{I_b=0} \tag{8-12}$$

where the second equation is valid if sinusoidal (phasor) voltages and currents are under consideration.

The common-emitter (CE) h-parameter model is shown in Fig. 8-6a. The common-collector (CC) hybrid model is given in Fig. 8-6b. Note from the CC configuration of Fig. 5-19 that the input terminals are B and C whereas the output terminals are E and C. An analogous equivalent circuit (Fig. 8-10) can be drawn for the common-base (CB) circuit, with input terminals E and B and output terminals C and B. For each configuration, we note from Kirchhoff's current law that

$$i_b + i_e + i_c = 0 \tag{8-13}$$

The circuit models and equations are valid for either an n-p-n or a p-n-p transistor and are independent of the type of load or the method of biasing.

8-4   THE h PARAMETERS[2]

Equations (8-4) and (8-5) give the form of the functional relationships for the common-emitter connection of total instantaneous collector current and base voltage in terms of two variables, namely, base current and collector voltage. Such functional relationships are represented in Chap. 5 by families of characteristic curves. Two families of curves are usually specified for transistors. The *output characteristic curves* depict the relationship between output current and voltage, with input current as the parameter. Figures 5-6 and 5-10 show typical output characteristic curves for the common-base and common-emitter transistor configurations. The *input characteristics* depict the relationship between input voltage and current with output voltage as the parameter. Typical input characteristic curves for the common-base and common-emitter transistor connections are shown in Figs. 5-7 and 5-11. If the input and output characteristics of a particular connection are given, the h parameters can be determined graphically.

The Parameter h_{fe}   For a common-emitter connection the output characteristics are shown in Fig. 8-7. From the definition of h_{fe} given in Eq. (8-11) and from Fig. 8-7, we have

$$h_{fe} = \frac{\partial i_C}{\partial i_B} \approx \frac{\Delta i_C}{\Delta i_B}\bigg|_{V_C} = \frac{i_{C2} - i_{C1}}{i_{B2} - i_{B1}} \tag{8-14}$$

The current increments are taken around the quiescent point Q, which corresponds to the base current $i_B = I_B$ and to the collector voltage $v_{CE} = V_C$ (a vertical line in Fig. 8-7).

The parameter h_{fe} is the most important small-signal parameter of the transistor. This common-emitter current transfer ratio, or CE alpha, is also written α_e, or β', and called the *small-signal beta* of the transistor. The relationship between $\beta' = h_{fe}$ and the *large-signal beta*, $\beta \approx h_{FE}$, is given in Eq. (5-23).

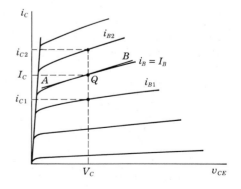

Fig. 8-7   Determination of h_{fe} and h_{oe} from the CE output characteristics.

The Parameter h_{oe} From Eq. (8-11)

$$h_{oe} = \frac{\partial i_C}{\partial v_C} \approx \frac{\Delta i_C}{\Delta v_C}\bigg|_{I_B} \tag{8-15}$$

The value of h_{oe} at the quiescent point Q is given by the slope of the output characteristic curve at that point. This slope can be evaluated by drawing the line AB in Fig. 8-7a tangent to the characteristic curve at the point Q.

The parameters h_{ie} and h_{re} may be obtained graphically from the input CE characteristics in a manner analogous to that illustrated in Fig. 8-7 (Prob. 8-2).

Based upon the definitions given in Secs. 8-3 and 8-9, simple experiments may be carried out for the direct determination of the hybrid parameters.[3]

Hybrid-parameter Variations From the discussion in this section we have seen that once a quiescent point Q is specified, the h parameters can be obtained from the slopes and spacing between curves at this point. Since the characteristic curves are not in general straight lines, equally spaced for equal changes in I_B (Fig. 8-2a) or V_{CE} (Fig. 8-2b), it is clear that the values of the h parameters depend upon the position of the quiescent point on the curves. Moreover, from our discussion in Chap. 5, we know that the shape

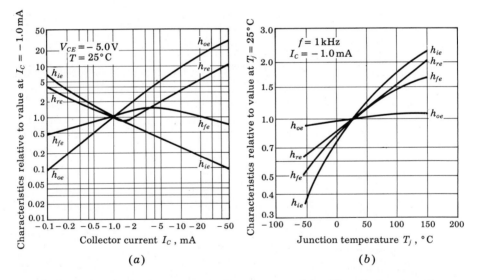

Fig. 8-8 Variation of common-emitter h parameters (a) with collector current normalized to unity at $V_{CE} = -5.0$ V and $I_C = -1.0$ mA for the type 2N996 diffused-silicon planar epitaxial transistor; (b) with junction temperature, normalized to unity at $T_j = 25°C$. (Courtesy of Fairchild Semiconductor.)

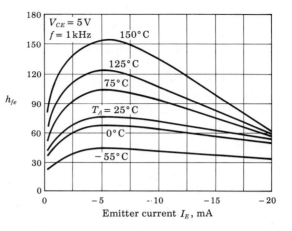

Fig. 8-9   Variation of h_{fe} with emitter current for the type 2N1573 silicon transistor. (Courtesy of Texas Instruments, Inc.)

and actual numerical values of the characteristic curves depend on the junction temperature. Hence the h parameters also will depend on temperature. Most transistor specification sheets include curves of the variation of the h parameters with the quiescent point and temperature. Such curves are shown for a typical silicon p-n-p transistor in Fig. 8-8a and b. These curves are plotted with respect to the values of a specific operating point, say, -5V collector-to-emitter voltage and -1 mA collector current. The variation in h parameters as shown in Fig. 8-8a is for a constant junction temperature of 25°C and a frequency of 1 kHz. Manufacturers usually also provide curves of h parameters versus V_{CE}, although this variation with V_{CE} is often not significant. Specifically, h_{fe} is more sensitive to I_C than to V_{CE}. Most transistors exhibit a well-defined maximum in the value of h_{fe} as a function of collector or emitter current. Such a maximum in the variation of h_{fe} with emitter current and temperature is shown in Fig. 8-9 for an n-p-n double-diffused silicon mesa transistor.

Table 8-2 shows values of h parameters for the three different transistor configurations of a typical junction transistor.

TABLE 8-2   Typical h-parameter values for a transistor (at $I_E = 1.3$ mA)

Parameter	CE	CC	CB
$h_{11} = h_i$	1,100 Ω	1,100 Ω	21.6 Ω
$h_{12} = h_r$	2.5×10^{-4}	~ 1	2.9×10^{-4}
$h_{21} = h_f$	50	-51	-0.98
$h_{22} = h_o$	24 μA/V	25 μA/V	0.49 μA/V
$1/h_o$	40 K	40 K	2.04 M

8-5   CONVERSION FORMULAS FOR THE PARAMETERS OF THE THREE TRANSISTOR CONFIGURATIONS

It may be necessary to convert from one set of transistor parameters to another set. Some transistor manufacturers specify all four common-emitter h parameters; others specify h_{fe}, h_{ib}, h_{ob}, and h_{rb}. In Table 8-3 we give approximate conversion formulas between the CE, CC, and CB h parameters. These equations can be verified by using the definitions of the parameters involved and Kirchhoff's laws. The general procedure is illustrated in the following example.

EXAMPLE Find h_{re} in terms of the CB h parameters.

Solution The CB h-parameter circuit is shown in Fig. 8-10, where capital letters are used to represent the rms value of the sinusoidal voltage or current. By definition,

$$h_{re} = \frac{V_{be}}{V_{ce}}\bigg|_{I_b = 0} = \frac{V_{bc} + V_{ce}}{V_{cc}}\bigg|_{I_b = 0} = \left(1 + \frac{V_{bc}}{V_{ce}}\right)\bigg|_{I_b = 0}$$

If $I_b = 0$, then $I_c = -I_e$, and the current I in h_{ob} in Fig. 8-10 is $I = (1 + h_{fb})I_e$. Since h_{ob} represents a conductance,

$$I = h_{ob}V_{bc} = (1 + h_{fb})I_e$$

Applying KVL to the path $EBCE$,

$$h_{ib}I_e + h_{rb}V_{cb} + V_{bc} + V_{ce} = 0$$

Combining the last two equations yields

$$\frac{h_{ib}h_{ob}}{1 + h_{fb}}V_{bc} - h_{rb}V_{bc} + V_{bc} + V_{ce} = 0$$

or

$$\frac{V_{bc}}{V_{ce}} = \frac{-(1 + h_{fb})}{h_{ib}h_{ob} + (1 - h_{rb})(1 + h_{fb})}$$

Hence

$$h_{re} = 1 + \frac{V_{bc}}{V_{ce}} = \frac{h_{ib}h_{ob} - (1 + h_{fb})h_{rb}}{h_{ib}h_{ob} + (1 - h_{rb})(1 + h_{fb})}$$

This is an exact expression. The simpler approximate formula is obtained by noting that, for the typical values given in Table 8-2,

$$h_{rb} \ll 1 \qquad \text{and} \qquad h_{ob}h_{ib} \ll 1 + h_{fb}$$

Hence

$$h_{re} \approx \frac{h_{ib}h_{ob}}{1 + h_{fb}} - h_{rb}$$

which is the formula given in Table 8-3.

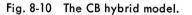

Fig. 8-10   The CB hybrid model.

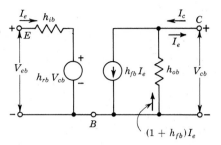

TABLE 8-3   Approximate conversion formulas for hybrid parameters

$h_{ic} = h_{ie}$	$h_{rc} = 1$
$h_{fc} = -(1 + h_{fe})$	$h_{oc} = h_{oe}$

$h_{ib} = \dfrac{h_{ie}}{1 + h_{fe}}$ †	$h_{rb} = \dfrac{h_{ie}h_{oe}}{1 + h_{fe}} - h_{re}$
$h_{fb} = -\dfrac{h_{fe}}{1 + h_{fe}}$	$h_{ob} = \dfrac{h_{oe}}{1 + h_{fe}}$

† The CE parameters in terms of the CB parameters are obtained by interchanging the subscripts b and e (Prob. 8-4).

8-6   ANALYSIS OF A TRANSISTOR AMPLIFIER CIRCUIT USING h PARAMETERS

To form a transistor amplifier it is only necessary to connect an external load and signal source as indicated in Fig. 8-11 and to bias the transistor properly. The two-port active network of Fig. 8-11 represents a transistor in any one of the three possible configurations. In Fig. 8-12 we treat the general case (connection not specified) by replacing the transistor with its small-signal hybrid model. The circuit used in Fig. 8-12 is valid for any type of load, whether it be a pure resistance, an impedance, or another transistor. This is true because the transistor hybrid model was derived without any regard

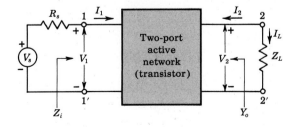

Fig. 8-11   A basic amplifier circuit.

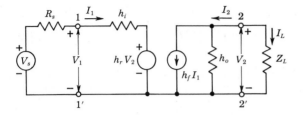

Fig. 8-12 The transistor in Fig. 8-11 is replaced by its h-parameter model.

to the external circuit in which the transistor is incorporated. The only restriction is the requirement that the h parameters remain substantially constant over the operating range.

Assuming sinusoidally varying voltages and currents, we can proceed with the analysis of the circuit of Fig. 8-12, using the phasor (sinor) notation to represent the sinusoidally varying quantities. The quantities of interest are *the current gain, the input impedance, the voltage gain,* and *the output impedance.*

The Current Gain, or Current Amplification, A_I For the transistor amplifier stage, A_I is defined as the ratio of output to input currents, or

$$A_I \equiv \frac{I_L}{I_1} = -\frac{I_2}{I_1} \tag{8-16}$$

From the circuit of Fig. 8-12, we have

$$I_2 = h_f I_1 + h_o V_2 \tag{8-17}$$

Substituting $V_2 = -I_2 Z_L$ in Eq. (8-17), we obtain

$$A_I = -\frac{I_2}{I_1} = -\frac{h_f}{1 + h_o Z_L} \tag{8-18}$$

The Input Impedance Z_i The resistance R_s in Figs. 8-11 and 8-12 represents the signal-source resistance. The impedance we see looking into the amplifier input terminals (1, 1') is the amplifier *input impedance Z_i,* or

$$Z_i \equiv \frac{V_1}{I_1} \tag{8-19}$$

From the input circuit of Fig. 8-12, we have

$$V_1 = h_i I_1 + h_r V_2 \tag{8-20}$$

Hence

$$Z_i = \frac{h_i I_1 + h_r V_2}{I_1} = h_i + h_r \frac{V_2}{I_1} \tag{8-21}$$

Substituting,

$$V_2 = -I_2 Z_L = A_I I_1 Z_L \tag{8-22}$$

in Eq. (8-21), we obtain

$$Z_i = h_i + h_r A_I Z_L = h_i - \frac{h_f h_r}{Y_L + h_o} \tag{8-23}$$

where use has been made of Eq. (8-18) and the fact that the load admittance is $Y_L \equiv 1/Z_L$. Note that *the input impedance is a function of the load impedance.*

The Voltage Gain, or Voltage Amplification, A_V The ratio of output voltage V_2 to input voltage V_1 gives the voltage gain of the transistor, or

$$A_V \equiv \frac{V_2}{V_1} \tag{8-24}$$

From Eq. (8-22) we have

$$A_V = \frac{A_I I_1 Z_L}{V_1} = \frac{A_I Z_L}{Z_i} \tag{8-25}$$

The Voltage Amplification A_{Vs}, Taking into Account the Resistance R_s of the Source This overall voltage gain A_{Vs} is defined by

$$A_{Vs} \equiv \frac{V_2}{V_s} = \frac{V_2}{V_1} \frac{V_1}{V_s} = A_V \frac{V_1}{V_s} \tag{8-26}$$

From the equivalent input circuit of the amplifier, shown in Fig. 8-13a,

$$V_1 = \frac{V_s Z_i}{Z_i + R_s}$$

Then

$$A_{Vs} = \frac{A_V Z_i}{Z_i + R_s} = \frac{A_I Z_L}{Z_i + R_s} \tag{8-27}$$

where use has been made of Eq. (8-25). Note that, if $R_s = 0$, then $A_{Vs} = A_V$. Hence A_V *is the voltage gain for an ideal voltage source* (one with zero internal resistance). In practice, the quantity A_{Vs} is more meaningful than A_V since, usually, the source resistance has an appreciable effect on the overall voltage amplification. For example, if Z_i is resistive and equal in magnitude to R_s, then $A_{Vs} = 0.5 A_V$.

Fig. 8-13 Input circuit of a transistor amplifier using (a) a Thévenin's equivalent for the source and (b) a Norton's equivalent for the source.

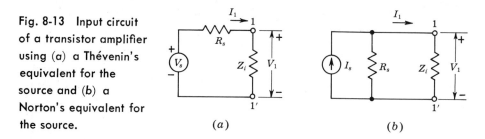

(a)   (b)

The Current Amplification A_{Is}, Taking into Account the Source Resistance R_s   If the input source is a current generator I_s in parallel with a resistance R_s, as indicated in Fig. 8-13b, then this *overall current gain* A_{Is} is defined by

$$A_{Is} \equiv \frac{-I_2}{I_s} = \frac{-I_2}{I_1}\frac{I_1}{I_s} = A_I \frac{I_1}{I_s} \tag{8-28}$$

From Fig. 8-13b

$$I_1 = \frac{I_s R_s}{Z_i + R_s}$$

and hence

$$A_{Is} = \frac{A_I R_s}{Z_i + R_s} \tag{8-29}$$

Note that, if $R_s = \infty$, then $A_{Is} = A_I$. Hence A_I is the current gain for an *ideal current source* (one with infinite source resistance).

Independent of the transistor characteristics, the voltage and current gains, taking source impedance into account, are related by

$$A_{Vs} = A_{Is}\frac{Z_L}{R_s} \tag{8-30}$$

This relationship is obtained by dividing Eq. (8-27) by Eq. (8-29), and is valid provided that the current and voltage generators have the *same* source resistance R_s.

The Output Admittance   By definition, the output impedance $Z_o = 1/Y_o$ is obtained by setting the source voltage V_s to zero and the load impedance Z_L to infinity and by driving the output terminals from a generator V_2. If the current drawn from V_2 is I_2, then

$$Y_o \equiv \frac{I_2}{V_2} \quad \text{with} \quad V_s = 0 \text{ and } R_L = \infty \tag{8-31}$$

From Eq. (8-17)

$$Y_o = h_f \frac{I_1}{V_2} + h_o \tag{8-32}$$

From Fig. 8-12, with $V_s = 0$,

$$R_s I_1 + h_i I_1 + h_r V_2 = 0 \tag{8-33}$$

or

$$\frac{I_1}{V_2} = -\frac{h_r}{h_i + R_s} \tag{8-34}$$

Substituting the expression for I_1/V_2 from Eq. (8-34) in Eq. (8-32), we obtain

$$Y_o = h_o - \frac{h_f h_r}{h_i + R_s} \tag{8-35}$$

Note that *the output admittance is a function of the source resistance.* If the source impedance is resistive, as we have assumed, then Y_o is real (a conductance).

In the above definition of $Y_o = 1/Z_o$, we have considered the load Z_L external to the amplifier. If the output impedance of the amplifier stage with Z_L included is desired, this loaded impedance can be calculated as the parallel combination of Z_L and Z_o.

Summary   The important formulas derived above are summarized for ready reference in Table 8-4. Note that the expressions for A_V, A_{Vs}, and A_{Is} are obtained without reference to the hybrid parameters, and hence are valid regardless of what equivalent circuit we use for the transistor. In particular, these expressions are valid at high frequencies, where the h parameters are functions of frequency or where we may prefer to use another model for the transistor (for example, the hybrid-Π model of Sec. 11-5).

TABLE 8-4   Small-signal analysis of a transistor amplifier

$$A_I = -\frac{h_f}{1 + h_o Z_L} \qquad\qquad Y_o = h_o - \frac{h_f h_r}{h_i + R_s} = \frac{1}{Z_o}$$

$$Z_i = h_i + h_r A_I Z_L \qquad\qquad A_{Vs} = \frac{A_V Z_i}{Z_i + R_s} = \frac{A_I Z_L}{Z_i + R_s}$$

$$A_V = \frac{A_I Z_L}{Z_i} \qquad\qquad\qquad A_{Is} = \frac{A_I R_s}{Z_i + R_s} = A_{Vs}\frac{R_s}{Z_L}$$

EXAMPLE   The transistor of Fig. 8-11 is connected as a common-emitter amplifier, and the h parameters are those given in Table 8-2. If $R_L = 10$ K and $R_s = 1$ K, find the various gains and the input and output impedances.

Solution   In making the small-signal analysis of this circuit it is convenient, first, to calculate A_I, then obtain R_i from A_I, and A_V from both these quantities. Using the expressions in Table 8-4 and the h parameters from Table 8-2,

$$A_I = -\frac{h_{fe}}{1 + h_{oe}R_L} = -\frac{50}{1 + 25 \times 10^{-6} \times 10^4} = -40.0$$

$$R_i = h_{ie} + h_{re}A_I R_L = 1,100 - 2.5 \times 10^{-4} \times 40.0 \times 10^4 = 1,000\ \Omega = 1\ K$$

$$A_V = \frac{A_I R_L}{R_i} = -\frac{40 \times 10}{1} = -400$$

$$A_{Vs} = \frac{A_V R_i}{R_i + R_s} = -\frac{400 \times 1}{1 + 1} = -200$$

$$A_{Is} = \frac{A_I R_s}{R_i + R_s} = -\frac{40.0 \times 1}{1 + 1} = -20.0$$

Note that $A_{Vs} = A_{Is}R_L/R_s$.

$$Y_o = h_{oe} - \frac{h_{fe}h_{re}}{h_{ie} + R_s} = 25 \times 10^{-6} - \frac{50 \times 2.5 \times 10^{-4}}{2,100} = 19.0 \times 10^{-6} \; \mho$$

$$= 19.0 \; \mu\text{A/V}$$

or

$$Z_o = \frac{1}{Y_o} = \frac{10^6}{19.0} \; \Omega = 52.6 \text{ K}$$

8-7   THÉVENIN'S AND NORTON'S THEOREMS AND COROLLARIES

The input source to an amplifier may be represented either by a series circuit, as in Fig. 8-13a, or by a parallel network, as in Fig. 8-13b. This result is a special case of Thévenin's and Norton's theorems. Thévenin's theorem states that *any two-terminal linear network may be replaced by a voltage source equal to the open-circuit voltage between the terminals in series with the output impedance seen at this port.* In Fig. 8-14a, V represents the open-circuit voltage and Z is the impedance between terminals 1 and 2. To find Z, all *independent* voltage sources are short-circuited and all *independent* current sources are open-circuited.

The dual of Thévenin's theorem is Norton's theorem, which states that *any two-terminal linear network may be replaced by a current source equal to the short-circuited current between the terminals in parallel with the output impedance seen at this port.* In Fig. 8-14b, I represents the short-circuit current and Z is the impedance between terminals 1 and 2. In other words, a voltage source V in series with an impedance Z is equivalent to a current source $I \equiv V/Z$ in parallel with the impedance Z.

Corollaries   As extensions of Thévenin's and Norton's theorems we have the following relationships: If V represents the open-circuit voltage, I the short-circuit current, and $Z(Y)$ the impedance (admittance) between two terminals in a network, then

$$Z = \frac{V}{I} \qquad I = \frac{V}{Z} = VY \qquad V = IZ = \frac{I}{Y} \qquad\qquad (8\text{-}36)$$

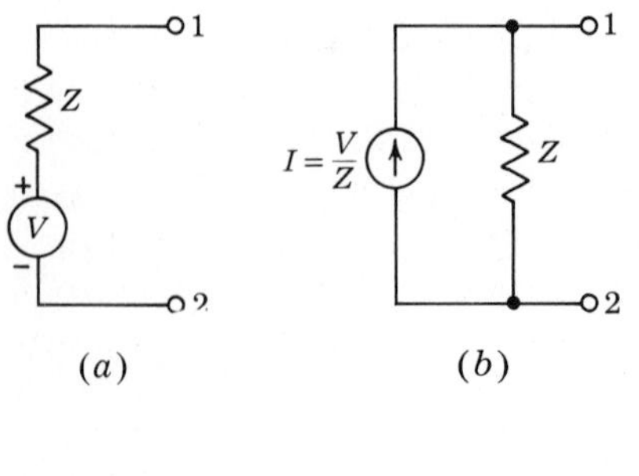

Fig. 8-14   As viewed from terminals 1 and 2, the Thévenin's circuit in (a) is equivalent to the Norton's circuit in (b).

(a)   (b)

The first relationship states that "the impedance between two nodes equals the open-circuit voltage divided by the short-circuit current." This method is one of the simplest for finding the output impedance Z_o.

The last relationship of Eqs. (8-36) is often the quickest way to calculate the voltage between two points in a network. This equation states that "the voltage equals the short-circuit current divided by the admittance."

The use of Thévenin's and Norton's theorems and the corollaries in Eq. (8-36) are used frequently in this text to simplify network analyses.

EXAMPLE Derive the output impedance in Fig. 8-12 using *the open-circuit-voltage short-circuit-current theorem.*

Solution From Eq. (8-36)

$$Y_o = \frac{I}{V}$$

where I is the current in a short circuit placed across the output terminals and V is the open-circuit output voltage ($Z_L = \infty$). If node 2 in Fig. 8-12 is connected to 2', then $V_2 = 0$ and

$$I = -h_f I_1 = -\frac{h_f V_s}{R_s + h_i} \tag{8-37}$$

With $Z_L = \infty$, $V_2 = V$ is given by $h_o V = -h_f I_1$. From KVL applied to the input circuit and with this value of I_1, we find

$$V_s = I_1(R_s + h_i) + h_r V = -\frac{h_o V}{h_f}(R_s + h_i) + h_r V \tag{8-38}$$

or

$$V = -\frac{h_f V_s/(R_s + h_i)}{h_o - h_f h_r/(R_s + h_i)} \tag{8-39}$$

Substituting I from Eq. (8-37) and V from Eq. (8-39), we obtain for $Y_o = I/V$ the expression in Eq. (8-35).

8-8   THE EMITTER FOLLOWER

The circuit diagram of a common-collector transistor amplifier, given in Fig. 5-19, is repeated in Fig. 8-15 for convenience. This configuration is also called the *emitter follower*, because its voltage gain is close to unity [Eq. (8-42)], and hence a change in base voltage appears as an equal change across the load at the emitter. In other words, the emitter *follows* the input signal. It is shown below that the input resistance R_i of an emitter follower is very high (hundreds of kilohms) and the output resistance R_o is very low (tens of ohms). Hence the most common use for the CC circuit is as a buffer stage which per-

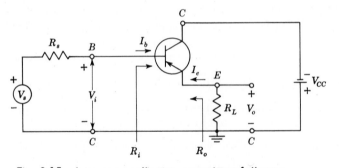

Fig. 8-15 A common-collector, or emitter-follower, configuration.

forms the function of resistance transformation (from high to low resistance) over a wide range of frequencies, with voltage gain close to unity. In addition, the emitter follower increases the power level of the signal.

From Tables 8-4 and 8-3 we obtain the following expressions for current gain, input resistance, voltage gain, and output resistance:

$$A_I = \frac{-I_e}{I_b} = \frac{-h_{fc}}{1 + h_{oc}R_L} = \frac{1 + h_{fe}}{1 + h_{oe}R_L} \tag{8-40}$$

$$R_i = \frac{V_i}{I_b} = h_{ic} + h_{rc}A_I R_L = h_{ie} + A_I R_L \tag{8-41}$$

$$A_V = \frac{V_o}{V_i} = \frac{A_I R_L}{R_i} = \frac{R_i - h_{ie}}{R_i} = 1 - \frac{h_{ie}}{R_i} \tag{8-42}$$

where the expression for $A_I R_L$ is taken from Eq. (8-41),

$$Y_o = h_{oc} - \frac{h_{fc}h_{rc}}{h_{ic} + R_s} = h_{oe} + \frac{1 + h_{fe}}{h_{ie} + R_s} \tag{8-43}$$

Using the same parameter values as in the illustrative example of Sec. 8-6, we find

$$A_I = 40.75 \qquad R_i = 409 \text{ K}$$

$$A_V = 0.997 \qquad R_o = \frac{1}{Y_o} = 41.2 \ \Omega$$

The overall voltage gain, taking the source resistance into account, is, from Table 8-4,

$$A_{Vs} = \frac{A_V R_i}{R_i + R_s} = 0.997 \times \frac{409}{410} = 0.995$$

These numerical values confirm that the emitter follower has the important characteristics mentioned above: a voltage gain close to unity, high input resistance, and low output resistance. It should also be observed that there is no phase shift between output and input in either voltage or current (since both A_V and A_I are positive and real).

8-9   COMPARISON OF TRANSISTOR AMPLIFIER CONFIGURATIONS

From Table 8-4 the values of current gain, voltage gain, input impedance, and output impedance are calculated as a function of load and source imped-ances.   These are plotted in Fig. 8-16 for each of the three configurations. A study of the shapes and relative amplitudes of these curves is instructive.

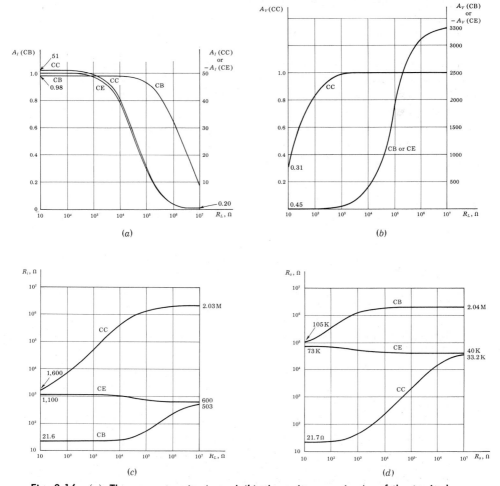

Fig. 8-16   (a) The current gain A_I and (b) the voltage gain A_V of the typical transistor of Table 8-2 as a function of its load resistance.   (c) The input resis-tance R_i as a function of its load resistance and (d) the output resistance R_o as a function of the source resistance of the typical transistor of Table 8-2.

The CE Configuration   From the curves and Table 8-5, it is observed that only the common-emitter stage is capable of both a voltage gain and a current gain greater than unity.   This configuration is the most versatile and useful of the three connections.

Note that R_i and R_o vary least with R_L and R_s, respectively, for the CE circuit.   Also observe that the magnitudes of R_i and R_o lie between those for the CB and CC configurations.

To realize a gain nominally equal to $(A_{Vs})_{max}$ would require not only that a zero-impedance voltage source be used, but also that R_L be many times larger than the output impedance.   Normally, however, so large a value of R_L is not feasible.   Suppose, for example, that a manufacturer specifies a maximum collector voltage of, say, 30 V.   Then we should not be inclined to use a collector supply voltage in excess of this maximum voltage, since in such a case the collector voltage would be exceeded if the transistor were driven to cutoff.   Suppose, further, that the transistor is designed to carry a collector current of, say, 5 mA when biased in the middle of its active region.   Then the load resistor should be selected to have a resistance of about $\frac{15}{5} = 3$ K.   We compute for the CE configuration a voltage gain under load of $A_V = -129$ (for $R_s = 0$).   Of course, the load resistance may be smaller than 3 K, as, for example, when a transistor is used to drive another transistor.   Or in some applications a higher value of R_L may be acceptable although load resistances in excess of 10 K are unusual.

The CB Configuration   For the common-base stage, A_I is less than unity, A_V is high (approximately equal to that of the CE stage), R_i is the lowest, and R_o is the highest of the three configurations.   The CB stage has few applications.   It is sometimes used to match a very low impedance source, to drive a high-impedance load, or as a noninverting amplifier with a voltage gain greater than unity.   It is also used as a constant-current source (for example, as a sweep circuit to charge a capacitor linearly[4]).

The CC Configuration   For the common-collector stage, A_I is high (approximately equal to that of the CE stage), A_V is less than unity, R_i is the highest, and R_o is the lowest of the three configurations.   This circuit finds wide application as a buffer stage between a high-impedance source and a low-impedance load.

TABLE 8-5   Comparison of transistor configurations ($R_L = 3$ K)

Quantity	CE	CC	CB
A_I...................	High (-46.5)	High (47.5)	Low (0.98)
A_V...................	High (-131)	Low (0.99)	High (131)
R_i....................	Medium (1,065 Ω)	High (144 K)	Low (22.5 Ω)
R_o ($R_s = 3$ K)........	Medium high (45.5 K)	Low (80.5 Ω)	High (1.72 M)

Summary The foregoing characteristics are summarized in Table 8-5, where the various quantities are calculated for $R_L = 3$ K and for the h parameters in Table 8-2.

8-10   LINEAR ANALYSIS OF A TRANSISTOR CIRCUIT

There are many transistor circuits which do not consist simply of the CE, CB, or CC configurations discussed above. For example, a CE amplifier may have a feedback resistor from collector to base, or it may have an emitter resistor. Furthermore, a circuit may consist of several transistors which are interconnected in some manner. An analytic determination of the small-signal behavior of even relatively complicated amplifier circuits may be made by following these simple rules:

1. Draw the actual wiring diagram of the circuit neatly.

2. Mark the points B (base), C (collector), and E (emitter) on this circuit diagram. Locate these points as the start of the equivalent circuit. Maintain the same relative positions as in the original circuit.

3. Replace each transistor by its h-parameter model (Fig. 8-6).

4. Transfer all circuit elements from the actual circuit to the equivalent circuit of the amplifier. Keep the relative positions of these elements intact.

5. Replace each independent dc source by its internal resistance. The ideal voltage source is replaced by a short circuit, and the ideal current source by an open circuit.

6. Solve the resultant linear circuit for mesh or branch currents and node voltages by applying Kirchhoff's current and voltage laws (KCL and KVL).

It should be emphasized that it is not necessary to use the foregoing general approach for a circuit consisting of a cascade of CE, CB, and/or CC stages. Such configurations are analyzed very simply by direct applications of the formulas in Table 8-4.

8-11   MILLER'S THEOREM AND ITS DUAL

Whereas the rules given in the preceding section are completely general, certain configurations can be analyzed more simply by using Miller's theorem.[5] Hence we digress briefly to discuss this important theorem. It is invoked often in connection with many topics in this book. Consider an arbitrary circuit configuration with N distinct nodes 1, 2, 3, . . . , N, as indicated in Fig. 8-17. Let the node voltages be V_1, V_2, V_3, . . . , V_N, where $V_N = 0$ since N is the reference, or ground, node. Nodes 1 and 2 (referred to as N_1 and N_2) are interconnected with an impedance Z'. We postulate that we know the ratio V_2/V_1. Designate the ratio V_2/V_1 by K, which in the sinus-

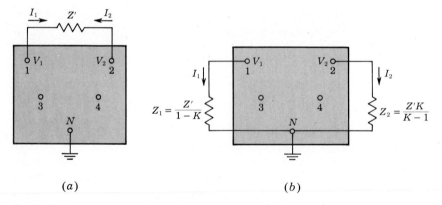

(a)   (b)

Fig. 8-17 Pertaining to Miller's theorem. By definition, $K \equiv V_2/V_1$. The networks in (a) and (b) have identical node voltages. Note that $I_1 = -I_2$.

oidal steady state will be a complex number and, more generally, will be a function of the Laplace transform variable s. We shall now show that the current I_1 drawn from N_1 through Z' can be obtained by disconnecting terminal 1 from Z' and by bridging an impedance $Z'/(1 - K)$ from N_1 to ground, as indicated in Fig. 8-17b.

The current I_1 is given by

$$I_1 = \frac{V_1 - V_2}{Z'} = \frac{V_1(1 - K)}{Z'} = \frac{V_1}{Z'/(1 - K)} = \frac{V_1}{Z_1} \tag{8-44}$$

Therefore, if $Z_1 \equiv Z'/(1 - K)$ were shunted across terminals N_1-N, the current I_1 drawn from N_1 would be the same as that from the original circuit. Hence the same expression is obtained for I_1 in terms of the node voltages for the two configurations (Figs. 8-17a and b).

In a similar way, it may be established that the correct current I_2 drawn from N_2 may be calculated by removing Z' and by connecting between N_2 and ground an impedance Z_2, given by

$$Z_2 \equiv \frac{Z'}{1 - 1/K} = \frac{Z'K}{K - 1} \tag{8-45}$$

Since identical nodal equations (KCL) are obtained from the configurations of Fig. 8-17a and b, these two networks are equivalent. It must be emphasized that this theorem will be useful in making calculations only if it is possible to find the value of K by some independent means.

EXAMPLE For the amplifier shown in Fig. 8-18a calculate R_i, R_i', A_V, A_{Vs}, and $A_I' = -I_2/I_1$. The transistor parameters are specified in Table 8-2.

Solution   Miller's theorem is applied to the 200-K resistance R' so that Fig. 8-18b is obtained, where $K = A_V$ is the voltage gain from base to collector.   Assuming that this gain is much larger than unity $(-A_V \gg 1)$, then

$$\frac{200}{1 - 1/A_V} \approx 200$$

The effective load resistance is $R'_L = 10 \| 200$, or

$$R'_L = \frac{10 \times 200}{210} = 9.52 \text{ K}$$

From Table 8-4

$$A_I = \frac{-h_{fe}}{1 + h_{oe}R'_L} = \frac{-50}{1 + (9.52/40)} = -40.3$$

$$R_i = h_{ie} + h_{re}A_I R'_L = 1.1 - 2.5 \times 10^{-4} \times 40.3 \times 9.52 = 1.00 \text{ K}$$

$$A_V = \frac{A_I R'_L}{R_i} = \frac{-40.3 \times 9.52}{1.00} = -384$$

We have thus verified the assumption that $-A_V \gg 1$

Since $200/(1 - A_V) = 200/385 = 0.520$ K, then $R'_i = R_i \| (0.520)$ K, or

$$R'_i = \frac{1,000 \times 520}{1,520} = 343 \ \Omega$$

$$A_{Vs} = A_V \frac{R'_i}{R'_i + R_s} = \frac{-384 \times 0.343}{0.343 + 10} = -12.7$$

$$A'_I = -\frac{I_2}{I_1} = \frac{V_o}{R_L} \frac{R'_i + R_s}{V_s} = A_{Vs} \frac{R'_i + R_s}{R_L} = -(12.7)\left(\frac{10.343}{10}\right) = -13.15$$

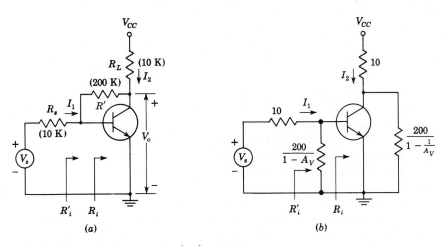

Fig. 8-18   Example using Miller's theorem.

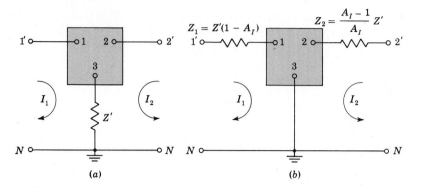

Fig. 8-19 Pertaining to the dual of Miller's theorem. By definition $A_I \equiv -I_2/I_1$. The networks in (a) and (b) have the same currents I_1 and I_2 if excited by the same voltages $V_{1'N}$ and $V_{2'N}$.

Dual of Miller's Theorem Consider the network of Fig. 8-19a with arbitrary active or passive linear elements between nodes 1, 2, and 3 and with an impedance Z' between node 3 and ground N. The two loops indicated are coupled by means of the common element Z'. We postulate that we know the current ratio $A_I \equiv -I_2/I_1$.

The dual of Miller's theorem is indicated in Fig. 8-19b, where node 3 is grounded, an impedance Z_1 is placed in mesh 1, and Z_2 is added to mesh 2. It is readily verified that the voltage $I_1 Z_1$ equals the drop $(I_1 + I_2)Z'$ across Z' if $Z_1 = Z'(1 - A_I)$. Hence $V_{1'N}$ is the same in the two circuits in Fig. 8-19a and b (for the same currents I_1 and I_2). In a similar manner $V_{2'N}$ has the same value in the two circuits if $Z_2 = [(A_I - 1)/A_I]Z'$. The two networks are therefore identical, in the sense that if the same voltages $V_{1'N}$ and $V_{2'N}$ are impressed on both, the same current I_1 will flow in mesh 1, and I_2 in mesh 2.

This transformation is useful in circuit analysis and is employed in Sec. 8-15.

8-12 CASCADING TRANSISTOR AMPLIFIERS

When the amplification of a single transistor is not sufficient for a particular purpose, or when the input or output impedance is not of the correct magnitude for the intended application, two or more stages may be connected in cascade; i.e., the output of a given stage is connected to the input of the next stage, as shown in Fig. 8-20. In the circuit of Fig. 8-21a the first stage is connected common-emitter, and the second is a common-collector stage. Figure 8-21b shows the small-signal circuit of the two-stage amplifier, with the biasing batteries omitted for simplicity, since these do not affect the small-signal calculations.

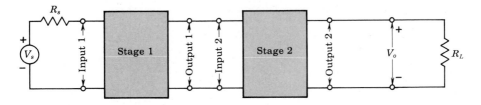

Fig. 8-20 Two cascaded stages.

To analyze a circuit such as the one of Fig. 8-21, we make use of the general expressions for A_I, Z_i, A_V, and Y_o from Table 8-4. It is necessary that we have available the h parameters for the specific transistors used in the circuit. The h-parameter values for a specific transistor are usually obtained from the manufacturer's data sheet.

EXAMPLE Shown in Fig. 8-21 is a two-stage amplifier circuit in a CE-CC configuration. The transistor parameters at the corresponding quiescent points are

$h_{ie} = 2\text{ K}$ $h_{fe} = 50$ $h_{re} = 6 \times 10^{-4}$ $h_{oe} = 25\ \mu\text{A/V}$

$h_{ic} = 2\text{ K}$ $h_{fc} = -51$ $h_{rc} = 1$ $h_{oc} = 25\ \mu\text{A/V}$

Find the input and output impedances and individual, as well as overall, voltage and current gains.

Solution We note that, in a cascade of stages, the collector resistance of one stage is shunted by the input impedance of the next stage. Hence it is advantageous to start the analysis with the last stage. In addition, it is convenient (as already noted in Sec. 8-6) to compute, first, the current gain, then the input impedance and the voltage gain. Finally, the output impedance may be calculated if desired by starting this analysis with the first stage and proceeding toward the output stage.

The second stage. From Table 8-4, with $R_L = R_{e2}$, the current gain of the last stage is

$$A_{I2} = -\frac{I_{e2}}{I_{b2}} = \frac{-h_{fc}}{1 + h_{oc}R_{e2}} = \frac{51}{1 + 25 \times 10^{-6} \times 5 \times 10^3} = 45.3$$

The input impedance R_{i2} is

$$R_{i2} = h_{ic} + h_{rc}A_{I2}R_{e2} = 2 + 45.3 \times 5 = 228.5\text{ K}$$

Note the high input impedance of the CC stage. The voltage gain of the second stage is

$$A_{V2} = \frac{V_o}{V_2} = A_{I2}\frac{R_{e2}}{R_{i2}} = \frac{45.3 \times 5}{228.5} = 0.991$$

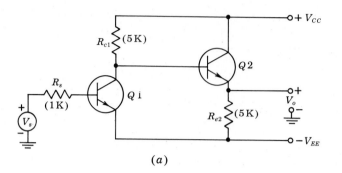

(a)

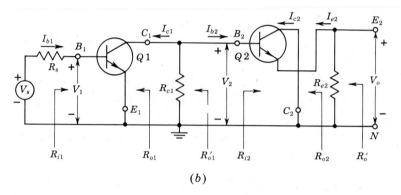

(b)

Fig. 8-21   (a) Common-emitter–common-collector amplifier.   (b) Small-signal circuit of the CE-CC amplifier.

or alternatively, from Eq. (8-42)

$$A_{V2} = 1 - \frac{h_{ie2}}{R_{i2}} = 1 - \frac{2}{228.5} = 0.991$$

The first stage.   We observe that the net load resistance R_{L1} of this stage is the parallel combination of R_{c1} and R_{i2} (written in symbolic form, $R_{L1} = R_{c1} \| R_{i2}$), or

$$R_{L1} = \frac{R_{c1}R_{i2}}{R_{c1} + R_{i2}} = \frac{5 \times 228.5}{233.5} = 4.9 \text{ K}$$

Hence

$$A_{I1} = -\frac{I_{c1}}{I_{b1}} = \frac{-h_{fe}}{1 + h_{oe}R_{L1}} = \frac{-50}{1 + 25 \times 10^{-6} \times 4.9 \times 10^{3}} = -44.5$$

The input impedance of the first stage, which is also the input impedance of the cascaded amplifier, is given by

$$R_{i1} = h_{ie} + h_{re}A_{I1}R_{L1} = 2 - 6 \times 10^{-4} \times 44.5 \times 4.9 = 1.87 \text{ K}$$

The voltage gain of the first stage is

$$A_{V1} = \frac{V_2}{V_1} = \frac{A_{I1}R_{L1}}{R_{i1}} = \frac{-44.5 \times 4.9}{1.87} = -116.6$$

The output admittance of the transistor is, from Eq. (8-35) or Table 8-4,

$$Y_{o1} = h_{oe} - \frac{h_{fe}h_{re}}{h_{ie} + R_s} = 25 \times 10^{-6} - \frac{50 \times 6 \times 10^{-4}}{2 \times 10^3 + 1 \times 10^3} = 15 \times 10^{-6}\ \mho$$

$$= 15\ \mu A/V$$

Hence

$$R_{o1} = \frac{1}{Y_{o1}} = \frac{10^6}{15}\ \Omega = 66.7\ K$$

The output impedance of the *first stage, taking R_{c1} into account,* is $R_{o1}\|R_{c1}$, or

$$R'_{o1} = \frac{R_{c1}R_{o1}}{R_{c1} + R_{o1}} = \frac{5 \times 66.7}{5 + 66.7} = 4.65\ K$$

The output resistance of the last stage. The effective source resistance R'_{s2} for the second transistor $Q2$ is $R_{o1}\|R_{c1}$. Thus $R'_{s2} = R'_{o1} = 4.65\ K$, and

$$Y_{o2} = h_{oc} - \frac{h_{fc}h_{rc}}{h_{ic} + R'_{s2}} = 25 \times 10^{-6} - \frac{-51 \times 1}{2 \times 10^3 + 4.65 \times 10^3} = 7.70 \times 10^{-3}\ A/V$$

Hence $R_{o2} = 1/Y_{o2} = 130\ \Omega$, where R_{o2} is the output impedance of transistor $Q2$ under open-circuit conditions. The output impedance R'_o of the amplifier, taking R_{e2} into account, is $R_{o2}\|R_{e2}$, or

$$R'_o = \frac{R_{o2}R_{e2}}{R_{o2} + R_{e2}} = \frac{130 \times 5,000}{130 + 5,000} = 127\ \Omega$$

The overall current and voltage gains. The total current gain of both stages is

$$A_I = -\frac{I_{e2}}{I_{b1}} = -\frac{I_{e2}}{I_{b2}}\frac{I_{b2}}{I_{c1}}\frac{I_{c1}}{I_{b1}} = -A_{I2}\frac{I_{b2}}{I_{c1}}A_{I1}$$

From Fig. 8-22 we have

$$\frac{I_{b2}}{I_{c1}} = -\frac{R_{c1}}{R_{i2} + R_{c1}} \tag{8-46}$$

Hence

$$A_I = A_{I2}A_{I1}\frac{R_{c1}}{R_{i2} + R_{c1}} = 45.3 \times (-44.5) \times \frac{5}{228.5 + 5} = -43.2 \tag{8-47}$$

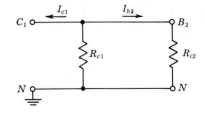

Fig. 8-22 Relating to the calculation of over-all current gain.

For the voltage gain of the amplifier we have

$$A_V = \frac{V_o}{V_1} = \frac{V_o}{V_2}\frac{V_2}{V_1} = A_{V2}A_{V1} \tag{8-48}$$

or

$$A_V = 0.99 \times (-116.6) = -115$$

The voltage gain can also be obtained from

$$A_V = A_I \frac{R_{e2}}{R_{i1}} = -43.2 \times \frac{5}{1.87} = -115$$

The overall voltage gain, taking the source impedance into account, is given by

$$A_{Vs} = \frac{V_o}{V_s} = A_V \frac{R_{i1}}{R_{i1} + R_s} = -115 \times \frac{1.87}{1.87 + 1} = -75.3$$

Table 8-6 summarizes the results obtained in the solution of this problem.

TABLE 8-6   Results of the example on page 259

	Transistor Q2 CC	Transistor Q1 CE	Both stages CE-CC
A_I	45.3	-44.5	-43.2
R_i	228.5 K	1.87 K	1.87 K
A_V	0.99	-116.6	-115
R'_o	127 Ω	4.65 K	127 Ω

Choice of the Transistor Configuration in a Cascade   It is important to note that the previous calculations of input and output impedances and voltage and current gains are applicable for any connection of the cascaded stages. They could be CC, CB, or combinations of all three possible connections.

Consider the following question: Which of the three possible connections must be used in cascade if maximum voltage gain is to be realized?   For the intermediate stages, the common-collector connection is not used because the voltage gain of such a stage is less than unity.   Hence it is not possible (without a transformer) to increase the overall voltage amplification by cascading common-collector stages.

Grounded-base RC-coupled stages also are seldom cascaded because the voltage gain of such an arrangement is approximately the same as that of the output stage alone.   This statement may be verified as follows: The voltage gain of a stage equals its current gain times the effective load resistance R_L divided by the input resistance R_i.   The effective load resistance R_L is the parallel combination of the actual collector resistance R_c and (except for the last stage) the input resistance R_i of the following stage.   This parallel combination is certainly less than R_i, and hence, for identical stages, the effective load resistance is less than R_i. The maximum current gain is h_{fb}, which is less than unity (but approximately equal to unity). Hence the voltage gain of any stage (except the last, or output, stage) is less than unity. (This analysis is not strictly correct because the R_i is a function of the effective load resistance and hence will vary somewhat from stage to stage.)

Since the short-circuit current gain h_{fe} of a common-emitter stage is much greater than unity, it is possible to increase the voltage amplification by cascading such stages. We may now state that *in a cascade the intermediate transistors should be connected in a common-emitter configuration.*

The choice of the input stage may be decided by criteria other than the maximization of voltage gain. For example, the amplitude or the frequency response of the transducer V_s may depend upon the impedance into which it operates. Some transducers require essentially open-circuit or short-circuit operation. In many cases the common-collector or common-base stage is used at the input because of impedance considerations, even at the expense of voltage or current gain. Noise is another important consideration which may determine the selection of a particular configuration of the input stage.

The output stage is selected also on the basis of impedance considerations. Since a CC stage has a very low output resistance, it is often used for the last stage if it is required to drive a low impedance (perhaps capacitive) load.

8-13   SIMPLIFIED COMMON–EMITTER HYBRID MODEL[6]

In the preceding sections, we carried out detailed calculations of current gain, voltage gain, input, and output impedances of illustrative transistor amplifier circuits.

In most practical cases it is appropriate to obtain approximate values of A_I, A_V, R_i, and R_o rather than to carry out the more lengthy exact calculations. We are justified in making such approximations because the h parameters themselves usually vary widely for the same type of transistor. Also, a better "physical feeling" for the behavior of a transistor circuit can be obtained from a simple approximate solution than from a more laborious exact calculation.

We show below that two of the four h parameters, h_{ie} and h_{fe}, are sufficient for the approximate analysis of low-frequency transistor circuits, provided that the load resistance is small enough to satisfy the condition $h_{oe}R_L < 0.1$. The simplified model is shown in Fig. 8-23. This equivalent circuit may be used for any configuration by grounding the appropriate node. The signal is connected between the input terminal and ground, and the load is placed between the output node and ground. By examining in detail the errors introduced in the calculation using this simplified model, it is found that, if R_L is no larger than $0.1/h_{oe}$, the error in A_I, R_i, A_V, or R_o' is less than 10 percent.

Consider first the common-emitter configuration. Figure 8-24 shows the

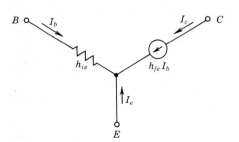

Fig. 8-23 Approximate hybrid model which may be used for all three configurations CE, CC, and CB.

CE stage with the transistor replaced by the approximate model of Fig. 8-23. We now compare the exact results with those obtained from the simplified analysis.

Current Gain   From Table 8-4 the CE current gain is given by

$$A_I = \frac{-h_{fe}}{1 + h_{oe}R_L} \tag{8-49}$$

Hence we immediately see that the approximation (Fig. 8-24)

$$A_I \approx -h_{fe}$$

overestimates the magnitude of the current gain by less than 10 percent if $h_{oe}R_L < 0.1$.

Input Impedance   From Table 8-4 the input resistance is given by

$$R_i = h_{ie} + h_{re}A_I R_L$$

which may be put in the form

$$R_i = h_{ie}\left(1 - \frac{h_{re}h_{fe}}{h_{ie}h_{oe}}\frac{|A_I|}{h_{fe}}h_{oe}R_L\right) \tag{8-50}$$

Using the typical h-parameter values in Table 8-2, we find $h_{re}h_{fe}/h_{ie}h_{oe} \approx 0.5$. From Eq. (8-49), we see that $|A_I| < h_{fe}$. Hence, if $h_{oe}R_L < 0.1$, it follows from Eq. (8-50) that the approximation obtained from Fig. 8-24, namely,

$$R_i = \frac{V_b}{I_b} \approx h_{ie} \tag{8-51}$$

overestimates the input resistance by less than 5 percent.

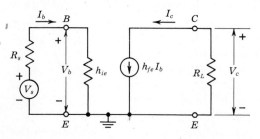

Fig. 8-24   Approximate CE model.

Voltage Gain From Table 8-4 the voltage gain is given by

$$A_V = A_I \frac{R_L}{R_i} = - \frac{h_{fe}R_L}{h_{ie}} \tag{8-52}$$

If we take the logarithm of this equation and then the differential, we obtain

$$\frac{dA_V}{A_V} = \frac{dA_I}{A_I} - \frac{dR_i}{R_i} \tag{8-53}$$

From the preceding discussion the maximum errors for $h_{oe}R_L < 0.1$ are

$$\frac{dA_I}{A_I} = +0.1 \quad \text{and} \quad \frac{dR_i}{R_i} = +0.05$$

Hence the maximum error in voltage gain is 5 percent, and the magnitude of A_V is overestimated by this amount.

Output Impedance The simplified circuit of Fig. 8-24 has infinite output resistance because, with $V_s = 0$ and an external voltage source applied at the output, we find $I_b = 0$, and hence $I_c = 0$. However, the true value depends upon the source resistance R_s and lies between 40 and 80 K (Fig. 8-16d). For a maximum load resistance of $R_L = 4$ K, the output resistance of the stage, taking R_L into account, is 4 K, if the simplified model is used, and the parallel combination of 4 K with 40 K (under the worst case), if the exact solution is used. Hence, using the approximate model leads to a value of output resistance under load which is too large, but by no more than 10 percent.

The approximate solution for the CE configuration is summarized in the first column of Table 8-7. In summary, two of the four h parameters, h_{ie} and h_{fe}, are sufficient for the approximate analysis of low-frequency transistor circuits, provided the load resistance R_L is no larger than $0.1/h_{oe}$. For the value of h_{oe} given in Table 8-2, R_L must be less than 4 K. The approximate circuit is always valid when CE transistors are operated in cascade because the low input impedance of a CE stage shunts the output of the previous stage so that the effective load resistance R_L' satisfies the condition $h_{oe}R_L' \leq 0.1$.

8-14 SIMPLIFIED CALCULATIONS FOR THE COMMON–COLLECTOR CONFIGURATION

Figure 8-25 shows the simplified circuit of Fig. 8-23 with the collector grounded (with respect to the signal) and a load R_L connected between emitter and ground.

Current Gain From Fig. 8-25 we see that

$$A_I = - \frac{I_e}{I_b} = 1 + h_{fe} \tag{8-54}$$

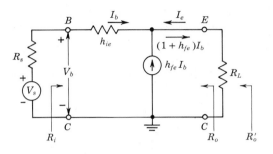

Fig. 8-25 Simplified hybrid model for the CC circuit.

Comparing this equation with the exact expression (8-40), we conclude that when the simplified equivalent circuit of Fig. 8-25 is used, the current gain is overestimated by less than 10 percent if $h_{oe}R_L < 0.1$.

Input Resistance From Fig. 8-25 we obtain

$$R_i = \frac{V_b}{I_b} = h_{ie} + (1 + h_{fe})R_L \tag{8-55}$$

Note that $R_i \gg h_{ie} \approx 1$ K even if R_L is as small as 0.5 K, because $h_{fe} \gg 1$. If we substitute from Eq. (8-54) into the exact expression (8-41), we obtain Eq. (8-55). However, we have just concluded that Eq. (8-54) gives too high a value of A_I by at most 10 percent. Hence it follows that R_i, as calculated from Eq. (8-55) or Fig. 8-25, is also overestimated by less than 10 percent.

Voltage Gain If Eq. (8-53) is used for the voltage gain, it follows from the same arguments as used in the CE case that there will be very little error in the value of A_V. An alternative proof is now given. The voltage gain of the emitter follower is close to unity, and is given by Eq. (8-42), or $1 - A_V = h_{ie}/R_i$. If, for example, $R_i = 10h_{ie}$, then $A_V = 0.9$. If, however, we use an approximate value of R_i which is 10 percent too high, then $h_{ie}/R_i = \frac{1}{11} = 0.09$ and $A_V = 0.91$. Hence the approximate calculation for A_V gives a value which is only 1 percent too high.

Output Impedance In Fig. 8-25 the open-circuit output voltage is V_s and the short-circuit output current is

$$I = (1 + h_{fe})I_b = \frac{(1 + h_{fe})V_s}{h_{ie} + R_s}$$

Hence the output admittance of the transistor alone is, from Eq. (8-36),

$$Y_o = \frac{I}{V_s} = \frac{1 + h_{fe}}{h_{ie} + R_s} \tag{8-56}$$

From Eq. (8-43), the exact expression for Y_o is

$$Y_o = h_{oe} + \frac{1 + h_{fe}}{h_{ie} + R_s} \tag{8-57}$$

Even if we choose an abnormally large value of source resistance, say, $R_s = 100$ K, then (using the typical h-parameter values in Table 8-2) we find that the second term in Eq. (8-57) is large ($500~\mu A/V$) compared with the first term ($25~\mu A/V$). Hence the value of the approximate output admittance given by Eq. (8-56) is smaller than the value given by Eq. (8-57) by less than 5 percent. The output resistance R_o of the transistor, calculated from the simplified model, namely,

$$R_o = \frac{h_{ie} + R_s}{1 + h_{fe}}$$

is an overestimation by less than 5 percent. The output resistance R_o' of the stage, taking the load into account, is R_o in parallel with R_L.

The approximate solution for the CC configuration is summarized in the third column of Table 8-7.

TABLE 8-7   Summary of approximate equations for $h_{oe}(R_e + R_L) \leq 0.1$†

	CE	CE with R_e	CC	CB
A_I	$-h_{fe}$	$-h_{fe}$	$1 + h_{fe}$	$-h_{fb} = \dfrac{h_{fe}}{1 + h_{fe}}$
R_i	h_{ie}	$h_{ie} + (1 + h_{fe})R_e$	$h_{ie} + (1 + h_{fe})R_L$	$h_{ib} = \dfrac{h_{ie}}{1 + h_{fe}}$
A_V	$-\dfrac{h_{fe}R_L}{h_{ie}}$	$-\dfrac{h_{fe}R_L}{R_i}$	$1 - \dfrac{h_{ie}}{R_i}$	$h_{fe}\dfrac{R_L}{h_{ie}}$
R_o	∞	∞	$\dfrac{R_s + h_{ie}}{1 + h_{fe}}$	∞
R_o'	R_L	R_L	$R_o \| R_L$	R_L

† $(R_i)_{CB}$ is an underestimation by less than 10 percent. All other quantities except R_o are too large in magnitude by less than 10 percent.

EXAMPLE Carry out the calculations for the two-stage amplifier of Fig. 8-21 using the simplified model of Fig. 8-23.

Solution First note that, since $h_{oe}R_L = 25 \times 10^{-6} \times 5 \times 10^3 = 0.125$, which is slightly larger than 0.1, we may expect errors in our approximation somewhat larger than 10 percent.

For the CC output stage we have, from Table 8-7

$$A_{I2} = 1 + h_{fe} = 51$$

$$R_{i2} = h_{ie} + (1 + h_{fe})R_L = 2 + (51)(5) = 257 \text{ K}$$

$$A_{V2} = 1 - \frac{h_{ie}}{R_{i2}} = 1 - \frac{2}{257} = 0.992$$

For the CE input stage, we find, from Table 8-7

$$A_{I1} = -h_{fe} = -50 \qquad R_{i1} = h_{ie} = 2 \text{ K}$$

The effective load on the first stage, its voltage gain, and output impedance are

$$R_{L1} = \frac{R_{c1}R_{i2}}{R_{c1} + R_{i2}} = \frac{(5)(257)}{262} = 4.9 \text{ K}$$

$$A_{V1} = \frac{A_{I1}R_{L1}}{R_{i1}} = \frac{-(50)(4.9)}{2} = -123$$

$$R'_{o1} = R_{c1} = 5 \text{ K}$$

Since R'_{o1} is the effective source impedance for Q2, then, from Table 8-7,

$$R_{o2} = \frac{h_{ie} + R_s}{1 + h_{fe}} = \frac{2,000 + 5,000}{51} = 137 \ \Omega$$

$$R'_{o2} = \frac{R_{o2}R_{L2}}{R_{o2} + R_{L2}} = \frac{(137)(5,000)}{5,137} = 134 \ \Omega$$

Finally, the overall voltage and current gains of the cascade are

$$A_V = A_{V1}A_{V2} = (-123)(0.992) = -122$$

$$A_I = A_{I1}A_{I2}\frac{R_{c1}}{R_{c1} + R_{i2}} = (-50)(51)\left(\frac{5}{5 + 257}\right) = -48.7$$

Alternatively, A_V may be computed from

$$A_V = A_I\frac{R_{L2}}{R_{i1}} = -\frac{(48.7)(5)}{2} = -122$$

Table 8-8 summarizes this solution, and should be compared with the exact values in Table 8-6.   We find that the maximum errors are just slightly above 10 percent, as anticipated.   It should also be noted that all the approximate values are numerically too large, as predicted.

TABLE 8-8   Approximate results of the example on page 267

	Q2, CC	Q1, CE	Both stages
A_I	51	−50	−48.7
R_i	257 K	2 K	2 K
A_V	0.992	−123	−122
R'_o	134 Ω	5 K	134 Ω

Following procedures exactly analogous to those explained in Secs. 8-13 and 8-14 for the CE and CC configurations, respectively, the approximate formulas given in the fourth column of Table 8-7 may be obtained for the CB stage.

8-15   THE COMMON–EMITTER AMPLIFIER WITH AN EMITTER RESISTANCE

Very often a transistor amplifier consists of a number of CE stages in cascade. Since the voltage gain of the amplifier is equal to the product of the voltage gains of each stage, it becomes important to stabilize the voltage amplification of each stage.   By stabilization of voltage or current gain, we mean that the amplification becomes essentially independent of the h parameters of the transistor.   From our discussion in Sec. 8-4, we know that the transistor parameters depend on temperature, aging, and the operating point.   Moreover, these parameters vary widely from device to device even for the same type of transistor.

The necessity for voltage stabilization is seen from the following example: Two commercially built six-stage amplifiers are to be compared.   If each stage of the first has a gain which is only 10 percent below that of the second, the overall amplification of the latter is $(0.9)^6 = 0.53$ (or about one-half that of the former).   And this value may be below the required specification.   A simple and effective way to obtain voltage-gain stabilization is to add an emitter resistor R_e to a CE stage, as indicated in the circuit of Fig. 8-26. This stabilization is a result of the feedback provided by the emitter resistor. The general concept of feedback is discussed in Chap. 13.

We show in this section that the presence of R_e has the following effects on the amplifier performance, in addition to the beneficial effect on bias

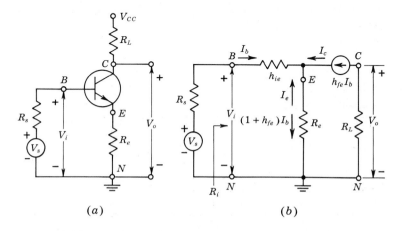

(a)   (b)

Fig. 8-26   (a) Common-emitter amplifier with an emitter resistor.
(b) Approximate small-signal equivalent circuit.

stability discussed in Chap. 9: It leaves the current gain A_I essentially unchanged; it increases the input impedance by $(1 + h_{fe})R_e$; it increases the output impedance; and under the condition $(1 + h_{fe})R_e \gg h_{ie}$, it stabilizes the voltage gain, which becomes essentially equal to $-R_L/R_e$ (and thus is independent of the transistor).

The Approximate Solution An approximate analysis of the circuit of Fig. 8-26a can be made using the simplified model of Fig. 8-25 as shown in Fig. 8-26b.

The current gain is, from Fig. 8-26b,

$$A_I = \frac{-I_c}{I_b} = \frac{-h_{fe}I_b}{I_b} = -h_{fe} \qquad (8\text{-}58)$$

The current gain equals the short-circuit value, and is unaffected by the addition of R_c.

The input resistance, as obtained from inspection of Fig. 8-26b, is

$$R_i = \frac{V_i}{I_b} = h_{ie} + (1 + h_{fe})R_e \qquad (8\text{-}59)$$

The input resistance is augmented by $(1 + h_{fe})R_e$, and may be very much larger than h_{ie}. For example, if $R_e = 1$ K and $h_{fe} = 50$, then

$$(1 + h_{fe})R_e = 51 \text{ K} \gg h_{ie} \approx 1 \text{ K}$$

Hence an emitter resistance greatly increases the input resistance.

The voltage gain is

$$A_V = \frac{A_I R_L}{R_i} = \frac{-h_{fe}R_L}{h_{ie} + (1 + h_{fe})R_e} \qquad (8\text{-}60)$$

Clearly, the addition of an emitter resistance greatly reduces the voltage amplification. This reduction in gain is often a reasonable price to pay for the improvement in stability. We note that, if $(1 + h_{fe})R_e \gg h_{ie}$, and since $h_{fe} \gg 1$, then

$$A_V \approx \frac{-h_{fe}}{1 + h_{fe}} \frac{R_L}{R_e} \approx \frac{-R_L}{R_e} \qquad (8\text{-}61)$$

Subject to the above approximations, A_V is completely stable (if stable resistances are used for R_L and R_e), since it is independent of all transistor parameters.

The output resistance of the transistor alone (with R_L considered external) is infinite for the approximate circuit of Fig. 8-26, just as it was for the CE amplifier of Sec. 8-13 with $R_e = 0$. Hence the output impedance of the stage, including the load, is R_L.

Looking into the Base and Emitter of a Transistor On the basis of Eq. (8-59), we draw the equivalent circuit of Fig. 8-27a from which to calculate

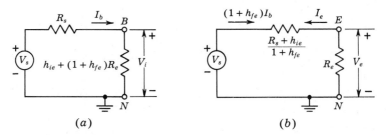

Fig. 8-27 (a) Equivalent circuit "looking into the base" of Fig. 8-26. This circuit gives (approximately) the correct base current. (b) Equivalent circuit "looking into the emitter" of Fig. 8-26. This circuit gives (approximately) the correct emitter voltage V_e and the correct emitter and base currents.

the base current with the signal source applied.   This network is the equivalent circuit "looking into the base."   From it we obtain

$$I_b = \frac{V_s}{R_s + h_{ie} + (1 + h_{fe})R_e} \tag{8-62}$$

From Fig. 8-26 and Eq. (8-62) we find the emitter-to-ground voltage to be

$$V_{en} \equiv V_e = (1 + h_{fe})I_b R_e = \frac{V_s R_e}{(R_s + h_{ie})/(1 + h_{fe}) + R_e} \tag{8-63}$$

This same expression may be obtained from Fig. 8-27b, which therefore represents the equivalent circuit "looking into the emitter."   From Fig. 8-27b we can obtain the emitter current $I_e = -V_e/R_e$ and also the base current $I_b = -I_e/(1 + h_{fe})$.

Note that I_b and $I_c = h_{fe}I_b$ are independent of R_c, and hence the approximate equivalent circuits in Fig. 8-27 are also valid for the emitter follower.

Validity of the Approximations   For the CE case, with $R_e = 0$, the approximate equivalent circuit of Fig. 8-24 is valid if $h_{oe}R_L \leq 0.1$.   What is the corresponding restriction for the circuit with $R_e \neq 0$?   We can answer this question and, at the same time, obtain an exact solution, if desired, by proceeding as indicated in Fig. 8-28.   The exact value of the current gain of Fig. 8-28a (which is the same as that of Fig. 8-26a) is $A_I = -I_c/I_b$.   The configuration in Fig. 8-28a corresponds to that in Fig. 8-19a with $Z' = R_e$, $I_1 = I_b$, and $I_2 = I_c$.   Hence applying the dual of Miller's theorem (Fig. 8-19b), we obtain the circuit of Fig. 8-28b.   The two amplifiers of Fig. 8-28a and b are equivalent in the sense that the base and collector currents are the same in the two circuits.

The effective load impedance R'_L is, from Fig. 8-28b,

$$R'_L = R_L + \frac{A_I - 1}{A_I} R_e \tag{8-64}$$

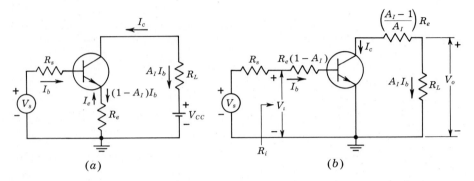

Fig. 8-28 (a) Transistor amplifier stage with unbypassed emitter resistor R_e. (b) Small-signal equivalent circuit.

We know from the above approximate solution that $A_I \approx -h_{fe}$, and since $h_{fe} \gg 1$, then $R'_L \approx R_{L.} + R_e$. Since in Fig. 8-28b the emitter is grounded and the collector resistance is R'_L, the approximate two-parameter (h_{ie} and h_{fe}) circuit is valid, provided that

$$h_{oe}R'_L = h_{oe}(R_L + R_e) \leq 0.1 \tag{8-65}$$

This condition means that the sum of R_L and R_e is no more than a few thousand ohms, say, 4 K for $1/h_{oe} = 40$ K. Furthermore, R_e is usually several times smaller than R_L in order to have an appreciable voltage gain [Eq. (8-61)].

The approximate solution for the CE amplifier with an emitter resistor R_e is summarized in the second column of Table 8-7.

The Exact Solution If the above inequality (8-65) is not satisfied for a particular amplifier, an exact solution can readily be obtained by referring to Fig. 8-28b and to Table 8-4. For example, the current gain is

$$A_I = \frac{-h_{fe}}{1 + h_{oe}R'_L} = \frac{-h_{fe}}{1 + h_{oe}\left(R_L + \dfrac{A_I - 1}{A_I} R_e\right)} \tag{8-66}$$

From the above equation we can solve explicitly for A_I, and we obtain the following:

$$A_I = \frac{h_{oe}R_e - h_{fe}}{1 + h_{oe}(R_L + R_e)} \tag{8-67}$$

If the inequality (8-65) is satisfied, then $h_{oe}R_e \ll h_{fe}$, and the exact expression (8-67) reduces to $A_I \approx -h_{fe}$, in agreement with Eq. (8-58).

The exact expression for the input resistance is, from Fig. 8-28b and Table 8-7,

$$R_i = \frac{V_i}{I_b} = (1 - A_I)R_e + h_{ie} + h_{re}A_IR_L' \tag{8-68}$$

where R_L' is given by Eq. (8-64). Usually, the third term on the right-hand side can be neglected, compared with the other two terms. The exact expression for the voltage amplification is

$$A_V = \frac{A_IR_L}{R_i} \tag{8-69}$$

where the exact values for A_I and R_i from Eqs. (8-67) and (8-68) must be used.

The exact expression for the output impedance (with R_L considered external to the amplifier) is found, as outlined in Prob. 8-47, to be

$$R_o = \frac{1}{h_{oe}} \frac{(1 + h_{fe})R_e + (R_s + h_{ie})(1 + h_{oe}R_e)}{R_e + R_s + h_{ie} - h_{re}h_{fe}/h_{oe}} \tag{8-70}$$

Note that, if $R_e \gg R_s + h_{ie}$, then

$$R_o \approx \frac{1 + h_{fe}}{h_{oe}} + \frac{(R_s + h_{ie})(1 + h_{oe}R_e)}{h_{oe}R_e}$$

$$= \frac{1}{h_{ob}} + (R_s + h_{ie})\left(1 + \frac{1}{h_{oe}R_e}\right) \tag{8-71}$$

where the conversion formula (Table 8-3) from the CE to the CB h parameters is used. Since $1/h_{ob} \approx 2$ M, we see that the addition of an emitter resistor greatly increases the output resistance of a CE stage. This statement is true even if R_e is of the same order of magnitude as R_s and h_{ie}. For example, for $R_e = R_s = 1$ K, and using the h-parameter values in Table 8-3, we find from Eq. (8-70) that $R_o = 817$ K, which is at least ten times the output resistance for an amplifier with $R_e = 0$ (Fig. 8-16d).

The Effect of a Collector-circuit Resistor in an Emitter Follower Consider a CC stage with an emitter resistor R_e and also a collector resistor R_c. The resistor R_c is frequently added in the circuit to protect the transistor against an accidental short circuit across R_e or against a large input-voltage swing.

From Fig. 8-28 we see that the relationship between the CE current gain A_{Ie} (designated simply A_I in the figure) and the CC current gain A_{Ic} is

$$A_{Ic} = 1 - A_{Ie} \tag{8-72}$$

where

$$A_{Ic} = -\frac{I_e}{I_b} \quad \text{and} \quad A_{Ie} = -\frac{I_c}{I_b}$$

Substituting Eq. (8-67) in Eq. (8-72), with R_L replaced by R_c, we obtain the exact expression

$$A_{Ic} = \frac{1 + h_{oe}R_c + h_{fe}}{1 + h_{oe}(R_c + R_e)} \qquad (8\text{-}73)$$

The value of R_i is obtained from Eq. (8-68), with A_I replaced by A_{Ie} and R_L by R_c. The voltage gain of the emitter follower with R_c present in the collector circuit is obtained as follows:

$$A_V = \frac{V_o}{V_i} = A_{Ic}\frac{R_e}{R_i} \qquad (8\text{-}74)$$

Subject to the restriction $h_{oe}(R_c + R_e) \ll 1$, the approximate formulas given in the third column of Table 8-7 are valid, and the protection resistor R_c has no effect on the small-signal operation of the emitter follower.

8-16   HIGH–INPUT–RESISTANCE TRANSISTOR CIRCUITS[7]

In some applications the need arises for an amplifier with a high input imped-ance. For input resistances smaller than about 500 K, the emitter follower discussed in Sec. 8-8 is satisfactory. To achieve larger input impedances, the circuit shown in Fig. 8-29a, called the *Darlington connection*, is used.†
Note that two transistors form a composite pair, the input resistance of the second transistor constituting the emitter load for the first. More specifically, the Darlington circuit consists of two cascaded emitter followers with infinite emitter resistance in the first stage, as shown in Fig. 8-29b.

The Darlington composite emitter follower will be analyzed by referring

† For many applications the field-effect transistor (Chap. 10), with its extremely high input impedance, would be preferred to the Darlington pair.

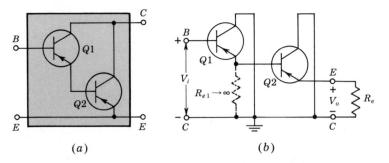

(a)   (b)

Fig. 8-29 (a) Darlington pair. Some vendors package this device as a single composite transistor with only three external leads. (b) The Darlington circuit drawn as two cascaded CC stages.

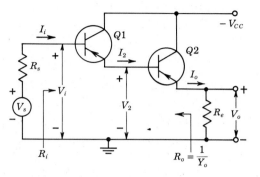

Fig. 8-30   Darlington emitter
follower.

to Fig. 8-30.   Assuming that $h_{oe}R_e \leq 0.1$ and $h_{fe}R_e \gg h_{ie}$, we have, from Table 8-7, for the current gain and the input impedance of the second stage,

$$A_{I2} = \frac{I_0}{I_2} \approx 1 + h_{fe} \qquad R_{i2} \approx (1 + h_{fe})R_e \qquad (8\text{-}75)$$

Since the effective load for transistor $Q1$ is R_{i2}, which usually does not meet the requirement $h_{oe}R_{i2} \leq 0.1$, we must use the exact expression of Eq. (8-40) for the current gain of the first transistor:

$$A_{I1} = \frac{I_2}{I_i} = \frac{1 + h_{fe}}{1 + h_{oe}R_{i2}} = \frac{1 + h_{fe}}{1 + h_{oe}(1 + h_{fe})R_e} \qquad (8\text{-}76)$$

and since $h_{oe}R_e \leq 0.1$, we have

$$A_{I1} \approx \frac{1 + h_{fe}}{1 + h_{oe}h_{fe}R_e} \qquad (8\text{-}77)$$

The overall current gain for Fig. 8-30 is

$$A_I = \frac{I_0}{I_i} = \frac{I_0}{I_2}\frac{I_2}{I_i} = A_{I2}A_{I1}$$

or

$$A_I \approx \frac{(1 + h_{fe})^2}{1 + h_{oe}h_{fe}R_e} \qquad (8\text{-}78)$$

Similarly, for the input resistance of $Q1$, we must use Eq. (8-41):

$$R_{i1} = h_{ie} + A_{I1}R_{i2} \approx \frac{(1 + h_{fe})^2 R_e}{1 + h_{oe}h_{fe}R_e} \qquad (8\text{-}79)$$

This equation for the input resistance of the Darlington circuit is valid for $h_{oe}R_e \leq 0.1$, and should be compared with the input resistance of the single-stage emitter follower given by Eq. (8-55).   If $R_e = 4$ K, and using the h parameters of Table 8-2, we obtain $R_{i2} = 205$ K for the emitter follower and $R_{i1} = 1.73$ M for the Darlington circuit.   We also find $A_I = 427$, which is much higher than the current gain of the emitter follower ($= 51$).

We have assumed in the above computations that the h parameters of $Q1$ and $Q2$ are identical. In reality, this is usually not the case, because the h parameters depend on the quiescent conditions of $Q1$ and $Q2$. Since the emitter current of $Q1$ is the base current of $Q2$, the quiescent current of the first stage is much smaller than that of the second. From Fig. 8-8 we see that h_{fe} does not vary drastically with current, and hence we have assumed

$$h_{fe1} = h_{fe2} = h_{fe}$$

in the above equations. The symbol h_{oe}, which increases rapidly with current, refers to $Q1$ in these equations. In Chap. 11 we show that the short-circuit input resistance varies approximately inversely with collector current [Eq. (11-11)]. Since the current in $Q2$ is $1 + h_{fe}$ times the current in $Q1$, then $h_{ie1} \approx (1 + h_{fe})h_{ie2}$. Using this relationship, we find (Prob. 8-50) that the voltage gain of the Darlington emitter follower is

$$A_V \approx 1 - \frac{h_{ie2}}{R_{i2}} (2 + h_{oe}h_{fe}R_e) \qquad (8\text{-}80)$$

and the output impedance is

$$R_o \approx \frac{R_s}{(1 + h_{fe})^2} + \frac{2h_{ie2}}{1 + h_{fe}} \qquad (8\text{-}81)$$

We conclude from the above equations that the Darlington emitter follower has a higher input resistance and a voltage gain less close to unity than does a single-stage emitter follower. The output impedance of the Darlington circuit may be greater or smaller than that of a single-transistor emitter follower, depending upon the value of R_s relative to h_{ie2}. If $R_s = 0$, then R_o for the Darlington combination is twice R_o for a single-stage follower, as can be seen by comparing Eq. (8-81) with Eq. (8-57).

A major drawback of the Darlington transistor pair is that the leakage current of the first transistor is amplified by the second. Hence the overall leakage current may be high and a Darlington connection of three or more transistors is usually impractical.

The composite transistor pair of Fig. 8-29a can, of course, be used as a common-emitter amplifier. The advantage of this pair would be a very high overall h_{fe}, nominally equal to the product of the CE short-circuit current gains of the two transistors. In fact, Darlington integrated transistor pairs are commercially available with h_{fe} as high as 30,000.

The Biasing Problem In discussing the Darlington transistor pair, we have emphasized its value in providing high-input impedance. However, we have oversimplified the problem by disregarding the effect of the biasing arrangement used in the circuit. Figure 8-31a shows a typical biasing network of resistors R_1, R_2, and R_e. The input resistance R_i' of this stage (discussed in detail in Sec. 9-3) consists of $R_i \| R'$, where $R' \equiv R_1 \| R_2$. Assume

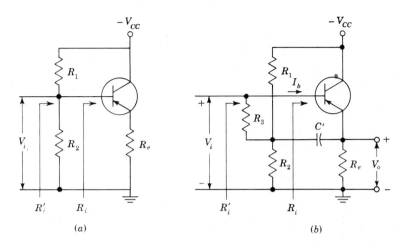

Fig. 8-31 (a) A self-biasing circuit. (b) The bootstrap principle increases the effective value of R_3.

that the input circuit is modified as in Fig. 8-31b by the addition of R_3 but with $C' = 0$ (that is, for the moment, ignore the presence of C').   Now R' is increased to $R_3 + R_1\|R_2$.   However, since R_i is usually much greater than R', it is seen that $R_i' \approx R'$, which may be a few hundred kilohms at most.

To overcome the decrease in the input resistance due to the biasing network, the input circuit of Fig. 8-31b is modified by the addition of C' between the emitter and the junction of R_1 and R_2.   The capacitance C' is chosen large enough to act as a short circuit at the lowest frequency under consideration. Hence the bottom of R_3 is effectively connected to the output (the emitter), whereas the top of R_3 is at the input (the base).   Since the input voltage is V_i and the output voltage is $V_o = A_V V_i$, the circuit of Fig. 8-17 and Miller's theorem can be used to calculate the current drawn by R_3 from the input signal. We can then see that the biasing arrangement R_1, R_2, and R_3 represents an effective input resistance of

$$R_\text{eff} = \frac{R_3}{1 - A_V} \tag{8-82}$$

Since, for an emitter follower, A_V approaches unity, R_eff becomes extremely large.   For example, with $A_V = 0.995$ and $R_3 = 100$ K, we find $R_\text{eff} = 20$ M. Note that the quiescent base current passes through R_3, and hence that a few hundred kilohms is probably an upper limit for R_3.

The above effect, when $A_V \to +1$, is called *bootstrapping*.   The term arises from the fact that, if one end of the resistor R_3 changes in voltage, the other end of R_3 moves through the same potential difference; it is as if R_3 were "pulling itself up by its bootstraps."   The input resistance of the CC amplifier as given by Eq. (8-42) is $R_i = h_{ie}/(1 - A_V)$.   Since this expression is of the form of Eq. (8-82), here is an example of bootstrapping of the resistance h_{ie} which appears between base and emitter.

In making calculations of A_I, R_i, and A_V, we should, in principle, take into account that the emitter follower is loaded, not only by R_e and $R_1 \| R_2$, but also by R_3. The extent to which R_3 loads the emitter follower is calculated as follows: The emitter end of R_3 is at a voltage A_V times as large as the base end of R_3. From Fig. 8-17, illustrating Miller's theorem, the effective resistance seen looking from the emitter to ground is not R_3 but, exaggerated by the Miller effect, is

$$R_{3M} = \frac{A_V R_3}{A_V - 1} \tag{8-83}$$

Since A_V is positive and slightly less than unity, R_{3M} is a (negative) resistance of large magnitude. Since R_{3M} is paralleled with the appreciably smaller (positive) resistors R_e and $R_1 \| R_2$, the effect of R_3 will usually be negligible.

Bootstrapped Darlington Circuit We find in Prob. 8-16 that, even neglecting the effect of the resistors R_1, R_2, and R_3 and assuming infinite emitter resistance, the maximum input resistance is limited to $1/h_{ob} \approx 2$ M. Since $1/h_{ob}$ is the resistance between base and collector, the input resistance can be greatly increased by bootstrapping the Darlington circuit through the addition of C_o between the first collector C_1 and the second emitter E_2, as indicated in Fig. 8-32a. Note that the collector resistor R_{c1} is essential because, without it, R_{e2} would be shorted to ground. If the input signal changes by V_i, then E_2 changes by $A_V V_i$ and (assuming that the reactance of C_o is negligible) the collector changes by the same amount. Hence $1/h_{ob}$ is now effectively increased to $1/(h_{ob})(1 - A_V) \approx 400$ M, for a voltage gain of 0.995.

An expression for the input resistance R_i of the bootstrapped Darlington pair can be obtained using the equivalent circuit of Fig. 8-32b. The effective resistance R_e between E_2 and ground is $R_e = R_{c1} \| R_{e2}$. If $h_{oe} R_e \leq 0.1$, then $Q2$ may be represented by the approximate h-parameter model. However, the exact hybrid model as indicated in Fig. 8-32b must be used for $Q1$. Since $1/h_{oe1} \gg h_{ie2}$, then h_{oe1} may be omitted from this figure. Solving for V_i/I_{b1}, we obtain (Prob. 8-51)

$$R_i \approx h_{fe1} h_{fe2} R_e \tag{8-84}$$

This equation shows that the input resistance of the bootstrapped Darlington emitter follower is essentially equal to the product of the short-circuit current gains and the effective emitter resistance. If $h_{fe1} = h_{fe2} = 50$ and $R_e = 4$ K, then $R_i \approx 10$ M. If transistors with current gains of the order of magnitude of 100 instead of 50 were used, an input resistance of 40 M would be obtained.

The biasing arrangement of Fig. 8-31b would also be used in the circuit of Fig. 8-32. Hence the input resistance, taking into account the bootstrapping both at the base and at the collector of $Q1$, would be $R_{\text{eff}} \| h_{fe1} h_{fe2} R_e$, where R_{eff} is given in Eq. (8-82).

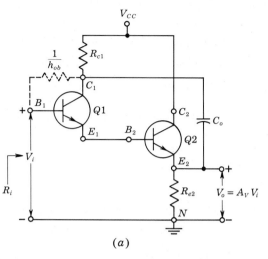

Fig. 8-32 (a) The boot-
strapped Darlington cir-
cuit. (b) The equivalent
circuit.

(a)

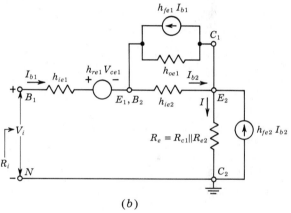

(b)

REFERENCES

1. IRE Standards on Semiconductor Symbols, *Proc. IRE*, vol. 44, pp. 935–937, July, 1956.

2. "Transistor Manual," 7th ed., pp. 52–55, General Electric Co., Syracuse, N.Y., 1964.

3. Ref. 2, pp. 477–482.

4. Millman, J., and H. Taub: "Pulse, Digital, and Switching Waveforms," pp. 528–532, McGraw-Hill Book Company, New York, 1965.

5. Miller, J. M.: Dependence of the Input Impedance of a Three-electrode Vacuum Tube upon the Load in the Plate Circuit, *Natl. Bur. Std. (U.S.) Res. Papers*, vol. 15, no. 351, pp. 367–385, 1919.

6. Dion, D. F.: Common Emitter Transistor Amplifiers, *Proc. IRE*. vol. 46, p. 920, May, 1958.

7. Levine, I.: High Input Impedance Transistor Circuits, *Electronics*, vol. 33, pp. 50–54, September, 1960.

REVIEW QUESTIONS

8-1 A transistor is excited by a large sinusoidal base current whose magnitude exceeds the quiescent value I_B for $0 < \omega t < \pi$ and is less than I_B for $\pi < \omega t < 2\pi$. Is the magnitude of the collector-current variation from the quiescent current greater at $\omega t = \pi/2$ or $3\pi/2$? Explain your answer with the aid of a graphical construction.

8-2 Is nonlinear distortion greater for a sinusoidal-input-base current or for a sinusoidal-input-base voltage? Explain with the aid of the input and output transistor characteristics.

8-3 Define in words and also as a partial derivative (a) h_{ie}; (b) h_{fe}; (c) h_{re}; (d) h_{oe}. Indicate what variable is held constant and give the dimensions of each h parameter.

8-4 Repeat Rev. 8-3 for (a) h_{ic}; (b) h_{fc}; (c) h_{rc}; (d) h_{oc}.

8-5 Repeat Rev. 8-3 for (a) h_{ib}; (b) h_{fb}; (c) h_{rb}; (d) h_{ob}.

8-6 Draw the circuit of a CE transistor configuration and give its h-parameter model.

8-7 Repeat Rev. 8-6 for the CC configuration.

8-8 Repeat Rev. 8-6 for the CB configuration.

8-9 Explain how to obtain from the output characteristic (a) h_{fe}; (b) h_{oe}.

8-10 Prove that (a) $h_{ic} = h_{ie}$; (b) $h_{fc} = -(h_{fe} + 1)$; (c) $h_{oc} = h_{oe}$; (d) $h_{rc} = 1 - h_{re}$.

8-11 Give (for $I_E \approx 1$ mA) the order of magnitude (including the sign) of (a) h_{ib}; (b) h_{ie}; (c) h_{re}; (d) h_{rc}; (e) h_{fb}; (f) h_{fe}; (g) h_{fc}; (h) $1/h_{oe}$; (i) $1/h_{ob}$.

8-12 In terms of the h parameters and the load impedance, derive the expressions for (a) A_I and (b) R_i.

8-13 Derive the expression for A_V in terms of A_I.

8-14 In terms of the h parameters and the source resistance, derive the equation for the output admittance.

8-15 Find (a) A_{Vs} in terms of A_V; (b) A_{Is} in terms of A_I.

8-16 Which of the configurations (CB, CE, CC) has the (a) highest R_i; (b) lowest R_i; (c) highest R_o; (d) lowest R_o; (e) lowest A_V; (f) lowest A_I.

8-17 State (a) Thévenin's theorem; (b) Norton's theorem; (c) the corollaries to these theorems.

8-18 (a) Draw the circuit of an emitter follower. (b) List its three most important characteristics.

8-19 (a) State Miller's theorem with the aid of a circuit diagram. (b) Repeat for the dual of Miller's theorem.

8-20 Draw a CE (first) stage cascaded with a CC (second) stage. In terms of A_{V1}, A_{V2}, A_{I1}, and A_{I2} derive the expression for (a) the resultant voltage gain A_V; (b) the resultant current gain A_I.

8-21 It is desired to have a high-gain amplifier with high input impedance and low output impedance.   If a cascade of four stages is used, what configuration should be used for each stage?

8-22   Using the approximate h-parameter model, obtain the expression for a CE circuit for (*a*) A_I; (*b*) R_i; (*c*) A_V; (*d*) R_o.

8-23   Repeat Rev. 8-22 for the emitter-follower circuit.

8-24   Repeat Rev. 8-22 for the CE circuit with an emitter resistor.

8-25   Repeat Rev. 8-22 for the emitter follower with a collector resistor.

8-26   Draw the equivalent circuit of a CE circuit with an emitter resistor (or of an emitter follower), looking into (*a*) the base and (*b*) the emitter.

8-27   Find a circuit with a grounded emitter which is equivalent to a CE circuit having an emitter resistor.   The new circuit will have resistors added into the base and collector, and these new resistors will depend upon A_I.

8-28   (*a*) Draw a Darlington emitter follower.   (*b*) Explain why the input impedance is higher than that of a single-stage emitter follower.

8-29   (*a*) Indicate the circuit of an emitter follower with biasing resistors R_1 and R_2.   Show that the input resistance is reduced because of these biasing resistors. (*b*) Add a bootstrapped resistor R_3 and explain how this increases the input resistance.

9 / TRANSISTOR BIASING AND THERMAL STABILIZATION

This chapter presents methods for establishing the quiescent operating point of a transistor amplifier in the active region of the characteristics. The operating point shifts with changes in temperature T because the transistor parameters (β, I_{CO}, and V_{BE}) are functions of T. A criterion is established for comparing the stability of different biasing circuits. Compensation techniques are also presented for quiescent-point stabilization.

9-1   THE OPERATING POINT

From our discussion of transistor characteristics in Secs. 5-6 to 5-8, it is clear that the transistor functions most linearly when it is constrained to operate in its active region. To establish an operating point in this region it is necessary to provide appropriate direct potentials and currents, using external sources. Once an operating point Q is established, such as the one shown in Fig. 8-2a, time-varying excursions of the input signal (base current, for example) should cause an output signal (collector voltage or collector current) of the same waveform. If the output signal is not a faithful reproduction of the input signal, for example, if it is clipped on one side, the operating point is unsatisfactory and should be relocated on the collector characteristics. The question now naturally arises as to how to choose the operating point. In Fig. 9-1 we show a common-emitter circuit. Figure 9-2 gives the output characteristics of the transistor used in Fig. 9-1. Note that even if we are free to choose R_c, R_L, R_b, and V_{CC}, we may not operate the transistor everywhere in the active region because the various transistor ratings limit the range of useful operation. These ratings (listed in the manufacturer's specification sheets) are maximum

282

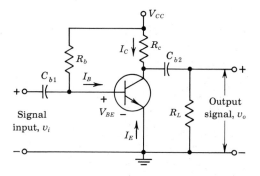

Fig. 9-1   The fixed-bias circuit.

collector dissipation $P_{C,\max}$, maximum collector voltage $V_{C,\max}$, maximum collector current $I_{C,\max}$, and maximum emitter-to-base voltage $V_{EB,\max}$. Figure 9-2 shows three of these bounds on typical collector characteristics.

Capacitive Coupling   Note that in the circuit of Fig. 8-1, neither side of the signal generator is grounded, and also that an auxiliary biasing supply V_{BB} is used.   Both of these difficulties are avoided by using a capacitor C_{b1} to couple the input signal to the transistor, as indicated in Fig. 9-1.   In this diagram one end of v_i is at ground potential, and the collector supply V_{CC} also provides the biasing base current I_B.   Under quiescent conditions (no input signal), C_{b1} (called a *blocking* capacitor) acts as an open circuit, because the reactance of a capacitor is infinite at zero frequency (dc).   The capacitances C_{b1} and C_{b2} are chosen large enough so that, at the lowest frequency of excitation, their reactances are small enough so that they can be considered to be short circuits.   These coupling capacitors block dc voltages but freely pass signal voltages.   For example, the quiescent collector voltage does not appear at the output, but v_o is an amplified replica of the input signal v_i.   The (ac or incremental) output signal voltage may be applied to the input of another amplifier without affecting its bias, because of the blocking capacitor C_{b2}.   The effect of the finite size of a blocking capacitor on the frequency response of an amplifier is considered in Sec. 12-8.

The Static and Dynamic Load Lines   We noted above that under dc conditions C_{b2} acts as an open circuit.   Hence the quiescent collector current and voltage are obtained by drawing a static (dc) load line corresponding to the resistance R_c through the point $i_C = 0$, $v_{CE} = V_{CC}$, as indicated in Fig. 9-2. If $R_L = \infty$ and if the input signal (base current) is large and symmetrical, we must locate the operating point Q_1 at the center of the load line.   In this way the collector voltage and current may vary approximately symmetrically around the quiescent values V_C and I_C, respectively.   If $R_L \neq \infty$, however, a *dynamic* (ac) load line must be drawn.   Since we have assumed that, at the signal frequency, C_{b2} acts as a short circuit, the effective load R'_L at the collector is R_c in parallel with R_L.   The dynamic load line must be drawn through

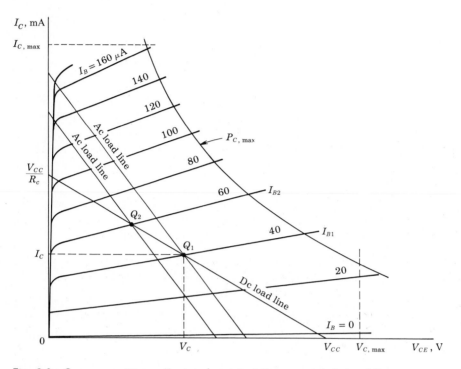

Fig. 9-2 Common-emitter collector characteristics; ac and dc load lines.

the operating point Q_1 and must have a slope corresponding to $R'_L = R_c \| R_L$. This ac load line is indicated in Fig. 9-2, where we observe that the input signal may swing a maximum of approximately 40 μA, around Q_1 because, if the base current decreases by more than 40 μA, the transistor is driven off.

If a larger input swing is available, then in order to avoid cutoff during a part of the cycle, the quiescent point must be located at a higher current. For example, by simple trial and error we locate Q_2 *on the dc load line* such that a line with a slope corresponding to the ac resistance R'_L and drawn through Q_2 gives as large an output as possible without too much distortion. In Fig. 9-2 the choice of Q_2 allows an input peak current swing of about 60 μA.

The Fixed-bias Circuit The point Q_2 can be established by noting the required current I_{B2} in Fig. 9-2 and choosing the resistance R_b in Fig. 9-1 so that the base current is equal to I_{B2}. Therefore

$$I_B = \frac{V_{CC} - V_{BE}}{R_b} = I_{B2} \tag{9-1}$$

The voltage V_{BE} across the forward-biased emitter junction is (Table 5-1, page 142) approximately 0.2 V for a germanium transistor and 0.7 V for a silicon transistor in the active region.   Since V_{CC} is usually much larger than V_{BE}, we have

$$I_B \approx \frac{V_{CC}}{R_b} \qquad (9\text{-}2)$$

The current I_B is constant, and the network of Fig. 9-1 is called the *fixed-bias circuit*.   In summary, we see that the selection of an operating point Q depends upon a number of factors.   Among these factors are the ac and dc loads of the stage, the available power supply, the maximum transistor ratings, the peak signal excursions to be handled by the stage, and the tolerable distortion.

9-2   BIAS STABILITY

In the preceding section we examined the problem of selecting an operating point Q on the load line of the transistor.   We now consider some of the problems of maintaining the operating point stable.

Let us refer to the biasing circuit of Fig. 9-1.   In this circuit the base current I_B is kept constant since $I_B \approx V_{CC}/R_b$.   Let us assume that the transistor of Fig. 9-1 is replaced by another of the same type.   In spite of the tremendous strides that have been made in the technology of the manufacture of semiconductor devices, transistors of a particular type still come out of production with a wide spread in the values of some parameters.   For example, Fig. 5-15 shows a range of $h_{FE} \approx \beta$ of about 3 to 1.   To provide information about this variability, a transistor data sheet, in tabulating parameter values, often provides columns headed minimum, typical, and maximum.

In Sec. 5-6 we see that the spacing of the output characteristics will increase or decrease (for equal changes in I_B) as β increases or decreases.   In Fig. 9-3 we have assumed that β is greater for the replacement transistor of Fig. 9-1, and since I_B is maintained constant at I_{B2} by the external biasing circuit, it follows that the operating point will move to Q_2.   This new operating point may be completely unsatisfactory.   Specifically, it is possible for the transistor to find itself in the saturation region.   We now conclude that maintaining I_B constant will not provide operating-point stability as β changes.   On the contrary, I_B should be allowed to change so as to maintain I_C and V_{CE} constant as β changes.

Thermal Instability   A second very important cause for bias instability is a variation in temperature.   In Sec. 5-7 we note that the reverse saturation current I_{CO}† changes greatly with temperature.   Specifically, I_{CO} doubles for every 10°C rise in temperature.   This fact may cause considerable practical difficulty in using a transistor as a circuit element.   For example, the collector

---

† Throughout this chapter I_{CBO} is abbreviated I_{CO} (Sec. 5-7).

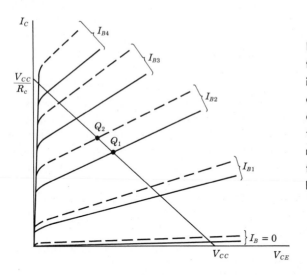

Fig. 9-3 Graphs showing the collector character-istics for two transistors of the same type. The dashed characteristics are for a transistor whose β is much larger than that of the transistor represented by the solid curves.

current I_C causes the collector-junction temperature to rise, which in turn increases I_{CO}. As a result of this growth of I_{CO}, I_C will increase [Eq. (5-16)], which may further increase the junction temperature, and consequently I_{CO}. It is possible for this succession of events to become cumulative, so that the ratings of the transistor are exceeded and the device burns out.

Even if the drastic state of affairs described above does not take place, it is possible for a transistor which was biased in the active region to find itself in

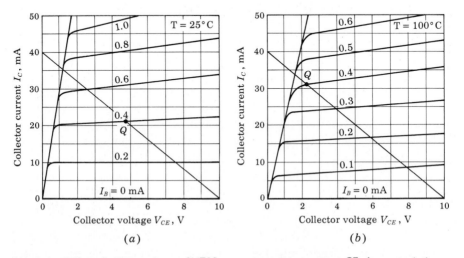

Fig. 9-4 Diffused silicon planar 2N708 n-p-n transistor output CE characteristics for (a) 25°C and (b) 100°C. (Courtesy of Fairchild Semiconductor.)

the saturation region as a result of this operating-point instability (Sec. 9-9). To see how this may happen, we note that if $I_B = 0$, then, from Eq. (5-16), $I_C = I_{CO}(1 + \beta)$. As the temperature increases, I_{CO} increases, and even if we assume that β remains constant (actually, it also increases), it is clear that the $I_B = 0$ line in the CE output characteristics will move upward. The characteristics for other values of I_B will also move upward by the same amount (provided that β remains constant), and consequently the operating point will move if I_B is forced to remain constant. In Fig. 9-4 we show the output characteristics of the 2N708 transistor at temperatures of $+25$ and $+100°C$. This transistor, used in the circuit of Fig. 9-1 with $V_{CC} = 10$ V, $R_c = 250$ Ω, $R_b = 24$ K, operates at Q with $I_B = (10 - 0.7)/24 \approx 0.4$ mA. Hence it would find itself almost in saturation at a temperature of $+100°C$, even though it would be biased in the middle of its active region at $+25°C$.

9-3   SELF–BIAS, OR EMITTER BIAS

A circuit which is used to establish a stable operating point is the self-biasing configuration of Fig. 9-5a. The current in the resistance R_e in the emitter lead causes a voltage drop which is in the direction to reverse-bias the emitter junction. Since this junction must be forward-biased, the base voltage is obtained from the supply through the R_1R_2 network.

    The physical reason for an improvement in stability with this circuit is the following: If I_C tends to increase, say, because I_{CO} has risen as a result of an elevated temperature, the current in R_e increases. As a consequence of the increase in voltage drop across R_e, the base current is decreased. Hence I_C will increase less than it would have, had there been no self-biasing resistor R_e.

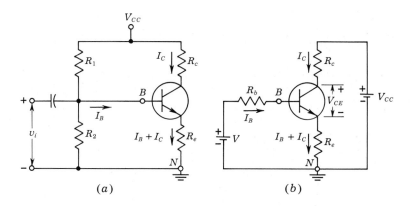

                              (a)   (b)

Fig. 9-5   (a) A self-biasing circuit.   (b) Simplification of the base circuit in (a) by the use of Thévenin's theorem.

Analysis of the Self-bias Circuit If the circuit component values in Fig. 9-5a are specified, the quiescent point is found as follows: Kirchhoff's voltage law around the collector circuit yields

$$-V_{CC} + I_C(R_c + R_e) + I_B R_e + V_{CE} = 0 \qquad (9\text{-}3)$$

If the drop in R_e due to I_B is neglected compared with that due to I_C, then this relationship between I_C and V_{CE} is a straight line whose slope corresponds to $R_c + R_e$ and whose intercept at $I_C = 0$ is $V_{CE} = V_{CC}$. This load line is drawn on the collector characteristics.

If the circuit to the left between the base B and ground N terminals in Fig. 9-5a is replaced by its Thévenin equivalent, the two-mesh circuit of Fig. 9-5b is obtained, where

$$V \equiv \frac{R_2 V_{CC}}{R_2 + R_1} \qquad R_b \equiv \frac{R_2 R_1}{R_2 + R_1} \qquad (9\text{-}4)$$

Obviously, R_b is the effective resistance seen looking back from the base terminal. Kirchhoff's voltage law around the base circuit yields

$$V = I_B R_b + V_{BE} + (I_B + I_C)R_e \qquad (9\text{-}5)$$

If I_C from Eq. (9-5) is substituted into Eq. (9-3), a relationship between I_B and V_{CE} results. For each value of I_B given on the collector curves, V_{CE} is calculated. The locus of these corresponding points V_{CE} and I_B plotted on the CE output characteristics is called the *bias curve*. The intersection of the load line and the bias curve gives the quiescent point.

In many cases transistor characteristics are not available but β is known. Then the calculation of the Q point may be carried out analytically as follows: In the active region the collector current is given by Eq. (5-16), namely,

$$I_C = \beta I_B + (1 + \beta)I_{CO} \qquad (9\text{-}6)$$

Equations (9-5) and (9-6) can now be solved for I_B and I_C (since V_{BE} is known in the active region). Note that with this method the currents (in the active region) are determined by the base circuit and the values of β and I_{CO}.

EXAMPLE A silicon transistor whose common-emitter output characteristics are shown in Fig. 9-6b is used in the circuit of Fig. 9-5a, with $V_{CC} = 22.5$ V, $R_c = 5.6$ K, $R_e = 1$ K, $R_2 = 10$ K, and $R_1 = 90$ K. For this transistor, $\beta = 55$. Find the Q point (a) graphically and (b) from the known value of β.

Solution   a. From Eqs. (9-4) we have

$$V = \frac{10 \times 22.5}{100} = 2.25 \text{ V} \qquad R_b = \frac{10 \times 90}{100} = 9.0 \text{ K}$$

The equivalent circuit is shown in Fig. 9-6a. The load line corresponding to a total resistance of 6.6 K and a supply of 22.5 V is drawn on the collector charac-

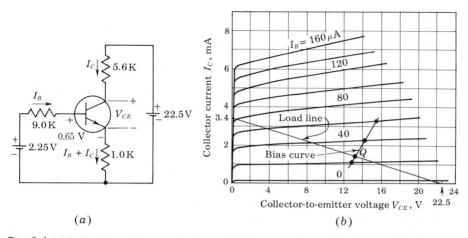

Fig. 9-6 (a) An illustrative example. (b) The intersection of the load line and the bias curve determines the Q point.

teristics of Fig. 9-6b. Kirchhoff's voltage law applied to the collector and base circuits, respectively, yields (with $V_{BE} = 0.65$)

$$-22.5 + 6.6I_C + I_B + V_{CE} = 0 \tag{9-7}$$

$$0.65 - 2.25 + I_C + 10.0I_B = 0 \tag{9-8}$$

Eliminating I_C from these two equations, we find the bias curve equation

$$V_{CE} = 65.0I_B + 11.9$$

Values of V_{CE} corresponding to $I_B = 20$, 40, and 60 μA are obtained from this equation and are plotted in Fig. 9-6b. We see that the intersection of the bias curve and the load line occurs at $V_{CE} = 13.3$ V, $I_C = 1.4$ mA, and from the bias-curve equation, $I_B = 26$ μA.

b. In many cases transistor characteristics are not available but β is known. Then the calculation of the Q point may be carried out as follows: For base currents large compared with the reverse saturation current ($I_B \gg I_{CO}$), it follows from Eq. (9-6) that

$$I_C = \beta I_B \tag{9-9}$$

This equation can now be used in place of the collector characteristics. Since $\beta = 55$ for the transistor used in this example, substituting $I_B = I_C/55$ in Eq. (9-8) for the base circuit yields

$$-1.60 + I_C + \tfrac{10}{55}I_C = 0$$

or

$$I_C = 1.36 \text{ mA} \quad \text{and} \quad I_B = \frac{I_C}{55} = \frac{1.36}{55} \text{ mA} = 24.8 \text{ } \mu\text{A}$$

These values are very close to those found from the characteristics.

The collector-to-emitter voltage can be found from Eq. (9-7) and the known values of I_B and I_C:

$$-22.5 + 6.6 \times 1.36 + 0.025 + V_{CE} = 0$$

or

$$V_{CE} = 13.5 \text{ V}$$

9-4   STABILIZATION AGAINST VARIATIONS IN I_{CO}, V_{BE}, AND β

The sources of instability of I_C are essentially three.[1] These are the reverse saturation current I_{CO}, which doubles for every 10°C increase in temperature; the base-to-emitter voltage V_{BE}, which decreases at the rate of 2.5 mV/°C for both Ge and Si transistors; and β, which increases with temperature (Tables 9-1 and 9-2).

We shall neglect the effect of the change of V_{CE} with temperature, because this variation is very small and we assume that the transistor operates in the active region, where I_C is approximately independent of V_{CE}.

The Transfer Characteristic The output current I_C is plotted in Fig. 9-7 as a function of input voltage for the germanium transistor, type 2N1631. This transfer characteristic for a silicon transistor is given in Fig. 5-17. Each curve shifts to the left at the rate of 2.5 mV/°C (at constant I_C) for increasing temperature. We now examine in detail the effect of the shift in transfer characteristics and the variation of β and I_{CO} with temperature. If Eq. (9-5), obtained by applying KVL around the base circuit of the self-bias circuit of

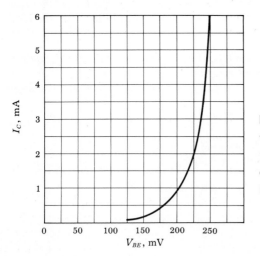

Fig. 9-7 Transfer characteristic for the 2N1631 germanium p-n-p alloy-type transistor at $V_{CE} = -9$ V and $T_A = 25°C$. (Courtesy of Radio Corp. of America.)

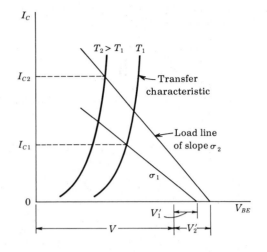

Fig. 9-8   Illustrating that the collector current varies with temperature T because V_{BE}, I_{CO}, and β change with T.

Fig. 9-5b, is combined with Eq. (9-6), which represents the collector characteristics in the active region, the result is

$$V_{BE} = V + (R_b + R_e)\frac{\beta + 1}{\beta}I_{CO} - \frac{R_b + R_e(1 + \beta)}{\beta}I_C \qquad (9\text{-}10)$$

Equation (9-10) represents a load line in the I_C-V_{BE} plane, and is indicated in Fig. 9-8. The intercept on the V_{BE} axis is $V + V'$, where

$$V' = (R_b + R_e)\frac{\beta + 1}{\beta}I_{CO} \approx (R_b + R_e)I_{CO} \qquad (9\text{-}11)$$

since $\beta \gg 1$. If at $T = T_1(T_2)$, $I_{CO} = I_{CO1}(I_{CO2})$ and $\beta = \beta_1(\beta_2)$, then $V_1' \approx (R_b + R_e)I_{CO1}$ and $V_2' \approx (R_b + R_e)I_{CO2}$. Hence the intercept of the load line on the V_{BE} axis is a function of temperature because I_{CO} increases with T. The slope of the load line is

$$\sigma = \frac{-\beta}{R_b + R_e(1 + \beta)}$$

and hence $|\sigma|$ increases with T because β increases with T. The transfer characteristic for $T = T_2 > T_1$ shifts to the left of the corresponding curve for $T = T_1$ because V_{BE} (at constant I_C) varies with T, as indicated above. The intersection of the load line with the transfer characteristic gives the collector current I_C. We see that $I_{C2} > I_{C1}$ because I_{CO}, β, and V_{BE} all vary with temperature.

Since, from Eq. (9-10), I_C is a function of I_{CO}, V_{BE}, and β, it is convenient to introduce the three partial derivatives of I_C with respect to these variables. These derivatives are called the *stability factors* S, S', and S'' and are defined as follows.

The Stabilization Factor S We define S as the rate of change of collector current with respect to the reverse saturation current, keeping β and V_{BE} constant, or

$$S \equiv \frac{\partial I_C}{\partial I_{CO}} \approx \frac{\Delta I_C}{\Delta I_{CO}} \tag{9-12}$$

The larger the value of S, the more likely the circuit is to exhibit thermal instability.†   Using the above definition and Eq. (9-10), we find

$$S = (1 + \beta) \frac{1 + R_b/R_e}{1 + \beta + R_b/R_e} \tag{9-13}$$

Note that S varies between 1 for small R_b/R_e and $1 + \beta$ for $R_b/R_e \to \infty$.   If $\beta + 1 \gg R_b/R_e$, Eq. (9-13) reduces to

$$S \approx 1 + \frac{R_b}{R_e} \tag{9-14}$$

Thus, for constant β, V_{BE}, and small S, we have

$$\frac{\Delta I_C}{I_C} \approx S \frac{\Delta I_{CO}}{I_C} \approx \frac{\Delta I_{CO}}{I_C} + \frac{R_b}{R_e} \frac{\Delta I_{CO}}{I_C} \tag{9-15}$$

For the typical design, $R_b/R_e > 1$, making the second term in Eq. (9-15) larger than the first.   The denominator of the second term is the dc voltage drop across R_e (since $|I_C| \approx |I_E|$) and is under the circuit designer's control.

The Stability Factor S' The variation of I_C with V_{BE} is given by the stability factor S', defined by

$$S' \equiv \frac{\partial I_C}{\partial V_{BE}} \approx \frac{\Delta I_C}{\Delta V_{BE}} \tag{9-16}$$

where both I_{CO} and β are considered constant.   From Eq. (9-10) we find

$$S' = \frac{-\beta}{R_b + R_e(1 + \beta)} = \frac{-\beta/R_e}{1 + \beta + R_b/R_e} \tag{9-17}$$

If we again assume that $\beta + 1 \gg R_b/R_e$, and also that $\beta \gg 1$, then Eq. (9-17) reduces to

$$S' \approx \frac{\Delta I_C}{\Delta V_{BE}} \approx -\frac{1}{R_e} \tag{9-18}$$

or

$$\frac{\Delta I_C}{I_C} \approx \frac{S' \Delta V_{BE}}{I_C} \approx -\frac{\Delta V_{BE}}{I_C R_e} \tag{9-19}$$

From the above equation and Eq. (9-15) we see that the dominant factor in stabilizing against I_{CO} and V_{BE} is the quiescent voltage drop across the emitter resistance R_e.   The larger this drop, the smaller is the percentage change in collector current due to ΔI_{CO} and ΔV_{BE}.

† In this sense, S should more properly be called an *instability factor*.

The Stability Factor S''   The variation of I_C with respect to β is given by the stability factor S'', defined by

$$S'' \equiv \frac{\partial I_C}{\partial \beta} \approx \frac{\Delta I_c}{\Delta \beta} \tag{9-20}$$

where both I_{CO} and V_{BE} are considered constant.   From Eq. (9-10)

$$I_C = \frac{\beta(V + V' - V_{BE})}{R_b + R_e(1 + \beta)} \tag{9-21}$$

where, from Eq. (9-11), V' may be taken to be independent of β.   We obtain, after differentiation and some algebraic manipulation,

$$S'' = \frac{\partial I_C}{\partial \beta} = \frac{I_C S}{\beta(1 + \beta)} \tag{9-22}$$

The change in collector current due to a change in β is

$$\Delta I_C \approx S'' \, \Delta \beta = \frac{I_C S}{\beta(1 + \beta)} \Delta \beta \tag{9-23}$$

where $\Delta \beta = \beta_2 - \beta_1$ may represent a large change in β.   Hence it is not clear whether to use β_1, β_2, or perhaps some average value of β in the expressions for S''.   (This problem does not arise for S or S' because these factors are almost independent of β.)   This difficulty is avoided if S'' is obtained by taking finite differences rather than by evaluating a derivative.   Thus

$$S'' \approx \frac{I_{C2} - I_{C1}}{\beta_2 - \beta_1} = \frac{\Delta I_C}{\Delta \beta} \tag{9-24}$$

From Eq. (9-21) we have

$$\frac{I_{C2}}{I_{C1}} = \frac{\beta_2}{\beta_1} \frac{R_b + R_e(1 + \beta_1)}{R_b + R_e(1 + \beta_2)} \tag{9-25}$$

Subtracting unity from both sides of Eq. (9-25) yields

$$\frac{I_{C2}}{I_{C1}} - 1 = \left(\frac{\beta_2}{\beta_1} - 1\right) \frac{R_b + R_e}{R_b + R_e(1 + \beta_2)} \tag{9-26}$$

or

$$S'' = \frac{\Delta I_C}{\Delta \beta} = \frac{I_{C1} S_2}{\beta_1(1 + \beta_2)} \tag{9-27}$$

where S_2 is the value of the stabilizing factor S when $\beta = \beta_2$ as given by Eq. (9-13).   Note that this equation reduces to Eq. (9-22) as $\Delta \beta = \beta_2 - \beta_1 \to 0$.

If we assume that S_2 is small so that the approximate value given in Eq. (9-14) is valid, then from Eq. (9-27) with $\beta \gg 1$ we find

$$\frac{\Delta I_C}{I_{C1}} \approx \left(1 + \frac{R_b}{R_e}\right) \frac{\Delta \beta}{\beta_1 \beta_2} = \left(1 + \frac{R_b}{R_e}\right) \frac{\beta_2/\beta_1 - 1}{\beta_2} \tag{9-28}$$

It is clear that R_b/R_e should be kept small. Also, for a given spread in the value of β (say, $\beta_2/\beta_1 = 3$), a high-β circuit will be more stable than one using a lower-β transistor.

Equation (9-27) is of prime importance because it allows us to determine the maximum value of S_2 for a given spread in β and a given I_{C1}. This variation in β may be due to any cause such as a temperature change, a transistor replacement, etc.

EXAMPLE Transistor type 2N335, used in the circuit of Fig. 9-5a, may have any value of β between 36 and 90 at a temperature of 25°C, and the leakage current I_{CO} has negligible effect on I_C at room temperature. Find R_e, R_1, and R_2 subject to the following specifications: $R_c = 4$ K, $V_{CC} = 20$ V; the nominal bias point is to be at $V_{CE} = 10$ V, $I_C = 2$ mA; and I_C should be in the range 1.75 to 2.25 mA as β varies from 36 to 90.

Solution   From the collector circuit (with $I_C \gg I_B$)

$$R_c + R_e = \frac{V_{CC} - V_{CE}}{I_C} = \frac{20 - 10}{2} = 5 \text{ K}$$

Hence $R_e = 5 - 4 = 1$ K.

From Eq. (9-27) we can solve for S_2. Hence, with $\Delta I_C = 2.25 - 1.75 = 0.5$ mA, $I_{C1} = 1.75$ mA, $\beta_1 = 36$, $\beta_2 = 90$, and $\Delta\beta = 54$, we obtain

$$\frac{0.5}{54} = \frac{1.75}{36} \frac{S_2}{1 + 90}$$

or

$$S_2 = 17.3$$

Substituting $S_2 = 17.3$, $R_e = 1$ K, and $\beta_2 = 90$ in Eq. (9-13) yields

$$(17.3)(91 + R_b) = 91(1 + R_b)$$

or

$$R_b = 20.1 \text{ K}$$

From Eq. (9-10), with $I_C = 1.75$ mA, $\beta = 36$, $R_b = 20.1$ K, $R_e = 1$ K, $V_{BE} = 0.65$ V, and $I_{CO} = 0$, we obtain

$$V = V_{BE} + \frac{R_b + R_e(1 + \beta)}{\beta} I_C = 0.65 + \left(\frac{20.1 + 37}{36}\right)(1.75) = 3.43 \text{ V}$$

From Eqs. (9-4), solving for R_1 and R_2, we find

$$R_1 = R_b \frac{V_{CC}}{V} = (20.1)\left(\frac{20}{3.43}\right) = 117 \text{ K}$$

$$R_2 = \frac{R_1 V}{V_{CC} - V} = \frac{(117)(3.43)}{20 - 3.43} = 24.2 \text{ K}$$

9-5   GENERAL REMARKS ON COLLECTOR–CURRENT STABILITY[2]

Stability factors were defined in the preceding section, which considered the change in collector current with respect to I_{CO}, V_{BE}, and β. These stability factors are repeated here for convenience:

$$S = \frac{\Delta I_C}{\Delta I_{CO}} \qquad S' = \frac{\Delta I_C}{\Delta V_{BE}} \qquad S'' = \frac{\Delta I_C}{\Delta \beta} \tag{9-29}$$

Each differential quotient (partial derivative) is calculated with all other parameters maintained constant.

If we desire to obtain the total change in collector current over a specified temperature range, we can do so by expressing this change as the sum of the individual changes due to the three stability factors. Specifically, by taking the total differential of $I_C = f(I_{CO}, V_{BE}, \beta)$, we obtain

$$\Delta I_C = \frac{\partial I_C}{\partial I_{CO}} \Delta I_{CO} + \frac{\partial I_C}{\partial V_{BE}} \Delta V_{BE} + \frac{\partial I_C}{\partial \beta} \Delta \beta$$

$$= S \, \Delta I_{CO} + S' \, \Delta V_{BE} + S'' \, \Delta \beta \tag{9-30}$$

The stability factors may be expressed in terms of the parameter M defined by

$$M \equiv \frac{1}{1 + R_b/[R_e(1 + \beta)]} \approx \frac{1}{1 + R_b/\beta R_e} \tag{9-31}$$

where we assume $\beta \gg 1$. Note that if $\beta R_e \gg R_b$, then $M \approx 1$. Substituting Eqs. (9-13), (9-17), and (9-28) into Eq. (9-30), we find for the fractional change in collector current

$$\frac{\Delta I_C}{I_{C1}} = \left(1 + \frac{R_b}{R_e}\right) \frac{M_1 \Delta I_{CO}}{I_{C1}} - \frac{M_1 \Delta V_{BE}}{I_{C1} R_e} + \left(1 + \frac{R_b}{R_e}\right) \frac{M_2 \Delta \beta}{\beta_1 \beta_2} \tag{9-32}$$

where $M_1(M_2)$ corresponds to $\beta_1(\beta_2)$. Note that as T increases, $\Delta I_{CO}/I_{C1}$ and $\Delta \beta$ increase, whereas $\Delta V_{BE}/I_{C1}$ decreases. Hence all terms in Eq. (9-32) are positive for increasing T and negative for decreasing T.

We now examine in detail the order of magnitude of the terms of Eq. (9-32) for both silicon and germanium transistors over their entire range of temperature operation as specified by transistor manufacturers. This range usually is -65 to $+75°C$ for germanium transistors and -65 to $+175°C$ for silicon transistors.

TABLE 9-1   Typical silicon transistor parameters

T, °C..........	-65	$+25$	$+175$
I_{CO}, nA.......	1.95×10^{-3}	1.0	$33,000$
β............	25	55	100
V_{BE}, V........	0.78	0.60	0.225

TABLE 9-2 Typical germanium transistor parameters

T, °C	-65	$+25$	$+75$
I_{CO}, μA	1.95×10^{-3}	1.0	32
β	20	55	90
V_{BE}, V	0.38	0.20	0.10

Tables 9-1 and 9-2 show typical parameters of silicon and germanium transistors, each having the same β (55) at room temperature. For Si, I_{CO} is much smaller than for Ge. Note that I_{CO} doubles approximately every 10°C and $|V_{BE}|$ decreases by approximately 2.5 mV/°C.

EXAMPLE For the self-bias circuit of Fig. 9-5a, $R_e = 4.7$ K, $R_b = 7.75$ K, and $R_b/R_e = 1.65$. The collector supply voltage and R_c are adjusted to establish a collector current of 1.5 mA at 25°C. (a) Determine the variation of I_C in the temperature range of -65 to $+175$°C when the silicon transistor of Table 9-1 is used. (b) Repeat (a) for the range -65 to $+75$°C when the germanium transistor of Table 9-2 is used.

Solution a. Since R_b/R_e is known, we can find the percentage change in I_C using Eq. (9-32). At room temperature

$$M_1 = \frac{1}{1 + R_b/\beta_1 R_e} = \frac{1}{1 + 1.65/55} = 0.97 \approx 1$$

Since at 175°C, $\beta_2 = 100$, then M_2 is even closer to unity than M_1. Hence at $T = +175$°C, we shall assume $M_1 = M_2 = 1$. From Eq. (9-32) we have

$$\frac{\Delta I_C(+175°C)}{I_{C1}} = (1 + 1.65) \times \frac{33,000 \times 10^{-9}}{1.5 \times 10^{-3}} + \frac{0.6 - 0.225}{1.5 \times 4.7}$$

$$+ (1 + 1.65) \times \frac{100 - 55}{55 \times 100} = 5.82 + 5.32 + 2.17\,\%$$

or the change in collector current is

$$\Delta I_C(+175°C) = 0.087 + 0.080 + 0.032 = 0.199 \text{ mA}$$

At -65°C, $M_2 = 1/(1 + 1.65/25) = 0.94$, and we shall take this small correction factor into account. From Eq. (9-32) we find

$$\frac{\Delta I_C(-65°C)}{I_{C1}} = -\frac{2.65 \times 10^{-9}}{1.5 \times 10^{-3}} - \frac{0.78 - 0.60}{1.5 \times 4.7} - \frac{2.65 \times (55 - 25) \times 0.94}{25 \times 55}$$

$$= 0 - 2.55 - 5.34\,\%$$

or

$$\Delta I_C(-65°C) = 0 - 0.038 - 0.080 = -0.118 \text{ mA}$$

Therefore, for the silicon transistor, the collector current will be approximately 1.70 mA at $+175$°C and 1.38 mA at -65°C.

b. Similarly for the germanium transistor at $+75°C$, we find with $M_1 \approx M_2 \approx 1$

$$\frac{\Delta I_C(+75°C)}{I_{C1}} = 2.65 \times \frac{31 \times 10^{-6}}{1.5 \times 10^{-3}} + \frac{0.10}{1.5 \times 4.7} + 2.65 \times \frac{35}{55 \times 90}$$

$$= 5.48 + 1.42 + 1.87\%$$

or the change in collector current is

$$\Delta I_C(+75°C) = 0.082 + 0.021 + 0.028 = 0.131 \text{ mA}$$

At $-65°C$ we find, with $M_1 \approx 1$ and $M_2 = 1/(1 + 1.65/20) = 0.93$,

$$\frac{\Delta I_C(-65°C)}{I_{C1}} = -2.65 \times \frac{10^{-6}}{1.5 \times 10^{-3}} - \frac{0.18}{1.5 \times 4.7} - 2.65 \times \frac{35}{20 \times 55} \times 0.93$$

$$= -0.18 - 2.56 - 7.85\%$$

or

$$\Delta I_C(-65°C) = -0.003 - 0.038 - 0.118 = -0.159 \text{ mA}$$

Therefore, for the germanium transistor, the collector current will be approximately 1.63 mA at $+75°C$ and 1.34 mA at $-65°C$.

Practical Considerations The foregoing example illustrates the superiority of silicon over germanium transistors because, approximately, the same change in collector current is obtained for a much higher temperature change in the silicon transistor. In the above example, with $S \approx 1 + R_b/R_e = 2.65$ and $R_e = 4.7$ K, the current change at the extremes of temperature is only about 10 percent. Hence this circuit could be used at temperatures in excess of 75°C for germanium and 175°C for silicon. If S is larger and R_e smaller, the current instability is greater. For example, in Prob. 9-23, we find for $R_e = 1$ K and $S = 7.70$ that the 25°C collector current varies about 30 percent at $-65°C$ and $+75°C$ (Ge) or at $-65°C$ and $+175°C$ (Si). These numerical values illustrate why a germanium transistor is seldom used above 75°C, and a silicon device above 175°C. The importance of keeping S and S' small and β large is clear.

The change in collector current that can be tolerated in any specific application depends on design requirements, such as peak signal voltage required across R_c. We should also point out that the tolerance in bias resistors and supply voltages must be taken into account, in addition to the variation of β, I_{CO}, and V_{BE}.

Our discussion of stability and the results obtained are independent of R_c, and hence they remain valid for $R_c = 0$. If the output is taken across R_e, such a circuit is called an *emitter follower* (Sec. 8-8). If we have a direct-coupled emitter follower *driven from an ideal voltage source*, then $R_b = 0$ and this circuit can be used to a higher temperature than a similar circuit with

$R_b \neq 0$. For the emitter follower, Eq. (9-32) reduces to

$$\frac{\Delta I_C}{I_{C1}} = \frac{\Delta I_{CO}}{I_{C1}} - \frac{\Delta V_{BE}}{I_{C1}R_e} + \frac{\Delta \beta}{\beta_1 \beta_2} \tag{9-33}$$

In the above example the increase in collector current from 25 to 75°C for a germanium transistor is 0.08 mA due to I_{CO}, 0.02 mA due to V_{BE}, and 0.03 mA due to β. Hence, for Ge, the effect of I_{CO} has the dominant influence on the collector current. On the other hand, the increase in I_C for a silicon transistor over the range from 25 to 175°C due to I_{CO} is approximately the same as that due to V_{BE}, but is much smaller due to β. However, if the temperature range is restricted somewhat, say, from 25 to 145°C, then $\Delta I_C = 0.01$ mA due to I_{CO} and $\Delta I_C = 0.06$ mA due to V_{BE}. These numbers are computed as follows: If T_{max} is reduced from 175 to 145°C, or by 30°, then I_{CO} is divided by $2^{\Delta T/10} = 2^3 = 8$. Hence $S \Delta I_{CO} = 0.087/8 \approx 0.01$ mA. Also, ΔV_{BE} is increased by $(30)(2.5) = 75$ mV, or ΔV_{BE} goes from -0.375 to -0.30 V and $S' \Delta V_{BE} \approx -\Delta V_{BE}/R_e = -0.30/4.7 = 0.06$ mA. Hence, for Si, the effect of V_{BE} is more important than that of I_{CO} on the collector current. However, for a transistor with a small β, a large spread in β may have the dominant influence on I_C.

EXAMPLE Design the self-bias circuit of Fig. 9-5a using a Si transistor type 2N3565 to meet the following specifications over the temperature range 25 to 65°C:

$$\frac{\Delta I_C}{I_C} \leq 15 \text{ percent}$$

V_{BE} at 25°C = 650 $\pm$ 50 mV V_{CC} = 20 V

β spread 150 to 600 at I_C = 1 mA and T = 25°C

Lowest β at 25°C = 150   highest β at 65°C = 1,200

I_{CO} at 25°C = 50 nA max I_{CO} at 65°C = 3.0 μA max

Solution We shall assume for our design that each factor I_{CO}, β, and V_{BE} causes the same percentage change (5 percent) in I_C. We now proceed in the following steps:

1. Select R_b/R_e using the $\Delta \beta$ term of Eq. (9-32) and assuming $M \approx 1$.

$$\left(1 + \frac{R_b}{R_e}\right) \frac{\Delta \beta}{\beta_1 \beta_2} = \left(1 + \frac{R_b}{R_e}\right) \times \frac{1,200 - 150}{1,200 \times 150} = 0.05$$

or

$$\frac{R_b}{R_e} = 7.56$$

Since the smallest β is 150, then

$$1 > M > \frac{1}{1 + R_b/\beta R_e} = \frac{1}{1 + 7.56/150} = 0.95$$

which justifies our assumption $M \approx 1$.

2. Select I_{C1}, considering the ΔI_{CO} term in Eq. (9-32).

$$\left(1 + \frac{R_b}{R_e}\right)\frac{\Delta I_{CO}}{I_{C1}} = 8.56 \frac{3 \times 10^{-6}}{I_{C1}} = 0.05$$

or $I_{C1} = 0.515$ mA. Use $I_C = 0.6$ mA.

3. Select $I_{C1}R_e$, considering the ΔV_{BE} term only in Eq. (9-32). Since V_{BE} changes -2.5 mV/°C, then $\Delta V_{BE} = -2.5 \times 40 = -100$ mV due to the temperature range. Since there is an uncertainty in V_{BE} at 25°C of ± 50 mV, the total increment is $\Delta V_{BE} = -100 - 100 = -200$ mV $= -0.2$ V. Hence

$$\frac{0.20}{I_{C1}R_e} = 0.05 \quad \text{or} \quad I_{C1}R_e = 4 \text{ V}$$

4. Since $I_{C1} = 0.6$ mA, then $R_e = 4/0.6 = 6.65$ K. Also $R_b = (R_b/R_e)R_e = 7.56 \times 6.65 = 50$ K.

5. To determine the value of the biasing resistors R_1 and R_2, we must first find V. From Fig. 9-5b or Eq. (9-5), at 25°C and using an average value of $\beta = \frac{1}{2}(150 + 600) = 375$,

$$V = I_B R_b + V_{BE} + (I_B + I_C)R_e = \frac{0.6 \times 50}{375} + 0.65 + 4 = 4.73 \text{ V}$$

Solving Eqs. (9-4) for R_1 and R_2, we obtain

$$R_1 = R_b \frac{V_{CC}}{V} = 50 \times \frac{20}{4.73} = 211 \text{ K}$$

$$R_2 = \frac{R_1 V}{V_{CC} - V} = \frac{211 \times 4.73}{20 - 4.73} = 66 \text{ K}$$

The value of R_c is selected on the basis of the required small-signal gain and symmetric operation of the circuit.

9-6 BIAS COMPENSATION[2]

From our discussion so far we see that in biasing a transistor in the active region we should strive to maintain the operating point stable by keeping I_C and V_{CE} constant. The techniques normally used to do so may be classified in two categories: *stabilization techniques* and *compensation techniques*. Stabilization techniques refer to the use of resistive biasing circuits (such as Fig. 9-5) which allow I_B to vary so as to keep I_C relatively constant, with variations in I_{CO}, β, and V_{BE}. Compensation techniques refer to the use of temperature-sensitive devices such as diodes, transistors, thermistors, etc. Two circuits using diode compensation will now be discussed.

Diode Compensation for V_{BE} A circuit utilizing the self-bias stabilization technique and diode compensation is shown in Fig. 9-9. The diode is kept biased in the forward direction by the source V_{DD} and resistance R_d.

If the diode is of the same material and type as the transistor, the voltage V_o across the diode will have the same temperature coefficient (-2.5 mV/°C) as the base-to-emitter voltage V_{BE}. If we write KVL around the base circuit of Fig. 9-9, then Eq. (9-21) becomes

$$I_C = \frac{\beta[V - (V_{BE} - V_o)] + (R_b + R_e)(\beta + 1)I_{CO}}{R_b + R_e(1 + \beta)} \tag{9-34}$$

Since V_{BE} tracks V_o with respect to temperature, it is clear from Eq. (9-34) that I_C will be insensitive to variations in V_{BE}. In practice, the compensation of V_{BE} as explained above is not exact, but it is sufficiently effective to take care of a great part of transistor drift due to variations in V_{BE}.

Diode Compensation for I_{CO} We demonstrate in Sec. 9-5 that changes of V_{BE} with temperature contribute significantly to changes in collector current of silicon transistors. On the other hand, for germanium transistors, changes in I_{CO} with temperature play the more important role in collector-current stability. The diode compensation circuit shown in Fig. 9-10 offers stabilization against variations in I_{CO}, and is therefore useful for stabilizing germanium transistors.

If the diode and the transistor are of the same type and material, the reverse saturation current I_o of the diode will increase with temperature at the same rate as the transistor collector saturation current I_{CO}. From Fig. 9-10 we have

$$I = \frac{V_{CC} - V_{BE}}{R_1} \approx \frac{V_{CC}}{R_1} = \text{constant}$$

Since the diode is reverse-biased by an amount $V_{BE} \approx 0.2$ V for germanium devices, it follows that the current through D is I_o. The base current is $I_B = I - I_o$. Substituting this expression for I_B in Eq. (9-6), we obtain

$$I_C = \beta I - \beta I_o + (1 + \beta)I_{CO} \tag{9-35}$$

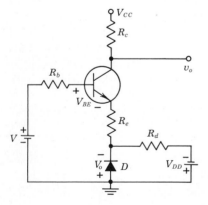

Fig. 9-9 Stabilization by means of self-bias and diode-compensation techniques.

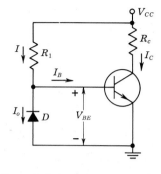

Fig. 9-10 Diode compensation for a germanium
transistor.

We see from Eq. (9-35) that if $\beta \gg 1$ and if I_o of D and I_{CO} of Q track each
other over the desired temperature range, then I_C remains essentially constant.

9-7   BIASING TECHNIQUES FOR LINEAR INTEGRATED CIRCUITS[3]

The self-bias circuit of Fig. 9-5a often requires a capacitor across R_e since
otherwise the negative feedback, due to R_e, reduces the signal gain drastically
(Sec. 8-15).   This bypass capacitance is much too large (of the order of microfarads, Sec. 12-8) to be fabricated by integrated-circuit technology.   Hence
the biasing technique shown in Fig. 9-11 has been developed for monolithic
circuits.   In Fig. 9-11a the transistor $Q1$ is connected as a diode across the
base-to-emitter junction of $Q2$ whose collector current is to be temperature-
stabilized.   The collector current of $Q1$ is given by

$$I_{C1} = \frac{V_{CC} - V_{BE}}{R_1} - I_{B1} - I_{B2} \tag{9-36}$$

For $V_{BE} \ll V_{CC}$ and $(I_{B1} + I_{B2}) \ll I_{C1}$, Eq. (9-36) becomes

$$I_{C1} \approx \frac{V_{CC}}{R_1} = \text{constant} \tag{9-37}$$

If transistors $Q1$ and $Q2$ are identical and have the same V_{BE}, their collector
currents will be equal.   Hence $I_{C2} = I_{C1} = \text{constant}$.   Even if the two transistors are not identical, experiments[3] have shown that this biasing scheme
gives collector-current matching between the biasing and operating transistors
typically better than 5 percent and is stable over a wide temperature range.
   The circuit of Fig. 9-11a is modified as indicated in Fig. 9-11b so that
the transistors are driven by equal base currents rather than the same base
voltage.   Since the collector current in the active region varies linearly with
I_B, but exponentially with V_{BE}, improved matching of collector currents
results.   The resistors R_2 and R_3 are fabricated in an identical manner, so
that $R_3 = R_2$.   Since the two bases are driven from a common voltage node

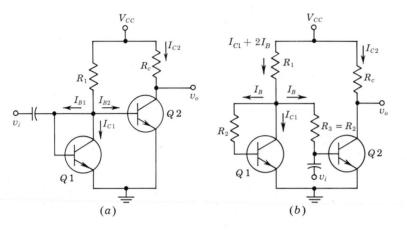

Fig. 9-11 Biasing techniques for linear integrated circuits.

through equal resistances, then $I_{B1} = I_{B2} \equiv I_B$, and the collector currents are well matched for identically constructed transistors.

From Fig. 9-11, the collector current of $Q1$ is given by

$$I_{C1} = \frac{V_{CC} - V_{BE}}{R_1} - \left(2 + \frac{R_2}{R_1}\right) I_B \tag{9-38}$$

Under the assumptions that $V_{BE} \ll V_{CC}$, and $(2 + R_2/R_1)I_B \ll V_{CC}/R_1$, Eq. (9-38) becomes

$$I_{C1} = I_{C2} = \frac{V_{CC}}{R_1}$$

If $R_c = \frac{1}{2}R_1$, then $V_{CE} = V_{CC} - I_{C2}R_c \approx V_{CC}/2$, which means that the amplifier will be biased at one-half the supply voltage V_{CC}, independent of the supply voltage as well as temperature, and dependent only on the matching of components within the integrated circuit. An evaluation of the effects of mismatch in this circuit on bias stability is given in Ref. 3.

9-8 THERMISTOR AND SENSISTOR COMPENSATION[2]

There is a method of transistor compensation which involves the use of temperature-sensitive resistive elements rather than diodes or transistors. The *thermistor* (Sec. 2-7) has a negative temperature coefficient, its resistance decreasing exponentially with increasing T. The circuit of Fig. 9-12 uses a thermistor R_T to minimize the increase in collector current due to changes in I_{CO}, V_{BE}, or β with T. As T rises, R_T decreases, and the current fed through R_T into R_e increases. Since the voltage drop across R_e is in the direction to reverse-bias the transistor, the temperature sensitivity of R_T acts so as to tend to compensate the increase in I_C due to T.

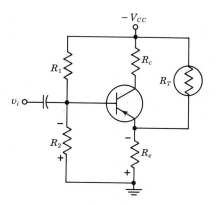

Fig. 9-12 Thermistor compensation of the increase in I_C with T.

An alternative configuration using thermistor compensation is to move R_T from its position in Fig. 9-12 and place it across R_2. As T increases, the drop across R_T decreases, and hence the forward-biasing base voltage is reduced. This behavior will tend to offset the increase in collector current with temperature.

Instead of a thermistor, it is possible to use a temperature-sensitive resistor with a positive temperature coefficient such as a metal, or the *sensistor* (Sec. 2-7). In the circuit of Fig. 9-12 (with R_T removed), temperature compensation may be obtained by placing a sensistor either in parallel with R_1 or in parallel with (or in place of) R_e. Why?

In practice it is often necessary to use silicon resistors and carbon resistors in series or parallel combinations to form the proper shaping network.[4] The characteristics required to eliminate the temperature effects can be determined experimentally as follows: A variable resistance is substituted for the shaping network and is adjusted to maintain constant collector current as the operating temperature changes. The resistance vs. temperature can then be plotted to indicate the required characteristics of the shaping network. The problem now is reduced to that of synthesizing a network with this measured temperature characteristic by using thermistors or sensistors padded with temperature-insensitive resistors.

9-9   THERMAL RUNAWAY

The maximum average power $P_{D,\max}$ which a transistor can dissipate depends upon the transistor construction and may lie in the range from a few milliwatts to 200 W. This maximum power is limited by the temperature that the collector-to-base junction can withstand. For silicon transistors this temperature is in the range 150 to 225°C, and for germanium it is between 60 and 100°C. The junction temperature may rise either because the ambient temperature rises or because of self-heating. The maximum power dissipation is usually specified for the transistor enclosure (case) or ambient temperature of 25°C. The problem of self-heating, which is mentioned in Sec. 9-2, results from the power dissipated at the collector junction. As a consequence of the junction power dissipation, the junction temperature rises, and this in turn increases the collector current, with a subsequent increase in power dissipation. If this phenomenon, referred to as *thermal runaway*, continues, it may result in permanently damaging the transistor.

Thermal Resistance It is found experimentally that the *steady-state* temperature rise at the collector junction is proportional to the power dissipated at the junction, or

$$T_j - T_A = \Theta P_D \qquad (9\text{-}39)$$

where T_j and T_A are the junction and ambient temperatures, respectively, in degrees centigrade, and P_D is the power in watts dissipated at the collector junction. The constant of proportionality Θ is called the *thermal resistance*. Its value depends on the size of the transistor, on convection or radiation to the surroundings, on forced-air cooling (if used), and on the thermal connection of the device to a metal chassis or to a heat sink. Typical values for various transistor designs vary from 0.2°C/W for a high-power transistor with an efficient heat sink to 1000°C/W for a low-power transistor in free air. The temperature rate at which power is dissipated under steady-state conditions is obtained by differentiating Eq. (9-39) with respect to T_j, or

$$\frac{\partial P_D}{\partial T_j} = \frac{1}{\Theta} \qquad (9\text{-}40)$$

The maximum collector power P_C allowed for safe operation is specified at 25°C. For ambient temperatures above this value, P_C must be decreased, and at the extreme temperature at which the transistor may operate, P_C is reduced to zero. A typical power-temperature derating curve, supplied in a manufacturer's specification sheet, is indicated in Fig. 9-13. The thermal

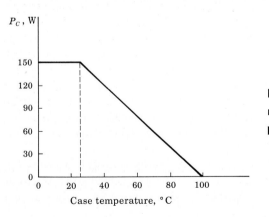

Fig. 9-13 Power-temperature derating curve for a germanium power transistor.

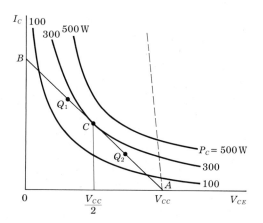

Fig. 9-14   Concerning transistor self-heating.   The dashed load line corresponds to a very small dc resistance.

resistance of this transistor is given by the reciprocal of the slope of the derating line in Fig. 9-13, or $\Theta = \frac{7.5}{150} = 0.5°C/W$.

Operating-point Considerations   The effects of self-heating may be appreciated by referring to Fig. 9-14, which shows three constant-power hyperbolas and a dc load line tangent to one of them.   It can be shown (Prob. 9-32) that the point of tangency C bisects the load line AB.   Consider that the quiescent point is above the point of tangency, say at Q_1.   If now the collector current increases, the result is a *lower* collector dissipation because Q_1 moves along the load line in the direction away from the 300-W toward the 100-hyperbola.   The opposite is true if the quiescent point is below the point of tangency, such as at Q_2.   We can conclude that if V_{CE} is less than $V_{CC}/2$, the quiescent point lies in a safe region, where an increase in collector current results in a decreased dissipation.   If, on the other hand, the operating point is located so that $V_{CE} > V_{CC}/2$, the self-heating results in even more collector dissipation, and the effect is cumulative.

It is not always possible to select an operating point which satisfies the restriction $V_{CE} < \frac{1}{2}V_{CC}$.   For example, if the load R_L is transformer-coupled to the collector, as in Fig. 9-15, then R_c represents the small primary dc resistance, and hence the load line is almost vertical, as indicated by the dashed line in Fig. 9-14.   Clearly, V_{CE} can be less than $\frac{1}{2}V_{CC}$ only for excessively large collector currents.   Hence thermal runaway can easily occur with a transformer-coupled load or with a power amplifier which has small collector and emitter resistances.   For such circuits it is particularly important to take precautions to keep the stability factors (discussed in the preceding sections) so small as to maintain essentially constant collector current.

The Condition for Thermal Stability   We now obtain the restrictions to be met if thermal runaway is to be avoided.   The required condition is that the rate at which heat is released at the collector junction must not exceed the rate at which the heat can be dissipated under steady-state conditions. From Eq. (9-40) it follows that

$$\frac{\partial P_C}{\partial T_j} < \frac{1}{\Theta} \tag{9-41}$$

is the condition which must be satisfied to prevent thermal runaway. By suitable circuit design it is possible to ensure that the transistor cannot run away below a specified ambient temperature or even under any conditions. Such an analysis is made in the next section.

9-10   THERMAL STABILITY

Let us refer to Fig. 9-5a and assume that the transistor is biased in the active region. The power generated at the collector junction with no signal is

$$P_C = I_C V_{CB} \approx I_C V_{CE} \tag{9-42}$$

If we assume that the quiescent collector and emitter currents are essentially equal, Eq. (9-42) becomes

$$P_C = I_C V_{CC} - I_C^2 (R_e + R_c) \tag{9-43}$$

Equation (9-41), the condition to avoid thermal runaway, can be rewritten as follows:

$$\frac{\partial P_C}{\partial I_C} \frac{\partial I_C}{\partial T_j} < \frac{1}{\Theta} \tag{9-44}$$

Since Θ and $\partial I_C / \partial T_j$ are positive, Eq. (9-44) is always satisfied if $\partial P_C / \partial I_C$ is negative. From Eq. (9-43)

$$\frac{\partial P_C}{\partial I_C} = V_{CC} - 2 I_C (R_e + R_c) \tag{9-45}$$

Hence, to avoid thermal runaway, it is necessary that

$$I_C > \frac{V_{CC}}{2(R_e + R_c)} \tag{9-46}$$

Since $V_{CE} = V_{CC} - I_C(R_e + R_c)$, then Eq. (9-46) implies that $V_{CE} < V_{CC}/2$, and this checks with our previous conclusion from Fig. 9-14. If the inequality of Eq. (9-46) is not satisfied and $V_{CE} > V_{CC}/2$, then from Eq. (9-45) we see that $\partial P_C / \partial I_C$ is positive, and the designer must ensure that Eq. (9-44) will be satisfied, or else thermal runaway will occur.

EXAMPLE   Find the value of Θ required for the Ge transistor of the example on page 296 in order for the circuit to be thermally stable. Assume $V_{CC} = 30$ V and $R_c = 2.0$ K and $R_e = 4.7$ K.

Solution   From Eq. (9-45), since $I_C = 1.5$ mA and $R_e = 4.7$ K,

$$\frac{\partial P_C}{\partial I_C} = 30 - (2)(1.5)(4.7 + 2.0) = 9.9 \text{ V}$$

From page 297 I_C increases by 0.131 mA over a temperature range of 25 to 75°C. Hence

$$\frac{\partial I_C}{\partial T_j} = \frac{0.131 \times 10^{-3}}{75 - 25} = 2.62 \times 10^{-6} \text{ A/°C}$$

From Eq. (9-44)

$$9.9 \times 2.62 \times 10^{-6} < \frac{1}{\Theta}$$

or

$$\Theta < 3.85 \times 10^4 \text{ °C}/W$$

This upper bound on Θ is so high that no transistor would violate it, and therefore this circuit will be safe from thermal runaway.

In some practical problems the effect of I_{CO} dominates, and we present an analysis of the thermal-runaway problem for this case.   From Eqs. (9-44) and (9-30)

$$\frac{\partial P_C}{\partial I_C}\left(S \frac{\partial I_{CO}}{\partial T_j}\right) < \frac{1}{\Theta} \tag{9-47}$$

In Sec. 5-7 it is noted that the reverse saturation current for either silicon or germanium increases about 7 percent/°C, or

$$\frac{\partial I_{CO}}{\partial T_j} = 0.07 I_{CO} \tag{9-48}$$

Substituting Eqs. (9-45) and (9-48) in Eq. (9-47) results in

$$[V_{CC} - 2I_C(R_e + R_c)](S)(0.07 I_{CO}) < \frac{1}{\Theta} \tag{9-49}$$

Equation (9-49) remains valid for a *p-n-p* transistor provided that I_C (and I_{CO}) are understood to represent the magnitude of the current.

   The above equation illustrates that amplifier circuits operated at low current and designed with low values of stability factor ($S < 10$) are rarely susceptible to thermal runaway.   In contrast, power amplifiers operate at high power levels.   In addition, in such circuits, R_e is a small resistance for power efficiency, and this results in a high stability factor S.   As a result, thermal runaway in power stages is a major consideration, and the designer must guard against it by attaching the collector to a heat sink.

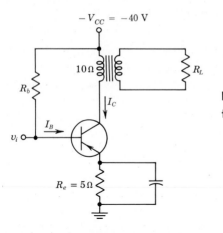

$-V_{CC} = -40$ V

Fig. 9-15　Power amplifier with a transformer-coupled load.

EXAMPLE Figure 9-15 shows a power amplifier using a *p-n-p* germanium transistor with $\beta = 100$ and $I_{co} = -5$ mA. The quiescent collector current is $I_C = -1$ A. Find (a) the value of resistor R_b; (b) the largest value of Θ that can result in a thermally stable circuit. Assume that the effect of I_{co} dominates.

Solution　a. The collector current is given by Eq. (9-6), or

$$I_C = \beta I_B + (1 + \beta)I_{co} \approx \beta(I_B + I_{co})$$

and

$$I_B = -\frac{1 - 5 \times 10^{-3} \times 100}{100} \text{ A} = -5 \text{ mA}$$

If we neglect V_{BE}, we have

$$5 \times 10^{-3} R_b = 40 - 5 \quad \text{or} \quad R_b = 7,000 \ \Omega$$

b. Since $|V_{CE}| = 40 - 15 = 25 > \frac{1}{2}|V_{CC}| = 20$ V, the circuit of Fig. 9-15 is not inherently stable. The stability factor S is obtained from Eq. (9-13).

$$S = 101 \frac{1 + 7,000/5}{101 + 7,000/5} = 94.3$$

Substituting in Eq. (9-49), we obtain

$$(40 - 2 \times 1 \times 15)(94.3)(0.07 \times 5 \times 10^{-3}) < \frac{1}{\Theta}$$

or

$$\Theta < 3.03°\text{C/W}$$

REFERENCES

1. Brown, W. L., and D. E. Perrine: Don't Guess at Bias Circuit Design, *Electron. Design*, May 9, 1968, pp. 80–86.

2. Hunter, L. P.: "Handbook of Semiconductor Electronics," McGraw-Hill Book Company, New York, 1970.

"Transistor Manual," 7th ed., General Electric Co., Syracuse, N.Y., 1964.

"Motorola Power Transistor Handbook," Phoenix, Ariz., 1961.

3. Widlar, R. I.: Some Circuit Design Techniques for Linear Integrated Circuits, *IEEE Trans. Circuit Theory*, vol. CT-12, no. 4, pp. 586–590, December, 1965.

4. Konjian, E., and J. S. Schaffner: Shaping of the Characteristics of Temperature-sensitive Elements, *Commun. and Electron.*, vol. 14, pp. 396–400, September, 1954.

REVIEW QUESTIONS

9-1   What ratings limit the range of operation of a transistor?

9-2   Why is capacitive coupling used to connect a signal source to an amplifier?

9-3   For a capacitively coupled load, is the dc load larger or smaller than the ac load? Explain.

9-4   (*a*) Draw a fixed-bias circuit. (*b*) Explain why the circuit is unsatisfactory if the transistor is replaced by another of the same type.

9-5   Discuss *thermal instability*.

9-6   (*a*) Draw a self-bias circuit. (*b*) Explain qualitatively why such a circuit is an improvement on the fixed-bias circuit, as far as stability is concerned.

9-7   (*a*) How is the load line drawn for a self-bias circuit?   Justify your answer. (*b*) Define *bias curve*.   (*c*) Explain how the bias curve is used to obtain the quiescent point for this circuit.

9-8   (*a*) List the three sources of instability of collector current. (*b*) Define the three stability factors.

9-9   How does the designer minimize the percentage variations in I_C (*a*) due to variations in I_{CO} and V_{BE} and (*b*) due to variations in β?

9-10   Over what temperature range can a transistor be used if it is (*a*) silicon and (*b*) germanium?

9-11   The collector-current variation is usually greater due to which parameter change (I_{CO} or V_{BE}) for (*a*) silicon and (*b*) germanium?

9-12   Define (*a*) *stabilization techniques* and (*b*) *compensation techniques*.

9-13   Draw a circuit which uses a diode to compensate for changes (*a*) in V_{BE} and (*b*) in I_{CO}.

9-14   (*a*) Draw a properly biased integrated-circuit linear amplifier. (*b*) How are the parameter values chosen so that the quiescent output voltage is $\frac{1}{2}V_{CC}$?

9-15   Draw a circuit employing (*a*) thermistor compensation and (*b*) sensitor compensation.

9-16   Discuss *thermal runaway*.

9-17   Define *thermal resistance*.   Give its dimensions and order of magnitude for a transistor.

9-18   (*a*) Draw a *derating curve* for a power transistor. (*b*) How is the thermal resistance obtained from this plot?

9-19   Show graphically that thermal runaway cannot take place if the quiescent point is located at $V_{CE} < \frac{1}{2}V_{CC}$.

9-20   What is the condition for thermal stability?   Explain.

10 / FIELD-EFFECT TRANSISTORS

The field-effect transistor[1] is a semiconductor device which depends for its operation on the control of current by an electric field. There are two types of field-effect transistors, the *junction field-effect transistor* (abbreviated JFET, or simply FET) and the *insulated-gate field-effect transistor* (IGFET), more commonly called the *metal-oxide-semiconductor (MOS) transistor* (MOST, or MOSFET).

The principles on which these devices operate, as well as the differences in their characteristics, are examined in this chapter. Representative circuits making use of FET transistors are also presented.

The field-effect transistor differs from the bipolar junction transistor in the following important characteristics:

1. Its operation depends upon the flow of majority carriers only. It is therefore a *unipolar* (one type of carrier) device.
2. It is simpler to fabricate and occupies less space in integrated form.
3. It exhibits a high input resistance, typically many megohms.
4. It is less noisy than a bipolar transistor.
5. It exhibits no offset voltage at zero drain current, and hence makes an excellent signal chopper.[2]

The main disadvantage of the FET is its relatively small gain-bandwidth product in comparison with that which can be obtained with a conventional transistor. The principal applications of MOSFETs are as LSI digital arrays.

10-1   THE JUNCTION FIELD–EFFECT TRANSISTOR

The structure of an *n-channel* field-effect transistor is shown in Fig. 10-1. Ohmic contacts are made to the two ends of a semiconductor bar of *n*-type material (if *p*-type silicon is used, the device is referred to as a *p-channel* FET). Current is caused to flow along the length of the bar because of the voltage supply connected between the ends.   This current consists of majority carriers, which in this case are electrons.   A simple side view of a JFET is indicated in Fig. 10-1*a* and a more detailed sketch is shown in Fig. 10-1*b*.   The circuit symbol with current and voltage polarities marked is given in Fig. 10-2.   The following FET notation is standard.

Source   The *source S* is the terminal through which the majority carriers enter the bar.   Conventional current entering the bar at *S* is designated by I_S.

Drain   The *drain D* is the terminal through which the majority carriers leave the bar.   Conventional current entering the bar at *D* is designated by I_D. The drain-to-source voltage is called V_{DS}, and is positive if *D* is more positive than *S*.   In Fig. 10-1, $V_{DS} = V_{DD}$ = drain supply voltage.

Gate   On both sides of the *n*-type bar of Fig. 10-1, heavily doped (p^+) regions of acceptor impurities have been formed by alloying, by diffusion, or

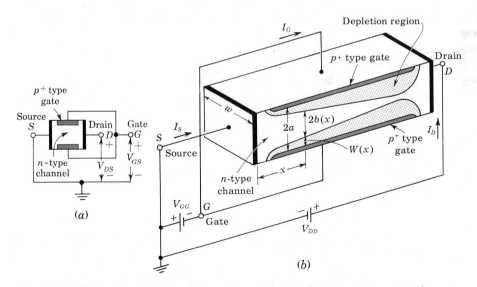

Fig. 10-1 The basic structure of an *n*-channel field-effect transistor. (*a*) Simplified view. (*b*) More detailed drawing. The normal polarities of the drain-to-source (V_{DD}) and gate-to-source (V_{GG}) supply voltages are shown. In a *p*-channel FET the voltages would be reversed.

by any other procedure available for creating p-n junctions. These impurity regions are called the *gate G*. Between the gate and source a voltage $V_{GS} = -V_{GG}$ is applied in the direction to reverse-bias the p-n junction. Conventional current entering the bar at G is designated I_G.

Channel The region in Fig. 10-1 of n-type material between the two gate regions is the *channel* through which the majority carriers move from source to drain.

FET Operation It is necessary to recall that on the two sides of the reverse-biased p-n junction (the transition region) there are space-charge regions (Sec. 3-7). The current carriers have diffused across the junction, leaving only uncovered positive ions on the n side and negative ions on the p side. The electric lines of field intensity which now originate on the positive ions and terminate on the negative ions are precisely the source of the voltage drop across the junction. As the reverse bias across the junction increases, so also does the thickness of the region of immobile uncovered charges. The conductivity of this region is nominally zero because of the unavailability of current carriers. Hence we see that the effective width of the *channel* in Fig. 10-1 will become progressively decreased with increasing reverse bias. Accordingly, for a fixed drain-to-source voltage, the drain current will be a function of the reverse-biasing voltage across the gate junction. The term *field effect* is used to describe this device because the mechanism of current control is the *effect* of the extension, with increasing reverse bias, of the *field* associated with the region of uncovered charges.

FET Static Characteristics The circuit, symbol, and polarity conventions for an FET are indicated in Fig. 10-2. The direction of the arrow at the gate of the junction FET in Fig. 10-2 indicates the direction in which gate current would flow if the gate junction were forward-biased. The common-source drain characteristics for a typical n-channel FET shown in Fig. 10-3 give I_D against V_{DS}, with V_{GS} as a parameter. To see qualitatively why the characteristics have the form shown, consider, say, the case for which $V_{GS} = 0$. For $I_D = 0$, the channel between the gate junctions is entirely open. In response

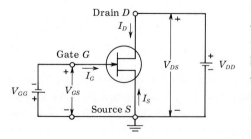

Fig. 10-2 Circuit symbol for an n-channel FET. (For a p-channel FET the arrow at the gate junction points in the opposite direction.) For an n-channel FET, I_D and V_{DS} are positive and V_{GS} is negative. For a p-channel FET, I_D and V_{DS} are negative and V_{GS} is positive.

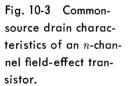

Fig. 10-3 Common-
source drain charac-
teristics of an n-chan-
nel field-effect tran-
sistor.

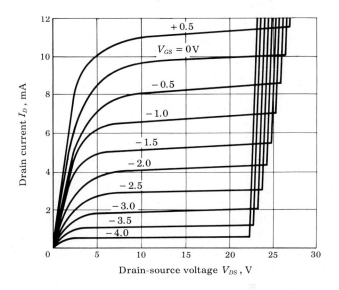

to a small applied voltage V_{DS}, the n-type bar acts as a simple semiconductor resistor, and the current I_D increases linearly with V_{DS}. With increasing current, the ohmic voltage drop between the source and the channel region reverse-biases the junction, and the conducting portion of the channel begins to constrict. Because of the ohmic drop along the length of the channel itself, the constriction is not uniform, but is more pronounced at distances farther from the source, as indicated in Fig. 10-1. Eventually, a voltage V_{DS} is reached at which the channel is "pinched off." This is the voltage, not too sharply defined in Fig. 10-3, where the current I_D begins to level off and approach a constant value. It is, of course, in principle not possible for the channel to close completely and thereby reduce the current I_D to zero. For if such, indeed, could be the case, the ohmic drop required to provide the necessary back bias would itself be lacking. Note that each characteristic curve has an ohmic region for small values of V_{DS}, where I_D is proportional to V_{DS}. Each also has a constant-current region for large values of V_{DS}, where I_D responds very slightly to V_{DS}.

If now a gate voltage V_{GS} is applied in the direction to provide additional reverse bias, pinch-off will occur for smaller values of $|V_{DS}|$, and the maximum drain current will be smaller. This feature is brought out in Fig. 10-3. Note that a plot for a silicon FET is given even for $V_{GS} = +0.5$ V, which is in the direction of forward bias. We note from Table 5-1 that, actually, the gate current will be very small, because at this gate voltage the Si junction is barely at the cutin voltage V_γ.

The maximum voltage that can be applied between any two terminals of the FET is the lowest voltage that will cause avalanche breakdown (Sec. 3-11) across the gate junction. From Fig. 10-3 it is seen that avalanche occurs at a lower value of $|V_{DS}|$ when the gate is reverse-biased than for $V_{GS} = 0$. This

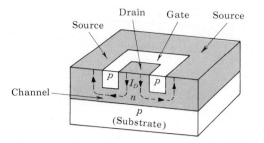

Fig. 10-4 Single-ended-geometry junction FET.

is caused by the fact that the reverse-bias gate voltage adds to the drain voltage, and hence increases the effective voltage across the gate junction.

We note from Fig. 10-2 that the n-channel FET requires zero or negative gate bias and positive drain voltage. The p-channel FET requires opposite voltage polarities. Either end of the channel may be used as a source. We can remember supply polarities by using the channel type, p or n, to designate the polarity of the *source* side of the drain supply.

A Practical FET Structure   The structure shown in Fig. 10-1 is not practical because of the difficulties involved in diffusing impurities into both sides of a semiconductor wafer. Figure 10-4 shows a single-ended-geometry junction FET where diffusion is from one side only. The substrate is of p-type material onto which an n-type channel is epitaxially grown (Sec. 7-3). A p-type gate is then diffused into the n-type channel. The substrate which may function as a second gate is of relatively low resistivity material. The diffused gate is also of very low resistivity material, allowing the depletion region to spread mostly into the n-type channel.

10-2   THE PINCH–OFF VOLTAGE V_P

We derive an expression for the gate reverse voltage V_P that removes all the free charge from the channel using the physical model described in the preceding section. This analysis was first made by Shockley,[1] using the structure of Fig. 10-1. In this device a slab of n-type semiconductor is sandwiched between two layers of p-type material, forming two p-n junctions.

Assume that the p-type region is doped with N_A acceptors per cubic meter, that the n-type region is doped with N_D donors per cubic meter, and that the junction formed is abrupt. The assumption of an abrupt junction is the same as that made in Sec. 3-7 and Fig. 3-10, and is chosen for simplicity. Moreover, if $N_A \gg N_D$, we see from Eq. (3-17) that $W_p \ll W_n$, and using Eq. (3-21), we have, for the space-charge width, $W_n(x) = W(x)$ at a distance x along the channel in Fig. 10-1:

$$W(x) = a - b(x) = \left\{ \frac{2\epsilon}{qN_D} [V_o - V(x)] \right\}^{\frac{1}{2}} \tag{10-1}$$

where ϵ = dielectric constant of channel material

$\qquad$ q = magnitude of electronic charge

$\qquad$ V_o = junction contact potential at x (Fig. 3-1d)

$\qquad$ $V(x)$ = applied potential across space-charge region at x and is a negative number for an applied reverse bias

$a - b(x)$ = penetration $W(x)$ of depletion region into channel at a point x along channel (Fig. 10-1)

If the drain current is zero, $b(x)$ and $V(x)$ are independent of x and $b(x) = b$. If in Eq. (10-1) we substitute $b(x) = b = 0$ and solve for V, on the assumption that $|V_o| \ll |V|$, we obtain the pinch-off voltage V_P, the diode reverse voltage that removes all the free charge from the channel. Hence

$$|V_P| = \frac{qN_D}{2\epsilon} a^2 \qquad\qquad (10\text{-}2)$$

If we substitute V_{GS} for $V_o - V(x)$ in Eq. (10-1), we obtain, using Eq. (10-2),

$$V_{GS} = \left(1 - \frac{b}{a}\right)^2 V_P \qquad\qquad (10\text{-}3)$$

The voltage V_{GS} in Eq. (10-3) represents the reverse bias across the gate junction and is independent of distance along the channel if $I_D = 0$.

EXAMPLE   For an n-channel silicon FET with $a = 3 \times 10^{-4}$ cm and $N_D = 10^{15}$ electrons/cm³, find (*a*) the pinch-off voltage and (*b*) the channel half-width for $V_{GS} = \frac{1}{2}V_P$ and $I_D = 0$.

Solution   *a.* The relative dielectric constant of silicon is given in Table 2-1 as 12, and hence $\epsilon = 12\epsilon_o$. Using the values of q and ϵ_o from Appendix A, we have, from Eq. (10-2), expressed in mks units,

$$|V_P| = \frac{1.60 \times 10^{-19} \times 10^{21} \times (3 \times 10^{-6})^2}{2 \times 12 \times (36\pi \times 10^9)^{-1}} = 6.8 \text{ V}$$

b. Solving Eq. (10-3) for b, we obtain for $V_{GS} = \frac{1}{2}V_P$

$$b = a\left[1 - \left(\frac{V_{GS}}{V_P}\right)^{\frac{1}{2}}\right] = (3 \times 10^{-4})[1 - (\tfrac{1}{2})^{\frac{1}{2}}] = 0.87 \times 10^{-4} \text{ cm}$$

Hence the channel width has been reduced to about one-third its value for $V_{GS} = 0$.

10-3   THE JFET VOLT–AMPERE CHARACTERISTICS

Assume, first, that a small voltage V_{DS} is applied between drain and source. The resulting small drain current I_D will then have no appreciable effect on the channel profile. Under these conditions we may consider the effective channel cross section A to be constant throughout its length. Hence $A = 2bw$, where $2b$ is the channel width corresponding to zero drain current as given by Eq. (10-3) for a specified V_{GS}, and w is the channel dimension perpendicular to the b direction, as indicated in Fig. 10-1.

Since no current flows in the depletion region, then, using Ohm's law [Eq. (2-7)], we obtain for the drain current

$$I_D = AqN_D\mu_n\mathcal{E} = 2bwqN_D\mu_n \frac{V_{DS}}{L} \tag{10-4}$$

where L is the length of the channel.

Substituting b from Eq. (10-3) in Eq. (10-4), we have, for small I_D,

$$I_D = \frac{2awqN_D\mu_n}{L} \left[1 - \left(\frac{V_{GS}}{V_P} \right)^{\frac{1}{2}} \right] V_{DS} \tag{10-5}$$

The ON Resistance $r_{d,\text{ON}}$ Equation (10-5) describes the volt-ampere characteristics of Fig. 10-3 for very small V_{DS}, and it suggests that under these conditions the FET behaves like an ohmic resistance whose value is determined by V_{GS}. The ratio V_{DS}/I_D at the origin is called the ON *drain resistance* $r_{d,\text{ON}}$. For a JFET we obtain from Eq. (10-5), with $V_{GS} = 0$,

$$r_{d,\text{ON}} = \frac{L}{2awqN_D\mu_n} \tag{10-6}$$

For the device values given in the illustrative example in this section and with $L/w = 1$, we find that $r_{d,\text{ON}} = 3.3$ K. For the dimensions and concentration used in commercially available FETs and MOSFETs (Sec. 10-5), values of $r_{d,\text{ON}}$ ranging from about 100 Ω to 100 K are measured. This parameter is important in switching applications where the FET is driven heavily ON. The bipolar transistor has the advantage over the field-effect device in that R_{CS} is usually only a few ohms, and hence is much smaller than $r_{d,\text{ON}}$. However, a bipolar transistor has the disadvantage for chopper applications[2] of possessing an offset voltage (Sec. 5-12), whereas the FET characteristics pass through the origin, $I_D = 0$ and $V_{DS} = 0$.

The Pinch-off Region   We now consider the situation where an electric field $\mathcal{E}_x$ appears along the x axis. If a substantial drain current I_D flows, the drain end of the gate is more reverse-biased than the source end, and hence the boundaries of the depletion region are not parallel to the longitudinal axis of the channel, but converge as shown in Fig. 10-1. If the convergence of the depletion region is gradual, the previous one-dimensional analysis is valid[1] in a thin slice of the channel of thickness Δx and at a distance x from the source. Subject to this condition of the "gradual" channel, the current may be written by inspection of Fig. 10-1 as

$$I_D = 2b(x)wqN_D\mu_n\mathcal{E}_x \tag{10-7}$$

As V_{DS} increases, $\mathcal{E}_x$ and I_D increase, whereas $b(x)$ decreases because the channel narrows, and hence the current density $J = I_D/2b(x)w$ increases. We now see that complete pinch-off ($b = 0$) cannot take place because, if it did, J would become infinite, which is a physically impossible condition. If J were to increase without limit, then, from Eq. (10-7), so also would $\mathcal{E}_x$, provided that μ_n remains constant. It is found experimentally,[3,4] however, that the mobility is a function of electric field intensity and remains constant only for $\mathcal{E}_x < 10^3$ V/cm in n-type silicon. For moderate fields, 10^3 to 10^4 V/cm, the mobility is approximately inversely proportional to the square root of the applied field. For still higher fields, such as are encountered at pinch-off, μ_n is inversely proportional to $\mathcal{E}_x$. In this region the drift velocity of the electrons ($v_x = \mu_n \mathcal{E}_x$) remains constant, and Ohm's law is no longer valid. From Eq. (10-7) we now see that both I_D and b remain constant, thus explaining the constant-current portion of the V-I characteristic of Fig. 10-3.

What happens[4] if V_{DS} is increased beyond pinch-off, with V_{GS} held constant? As explained above, the minimum channel width $b_{min} = \delta$ has a small nonzero constant value. This minimum width occurs at the drain end of the bar. As V_{DS} is increased, this increment in potential causes an increase in $\mathcal{E}_x$ in an adjacent channel section toward the source. Referring to Fig. 10-5, the velocity-limited region L' increases with V_{DS}, whereas δ remains at a fixed value.

The Region before Pinch-off   We have verified that the FET behaves as an ohmic resistance for small V_{DS} and as a constant-current device for large V_{DS}. An analysis giving the shape of the volt-ampere characteristic between these two extremes is complicated. It has already been mentioned that in this region the mobility is at first independent of electric field and then μ varies with $\mathcal{E}_x^{-\frac{1}{2}}$ for larger values of $\mathcal{E}_x$ (before pinch-off). Taking this relationship into account, it is possible[3–5] to obtain an expression for I_D as a

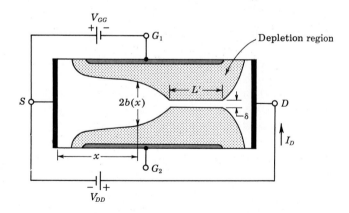

Fig. 10-5   After pinch-off, as V_{DS} is increased, then L' increases but δ and I_D remain essentially constant. (G_1 and G_2 are tied together.)

function of V_{DS} and V_{GS} which agrees quite well with experimentally determined curves.

The Transfer Characteristic   In amplifier applications the FET is almost always used in the region beyond pinch-off (also called the *constant-current, pentode,* or *current-saturation region*).   Let the saturation drain current be designated by I_{DS}, and its value with the gate shorted to the source ($V_{GS} = 0$) by I_{DSS}.   It has been found[6] that the transfer characteristic, giving the relationship between I_{DS} and V_{GS}, can be approximated by the parabola

$$I_{DS} = I_{DSS}\left(1 - \frac{V_{GS}}{V_P}\right)^2 \tag{10-8}$$

This simple parabolic approximation gives an excellent fit, with the experimentally determined transfer characteristics for FETs made by the diffusion process.

Cutoff   Consider an FET operating at a fixed value of V_{DS} in the constant-current region.   As V_{GS} is increased in the direction to reverse-bias the gate junction, the conducting channel will narrow.   When $V_{GS} = V_P$, the channel width is reduced to zero, and from Eq. (10-8), $I_{DS} = 0$.   With a physical device some leakage current $I_{D,\text{OFF}}$ still flows even under the cutoff condition $|V_{GS}| > |V_P|$.   A manufacturer usually specifies a maximum value of $I_{D,\text{OFF}}$ at a given value of V_{GS} and V_{DS}.   Typically, a value of a few nanoamperes may be expected for $I_{D,\text{OFF}}$ for a silicon FET.

The gate reverse current, also called *the gate cutoff current*, designated by I_{GSS}, gives the gate-to-source current, with the drain shorted to the source for $|V_{GS}| > |V_P|$.   Typically, I_{GSS} is of the order of a few nanoamperes for a silicon device.

10-4   THE FET SMALL–SIGNAL MODEL

The linear small-signal equivalent circuit for the FET can be obtained in a manner analogous to that used to derive the corresponding model for a transistor.   We employ the same notation in labeling time-varying and dc currents and voltages as used in Secs. 8-1 and 8-2 for the transistor.   We can formally express the drain current i_D as a function f of the gate voltage v_{GS} and drain voltage v_{DS} by

$$i_D = f(v_{GS}, v_{DS}) \tag{10-9}$$

The Transconductance g_m and Drain Resistance r_d   If both the gate and drain voltages are varied, the change in drain current is given approximately by the first two terms in the Taylor's series expansion of Eq. (10-9), or

$$\Delta i_D = \left.\frac{\partial i_D}{\partial v_{GS}}\right|_{V_{DS}} \Delta v_{GS} + \left.\frac{\partial i_D}{\partial v_{DS}}\right|_{V_{GS}} \Delta v_{DS} \tag{10-10}$$

In the small-signal notation of Sec. 8-1, $\Delta i_D = i_d$, $\Delta v_{GS} = v_{gs}$, and $\Delta v_{DS} = v_{ds}$, so that Eq. (10-10) becomes

$$i_d = g_m v_{gs} + \frac{1}{r_d} v_{ds} \tag{10-11}$$

where

$$g_m \equiv \frac{\partial i_D}{\partial v_{GS}}\bigg|_{V_{DS}} \approx \frac{\Delta i_D}{\Delta v_{GS}}\bigg|_{V_{DS}} = \frac{i_d}{v_{gs}}\bigg|_{V_{DS}} \tag{10-12}$$

is the *mutual conductance,* or *transconductance.* It is also often designated by y_{fs} or g_{fs} and called the (*common-source*) *forward transadmittance.* The second parameter r_d in Eq. (10-11) is the *drain* (or *output*) *resistance,* and is defined by

$$r_d \equiv \frac{\partial v_{DS}}{\partial i_D}\bigg|_{V_{GS}} \approx \frac{\Delta v_{DS}}{\Delta i_D}\bigg|_{V_{GS}} = \frac{v_{ds}}{i_d}\bigg|_{V_{GS}} \tag{10-13}$$

The reciprocal of r_d is the drain conductance g_d. It is also designated by y_{os} and g_{os} and called the (common-source) output conductance.

An *amplification factor* μ for an FET may be defined by

$$\mu \equiv -\frac{\partial v_{DS}}{\partial v_{GS}}\bigg|_{I_D} = -\frac{\Delta v_{DS}}{\Delta v_{GS}}\bigg|_{I_D} = -\frac{v_{ds}}{v_{gs}}\bigg|_{i_d=0} \tag{10-14}$$

We can verify that μ, r_d, and g_m are related by

$$\mu = r_d g_m \tag{10-15}$$

by setting $i_d = 0$ in Eq. (10-11).

An expression for g_m is obtained by applying the definition of Eq. (10-12) to Eq. (10-8). The result is

$$g_m = g_{mo}\left(1 - \frac{V_{GS}}{V_P}\right) = \frac{2}{|V_P|}(I_{DSS}I_{DS})^{\frac{1}{2}} \tag{10-16}$$

where g_{mo} is the value of g_m for $V_{GS} = 0$, and is given by

$$g_{mo} = \frac{-2I_{DSS}}{V_P} \tag{10-17}$$

Since I_{DSS} and V_P are of opposite sign, g_{mo} is always positive. Note that the transconductance varies as the square root of the drain current. The relationship connecting g_{mo}, I_{DSS}, and V_P has been verified experimentally.[7] Since g_{mo} can be measured and I_{DSS} can be read on a dc milliammeter placed in the drain lead (with zero gate excitation), Eq. (10-17) gives a method for obtaining V_P.

The dependence of g_m upon V_{GS} is indicated in Fig. 10-6 for the 2N3277 FET (with $V_P \approx 4.5$ V) and the 2N3278 FET (with $V_P \approx 7$ V). The linear relationship predicted by Eq. (10-16) is seen to be only approximately valid.

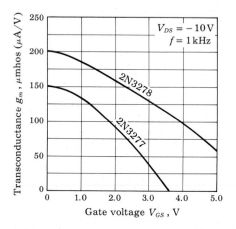

Fig. 10-6 Transconductance g_m versus gate voltage for types 2N3277 and 2N3278 FETs. (Courtesy of Fairchild Semiconductor Company.)

Temperature Dependence Curves of g_m and r_d versus temperature are given in Fig. 10-7.   The drain current I_{DS} has the same temperature variation as does g_m.   The principal reason for the negative temperature coefficient of I_{DS} is that the mobility decreases with increasing temperature.[8]   Since this majority-carrier current decreases with temperature (unlike the bipolar transistor whose minority-carrier current increases with temperature), the troublesome phenomenon of *thermal runaway* (Sec. 9-9) is not encountered with field-effect transistors.

The FET Model A circuit which satisfies Eq. (10-11) is indicated in Fig. 10-8a.   This low-frequency small-signal model has a Norton's output circuit with a dependent current generator whose current is proportional to the gate-to-source voltage.   The proportionality factor is the transconductance g_m, which is consistent with the definition of g_m in Eq. (10-12).   The output resistance is r_d, which is consistent with the definition in Eq. (10-13).   The input resistance between gate and source is infinite, since it is assumed that

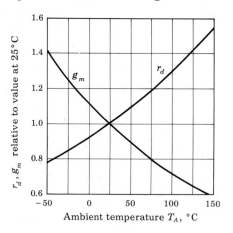

Fig. 10-7 Normalized g_m and normalized r_d versus T_A (for the 2N3277 and the 2N3278 FETs with $V_{DS} = -10$ V, $V_{GS} = 0$ V, and $f = 1$ kHz). (Courtesy of Fairchild Semiconductor Company.)

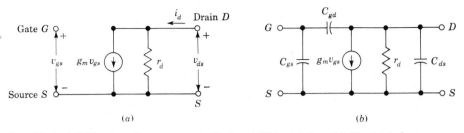

Fig. 10-8 (a) The low-frequency small-signal FET model. (b) The high-frequency model, taking node capacitors into account.

the reverse-biased gate takes no current. For the same reason the resistance between gate and drain is assumed to be infinite.

The FET model of Fig. 10-8a should be compared with the h-parameter model of the bipolar junction transistor of Fig. 8-6. The latter also has a Norton's output circuit, but the current generated depends upon the input *current*, whereas in the FET model the generator current depends upon the input *voltage*. Note that there is no feedback at low frequencies from output to input in the FET, whereas such feedback exists in the bipolar transistor through the parameter h_{re}. Finally, observe that the high (almost infinite) input resistance of the FET is replaced by an input resistance of about 1 K for a CE amplifier. In summary, the field-effect transistor is a much more ideal amplifier than the conventional transistor *at low frequencies*. Unfortunately, this is not true beyond the audio range, as we now indicate.

The high-frequency model given in Fig. 10-8b is identical with Fig. 10-8a except that the capacitances between pairs of nodes have been added. The capacitor C_{gs} represents the barrier capacitance between gate and source, and C_{gd} is the barrier capacitance between gate and drain. The element C_{ds} represents the drain-to-source capacitance of the channel. Because of these internal capacitances, feedback exists between the input and output circuits, and the voltage amplification drops rapidly as the frequency is increased (Sec. 10-11). The order of magnitudes of the parameters in the model for a diffused-junction FET is given in Table 10-1.

TABLE 10-1   Range of parameter values for an FET

Parameter	JFET	MOSFET†
g_m	0.1–10 mA/V	0.1–20 mA/V or more
r_d	0.1–1 M	1–50 K
C_{ds}	0.1–1 pF	0.1–1 pF
C_{gs}, C_{gd}	1–10 pF	1–10 pF
r_{gs}	$>10^8\ \Omega$	$>10^{10}\ \Omega$
r_{gd}	$>10^8\ \Omega$	$>10^{14}\ \Omega$

† Discussed in Sec. 10-5.

10-5   THE METAL–OXIDE–SEMICONDUCTOR FET (MOSFET)

In preceding sections we developed the volt-ampere characteristics and small-signal properties of the junction field-effect transistor. We now turn our attention to the insulated-gate FET, or metal-oxide-semiconductor FET,[9] which is of much greater commercial importance than the junction FET.

The p-channel MOSFET consists of a lightly doped n-type substrate into which two highly doped p^+ regions are diffused, as shown in Fig. 10-9. These p^+ sections, which will act as the source and drain, are separated by about 5 to 10 μm. A thin (1,000 to 2,000 Å) layer of insulating silicon dioxide (SiO$_2$) is grown over the surface of the structure, and holes are cut into the oxide layer, allowing contact with the source and drain. Then the gate-metal area is overlaid on the oxide, covering the entire channel region. Simultaneously, metal contacts are made to the drain and source, as shown in Fig. 10-9. The contact to the metal over the channel area is the gate terminal. The chip area of a MOSFET is 3 square mils or less, which is only about 5 percent of that required by a bipolar junction transistor.

The metal area of the gate, in conjunction with the insulating dielectric oxide layer and the semiconductor channel, form a parallel-plate capacitor. The insulating layer of silicon dioxide is the reason why this device is called the insulated-gate field-effect transistor. This layer results in an extremely high input resistance (10^{10} to 10^{15} Ω) for the MOSFET. The p-channel enhancement MOSFET is the most commonly available field-effect device (1972), and its characteristics will now be described.

The Enhancement MOSFET   If we ground the substrate for the structure of Fig. 10-9 and apply a negative voltage at the gate, an electric field will be directed perpendicularly through the oxide. This field will end on "induced" positive charges on the semiconductor site, as shown in Fig. 10-9. The positive charges, which are minority carriers in the n-type substrate, form an "inversion layer." As the magnitude of the negative voltage on the gate increases, the induced positive charge in the semiconductor increases. The region beneath the oxide now has p-type carriers, the conductivity increases, and current flows from source to drain through the induced channel. Thus

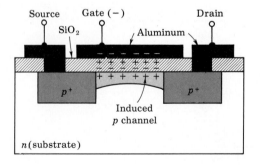

Fig. 10-9   Enhancement in a p-channel MOSFET.   (Courtesy of Motorola Semiconductor Products, Inc.)

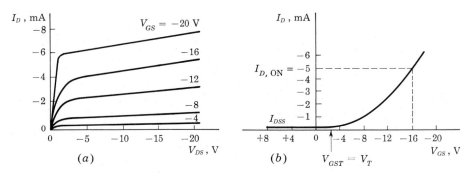

Fig. 10-10 (a) The drain characteristics and (b) the transfer curve (for $V_{DS} =$ 10 V) of a p-channel enhancement-type MOSFET.

the drain current is "enhanced" by the negative gate voltage, and such a device is called an *enhancement-type* MOS.

Threshold Voltage The volt-ampere drain characteristics of a p-channel enhancement-mode MOSFET are given in Fig. 10-10a, and its transfer curve in Fig. 10-10b. The current I_{DSS} at $V_{GS} \geq 0$ is very small, of the order of a few nanoamperes. As V_{GS} is made negative, the current $|I_D|$ increases slowly at first, and then much more rapidly with an increase in $|V_{GS}|$. The manufacturer often indicates the *gate-source* threshold voltage V_{GST}, or V_T,† at which $|I_D|$ reaches some defined small value, say 10 μA. A current $I_{D,ON}$ corresponding approximately to the maximum value given on the drain characteristics, and the value of V_{GS} needed to obtain this current are also usually given on the manufacturer's specification sheets.

The value of V_T for the p-channel standard MOSFET is typically -4 V, and it is common to use a power-supply voltage of -12 V for the drain supply. This large voltage is incompatible with the power-supply voltage of typically 5 V used in bipolar integrated circuits. Thus various manufacturing techniques[10] have been developed to reduce V_T. In general, a low threshold voltage allows (1) the use of a small power-supply voltage, (2) compatible operation with bipolar devices, (3) smaller switching time due to the smaller voltage swing during switching and (4) higher packing densities.

Three methods are used to lower the magnitude of V_T.

1. The high-threshold MOSFET described above uses a silicon crystal with ⟨111⟩ orientation. If a crystal is utilized in the ⟨100⟩ direction it is found that a value of V_T results which is about one-half that obtained with ⟨111⟩ orientation.

2. The silicon nitride approach makes use of a layer of Si_3N_4 and SiO_2,

---

† In this chapter the threshold voltage should not be confused with the volt equivalent of temperature V_T of Sec. 2-9.

whose dielectric constant is about twice that of SiO_2 alone. A FET constructed in this manner (designated an MNOS device) decreases V_T to approximately 2 V.

3. Polycrystalline silicon doped with boron is used as the gate electrode instead of aluminum. This reduction in the difference in contact potential between the gate electrode and the gate dielectric reduces V_T. Such devices are called *silicon gate* MOS transistors. All three of the fabrication methods described above result in a low-threshold device with V_T in the range 1.5 to 2.5 V, whereas the standard high-threshold MOS has a V_T of approximately 4 to 6 V.

Power Supply Requirements   Table 10-2 gives the voltages customarily used with high-threshold and low-threshold p-channel MOSFETs. Note that V_{SS} refers to the source, V_{DD} to the drain, and V_{GG} to the gate supply voltages. The subscript 1 denotes that the source is grounded and the subscript 2 designates that the drain is at ground potential.

The low-threshold MOS circuits require lower power supply voltages and this means less expensive system power supplies. In addition, the input voltage swing for turning the device ON and OFF is smaller for the lower-threshold voltage, and this means faster operation. Another very desirable feature of low-threshold MOS circuits is that they are directly compatible with bipolar ICs. They require and produce essentially the same input and output signal swings and the system designer has the flexibility of using MOS and bipolar circuits in the same system.

TABLE 10-2   Power supply voltages for p-channel MOSFETs, in volts

	V_{SS1}	$-V_{DD1}$	$-V_{GG1}$	V_{SS2}	V_{DD2}	$-V_{GG2}$
High-threshold	0	-12	-24	$+12$	0	-12
Low-threshold	0	-5	-17	$+5$	0	-12

Ion Implantation[10]   The ion-implantation technique demonstrated in Fig. 10-11 provides very precise control of doping. Ions of the proper dopant such as phosphorus or boron are accelerated to a high energy of up to 300,000 eV and are used to bombard the silicon wafer target. The energy of the ions determines the depth of penetration into the target. In those areas where ion implantation is not desired, an aluminum mask or a thick (12,000 Å) oxide layer absorbs the ion. Virtually any value of V_T can be obtained using ion implantation. In addition, we see from Fig. 10-11 that there is no overlap between the gate and drain or gate and source electrodes (compare Fig. 10-11 with Fig. 10-9). Consequently, due to ion implantation, there is a drastic reduction in C_{gd} and C_{gs}.

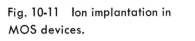

Fig. 10-11   Ion implantation in MOS devices.

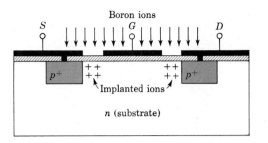

The Depletion MOSFET   A second type of MOSFET can be made if, to the basic structure of Fig. 10-9, a channel is diffused between the source and the drain, with the same type of impurity as used for the source and drain diffusion.   Let us now consider such an n-channel structure, shown in Fig. 10-12a.   With this device an appreciable drain current I_{DSS} flows for zero gate-to-source voltage $V_{GS} = 0$.   If the gate voltage is made negative, positive charges are induced in the channel through the SiO_2 of the gate capacitor. Since the current in an FET is due to majority carriers (electrons for an n-type material), the induced positive charges make the channel less conductive, and the drain current drops as V_{GS} is made more negative.   The redistribution of charge in the channel causes an effective depletion of majority carriers, which accounts for the designation *depletion* MOSFET.   Note in Fig. 10-12b that, because of the voltage drop due to the drain current, the channel region nearest the drain is more depleted than is the volume near the source.   This phenomenon is analogous to that of pinch-off occurring in a JFET at the drain end of the channel (Fig. 10-1).   As a matter of fact, the volt-ampere characteristics of the depletion-mode MOS and the JFET are quite similar.

A MOSFET of the depletion type just described may also be operated

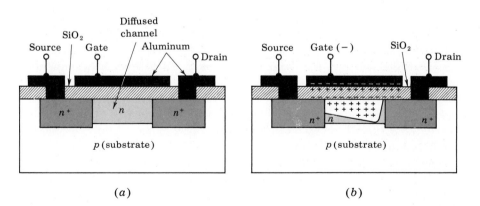

(a)   (b)

Fig. 10-12   (a) An n-channel depletion-type MOSFET.   (b) Channel depletion with the application of a negative gate voltage.   (Courtesy of Motorola Semiconductor Products, Inc.)

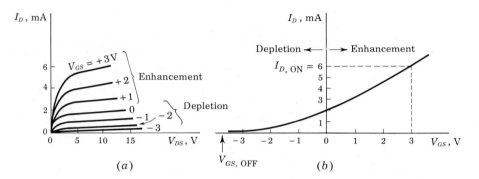

Fig. 10-13   (a) The drain characteristics and (b) the transfer curve (for $V_{DS} = $ 10 V) for an n-channel MOSFET which may be used in either the enhancement or the depletion mode.

in an enhancement mode.   It is only necessary to apply a positive gate voltage so that negative charges are induced into the n-type channel.   In this manner the conductivity of the channel increases and the current rises above I_{DSS}. The volt-ampere characteristics of this device are indicated in Fig. 10-13a, and the transfer curve is given in Fig. 10-13b.   The depletion and enhancement regions, corresponding to V_{GS} negative and positive, respectively, should be noted.   The manufacturer sometimes indicates the *gate-source cutoff voltage* $V_{GS,\text{OFF}}$, at which I_D is reduced to some specified negligible value at a recommended V_{DS}.   This gate voltage corresponds to the pinch-off voltage V_P of a JFET.

The foregoing discussion is applicable in principle also to the p-channel MOSFET.   For such a device the signs of all currents and voltages in the volt-ampere characteristics of Fig. 10-13 must be reversed.

Comparison of p- with n-Channel FETs   The p-channel enhancement FET, shown in Fig. 10-9, is very popular in MOS systems because it is much easier to produce than the n-channel device.   Most of the contaminants in MOS fabrication are mobile ions which are positively charged and are trapped in the oxide layer between gate and substrate.   In an n-channel enhancement device the gate is normally positive with respect to the substrate and, hence, the positively charged contaminants collect along the interface between the SiO_2 and the silicon substrate.   The positive charge from this layer of ions attracts free electrons in the channel which tends to make the transistor turn on prematurely.   In p-channel devices the positive contaminant ions are pulled to the opposite side of the oxide layer (to the aluminum-SiO_2 interface) by the negative gate voltage and there they cannot affect the channel. The hole mobility in silicon and at normal field intensities is approximately 500 cm²/V-s.   On the other hand, electron mobility is about 1,300 cm²/V-s.   Thus the p-channel device will have more than twice the ON resistance of an equivalent n-channel of the same geometry and under the same operating conditions. In other words, the p-channel device must have more than twice the area of the n-channel device to achieve the same resistance. Therefore n-channel MOS circuits can be smaller for the same complexity than p-channel devices. The higher packing density of the n-channel MOS also makes it faster in switching applications due to the smaller junction areas. The operating speed is limited primarily by the internal RC time constants, and the capacitance is directly proportional to the junction cross sections. For all the above reasons it is clear that n-channel MOS circuits are more desirable than p-channel circuits. However, the more extensive process control needed for n-channel fabrication makes them expensive and unable to compete economically with p-channel devices at this time (1972).

MOSFET Gate Protection Since the SiO_2 layer of the gate is extremely thin, it may easily be damaged by excessive voltage. An accumulation of charge on an open-circuited gate may result in a large enough field to punch through the dielectric. To prevent this damage some MOS devices are fabricated with a Zener diode between gate and substrate. In normal operation this diode is open and has no effect upon the circuit. However, if the voltage at the gate becomes excessive, then the diode breaks down and the gate potential is limited to a maximum value equal to the Zener voltage.

Circuit Symbols It is possible to bring out the connection to the substrate externally so as to have a tetrode device. Most MOSFETs, however, are triodes, with the substrate internally connected to the source. The circuit symbols used by several manufacturers are indicated in Fig. 10-14. Often the substrate lead is omitted from the symbol as in (a), and is then under-

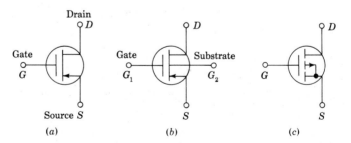

Fig. 10-14 Three circuit symbols for a p-channel MOSFET. (a) and (b) can be either depletion or enhancement types, whereas (c) represents specifically an enhancement device. In (a) the substrate is understood to be connected internally to the source. For an n-channel MOSFET the direction of the arrow is reversed.

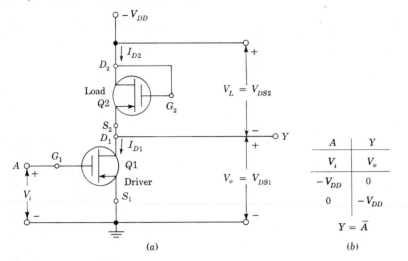

A	Y
V_i	V_o
$-V_{DD}$	0
0	$-V_{DD}$

$$Y = \bar{A}$$

(a)

(b)

Fig. 10-15 (a) MOS inverter (NOT circuit). (b) The voltage truth table and Boolean expression.

stood to be connected to the source internally. For the enhancement-type MOSFET of Fig. 10-14c, G_2 is shown to be internally connected to S.

Small-signal MOSFET Circuit Model[11] If the small resistances of the source and drain regions are neglected, the small-signal equivalent circuit of the MOSFET between terminals $G (= G_1)$, S, and D is identical with that given in Fig. 10-8 for the JFET. The transconductance g_m and the interelectrode capacitances have comparable values for the two types of devices. However, as noted in Table 10-1 on page 321, the drain resistance r_d of the MOSFET is very much smaller than that of the JFET. It should also be noted in Table 10-1 that the input resistance r_{gs} and the feedback resistance r_{gd} are very much larger for the MOSFET than for the JFET.

If the substrate terminal G_2 is not connected to the source, the model of Fig. 10-8 must be generalized as follows: Between node G_2 and S, a diode $D1$ is added to represent the p-n junction between the substrate and the source. Similarly, a second diode $D2$ is included between G_2 and D to account for the p-n junction formed by the substrate and the drain.

10-6   DIGITAL MOSFET CIRCUITS[12]

The most common applications of MOS devices are digital, such as logic gates (discussed in this section) and registers, or memory arrays (Chap. 17). Because of the gate-to-drain and gate-to-source and substrate parasitic capacitances, MOSFET circuits are slower than corresponding bipolar circuits.

However, the lower power dissipation and higher density of fabrication make MOS devices attractive and economical for many low-speed applications.

Inverter   MOSFET digital circuits consist *entirely* of FETs and no other devices such as diodes, resistors, or capacitors (except for parasitic capacitances).   For example, consider the MOSFET inverter of Fig.10-15*a*.   Device *Q*1 is the *driver FET*, whereas *Q*2 acts as its load resistance and is called the *load FET*.   The nonlinear character of the load is brought into evidence as follows: Since the gate is tied to the drain, $V_{GS2} = V_{DS2}$.   The drain characteristics of Fig. 10-10 are reproduced in Fig. 10-16*a*, and the shaded curve represents the locus of the points $V_{GS2} = V_{DS2} = V_L$.   This curve also gives I_{D2} versus V_L (for $V_{GS2} = V_{DS2}$), and its slope gives the incremental load conductance g_L of *Q*2 as a load.   Clearly, the load resistance is nonlinear.   Note that *Q*2 is always conducting, (for $|V_{DS2}| > |V_T|$), regardless of whether *Q*1 is ON or OFF.

An analytical expression for the load curve is given by Eq. (10-8) with $V_{GS} = V_{DS} = V_L$ and with V_P replaced by the threshold voltage $V_{GST} = V_T$.

$$I_{DS} = I_{DSS} \left(1 - \frac{V_{DS}}{V_T}\right)^2 \quad \text{for } |V_{DS}| \geq |V_T| \tag{10-18}$$

and we see that this is a quadratic, rather than a linear, relationship.   From Eq.(10-18) we find (Prob. 10-9) that the load conductance is equal to the transconductance of the FET, $g_L = g_m$.   The same result is obtained in Sec. 10-7.

The incremental resistance is not a very useful parameter when considering large-signal (ON-OFF) digital operation.   It is necessary to draw the *load curve* (corresponding to a *load line* with a constant resistance) on the volt-ampere characteristics of the driver FET *Q*1.   The *load curve* is a plot of

$$I_D = I_{D1} \quad \text{versus} \quad V_{DS1} = V_o = -V_{DD} - V_L = -20 - V_{DS2}$$

where we have assumed a 20-V power supply.   For a given value of $I_{D2} = I_{D1}$, we find $V_{DS2} = V_L$ from the shaded curve in Fig. 10-16*a* and then plot the locus of the values I_{D1} versus $V_o = V_{DS1}$ in Fig. 10-16*b*.   For example, from Fig. 10-16*a* for $I_{D2} = 4$ mA, we find $V_{DS2} = -14$ V.   Hence $I_{D1} = 4$ mA is located at $V_{DS1} = -20 + 14 = -6$ V in Fig. 10-16*b*.

We now confirm that the circuit of Fig. 10-15 is an inverter, or NOT circuit. Let us assume *negative* logic (Sec. 6-1) with the 1, or low state, given by $V(1) \approx -V_{DD} = -20$ V and the 0, or high state, given by $V(0) \approx 0$.   If $V_i = V_{GS1} = -20$ V, then from Fig. 10-16*b*, $V_o = V_{ON} \approx -2$ V.   Hence, $V_i = V(1)$ gives $V_o = V(0)$.   Similarly from Fig. 10-16*b*, if $V_i = 0$ V, then $V_o = -V_{DD} - V_T = -17$ V for $V_T \approx -3$ V.   Hence, $V_i = V(0)$ gives $V_o = V(1)$, thus confirming the truth table of Fig. 10-15*b*.

We shall simplify the remainder of the discussion in this section by assuming that $|V_{ON}|$ and $|V_T|$ are small compared with $|V_{DD}|$ and shall take $V_{ON} = 0$ and $V_T = 0$.   Hence, to a first approximation the load FET may be con-

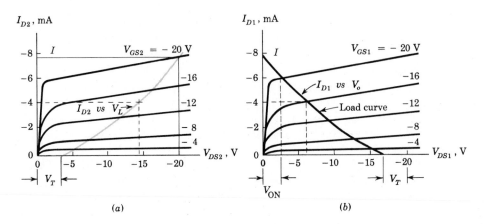

Fig. 10-16   (a) The load current I_{D2} versus $V_L = V_{DS2}$.   (b) The load curve $I_{D1} = I_{D2}$ versus $V_o = V_{DS1}$.

sidered to be a constant resistance R_L and may be represented by a load line passing through $I_D = 0$, $V_{DS} = -V_{DD}$ and $I_D = I$, $V_{DS} = 0$, where I is the drain current for $V_{DS} = V_{GS} = -V_{DD}$. In other words, $R_L = -V_{DD}/I$. Most MOSFETs are p-channel enhancement-type devices, and negative logic is used with $V(0) = 0$ and $V(1) = -V_{DD}$.

NAND **Gate**   The operation of the negative NAND gate of Fig. 10-17 can be understood if we realize that if either input V_1 or V_2 is at 0 V (the 0 state), the corresponding FET is OFF and the current is zero. Hence the voltage drop across the load FET is zero and the output $V_o = -V_{DD}$ (the 1 state). If both V_1 and V_2 are in the 1 state ($V_1 = V_2 = -V_{DD}$), then both $Q1$ and $Q2$ are ON and the output is 0 V, or at the 0 state. These values are in agreement with the voltage truth table of Fig. 10-17b. If 1 is substituted for $-V_{DD}$ in Fig. 10-17b, then this logic agrees with the truth table for a NAND gate, given in Fig. 6-18. We note that only during one of the four possible input states is power delivered by the power supply.

NOR **Gate**   The circuit of Fig. 10-18a is a negative NOR gate. When either one of the two inputs (or both) is at $-V_{DD}$, the corresponding FET is ON and the output is at 0 V. If both inputs are at 0 V, both transistors $Q1$ and $Q2$ are OFF and the output is at $-V_{DD}$. These values agree with the truth table of Fig. 10-18b. Note that power is drawn from the supply during three of the four possible input states. Because of the high density of MOS devices on the same chip, it is important to minimize power consumption in LSI MOSFET systems (Sec. 17-17).
   The circuit of Fig. 10-17 may be considered to be a *positive* NOR gate, and that of Fig. 10-18 to be a *positive* NAND gate (Sec. 6-9). These MOSFET circuits are examples of direct-coupled transistor logic (DCTL), mentioned

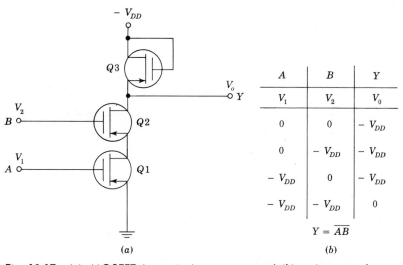

A	B	Y
V_1	V_2	V_0
0	0	$-V_{DD}$
0	$-V_{DD}$	$-V_{DD}$
$-V_{DD}$	0	$-V_{DD}$
$-V_{DD}$	$-V_{DD}$	0

$$Y = \overline{AB}$$

(a)   (b)

Fig. 10-17 (a) MOSFET (negative) NAND gate and (b) voltage truth table and Boolean expression. (Remember that 0 V is the zero state and $-V_{DD}$ is the 1 state.)

in Sec. 6-14.   However, MOSFET DCTL circuits have none of the disadvantages (such as base-current "hogging") of bipolar DCTL gates.   A FLIP-FLOP constructed from MOSFETs is indicated in Prob. 10-11.   An AND (OR) gate is obtained by cascading a NAND (NOR) gate with a NOT gate.   Typically, a three-input NAND gate uses about 16 mils² of chip area, whereas a single bipolar junction transistor may need about three times this area.

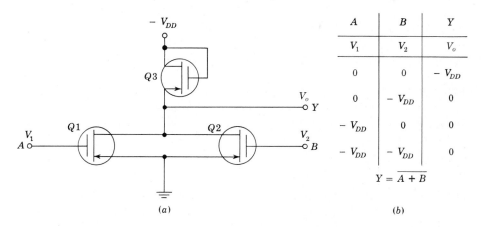

A	B	Y
V_1	V_2	V_0
0	0	$-V_{DD}$
0	$-V_{DD}$	0
$-V_{DD}$	0	0
$-V_{DD}$	$-V_{DD}$	0

$$Y = \overline{A + B}$$

(a)   (b)

Fig. 10-18 (a) MOSFET (negative) NOR gate and (b) voltage truth table and Boolean equation.

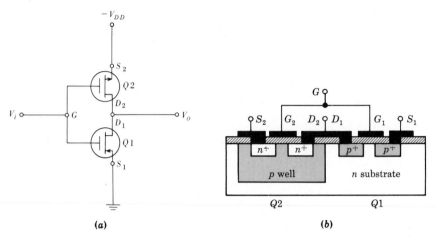

Fig. 10-19 (a) Complementary MOS inverter. (b) Cross section of complementary MOSFETS. Note that the p-type well is diffused into the n-type substrate and that the n-channel MOS $Q2$ is formed in this region.

Complementary MOS (CMOS)[12] It is possible to reduce the power dissipation to very small (50 nW) levels by using complementary p-channel and n-channel enhancement MOS devices on the same chip. The basic complementary MOS inverter circuit is shown in Fig. 10-19. Transistor $Q1$ is the p-channel unit, and transistor $Q2$ is n-channel. The two devices are in series, with their drains tied together and their gates also connected together. The logic swing gate voltage V_i varies from 0 V to the power supply $-V_{DD}$. When $V_i = -V_{DD}$ (logic 1) transistor $Q1$ is turned ON (but draws no appreciable steady-state current) and $Q2$ is turned OFF, the output V_o is then at 0 V (logic 0), and inversion has been accomplished. When zero voltage (logic 0) is applied at the input, the n-channel $Q2$ is turned ON (at no steady-state current) and $Q1$ is turned OFF. Thus the output is at $-V_{DD}$ (logic 1). In either logic state, $Q1$ or $Q2$ is OFF and the quiescent power dissipation for this simple inverter is the product of the OFF leakage current and $-V_{DD}$.

More complicated digital CMOS circuits (NAND, NOR, and FLIP-FLOPS) can be formed by combining simple inverter circuits (Probs. 10-13 and 10-14).

The remainder of the chapter considers the FET under small-signal operation. We first discuss low-frequency gain, then methods of biasing the device in the linear range, and finally the high-frequency limitations of the FET.

10-7   THE LOW–FREQUENCY COMMON–SOURCE AND COMMON–DRAIN AMPLIFIERS

The common-source (CS) stage is indicated in Fig. 10-20a, and the common-drain (CD) configuration in Fig. 10-20b. The former is analogous to the

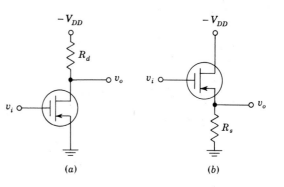

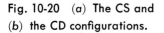

Fig. 10-20 (a) The CS and
(b) the CD configurations.

(a)   (b)

bipolar transistor CE amplifier, and the latter to the CC stage. We shall
analyze both of these circuits simultaneously by considering the generalized
configuration in Fig. 10-21a. For the CS stage the output is v_{o1} taken at the
drain and $R_s = 0$. For the CD stage the output is v_{o2} taken at the source and
$R_d = 0$. The signal-source resistance is unimportant since it is in series with
the gate, which draws negligible current. No biasing arrangements are indicated (Sec. 10-8), but it is assumed that the stage is properly biased for linear
operation.

Replacing the FET by its low-frequency small-signal model of Fig. 10-8,
the equivalent circuit of Fig. 10-21b is obtained. Applying KVL to the output
circuit yields

$$i_d R_d + (i_d - g_m v_{gs})r_d + i_d R_s = 0 \tag{10-19}$$

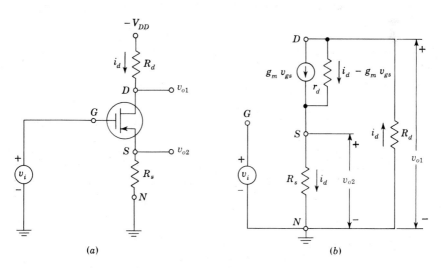

(a)   (b)

Fig. 10-21 (a) A generalized FET amplifier configuration. (b) The small-
signal equivalent circuit.

From Fig. 10-21b the voltage from G to S is given by

$$v_{gs} = v_i - i_d R_s \tag{10-20}$$

Combining Eqs. (10-19) and (10-20) and remembering that $\mu = r_d g_m$ [Eq. (10-15)], we find

$$i_d = \frac{\mu v_i}{r_d + R_d + (\mu + 1) R_s} \tag{10-21}$$

The CS Amplifier with an Unbypassed Source Resistance Since $v_{o1} = -i_d R_d$, then

$$v_{o1} = \frac{-\mu v_i R_d}{r_d + R_d + (\mu + 1) R_s} \tag{10-22}$$

From Eq. (10-22) we obtain the Thévenin's equivalent circuit of Fig. 10-22a "looking into" the drain node (to ground). The open-circuit voltage is $-\mu v_i$, and the output resistance is $R_o = r_d + (\mu + 1) R_s$. The voltage gain is $A_V = v_{o1}/v_i$. The minus sign in Eq. (10-22) indicates that the output is 180° out of phase with the input. If R_s is bypassed with a large capacitance or if the source is grounded, the above equations are valid with $R_s = 0$. Under these circumstances,

$$A_V = \frac{v_{o1}}{v_i} = \frac{-\mu R_d}{r_d + R_d} = -g_m R_d' \tag{10-23}$$

where $\mu = r_d g_m$ [Eq. (10-15)] and $R_d' = R_d \| r_d$.

The CD Amplifier with a Drain Resistance Since $v_{o2} = i_d R_s$, then from Eq. (10-21)

$$v_{o2} = \frac{\mu v_i R_s}{r_d + R_d + (\mu + 1) R_s} = \frac{[\mu v_i/(\mu + 1)] R_s}{(r_d + R_d)/(\mu + 1) + R_s} \tag{10-24}$$

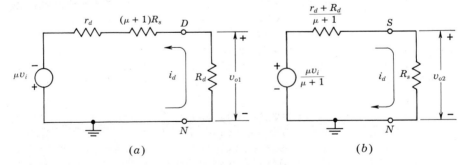

(a)   (b)

Fig. 10-22 The equivalent circuits for the generalized amplifier of Fig. 10-21 "looking into" (a) the drain and (b) the source. Note that $\mu = r_d g_m$.

From Eq. (10-24) we obtain the Thévenin's equivalent circuit of Fig. 10-22b "looking into" the source node (to ground). The open-circuit voltage is $\mu v_i/(\mu + 1)$, and the output resistance is $R_o = (r_d + R_d)/(\mu + 1)$. The voltage gain is $A_V = v_{o2}/v_i$. Note that there is no phase shift between input and output. If $R_d = 0$ and if $(\mu + 1)R_s \gg r_d$, then $A_V \approx \mu/(\mu + 1) \approx 1$ for $\mu \gg 1$. A voltage gain of unity means that the output (at the source) follows the input (at the gate). Hence the CD configuration is called a *source follower* (analogous to the *emitter* follower for a bipolar junction transistor).

Note that the open-circuit voltage and the output impedance in either Fig. 10-22a or b are independent of the load (R_d in Fig. 10-22a and R_s in Fig. 10-22b). These restrictions must be satisfied if the networks in Fig. 10-22 are to represent the true Thévenin equivalents of the amplifier in Fig. 10-21.

For the source follower ($R_d = 0$) with $\mu \gg 1$, the output conductance is

$$g_o = \frac{1}{R_o} = \frac{\mu + 1}{r_d} \approx \frac{\mu}{r_d} = g_m \tag{10-25}$$

which agrees with the result obtained in Sec. 10-6 for the conductance looking into the source of a MOSFET with the gate at a constant voltage. In the discussion of diffused resistors in Sec. 7-8, it is indicated that 30 K is about the maximum resistance that can be fabricated. Larger values may be obtained by using the MOS structure as a load with gate connected to drain and tied to a fixed voltage such as $Q2$ in Fig. 10-15. By using a low g_m FET, a high value of effective resistance is obtained. For example, for $g_m = 10 \ \mu A/V$, we obtain $R_o = 1/g_m = 100$ K. This value of effective resistance requires approximately 5 mil² of chip area compared with 300 mil² to yield a diffused 20-K resistance.

10-8   BIASING THE FET

The selection of an appropriate operating point (I_D, V_{GS}, V_{DS}) for an FET amplifier stage is determined by considerations similar to those given to transistors, as discussed in Chap. 9. These considerations are output-voltage swing, distortion, power dissipation, voltage gain, and drift of drain current. In most cases it is not possible to satisfy all desired specifications simultaneously. In this section we examine several biasing circuits for field-effect devices.

Source Self-bias   The configuration shown in Fig. 10-23 can be used to bias junction FET devices or depletion-mode MOS transistors. For a specified drain current I_D, the corresponding gate-to-source voltage V_{GS} can be obtained applying either Eq. (10-8) or the plotted drain or transfer characteristics. Since the gate current (and, hence, the voltage drop across R_g) is negligible, the source resistance R_s can be found as the ratio of V_{GS} to the desired I_D.

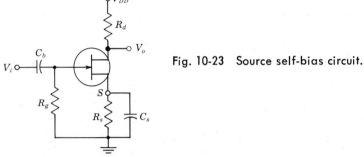

Fig. 10-23   Source self-bias circuit.

EXAMPLE The amplifier of Fig. 10-23 utilizes an n-channel FET for which $V_P = -2.0$ V and $I_{DSS} = 1.65$ mA. It is desired to bias the circuit at $I_D = 0.8$ mA, using $V_{DD} = 24$ V. Assume $r_d \gg R_d$. Find (a) V_{GS}, (b) g_m, (c) R_s, (d) R_d, such that the voltage gain is at least 20 dB, with R_s bypassed with a very large capacitance C_s.

Solution a. Using Eq. (10-8), we have $0.8 = 1.65(1 + V_{GS}/2.0)^2$. Solving, $V_{GS} = -0.62$ V.

b. Equation (10-17) now yields

$$g_{mo} = -\frac{2I_{DSS}}{V_P} = \frac{(2)(1.65)}{2} = 1.65 \text{ mA/V}$$

and from Eq. (10-16)

$$g_m = g_{mo}\left(1 - \frac{V_{GS}}{V_P}\right) = (1.65)\left(1 - \frac{0.62}{2.0}\right) = 1.14 \text{ mA/V}$$

c. $R_s = -\dfrac{V_{GS}}{I_D} = \dfrac{0.62}{0.8} = 0.77 \text{ K} = 770 \ \Omega$

d. Since 20 dB corresponds to a voltage gain of 10, then from Eq. (10-23), with $r_d \gg R_d$, $|A_V| = g_m R_d \geq 10$, or $R_d \geq 10/1.14 = 8.76$ K.

Biasing against Device Variation   FET manufacturers usually supply information on the maximum and minimum values of I_{DSS} and V_P at room temperature. They also supply data to correct these quantities for temperature variations. The transfer characteristics for a given type of n-channel FET may appear as in Fig. 10-24a, where the top and bottom curves are for extreme values of temperature and device variation. Assume that, on the basis of considerations previously discussed, it is necessary to bias the device at a drain current which will not drift outside of $I_D = I_A$ and $I_D = I_B$. Then the bias line $V_{GS} = -I_D R_s$ must intersect the transfer characteristics between the points A and B, as indicated in Fig. 10-24a. The slope of the bias line is determined by the source resistance R_s. For any transfer characteristic between the two extremes indicated, the current I_Q is such that $I_A < I_Q < I_B$, as desired.

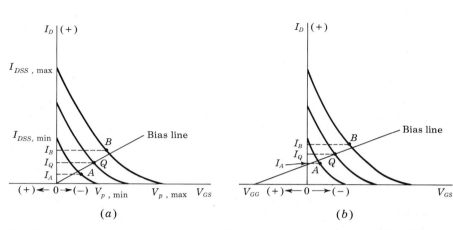

(a)   *(b)*

Fig. 10-24   Maximum and minimum transfer curves for an *n*-channel FET.   The
drain current must lie between I_A and I_B.   The bias line can be drawn through the
origin for the current limits indicated in (*a*), but this is not possible for the currents
specified in (*b*).

Consider the situation indicated in Fig. 10-24*b*, where a line drawn to pass
between points *A* and *B* does not pass through the origin.   This bias line
satisfies the equation

$$V_{GS} = V_{GG} - I_D R_s \qquad (10\text{-}26)$$

Such a bias relationship may be obtained by adding a fixed bias to the gate
in addition to the source self-bias, as indicated in Fig. 10-25*a*.   A circuit

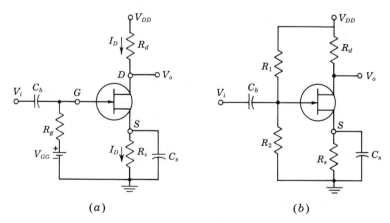

(a)   *(b)*

Fig. 10-25   (*a*) Biasing an FET with a fixed-bias V_{GG} in addition to
self-bias through R_s.   (*b*) A single power-supply configuration
which is equivalent to the circuit in (*a*).

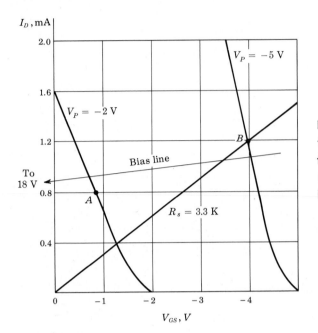

Fig. 10-26   Extreme transfer curves for the 2N3684 field-effect transistor. (Courtesy of Union Carbide Corporation.)

requiring only one power supply and which can satisfy Eq. (10-26) is shown in Fig. 10-25b.   For this circuit

$$V_{GG} = \frac{R_2 V_{DD}}{R_1 + R_2} \qquad R_g = \frac{R_1 R_2}{R_1 + R_2}$$

We have assumed that the gate current is negligible.   It is also possible for V_{GG} to fall in the reverse-biased region so that the line in Fig. 10-24b intersects the axis of abscissa to the right of the origin.   Under these circumstances two separate supply voltages must be used.

EXAMPLE   FET 2N3684 is used in the circuit of Fig. 10-25b.   For this n-channel device the manufacturer specifies $V_{P,\mathrm{min}} = -2$ V, $V_{P,\mathrm{max}} = -5$ V, $I_{DSS,\mathrm{min}} = 1.6$ mA, and $I_{DSS,\mathrm{max}} = 7.05$ mA.   The extreme transfer curves are plotted in Fig. 10-26.   It is desired to bias the circuit so that $I_{D,\mathrm{min}} = 0.8$ mA $= I_A$ and $I_{D,\mathrm{max}} = 1.2$ mA $= I_B$ for $V_{DD} = 24$ V.   Find (a) V_{GG} and R_s, and (b) the range of possible values in I_D if $R_s = 3.3$ K and $V_{GG} = 0$.

Solution   a. The bias line will lie between A and B, as indicated, if it is drawn to pass through the two points $V_{GS} = 0$, $I_D = 0.9$ mA, and $V_{GS} = -4$ V, $I_D = 1.1$ mA.   The slope of this line determines R_s, or

$$R_s = \frac{4 - 0}{1.1 - 0.9} = 20 \text{ K}$$

Then, from the first point and Eq. (10-26), we find

$$V_{GG} = I_D R_s = (0.9)(20) = 18 \text{ V}$$

Fig. 10-27 (a) Drain-to-
gate bias circuit for en-
hancement-mode MOS
transistors; (b) improved
version of (a).

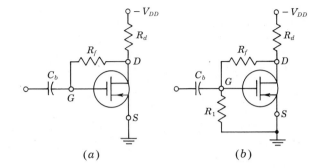

(a)   (b)

  b. If $R_s = 3.3$ K, we see from the curves that $I_{D,\min} = 0.4$ mA and $I_{D,\max} = 1.2$ mA. The minimum current is far below the specified value of 0.8 mA.

  Biasing the Enhancement MOSFET   The self-bias technique of Fig. 10-23 cannot be used to establish an operating point for the enhancement-type MOSFET because the voltage drop across R_s is in a direction to reverse-bias the gate, and a forward gate bias is required.   The circuit of Fig. 10-27a can be used, and for this case we have $V_{GS} = V_{DS}$, since no current flows through R_f.   If for reasons of linearity in device operation or maximum output voltage it is desired that $V_{GS} \neq V_{DS}$, then the circuit of Fig. 10-27b is suitable.   We note that $V_{GS} = [R_1/(R_1 + R_f)]V_{DS}$.   Both circuits discussed here offer the advantages of dc stabilization through the feedback introduced with R_f. However, the input impedance is reduced because, by Miller's theorem (Sec. 8-11), R_f corresponds to an equivalent resistance $R_i = R_f/(1 - A_V)$ shunting the amplifier input.
  Finally, note that the circuit of Fig. 10-25b is often used with the enhancement MOSFET.   The dc stability introduced in Fig. 10-27 through the feedback resistor R_f is then missing, and is replaced by the dc feedback through R_s.

10-9   THE FET AS A VOLTAGE–VARIABLE RESISTOR[13] (VVR)

In most linear applications of field-effect transistors the device is operated in the constant-current portion of its output characteristics.   We now consider FET transistor operation in the region before pinch-off, where V_{DS} is small. In this region the FET is useful as a voltage-controlled resistor; i.e., the drain-to-source resistance is controlled by the bias voltage V_{GS}.   In such an application the FET is also referred to as a *voltage-variable resistor* (VVR), or *voltage-dependent resistor* (VDR).
  Figure 10-28a shows the low-level bidirectional characteristics of an FET. The slope of these characteristics gives r_d as a function of V_{GS}.   Figure 10-28a has been extended into the third quadrant to give an idea of device linearity around $V_{DS} = 0$.

In our treatment of the junction FET characteristics in Sec. 10-3, we derive Eq. (10-5), which gives the drain-to-source conductance $g_d = I_D/V_{DS}$ for small values of V_{DS}. From this equation we have

$$g_d = g_{do}\left[1 - \left(\frac{V_{GS}}{V_P}\right)^{\frac{1}{2}}\right] \qquad (10\text{-}27)$$

where g_{do} is the value of the drain conductance when the bias is zero. Variation of r_d with V_{GS} is plotted in Fig. 10-28b for the 2N3277 and 2N3278 FETs. The variation of r_d with V_{GS} can be closely approximated by the empirical expression

$$r_d = \frac{r_o}{1 - KV_{GS}} \qquad (10\text{-}28)$$

where r_o = drain resistance at zero gate bias
K = a constant, dependent upon FET type
V_{GS} = gate-to-source voltage

Applications of the VVR Since the FET operated as described above acts like a variable passive resistor, it finds applications in many areas where this property is useful. The VVR, for example, can be used to vary the voltage gain of a multistage amplifier A as the signal level is increased. This action is called *automatic gain control* (AGC). A typical arrangement is shown in Fig. 10-29. The signal is taken at a high-level point, rectified, and filtered to produce a dc voltage proportional to the output-signal level. This voltage is applied to the gate of $Q2$, thus causing the ac resistance between the drain and source to change, as shown in Fig. 10-28b. We thus may cause the gain of transistor $Q1$ to decrease as the output-signal level increases. The dc bias

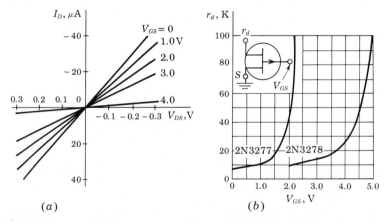

(a)   (b)

Fig. 10-28 (a) FET low-level drain characteristics for 2N3278.
(b) Small-signal FET resistance variation with applied gate voltage.
(Courtesy of Fairchild Semiconductor Company.)

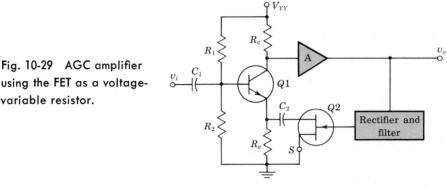

Fig. 10-29 AGC amplifier
using the FET as a voltage-
variable resistor.

conditions of $Q1$ are not affected by $Q2$ since $Q2$ is isolated from $Q1$ by means
of capacitor C_2.

10-10   THE COMMON–SOURCE AMPLIFIER AT HIGH FREQUENCIES

The circuits discussed in this and the following section apply equally well to
either JFETs or MOSFETs (except for the method of biasing). The low-
frequency analysis of Sec. 10-8 is now modified to take into account the effect
of the internal node capacitances.

Voltage Gain   The circuit of Fig. 10-30a is the basic CS amplifier configuration. If the FET is replaced by the circuit model of Fig. 10-8b, we
obtain the network in Fig. 10-30b. The output voltage V_o between D and S
is easily found with the aid of the theorem of Sec. 8-7, namely, $V_o = IZ$, where
I is the short-circuit current and Z is the impedance seen between the terminals.

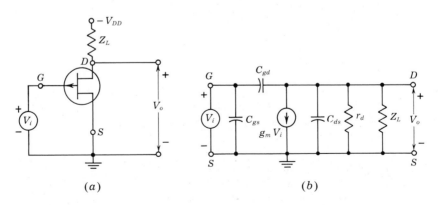

Fig. 10-30 (a) The common-source amplifier circuit; (b) small-signal
equivalent circuit at high frequencies. (The biasing network is not
indicated.)

To find Z, the independent generator V_i is (imagined) short-circuited, so that $V_i = 0$, and hence there is no current in the dependent generator $g_m V_i$. We then note that Z is the parallel combination of the impedances corresponding to Z_L, C_{ds}, r_d, and C_{gd}. Hence

$$Y = \frac{1}{Z} = Y_L + Y_{ds} + g_d + Y_{gd} \tag{10-29}$$

where
$Y_L = 1/Z_L$ = admittance corresponding to Z_L
$Y_{ds} = j\omega C_{ds}$ = admittance corresponding to C_{ds}
$g_d = 1/r_d$ = conductance corresponding to r_d
$Y_{gd} = j\omega C_{gd}$ = admittance corresponding to C_{gd}

The current in the direction from D to S in a zero-resistance wire connecting the output terminals is

$$I = -g_m V_i + V_i Y_{gd} \tag{10-30}$$

The amplification A_V with the load Z_L in place is given by

$$A_V = \frac{V_o}{V_i} = \frac{IZ}{V_i} = \frac{I}{V_i Y} \tag{10-31}$$

or from Eqs. (10-29) and (10-30)

$$A_V = \frac{-g_m + Y_{gd}}{Y_L + Y_{ds} + g_d + Y_{gd}} \tag{10-32}$$

At low frequencies the FET capacitances can be neglected and hence

$$Y_{ds} = Y_{gd} = 0$$

Under these conditions Eq. (10-32) reduces to

$$A_V = \frac{-g_m}{Y_L + g_d} = \frac{-g_m r_d Z_L}{r_d + Z_L} = -g_m Z_L' \tag{10-33}$$

where $Z_L' \equiv Z_L \| r_d$. This equation agrees with Eq. (10-23), with Z_L replaced by R_d.

Input Admittance An inspection of Fig. 10-30b reveals that the gate circuit is not isolated from the drain circuit, but rather that they are connected by C_{gd}. From Miller's theorem (Sec. 8-11), this admittance may be replaced by $Y_{gd}(1 - K)$ between G and S, and by $Y_{gd}(1 - 1/K)$ between D and S, where $K = A_V$. Hence the input admittance is given by

$$Y_i = Y_{gs} + (1 - A_V)Y_{gd} \tag{10-34}$$

This expression indicates that for an FET to possess negligible input admittance over a wide range of frequencies, the gate-source and gate-drain capacitances must be negligible.

Input Capacitance (Miller Effect) Consider an FET with a drain-circuit resistance R_d. From the previous discussion it follows that within the audio-frequency range, the gain is given by the simple expression $A_V = -g_m R'_d$, where R'_d is $R_d \| r_d$. In this case, Eq. (10-34) becomes

$$\frac{Y_i}{j\omega} \equiv C_i = C_{gs} + (1 + g_m R'_d)C_{gd} \qquad (10\text{-}35)$$

This increase in input capacitance C_i over the capacitance from gate to source is called the *Miller effect*.

This input capacitance is important in the operation of cascaded amplifiers. In such a system the output from one stage is used as the input to a second amplifier. Hence the input impedance of the second stage acts as a shunt across the output of the first stage and R_d is shunted by the capacitance C_i. Since the reactance of a capacitor decreases with increasing frequencies, the resultant output impedance of the first stage will be correspondingly low for the high frequencies. This will result in a decreasing gain at the higher frequencies.

EXAMPLE A MOSFET has a drain-circuit resistance R_d of 100 K and operates at 20 kHz. Calculate the voltage gain of this device as a single stage, and then as the first transistor in a cascaded amplifier consisting of two identical stages. The MOSFET parameters are $g_m = 1.6$ mA/V, $r_d = 44$ K, $C_{gs} = 3.0$ pF, $C_{ds} = 1.0$ pF, and $C_{gd} = 2.8$ pF.

Solution

$$Y_{gs} = j\omega C_{gs} = j2\pi \times 2 \times 10^4 \times 3.0 \times 10^{-12} = j3.76 \times 10^{-7} \ \mho$$

$$Y_{ds} = j\omega C_{ds} = j1.26 \times 10^{-7} \ \mho$$

$$Y_{gd} = j\omega C_{gd} = j3.52 \times 10^{-7} \ \mho$$

$$g_d = \frac{1}{r_d} = 2.27 \times 10^{-5} \ \mho$$

$$Y_d = \frac{1}{R_d} = 10^{-5} \ \mho$$

$$g_m = 1.60 \times 10^{-3} \ \mho$$

The gain of a one-stage amplifier is given by Eq. (10-32):

$$A_V = \frac{-g_m + Y_{gd}}{g_d + Y_d + Y_{ds} + Y_{gd}} = \frac{-1.60 \times 10^{-3} + j3.52 \times 10^{-7}}{3.27 \times 10^{-5} + j4.78 \times 10^{-7}}$$

It is seen that the j terms (arising from the interelectrode capacitances) are negligible in comparison with the real terms. If these are neglected, then $A_V = -48.8$. This value can be checked by using Eq. (10-23), which neglects interelectrode capacitances. Thus

$$A_V = \frac{-\mu R_d}{R_d + r_d} = \frac{-1.6 \times 44 \times 100}{100 + 44} = -48.8 = -g_m R'_d$$

Since the gain is a real number, the input impedance consists of a capacitor whose value is given by Eq. (10-35):

$$C_i = C_{gs} + (1 + g_m R'_d)C_{gd} = 3.0 + (1 + 49)(2.8) = 143 \text{ pF}$$

Consider now a two-stage amplifier, each stage consisting of an FET operating as above. The gain of the second stage is that just calculated. However, in calculating the gain of the first stage, it must be remembered that *the input impedance of the second stage acts as a shunt on the output of the first stage.* Thus the drain load now consists of a 100-K resistance in parallel with 143 pF. To this must be added the capacitance from drain to source of the first stage since this is also in shunt with the drain load. Furthermore, any stray capacitances due to wiring should be taken into account. For example, for every 1-pF capacitance between the leads going to the drain and gate of the second stage, 50 pF is effectively added across the load resistor of the first stage! This clearly indicates the importance of making connections with very short direct leads in high-frequency amplifiers. Let it be assumed that the input capacitance, taking into account the various factors just discussed, is 200 pF. Then the load admittance is

$$Y_L = \frac{1}{R_d} + j\omega C_i = 10^{-5} + j2\pi \times 2 \times 10^4 \times 200 \times 10^{-12}$$

$$= 10^{-5} + j2.52 \times 10^{-5} \; \mho$$

The gain is given by Eq. (10-33):

$$A_V = \frac{-g_m}{g_d + Y_L} = \frac{-1.6 \times 10^{-3}}{2.27 \times 10^{-5} + 10^{-5} + j2.52 \times 10^{-5}}$$

$$= -30.7 + j23.7 = 38.8 \underline{/143.3°}$$

Thus the effect of the capacitances has been to reduce the magnitude of the amplification from 48.8 to 38.8 and to change the phase angle between the output and input from 180 to 143.3°.

If the frequency were higher, the gain would be reduced still further. For example, this circuit would be useless as a video amplifier, say, to a few megahertz, since the gain would then be less than unity. This variation of gain with frequency is called *frequency distortion.* Cascaded amplifiers and frequency distortion are discussed in detail in Chap. 12.

Output Admittance   For the common-source amplifier of Fig. 10-30 the output impedance is obtained by "looking into the drain" with the input voltage set equal to zero. If $V_i = 0$ in Fig. 10-30b, we see r_d, C_{ds}, and C_{gd} in parallel. Hence the output admittance with Z_L considered external to the amplifier is given by

$$Y_o = g_d + Y_{ds} + Y_{gd} \qquad\qquad (10\text{-}36)$$

10-11   THE COMMON–DRAIN AMPLIFIER AT HIGH FREQUENCIES

The source-follower configuration is given in Fig. 10-20b, with $R_d = 0$, and is repeated in Fig. 10-31a. Its equivalent circuit with the FET replaced by its high-frequency model of Fig. 10-8b is shown in Fig. 10-31b.

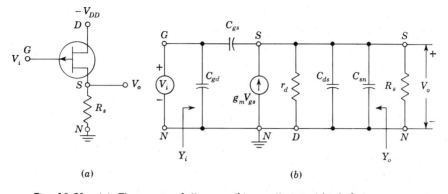

Fig. 10-31 (a) The source-follower; (b) small-signal high-frequency equivalent circuit. (The biasing network is not indicated.)

Voltage Gain The output voltage V_o can be found from the product of the short-circuit current and the impedance between terminals S and N. We now find for the voltage gain

$$A_V = \frac{(g_m + j\omega C_{gs})R_s}{1 + (g_m + g_d + j\omega C_T)R_s} \tag{10-37}$$

$$C_T \equiv C_{gs} + C_{ds} + C_{sn} \tag{10-38}$$

where C_{sn} represents the capacitance from source to ground. At low frequencies the gain reduces to

$$A_V \approx \frac{g_m R_s}{1 + (g_m + g_d)R_s} \tag{10-39}$$

Note that the amplification is positive and has a value less than unity. If $g_m R_s \gg 1$, then $A_V \approx g_m/(g_m + g_d) = \mu/(\mu + 1)$.

Input Admittance The source follower offers the important advantage of lower input capacitance than the CS amplifier. The input admittance Y_i is obtained by applying Miller's theorem to C_{gs}. We find

$$Y_i = j\omega C_{gd} + j\omega C_{gs}(1 - A_V) \approx j\omega C_{gd} \tag{10-40}$$

because $A_V \approx 1$.

Output Admittance The output admittance Y_o, with R_s considered external to the amplifier, is given by

$$Y_o = g_m + g_d + j\omega C_T \tag{10-41}$$

where C_T is given by Eq. (10-38). At low frequencies the output resistance R_o is

$$R_o = \frac{1}{g_m + g_d} \approx \frac{1}{g_m} \tag{10-42}$$

since $g_m \gg g_d$. For $g_m = 2$ mA/V, then $R_o = 500$ Ω.

The source follower is used for the same applications as the emitter follower, those requiring high input impedance and low output impedance.

REFERENCES

1. Shockley, W.: A Unipolar Field-effect Transistor, *Proc. IRE*, vol. 40, pp. 1365–1376, November, 1952.

   Dacey, G. C., and I. M. Ross: The Field Effect Transistor, *Bell System Tech. J.*, vol. 34, pp. 1149–1189, November, 1955.

   Wallmark, J. T., and H. Johnson: "Field-effect Transistors," Prentice-Hall, Inc., Englewood Cliffs, N.J., 1966.

   Sevin, L. J.: "Field-effect Transistors," McGraw-Hill Book Company, New York, 1965.

2. Millman, J., and H. Taub: "Pulse, Digital, and Switching Waveforms," sec. 17-20, McGraw-Hill Book Company, New York, 1965.

3. Wallmark, J. T., and H. Johnson: "Field-effect Transistors," p. 115, Prentice-Hall, Inc., Englewood Cliffs, N.J., 1966.

4. Sevin, L. J., Ref. 1, pp. 13–17.

5. Halladay, H. E., and A. Van der Ziel: DC Characteristics of Junction Gate Field-effect Transistors, *IEEE Trans. Electron. Devices*, vol. ED-13, no. 6, pp. 531–532, June, 1966.

6. Sevin, L. J., Ref. 1, p. 21.

7. Sevin, L. J., Ref. 1, p. 23.

8. Sevin, L. J., Ref. 1, p. 34.

9. Ref. 3, pp. 187–215.

10. Macdougall, J., and K. Manchester: Ion Implantation, *Electronics*, vol. 43, no. 13, no. 13, pp. 86–90, June 22, 1970.

11. Ref. 3, pp. 256–259.

12. Garrett, L.: Integrated-circuit Digital Logic Families, *Spectrum*, vol. 7, no. 12, pp. 30–42, December, 1970.

13. Bilotti, A.: Operation of a MOS Transistor as a Voltage Variable Resistor, *Proc. IEEE*, vol. 54, pp. 1093–1094, August, 1966.

REVIEW QUESTIONS

10-1 (*a*) Sketch the basic structure of an *n*-channel junction field-effect transistor. (*b*) Show the circuit symbol for the JFET.

10-2 (a) Draw a family of CS drain characteristics of an n-channel JFET. (b) Explain the shape of these curves qualitatively.

10-3 How does the FET behave (a) for small values of $|V_{DS}|$? (b) For large $|V_{DS}|$?

10-4 (a) Define the *pinch-off voltage* V_P. (b) Sketch the depletion region before and after pinch-off.

10-5 Sketch the geometry of a JFET in integrated form.

10-6 (a) How does the drain current vary with gate voltage in the saturation region? (b) How does the transconductance vary with drain current?

10-7 Define (a) *transconductance* g_m, (b) *drain resistance* r_d, and (c) *amplification factor* μ of an FET.

10-8 Give the order of magnitude of g_m, r_d, and μ for a MOSFET.

10-9 Show the small-signal model of an FET (a) at low frequencies and (b) at high frequencies.

10-10 (a) Sketch the cross section of a p-channel enhancement MOSFET. (b) Show two circuit symbols for this MOSFET.

10-11 For the MOSFET in Rev. 10-10 draw (a) the drain characteristics and (b) the transfer curve.

10-12 Repeat Rev. 10-10 for an n-channel depletion MOSFET.

10-13 (a) Draw the circuit of a MOSFET NOT circuit. (b) Explain how it functions as an inverter.

10-14 (a) Explain how a MOSFET is used as a load. (b) Obtain the volt-ampere characteristic of this load graphically.

10-15 Sketch a two-input NAND gate and verify that it satisfies the Boolean NAND equation.

10-16 Repeat Rev. 10-15 for a two-input NOR gate.

10-17 Sketch a CMOS inverter and explain its operation.

10-18 (a) Draw the circuit of an FET amplifier with a source resistance R_s and a drain resistance R_d. (b) What is the Thévenin's equivalent circuit looking into the drain at low frequencies?

10-19 Repeat Rev. 10-18 looking into the source.

10-20 (a) Sketch the circuit of a source-follower. At low frequencies what is (b) the maximum value of the voltage gain? (c) The order of magnitude of the output impedance?

10-21 (a) Sketch the circuit of a CS amplifier. (b) Derive the expression for the voltage gain at low frequencies. (c) What is the maximum value of A_V?

10-22 (a) Draw two biasing circuits for a JFET or a depletion-type MOSFET. (b) Explain under what circumstances each of these two arrangements should be used.

10-23 Draw two biasing circuits for an enhancement-type MOSFET.

10-24 (a) How is an FET used as a voltage-variable resistance? (b) Explain.

10-25 (a) Sketch the small-signal high-frequency circuit of a CS amplifier. (b) Derive the expression for the voltage gain.

10-26 (a) From the circuit of Rev. 10-25, derive the input admittance. (b) What is the expression for the input capacitance in the audio range?

10-27 What specific capacitance has the greatest effect on the high-frequency response of a cascade of FET amplifiers? Explain.

10-28 Repeat Rev. 10-25 for a source-follower circuit.

10-29 Repeat Rev. 10-26 for a CD amplifier.

11 / THE TRANSISTOR AT HIGH FREQUENCIES

At low frequencies it is assumed that the transistor responds instantly to changes of input voltage or current. Actually, such is not the case, because the mechanism of the transport of charge carriers from emitter to collector is essentially one of diffusion. Hence, to find out how the transistor behaves at high frequencies, it is necessary to examine this diffusion mechanism in more detail. Such an analysis[1] is complicated, and the resulting equations are suggestive of those encountered in connection with a lossy transmission line. A model based upon the transmission-line equations would be quite accurate, but unfortunately, the resulting equivalent circuit is too complicated to be of practical use. Hence it is necessary to make approximations. Of course, the cruder the approximation, the simpler the circuit becomes. The hybrid-pi model developed in this chapter gives a reasonable compromise between accuracy and simplicity. Using this model, a detailed analysis of a single-stage CE transistor amplifier is made.

11-1   THE HYBRID–PI (Π) COMMON–EMITTER TRANSISTOR MODEL[2]

In Chap. 8 it is emphasized that the common-emitter circuit is the most important practical configuration. Hence we now seek a CE model which will be valid at high frequencies. Such a circuit, called *the hybrid-π, or Giacoletto, model*, is indicated in Fig. 11-1. Analyses of circuits using this model are not too difficult and give results which are in excellent agreement with experiment at all frequencies for which the transistor gives reasonable amplification. Furthermore, the resistive components in this circuit can be obtained from the low-frequency h parameters. All parameters (resistances and capacitances) in the

Fig. 11-1 The hybrid-π
model for a transistor
in the CE configuration.

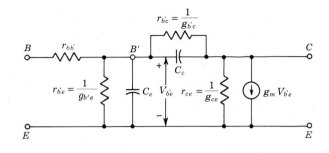

model are assumed to be independent of frequency.   They may vary with the
quiescent operating point, but under given bias conditions are reasonably
constant for small-signal swings.

Discussion of Circuit Components   The internal node B' is not physically
accessible.   The ohmic base-spreading resistance $r_{bb'}$ is represented as a lumped
parameter between the external base terminal and B'.

For small changes in the voltage $V_{b'e}$ across the emitter junction, the
excess-minority-carrier concentration injected into the base is proportional to
$V_{b'e}$, and therefore the resulting small-signal collector current, with the collector
shorted to the emitter, is proportional to $V_{b'e}$.   This effect accounts for the
current generator $g_m V_{b'e}$ in Fig. 11-1.

The increase in minority carriers in the base results in increased recombination base current, and this effect is taken into account by inserting a
conductance $g_{b'e}$ between B' and E.   The excess-minority-carrier storage in
the base is accounted for by the diffusion capacitance C_e connected between
B' and E.

The Early effect (Sec. 5-5) indicates that the varying voltage across the
collector-to-emitter junction results in *base-width modulation*.   A change in
the effective base width causes the emitter (and hence collector) current to
change because the slope of the minority-carrier distribution in the base
changes.   This feedback effect between output and input is taken into account
by connecting $g_{b'c}$ between B' and C.   The conductance between C and E is g_{ce}.

Finally, the collector-junction barrier capacitance is included in C_c.   Sometimes it is necessary to split the collector-barrier capacitance in two parts
and connect one capacitance between C and B' and another between C and B.
The last component is known as the overlap-diode capacitance.

Hybrid-II Parameter Values   Typical magnitudes for the elements of
the hybrid-π model at room temperature and for $I_C = 1.3$ mA are

$$g_m = 50 \text{ mA/V} \qquad r_{bb'} = 100 \ \Omega \qquad r_{b'e} = 1 \text{ K} \qquad r_{b'c} = 4 \text{ M}$$
$$r_{ce} = 80 \text{ K} \qquad C_c = 3 \text{ pF} \qquad C_e = 100 \text{ pF}$$

That these values are taken as reasonable is justified in the following section.

11-2   HYBRID–II CONDUCTANCES

We now demonstrate that all the resistive components in the hybrid-π model can be obtained from the h parameters in the CE configuration.

Transistor Transconductance g_m Figure 11-2 shows a p-n-p transistor in the CE configuration with the collector shorted to the emitter for time-varying signals.   In the active region the collector current is given by Eq. (5-3), repeated here for convenience, with $\alpha_N = \alpha_o$:

$$I_C = I_{CO} - \alpha_o I_E$$

Since the short-circuit current in Fig. 11-1 is $g_m V_{b'e}$, the transconductance g_m is defined by

$$g_m \equiv \frac{\partial I_C}{\partial V_{B'E}}\bigg|_{V_{CE}} = -\alpha_o \frac{\partial I_E}{\partial V_{B'E}} = \alpha_o \frac{\partial I_E}{\partial V_E} \qquad (11\text{-}1)$$

In the above we have assumed that α_N is independent of V_E.   For a p-n-p transistor, $V_E = -V_{B'E}$, as shown in Fig. 11-2.   If the emitter diode resistance is r_e, then $r_e = \partial V_E/\partial I_E$, and hence

$$g_m = \frac{\alpha_o}{r_e} \qquad (11\text{-}2)$$

The dynamic resistance of a forward-biased diode is given in Eq. (3-14) as V_T/I_E,† where $V_T = \bar{k}T/q$, and hence

$$g_m = \frac{\alpha_o I_E}{V_T} = \frac{I_{CO} - I_C}{V_T} \qquad (11\text{-}3)$$

For a p-n-p transistor I_C is negative.   For an n-p-n transistor I_C is positive, but the foregoing analysis (with $V_E = +V_{B'E}$) leads to $g_m = (I_C - I_{CO})/V_T$. Hence, for either type of transistor, g_m is positive.   Since $|I_C| \gg |I_{CO}|$, then g_m is given by

$$g_m \approx \frac{|I_C|}{V_T} \qquad (11\text{-}4)$$

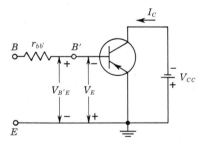

Fig. 11-2   Pertaining to the derivation of g_m.

† Since the recombination current in the emitter space-charge region does not reach the collector, the factor η in Eq. (3-14) is taken as unity in the calculation of g_m.

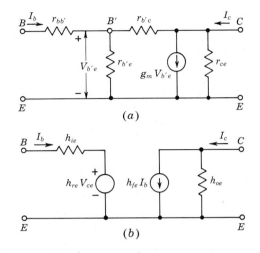

Fig. 11-3 (a) The hybrid-π model
at low frequencies. (b) the *h*-param-
eter model at low frequencies.

where, from Eq. (2-38), $V_T = T/11,600$. Note that g_m *is directly proportional
to current and inversely proportional to temperature.* At room temperature

$$g_m = \frac{|I_C|(\text{mA})}{26}$$ (11-5)

For $I_C = 1.3$ mA, $g_m = 0.05\,\mho = 50$ mA/V. For $I_C = 10$ mA, $g_m \approx 400$ mA/V.
These values are much larger than the transconductances obtained with FETs.

The Input Conductance $g_{b'e}$ In Fig. 11-3a we show the hybrid-π model
valid at low frequencies, where all capacitances are negligible. Figure 11-3b
represents the same transistor, using the *h*-parameter equivalent circuit.
    From the component values given in Sec. 11-1, we see that $r_{b'c} \gg r_{b'e}$.
Hence I_b flows into $r_{b'e}$ and $V_{b'e} \approx I_b r_{b'e}$. The short-circuit collector current
is given by

$$I_c = g_m V_{b'e} \approx g_m I_b r_{b'e}$$

The short-circuit current gain h_{fe} is defined by

$$h_{fe} = \frac{I_c}{I_b}\bigg|_{V_{CE}} = g_m r_{b'e}$$

or

$$r_{b'e} = \frac{h_{fe}}{g_m} = \frac{h_{fe}V_T}{|I_C|} \quad \text{or} \quad g_{b'e} = \frac{g_m}{h_{fe}}$$ (11-6)

Note that, over the range of currents for which h_{fe} remains fairly constant,
$r_{b'e}$ *is directly proportional to temperature and inversely proportional to current.*
Observe in Fig. 8-8a that at both very low and very high currents, h_{fe} decreases.

The Feedback Conductance $g_{b'c}$ With the input open-circuited, h_{re} is defined as the reverse voltage gain, or from Fig. 11-3a with $I_b = 0$,

$$h_{re} = \frac{V_{b'e}}{V_{ce}} = \frac{r_{b'e}}{r_{b'e} + r_{b'c}} \tag{11-7}$$

or

$$r_{b'e}(1 - h_{re}) = h_{re}r_{b'c}$$

Since $h_{re} \ll 1$, then to a good approximation

$$r_{b'e} = h_{re}r_{b'c} \qquad \text{or} \qquad g_{b'c} = h_{re}g_{b'e} \tag{11-8}$$

Since $h_{re} \approx 10^{-4}$, Eq. (11-8) verifies that $r_{b'c} \gg r_{b'e}$.

The Base-spreading Resistance $r_{bb'}$ The input resistance with the output shorted is h_{ie}. Under these conditions $r_{b'e}$ is in parallel with $r_{b'c}$. Using Eq. (11-8), we have $r_{b'e}\|r_{b'c} \approx r_{b'e}$, and hence

$$h_{ie} = r_{bb'} + r_{b'e} \tag{11-9}$$
or

$$r_{bb'} = h_{ie} - r_{b'e} \tag{11-10}$$

Incidentally, note from Eqs. (11-6) and (11-9) that the short-circuit input impedance h_{ie} varies with current and temperature in the following manner:

$$h_{ie} = r_{bb'} + \frac{h_{fe}V_T}{|I_C|} \approx \frac{h_{fe}V_T}{|I_C|} \tag{11-11}$$

The Output Conductance g_{ce} With the input open-circuited, this conductance is defined as h_{oe}. For $I_b = 0$, we have

$$I_c = \frac{V_{ce}}{r_{ce}} + \frac{V_{ce}}{r_{b'c} + r_{b'e}} + g_m V_{b'e} \tag{11-12}$$

With $I_b = 0$, we have, from Eq. (11-7), $V_{b'e} = h_{re}V_{ce}$, and from Eq. (11-12), we find

$$h_{oe} \equiv \frac{I_c}{V_{ce}} = \frac{1}{r_{ce}} + \frac{1}{r_{b'c}} + g_m h_{re} \tag{11-13}$$

where we made use of the fact that $r_{b'c} \gg r_{b'e}$. If we substitute Eqs. (11-6) and (11-8) in Eq. (11-13), we have

$$h_{oe} = g_{ce} + g_{b'c} + g_{b'e}h_{fe}\frac{g_{b'c}}{g_{b'e}}$$

or

$$g_{ce} = h_{oe} - (1 + h_{fe})g_{b'c} \tag{11-14}$$

If $h_{fe} \gg 1$, this equation may be put in the form [using Eq. (11-8)]

$$g_{ce} \approx h_{oe} - g_m h_{re} \tag{11-15}$$

Summary If the CE h parameters at low frequencies are known at a given collector current I_C, the conductances or resistances in the hybrid-π circuit are calculable from the following five equations in the order given:

$$g_m = \frac{|I_C|}{V_T}$$

$$r_{b'e} = \frac{h_{fe}}{g_m} = \frac{h_{fe} V_T}{|I_C|} \qquad \text{or} \qquad g_{b'e} = \frac{g_m}{h_{fe}}$$

$$r_{bb'} = h_{ie} - r_{b'e} \tag{11-16}$$

$$r_{b'c} = \frac{r_{b'e}}{h_{re}} \qquad \text{or} \qquad g_{b'c} = \frac{h_{re}}{r_{b'e}}$$

$$g_{ce} = h_{oe} - (1 + h_{fe})g_{b'c} = \frac{1}{r_{ce}}$$

For the typical h parameters in Table 8-2, at $I_C = 1.3$ mA and room temperature, we obtain the component values listed on page 349.

11-3   THE HYBRID-Π CAPACITANCES

The hybrid-π model for a transistor as shown in Fig. 11-1 includes two capacitances. The collector-junction capacitance $C_c = C_{b'c}$ is the measured CB output capacitance with the input open ($I_E = 0$), and is usually specified by manufacturers as C_{ob}. Since in the active region the collector junction is reverse-biased, then C_c is the transition capacitance and varies as V_{CB}^{-n}, where n is $\frac{1}{2}$ or $\frac{1}{3}$ for an abrupt or graded junction, respectively.

The capacitance C_e represents the sum of the emitter diffusion capacitance C_{De} and the emitter junction capacitance C_{Te}. For a forward-biased emitter junction, C_{De} is usually much larger than C_{Te}, and hence

$$C_e = C_{De} + C_{Te} \approx C_{De} \tag{11-17}$$

We shall now show that C_{De} is proportional to the emitter bias current I_E and is almost independent of temperature.

The Diffusion Capacitance Refer to Fig. 11-4, which represents the injected hole concentration vs. distance in the base region of a p-n-p transistor. The base width W is assumed to be small compared with the diffusion length L_B of the minority carriers. Since the collector is reverse-biased, the injected charge concentration p' at the collector junction is essentially zero. If $W \ll L_B$, then p' varies almost linearly from the value $p'(0)$ at the emitter to zero at the collector, as indicated in Fig. 11-4. The stored base charge Q_B is the average concentration $p'(0)/2$ times the volume of the base WA (where A is the base cross-sectional area) times the electronic charge q; that is,

$$Q_B = \tfrac{1}{2}p'(0) A W q \tag{11-18}$$

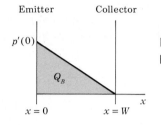

Emitter   Collector

Fig. 11-4   Minority-carrier charge distribution in the base region.

The diffusion current is [from Eq. 2-36)]

$$I = -AqD_B \frac{dp'}{dx} = AqD_B \frac{p'(0)}{W} \tag{11-19}$$

where D_B is the diffusion constant for minority carriers in the base. Combining Eqs. (11-18) and (11-19),

$$Q_B = \frac{IW^2}{2D_B} \tag{11-20}$$

The static emitter diffusion capacitance C_{De} is given by the rate of change of Q_B with respect to emitter voltage V, or

$$C_{De} = \frac{dQ_B}{dV} = \frac{W^2}{2D_B} \frac{dI}{dV} = \frac{W^2}{2D_B} \frac{1}{r_e} \tag{11-21}$$

where $r_e \equiv dV/dI = V_T/I_E$ is the emitter-junction incremental resistance.

$$C_{De} = \frac{W^2 I_E}{2D_B V_T} = g_m \frac{W^2}{2D_B} \tag{11-22}$$

which indicates that *the diffusion capacitance is proportional to the emitter bias current* I_E. From the Einstein relationship $D = \mu V_T$ [Eq. (2-37)] and the discussion in Sec. 2-5, it follows that the diffusion constant D varies as T^{-m} over a temperature range of 100 to 400°K. For silicon, $m = 1.5$ (1.7) for electrons (holes), and for germanium, $m = 0.66$ (1.33) for electrons (holes). Thus C_{DE} varies as T^{+n}, where for silicon $n = +0.5$ (+0.7) for electrons (holes) and for germanium $n = -0.34$ (+0.33) for electrons (holes).

It is pointed out in Sec. 3-9 that the dynamic capacitance of a diode differs from the static capacitance (by a factor of 2). Proceeding as in Sec. 19-12, it can be shown that for a transistor the capacitance for a sinusoidal input equals two-thirds the static capacitance given in Eq. (11-21).

Experimentally, C_e is determined from a measurement of f_T, the frequency at which the CE short-circuit current gain drops to unity. We verify in Sec. 11-6 that

$$C_e \approx \frac{g_m}{2\pi f_T} \tag{11-23}$$

11-4   VALIDITY OF HYBRID–Π MODEL[2,4]

In the derivation of Eqs. (11-1) and (11-21) for g_m and C_e we have assumed that V_{BE} changes slowly enough so that the minority-carrier charge distribution in the base region is always triangular, as shown in Fig. 11-4. If the distribution remains triangular under varying V_{BE}, then the slope at $x = 0$ is the same as at $x = W$, and hence the emitter and collector currents are equal. Consequently, the hybrid-π model is valid under dynamic conditions when the rate of change of V_{BE} is small enough so that the base incremental current I_b is small compared with the collector incremental current I_c. Giacolletto,[2] in his original paper on the hybrid-π equivalent circuit, has shown that the network elements of Fig. 11-1 are frequency-independent provided that

$$2\pi f \frac{W^2}{6D_B} \ll 1 \tag{11-24}$$

From Eqs. (11-22) and (11-23) we have

$$\frac{W^2}{6D_B} = \frac{C_e}{3g_m} = \frac{1}{6\pi f_T}$$

Thus Eq. (11-24) becomes

$$f \ll \frac{6\pi f_T}{2\pi} = 3f_T \tag{11-25}$$

It follows that the hybrid-π model is valid for frequencies up to approximately $f_T/3$.

11-5   VARIATION OF HYBRID–Π PARAMETERS[4]

In the preceding two sections we have derived expressions for the hybrid-π conductances and capacitances in terms of the low-frequency h parameters and other transistor parameters, such as the base width or the diffusion constant for minority carriers in the base.

Table 11-1 summarizes the dependence of g_m, $r_{b'e'}$, $r_{bb'}$, C_e, C_c, h_{fe}, and h_{ie} on the collector current magnitude $|I_C|$, the collector-to-emitter voltage magnitude $|V_{CE}|$, and the temperature. The conclusions in the table are based on Eqs. (11-16), (11-17), and (11-22). We must also recall that increasing $|V_{CE}|$ decreases the effective base width (Fig. 5-8). The dependence of $r_{bb'}$ on $|I_C|$ and temperature requires some explanation. The decrease of $r_{bb'}$ with $|I_C|$ is due to conductivity modulation of the base with increasing collector current. On the contrary, $r_{bb'}$ increases with increasing temperature because the mobility of majority and minority carriers decreases, and this results in reduced conductivity. The increase of h_{fe} with temperature has been determined experimentally, whereas the increase with $|V_{CE}|$ is due to the decrease of the base width and the reduction in recombination which increases the transistor alpha (Sec. 5-6).

No entry in Table 11-1 means that the particular dependence varies with the absolute value of $|I_C|$, $|V_{CE}|$, or T in a complicated fashion.

TABLE 11-1 Dependence of parameters upon current, voltage, and temperature

Parameter	Variation with increasing:		
	$|I_C|$	$|V_{CE}|$	T
g_m	$|I_C|$	Independent	$1/T$
$r_{bb'}$	Decreases		Increases
$r_{b'e}$	$1/|I_C|$	Increases	Increases
C_e	$|I_C|$	Decreases	
C_c	Independent	Decreases	Independent
h_{fe}	See Fig. 8-8	Increases	Increases
h_{ie}	$1/|I_C|$	Increases	Increases

11-6   THE CE SHORT–CIRCUIT CURRENT GAIN

Consider a single-stage CE transistor amplifier, or the last stage of a cascade. The load R_L on this stage is the collector-circuit resistor, so that $R_c = R_L$. In this section we assume that $R_L = 0$, whereas the circuit with $R_L \neq 0$ is analyzed in the next section. To obtain the *frequency response* (the gain as a function of frequency) of the transistor amplifier, we use the hybrid-Π model of Fig. 11-1, which is repeated for convenience in Fig. 11-5. Representative values of the circuit components are specified on page 349 for a transistor intended for use at high frequencies. We use these values as a guide in making simplifying assumptions.

The approximate equivalent circuit from which to calculate the short-

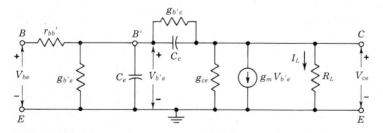

Fig. 11-5 The hybrid-π circuit for a single transistor with a resistive load R_L.

Fig. 11-6 Approximate equivalent circuit for the calculation of the short-circuit CE current gain.

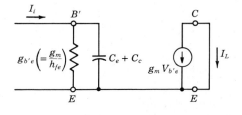

circuit current gain is shown in Fig. 11-6. A current source furnishes a sinusoidal input current of magnitude I_i, and the load current is I_L. We have neglected $g_{b'c}$, which should appear across terminals $B'C$, because $g_{b'c} \ll g_{b'e}$. And of course g_{ce} disappears, because it is in shunt with a short circuit. An additional approximation is involved, in that we have neglected the current delivered directly to the output through $g_{b'c}$ and C_c. We see shortly that this approximation is justified.

The load current is $I_L = -g_m V_{b'e}$, where

$$V_{b'e} = \frac{I_i}{g_{b'e} + j\omega(C_e + C_c)} \tag{11-26}$$

The current amplification under short-circuited conditions is

$$A_i = \frac{I_L}{I_i} = \frac{-g_m}{g_{b'e} + j\omega(C_e + C_c)} \tag{11-27}$$

Using the results given in Eqs. (11-16), we have

$$A_i = \frac{-h_{fe}}{1 + j(f/f_\beta)} \qquad |A_i| = \frac{h_{fe}}{[1 + (f/f_\beta)^2]^{\frac{1}{2}}} \tag{11-28}$$

where

$$f_\beta = \frac{g_{b'e}}{2\pi(C_e + C_c)} = \frac{1}{h_{fe}} \frac{g_m}{2\pi(C_e + C_c)} \tag{11-29}$$

At $f = f_\beta$, $|A_i|$ is equal to $1/\sqrt{2} = 0.707$ of its low-frequency value h_{fe}. The frequency range up to f_β is referred to as the *bandwidth* of the circuit. Note that the value of A_i at $\omega = 0$ is $-h_{fe}$, in agreement with the definition of $-h_{fe}$ as the low-frequency short-circuit CE current gain.

The Parameter f_T We now introduce f_T, which is defined as the *frequency at which the short-circuit common-emitter current gain attains unit magnitude.* Since $h_{fe} \gg 1$, we have, from Eqs. (11-28) and (11-29), that f_T is given by

$$f_T \approx h_{fe}f_\beta = \frac{g_m}{2\pi(C_e + C_c)} \approx \frac{g_m}{2\pi C_e} \tag{11-30}$$

since $C_e \gg C_c$. Hence, from Eq. (11-28),

$$A_i \approx \frac{-h_{fe}}{1 + jh_{fe}(f/f_T)} \tag{11-31}$$

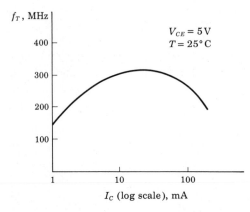

Fig. 11-7 Variation of f_T with collector current.

The parameter f_T is an important high-frequency characteristic of a transistor. Like other transistor parameters, its value depends on the operating conditions of the device. Typically, the dependence of f_T on collector current is as shown in Fig. 11-7.

Since $f_T \approx h_{fe}f_\beta$, this parameter may be given a second interpretation. It represents the *short-circuit current-gain–bandwidth product;* that is, for the CE configuration with the output shorted, f_T is the product of the low-frequency current gain and the upper 3-dB frequency. For our typical transistor (page 349), $f_T = 80$ MHz and $f_\beta = 1.6$ MHz. It is to be noted from Eq. (11-30) that there is a sense in which gain may be sacrificed for bandwidth, and vice versa. Thus, if two transistors are available with equal f_T, the transistor with lower h_{fe} will have a correspondingly larger bandwidth.

In Fig. 11-8, A_i expressed in decibels (i.e., 20 log $|A_i|$) is plotted against

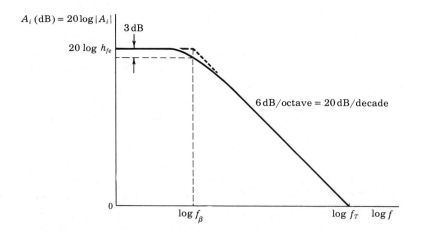

Fig. 11-8 The short-circuit CE current gain vs. frequency (plotted on a log-log scale).

frequency on a logarithmic frequency scale. When $f \ll f_\beta$, $A_i \approx -h_{fe}$, and A_i (dB) approaches asymptotically the horizontal line A_i (dB) = 20 log h_{fe}. When $f \gg f_\beta$, $|A_i| \approx h_{fe}f_\beta/f = f_T/f$, so that A_i (dB) = 20 log f_T − 20 log f. Accordingly, A_i (dB) = 0 dB at $f = f_T$. And for $f \gg f_\beta$, the plot approaches as an asymptote a straight line passing through the point $(f_T, 0)$ and having a slope which causes a decrease in A_i (dB) of 6 dB per octave (f is multiplied by a factor of 2, and 20 log 2 = 6 dB), or 20 dB per decade. The intersection of the two asymptotes occurs at the "corner" frequency $f = f_\beta$, where A_i is down by 3 dB. Hence f_β is also called the 3-dB frequency.

Earlier, we neglected the current delivered directly to the output through $g_{b'c}$ and C_c. Now we may see that this approximation is justified. Consider, say, the current through C_c. The magnitude of this current is $\omega C_c V_{b'e}$, whereas the current due to the controlled generator is $g_m V_{b'e}$. The ratio of currents is $\omega C_c / g_m$. At the highest frequency of interest f_T, we have, from Eq. (11-30), using the typical values from page 349,

$$\frac{\omega C_c}{g_m} = \frac{2\pi f_T C_c}{g_m} = \frac{C_c}{C_e + C_c} \approx 0.03$$

In a similar way the current delivered to the output through $g_{b'c}$ may be shown to be negligible.

Measurement of f_T   The frequency f_T is often inconveniently high to allow a direct experimental determination of f_T. However, a procedure is available which allows a measurement of f_T at an appreciably lower frequency. We note from Eq. (11-28) that, for $f \gg f_\beta$, we may neglect the unity in the denominator and write $|A_i|f \approx h_{fe}f_\beta = f_T$ from Eq. (11-30). Accordingly, at some particular frequency f_1 (say, f_1 is five or ten times f_β), we measure the gain $|A_{i1}|$. The parameter f_T may be calculated now from $f_T = f_1|A_{i1}|$. In the case of our typical transistor, for which f_T = 80 MHz and f_β = 1.6 MHz, the frequency f_1 may be $f_1 = 5 \times 1.6 = 8.0$ MHz, a much more convenient frequency than 80 MHz.

The experimentally determined value of f_T is used to calculate the value of C_e in the hybrid-π circuit. From Eq. (11-30)

$$C_e = \frac{g_m}{2\pi f_T} \tag{11-32}$$

11-7   CURRENT GAIN WITH RESISTIVE LOAD

To minimize the complications which result when the load resistor R_L in Fig. 11-5 is not zero, we find it convenient to deal with the parallel combination of $g_{b'c}$ and C_c, using Miller's theorem of Sec. 8-11. We identify $V_{b'e}$ with V_1 in Fig. 8-17 and V_{ce} with V_2. On this basis the circuit of Fig. 11-5 may be

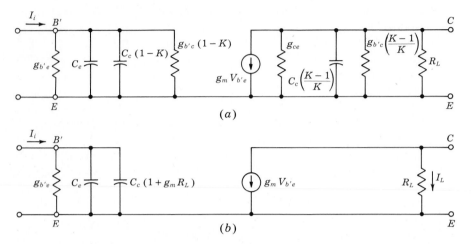

Fig. 11-9 (a) Approximate equivalent circuit for calculation of response of a transistor amplifier stage with a resistive load; (b) further simplification of the equivalent circuit.

replaced by the circuit of Fig. 11-9a. Hence $K \equiv V_{ce}/V_{b'e}$. This circuit is still rather complicated because it has two independent time constants, one associated with the input circuit and one associated with the output. We now show that in a practical situation the output time constant is negligible in comparison with the input time constant, and may be ignored. Let us therefore delete the output capacitance $C_c(K-1)/K$, consider the resultant circuit, and then show that the reintroduction of the output capacitance makes no significant change in the performance of the circuit. We assume here that K is independent of frequency.

Since $K \equiv V_{ce}/V_{b'e}$ is (approximately) the voltage gain, we normally have $|K| \gg 1$. Hence $g_{b'c}(K-1)/K \approx g_{b'c}$. Since $g_{b'c} \ll g_{ce}$ ($r_{b'c} \approx 4$ M and $r_{ce} \approx 80$ K), we may omit $g_{b'c}$ from Fig. 11-9a. In a wideband amplifier, R_L seldom exceeds 2 K. The conductance g_{ce} may be neglected compared with $1/R_L$, and the output circuit consists of the current generator $g_m V_{b'e}$ feeding the load R_L, as indicated in Fig. 11-9b. Even if the above approximations were not valid for some particular transistor or load, the analysis to follow is still valid provided that R_L is interpreted as the parallel combination of the collector-circuit resistor, r_{ce} and $r_{b'c}$.

By inspection of Fig. 11-9b, $K = V_{ce}/V_{b'e} = -g_m R_L$. For $g_m = 50$ mA/V and $R_L = 2,000\ \Omega$, $K = -100$. For this maximum value of K, conductance $g_{b'c}(1-K) \approx 0.025$ mA/V is negligible compared with $g_{b'e} \approx 1$ mA/V. Hence the circuit of Fig. 11-9a is reduced to that shown in Fig. 11-9b. The load resistance R_L has been restricted to a maximum value of 2 K because, at values of R_L much above 2,000 Ω, the capacitance $C_c(1 + g_m R_L)$ becomes excessively large and the bandpass correspondingly small.

Now let us return to the capacitance $C_c(K - 1)/K \approx C_c$, which we neglected above. For $R_L = 2,000\ \Omega$,

$$R_L C_c = 2 \times 10^3 \times 3 \times 10^{-12} = 6 \times 10^{-9}\ \text{s} = 6\ \text{ns}$$

The input time constant is

$$r_{b'e}[C_e + C_c(1 + g_m R_L)] = 10^3(100 + 3 \times 101)10^{-12}\ \text{s} = 403\ \text{ns}$$

It is therefore apparent that the bandpass of the amplifier will be determined by the time constant of the input circuit and that, in the useful frequency range of the stage, the capacitance C_c will not make itself felt in the output circuit. Of course, if the transistor works into a highly capacitive load, this capacitance will have to be taken into account, and it then might happen that the output time constant will predominate.

The circuit of Fig. 11-9b is different from the circuit of Fig. 11-6 only in that a load R_L has been included and C_c has been augmented by $g_m R_L C_c$. To the accuracy of our approximations, the low-frequency current gain A_{I_o} under load is equal to the low-frequency gain $A_{io} = -h_{fe}$ with output shorted. However, the 3-dB frequency is now f_H (rather than f_β), where

$$f_H = \frac{1}{2\pi r_{b'e}C} = \frac{g_{b'e}}{2\pi C} \tag{11-33}$$

where

$$C \equiv C_e + C_c(1 + g_m R_L) \tag{11-34}$$

Therefore A_I is given by Eq. (11-28), and f_β replaced by f_H.

11-8   SINGLE–STAGE CE TRANSISTOR AMPLIFIER RESPONSE

In the preceding sections we assume that the transistor is driven from an ideal current source, that is, a source of infinite resistance. We now remove this restriction and consider that the source V_s has a finite resistance R_s. The equivalent circuit from which to obtain the response is shown in Fig. 11-10. This circuit is obtained from Fig. 11-5 by adding V_s in series with R_s between B and E and by omitting $r_{ce} = 1/g_{ce}$ and $r_{b'c} = 1/g_{b'c}$, since they are much greater than R_L.

The Transfer Function   We wish to calculate the transfer function V_o/V_s as a function of the complex-frequency (or Laplace transform) variable s. Introducing $R_s' \equiv R_s + r_{bb'} = 1/G_s'$, and noting that the admittance of a capacitor C is sC, we obtain the following KCL equations at nodes B' and C, respectively (the so-called nodal equations):

$$G_s' V_s = [G_s' + g_{b'e} + s(C_e + C_c)]V_{b'e} - sC_c V_o \tag{11-35}$$

$$0 = (g_m - sC_c)V_{b'e} + \left(\frac{1}{R_L} + sC_c\right)V_o \tag{11-36}$$

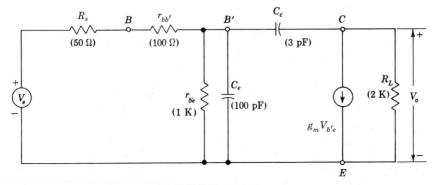

Fig. 11-10 Equivalent circuit for frequency analysis of CE amplifier stage driven from a voltage source.

Solving Eqs. (11-35) and (11-36), we find

$$\frac{V_o}{V_s} = \frac{-G'_s R_L (g_m - sC_c)}{s^2 C_e C_c R_L + s[C_e + C_c + C_c R_L (g_m + g_{b'e} + G'_s)] + G'_s + g_{b'e}}$$
(11-37)

The above equation is of the form

$$A_{Vs} \equiv \frac{V_o}{V_s} = \frac{K_1(s - s_o)}{(s - s_1)(s - s_2)}$$
(11-38)

We thus see that the transfer function of the single CE transistor at high frequencies has one zero $s_o = g_m/C_c$ and two poles s_1 and s_2. These poles are calculated by finding the roots of the denominator of Eq. (11-37). Also from this equation it follows that $K_1 = G'_s R_L C_c / C_e C_c R_L = G'_s / C_e$.

For the numerical values indicated in Fig. 11-10 and with $g_m = 50$ mA/V, we find

$$K_1 = 6.67 \times 10^7 \qquad\qquad s_o = 1.67 \times 10^9 \text{ rad/s}$$

$$s_1 = -1.75 \times 10^7 \text{ rad/s} \qquad s_2 = -7.30 \times 10^8 \text{ rad/s}$$

The magnitude and the phase of the transfer function [obtained from Eq. (11-37) with $s = j\omega = j2\pi f$] are plotted in Fig. 11-11. The 3-dB frequency is found to be 2.8 MHz.

Approximate Analysis We can obtain a very simple approximate expression for the transfer function by applying Miller's theorem to the circuit of Fig. 11-10. Proceeding as in Sec. 11-7, we obtain the circuit of Fig. 11-12, with $K \equiv V_{ce}/V_{b'e}$. Since $|K| \gg 1$, the output capacitance is C_c and the output time constant is $C_c R_L = 6$ ns, as in Sec. 11-7. Neglecting C_c, it follows that $K = -g_m R_L$ and the-input capacitance is

$$C \equiv C_e + C_c(1 + g_m R_L) = 100 + (3)(101) = 403 \text{ pF}$$
(11-39)

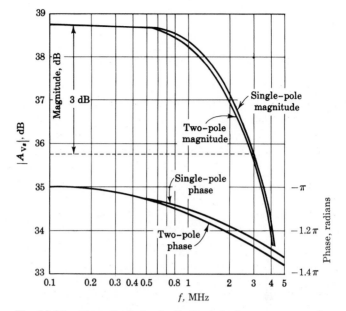

Fig. 11-11 Magnitude in decibels and phase response for two-pole and single-pole transfer functions.

The input-loop resistance is

$$R \equiv R'_s \| r_{b'e} = \frac{1}{G'_s + g_{b'e}} = \frac{1}{1/150 + 1/1,000} = 130 \ \Omega \qquad (11\text{-}40)$$

and the input time constant is $(130)(403)$ ps $= 53$ ns. Since this time constant is almost nine times as large as the output time constant, we assume, as in Sec. 11-7, that the bandpass of the amplifier is determined by the input time constant alone. Neglecting the output time constant and using $K = -g_m R_L$, the transfer function obtained from Fig. 11-12 is

$$A_{Vs} = \frac{V_o}{V_s} = \frac{-g_m R_L G'_s}{G'_s + g_{b'e} + sC} \qquad (11\text{-}41)$$

which is of the form

$$A_{Vs} = \frac{K_2}{s - s_1} \qquad (11\text{-}42)$$

and therefore is a one-pole approximation for the transfer function. The pole is given by $s = s_1$, where

$$s_1 = -\frac{G'_s + g_{b'e}}{C} = \frac{-1}{RC} \qquad (11\text{-}43)$$

or

$$s_1 = \frac{-1}{(130)(403 \times 10^{-12})} = -1.90 \times 10^7 \ \text{rad/s}$$

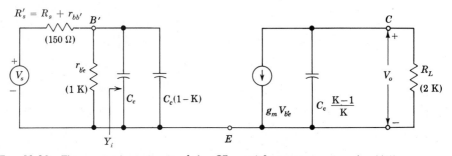

Fig. 11-12 The equivalent circuit of the CE amplifier stage, using the Miller effect.

To find the frequency response, let $s = j2\pi f$, and Eq. (11-42) becomes

$$A_{Vs} = \frac{K_2}{j2\pi f - s_1} = \frac{-K_2}{s_1} \frac{1}{1 - j2\pi f/s_1} \qquad (11\text{-}44)$$

The magnitude of voltage gain as a function of frequency is

$$|A_{Vs}| = \left| \frac{K_2}{s_1} \right| \frac{1}{[1 + (2\pi f/s_1)^2]^{\frac{1}{2}}} \qquad (11\text{-}45)$$

and the phase lead angle is

$$\theta_1 = -\pi - \arctan \frac{2\pi f}{|s_1|} \qquad (11\text{-}46)$$

The phase angle $-\pi$ results from the fact that both K_2 and s_1 are negative, and hence that $-K_2/s_1$ is negative.

The upper 3-dB frequency f_H occurs when the gain drops to $1/\sqrt{2}$ of its low-frequency value. Hence

$$f_H = \frac{|s_1|}{2\pi} = \frac{1}{2\pi RC} = \frac{1.90 \times 10^7}{2\pi} \text{ Hz} = 3.0 \text{ MHz} \qquad (11\text{-}47)$$

Note that *the upper 3-dB frequency of a single-pole circuit is given by* $1/(2\pi\tau)$, *where* τ *is the time constant of the circuit.* The magnitude and phase as given by Eq. (11-44) are plotted in Fig. 11-11. A comparison of the two-pole with the one-pole curves shows that the value obtained with the more exact two-pole model is $f_H = 2.8$ MHz. The error in the single-pole value of 3.0 MHz is about 7 percent, and hence it is not necessary to use the more complicated two-pole solution.

The Miller Input Impedance[5] In the Miller capacitance C we used the low-frequency value of $|K| = g_m R_L$. Since this is the maximum value of K, too large a value of C is used, and we should expect to obtain too small a value of f_H. However, as noted above, the single-pole approximation gives too large a value of f_H (3.0 MHz instead of the correct value of 2.8 MHz). This apparent anomaly is resolved if we take into account the frequency dependence of K in calculating the input admittance Y_i. From Fig. 11-12

$$Y_{b'e} = Y_i = j\omega[C_e + C_c(1 - K)] \tag{11-48}$$

Since $|K| \gg 1$ even at the 3-dB frequency, the output circuit consists of a capacitor C_c in parallel with R_L. Hence

$$K = \frac{V_o}{V_{b'e}} = \frac{-g_m}{j\omega C_c + 1/R_L} = \frac{-g_m R_L}{1 + j\omega C_c R_L} \tag{11-49}$$

and

$$Y_i = j\omega \left[C_e + C_c \left(1 + \frac{g_m R_L}{1 + j\omega C_c R_L} \right) \right] \tag{11-50}$$

If we consider the input to consist of a capacitor C_i in parallel with a resistor R_i, then

$$Y_i = j\omega C_i + \frac{1}{R_i} \tag{11-51}$$

From Eq. (11-50) it follows that

$$C_i = C_e + C_c + \frac{g_m R_L C_c}{1 + \omega^2 C_c^2 R_L^2} \tag{11-52}$$

and

$$R_i = \frac{1}{g_m} \left(1 + \frac{1}{\omega^2 C_c^2 R_L^2} \right) \tag{11-53}$$

At the frequency $f_H = 3.0 \times 10^6$ Hz,

$$\omega^2 C_c^2 R_L^2 = [(2\pi)(3 \times 10^6)(3 \times 10^{-12})(2 \times 10^3)]^2 = 0.0128$$

Hence, for $0 \leq f \leq f_H$, C_i remains essentially constant at its zero frequency value of $C = C_e + C_c(1 + g_m R_L)$, whereas R_i decreases from infinity to $[1/(50 \times 10^{-3})](1 + 1/0.0128) = 1,590$ Ω. This value is comparable with $r_{b'e} = 1,000$ Ω, and hence the Miller input resistance reduces the value of $V_{b'e}/V_s$, and hence also V_o/V_s. This explains why the bandwidth obtained with the two-pole exact transfer function is more conservative (smaller by approximately 7 percent, as seen from Fig. 11-11) than the value obtained from the single-pole approximation. For most applications the approximate single-pole transfer function is sufficiently accurate for bandwidth calculations.

11-9   THE GAIN–BANDWIDTH PRODUCT

Using the approximate single-pole transfer function obtained from Fig. 11-9b (with $R_s + r_{bb'}$ added to the input circuit), it is found in Prob. 11-17 that the gain-bandwidth product for voltage and current is, respectively,

$$|A_{Vso}f_H| = \frac{g_m}{2\pi C} \frac{R_L}{R_s + r_{bb'}} = \frac{f_T}{1 + 2\pi f_T C_c R_L} \frac{R_L}{R_s + r_{bb'}}$$

$$|A_{Iso}f_H| = \frac{f_T}{1 + 2\pi f_T C_c R_L} \frac{R_s}{R_s + r_{bb'}}$$

(11-54)

The quantities f_H, A_{Iso}, and A_{Vso}, which characterize the transistor stage, depend on both R_L and R_s. The form of this dependence, as well as the order of magnitude of these quantities, may be seen in Fig. 11-13. Here f_H has been plotted as a function of R_L, up to $R_L = 2,000\ \Omega$, for several values of R_s. The topmost f_H curve in Fig. 11-13 for $R_s = 0$ corresponds to ideal-voltage-source drive. The voltage gain ranges from zero at $R_L = 0$ to 90.9 at $R_L = 2,000\ \Omega$. Note that a source impedance of only 100 Ω reduces the bandwidth by a factor of about 1.8. The bottom curve has $R_s = \infty$ and corresponds to the ideal current source. The voltage gain is zero for all R_L if $R_s = \infty$. For any R_L the bandwidth is highest for lowest R_s. The voltage-gain–bandwidth

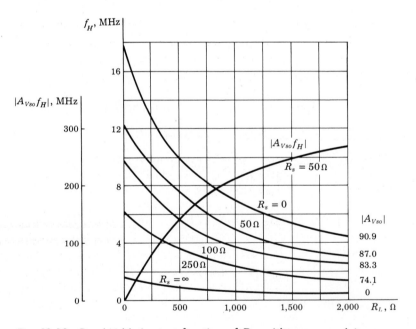

Fig. 11-13 Bandwidth f_H as a function of R_L, with source resistance as a parameter, for an amplifier consisting of one CE transistor whose parameters are given in Sec. 11-1. Also, the gain-bandwidth product for a 50-Ω source is plotted. The tabulated values of $|A_{Vso}|$ correspond to $R_L = 2,000\ \Omega$ and to the values of R_s on the curves.

product increases with increasing R_L and decreases with increasing R_s. Even if we know the gain-bandwidth product at a particular R_s and R_L, we cannot use the product to determine the improvement, say, in bandwidth corresponding to a sacrifice in gain. For if we change the gain by changing R_s or R_L or both, generally, the gain-bandwidth product will no longer be the same as it had been.

Summary The high-frequency response of a transistor amplifier is obtained in this chapter in terms of the transistor parameters g_m, $r_{b'e}$, $r_{bb'}$, C_e, and C_c. We shall now show that these may be obtained from the four independent parameters h_{fe}, f_T, h_{ie}, and $C_o = C_{ob}$.

From the operating current I_C and the temperature T, the transconductance is obtained [Eqs. (11-16)] as $g_m = |I_C|/V_T$ and is independent of the particular device under consideration. Knowing g_m, we can find, from Eqs. (11-16) and (11-23),

$$r_{b'e} = \frac{h_{fe}}{g_m} \qquad r_{bb'} = h_{ie} - r_{b'e} \qquad C_e \approx \frac{g_m}{2\pi f_T}$$

If R_s and R_L are given, all quantities in Eq. (11-37) or (11-41) are known. We have therefore verified that the frequency response may be determined from the four parameters h_{fe}, f_T, h_{ie}, and C_c. Hence these four are usually specified by manufacturers of high-frequency transistors.

11-10   EMITTER FOLLOWER AT HIGH FREQUENCIES

In this section we examine the high-frequency response of the emitter follower shown in Fig. 11-14a. A capacitance C_L is included across the load because the emitter follower (due to its low output resistance) is often used to drive capacitive loads.

Writing nodal equations at the nodes B' and E, respectively, we have

$$G_s' V_s = [G_s' + g_{b'e} + s(C_c + C_e)]V_i' - (g_{b'e} + sC_e)V_e \qquad (11\text{-}55)$$

$$0 = -(g + sC_e)V_i' + \left[g + \frac{1}{R_L} + s(C_e + C_L)\right]V_e \qquad (11\text{-}56)$$

where

$$G_s' \equiv \frac{1}{R_s + r_{bb'}} \qquad \text{and} \qquad g \equiv g_m + g_{b'e} \qquad (11\text{-}57)$$

If V_i' is eliminated from these equations, the voltage gain V_e/V_s as a function of s is obtained. The result, of the form given in Eq. (11-38), has one zero and two poles. The exact solution can be found by proceeding as in Sec. 11-8 (Prob. 11-18).

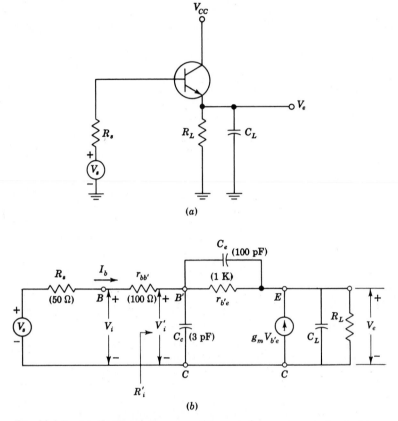

Fig. 11-14 (a) Emitter follower. (b) High-frequency equivalent circuit of emitter follower.

Single-pole Solution We can obtain a very simple approximate expression for the transfer function by applying Miller's theorem to the circuit of Fig. 11-14b. With $K \equiv V_e/V_i'$ we obtain the circuit of Fig. 11-15.

The low-frequency gain of an emitter follower is close to unity: $K \approx 1$ and $1 - K \approx 0$. Hence the input time constant $\tau_i \approx (R_s + r_{bb'})C_c$. The output time constant τ_o is proportional to C_L, and since we have assumed that the load is highly capacitive, then $\tau_o \gg \tau_i$. Hence the upper 3-dB frequency is determined, to a good approximation, by the output circuit alone. Using $K = 1$, we obtain

$$V_e = \frac{g_m V_{b'e}}{1/R_L + j\omega C_L} = \frac{g_m R_L(V_i' - V_e)}{1 + j\omega C_L R_L} \tag{11-58}$$

Solving for $V_e/V_i' = K$, we obtain

$$K = \frac{g_m R_L}{1 + g_m R_L}\frac{1}{1 + jf/f_H} = \frac{K_o}{1 + jf/f_H} \tag{11-59}$$

where

$$K_o \equiv \frac{g_m R_L}{1 + g_m R_L} \approx 1 \tag{11-60}$$

and

$$f_H \equiv \frac{1 + g_m R_L}{2\pi C_L R_L} \approx \frac{g_m}{2\pi C_L} = \frac{f_T C_e}{C_L} \tag{11-61}$$

and f_T is given by Eq. (11-30). Since $f_H = 1/2\pi\tau_o$, we see that $\tau_o = C_L/g_m$, and the condition $\tau_o \gg \tau_i$ requires

$$C_L \gg g_m (R_s + r_{bb'}) C_c \tag{11-62}$$

For the parameter values in Fig. 11-14 and $g_m = 50$ mA/V, this condition is $C_L \gg (50 \times 10^{-3})(150)(3) = 23$ pF. Note that $f_H \sim f_T$.

Since the input impedance between terminals B' and C is very large compared with $R_s + r_{bb'}$, then K also represents the overall voltage gain $A_{Vs} \equiv V_e/V_s$. Incidentally, a somewhat better approximation for f_H is given in Prob. 11-20, where we find

$$f_H = \frac{g_m + g_{b'e}}{2\pi(C_L + C_e)} \tag{11-63}$$

Input Admittance We can find the input admittance (excluding $r_{bb'}$) by referring to Fig. 11-15.

$$Y'_i = \frac{I_b}{V'_i} = j\omega[C_c + (1 - K)C_e] + (1 - K)g_{b'e}$$

Substituting K from Eq. (11-59) in this equation, we find

$$Y'_i = j2\pi f C_c + (g_{b'e} + j2\pi f C_e)\frac{1 - K_o + jf/f_H}{1 + jf/f_H} \tag{11-64}$$

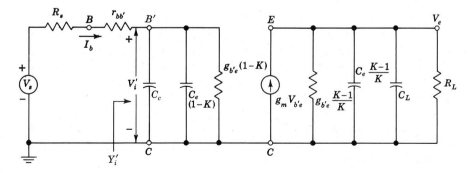

Fig. 11-15 The equivalent circuit of the emitter follower, using Miller's theorem.

Since $K_o \approx 1$, the numerator of Eq. (11-64) is affected by frequency at a much lower value of f than is the denominator. Hence, for $f < f_H$, Eq. (11-64) can be written

$$Y_i' \approx j2\pi f C_c + (g_{b'e} + j2\pi f C_e)\left(1 - K_o + \frac{jf}{f_H}\right)$$

$$\approx j2\pi f[C_c + (1 - K_o)C_e] + g_{b'e}(1 - K_o) + jg_{b'e}\frac{f}{f_H} - 2\pi f^2 \frac{C_e}{f_H}$$
(11-65)

where the last term represents a negative resistance which is a function of frequency. Thus, the input impedance consists of a capacitance shunted by a negative resistance and if the source resistance R_s contains some inductance in series with it, it is possible for the circuit to sustain undesirable oscillations. One way to remedy this condition is to use a small resistance in series with R_s so that the net resistance seen by the source is positive.

REFERENCES

1. Phillips, A. B.: "Transistor Engineering," chaps. 13 and 14, McGraw-Hill Book Company, New York, 1962.

   Pritchard, R. L.: Electric-network Representations of Transistors: A Survey, *IRE Trans. Circuit Theory*, vol. CT-3, no. 1, pp. 5–21, March, 1956.

   Searle, C. L., A. R. Boothroyd, E. J. Angelo, Jr., P. E. Gray, and D. O. Pederson: "Elementary Circuit Properties of Transistors," vol. 3, Semiconductor Electronics Education Committee, John Wiley & Sons, Inc., New York, 1964.

2. Giacoletto, L. J.: Study of *p-n-p* Alloy Junction Transistors from DC through Medium Frequencies, *RCA Rev.*, vol. 15, no. 4, pp. 506–562, December, 1954.

   Searle, C. L., et al.: Ref. 1, vol. 3, chap. 3.

3. Phillips, A. B., Ref. 1, pp. 129–130.

4. Gray, P. E., and C. L. Searle: "Electronic Principles," pp. 373–380, 421–424, John Wiley & Sons, Inc., New York, 1969.

5. Cherry, E. M., and D. E. Hooper: "Amplifying Devices and Low-pass Amplifier Design," pp. 337–343, John Wiley & Son's, Inc., New York, 1968.

REVIEW QUESTIONS

11-1 Draw the small-signal high-frequency CE model of a transistor.

11-2 (a) What is the physical origin of the two capacitors in the hybrid-π model? (b) What is the order of magnitude of each capacitance?

11-3 What is the order of magnitude of each resistance in the hybrid-π model?

11-4 How does g_m vary with (a) $|I_C|$; (b) $|V_{CE}|$; (c) T?

11-5 Prove that $h_{fe} = g_m r_{b'e}$.

11-6 (a) Prove that $h_{ie} = r_{bb} + r_{b'e}$. (b) Assuming $r_{bb'} \ll r_{b'e}$, how does h_{ie} vary with $|I_C|$?

11-7 (a) How does C_e vary with $|I_C|$ and $|V_{CE}|$? (b) How does C_c vary with $|I_C|$ and $|V_{CE}|$?

11-8 Derive the expression for the CE short-circuit current gain A_i as a function of frequency.

11-9 (a) Define f_β. (b) Define f_T. (c) What is the relationship between f_β and f_T?

11-10 Consider a CE stage with a resistive load R_L. Using Miller's theorem, what is the midband input capacitance? (b) Assuming the output time constant is small compared with the input time constant, what is the high 3-dB frequency f_H for the current gain?

11-11 Explain why the 3-dB frequency for current gain is not the same as f_H for voltage gain.

11-12 In terms of what four parameters is the high-frequency response of a CE stage obtained?

11-13 Draw the small-signal equivalent circuit for an emitter-follower stage at high frequencies.

12 / MULTISTAGE AMPLIFIERS

Frequently, the need arises for amplifying a signal with a minimum of distortion. Under these circumstances the active devices involved must operate linearly. In the analysis of such circuits the first step is the replacement of the actual circuit by a linear model. Thereafter it becomes a matter of circuit analysis to determine the distortion produced by the transmission characteristics of the linear network.

The frequency range of the amplifiers discussed in this chapter extends from a few cycles per second (hertz), or possibly from zero, up to some tens of megahertz. The original impetus for the study of such wideband amplifiers was supplied because they were needed to amplify the pulses occurring in a television signal. Therefore such amplifiers are often referred to as *video amplifiers*. Basic amplifier circuits are discussed here.

In this chapter, then, we consider the following problem: Given a low-level input waveform which is not necessarily sinusoidal but may contain frequency components from a few hertz to a few megahertz, how can this voltage signal be amplified with a minimum of distortion?

We also discuss many topics associated with the general problem of amplification, such as the classification of amplifiers, hum and noise in amplifiers, etc.

12-1 CLASSIFICATION OF AMPLIFIERS

Amplifiers are described in many ways, according to their frequency range, the method of operation, the ultimate use, the type of load, the method of interstage coupling, etc. The frequency classification includes dc (from zero frequency), audio (20 Hz to 20 kHz), video or pulse (up to a few megahertz), radio-frequency (a few kilohertz to hundreds of megahertz), and ultrahigh-frequency (hundreds or thousands of megahertz) amplifiers.

The position of the quiescent point and the extent of the characteristic that is being used determine the method of operation. Whether the transistor or FET is operated as a Class A, Class AB, Class B, or Class C amplifier is determined from the following definitions.

Class A A Class A amplifier is one in which the operating point and the input signal are such that the current in the output circuit (in the collector, or drain electrode) flows at all times. A Class A amplifier operates essentially over a linear portion of its characteristic.

Class B A Class B amplifier is one in which the operating point is at an extreme end of its characteristic, so that the quiescent power is very small. Hence either the quiescent current or the quiescent voltage is approximately zero. If the signal voltage is sinusoidal, amplification takes place for only one-half a cycle. For example, if the quiescent output-circuit current is zero, this current will remain zero for one-half a cycle.

Class AB A Class AB amplifier is one operating between the two extremes defined for Class A and Class B. Hence the output signal is zero for part but less than one-half of an input sinusoidal signal cycle.

Class C A Class C amplifier is one in which the operating point is chosen so that the output current (or voltage) is zero for more than one-half of an input sinusoidal signal cycle.

Amplifier Applications The classification according to use includes voltage, power, current, or general-purpose amplifiers. In general, the load of an amplifier is an impedance. The two most important special cases are the idealized resistive load and the tuned circuit operating near its resonant frequency.

Class AB and Class B operation are used with untuned power amplifiers (Chap. 18), whereas Class C operation is used with tuned radio-frequency amplifiers. Many important waveshaping functions may be performed by Class B or C overdriven amplifiers. This chapter considers only the untuned audio or video voltage amplifier with a resistive load operated in Class A.

12-2   DISTORTION IN AMPLIFIERS

The application of a sinusoidal signal to the input of an ideal Class A amplifier will result in a sinusoidal output wave. Generally, the output waveform is not an exact replica of the input-signal waveform because of various types of distortion that may arise, either from the inherent nonlinearity in the characteristics of the transistors or FETs or from the influence of the associated circuit. The types of distortion that may exist either separately or simultaneously are called *nonlinear distortion, frequency distortion,* and *delay or phase-shift distortion.*

Nonlinear Distortion This type of distortion results from the production of new frequencies in the output which are not present in the input signal. These new frequencies, or harmonics, result from the existence of a nonlinear dynamic curve for the active device; they are considered in some detail in Secs. 18-2 and 18-3. This distortion is sometimes referred to as "amplitude distortion."

Frequency Distortion This type of distortion exists when the signal components of different frequencies are amplified differently. In either a transistor or a FET this distortion may be caused by the internal device capacitances, or it may arise because the associated circuit (for example, the coupling components or the load) is reactive. Under these circumstances, the gain A is a complex number whose magnitude and phase angle depend upon the frequency of the impressed signal. A plot of gain (magnitude) vs. frequency of an amplifier is called the *amplitude frequency-response characteristic.* Such a plot for a single-stage transistor amplifier with a resistive load is indicated in Fig. 11-11. If the frequency-response characteristic is not a horizontal straight line over the range of frequencies under consideration, the circuit is said to exhibit frequency distortion over this range.

Phase-shift Distortion Phase-shift distortion results from unequal phase shifts of signals of different frequencies. This distortion is due to the fact that the phase angle of the complex gain A depends upon the frequency.

12-3   FREQUENCY RESPONSE OF AN AMPLIFIER

A criterion which may be used to compare one amplifier with another with respect to fidelity of reproduction of the input signal is suggested by the following considerations: Any arbitrary waveform of engineering importance may be resolved into a Fourier spectrum. If the waveform is periodic, the spectrum will consist of a series of sines and cosines whose frequencies are all integral multiples of a fundamental frequency. The fundamental frequency is the reciprocal of the time which must elapse before the waveform repeats itself. If the waveform is not periodic, the fundamental period extends in a sense from a time $-\infty$ to a time $+\infty$. The fundamental frequency is then infinitesimally small; the frequencies of successive terms in the Fourier series differ by an infinitesimal amount rather than by a finite amount; and the

Fourier series becomes instead a Fourier integral.   In either case the spectrum includes terms whose frequencies extend, in the general case, from zero frequency to infinity.

Fidelity Considerations   Consider a sinusoidal signal of angular frequency ω represented by $V_m \sin(\omega t + \phi)$.   If the voltage gain of the amplifier has a magnitude A and if the signal suffers a phase change (lead angle) θ, then the output will be

$$AV_m \sin(\omega t + \phi + \theta) = AV_m \sin\left[\omega\left(t + \frac{\theta}{\omega}\right) + \phi\right]$$

Therefore, *if the amplification A is independent of frequency and if the phase shift θ is proportional to frequency (or is zero), then the amplifier will preserve the form of the input signal, although the signal will be shifted in time by an amount $D = \theta/\omega$.*

This discussion suggests that the extent to which an amplifier's amplitude response is not uniform, and its time delay is not constant with frequency, may serve as a measure of the lack of fidelity to be anticipated in it.   In principle, it is not necessary to specify both amplitude and delay response since, for most practical circuits, the two are related and, one having been specified, the other is uniquely determined.   However, in particular cases, it may well be that either the time-delay or amplitude response is the more sensitive indicator of frequency distortion.

For an amplifier stage the frequency characteristics may be divided into three regions: There is a range, called the *midband frequencies*, over which the amplification is reasonably constant and equal to A_o and over which the delay is also quite constant.   For the present discussion we assume that the midband gain is normalized to unity, $A_o = 1$.   In the second (low-frequency) region, below midband, an amplifier stage may behave like the simple high-pass circuit of Fig. 12-1.   The response decreases with decreasing frequency, and the output usually approaches zero at dc ($f = 0$).   In the third (high-frequency) region, above midband, the circuit often behaves like the simple low-pass network of Fig. 12-2, and the response decreases with increasing frequency.   The total frequency characteristic, indicated in Fig. 12-3 for all three regions, will now be discussed.

Low-frequency Response   From the circuit of Fig. 12-1, we find, using the complex variable s,

$$V_o(s) = \frac{V_i(s)R_1}{R_1 + 1/sC_1} = V_i(s)\frac{s}{s + 1/R_1C_1} \tag{12-1}$$

Thus the voltage transfer function at low frequencies, $A_L(s) \equiv V_o(s)/V_i(s)$, has one zero at $s = 0$ and one pole at $s = -1/R_1C_1$.   For real frequencies

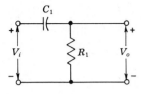

Fig. 12-1   A high-pass RC circuit used to calculate the low-frequency response of an amplifier.

$(s = j\omega = j2\pi f)$, Eq. (12-1) becomes

$$A_L(jf) = \frac{1}{1 - j(f_L/f)} \qquad (12\text{-}2)$$

where

$$f_L \equiv \frac{1}{2\pi R_1 C_1} \qquad (12\text{-}3)$$

The magnitude $|A_L|$ and the phase lead θ_L of the gain are given by

$$|A_L(jf)| = \frac{1}{\sqrt{1 + (f_L/f)^2}} \qquad \theta_L = \arctan \frac{f_L}{f} \qquad (12\text{-}4)$$

At the frequency $f = f_L$, $A_L = 1/\sqrt{2} = 0.707$, whereas in the midband region $(f \gg f_L)$, $A_L \to 1$. Hence f_L is that frequency at which the gain has fallen to 0.707 times its midband value A_o. This drop in signal level corresponds to a decibel reduction of 20 log $(1/\sqrt{2})$, or 3 dB. Accordingly, f_L is referred to as the *lower 3-dB frequency*. From Eq. (12-3) we see that f_L is that frequency for which the resistance R_1 equals the capacitive reactance $1/2\pi f_L C_1$.

High-frequency Response   In the high-frequency region, above the midband, the amplifier stage can often be approximated by the simple low-pass circuit of Fig. 12-2. Such is the case, for example, for the CE transistor stage if we use the Miller approximation (Sec. 11-8). In terms of the complex variable s, we find

$$V_o(s) = \frac{1/sC_2}{R_2 + 1/sC_2} V_i(s) = \frac{1}{1 + sR_2C_2} V_i(s) \qquad (12\text{-}5)$$

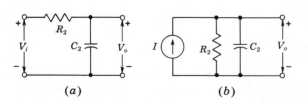

(a)   (b)

Fig. 12-2   (a) A low-pass RC circuit used to calculate the high-frequency response of an amplifier. (b) The Norton's equivalent of the circuit in (a), where $I = V_i/R_2$.

Hence the transfer function in this region has a single pole of $s = -1/R_2C_2$. For real frequencies $(s = j\omega = j2\pi f)$ we obtain for the magnitude

$$|A_H(jf)| = |V_o(s)/V_i(s)|_{s=j2\pi f}$$

and for the phase lead angle θ_H of the gain in the high-frequency region,

$$|A_H(jf)| = \frac{1}{\sqrt{1 + (f/f_H)^2}} \qquad \theta_H = -\arctan\frac{f}{f_H} \qquad (12\text{-}6)$$

where

$$f_H = \frac{1}{2\pi R_2 C_2} \qquad (12\text{-}7)$$

Since at $f = f_H$ the gain is reduced to $1/\sqrt{2}$ times its midband value, then f_H is called the *upper 3-dB frequency*. It also represents that frequency at which the resistance R_2 equals the capacitive reactance $1/2\pi f_H C_2$. In the above expressions θ_L and θ_H represent the angle by which the output *leads* the input, neglecting the initial 180° phase shift through the amplifier. The frequency dependence of the gains in the high- and low-frequency range is to be seen in Fig. 12-3. Such characteristics, called *Bode plots*, are discussed in detail in the following section.

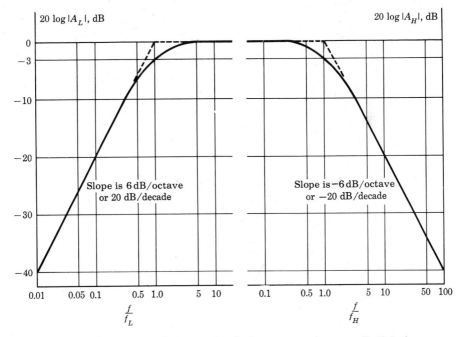

Fig. 12-3   A semi-log plot of the amplitude frequency-response (Bode) charac-teristic of an RC-coupled amplifier.   The dashed curve is the idealized Bode plot.

Bandwidth The frequency range from f_L to f_H is called the *bandwidth* of the amplifier stage. We may anticipate in a general way that a signal, all of whose Fourier components of appreciable amplitude lie well within the range f_L to f_H, will pass through the stage without excessive distortion. This criterion must be applied, however, with caution.

12-4   BODE PLOTS[1]

The frequency response of an amplifier (or any linear network) is indicated by plotting two curves: (1) the magnitude of the transfer gain and (2) the phase-lead angle as a function of frequency. These characteristics are called *Bode plots*. We shall now demonstrate that these curves may be approximated by piecewise linear regions. These interconnected straight-line characteristics are referred to as *idealized Bode plots*.

Single-pole Transfer Function A one-pole transfer gain is of the form given in Eq. (12-5), or with $s = j2\pi f$ and with f_p equal to the frequency of the pole,

$$A(jf) = \frac{A_o}{1 + jf/f_p} \qquad (12\text{-}8)$$

The magnitude in decibels is defined by

$$|A|(\text{dB}) \equiv 20 \log |A| = 20 \log |A_o| - 20 \log \sqrt{1 + \left(\frac{f}{f_p}\right)^2} \qquad (12\text{-}9)$$

or

$$|A|(\text{dB}) = \begin{cases} 20 \log |A_o| & \text{if } \dfrac{f}{f_p} \ll 1 \\[2ex] 20 \log |A_o| - 20 \log \dfrac{f}{f_p} & \text{if } \dfrac{f}{f_p} \gg 1 \end{cases} \qquad (12\text{-}10)$$

These equations are plotted on semi-log paper (that is, $20 \log |A|$ versus $\log f/f_p$) in Fig. 12-4. For frequencies below the pole frequency f_p ($f/f_p < 1$), the characteristic is asymptotic to the horizontal line $20 \log A_o$. For frequencies above f_p ($f/f_p > 1$), Eq. (12-10) indicates that the curve of the gain magnitude in decibels asymptotically approaches a straight line whose slope is -20 dB per decade of frequency ($f/f_p = 10$), or -6 dB per octave ($f/f_p = 2$). The Bode characteristic is drawn in Fig. 12-4, where the asymptotes are shown as shaded straight lines. Note that for $f = f_p$, both asymptotic relationships in Eq. (12-10) yield $|A|(dB) = 20 \log A_o$. Hence, the two lines intersect at $f = f_p$ as indicated in Fig. 12-4 and the 3-dB frequency f_p is also called the *corner frequency*.

Since A_o and f_p are known, the two asymptotes can be easily located. The true Bode characteristic can then be sketched in simply by noting the deviations listed in the table given in Fig. 12-4. For example, the true

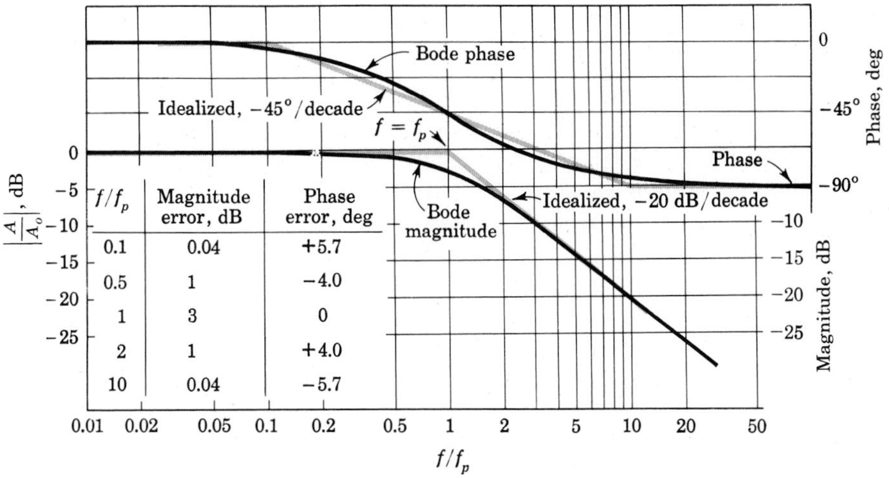

f/f_p	Magnitude error, dB	Phase error, deg
0.1	0.04	+5.7
0.5	1	−4.0
1	3	0
2	1	+4.0
10	0.04	−5.7

Fig. 12-4 Bode plots of a single-pole low-pass amplifier. The piecewise linear approximations are shown shaded.

response curve is 3 dB below the idealized Bode plot at $f = f_p$; is 1 dB below at $f = 0.5f_p$; etc.

The phase-shift angle θ of the single-pole transfer function is given by

$$\theta = -\arctan\frac{f}{f_p} \tag{12-11}$$

For $f \ll f_p$, $\theta \to 0°$, and for $f \gg f_p$, $\theta \to -90°$. At $f = f_p$, $\theta = -45°$. In view of these facts, a piecewise linear approximation to the phase characteristic is constructed as follows: The two asymptotes are the horizontal lines $\theta = 0°$ for $0 \le f/f_p \le 0.1$ and $\theta = -90°$ for $f/f_p \ge 10$. These are joined by a line of slope $-45°$ per decade passing through $\theta = -45°$ at $f = f_p$. In other words this line passes through the points $\theta = 0°$, $f = 0.1f_p$ and $\theta = -90°$, $f = 10f_p$. This broken-line Bode phase characteristic is indicated shaded in Fig. 12-4. The table shows that the idealized Bode plot differs from the true characteristic by less than 6° everywhere.

Single-zero Transfer Function For the transfer gain with one zero at f_z,

$$A(jf) = A_o\left(1 + j\frac{f}{f_z}\right) \tag{12-12}$$

The magnitude in decibels is

$$|A|(\text{dB}) = 20\log|A| = 20\log|A_o| + 20\log\sqrt{1 + \left(\frac{f}{f_z}\right)^2}$$

or

$$|A|(\mathrm{dB}) = \begin{cases} 20 \log |A_o| & \text{if } f \ll f_z \\ 20 \log |A_o| + 20 \log \dfrac{f}{f_z} & \text{if } f \gg f_z \end{cases} \qquad (12\text{-}14)$$

For frequencies below the zero frequency f_z ($f/f_z < 1$), the characteristic is asymptotic to the horizontal line $20 \log |A_o|$. For frequencies above f_z ($f/f_z > 1$), Eq. (12-14) indicates that the curve of $|A|$ (dB) asymptotically approaches a straight line whose slope is $+20$ dB per decade, or $+6$ dB per octave. The Bode characteristic is drawn in Fig. 12-5, where the asymptotes are shown as shaded straight lines. As indicated in connection with the one-pole function, we can quickly sketch the true Bode single-zero characteristic from the two asymptotes determined by A_o and f_z.

The phase angle θ of the single-zero transfer gain is given by

$$\theta = + \arctan \frac{f}{f_z} \qquad (12\text{-}15)$$

For $f \ll f_z$, $\theta \to 0$, and for $f \gg f_z$, $\theta \to +90°$. At $f = f_z$, $\theta = +45°$. Hence the piecewise linear phase approximation is obtained as follows: The two asymptotes are the horizontal lines $\theta = 0°$ for $0 \le f/f_z \le 0.1$ and $\theta = +90°$ for $f/f_z \ge 10$. These are joined by a line of slope $+45°$ per decade passing through $\theta = +45°$ at $f = f_z$. The broken-line Bode plot and the true phase characteristic are indicated in Fig. 12-5.

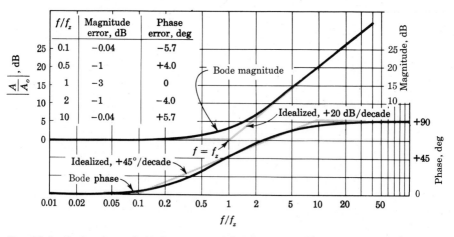

Fig. 12-5   Bode plots of single-zero transfer function.   The piecewise linear approximations are shown shaded.

The single-zero transfer gain is unrealistic because it indicates a response which increases without limit as $f \rightarrow \infty$. A practical circuit has a transfer characteristic consisting of one or more zeros and simultaneously one or more poles, but with the number of poles larger than the number of zeros. It is the poles which determine the upper and lower 3-dB frequencies.

Two-pole Transfer Function   A transfer gain with one pole at f_{p1} and a second at f_{p2} is given by

$$A(jf) = \frac{A_o}{[1 + j(f/f_{p1})][1 + j(f/f_{p2})]} \quad (12\text{-}16)$$

The magnitude in decibels is

$$|A|(\text{dB}) = 20 \log |A_o| - 20 \log \sqrt{1 + \left(\frac{f}{f_{p1}}\right)^2}$$

$$- 20 \log \sqrt{1 + \left(\frac{f}{f_{p2}}\right)^2} \quad (12\text{-}17)$$

Proceeding as above, we conclude that the first term is a horizontal straight line, the second term has an asymptote passing through $f = f_{p1}$ with a slope of -20 dB per decade, or -6 dB per octave, and the third term has an asymptote passing through $f = f_{p2}$ with the same slope. These lines are shown dashed in Fig. 12-6a, and the sum of the three asymptotes, given by the solid-broken-line continuous curve, is the resultant idealized Bode plot. Note that it has been assumed that $f_{p2} > f_{p1}$. For $f > f_{p2}$ the resultant slope is -40 dB per decade, or -12 dB per octave.

The phase response is given by

$$\theta = - \arctan \frac{f}{f_{p1}} - \arctan \frac{f}{f_{p2}} \quad (12\text{-}18)$$

The linearized Bode plot is obtained by considering each term in Eq. (12-18) separately and proceeding as in Fig. 12-6b. The contribution to the phase θ by each pole is indicated separately by the dashed lines. For example, the curve marked θ_2 corresponds to the pole at f_{p2}. Consistent with the above discussion $\theta_2 = 0$ for $f \leq 0.1f_{p2}$ and $\theta = -90°$ for $f \geq 10f_{p2}$, and θ_2 decreases by $45°$ per decade for $0.1f_{p2} \leq f \leq 10f_{p2}$. At $f = f_{p2}$, $\theta_2 = -45°$. The resultant phase at any frequency is the sum of the two dashed curves at that frequency and is indicated by the shaded-broken-line plot. (Figure 12-6 is drawn for the special case where $f_{p2} = 5f_{p1}$.)

The true (not linearized) Bode plot may be obtained by using the table in Fig. 12-4. The two individual curves corresponding to break points at f_{p1} and f_{p2} are drawn and then are added to obtain the resultant Bode amplitude (Prob. 12-5). The Bode phase response is obtained in an analogous manner by using the table in Fig. 12-4 to correct the piecewise linear phase lines.

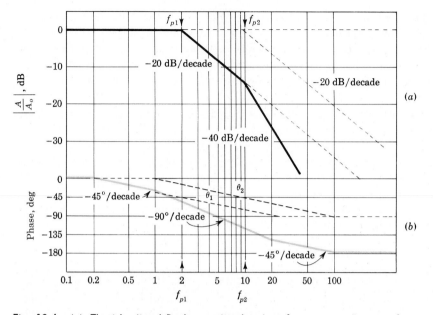

Fig. 12-6 (a) The idealized Bode amplitude plots for a two-pole transfer function. The individual asymptotes for each pole are shown dashed, and the resultant is drawn as a solid continuous broken-line plot. (b) Phase-response Bode plot for (a).

The Dominant Pole If in Eq. (12-16) f_{p1} is much smaller than f_{p2}, the above discussion in connection with Fig. 12-6 indicates that the upper 3-dB frequency is given approximately by f_{p1}. If $f_{p2} = 4f_{p1}$, an exact plot indicates (Prob. 12-7) that the 3-dB frequency is only 6 percent smaller than f_{p1}. We conclude that *if a transfer function has several poles determining the high-frequency response, if the smallest of these is f_{p1} and if each other pole is at least two octaves higher, then the amplifier behaves essentially as a single-time-constant circuit whose 3-dB frequency is f_{p1}.* The frequency f_{p1} is called the *dominant pole.*

12-5 STEP RESPONSE OF AN AMPLIFIER

An alternative criterion of amplifier fidelity is the response of the amplifier to a particular input waveform. Of all possible available waveforms, the most generally useful is the step voltage. In terms of a circuit's response to a step, the response to an arbitrary waveform may be written in the form of the superposition integral. Another feature which recommends the step voltage is the fact that this waveform is one which permits small distortions to stand out clearly. Additionally, from an experimental viewpoint, we note that excellent pulse (a short step) and square-wave (a repeated step) generators are available commercially.

As long as an amplifier can be represented by a dominant pole, the correlation between its frequency response and the output waveshape for a step input is that given below. Quite generally, even for more complicated amplifier circuits, there continues to be an intimate relationship between the distortion of the leading edge of a step and the high-frequency response. Similarly, there is a close relationship between the low-frequency response and the distortion of the flat portion of the step. We should, of course, expect such a relationship, since the high-frequency response measures essentially the ability of the amplifier to respond faithfully to rapid variations in signal, whereas the low-frequency response measures the fidelity of the amplifier for slowly varying signals. An important feature of a step is that it is a combination of the most abrupt voltage change possible and of the slowest possible voltage variation.

Rise Time The response of the low-pass circuit of Fig. 12-2 to a step input of amplitude V is exponential with a time constant R_2C_2. Since the capacitor voltage cannot change instantaneously, the output starts from zero and rises toward the steady-state value V, as shown in Fig. 12-7. The output is given by

$$v_o = V(1 - \epsilon^{-t/R_2C_2}) \qquad (12\text{-}19)$$

The time required for v_o to reach one-tenth of its final value is readily found to be $0.1R_2C_2$, and the time to reach nine-tenths its final value is $2.3R_2C_2$. The difference between these two values is called the *rise time t_r* of the circuit and is shown in Fig. 12-7. The time t_r is an indication of how fast the amplifier can respond to a discontinuity in the input voltage. We have, using Eq. (12-7),

$$t_r = 2.2R_2C_2 = \frac{2.2}{2\pi f_H} = \frac{0.35}{f_H} \qquad (12\text{-}20)$$

Note that the rise time is inversely proportional to the upper 3-dB frequency. For an amplifier with 1 MHz bandpass, $t_r = 0.35 \ \mu$s.

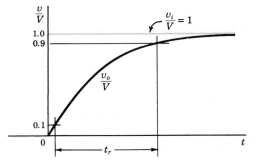

Fig. 12-7 Step-voltage response of the low-pass RC circuit. The rise time t_r is indicated.

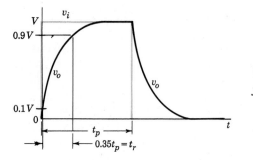

Fig. 12-8 Pulse response for the case $f_H = 1/t_p$.

Consider a pulse of width t_p. What must be the high 3-dB frequency f_H of an amplifier used to amplify this signal without excessive distortion? A reasonable answer to this question is: *Choose f_H equal to the reciprocal of the pulse width, $f_H = 1/t_p$.* From Eq. (12-20) we then have $t_r = 0.35t_p$. Using this relationship, the (shaded) pulse in Fig. 12-8 becomes distorted into the (solid) waveform, which is clearly recognized as a pulse.

Tilt or Sag If a step of amplitude V is impressed on the high-pass circuit of Fig. 12-1, the output is

$$v_o = V\epsilon^{-t/R_1C_1} \tag{12-21}$$

For times t which are small compared with the time constant R_1C_1, the response is given by

$$v_o \approx V\left(1 - \frac{t}{R_1C_1}\right) \tag{12-22}$$

From Fig. 12-9 we see that the output is tilted, and the percent tilt, or sag, in time t_1 is given by

$$P \equiv \frac{V - V'}{V} \times 100 = \frac{t_1}{R_1C_1} \times 100\% \tag{12-23}$$

It is found[6] that this same expression is valid for the tilt of each half cycle of a symmetrical square wave of peak-to-peak value V and period T provided

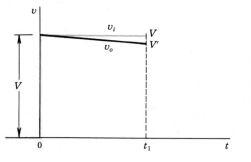

Fig. 12-9 The response v_o, when a step v_i is applied to a high-pass RC circuit, exhibits a tilt.

that we set $t_1 = T/2$. If $f = 1/T$ is the frequency of the square wave, then, using Eq. (12-3), we may express P in the form

$$P = \frac{T}{2R_1C_1} \times 100 = \frac{1}{2fR_1C_1} \times 100 = \frac{\pi f_L}{f} \times 100\% \tag{12-24}$$

Note that the tilt is directly proportional to the lower 3-dB frequency. If we wish to pass a 50-Hz square wave with less than 10 percent sag, then f_L must not exceed 1.6 Hz.

Square-wave Testing An important experimental procedure (called *square-wave testing*) is to observe with an oscilloscope the output of an amplifier excited by a square-wave generator. It is possible to improve the response of an amplifier by adding to it certain circuit elements,[1] which then must be adjusted with precision. It is a great convenience to be able to adjust these elements and to see simultaneously the effect of such an adjustment on the amplifier output waveform. The alternative is to take data, after each successive adjustment, from which to plot the amplitude and phase responses. Aside from the extra time consumed in this latter procedure, we have the problem that it is usually not obvious which of the attainable amplitude and phase responses corresponds to optimum fidelity. On the other hand, the step response gives immediately useful information.

It is possible, by judicious selection of two square-wave frequencies, to examine individually the high-frequency and low-frequency distortion. For example, consider an amplifier which has a high-frequency time constant of 1 μs and a low-frequency time constant of 0.1 s. A square wave of half period equal to several microseconds, on an appropriately fast oscilloscope sweep, will display the rounding of the leading edge of the waveform and will not display the tilt. At the other extreme, a square wave of half period approximately 0.01 s on an appropriately slow sweep will display the tilt, and not the distortion of the leading edge. Such a waveform is indicated in Fig. 12-10.

It should *not* be inferred from the above comparison between steady-state and transient response that the phase and amplitude responses are of no importance at all in the study of amplifiers. The frequency characteristics are useful for the following reasons. In the first place, much more is known generally about the analysis and synthesis of circuits in the frequency domain than in the time domain, and for this reason the design of coupling networks is often done on a frequency-response basis. Second, it is often possible to arrive at least at a qualitative understanding of the properties of a circuit from a study of the steady-state-response in circumstances where transient calculations are extremely cumbersome. Third, compensating an amplifier against unwanted oscillations (Chap. 14) is accomplished in the frequency domain. Finally, it happens occasionally that an amplifier is required whose characteristics are specified on a frequency basis, the principal emphasis being to amplify a sine wave.

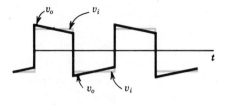

Fig. 12-10   A square-wave (shaded) input signal is distorted by an amplifier with a lower 3-dB frequency f_L. The output (solid) waveform shows a tilt where the input is horizontal.

12-6   BANDPASS OF CASCADED STAGES

The high 3-dB frequency for n cascaded stages is f_H^* and equals the frequency for which the overall voltage gain falls 3 dB to $1/\sqrt{2}$ of its midband value. To obtain the overall transfer function of *noninteracting* stages, the transfer gains of the individual stages are multiplied together.   Hence, if each stage has a dominant pole and if the high 3-dB frequency of the ith stage is f_{Hi}, where $i = 1, 2, \ldots, n$, then f_H^* can be calculated from the product

$$\frac{1}{\sqrt{1 + (f_H^*/f_{H1})^2}} \cdots \frac{1}{\sqrt{1 + (f_H^*/f_{Hi})^2}} \cdots \frac{1}{\sqrt{1 + (f_H^*/f_{Hn})^2}}$$
$$= \frac{1}{\sqrt{2}} \quad (12\text{-}25)$$

For n stages with identical upper 3-dB frequencies we have

$$f_{H1} = f_{H2} = \cdots = f_{Hi} = \cdots = f_{Hn} = f_H$$

Thus f_H^* is calculated from

$$\left[\frac{1}{\sqrt{1 + (f_H^*/f_H)^2}}\right]^n = \frac{1}{\sqrt{2}}$$

to be

$$\frac{f_H^*}{f_H} = \sqrt{2^{1/n} - 1} \qquad\qquad (12\text{-}26)$$

For example, for $n = 2$, $f_H^*/f_H = 0.64$.   Hence two cascaded stages, each with a bandwidth $f_H = 10$ kHz, have an overall bandwidth of 6.4 kHz.   Similarly, three cascaded 10-kHz stages give a resultant upper 3-dB frequency of 5.1 kHz, etc.

If the low 3-dB frequency for n identical noninteracting cascaded stages is f_L^*, then, corresponding to Eq. (12-26), we find

$$\frac{f_L^*}{f_L} = \frac{1}{\sqrt{2^{1/n} - 1}} \qquad\qquad (12\text{-}27)$$

We see that a cascade of stages has a lower f_H and a higher f_L than a single stage, resulting in a shrinkage in bandwidth.

If the amplitude response for a single stage is plotted on log-log paper, the resulting graph will approach a straight line whose slope is 6 dB per octave both at the low and at the high frequencies, as indicated in Fig. 12-3.   For an

n-stage amplifier it follows that the amplitude response falls $6n$ dB per octave, or equivalently, $20n$ dB per decade.

Interacting Stages   If in a cascade of stages the input impedance of one stage is low enough to act as an appreciable shunt on the output impedance of the preceding stage, then it is no longer possible to isolate stages.   Under these circumstances individual 3-dB frequencies for each stage cannot be defined.   However, when the overall transfer function of the cascade is obtained (Sec. 12-10), it is found to contain n poles (in addition to k zeros). If the pole frequencies are $f_1, \ldots, f_i, \ldots, f_n$, then the high 3-dB frequency of the entire cascade f_H^* is given by Eq. (12-25) (with f_{Hi} replaced by f_i), provided that the zero frequencies are much higher than the pole frequencies (Prob. 12-14).

If the cascade has a dominant pole f_D which is much smaller than all other poles, all terms in the product in Eq. (12-25) may be neglected except the first. It then follows that $f_H = f_D$, or the high 3-dB frequency equals the dominant-pole frequency.   (From here on we shall drop the asterisk on f_H^*.)

Consider now the situation discussed in Sec. 12-4, where there is a dominant frequency f_D, a second pole whose frequency is only two octaves away, and all other poles are at very much higher frequencies.   Then Eq. (12-25) becomes

$$\frac{1}{\sqrt{1 + (f_H/f_D)^2}} \frac{1}{\sqrt{1 + (f_H/4f_D)^2}} = \frac{1}{\sqrt{2}} \tag{12-28}$$

Since we expect the 3-dB frequency to be approximately equal to the dominant frequency, substitute $f_H = f_D$ into the second term in Eq. (12-28) to obtain

$$1 + \left(\frac{f_H}{f_D}\right)^2 = \frac{2}{1 + (\tfrac{1}{4})^2} \tag{12-29}$$

or

$$f_H = 0.94 f_D \tag{12-30}$$

This calculation verifies that the high 3-dB frequency is less than 6 percent smaller than the dominant frequency provided that the next higher pole frequency is at least two octaves away.

If the pole frequencies are not widely separated, the result of Prob. 12-15 indicates that f_H is given (to within 10 percent) by

$$\frac{1}{f_H} = 1.1 \sqrt{\frac{1}{f_1{}^2} + \frac{1}{f_2{}^2} + \cdots + \frac{1}{f_n{}^2}} \tag{12-31}$$

If this equation is applied to the case considered above, $f_1 = f_D$ and $f_2 = 4f_D$ and all other poles much higher, the result is $f_H = 0.89 f_D$, in close agreement with Eq. (12-30).   If Eq. (12-31) is applied to the case where $f_1 = f_2$ and all other poles are much higher, then $f_H = 0.65 f_1$ (instead of the exact value of $0.64 f_1$).   For three equal poles, Eq. (12-31) yields $f_H = 0.53 f_1$ (instead of the exact value of $0.51 f_1$).

Step Response If the rise time of isolated individual cascaded stages is $t_{r1}, t_{r2}, \ldots, t_{rn}$ and if the input waveform rise time is t_{ro}, then, corresponding to Eq. (12-31) for the resultant upper 3-dB frequency, we have that the output-signal rise time t_r is given (to within 10 percent) by

$$t_r = 1.1 \sqrt{t_{ro}^2 + t_{r1}^2 + t_{r2}^2 + \cdots + t_{rn}^2} \tag{12-32}$$

If, upon application of a voltage step, one circuit produces a tilt of P_1 percent and if a second stage gives a tilt of P_2 percent, the effect of cascading these two noninteracting circuits is to produce a tilt of $P_1 + P_2$ percent. This result applies only if the individual tilts and the combined tilt are small enough so that in each case the waveform falls approximately linearly with time.

12-7   THE *RC*-COUPLED AMPLIFIER

A cascaded arrangement of common-emitter (CE) transistor stages is shown in Fig. 12-11a, and of common-source (CS) FET stages is shown in Fig. 12-11b.

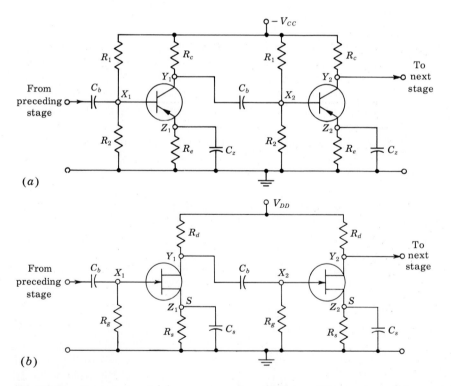

(a)

(b)

Fig. 12-11 A cascade of (a) common-emitter (CE) transistor stages; (b) common-source (CS) depletion-type, or JFET stages.

The output Y_1 of one stage is coupled to the input X_2 of the next stage via a blocking capacitor C_b which is used to keep the dc component of the output voltage at Y_1 from reaching the input X_2.   Resistor R_g is from gate to ground, and the collector (drain) circuit resistor is $R_c(R_d)$.   The source resistor R_s, the emitter resistor R_e, and the resistors R_1 and R_2 are used to establish the bias. The bypass capacitors, used to prevent loss of amplification due to negative feedback (Chap. 13), are C_z in the emitter circuit and C_s in the source circuit. Also present are junction capacitances, to be taken into account when we consider the high-frequency response, which is limited by their presence.   In any practical mechanical arrangement of the amplifier components, there are also capacitances associated with device sockets and the proximity to the chassis of components (for example, the body of C_b) and signal leads.   These stray capacitances are also considered later.   We assume that the active device operates linearly, so that small-signal models are used throughout this chapter.

12-8   LOW–FREQUENCY RESPONSE OF AN RC–COUPLED STAGE

The effect of the bypass capacitors C_z, and C_s on the low-frequency characteristics is discussed in the next section.   For the present we assume that these capacitances are arbitrarily large and act as ac (incremental) short circuits across R_e and R_s, respectively.   A single intermediate stage of any of the cascades in Fig. 12-11 may be represented schematically as in Fig. 12-12. The resistor R_b represents the gate resistor R_g for an FET, and equals R_1 in parallel with R_2 if a transistor stage is under consideration.   The resistor R_y represents R_c for a transistor, or R_d for an FET, and R_i represents the input resistance of the following stage.

The low-frequency equivalent circuit is obtained by neglecting all shunting capacitances and all junction capacitances by replacing amplifier A_1 by its Norton's equivalent, as indicated in Fig. 12-13a.   For a field-effect transistor, $R_i = \infty$, the output impedance is $R_o = r_d$ (the drain resistance), and $I = g_m V_i$ (transconductance times gate signal voltage).   For a transistor these quantities may be expressed in terms of the CE hybrid parameters as in Sec. 8-13: $R_i \approx h_{ie}$ (for small values of R_c); $R_o = 1/h_{oe}$ (for a current drive); and $I = h_{fe} I_b$, where I_b is the base signal current.   Let R_o' represent R_o in parallel with R_y, and let R_i' be R_i in parallel with R_b.   Then, replacing I and

Fig. 12-12   A schematic representation of either an FET or transistor stage.   Biasing arrangements and supply voltages are not indicated.

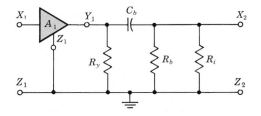

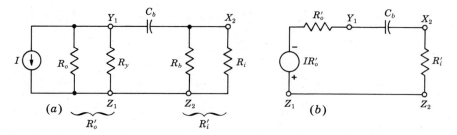

Fig. 12-13 (a) The low-frequency model of an RC-coupled amplifier; (b) an equivalent representation. For an **FET**, $I = g_m V_i$, $R_o = r_d$, $R_y = R_d$, $R_b = R_g$, and $R_i = \infty$. For a transistor, $I = h_{fe} I_b$, $R_o \approx 1/h_{oe}$, $R_b = R_1 \| R_2$, $R_y = R_c$, and $R_i \approx h_{ie}$. Also, $R'_i = R_i \| R_b$ and $R'_o = R_o \| R_y$.

R'_o by the Thévenin's equivalent, the single-time-constant high-pass circuit of Fig. 12-13b results. Hence, from Eq. (12-3), the low 3-dB frequency is

$$f_L = \frac{1}{2\pi(R'_o + R'_i)C_b} \tag{12-33}$$

This result is easy to remember since the time constant equals C_b multiplied by the sum of the effective resistances R'_o to the left of the blocking capacitor, and R'_i to the right of C_b. For an FET amplifier, $R'_i = R_g \gg R_d$. Since $R'_o < R_d$ because R'_o is R_d in parallel with R_o, then $R'_i = R_g \gg R'_o$ and $f_L \approx 1/2\pi C_b R_g$.

EXAMPLE It is desired to have a low 3-dB frequency of not more than 10 Hz for an RC-coupled amplifier for which $R_y = 1$ K. What minimum value of coupling capacitance is required if (a) FETs with $R_g = 1$ M are used; (b) transistors with $R_i = 1$ K and $1/h_{oe} = 40$ K are used?

Solution a. From Eq. (12-33) we have

$$f_L = \frac{1}{2\pi(R'_o + R'_i)C_b} \le 10$$

or

$$C_b \ge \frac{1}{62.8(R'_o + R'_i)}$$

Since $R'_i = 1$ M and $R'_o < R_y = 1$ K, then $R'_o + R'_i \approx 1$ M and $C_b \ge 0.016$ μF.

b. From Eq. (8-35) we find for a transistor $R_o \ge 1/h_{oe} \approx 40$ K, and hence $R'_o \approx R_c = 1$ K. If we assume that $R_b \gg R_i = 1$ K, then $R'_i \approx 1$ K. Hence

$$C_b \ge \frac{1}{(62.8)(2 \times 10^3)} \text{ F} = 8.0 \ \mu\text{F}$$

Note that, because the input impedance of a transistor is much smaller than that of an FET, a coupling capacitor is required with the transistor which is 500 times larger than that required with the FET. Fortunately, it is possible to obtain physically small electrolytic capacitors having such high capacitance values at the low voltages at which transistors operate. Since the coupling capacitances required for good low-frequency response are far larger than those obtainable in integrated form, cascaded integrated stages must be direct-coupled (Chap. 15).

12-9   EFFECT OF AN EMITTER BYPASS CAPACITOR ON LOW–FREQUENCY RESPONSE

If an emitter resistor R_e is used for self-bias in an amplifier and if it is desired to avoid the degeneration, and hence the loss of gain due to R_e, we might attempt to bypass this resistor with a very large capacitance C_z. The circuit is indicated in Fig. 12-11a. It is shown below that the effect of this capacitor is to affect adversely the low-frequency response.

Consider the single stage of Fig. 12-14a. To simplify the analysis we assume that $R_1 \| R_2 \gg R_s$ and that the load R_c is small enough so that the simplified hybrid model of Fig. 8-23 is valid. The equivalent circuit subject to these assumptions is shown in Fig. 12-14b. The blocking capacitor C_b is omitted from Fig. 12-14b; its effect is considered in Sec. 12-8.

The output voltage V_o is given by

$$V_o = -I_b h_{fe} R_c = -\frac{V_s h_{fe} R_c}{R_s + h_{ie} + Z'_e} \qquad (12\text{-}34)$$

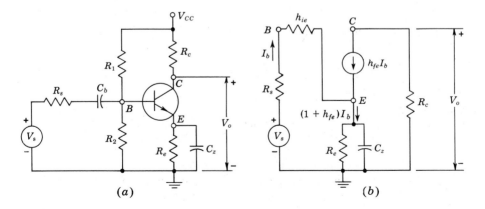

Fig. 12-14   (a) An amplifier with a bypassed emitter resistor; (b) the low-frequency simplified h-parameter model of the circuit in (a).

where

$$Z'_e \equiv (1 + h_{fe}) \frac{R_e}{1 + j\omega C_z R_e} \tag{12-35}$$

Substituting Eq. (12-35) in Eq. (12-34) and solving for the voltage gain A_V, we find

$$A_V = \frac{V_o}{V_s} = -\frac{h_{fe}R_c}{R + R'} \frac{1 + j\omega C_z R_e}{1 + j\omega C_z[R_e R/(R + R')]} \tag{12-36}$$

where

$$R \equiv R_s + h_{ie} \qquad \text{and} \qquad R' \equiv (1 + h_{fe})R_e \tag{12-37}$$

The midband gain A_o is obtained as $\omega \to \infty$, or

$$A_o = -\frac{h_{fe}R_c}{R} = \frac{-h_{fe}R_c}{R_s + h_{ie}} \tag{12-38}$$

Hence

$$\frac{A_V}{A_o} = \frac{1}{1 + R'/R} \frac{1 + jf/f_o}{1 + jf/f_p} \tag{12-39}$$

where

$$f_o \equiv \frac{1}{2\pi C_z R_e} \qquad f_p \equiv \frac{1 + R'/R}{2\pi C_z R_e} \tag{12-40}$$

Note that f_o determines the zero and f_p the pole of the gain A_V/A_o. Since, usually, $R'/R \gg 1$, then $f_p \gg f_o$, so that the pole and zero are widely separated. For example, assuming $R_s = 0$, $R_e = 1$ K, $C_z = 100$ μF, $h_{fe} = 50$, $h_{ie} = 1.1$ K, and $R_c = 2$ K, we find $f_o = 1.6$ Hz and $f_p = 76$ Hz.

The magnitude of $|A_V/A_o|$ in decibels is given by

$$20 \log \left| \frac{A_V}{A_o} \right| = -20 \log \left(1 + \frac{R'}{R} \right) + 20 \log \sqrt{1 + \left(\frac{f}{f_o} \right)^2}$$
$$- 20 \log \sqrt{1 + \left(\frac{f^2}{f_p} \right)} \tag{12-41}$$

From the discussion in Sec. 12-4, we conclude that the first term represents a horizontal line, the second term has an asymptote passing through $f = f_o$ with a positive slope of 6 dB per octave, and the third term has an asymptote passing through $f = f_p$, with a negative slope of the same magnitude. These lines are shown dashed in Fig. 12-15, and the idealized Bode plot is obtained by adding the three asymptotes together to form the shaded-broken-line continuous curve. Note that, for $f \geq f_p$, the result of adding a gain which increases 6 dB per octave to a gain which decreases 6 dB per octave gives a constant gain. For $f \gg f_p$, the gain must approach its midband value A_o, and hence the horizontal line for $f > f_p$ occurs at $20 \log |A_V/A_o| = 0$ dB, as indicated.

The Bode amplitude response curve for the amplifier of Fig. 12-14a is plotted in Fig. 12-16. The asymptotic behavior discussed in Fig. 12-4 is now included as the shaded-broken-line plot.

Fig. 12-15 The idealized
Bode amplitude plot for
a transfer function with
one zero $f = f_o$ and one
pole $f = f_p$. The individual asymptotes are
shown dashed, and the
resultant is drawn as a
shaded continuous
broken-line plot.

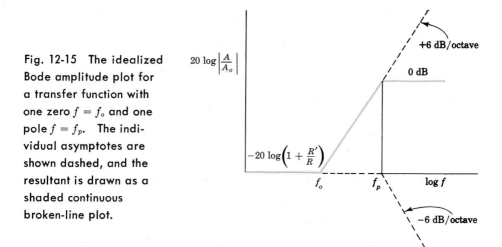

Remembering that $f_p \gg f_o$ and using Eqs. (12-39) and (12-40), the magnitude of A_V/A_o becomes, for $f = f_p$,

$$\left|\frac{A_V}{A_o}\right| = \frac{1}{1 + R'/R}\frac{f_p/f_o}{\sqrt{1+1}} = \frac{1}{\sqrt{2}}$$

Hence $f = f_p$ is that frequency at which the gain has dropped 3 dB.   Thus the low 3-dB frequency f_L is approximately equal to f_p.   If the condition

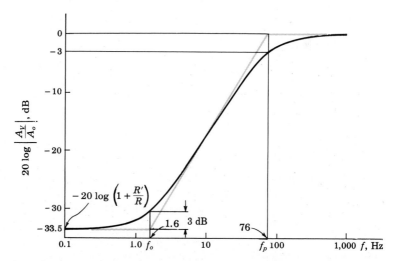

Fig. 12-16 The frequency response of an amplifier with a bypassed emitter resistor.   The numerical values correspond to the component values given below Eq. (12-40).

$f_p \gg f_o$ is *not* satisfied, then $f_L \neq f_p$. As a matter of fact, a 3-dB frequency may not exist (Prob. 12-24).

Square-wave Response   If $f_p \gg f_o$, the network in Fig. 12-14 behaves like a single-time-constant circuit and the percentage tilt to a square wave is given by Eq. (12-24), or

$$P = \frac{\pi f_p}{f} \times 100 = \frac{1 + R'/R}{2fC_zR_e} \times 100\% \tag{12-42}$$

Since $R'/R \gg 1$,

$$P \approx \frac{R' \times 100}{2fC_zRR_e} = \frac{1 + h_{fe}}{2fC_z(R_s + h_{ie})} \times 100\% \tag{12-43}$$

Practical Considerations   Let us calculate the size of C_z so that we may reproduce a 50-Hz square wave with a tilt of less than 10 percent. Using the parameters given above, we obtain

$$C_z = \frac{(51)(100)}{(2)(50)(1,100)(10)} \text{ F} = 4,600 \ \mu\text{F}$$

Such a large value of capacitance is impractical, and it must be concluded that if very small tilts are to be obtained for very low frequency signals, the emitter resistor must be left unbypassed. The flatness will then be obtained at the sacrifice of gain because of the degeneration caused by R_e. If the loss in amplification cannot be tolerated, R_e cannot be used.

Electrolytic capacitors are often used as emitter or source bypass capacitors because they offer the greatest capacitance per unit volume. It is important to note that these capacitors have a series resistance which arises from the conductive losses in the electrolyte. This resistance, typically 1 to 20 Ω, must be taken into account in computing the midband gain of the stage.

Response Due to Both Emitter and Coupling Capacitors   If in a given stage both C_z and the coupling capacitor C_b are present, we can assume, first, C_z to be infinite and compute the low 3-dB frequency due to C_b alone. We then calculate f_L due to C_z by assuming C_b to be infinite. If the two cutoff frequencies are significantly different (by two octaves or more), the higher of the two is approximately the low 3-dB frequency for the stage.

If a dominant pole does not exist at low frequencies, the response must be obtained by writing the network equations for a sinusoidal excitation ($s = j2\pi f$) and with both C_z and C_b finite. The transfer function V_o/V_s is plotted as a function of frequency, and the low 3-dB frequency is read from this plot. This general method of solution is illustrated (at the high-frequency end of the spectrum) in the following section.

The low-frequency analysis of an FET amplifier with a source resistor R_s bypassed with a capacitor C_s is considered in Prob. 12-27.

12-10   HIGH–FREQUENCY RESPONSE OF TWO CASCADED CE TRANSISTOR STAGES[2]

The high-frequency analysis of a single-stage CE transistor amplifier, or the last stage of a cascade, is given in detail in Secs. 11-7 and 11-8. We have since established that there is interaction between CE cascaded transistor stages. The analysis of a multistage amplifier is complicated and tedious. Fortunately, it is possible to make certain approximations in the analysis, and thus reduce the complexity of bandwidth calculations while keeping the error under approximately 20 percent.

Figure 12-11a shows two CE transistors in cascade. For high-frequency calculations each transistor is replaced by its small-signal hybrid-Π model, as indicated in Fig. 12-17. The elements $g_{b'c}$ and g_{ce} have been deleted, because, as demonstrated in Sec. 11-6, their omission introduces little error. We have included a voltage source V_s with $R_s = 50\ \Omega$ and have assumed that capacitors C_b and C_z represent short circuits for high frequencies. The base biasing resistors R_1 and R_2 in Fig. 12-11a are assumed to be large compared with R_s. The symbol R_{L1} represents the parallel combination of R_1, R_2, and collector circuit resistance R_c of the first stage. The symbol R_{L2} is the total load resistance of the second stage. A complete stage is included in each shaded block.

The network can be described by four nodal equations. If

$$R_s' \equiv R_s + r_{bb'} = 1/G_s',\ G_{L1} = 1/R_{L1},\ G_{L2} = 1/R_{L2},\ \text{and}\ g_{bb'} = 1/r_{bb'}$$

these equations are

$$
\begin{aligned}
G_s'V_s &= [G_s' + g_{b'e} + s(C_e + C_c)]V_1 - sC_cV_2 \\
0 &= (g_m - sC_c)V_1 + (G_{L1} + g_{bb'} + sC_c)V_2 - g_{bb'}V_3 \\
0 &= -g_{bb'}V_2 + [g_{b'e} + g_{bb'} + s(C_e + C_c)]V_3 - sC_cV_4 \\
0 &= (g_m - sC_c)V_3 + (G_{L2} + sC_c)V_4
\end{aligned}
\qquad (12\text{-}44)
$$

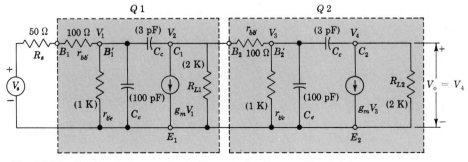

Fig. 12-17 Two-stage interacting CE amplifier ($g_m = 50$ mA/V).

We can find the transfer gain V_o/V_s from these equations by recalling Cramer's rule,

$$A_V \equiv \frac{V_4}{V_s} = \frac{G_s' \Delta_{14}}{\Delta} = \text{transfer function} \tag{12-45}$$

where Δ is the determinant of the set of equations, and Δ_{14} is the minor formed by removing the first row and fourth column of the complete determinant. Thus we see that the poles of the transfer function are given by the zeros of the determinant $\Delta = 0$, and the zeros of the transfer function are given by the zeros of the minor $\Delta_{14} = 0$. So much effort is required to find the poles and zeros that it is advisable to use a computer. There are several computer programs[3] for finding roots of determinants.

The transfer function V_4/V_s of Fig. 12-17 must have four poles since the network contains four independent energy storage elements (in the sense that there exists no loop formed exclusively of capacitors). In addition, the C_e capacitance for each of the two transistors provides a short circuit to ground as $s \to \infty$, and thus $V_4/V_s \to 1/s^2$ as $s \to \infty$. Hence we must have two zeros in addition to the poles. The values of the zeros for the circuit of Fig. 12-17 can be found by inspection. If, for some value of s, say s_5, V_4 is zero, the current fed through C_c must be equal to $g_m V_3$. Hence

$$s_5 C_c V_3 = g_m V_3 \tag{12-46}$$

and the zero is $s_5 = g_m/C_c$. Similarly, for $Q1$, the zero is $s_6 = g_m/C_c$. The transfer function now has the form

$$A_V = \frac{K(s - s_5)(s - s_6)}{(s - s_1)(s - s_2)(s - s_3)(s - s_4)} \tag{12-47}$$

Using one of the standard computer programs to solve $\Delta = 0$, we find for the numerical values given in Fig. 12-17 that the poles are given by

$$s_1 = -0.00342 \times 10^9 \text{ rad/s} \qquad s_2 = -0.0670 \times 10^9 \text{ rad/s}$$

$$s_3 = -0.680 \times 10^9 \text{ rad/s} \qquad s_4 = -4.21 \times 10^9 \text{ rad/s}$$

The zeros are

$$s_5 = s_6 = \frac{g_m}{C_c} = 16.65 \times 10^9 \text{ rad/s}$$

The program used to obtain the poles and zeros is CORNAP.[4] This computer program is a network-analysis routine that finds the state equations, the transfer function, and the frequency response of a general linear active network. The frequency response of the amplifier shown in Fig. 12-17 as obtained using CORNAP is plotted in Fig. 12-18. From the frequency-response curve we can read the high 3-dB frequency of the amplifier as $f_H = 540$ kHz.

The two transistors in Fig. 12-17 were assumed to have identical parameters. This condition simplifies the notation somewhat, but is not an essential restriction. The general method of analysis outlined above is equally valid for nonidentical transistors.

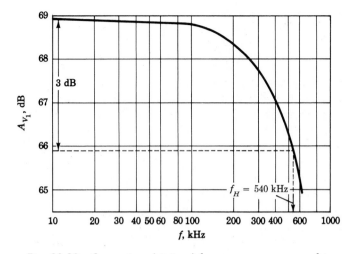

Fig. 12-18 Computer-obtained frequency response of two stage amplifier.

Dominant Pole In the above example we observe that one of the poles, $s_1 = -0.00342 \times 10^9$ rad/s, is much lower than the other poles and zeros, and hence is the dominant pole. From our discussion of Bode plots in Sec. 12-4, we recognize that if all other poles and zeros are at least two octaves away, the response is down 3 dB at the dominant pole. Hence the upper 3-dB frequency for the amplifier is given, approximately, by the dominant pole

$$f_H \approx f_1 = \frac{-s_1}{2\pi} = \frac{0.00342 \times 10^9}{2\pi} \text{ Hz} = 545 \text{ kHz}$$

which is essentially the same value read from the curve.

Two-stage-cascade Simplified Analysis If a computer is not available to help with the computations, we must make simplified assumptions in order to proceed with the analysis. We follow the method outlined in Sec. 11-8.

The effect of C_c is approximated using Miller's theorem and the midband value of the stage gain. Thus C_c of Q_2 is replaced by a capacitance

$$C_c(1 + g_m R_{L2}) = 3(1 + 50 \times 2) = 303 \text{ pF}$$

across the input of $Q2$. Similarly, C_c of $Q1$ is replaced by

$$C_c[1 + g_m R_{L1}||(r_{b'e} + r_{bb'})] = 3(1 + 50 \times 0.709) = 109 \text{ pF}$$

across the input of $Q1$. The circuit is now considerably simplified since, as shown in Fig. 12-19, there are only two independent capacitors in the network.

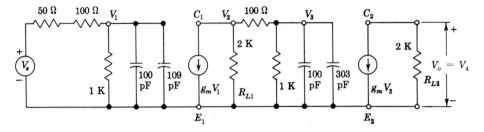

Fig. 12-19 Two-stage interacting CE amplifier using Miller approximation.

The transfer function can be obtained from Fig. 12-19 by writing the four nodal equations. These are quite simple (Prob. 12-28) and result in

$$A_V \equiv \frac{V_4}{V_s} = \frac{2{,}810}{\left(1 + j\dfrac{f}{5.85 \times 10^5}\right)\left(1 + j\dfrac{f}{5.85 \times 10^6}\right)} \qquad (12\text{-}48)$$

Clearly, we have a dominant pole at $f_1 = 585$ kHz, and thus this is the approximate high 3-dB frequency of the two-stage cascade.

We note that the simplified analysis yields a value for the bandwidth which is higher than the 540 kHz obtained using the exact method, by 8 percent.

12-11   MULTISTAGE CE AMPLIFIER CASCADE AT HIGH FREQUENCIES

Regardless of the number of stages in an amplifier chain, the general method of solution is that given in the preceding section. Of course, the larger the number of stages, the greater is the computational complexity. We shall now outline the analysis of the three-stage CE cascade. Figure 12-20a shows the network with each transistor replaced by its small-signal model for high-frequency analysis. We assume that all transistors are identical with $g_m = 50$ mA/V. We desire to calculate and plot the transfer function V_o/V_s as a function of frequency. If we write nodal equations for nodes V_1 through V_6, we obtain a system of six equations in six unknowns similar to Eqs. (12-44). Clearly, due to the six independent capacitors, the transfer function must have six poles. In addition, three zeros must also be present, because each one of the three emitter capacitors causes the output to vary as $1/s$ and hence $V_o \to 1/s^3$ as $s \to \infty$. Thus the transfer function is of the form

$$\frac{V_o}{V_s} = K \frac{(s - s_7)(s - s_8)(s - s_9)}{(s - s_1)(s - s_2)(s - s_3)(s - s_4)(s - s_5)(s - s_6)} \qquad (12\text{-}49)$$

The zeros of this equation can be obtained by inspection of this particular network, as we explained in the preceding section. To obtain the poles, we must solve for the zeros of the network determinant, and since this determinant is of order six by six, the amount of computational labor required is prohibitive without the aid of a digital computer.

Using the program CORNAP, not only are the poles, zeros, and the gain constant K obtained, but the magnitude, phase, and the derivative of the phase with respect to frequency (this is the delay introduced by the amplifier) are also computed.

Assuming identical stages and using the same parameter values as in the amplifier in Fig. 12-17, the computer gives the following values for the poles:

$$s_1 = -27.5 \times 10^5 \text{ rad/s} \qquad s_2 = -13.3 \times 10^6 \text{ rad/s}$$

$$s_3 = -7.75 \times 10^7 \text{ rad/s} \qquad s_4 = -6.80 \times 10^8 \text{ rad/s}$$

$$s_5 = -39.7 \times 10^8 \text{ rad/s} \qquad s_6 = -44.2 \times 10^8 \text{ rad/s}$$

and the zeros:

$$s_7 = s_8 = s_9 = 16.6 \times 10^9 \text{ rad/s}$$

In Fig. 12-21 is plotted the frequency response, from which we find that the high 3-dB frequency is 420 kHz.

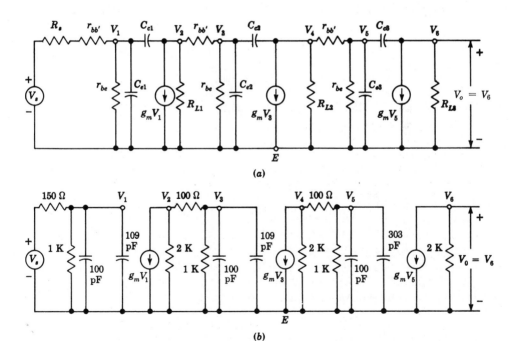

Fig. 12-20 (a) Three-stage interacting amplifier; (b) noninteracting Miller approximation of (a).

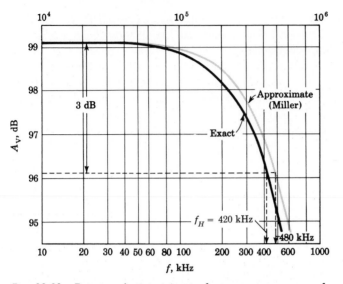

Fig. 12-21 Exact and approximate frequency response of three-stage CE amplifier.

Approximate Analysis We now perform an approximate analysis of the previous amplifier. We replace C_{c3} by the Miller capacitance

$$C_{c3}(1 + g_m R_{L3}) = 3(1 + 50 \times 2) = 303 \text{ pF}$$

shunting C_{e3}. Similarly, we replace C_{c2} and C_{c1} with Miller capacitances of 109 pF shunting C_{e2} and C_{e1}, respectively. Thus we obtain the simplified circuit of Fig. 12-20b, from which we obtain the transfer function (with f expressed in megahertz)

$$\frac{V_o}{V_s} = -\frac{90.5 \times 10^4}{\left(1 + j\,\dfrac{f}{0.582}\right)\left(1 + j\,\dfrac{f}{1.12}\right)\left(1 + j\,\dfrac{f}{5.86}\right)} \tag{12-50}$$

The plot of Eq. (12-50) is shown in Fig. 12-21. We find that two of the poles are closer together than four octaves, and hence this amplifier does not have a dominant pole. We use the plot to deduce the upper 3-dB frequency. From Fig. 12-21 we read $f_H = 480$ kHz as compared with $f_H = 420$ kHz, obtained using the exact six-pole transfer function, an overestimate of 14 percent.

We can avoid the effort of plotting Eq. (12-50) if we use the result of Eq. (12-31). In this case we have

$$\frac{1}{f_H} = 1.1 \sqrt{\left(\frac{1}{0.582}\right)^2 + \left(\frac{1}{1.12}\right)^2 + \left(\frac{1}{5.86}\right)^2}$$

or $f_H = 0.466$ MHz $= 466$ kHz, as compared with $f_H = 480$ kHz from Fig. 12-21. It is instructive to recall that in Sec. 11-8 we found that f_H of a single stage using the transistor of the above example and $R_s = 50\ \Omega$, $R_L = 2\ K$, is $f_H = 2.8\ \text{MHz}$. Two identical stages cascaded result in $f_H = 540\ \text{kHz}$. If these two stages were noninteracting, each with $f_H = 2.8\ \text{MHz}$, the overall bandwidth would be $f_H = 0.64 \times 2.8\ \text{MHz} = 1.79\ \text{MHz}$, using Eq. (12-26). Similarly, for the three-stage noninteracting amplifier, we would obtain $f_H = 0.51 \times 2.8\ \text{MHz} = 1.43\ \text{MHz}$, as compared with the value $f_H = 420\ \text{kHz}$ of the three-stage interacting amplifier. The capacitive loading of one stage on the preceding one reduces the upper 3-dB frequency drastically.

12-12   NOISE

It is found that there is an inherent limit to the amplification obtainable from an amplifier, because even when there is no signal impressed at the input, a small output, called *amplifier noise*,[5] is obtained. If, therefore, only a very small voltage is available, such as a weak radio, television, radar, etc., signal, it may be impossible to distinguish the signal from the background noise. The term *noise* arises from the fact that, with no input, the output from the loudspeaker of an audio amplifier with the gain control set at a maximum is an audible hiss or crackle. In the case of a video amplifier, the term *snow* is often used in place of noise, because of the snowlike appearance on a TV screen when the set is tuned to a weak station. The various noise sources in an amplifier are now considered.

Thermal, or Johnson, Noise[6]   The electrons in a conductor possess varying amounts of energy by virtue of the temperature of the conductor. The slight fluctuations in energy about the values specified by the most probable distribution are very small, but they are sufficient to produce small noise potentials within a conductor. These random fluctuations produced by the thermal agitation of the electrons are called the *thermal*, or Johnson, noise. The rms value of the thermal-resistance noise voltage V_n over a frequency range $f_H - f_L$ is given by the expression

$$V_n{}^2 = 4\bar{k}TRB \tag{12-51}$$

where $\bar{k}$ = Boltzmann constant, J/°K
   T = resistor temperature, °K
   R = resistance, Ω
   $B = f_H - f_L$ = bandwidth, Hz

It should be observed that the same noise power exists in a given bandwidth regardless of the center frequency. Such a distribution, which gives the same noise per unit bandwidth anywhere in the spectrum, is called *white noise*.

If the conductor under consideration is the input resistor to an ideal (noiseless) amplifier, the input noise voltage to the amplifier is given by Eq. (12-51). An idea of the order of magnitude of the voltage involved is obtained by calculating the noise voltage generated in a 1-M resistance at room temperature over a 10-kHz bandpass. Equation (12-51) yields for V_n the value 13 μV. Clearly, if the bandpass of an amplifier is wider, the input resistance must be smaller, if excessive noise is to be avoided. Thus, if the amplifier considered is 10 MHz wide, its input resistance cannot exceed 1,000 Ω if the fluctuation noise is not to exceed that of the 10-kHz audio amplifier.

It is obvious that the bandpass of an amplifier should be kept as narrow as possible (without introducing excessive frequency distortion) because the noise power is directly proportional to the bandwidth. The noise output voltage squared from the amplifier due to R_s only is given by Eq. (12-51) provided that the value of $V_n{}^2$ is multiplied by $|A_{Vo}|^2$ and that the noise bandwidth B_n is defined by

$$B_n \equiv \frac{1}{|A_{Vo}|^2} \int_0^\infty |A_V(f)|^2 \, df \qquad (12\text{-}52)$$

Shot, or Schottky, Noise[7] Shot noise is attributed to the discrete-particle nature of current carriers in semiconductors. Normally, one assumes that the current in a transistor or FET under dc conditions is a constant at every instant. Actually, however, the current from the emitter to the collector consists of a stream of individual electrons or holes, and it is only the time-average flow which is measured as the constant current. The fluctuation in the number of carriers is called the *shot noise*. The mean-square shot-noise current in any device is given by

$$I_n{}^2 = 2qI_{dc}B \qquad (12\text{-}53)$$

where q = electronic charge magnitude
$\quad I_{dc}$ = dc current
$\quad B$ = bandwidth
If the load resistor is R_L, a noise voltage of magnitude I_nR_L will appear across the load.

Noise Figure A *noise figure NF* has been introduced in order to be able to specify quantitatively how noisy a device is. By definition, NF is the ratio of the noise power output of the circuit under consideration to the noise power output which would be obtained in the same bandwidth if the only source of noise were the thermal noise in the internal resistance R_s of the signal source. Thus the noise figure is a quantity which compares the noise in an actual amplifier with that in an ideal (noiseless) amplifier. Usually, NF is expressed in decibels.

We define the following symbols:

$\quad S_{Pi}(S_{Vi})$ = signal power (voltage) input

$\quad N_{Pi}(N_{Vi})$ = noise power (voltage) input due to R_s

$S_{Po}(S_{Vo})$ = signal power (voltage) output

$N_{Po}(N_{Vo})$ = noise power (voltage) output due to R_s and any noise sources within the active device

From Eq. (12-51), $N_{Vi} = V_n = (4\bar{k}TR_sB)^{\frac{1}{2}}$.

From the definition of noise figure

$$NF \equiv 10 \log \frac{\text{total noise power output}}{\text{noise power output due to } R_s} = 10 \log \frac{N_{Po}}{A_P N_{Pi}} \qquad (12\text{-}54)$$

where the power gain of the active device is $A_P \equiv S_{Po}/S_{Pi}$. Hence

$$NF = 10 \log \frac{N_{Po}S_{Pi}}{S_{Po}N_{Pi}} = 10 \log \frac{S_{Pi}/N_{Pi}}{S_{Po}/N_{Po}} \qquad (12\text{-}55)$$

The quotient S_P/N_P is called the *signal-to-noise power ratio*. The noise figure is the input signal-to-noise power ratio divided by the output signal-to-noise power ratio. Expressed in decibels, the noise figure is given by the input signal-to-noise power ratio in decibels minus the output signal-to-noise power ratio in decibels. Since the signal and noise appear across the same load, Eq. (12-55) takes the form

$$NF = 20 \log \frac{S_{Vi}/N_{Vi}}{S_{Vo}/N_{Vo}} = 20 \log \frac{S_{Vi}}{N_{Vi}} - 20 \log \frac{S_{Vo}}{N_{Vo}} \qquad (12\text{-}56)$$

where S_V/N_V is called the *signal-to-noise voltage ratio*.

Measurement of Noise Figure A very simple method[7] for measuring the noise figure of an active device Q is indicated in Fig. 12-22. An audio sinusoidal generator V_s with source resistance R_s is connected to the input of Q. The active device is cascaded with a low-noise amplifier and a filter, and the output of this system is measured on a true rms reading voltmeter M. The experimental procedure for determining NF is as follows:

1. Measure R_s and calculate $N_{Vi} \equiv V_n$ from Eq. (12-51). The bandwidth B is set by the filter.

2. Adjust the audio signal voltage so that it is ten times the noise voltage: $V_s = 10V_n$ or $S_{Vi} = 10N_{Vi}$. Measure the output voltage with M. For such a

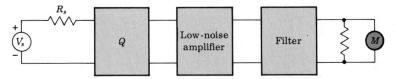

Fig. 12-22   A system used to measure the noise figure of an active device Q.

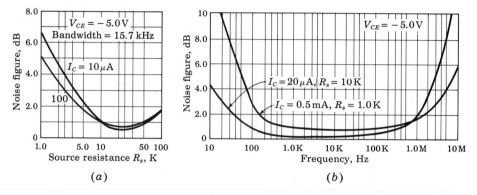

Fig. 12-23 Noise figure of a 2N3964 transistor. (a) Broadband noise figure as a function of source resistance. (b) Spot noise figure as a function of frequency. (Courtesy of Fairchild Semiconductor Corp.)

large signal-to-noise ratio (S_{Vi}/N_{Vi} = 20 dB), we may neglect the noise and assume that the voltmeter reading gives the signal output voltage S_{Vo}.

3. Set $V_s = 0$ and measure the output voltage N_{Vo} with M.

4. From Eq. (12-56) the noise figure is given by

$$NF = 20 - 20 \log \frac{S_{Vo}}{N_{Vo}} \tag{12-57}$$

where S_{Vo} and N_{Vo} are the meter readings obtained in measurements 2 and 3, respectively.

The low-noise amplifier is required only if the noise output of Q is too low to be detected with M. It should be pointed out that the amplifier-filter combination does not affect NF (for a given B) since the ratio S_{Vo}/N_{Vo} is used in Eq. (12-57).

The accuracy of the method described is based on the assumption that the output signal and noise can be measured separately. This is not strictly true since the noise cannot be turned off while measuring the output signal. It is found[7] that for a 20-dB input signal-to-noise ratio, transistor noise figures may be measured up to 10 dB with less than 0.5 dB error. The larger the S_{Vi}/N_{Vi}, the smaller is the error in this measurement. Usually, the output signal voltage is monitored on an oscilloscope to make certain that the system operates linearly so that no clipping takes place and no 60-Hz hum is present.

If a filter with a very narrow bandwidth (a few hertz) is used, the foregoing measurement gives the *spot, single-frequency*, or *incremental noise figure*. On the other hand, if the filter bandwidth is large (from $f_1 = 10$ Hz to $f_2 = 10$ kHz), the circuit of Fig. 12-22 gives the *broadband* or *integrated noise figure*. Other methods of measuring NF are available,[8,9] but these have the disadvantage of requiring a calibrated noise generator.

Transistor Noise[10]   In addition to thermal noise in a transistor, there is noise due to the random motion of the carriers crossing the emitter and collector

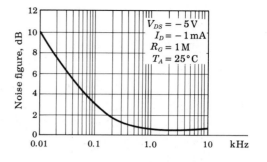

Fig. 12-24 The spot noise figure for a 2N2497 FET. (Courtesy of Texas Instruments, Inc.)

junctions and to the random recombination of holes and electrons in the base. There is also a partition effect arising from the random fluctuation in the division of current between the collector and base. It is found that a transistor does not generate white noise, except over a midband region. Also, the amount of noise generated depends upon the quiescent conditions and the source resistance. Hence, in specifying the noise in a transistor, the center frequency, the operating point, and R_s must be given.

Figure 12-23a and b show the noise figure vs. source resistance and frequency for the 2N3964 diffused planar transistor. There are three distinct regions in Fig. 12-23b. At low frequencies the noise varies approximately as $1/f$, and is called *excess*, or *flicker*, noise. The source of this noise is not clearly understood, but is thought to be caused by the recombination and generation of carriers on the surface of the crystal. In intermediate frequencies the noise is independent of frequency. This white noise is caused by the bulk resistance of the semiconductor material and the statistical variation of the currents (shot noise). The third region in Fig. 12-23b is characterized by an increase of the noise figure with frequency, and is essentially caused by a decrease in power gain with frequency.[9]

FET Noise[11] The field-effect transistor exhibits excellent noise characteristics. The main sources of noise in the FET are the thermal noise of the conducting channel, the shot noise caused by the gate leakage current, and the $1/f$ noise caused by surface effects.

The noise figure vs. frequency for the 2N2497 FET transistor is shown in Fig. 12-24. It should be pointed out that, unlike the bipolar transistor, the noise figure of the FET is essentially independent of the quiescent point (I_D and V_{DS}).

REFERENCES

1. Bode, Hendrik, W.: "Network Analysis and Feedback Amplifier Design," chap. 15, D. Van Nostrand Company, Inc., Princeton, N.J., 1945.

Friedland, B., O. Wing, and R. Ash: "Principles of Linear Networks," chap. 8, McGraw-Hill Book Company, New York, 1961.

2. Thornton, R. D., C. L. Searle, D. O. Pederson, R. B. Adler, and E. J. Angelo, Jr.: "Multistage Transistor Circuits," SEEC Series, vol. 5, chaps. 1, 2, John Wiley & Sons, Inc., New York, 1965.

3. Gray, P. E., and C. L. Searle: "Electronic Principles, Physics, Models and Circuits," appendix C, John Wiley & Sons, Inc., New York, 1969.

4. Pottle, C.: A Textbook Computerized State Space Network Analysis Algorithm, *Cornell Univ. Elec. Eng. Res. Lab. Rept.*, September, 1968.

5. Van der Ziel, A.: "Noise," Prentice-Hall, Inc., Englewood Cliffs, N.J., 1954.
Terman, F. E.: "Electronic and Radio Engineering," 4th ed., pp. 434–442, 796–798, McGraw-Hill Book Company, New York, 1952.
Seely, S.: "Radio Electronics," pp. 143–149, McGraw-Hill Book Company, New York, 1956.
Valley, G. E., Jr., and H. Wallman (eds.): "Vacuum Tube Amplifiers," MIT Radiation Laboratory Series, vol. 18, pp. 496–720, McGraw-Hill Book Company, New York, 1948.

6. Van der Ziel, A.: "Electronics," chap. 23, Allyn and Bacon, Inc., Boston, 1966.

7. Miller, J. R. (ed.): "Solid-state Communications," Texas Instruments Electronic Series, pp. 194–197, McGraw-Hill Book Company, New York, 1966.

8. Terman, F. E., and J. M. Pettit: "Electronic Measurements," 2d ed., pp. 362–379, McGraw-Hill Book Company, New York, 1952.

9. Dosse, J.: "The Transistor," 4th ed., pp. 144–152, D. Van Nostrand Company, Inc., Princeton, N.J., 1964.

10. Thornton, R. D., et al.: "Characteristics and Limitations of Transistors," vol. 4, Semiconductor Electronics Education Committee, John Wiley & Sons, Inc., New York, 1966.

11. Van der Ziel, A.: Thermal Noise in Field Effect Transistors, *Proc. IRE*, vol. 50, pp. 1808–1812, August, 1962.
Van der Ziel, A., Ref. 6, chap. 23.
Sevin, L. J.: "Field-effect Transistors," pp. 46–50, McGraw-Hill Book Company, New York, 1965.

REVIEW QUESTIONS

12-1 Define the following modes of operation of an amplifier: (*a*) Class A; (*b*) Class B; (*c*) Class AB; (*d*) Class C.

12-2 Define the following types of distortion: (*a*) *nonlinear;* (*b*) *frequency;* (*c*) *phase-shift distortion.*

12-3 Under what conditions does an amplifier preserve the form of the input signal?

12-4 (*a*) Define the *frequency-response magnitude characteristic* of an amplifier.

(b) Sketch a typical response curve.   (c) Indicate the high and low 3-dB frequencies. (d) Define *bandwidth*.

12-5   (a) For a low-pass single-pole amplifier, sketch the Bode magnitude plot and its piecewise linear approximation.   (b) Repeat for the Bode phase plot.   (c) What are the slopes of the idealized Bode plots?   (d) What are the corner frequencies, in both plots?

12-6   Repeat Rev. 12-5 for a single-zero transfer function.

12-7   Repeat Rev. 12-5 for a two-pole transfer function.

12-8   Define *dominant pole*.

12-9   (a) Sketch the high-frequency *step response* of a low-pass single-pole amplifier.   (b) Define the *rise time* t_r.   (c) What is the relationship between t_r and the high 3-dB frequency f_H?

12-10   (a) The input to a low-pass amplifier is a pulse of width t_p.   Sketch the output waveshape.   (b) What must be the relationship between t_p and f_H in order to amplify the pulse without excessive distortion?

12-11   (a) Sketch the response of an amplifier to a low-frequency square wave. (b) Define *tilt*.   (c) How is the tilt related to the low 3-dB frequency f_L?

12-12   Derive the expression for the high 3-dB frequency f_H^* of n identical noninteracting stages in terms of f_H for one stage.

12-13   (a) Is f_H^* for two stages greater or smaller than f_H for a single stage? Explain.   (b) Repeat for f_L^* versus f_L.

12-14   (a) Give an approximate expression relating f_H^* and the 3-dB frequencies of n nonidentical stages.   (b) For two identical stages, what is f_H^*/f_H?   Repeat for three stages.

12-15   Give an approximate relationship between the output rise time t_r, the rise time t_{ro} of an input signal, and the rise times of n nonidentical stages.

12-16   (a) Sketch two *RC*-coupled CE transistor stages.   (b) Show the low-frequency model for one stage.   (c) What is the expression for f_L?

12-17   Repeat Rev. 12-16 for CS JFET stages.

12-18   (a) An amplifier with a bypassed emitter resistor has a transfer function consisting of one pole and one zero.   Explain why, qualitatively (write no equations). (b) Sketch the Bode amplitude plot for the case where the pole and zero are widely separated.   (c) Does f_L always exist?   Explain.

12-19   (a) Outline the general method for obtaining the high-frequency response of two interacting transistor amplifier stages.   (b) Outline an approximate method of solution.

12-20   (a) For a cascade of n transistor CE stages, how many poles will the voltage transfer function have?   Explain.   (b) Repeat for the number of zeros.

12-21   (a) What is meant by *amplifier noise?*   (b) Define *white noise*.

12-22   (a) Define *thermal* or *Johnson noise*.   (b) Upon what factors does the noise voltage depend?

12-23   (a) Define *shot noise*.   (b) Upon what factors does the shot noise current depend?

12-24   Define (a) *noise figure;* (b) *signal-to-noise ratio*.

12-25   Derive the expression for the noise figure in terms of the input and output signal-to-noise ratios.

12-26   List 4 sources of noise in a transistor.

12-27   List 3 sources of noise in a FET.

13 / FEEDBACK AMPLIFIERS

In this chapter we introduce the concept of feedback and show how to modify the characteristics of an amplifier by combining a portion of the output signal with the external signal. Many advantages are to be gained from the use of negative (degenerative) feedback, and these are studied. Examples of feedback amplifier circuits at low frequencies are given, but the frequency response of feedback amplifiers is deferred to the following chapter.

13-1   CLASSIFICATION OF AMPLIFIERS

Before proceeding with the concept of feedback, it is useful to classify amplifiers into four broad categories,[1] as either *voltage, current, transconductance,* or *transresistance amplifiers* This classification is based on the magnitudes of the input and output impedances of an amplifier relative to the source and load impedances, respectively.

Voltage Amplifier Figure 13-1 shows a Thévenin's equivalent circuit of a two-port network which represents an amplifier. If the amplifier input resistance R_i is large compared with the source resistance R_s, then $V_i \approx V_s$. If the external load resistance R_L is large compared with the output resistance R_o of the amplifier, then $V_o \approx A_v V_i \approx A_v V_s$. This amplifier provides a voltage output proportional to the voltage input, and *the proportionality factor is independent of the magnitudes of the source and load resistances.* Such a circuit is called a *voltage amplifier.* An ideal voltage amplifier must have infinite input resistance R_i and zero output resistance R_o. The symbol A_v in Fig. 13-1 represents V_o/V_i, with $R_L = \infty$, and hence represents the open-circuit voltage amplification, or gain.

408

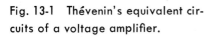
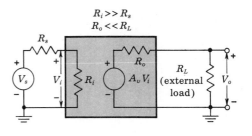

Fig. 13-1 Thévenin's equivalent circuits of a voltage amplifier.

Current Amplifier An ideal current amplifier[1] is defined as an amplifier
which provides an output current proportional to the signal current, and *the
proportionality factor is independent of R_s and R_L*. An ideal current amplifier
must have zero input resistance R_i and infinite output resistance R_o. In
practice, the amplifier has low input resistance and high output resistance.
It drives a low-resistance load ($R_o \gg R_L$), and is driven by a high-resistance
source ($R_i \ll R_s$). Figure 13-2 shows Norton's equivalent circuit of a current amplifier. Note that $A_i \equiv I_L/I_i$, with $R_L = 0$, representing the short-
circuit current amplification, or gain. We see that if $R_i \ll R_s$, $I_i \approx I_s$, and if
$R_o \gg R_L$, $I_L \approx A_iI_i \approx A_iI_s$. Hence the output current is proportional to the
signal current. The characteristics of the four ideal amplifier types are summarized in Table 13-1.

TABLE 13-1   Ideal amplifier characteristics

Parameter	Amplifier type			
	Voltage	Current	Transconductance	Transresistance
R_i..................	∞	0	∞	0
R_o..................	0	∞	∞	0
Transfer characteristic..	$V_o = A_vV_s$	$I_L = A_iI_s$	$I_L = G_mV_s$	$V_o = R_mI_s$
Reference..............	Fig. 13-1	Fig. 13-2	Fig. 13-3	Fig. 13-4

Transconductance Amplifier The ideal transconductance amplifier[1]
supplies an output current which is proportional to the signal voltage, inde-

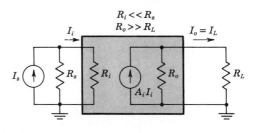

Fig. 13-2 Norton's equivalent circuits of a current amplifier.

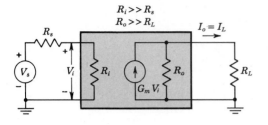

Fig. 13-3 A transconductance amplifier is represented by a Thévenin's equivalent in its input circuit and a Norton's equivalent in its output circuit.

pendently of the magnitudes of R_s and R_L. This amplifier must have an infinite input resistance R_i and infinite output resistance R_o. A practical transconductance amplifier has a large input resistance ($R_i \gg R_s$) and hence must be driven by a low-resistance source. It presents a high output resistance ($R_o \gg R_L$) and hence drives a low-resistance load. The equivalent circuit of a transconductance amplifier is shown in Fig. 13-3.

Transresistance Amplifier Finally, in Fig. 13-4, we show the equivalent circuit of an amplifier which ideally supplies an output voltage V_o in proportion to the signal current I_s independently of R_s and R_L. This amplifier is called a *transresistance amplifier*. For a practical transresistance amplifier we must have $R_i \ll R_s$ and $R_o \ll R_L$. Hence the input and output resistances are low relative to the source and load resistances. From Fig. 13-4 we see that if $R_s \gg R_i$, $I_i \approx I_s$, and if $R_o \ll R_L$, $V_o \approx R_m I_i \approx R_m I_s$. Note that $R_m \equiv V_o/I_i$, with $R_L = \infty$. In other words, R_m is the open-circuit mutual or transfer resistance.

13-2   THE FEEDBACK CONCEPT[2]

In the preceding section we summarize the properties of four basic amplifier types. In each one of these circuits we may sample the output voltage or current by means of a suitable sampling network and apply this signal to the input through a feedback two-port network, as shown in Fig. 13-5. At the input the feedback signal is combined with the external (source) signal through a mixer network and is fed into the amplifier proper.

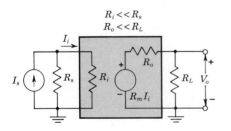

Fig. 13-4 A transresistance amplifier is represented by a Norton's equivalent in its input circuit and a Thévenin's equivalent in its output circuit.

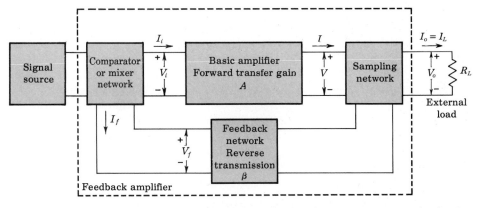

Fig. 13-5 Representation of any single-loop feedback connection around a basic amplifier. The transfer gain A may represent A_V, A_I, G_M, or R_M.

Signal Source This block in Fig. 13-5 is either a signal voltage V_s in series with a resistor R_s (a Thévenin's representation as in Fig. 13-1) or a signal current I_s in parallel with a resistor R_s (a Norton's representation as in Fig. 13-2).

Feedback Network This block in Fig. 13-5 is usually a passive two-port network which may contain resistors, capacitors, and inductors. Most often it is simply a resistive configuration.

Sampling Network Two sampling blocks are shown in Fig. 13-6. In Fig. 13-6*a* the output voltage is sampled by connecting the feedback network *in shunt* across the output. This type of connection is referred to as *voltage,* or *node, sampling.* Another feedback connection which samples the output

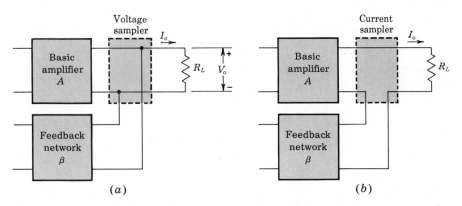

(*a*)   (*b*)

Fig. 13-6 Feedback connections at the output of a basic amplifier, sampling the output (*a*) voltage and (*b*) current.

current is shown in Fig. 13-6b, where the feedback network is connected *in series* with the output. This type of connection is referred to as *current*, or *loop, sampling*. Other sampling networks are possible.

Comparator, or Mixer, Network   Two mixing blocks are shown in Fig. 13-7. Figure 13-7a and b shows the simple and very common *series (loop) input* and *shunt (node) input* connections, respectively. A differential amplifier (Sec. 15-2) is often also used as the mixer. Such an amplifier has two inputs and gives an output proportional to the difference between the signals at the two inputs.

Transfer Ratio, or Gain   The symbol A in Fig. 13-5 represents the ratio of the output signal to the input signal of the basic amplifier. The transfer ratio V/V_i is the voltage amplification, or the *voltage gain*, A_V. Similarly, the transfer ratio I/I_i is the current amplification, or *current gain*, A_I for the amplifier. The ratio I/V_i of the basic amplifier is the transconductance G_M, and V/I_i is the transresistance R_M. Although G_M and R_M are defined as the ratio of two signals, one of these is a current and the other is a voltage waveform. Hence the symbol G_M or R_M does not represent an amplification in the usual sense of the word. Nevertheless, it is convenient to refer to each of the four quantities A_V, A_I, G_M, and R_M as a *transfer gain of the basic amplifier without feedback* and to use the symbol A to represent any one of these quantities.

The symbol A_f is defined as the ratio of the output signal to the input signal of the amplifier configuration of Fig. 13-5 and is called the *transfer gain of the amplifier with feedback*. Hence A_f is used to represent any one of the four ratios $V_o/V_s \equiv A_{Vf}$, $I_o/I_s \equiv A_{If}$, $I_o/V_s \equiv G_{Mf}$, and $V_o/I_s \equiv R_{Mf}$. The rela-

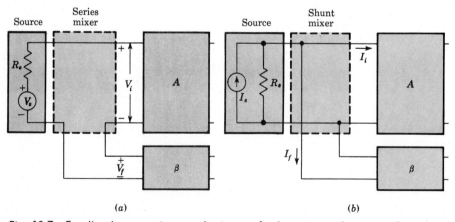

(a)   (b)

Fig. 13-7   Feedback connections at the input of a basic amplifier.   (a) Series comparison.   (b) Shunt mixing.

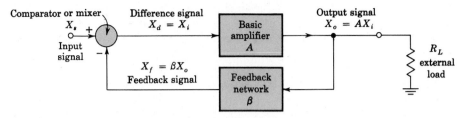

Fig. 13-8 Schematic representation of a single-loop feedback amplifier.

tionship between the transfer gain A_f with feedback and the gain A of the ideal amplifier without feedback is derived below [Eq. (13-4)].

Advantages of Negative Feedback   When any increase in the output signal results in a feedback signal into the input in such a way as to cause a decrease in the output signal, the amplifier is said to have *negative feedback*. The usefulness of negative feedback lies in the fact that, in general, any of the four basic amplifier types discussed in Sec. 13-1 may be improved by the proper use of negative feedback.   For example, the normally high input resistance of a voltage amplifier can be made higher, and its normally low output resistance can be lowered.   Also, the transfer gain A_f of the amplifier with feedback can be stabilized against variations of the h or hybrid-Π parameters of the transistors or the parameters of the other active devices used in the amplifier.   Another important advantage of the proper use of negative feedback is the significant improvement in the frequency response and in the linearity of operation of the feedback amplifier compared with that of the amplifier without feedback.

It should be pointed out that all the advantages mentioned above are obtained at the expense of the gain A_f with feedback, which is lowered in comparison with the transfer gain A of an amplifier without feedback.   Also, under certain circumstances, discussed in the next chapter, a negative-feedback amplifier may become unstable and break into oscillations.   The precautions which must be taken to avoid this undesirable effect are given in Chap. 14.

13-3   THE TRANSFER GAIN WITH FEEDBACK

Any one of the output connections of Fig. 13-6 may be combined with any of the input connections of Fig. 13-7 to form the feedback amplifier of Fig. 13-5. The analysis of the feedback amplifier can then be carried out by replacing each active element (transistor, FET, or vacuum tube) by its small-signal model and by writing Kirchhoff's loop, or nodal, equations.   That approach, however, does not place in evidence the main characteristics of feedback.

As a first step toward a method of analysis which emphasizes the benefits of feedback, consider Fig. 13-8, which represents a generalized feedback amplifier.   The basic amplifier of Fig. 13-8 may be a voltage, transconductance, current, or transresistance amplifier connected in a feedback configuration, as indicated in Fig. 13-9. The four topologies indicated in this figure are referred to as (1) *voltage-series feedback*, (2) *current-series feedback*, (3) *current-shunt feedback*, and (4) *voltage-shunt feedback*. In Fig. 13-8, the source resistance R_s is considered to be part of the amplifier, and the transfer gain $A(A_V, G_M, A_I, R_M)$ includes the effect of the loading of the β network (as well as R_L) upon the amplifier. The input signal X_s, the output signal X_o, the feedback signal X_f, and the difference signal X_d, each represents either a voltage or a current. These signals and also the ratios A and β are summarized in Table 13-2. The

TABLE 13-2　　　Voltage and current signals in feedback amplifiers

Signal or ratio	Type of feedback			
	Voltage-series Fig. 13-9a	Current-series Fig. 13-9b	Current-shunt Fig. 13-9c	Voltage-shunt Fig. 13-9d
X_o	Voltage	Current	Current	Voltage
X_s, X_f, X_d	Voltage	Voltage	Current	Current
A	A_V	G_M	A_I	R_M
β	V_f/V_o	V_f/I_o	I_f/I_o	I_f/V_o

symbol indicated by the circle in Fig. 13-8 represents a mixing, or comparison, network, whose output is the sum of the inputs, taking the sign shown at each input into account. Thus

$$X_d = X_s - X_f = X_i \tag{13-1}$$

Since X_d represents the difference between the applied signal and that fed back to the input, X_d is called the *difference, error,* or *comparison, signal.*

The reverse transmission factor β is defined by

$$\beta \equiv \frac{X_f}{X_o} \tag{13-2}$$

The factor β is often a positive or a negative real number, but in general, β is a complex function of the signal frequency. (This symbol should not be confused with the symbol β used previously for the CE short-circuit current gain.) The symbol X_o is the output voltage, or the output (load) current.

The transfer gain A is defined by

$$A \equiv \frac{X_o}{X_i} \tag{13-3}$$

By substituting Eqs. (13-1) and (13-2) into (13-3), we obtain for A_f the gain with feedback,

$$A_f \equiv \frac{X_o}{X_s} = \frac{A}{1 + \beta A} \tag{13-4}$$

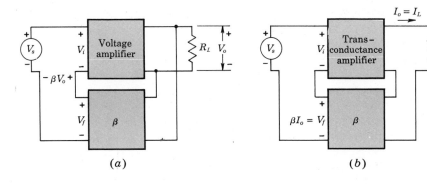

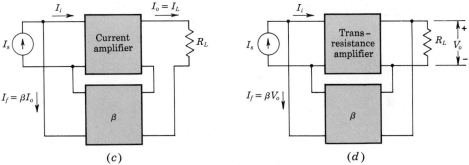

Fig. 13-9 Feedback-amplifier topologies. The source resistance is considered to be part of the amplifier. (a) Voltage amplifier with voltage-series feedback. (b) Transconductance amplifier with current-series feedback. (c) Current amplifier with current-shunt feedback. (d) Transresistance amplifier with voltage-shunt feedback.

The quantity A in Eqs. (13-3) and (13-4) represents the transfer gain of the corresponding amplifier without feedback, but *including the loading of the β network*, R_L and R_s. In the following section many of the desirable features of feedback are deduced, starting with the fundamental relationship given in Eq. (13-4).

If $|A_f| < |A|$, the feedback is termed *negative*, or *degenerative*. If $|A_f| > |A|$, the feedback is termed *positive*, or *regenerative*. From Eq. (13-4) we see that, in the case of negative feedback, the gain of the basic ideal amplifier with feedback is divided by the factor $|1 + \beta A|$, which exceeds unity.

Loop Gain The signal X_d in Fig. 13-8 is multiplied by A in passing through the amplifier, is multiplied by β in transmission through the feedback network, and is multiplied by -1 in the mixing or differencing network. Such a path takes us from the input terminals around the loop consisting of the amplifier and feedback network back to the input; the product $-A\beta$ is called the *loop gain*, or *return ratio*. The difference between unity and the loop gain is called the *return difference* $D = 1 + A\beta$. Also, the amount of feedback introduced into an amplifier is often expressed in decibels by the definition

$$N = \text{dB of feedback} = 20 \log \left| \frac{A_f}{A} \right| = 20 \log \left| \frac{1}{1 + A\beta} \right| \qquad (13\text{-}5)$$

If negative feedback is under consideration, N will be a negative number.

Fundamental Assumptions Three conditions must be satisfied for the feedback network of Fig. 13-8 in order that Eq. (13-4) be true and that the expressions for input and output resistances (derived in Secs. 13-5 and 13-6) be valid.

1. The input signal is transmitted to the output through the amplifier A and *not* through the β network. In other words, if the A is deactivated (say, set $A = 0$ by reducing h_{fe} or g_m for a transistor to zero), the output signal must drop to zero.

This first assumption is equivalent to the statement that the system has rendered the β block unilateral, so that it does *not* transmit a signal from input to output (but only in the reverse direction). This condition is often not satisfied exactly because β is a passive bilateral network. It is, however, approximately valid for practical feedback connections, as we shall verify in each of the feedback amplifiers to be considered.

2. The feedback signal is transmitted from the output to the input through the β block, and *not* through the amplifier. In other words, the basic amplifier is unilateral from input to output and the reverse transmission is zero. Note that the amplifiers in Figs. 13-1 to 13-4 satisfy this unilateral condition (such is not the case, for example, with a transistor amplifier at low frequencies if $h_{re} \neq 0$).

3. The reverse transmission factor β of the feedback network is independent of the load and the source resistances R_L and R_s.

For each topology studied we shall point out the approximations involved.

13-4 GENERAL CHARACTERISTICS OF NEGATIVE–FEEDBACK AMPLIFIERS[3]

Since negative feedback reduces the transfer gain, why is it used? The answer is that many desirable characteristics are obtained for the price of gain reduction. We now examine some of the advantages of negative feedback.

Desensitivity of Transfer Amplification The variation due to aging, temperature, replacement, etc., of the circuit components and transistor or FET characteristics is reflected in a corresponding lack of stability of the amplifier transfer gain. The fractional change in amplification with feedback divided by the fractional change without feedback is called the *sensitivity* of the transfer gain. If Eq. (13-4) is differentiated with respect to A, the absolute value of the resulting equation is

$$\left| \frac{dA_f}{A_f} \right| = \frac{1}{|1 + \beta A|} \left| \frac{dA}{A} \right| \tag{13-6}$$

Hence the sensitivity is $1/|1 + \beta A|$. If, for example, the sensitivity is 0.1, the percentage change in gain with feedback is one-tenth the percentage variation in amplification if no feedback is present. The reciprocal of the sensitivity is called the *desensitivity* D, or

$$D \equiv 1 + \beta A \tag{13-7}$$

The fractional change in gain without feedback is divided by the desensitivity D when feedback is added. [In passing, note that the desensitivity is another name for the *return difference*, and that the amount of feedback is $-20 \log D$ (Eq. 13-5).] For an amplifier with 20 dB of negative feedback, $D = 10$, and hence, for example, a 5 percent change in gain without feedback is reduced to a 0.5 percent variation after feedback is introduced.

Note from Eq. (13-4) that the transfer gain is divided by the desensitivity after feedback is added. Thus

$$A_f = \frac{A}{D} \tag{13-8}$$

In particular, if $|\beta A| \gg 1$, then

$$A_f = \frac{A}{1 + \beta A} \approx \frac{A}{\beta A} = \frac{1}{\beta} \tag{13-9}$$

and the gain may be made to depend entirely on the feedback network. The worst offenders with respect to stability are usually the active devices (transistors) involved. If the feedback network contains only stable passive elements, the improvement in stability may indeed be pronounced.

Since A represents either A_V, G_M, A_I, or R_M, then A_f represents the corresponding transfer gains with feedback: either A_{Vf}, G_{Mf}, A_{If}, or R_{Mf}. The topology determines which transfer ratio (Table 13-2) is stabilized. For example, for voltage-series feedback, Eq. (13-9) signifies that $A_{Vf} \approx 1/\beta$, and it is the voltage gain which is stabilized. For current-series feedback, Eq. (13-9) is $G_{Mf} \approx 1/\beta$, and hence, for this topology, it is the transconductance gain which is desensitized. Similarly, it follows from Eq. (13-9) that the current gain is stabilized for current-shunt feedback ($A_{If} \approx 1/\beta$) and the transresistance gain is desensitized for voltage-shunt feedback ($R_{Mf} \approx 1/\beta$).

Feedback is used to improve stability in the following way: Suppose an amplifier of gain A_1 is required. We start by building an amplifier of gain $A_2 = DA_1$, in which D is a large number. Feedback is now introduced to divide the gain by the factor D. The stability will be improved by the same factor D, since both gain and instability are divided by the desensitivity D.

If now the instability of the amplifier of gain A_2 is not appreciably larger than the instability of an amplifier of gain without feedback equal to A_1, this procedure will have been useful. It often happens as a matter of practice that amplifier gain may be increased appreciably without a corresponding loss of stability. For example, the voltage gain of a transistor may be increased by increasing the collector resistance R_c.

Frequency Distortion It follows from Eq. (13-9) that if the feedback network does not contain reactive elements, the overall gain is not a function of frequency. Under these circumstances a substantial reduction in frequency and phase distortion is obtained. The frequency response of feedback amplifiers is analyzed in the following chapter.

If a frequency-selective feedback network is used, so that β depends upon frequency, the amplification may depend markedly upon frequency. For example, it is possible to obtain an amplifier with a high-Q bandpass characteristic by using a feedback network which gives little feedback at the center of the band and a great deal of feedback on both sides of this frequency.

Nonlinear Distortion Suppose that a large amplitude signal is applied to a stage of an amplifier so that the operation of the device extends slightly beyond its range of linear operation, and as a consequence the output signal is slightly distorted. Negative feedback is now introduced, and the input signal is increased by the same amount by which the gain is reduced, so that the output-signal amplitude remains the same. For simplicity, let us consider that the input signal is sinusoidal and that the distortion consists, simply, of a second-harmonic signal generated within the active device. We assume that the second-harmonic component, in the absence of feedback, is equal to B_2. Because of the effects of feedback, a component B_{2f} actually appears in the output. To find the relationship that exists between B_{2f} and B_2, it is noted that the output will contain the term $-A\beta B_{2f}$, which arises from the component $-\beta B_{2f}$ that is fed back to the input. Thus the output contains two terms: B_2, generated in the transistor, and $-A\beta B_{2f}$, which represents the effect of the feedback. Hence

$$B_2 - A\beta B_{2f} = B_{2f}$$

or

$$B_{2f} = \frac{B_2}{1 + \beta A} = \frac{B_2}{D} \tag{13-10}$$

Since A and β are generally functions of the frequency, they must be evaluated at the second-harmonic frequency.

The signal X_s to the feedback amplifier may be the actual signal externally available, or it may be the output of an amplifier preceding the feedback stage or stages under consideration. To multiply the input to the feedback amplifier by the factor $|1 + A\beta|$, it is necessary either to increase the nominal gain of the preamplifying stages or to add a new stage. If the full benefit of the feedback amplifier in reducing nonlinear distortion is to be obtained, these preamplifying stages must not introduce additional distortion, because of the increased output demanded of them. Since, however, appreciable harmonics are introduced only when the output swing is large, most of the distortion arises in the last stage. The preamplifying stages are of smaller importance in considerations of harmonic generation.

It has been assumed in the derivation of Eq. (13-10) that the small amount of additional distortion that might arise from the second-harmonic component fed back from the output to the input is negligible. This assumption leads to little error. Further, it must be noted that the result given by Eq. (13-10) applies only in the case of small distortion. The principle of superposition has been used in the derivation, and for this reason it is required that the device operate approximately linearly.

Reduction of Noise By employing the same reasoning as that in the discussion of nonlinear distortion, it can be shown that the noise introduced in an amplifier is divided by the factor D if feedback is employed. If D is much larger than unity, this would seem to represent a considerable reduction in the output noise. However, as noted above, for a given output the amplification of the preamplifier for a specified overall gain must be increased by the factor D. Since the noise generated is independent of the signal amplitude, there may be as much noise generated in the preamplifying stage as in the output stage. Furthermore, this additional noise will be amplified, as well as the signal, by the feedback amplifier, so that the complete system may actually be noisier than the original amplifier without feedback. If the additional gain required to compensate what is lost because of the presence of inverse feedback can be obtained by a readjustment of the circuit parameters rather than by the addition of an extra stage, a definite reduction will result from the presence of the feedback. In particular, the hum introduced into the circuit by a poorly filtered power supply may be decreased appreciably.

13-5   INPUT RESISTANCE[4]

We now discuss qualitatively the effect of the topology of a feedback amplifier upon the input resistance. If the feedback signal is returned to the input in *series* with the applied voltage (regardless of whether the feedback is obtained by sampling the output current or voltage), it *increases the input resistance*. Since the feedback voltage V_f opposes V_s, the input current I_i is less than it would be if V_f were absent. Hence the input resistance with feedback $R_{if} \equiv V_s/I_i$ (Fig. 13-10) is greater than the input resistance without feedback R_i. We show below that, for this topology, $R_{if} = R_i(1 + \beta A) = R_i D$.

Negative feedback in which the feedback signal is returned to the input in *shunt* with the applied signal (regardless of whether the feedback is obtained

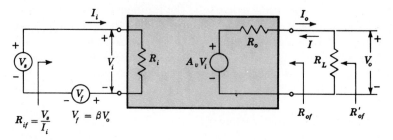

$$R_{if} = \frac{V_s}{I_i} \qquad V_f = \beta V_o$$

Fig. 13-10   Voltage-series feedback circuit used to calculate input and output resistances.

by sampling the output current or voltage) *decreases the input resistance*. Since $I_s = I_i + I_f$ [Eq. (13-1)], then the current I_i (for a fixed value of I_s) is decreased from what it would be if there were no feedback current. Hence $R_{if} \equiv V_i/I_s = I_iR_i/I_s$ (Fig. 13-11) is decreased because of this type of feedback. We show below that, for this topology, $R_{if} = R_i/(1 + \beta A) = R_i/D$.

Table 13-3 summarizes the characteristics of the four types of negative-feedback configurations: *For series comparison, $R_{if} > R_i$, whereas for shunt mixing, $R_{if} < R_i$.*

TABLE 13-3   Effect of negative feedback on amplifier characteristics

	Type of feedback			
	Voltage-series	Current-series	Current-shunt	Voltage-shunt
Reference.................	Fig. 13-9a	Fig. 13-9b	Fig. 13-9c	Fig. 13-9d
R_{of}.....................	Decreases	Increases	Increases	Decreases
R_{if}......................	Increases	Increases	Decreases	Decreases
Improves characteristics of.	Voltage amplifier	Transconductance amplifier	Current amplifier	Transresistance amplifier
Desensitizes...............	A_{Vf}	G_{Mf}	A_{If}	R_{Mf}
Bandwidth................	Increases	Increases	Increases	Increases
Nonlinear distortion........	Decreases	Decreases	Decreases	Decreases

Voltage-series Feedback   We now obtain R_{if} quantitatively. The topology of Fig. 13-9a is indicated in Fig. 13-10, with the amplifier replaced by its Thévenin's model. In this circuit A_v (corresponding to A_{vs} in Chap. 8) represents the open-circuit voltage gain *taking R_s into account*. Since throughout the discussion of feedback amplifiers we shall consider R_s to be part of the amplifier, we shall drop the subscript s on the transfer gain and input impedance (A_v instead of A_{vs}, R_i instead of R_{is}, R_{if} instead of R_{ifs}, G_m instead of G_{ms}, etc.).

From Fig. 13-10 the input impedance with feedback is $R_{if} = V_s/I_i$.   Also

$$V_s = I_iR_i + V_f = I_iR_i + \beta V_o \tag{13-11}$$

and

$$V_o = \frac{A_v V_i R_L}{R_o + R_L} = A_v I_i R_i \tag{13-12}$$

where

$$A_V \equiv \frac{V_o}{V_i} = \frac{A_v R_L}{R_o + R_L} \tag{13-13}$$

From Eqs. (13-11) and (13-12)

$$R_{if} = \frac{V_s}{I_i} = R_i(1 + \beta A_V) \qquad \cdot \tag{13-14}$$

Whereas A_v represents the open-circuit voltage gain without feedback, Eq. (13-13) indicates that A_V is the voltage gain without feedback taking the load R_L into account.   Therefore

$$A_v = \lim_{R_L \to \infty} A_V \tag{13-15}$$

Current-series Feedback   Proceeding in a similar manner for the topology of Fig. 13-9b, we obtain

$$R_{if} = R_i(1 + \beta G_M) \tag{13-16}$$

where

$$G_m = \lim_{R_L \to 0} G_M \tag{13-17}$$

and

$$G_M \equiv \frac{I_o}{V_i} = \frac{G_m R_o}{R_o + R_L} \tag{13-18}$$

Note that G_m is the short-circuit transconductance, whereas G_M is the transconductance without feedback taking the load into account.   Note that Eqs. (13-14) and (13-16) confirm that for series mixing $R_{if} > R_i$.

Current-shunt Feedback   The topology of Fig. 13-9c is indicated in Fig. 13-11, with the amplifier replaced by its Norton's model.   In this circuit

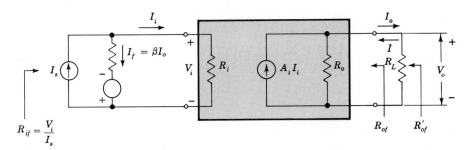

Fig. 13-11   Current-shunt feedback circuit used to calculate input and output resistances.

A_i represents the short-circuit current gain *taking R_s into account*. From Fig. 13-11

$$I_s = I_i + I_f = I_i + \beta I_o \tag{13-19}$$

and

$$I_o = \frac{A_i R_o I_i}{R_o + R_L} = A_I I_i \tag{13-20}$$

where

$$A_I \equiv \frac{I_o}{I_i} = \frac{A_i R_o}{R_o + R_L} \tag{13-21}$$

From Eqs. (13-19) and (13-20)

$$I_s = (1 + \beta A_I) I_i \tag{13-22}$$

From Fig. 13-11, $R_{if} = V_i/I_s$ and $R_i = V_i/I_i$. Using Eq. (13-22), we obtain

$$R_{if} = \frac{V_i}{(1 + \beta A_I) I_i} = \frac{R_i}{1 + \beta A_I} \tag{13-23}$$

Whereas A_i represents the short-circuit current gain, Eq. (13-21) indicates that A_I *is the current gain without feedback taking the load R_L into account.* Therefore

$$A_i = \lim_{R_L \to 0} A_I \tag{13-24}$$

Voltage-shunt Feedback Proceeding in a similar manner for the topology of Fig. 13-9d, we obtain

$$R_{if} = \frac{R_i}{1 + \beta R_M} \tag{13-25}$$

where

$$R_M \equiv \frac{V_o}{I_i} = \frac{R_m R_L}{R_o + R_L} \tag{13-26}$$

Note that R_m is the open-circuit transresistance, whereas R_M is the transresistance without feedback taking the load into account. Therefore

$$R_m = \lim_{R_L \to \infty} R_M \tag{13-27}$$

Note that Eqs. (13-23) and (13-25) confirm that for shunt comparison $R_{if} < R_i$. The expressions for R_{if} are summarized in Table 13-4.

13-6   OUTPUT RESISTANCE[4]

We now discuss qualitatively the effect of the topology of a feedback amplifier upon the output resistance. Negative feedback which samples the output *voltage*, regardless of how this output signal is returned to the input, tends to *decrease the output resistance.* For example, if R_L increases so that V_o increases, the effect of feeding this voltage back to the input in a degenerative manner (negative feedback) is to cause V_o to increase less than it would if there were no feedback. Hence the output voltage tends to remain constant as R_L changes, which means that $R_{of} \ll R_L$. This argument leads to the conclusion that this type of feedback (sampling the output voltage) reduces the output resistance.

By reasoning similar to that given above, negative feedback which samples the output *current* will tend to hold this current constant. Hence an output-current source is created ($R_{of} \gg R_L$), and we conclude that this type of sampling connection increases the output resistance.

In summary (Table 13-3): *For voltage sampling, $R_{of} < R_o$, whereas for current sampling, $R_{of} > R_o$.*

Voltage-series Feedback We now obtain quantitatively the resistance with feedback R_{of} looking into the output terminals but with R_L disconnected. To find R_{of} we must remove the external signal (set $V_s = 0$ or $I_s = 0$), let $R_L = \infty$, impress a voltage V across the output terminals, and calculate the current I delivered by V. Then $R_{of} \equiv V/I$. From Fig. 13-10 we find (with V_o replaced by V)

$$I = \frac{V - A_v V_i}{R_o} = \frac{V + \beta A_v V}{R_o} \tag{13-28}$$

because, with $V_s = 0$, $V_i = -V_f = -\beta V$. Hence

$$R_{of} \equiv \frac{V}{I} = \frac{R_o}{1 + \beta A_v} \tag{13-29}$$

Note that R_o is divided by the desensitivity factor $1 + \beta A_v$, which contains the open-circuit voltage gain A_v (*not* A_V).

The output resistance with feedback R'_{of} *which includes R_L as part of the amplifier* is given by R_{of} in parallel with R_L, or

$$R'_{of} = \frac{R_{of} R_L}{R_{of} + R_L} = \frac{R_o R_L}{1 + \beta A_v} \frac{1}{R_o/(1 + \beta A_v) + R_L} = \frac{R_o R_L}{R_o + R_L + \beta A_v R_L}$$

$$= \frac{R_o R_L/(R_o + R_L)}{1 + \beta A_v R_L/(R_o + R_L)} \tag{13-30}$$

Since $R'_o = R_o \| R_L$ is the output resistance without feedback but *with R_L considered as part of the amplifier*, and using Eq. (13-13) relating A_V to A_v, we obtain

$$R'_{of} = \frac{R'_o}{1 + \beta A_V} \tag{13-31}$$

Note that R'_o is now divided by the desensitivity factor $1 + \beta A_V$ which contains the voltage gain A_V that takes R_L into account.

Voltage-shunt Feedback   Proceeding as outlined above, we obtain for this topology

$$R_{of} = \frac{R_o}{1 + \beta R_m} \quad \text{and} \quad R'_{of} = \frac{R'_o}{1 + \beta R_M} \tag{13-32}$$

Note that Eqs. (13-31) and (13-32) confirm that for voltage sampling $R_{of} < R_o$.

Current-shunt Feedback   From Fig. 13-11 we find (with V_o replaced by V)

$$I = \frac{V}{R_o} - A_i I_i \tag{13-33}$$

With $I_s = 0$, $I_i = -I_f = -\beta I_o = +\beta I$.   Hence

$$I = \frac{V}{R_o} - \beta A_i I \quad \text{or} \quad I(1 + \beta A_i) = \frac{V}{R_o} \tag{13-34}$$

$$R_{of} = \frac{V}{I} = R_o(1 + \beta A_i) \tag{13-35}$$

Note that R_o is multiplied by the desensitivity factor $1 + \beta A_i$ which contains the short-circuit current gain A_i (*not* A_I).

The output resistance R'_{of} which includes R_L as part of the amplifier is not given by $R'_o(1 + \beta A_I)$, as one might thoughtlessly expect.   We shall now find the correct expression for R'_{of}.

$$R'_{of} = \frac{R_{of}R_L}{R_{of} + R_L} = \frac{R_o(1 + \beta A_i)R_L}{R_o(1 + \beta A_i) + R_L}$$

$$= \frac{R_o R_L}{R_o + R_L}\frac{1 + \beta A_i}{1 + \beta A_i R_o/(R_o + R_L)} \tag{13-36}$$

Using Eq. (13-21) and with $R'_o = R_o \| R_L$, we obtain

$$R'_{of} = R'_o\frac{1 + \beta A_i}{1 + \beta A_I} \tag{13-37}$$

For $R_L = \infty$, $A_I = 0$ and $R'_o = R_o$, so that Eq. (13-37) reduces to

$$R'_{of} = R_o(1 + \beta A_i)$$

in agreement with Eq. (13-35).

Current-series Feedback   Proceeding as outlined above, we obtain for this topology

$$R_{of} = R_o(1 + \beta G_m) \quad \text{and} \quad R'_{of} = R'_o\frac{1 + \beta G_m}{1 + \beta G_M} \tag{13-38}$$

Note that Eqs. (13-37) and (13-38) confirm that, for current sampling, $R_{of} > R_o$. The expressions for R_{of} and R'_{of} are summarized in Table 13-4.   The above derivations do *not* assume that the network is resistive.   Hence, if A or β is a function of frequency, R should be changed to Z in Table 13-4. Then $Z_{if}(Z_{of})$ gives the input (output) impedance with feedback.

13-7   METHOD OF ANALYSIS OF A FEEDBACK AMPLIFIER

It is desirable to separate the feedback amplifier into two blocks, the basic amplifier A and the feedback network β, because with a knowledge of A and β, we can calculate the important characteristics of the feedback system, namely, A_f, R_{if}, and R_{of}. The basic amplifier configuration *without feedback but taking the loading of the β network into account* is obtained[5] by applying the following rules:

To find the input circuit:
1. Set $V_o = 0$ for voltage sampling. In other words, short the output node.
2. Set $I_o = 0$ for current sampling. In other words, open the output loop.
To find the output circuit:
1. Set $V_i = 0$ for shunt comparison. In other words, short the input node.
2. Set $I_i = 0$ for series comparison. In other words, open the input loop.

These procedures ensure that the feedback is reduced to zero without altering the loading on the basic amplifier.

The complete analysis of a feedback amplifier is obtained by carrying out the following steps:

1. Identify the topology. (*a*) Is the feedback signal X_f a voltage or a current? In other words, is X_f applied in series or in shunt with the external excitation? (*b*) Is the sampled signal X_o a voltage or a current? In other words, is the sampled signal taken at the output node or from the output loop?
2. Draw the basic amplifier circuit without feedback, following the rules listed above.
3. Use a Thévenin's source if X_f is a voltage and a Norton's source if X_f is a current.
4. Replace each active device by the proper model (for example, the hybrid-Π model for a transistor at high frequencies or the h-parameter model at low frequencies).
5. Indicate X_f and X_o on the circuit obtained by carrying out steps 2, 3, and 4. Evaluate $\beta = X_f/X_o$.
6. Evaluate A by applying KVL and KCL to the equivalent circuit obtained after step 4.
7. From A and β, find D, A_f, R_{if}, R_{of}, and R'_{of}.

Table 13-4 summarizes the above procedure and should be referred to when carrying out the analyses of the feedback circuits discussed in the following sections.   We shall consider only the low-frequency response in this chapter and reserve the high-frequency analysis for the next chapter.

13-8   VOLTAGE–SERIES FEEDBACK

Two examples of the voltage-series topology are considered in this section: the FET common-drain amplifier (source follower) and the bipolar transistor common-collector amplifier (emitter follower).   A transistor two-stage voltage-series configuration is given in the following section.

The FET Source Follower   The circuit is given in Fig. 13-12a.   The feedback signal is the voltage V_f across R, and the sampled signal is the output voltage V_o across R.   Hence this is the case of voltage-series feedback, and we must refer to the first topology in Table 13-4.

TABLE 13-4   Feedback amplifier analysis

Characteristic Topology	(1) Voltage-series	(2) Current-series	(3) Current-shunt	(4) Voltage-shunt
Feedback signal X_f.....	Voltage	Voltage	Current	Current
Sampled signal X_o......	Voltage	Current	Current	Voltage
To find input loop, set†.	$V_o = 0$	$I_o = 0$	$I_o = 0$	$V_o = 0$
To find output loop, set†	$I_i = 0$	$I_i = 0$	$V_i = 0$	$V_i = 0$
Signal source..........	Thévenin	Thévenin	Norton	Norton
$\beta = X_f/X_o$.............	V_f/V_o	V_f/I_o	I_f/I_o	I_f/V_o
$A = X_o/X_i$.............	$A_V = V_o/V_i$	$G_M = I_o/V_i$	$A_I = I_o/I_i$	$R_M = V_o/I_i$
$D = 1 + \beta A$..........	$1 + \beta A_V$	$1 + \beta G_M$	$1 + \beta A_I$	$1 + \beta R_M$
A_f....................	A_V/D	G_M/D	A_I/D	R_M/D
R_{if}....................	$R_i D$	$R_i D$	R_i/D	R_i/D
R_{of}..................	$\dfrac{R_o}{1 + \beta A_v}$	$R_o(1 + \beta G_m)$	$R_o(1 + \beta A_i)$	$\dfrac{R_o}{1 + \beta R_m}$
$R'_{of} = R_{of} \| R_L$.........	$\dfrac{R'_o}{D}$	$R'_o \dfrac{1 + \beta G_m}{D}$	$R'_o \dfrac{1 + \beta A_i}{D}$	$\dfrac{R'_o}{D}$

† This procedure gives the basic amplifier circuit without feedback but taking the loading of β, R_L and R_s into account.

We must now draw the basic amplifier without feedback.   To find the input circuit, set $V_o = 0$, and hence V_s appears directly between G and S.   To find the output circuit, set $I_i = 0$ (the input loop is opened), and hence R appears only in the output loop.   Following these rules we obtain Fig. 13-12b. If the FET is replaced by its low-frequency model of Fig. 10-8, the result is Fig. 13-12c.   From this figure V_f and V_o are equal, and $\beta = V_f/V_o = 1$.

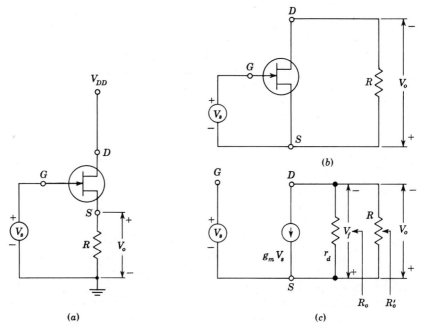

Fig. 13-12 (a) The source-follower. (b) The amplifier without feedback and (c) the FET replaced by its small-signal low-frequency model.

This topology stabilizes voltage gain. A_V is calculated by inspection of Fig. 13-12c. Since without feedback $V_i = V_s$, then

$$A_V = \frac{V_o}{V_i} = \frac{g_m V_s r_d R}{(r_d + R)V_s} = \frac{\mu R}{r_d + R} \tag{13-39}$$

where $\mu = g_m r_d$ from Eq. (10-15).

$$D = 1 + \beta A_V = 1 + \frac{\mu R}{r_d + R} = \frac{r_d + (1 + \mu)R}{r_d + R} \tag{13-40}$$

$$A_{Vf} = \frac{A_V}{D} = \frac{\mu R}{r_d + (1 + \mu)R} \tag{13-41}$$

The input impedance of an FET is infinite, $R_i = \infty$, and hence $R_{if} = R_i D = \infty$.

We are interested in finding the output resistance seen looking into the FET source S. Hence R is considered as an external load R_L. From Table 13-4

$$R_{of} = \frac{R_o}{1 + \beta A_v} = \frac{r_d}{1 + \mu} \tag{13-42}$$

because $R_o = r_d$ from Fig. 13-12c, $\beta = 1$, and $A_v = \lim_{R \to \infty} A_V = \mu$ from Eq. (13-15). Also, from Table 13-4

$$R'_{of} = \frac{R'_o}{D} = \frac{Rr_d}{R + r_d} \frac{r_d + R}{r_d + (\mu + 1)R} = \frac{Rr_d}{r_d + (\mu + 1)R} \tag{13-43}$$

Note that

$$R_{of} = \lim_{R \to \infty} R'_{of} = \frac{r_d}{\mu + 1}$$

which agrees with Eq. (13-42).

Since the three assumptions listed in Sec. 13-3 are satisfied, the above results are exact and agree with those obtained in Sec. 10-7 without the use of feedback formulas.

The Emitter Follower   The circuit is given in Fig. 13-13a.   The feedback signal is the voltage V_f across R_e, and the sampled signal is V_o across R_e. Hence this is a case of voltage-series feedback, and we must refer to the first topology in Table 13-4.

We now draw the basic amplifier without feedback.   To find the input circuit, set $V_o = 0$, and hence V_s in series with R_s appears between B and E.

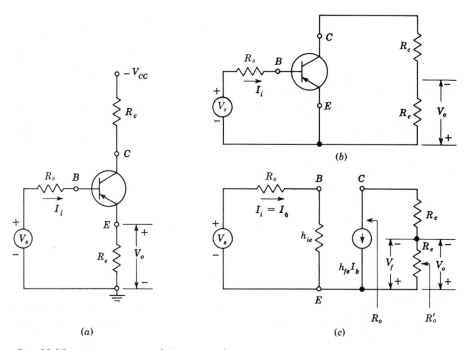

Fig. 13-13   (a) An emitter follower.   (b) The amplifier without feedback and (c) the transistor replaced by its approximate low-frequency model.

To find the output circuit, set $I_i = I_b = 0$ (the input loop is opened), and hence R_e appears only in the output loop. Following these rules, we obtain the circuit of Fig. 13-13b. If the transistor is replaced by its low-frequency approximate model of Fig. 8-23, the result is Fig. 13-13c. From this figure $V_o = V_f$ and $\beta = V_f/V_o = 1$.

This topology stabilizes the voltage gain. A_V is calculated by inspection of Fig. 13-13c. Since R_s is considered as part of the amplifier, then $V_i = V_s$, and

$$A_V = \frac{V_o}{V_i} = \frac{h_{fe}I_b R_e}{V_s} = \frac{h_{fe}R_e}{R_s + h_{ie}} \tag{13-44}$$

$$D = 1 + \beta A_V = 1 + \frac{h_{fe}R_e}{R_s + h_{ie}} = \frac{R_s + h_{ie} + h_{fe}R_e}{R_s + h_{ie}} \tag{13-45}$$

$$A_{Vf} = \frac{A_V}{D} = \frac{h_{fe}R_e}{R_s + h_{ie} + h_{fe}R_e} \tag{13-46}$$

For $h_{fe}R_e \gg R_s + h_{ie}$, $A_{Vf} \approx 1$, as it should be for an emitter follower.

The input resistance without feedback is $R_i = R_s + h_{ie}$ from Fig. 13-13c. Hence

$$R_{if} = R_i D = (R_s + h_{ie})\frac{R_s + h_{ie} + h_{fe}R_e}{R_s + h_{ie}} = R_s + h_{ie} + h_{fe}R_e \tag{13-47}$$

We are interested in the resistance seen looking into the emitter. Hence R_e is considered as an external load. From Table 13-4

$$R_{of} = \frac{R_o}{1 + \beta A_v} = \frac{\infty}{\infty} \tag{13-48}$$

because, from Fig. 13-13c, we are looking into a current source $R_o = \infty$ and $A_v = \lim\limits_{R_e \to \infty} A_V = \infty$ from Eq. (13-15). The indeterminacy in Eq. (13-48) may be resolved by first evaluating R'_{of} and then going to the limit $R_e \to \infty$. Thus, since $R'_o = R_e$,

$$R'_{of} = \frac{R'_o}{D} = \frac{R_e(R_s + h_{ie})}{R_s + h_{ie} + h_{fe}R_e} \tag{13-49}$$

and

$$R_{of} = \lim\limits_{R_e \to \infty} R'_{of} = \frac{R_s + h_{ie}}{h_{fe}} \tag{13-50}$$

Note that the feedback desensitizes voltage gain with respect to changes in h_{fe} and that it increases the input resistance and decreases the output resistance.

The foregoing expressions for A_{Vf}, R_{if}, and R_{of} are based on the assumption of zero forward transmission through the feedback network. Since there is such forward transmission because the input current passes through R_e in Fig. 13-13a, these expressions are only approximately true. In this example we have in effect neglected the base current which flows in R_e compared with the collector

current. The more exact answers are obtained in Sec. 8-8, and they differ
from those given above only in that h_{fe} must be replaced by $h_{fe} + 1$.

13-9   A VOLTAGE–SERIES FEEDBACK PAIR

Figure 13-14 shows two cascaded stages whose voltage gains are A_{V1} and A_{V2},
respectively.   The output of the second stage is returned through the feedback
network R_1R_2 in opposition to the input signal V_s.   Clearly, then, this is a
case of voltage-series negative feedback.   According to Table 13-3, we should
expect the input resistance to increase, the output resistance to decrease, and
the voltage gain to be stabilized (desensitized).
    The first basic assumption listed in Sec. 13-3 is not strictly satisfied for the
circuit of Fig. 13-14a because I' represents transmission through the feedback
network from input to the output.   We shall neglect I' compared with I on
the realistic assumption that the current gain of the second stage is much
larger than unity.   Under these circumstances very little error is made in using
the feedback formulas developed in this chapter.
    The input of the basic circuit without feedback is found by setting $V_o = 0$
(Table 13-4), and hence R_2 appears in parallel with R_1.   The output of the
basic amplifier without feedback is found by opening the input loop (set $I' = 0$),
and hence R_1 is placed in series with R_2.   Following these rules results in Fig.
13-14b, to which has been added the series feedback voltage V_f across R_1 in the
output circuit.   Clearly,

$$\beta = \frac{V_f}{V_o} = \frac{R_1}{R_1 + R_2} \tag{13-51}$$

Second-collector to First-emitter Feedback Pair   The circuit of Fig. 13-15
shows a two-stage amplifier which makes use of voltage-series feedback by

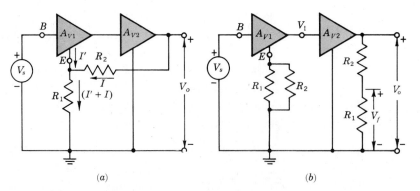

(a)   (b)

Fig. 13-14 (a) Voltage-series feedback pair. (b) Equivalent circuit,
without external feedback, but including the loading of R_2.

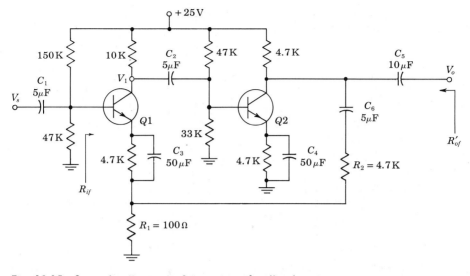

Fig. 13-15 Second-collector to first-emitter feedback pair.

connecting the second collector to the first emitter through the voltage divider R_1R_2. Capacitors C_1, C_2, C_5, and C_6 are dc blocking capacitors, and capacitors C_3 and C_4 are bypass capacitors for the emitter bias resistors. All these capacitances represent negligible reactances at the frequencies of operation of this circuit. For this amplifier the voltage gain A_{Vf} is given approximately by $1/\beta$, and is thus stabilized against temperature changes and transistor replacement. A more accurate determination of A_{Vf}, as well as a calculation of input and output resistance, is given in the following illustrative problem.

EXAMPLE Calculate A_{Vf}, R_{of}, and R_{if} for the amplifier of Fig. 13-15. Assume $R_s = 0$, $h_{fe} = 50$, $h_{ie} = 1.1$ K, $h_{re} = h_{oe} = 0$, and identical transistors.

Solution We first calculate the overall voltage gain without feedback from $A_V = A_{V1}A_{V2}$. The effective load R'_{L1} of transistor $Q1$ is

$$R'_{L1} = 10\|47\|33\|1.1 \text{ K} = 942 \text{ }\Omega$$

From Fig. 13-14b we see that the effective load R'_{L2} of transistor $Q2$ is the collector resistance $R_{c2} = 4.7$ K in parallel with $R_1 + R_2 = 4.8$ K,

$$R'_{L2} = 4.7\|4.8 = 2.37 \text{ K}$$

From Fig. 13-14b we see that the effective emitter impedance R_e of $Q1$ is $R_1\|R_2$ or

$$R_e = R_1\|R_2 = 0.1\|4.7 \text{ K} = 0.098 \text{ K} = 98 \text{ }\Omega$$

The voltage gain A_{V1} of $Q1$ is, from Eq. (8-60) and Fig. 13-14b with $V_i = V_s$,

$$A_{V1} \equiv \frac{V_1}{V_i} = \frac{-h_{fe}R'_{L1}}{h_{ie} + (1 + h_{fe})R_e} = \frac{-50 \times 0.942}{1.1 + 51 \times 0.098} = -7.72$$

The voltage gain A_{V2} of $Q2$ is, from Eq. (8-52),

$$A_{V2} \equiv \frac{V_o}{V_1} = -h_{fe}\frac{R'_{L2}}{h_{ie}} = -50 \times \frac{2.37}{1.1} = -108$$

Hence the voltage gain A_V of the two stages in cascade without feedback is

$$A_V \equiv \frac{V_o}{V_i} = A_{V1}A_{V2} = 7.72 \times 108 = 834$$

$$\beta = \frac{R_1}{R_1 + R_2} = \frac{100}{4,800} = \frac{1}{48} \quad \text{and} \quad A_V\beta = \frac{834}{48} = 17.4$$

$$D = 1 + \beta A_V = 18.4$$

$$A_{Vf} = \frac{A_V}{D} = \frac{834}{18.4} = 45.4$$

This value is to be compared with the approximate solution (based upon $A_V \to \infty$) given by $A_{Vf} = 1/\beta = 48$.

The input resistance without external feedback is, from Eq. (8-55),

$$R_i \equiv h_{ie} + (1 + h_{fe})R_e = 1.1 + 51 \times 0.098 = 6.1 \text{ K}$$

Hence, from Eq. (13-14),

$$R_{if} = R_i D = 6.1 \times 18.4 = 112 \text{ K}$$

The output resistance without feedback is $R'_o = R'_{L2} = 2.37$ K. Hence, from Table 13-4

$$R'_{of} = \frac{R'_o}{D} = \frac{2.37}{18.4} \text{ K} = 129 \text{ }\Omega$$

It is interesting to note that there is internal (local) feedback in the first stage of Fig. 13-14b because the R_1R_2 parallel combination acts as an emitter resistor. This first stage is an example of current-series feedback, which is analyzed in the next section.

13-10   CURRENT–SERIES FEEDBACK

Two examples of the current-series topology are considered in this section. The common-emitter transistor amplifier with a resistance R_e in the emitter is analyzed first. Then the FET common-source amplifier with a resistor R in the source lead is studied.

The Transistor Configuration   The circuit is given in Fig. 13-16a. The feedback signal is the voltage V_f across R_e and the sampled signal is the load current I_o. [For the present argument, we neglect base current compared with collector (load) current.] Hence this is a case of current-series feedback.

In passing, note that although I_o is proportional to V_o, it is *not* correct to conclude that this is a voltage-series feedback.   Thus, if the output signal is taken as the voltage V_o, then

$$\beta = \frac{V_f}{V_o} = \frac{-I_o R_e}{I_o R_L} = -\frac{R_e}{R_L}$$

Since β is now a function of the load R_L, the third basic assumption given in Sec. 13-3 is violated.

We must refer to the second topology in Table 13-4.   The input circuit of the amplifier without feedback is obtained by opening the output loop. Hence R_e must appear in the input side.   Similarly, the output circuit is obtained by opening the input loop, and this places R_e also in the output side. The resulting equivalent circuit is given in Fig. 13-16b.   No ground can be indicated in this circuit because to do so would again couple the input to the output via R_e; that is, it would reintroduce feedback.   And the circuit of Fig. 13-16b represents the basic amplifier without feedback, but taking the loading of the β network into account.

This topology stabilizes the transconductance G_M.   In Fig. 13-16c the transistor is replaced by its low-frequency approximate h-parameter model.

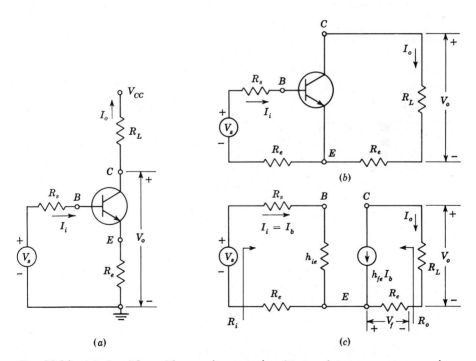

Fig. 13-16 (a) Amplifier with an unbypassed emitter resistance as an example of current-series feedback. (b) The amplifier without feedback, but including the loading of R_e. (c) The h-parameter model used for the transistor in (b).

Since the feedback voltage V_f appears across R_e in the output circuit, then, from Fig. 13-16c

$$\beta = \frac{V_f}{I_o} = \frac{-I_o R_e}{I_o} = -R_e \tag{13-52}$$

Since the input signal V_i without feedback is the V_s of Fig. 13-16c, then

$$G_M = \frac{I_o}{V_i} = \frac{-h_{fe}I_b}{V_s} = \frac{-h_{fe}}{R_s + h_{ie} + R_e} \tag{13-53}$$

$$D = 1 + \beta G_M = 1 + \frac{h_{fe}R_e}{R_s + h_{ie} + R_e} = \frac{R_s + h_{ie} + (1 + h_{fe})R_e}{R_s + h_{ie} + R_e} \tag{13-54}$$

$$G_{Mf} = \frac{G_M}{D} = \frac{-h_{fe}}{R_s + h_{ie} + (1 + h_{fe})R_e} \tag{13-55}$$

Note that if $(1 + h_{fe})R_e \gg R_s + h_{ie}$, and since $h_{fe} \gg 1$, then $G_{Mf} \approx -1/R_e$, in agreement with $G_{Mf} \approx 1/\beta$. If R_e is a stable resistor, the transconductance gain with feedback is stabilized (desensitized). The load current is given by

$$I_o = G_{Mf}V_s = \frac{-h_{fe}V_s}{R_s + h_{ie} + (1 + h_{fe})R_e} \approx -\frac{V_s}{R_e} \tag{(13-56)}$$

Under the conditions $(1 + h_{fe})R_e \gg R_s + h_{ie}$ and $h_{fe} \gg 1$, *the load current is directly proportional to the input voltage, and this current depends only upon R_e, and not upon any other circuit or transistor parameter.* As an example, consider that this circuit is used as the driver for the deflection current I_o in a magnetic cathode-ray oscilloscope. The load is then the deflection-yoke impedance, which is essentially an inductance whose reactance is proportional to frequency. Yet, from Eq. (13-56) the load current is independent of the characteristics of the yoke. If a deflection which varies linearly with time is desired, it is only necessary to generate a voltage waveform V_s which increases linearly with time (we are assuming that the deflection of the spot on the tube face is proportional to the yoke current).

The voltage gain is given by

$$A_{Vf} = \frac{I_o R_L}{V_s} = G_{Mf}R_L = \frac{-h_{fe}R_L}{R_s + h_{ie} + (1 + h_{fe})R_e} \tag{13-57}$$

Subject to the approximations made above, $A_{Vf} \approx -R_L/R_e$ and the voltage gain is stable if R_L and R_e are stable resistors.

From Fig. 13-16c, we see that $R_i = R_s + h_{ie} + R_e$. Hence, from Eq. (13-16),

$$R_{if} = R_i D = R_s + h_{ie} + (1 + h_{fe})R_e \tag{13-58}$$

Because R_s is considered to be part of the amplifier, it appears above as a component of the input resistance.

Since $R_o = \infty$, then $R_{of} = R_o(1 + \beta G_m) = \infty$. Hence $R'_{of} = R_L \| R_{of} = R_L$. An alternative derivation is to use the expression in Table 13-4, namely,

$$R'_{of} = R'_o \frac{1 + \beta G_m}{1 + \beta G_M}$$

Since G_m represents the short-circuit transconductance, then $G_m = \lim_{R_L \to 0} G_M$ [Eq. (13-17)]. However, from Eq. (13-53), G_M is independent of R_L, and hence $G_m = G_M$ and $R'_{of} = R'_o = R_L$.

The above results agree exactly with those derived in Sec. 8-15 because all three assumptions listed in Sec. 13-3 are satisfied. Note, in particular, that if the amplifier is deactivated (say, $h_{fe} = 0$), then $I_o = 0$, which means that none of the input signal appears at the output via the feedback block. Hence the first condition is satisfied, even though the β network is simply a resistor R_e. Note that for this topology it is *not* necessary to assume that the base current is negligible compared with the collector current.

EXAMPLE The circuit of Fig. 13-16a is to have an overall transconductance gain of -1 mA/V, a voltage gain of -4, and a desensitivity of 50. If $R_s = 1$ K, $h_{fe} = 150$, and $r_{bb'}$ is negligible, find (a) R_e, (b) R_L, (c) R_{if}, and (d) the quiescent collector current I_C at room temperature.

Solution   a. $G_{Mf} = \dfrac{G_M}{D} = \dfrac{G_M}{50} = -1$ mA/V

or

$G_M = -50$ mA/V

Since $\beta = -R_e$, then

$D = 1 + \beta G_M = 1 + 50R_e = 50$

or

$R_e = 0.98$ K ≈ 1 K

b. $A_{Vf} = G_{Mf}R_L$

or

$R_L = \dfrac{A_{Vf}}{G_{Mf}} = \dfrac{-4}{-1} = 4$ K

c. From Eq. (13-53)

$G_M = -50 = \dfrac{-h_{fe}}{R_s + h_{ie} + R_e} = \dfrac{-150}{1 + h_{ie} + 1}$

or

$h_{ie} = 1$ K

$R_i = R_s + h_{ie} + R_e = 3$ K

$$R_{if} = R_i D = (3)(50) = 150 \text{ K}$$

d. From Eqs. (11-9) and (11-6)

$$h_{ie} = r_{bb'} + r_{b'e} \approx \frac{h_{fe}}{g_m} = \frac{h_{fe} V_T}{I_C}$$

or

$$I_C = \frac{h_{fe} V_T}{h_{ie}} = \frac{(150)(0.026)}{1} = 3.9 \text{ mA}$$

The FET CS Stage with a Source Resistor R   The circuit of Fig. 13-17a is analogous to the transistor CE stage with an emitter resistor R_e.   Proceeding as we did for the transistor amplifier, we obtain the circuit of Fig. 13-17b. Replacing the FET by its low-frequency model results in Fig. 13-17c.   Without feedback $V_i = V_s$ and

$$G_M = \frac{I_o}{V_i} = \frac{I_o}{V_s} = \frac{-g_m r_d}{r_d + R_L + R} = \frac{-\mu}{r_d + R_L + R} \tag{13-59}$$

where $\mu = r_d g_m$ from Eq. (10-15).

$$\beta = \frac{V_f}{I_o} = -R \tag{13-60}$$

$$D = 1 + \beta G_M = 1 + \frac{\mu R}{r_d + R_L + R} = \frac{r_d + R_L + (\mu + 1)R}{r_d + R_L + R} \tag{13-61}$$

$$G_{Mf} = \frac{G_M}{D} = \frac{-\mu}{r_d + R_L + (\mu + 1)R} \tag{13-62}$$

Since $R_i = \infty$, then

$$R_{if} = R_i D = \infty \tag{13-63}$$

If R_L is considered to be an external load, then from Fig. 13-17c

$$R_o = r_d + R$$

To calculate R_{of} we need G_m, and from Eq. (13-17), $G_m = \lim_{R_L \to 0} G_M$.   Since β is independent of R_L, then using Eq. (13-61),

$$1 + \beta G_m = \lim_{R_L \to 0} D = \frac{r_d + (\mu + 1)R}{r_d + R} \tag{13-64}$$

$$R_{of} = R_o(1 + \beta G_m) = (r_d + R)\frac{r_d + (\mu + 1)R}{r_d + R} = r_d + (\mu + 1)R \tag{13-65}$$

The above results agree with those obtained in Sec. 10-7.

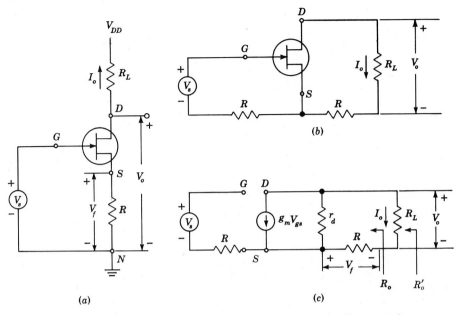

Fig. 13-17 (a) An FET amplifier with a source resistor R. (b) The amplifier without feedback, but including the loading of R. (c) The FET replaced by its small-signal low-frequency model.

R'_{of} is most easily calculated as $R_L \| R_{of}$. The same result may be obtained from the expression in Table 13-4, with $R'_o = R_o \| R_L$. Thus

$$R'_{of} = R'_o \frac{1 + \beta G_m}{D}$$

$$= \frac{(r_d + R)R_L}{r_d + R_L + R} \frac{r_d + (\mu + 1)R}{r_d + R} \frac{r_d + R_L + R}{r_d + R_L + (\mu + 1)R}$$

$$= \frac{R_L[r_d + (\mu + 1)R]}{r_d + R_L + (\mu + 1)R} \tag{13-66}$$

which is equivalent to R_L in parallel with R_{of}.

13-11   CURRENT–SHUNT FEEDBACK

Figure 13-18 shows two transistors in cascade with feedback from the second emitter to the first base through the resistor R'. We now verify that this connection produces negative feedback. The voltage V_{i2} is much larger than V_{i1} because of the voltage gain of $Q1$. Also, V_{i2} is $180°$ out of phase with V_{i1}. Because of emitter-follower action, V_{e2} is only slightly smaller than V_{i2}, and these voltages are in phase. Hence V_{e2} is larger in magnitude than V_{i1} and is 180° out of phase with V_{i1}. If the input signal increases so that I_s' increases, I_f also increases, and $I_i = I_s' - I_f$ is smaller than it would be if there were no feedback. This action is characteristic of *negative* feedback.

We now show that the configuration of Fig. 13-18 approximates a current-shunt feedback pair. Since $V_{e2} \gg V_{i1}$, and neglecting the base current of $Q2$ compared with the collector current,

$$I_f = \frac{V_{i1} - V_{e2}}{R'} \approx -\frac{V_{e2}}{R'} = \frac{(I_o - I_f)R_e}{R'} \tag{13-67}$$

or

$$I_f = \frac{R_e I_o}{R' + R_e} = \beta I_o \tag{13-68}$$

where $\beta = R_e/(R' + R_e)$. Since the feedback current is proportional to the output current, this circuit is an example of a current-shunt feedback amplifier. From Table 13-3 we expect the transfer (current) gain A_{If} to be stabilized.

$$A_{If} = \frac{I_o}{I_s} \approx \frac{1}{\beta} = \frac{R' + R_e}{R_e} \tag{13-69}$$

and hence we have verified that A_{If} is desensitized provided that R' and R_e are stable resistances. Note that $I_s \equiv V_s/R_s$.

From Table 13-3 we expect the input resistance to be low and the output resistance to be high.

The voltage gain with feedback is

$$A_{Vf} = \frac{V_o}{V_s} = \frac{I_o R_{c2}}{I_s R_s} \approx \frac{R' + R_e}{R_e} \frac{R_{c2}}{R_s} = \frac{R_{c2}}{\beta R_s} \tag{13-70}$$

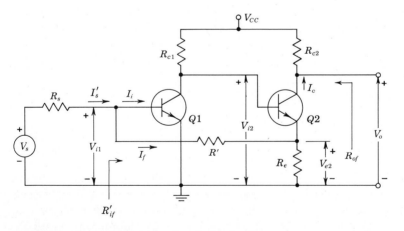

Fig. 13-18 Second-emitter to first-base feedback pair. (The input blocking capacitor and the biasing resistors are not indicated.)

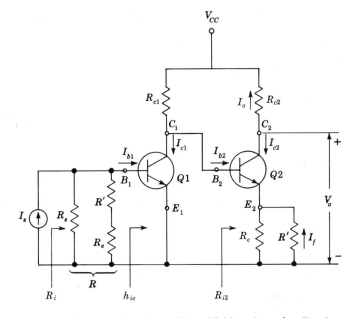

Fig. 13-19 The amplifier of Fig. 13-18 without feedback, but including the loading of R'.

Note that if R_e, R', R_{c2}, and R_s are stable elements, then A_{Vf} is stable (independent of the transistor parameters, the temperature, or supply-voltage variations).

The Amplifier without Feedback We must refer to the third topology in Table 13-4. The input circuit of the amplifier without feedback is obtained by opening the output loop at the emitter of $Q2$. This places R' in series with R_e from base to emitter of $Q1$. The output circuit is found by shorting the input node (the base of $Q1$). This places R' in parallel with R_e. The resultant equivalent circuit is given in Fig. 13-19. Since the feedback signal is a current, the source is represented by a Norton's equivalent circuit with $I_s = V_s/R_s$.

The feedback signal is the current I_f in the resistor R', which is in the output circuit. From Fig. 13-19, with $I_{b2} < I_{c2} = |I_o|$.

$$\beta = \frac{I_f}{I_o} = \frac{R_e}{R' + R_e} \tag{13-71}$$

in agreement with Eq. (13-68).

EXAMPLE The circuit of Fig. 13-18 has the following parameters: $R_{c1} = 3$ K, $R_{c2} = 500$ Ω, $R_e = 50$ Ω, $R' = R_s = 1.2$ K, $h_{fe} = 50$, $h_{ie} = 1.1$ K, and $h_{re} = h_{oe} = 0$. Find (a) A_{Vf}; (b) R_{if}; (c) the resistance seen by the voltage source; and (d) the output resistance.

Solution   a. Since the current gain is stabilized, we first calculate A_{If} from A_I. We can then obtain A_{Vf} from A_{If}. Referring to Fig. 13-19,

$$A_I = \frac{-I_{c2}}{I_s} = \frac{-I_{c2}}{I_{b2}}\frac{I_{b2}}{I_{c1}}\frac{I_{c1}}{I_{b1}}\frac{I_{b1}}{I_s} \tag{13-72}$$

Using the low-frequency approximate h-parameter models for $Q1$ and $Q2$,

$$\frac{-I_{c2}}{I_{b2}} = -h_{fe} = -50 \qquad \frac{I_{c1}}{I_{b1}} = +h_{fe} = +50 \tag{13-73}$$

$$\frac{I_{b2}}{I_{c1}} = \frac{-R_{c1}}{R_{c1}+R_{i2}} = \frac{-3}{3+3.55} = -0.458 \tag{13-74}$$

because, from Eq. (13-58),

$$R_{i2} = h_{ie} + (1+h_{fe})(R_e\|R') = 1.1 + (51)\left(\frac{0.05\times1.20}{1.25}\right) = 3.55\text{ K}$$

If R is defined by

$$R \equiv R_s\|(R'+R_e) = \frac{(1.2)(1.25)}{1.2+1.25} = 0.612\text{ K} \tag{13-75}$$

then from Fig. 13-19

$$\frac{I_{b1}}{I_s} = \frac{R}{R+h_{ie}} = \frac{0.61}{0.61+1.1} = 0.358 \tag{13-76}$$

Substituting the numerical values in Eqs. (13-73), (13-74), and (13-76) into Eq. (13-72) yields

$$A_I = (-50)(-0.458)(50)(0.358) = +410$$

$$\beta = \frac{R_e}{R'+R_e} = \frac{50}{1,250} = 0.040$$

$$D = 1 + \beta A_I = 1 + (0.040)(410) = 17.4$$

$$A_{If} = \frac{A_I}{D} = \frac{410}{17.4} = 23.6$$

$$A_{Vf} = \frac{V_o}{V_s} = \frac{-I_{c2}R_{c2}}{I_sR_s} = \frac{A_{If}R_{c2}}{R_s} = \frac{(23.6)(0.5)}{1.2} = 9.83$$

The approximate expression of Eq. (13-70) yields

$$A_{Vf} \approx \frac{R_{c2}}{\beta R_s} = \frac{0.5}{(0.040)(1.2)} = 10.4$$

which is in error by 6 percent.

b. From Fig. 13-19, the input impedance without feedback seen by the current source is, using Eq. (13-75),

$$R_i = R\|h_{ie} = \frac{(0.61)(1.1)}{1.71} = 0.394\text{ K}$$

and from Table 13-4 the resistance R_{if} with feedback seen by the current source is

$$R_{if} = \frac{R_i}{D} = \frac{394}{17.4} = 22.6 \ \Omega$$

c. Note that the input resistance is quite small, as predicted. If the resistance looking to the right of R_s (from base to emitter of $Q1$) in Fig. 13-18 is R'_{if}, then $R_{if} = R'_{if} \| R_s$, or

$$22.6 = \frac{1{,}200 R'_{if}}{1{,}200 + R'_{if}}$$

which yields $R'_{if} = 23.0 \ \Omega$. Hence from Fig. 13-18, the resistance with feedback seen by the voltage source V_s is

$$R_s + R'_{if} = 1{,}200 + 23.0 \ \Omega = 1.22 \ \mathrm{K}$$

d. If R_{c2} is considered as an external load, then R_o is the resistance seen looking into the collector of $Q2$. Since $h_{oe} = 0$, then $R_o = \infty$. From Table 13-4, $R_{of} = R_o(1 + \beta A_i) = \infty$.

From the calculations in part a, we note that A_I is independent of the load $R_L = R_{c2}$. Hence $A_i = \lim_{R_{c2} \to 0} A_I = A_I$. Since $R'_o = R_o \| R_{c2} = R_{c2}$, then from Table 13-4

$$R'_{of} = R'_o \frac{1 + \beta A_i}{1 + \beta A_I} = R'_o = R_{c2} = 500 \ \Omega$$

R'_{of} may also be calculated as the ratio of the open-circuit voltage V_o to the short-circuit output current I_o. Since for $h_{oe} = 0$, $I_o = -I_{c2}$ is independent of R_{c2}, then

$$R'_{of} = \frac{V_o}{I_o} = \frac{V_o}{V_s} \frac{V_s}{I_s} \frac{I_s}{I_o} = \frac{A_{Vf} R_s}{A_{If}} = \frac{(9.83) \times (1.2)}{23.6} = 0.50 \ \mathrm{K}$$

which agrees with the value found above using Table 13-4.

13-12   VOLTAGE–SHUNT FEEDBACK

Figure 13-20a shows a common-emitter stage with a resistor R' connected from the output to the input. We first show that this configuration conforms to voltage-shunt topology, and then obtain approximate expressions for transresistance and the voltage gain with feedback.

In the circuit of Fig. 13-20a, the output voltage V_o is much greater than the input voltage V_i and is 180° out of phase with V_i. Hence

$$I_f = \frac{V_i - V_o}{R'} \approx -\frac{V_o}{R'} = \beta V_o \tag{13-77}$$

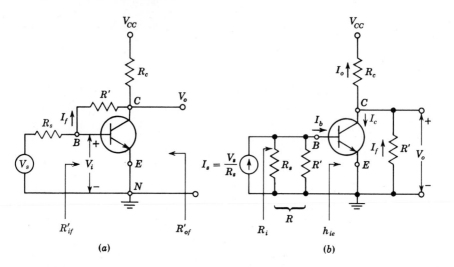

Fig. 13-20 (a) Voltage-shunt feedback. (b) The amplifier without feedback, but including the loading of R'.

where $\beta = -1/R'$. Since the feedback current is proportional to the output voltage, this circuit is an example of a *voltage-shunt feedback amplifier*. From Table 13-3 we expect the transfer gain (the transresistance) R_{Mf} to be desensitized. From Eq. (13-9)

$$R_{Mf} \equiv \frac{V_o}{I_s} \approx \frac{1}{\beta} = -R' \tag{13-78}$$

Note that the transresistance equals the negative of the feedback resistance from output to input of the transistor and is stable if R' is a stable resistance.

From Table 13-3 we expect both the input and output resistance to be low because of the voltage-shunt feedback. If we assume that $R'_{if} = 0$, then the voltage gain with feedback is

$$A_{Vf} = \frac{V_o}{V_s} = \frac{V_o}{I_s R_s} \approx \frac{1}{\beta R_s} = -\frac{R'}{R_s} \tag{13-79}$$

where use is made of Eq. (13-78). Note that if R' and R_s are stable elements, then A_{Vf} is stable (independent of the transistor parameters, the temperature, and supply-voltage variations).

The Amplifier without Feedback   We must refer to the fourth topology in Table 13-4. The input circuit of the amplifier without feedback is obtained by shorting the output node ($V_o = 0$). This places R' from base to emitter of the transistor. The output circuit is found by shorting the input node ($V_i = 0$), thus connecting R' from collector to emitter. The resultant equivalent circuit is given in Fig. 13-20b. Since the feedback signal is a current, the source is represented by a Norton's equivalent with $I_s = V_s/R_s$.

The feedback signal is the current I_f in the resistor R' which is in the output circuit. From Fig. 13-20b

$$\beta = \frac{I_f}{V_o} = -\frac{1}{R'} \tag{13-80}$$

in agreement with Eq. (13-78).

The first assumption in Sec. 13-3 is not satisfied exactly. If the amplifier is deactivated by reducing h_{fe} to zero, a current I_f passes through the β network (the resistor R') from input to output. This current is given by

$$I_f = \frac{V_s}{R_s + R' + R_c}$$

The output current I_o with the amplifier activated is

$$I_o = \frac{V_o}{R_c} = \frac{A_{Vf}V_s}{R_c}$$

Hence the condition that the forward transmission through the feedback network can be neglected is $|I_o| \gg |I_f|$, or

$$|A_{Vf}| \gg \frac{R_c}{R_s + R' + R_c} \tag{13-81}$$

Since the voltage gain is at least unity, this inequality is easily satisfied by selecting $R_s + R' \gg R_c$.

EXAMPLE The circuit of Fig. 13-20 has the following parameters: $R_c = 4$ K, $R' = 40$ K, $R_s = 10$ K, $h_{ie} = 1.1$ K, $h_{fe} = 50$, and $h_{re} = h_{oe} = 0$. Find (a) A_{Vf}, (b) R_{if}, and (c) R'_{of}.

Solution   a. Since the transresistance is stabilized, we first calculate R_{Mf} from R_M. Define R'_c and R by

$$R'_c \equiv R_c \| R' = \frac{(4)(40)}{44} = 3.64 \text{ K} \tag{13-82}$$

and

$$R \equiv R_s \| R' = \frac{(10)(40)}{50} = 8 \text{ K} \tag{13-83}$$

From Fig. 13-20b

$$R_M = \frac{V_o}{I_s} = \frac{-I_c R'_c}{I_s} = \frac{-h_{fe}I_b R'_c}{I_s} = \frac{-h_{fe}R'_c R}{R + h_{ie}} \tag{13-84}$$

$$R_M = \frac{(-50)(3.64)(8)}{8 + 1.1} = -160 \text{ K}$$

$$\beta = -\frac{1}{R'} = -\frac{1}{40} = -0.025 \text{ mA/V}$$

$$D = 1 + \beta R_M = 1 + 0.025 \times 160 = 5.00$$

$$R_{Mf} = \frac{R_M}{D} = \frac{-160}{5.00} = -32.0 \text{ K}$$

$$A_{Vf} = \frac{V_o}{V_s} = \frac{V_o}{I_s R_s} = \frac{R_{Mf}}{R_s} \tag{13-85}$$

or

$$A_{Vf} = \frac{-32.0}{10} = -3.20$$

b. From Fig. 13-20b

$$R_i = \frac{R h_{ie}}{R + h_{ie}} = \frac{(8)(1.1)}{9.1} = 0.968 \text{ K} = 968 \text{ } \Omega$$

From Table 13-4

$$R_{if} = \frac{R_i}{D} = \frac{968}{5.00} = 193 \text{ } \Omega$$

Note that the input resistance is quite small, as predicted.

If the input resistance looking to the right of R_s (from base to emitter in Fig. 13-20a) is R'_{if}, then $R_{if} = R'_{if} \| R_s$. Solving, we find $R'_{if} = 196 \text{ } \Omega$. The impedance seen by the voltage source V_s is $R_s + R'_{if} = 10.2 \text{ K}$.

c. If R_c is considered an external load, the output resistance, neglecting feedback, is $R_o = R' = 40 \text{ K}$. Since

$$R_m = \lim_{R_c \to \infty} R_M = \frac{-h_{fe} R' R}{R + h_{ie}} = \frac{(-50)(40)(8)}{8 + 1.1} = -1,760 \text{ K},$$

because in Eq. (13-83) $\lim_{R_c \to \infty} R'_c = R'$. From Table 13-4 (with $R_o = R'$)

$$R_{of} = \frac{R_o}{1 + \beta R_m} = \frac{40}{1 + (0.025)(1,760)} \text{ K} \approx 890 \text{ } \Omega$$

and

$$R'_{of} = R_{of} \| R_c = \frac{(890)(4,000)}{4,890} = 728 \text{ } \Omega$$

Alternatively, R'_{of} can be calculated from the formula in Table 13-4. The output resistance, taking R_c into account but neglecting feedback, is, from Fig. 13-20b, $R'_o = R_c \| R' = R'_c = 3.64 \text{ K}$. From Table 13-4

$$R'_{of} = \frac{R'_o}{D} = \frac{3.64}{5.00} \text{ K} = 728 \text{ } \Omega$$

in agreement with the value calculated above.

It is instructive to examine the approximate expression for the voltage gain given in Eq. (13-79).

$$A_{Vf} \approx \frac{-R'}{R_s} = -\frac{40}{10} = -4.00$$

which differs from the value of -3.20 by about 22 percent. This approximate formula leads to the erroneous conclusion that A_{Vf} increases without limit as $R_s \to 0$. The difficulty arises because Eq. (13-79) is valid only if $\beta R_M \gg 1$. However, from Eqs. (13-84) and (13-83)

$$\lim_{R_s \to 0} R_M = \lim_{R_s \to 0} \left(-\frac{h_{fe} R_c' R}{R + h_{ie}} \right) = \lim_{R_s \to 0} \left(-\frac{h_{fe} R_c' R_s}{R_s + h_{ie}} \right) \to 0$$

The correct result for A_{Vf} is obtained from Eq. (13-85), namely,

$$A_{Vf} = \lim_{R_s \to 0} \frac{R_{Mf}}{R_s} = \lim_{R_s \to 0} \frac{-h_{fe} R_c'}{R_s + h_{ie}} = -\frac{h_{fe} R_c'}{h_{ie}} \qquad (13\text{-}86)$$

This equation can be obtained by inspection of Fig. 13-20b. With $R_s = 0$, $A_{Vf} = V_o/V_s$ is the voltage gain of a CE amplifier with a load $R_c' = R_c \| R'$.

The circuit of Fig. 13-20 is the basic form of the *operational amplifier*, which is discussed in detail in Chap. 15.

REFERENCES

1. Jennings, R. R.: Negative Feedback in Voltage Amplifiers, *Electro-technol.* (*New York*), vol. 70, pp. 80–83, December, 1962.

   Jennings, R. R.: Negative Feedback in Current Amplifier, *ibid.*, vol. 72, pp. 100–103, July, 1963.

   Jennings, R. R.: Negative Feedback in Transconductance and Transresistance Amplifiers, *ibid.*, vol. 74, pp. 37–41, July, 1964.

2. Bode, H. W.: "Network Analysis and Feedback Amplifier Design," D. Van Nostrand Company, Inc., Princeton, N.J., 1945.

3. Uzunoglu, V.: "Semiconductor Network Analysis and Design," chap. 8, McGraw-Hill Book Company, New York, 1964.

   Ghausi, M. S.: "Principles and Design of Linear Active Circuits," chap. 4, McGraw-Hill Book Company, New York, 1965.

   Thornton, R. D., et al.: "Multistage Transistor Circuits," vol. 5, chap. 3, Semiconductor Electronics Education Committee, John Wiley & Sons, Inc., New York, 1965.

   Hakim, S. S.: "Junction Transistor Circuit Analysis," John Wiley & Sons, Inc., New York, 1962.

4. Uzunoglu, V.: Feedback and Impedance Levels in Transistor Circuits, *Electron. Equipment Eng.*, pp. 42–43, July, 1962.

   Blecher, F. H.: Design Principles for Single Loop Transistor Feedback Amplifiers, *IRE Trans. Circuit Theory*, vol. CT-4, p. 145, September, 1957.

   Blackman, R. B.: Effect of Feedback on Impedance, *Bell System Tech. J.*, vol. 22, no. 3, p. 269, October, 1943.

5. Gray, P. E., and Searle, C. L.: "Electronic Principles," chap. 18, John Wiley & Sons, Inc., New York, 1969.

REVIEW QUESTIONS

13-1 (a) Draw the equivalent circuit for a voltage amplifier. (b) For the ideal amplifier, what are the values of R_i and R_o? (c) What are the dimensions of the transfer gain?

13-2 Repeat Rev. 13-1 for a current amplifier.

13-3 Repeat Rev. 13-1 for a transconductance amplifier.

13-4 Repeat Rev. 13-1 for a transresistance amplifier.

13-5 Draw a feedback amplifier in block-diagram form. Identify each block, and state its function.

13-6 (a) What are the four possible topologies of a feedback amplifier? (b) Identify the output signal X_o and the feedback signal X_f for each topology (either as a current or voltage). (c) Identify the transfer gain A for each topology (for example, give its dimensions). (d) Define the feedback factor β.

13-7 (a) What is the relationship between the transfer gain with feedback A_f and that without feedback A? (b) Define *negative feedback*. (c) Define *positive feedback*. (d) Define the amount of feedback in decibels.

13-8 State the three fundamental assumptions which are made in order that the expression $A_f = A/(1 + A\beta)$ be satisfied exactly.

13-9 (a) Define *desensitivity D*. (b) For large values of D, what is A_f? What is the significance of this result?

13-10 List five characteristics of an amplifier which are modified by negative feedback.

13-11 State whether the input resistance R_{if} is increased or decreased for each topology.

13-12 Repeat Rev. 13-11 for the output resistance R_{of}.

13-13 List the procedures to follow to obtain the basic amplifier configuration without feedback but taking the loading of the β network into account.

13-14 List the steps required to carry out the analysis of a feedback amplifier.

13-15 Draw the circuit of a single-stage voltage-series feedback amplifier.

13-16 Repeat Rev. 13-15 for a two-stage amplifier.

13-17 Find A_f for a source follower using the feedback method of analysis.

13-18 Repeat Rev. 13-17 for an emitter follower.

13-19 Draw a circuit of a current-series feedback amplifier.

13-20 Repeat Rev. 13-17 for a CE stage with an unbypassed emitter resistor.

13-21 Repeat Rev. 13-17 for a CS stage with an unbypassed resistance in the source lead.

13-22 Draw the circuit of a feedback pair with current-shunt topology.

13-23 Draw the circuit of a voltage-shunt feedback amplifier.

14 / STABILITY AND OSCILLATORS

The frequency response characteristics of multipole feedback amplifiers are studied. It is shown that single-pole and double-pole closed-loop amplifiers are inherently stable. The step response of these amplifiers is obtained. A feedback amplifier with more than two poles can become unstable and break into oscillation if too much feedback is applied.

The problem of stability is examined. Compensation techniques which are employed to prevent a feedback amplifier from becoming unstable are studied. Sinusoidal oscillators are analyzed as special cases of feedback amplifiers which are intentionally rendered unstable.

14-1 EFFECT OF FEEDBACK ON AMPLIFIER BANDWIDTH

The transfer gain of an amplifier employing feedback is given by Eq. (13-4), namely,

$$A_f = \frac{A}{1 + \beta A} \tag{14-1}$$

If $|\beta A| \gg 1$, then

$$A_f \approx \frac{A}{\beta A} = \frac{1}{\beta}$$

and from this result we conclude that the transfer gain may be made to depend entirely on the feedback network β. However, it is now important to consider the fact that even if β is constant, the gain A is not, since it depends on frequency. This means that at certain high or low frequencies, $|\beta A|$ will not be much larger than unity. To study the effect of feedback on bandwidth we shall assume in this section that the transfer gain A without feedback is given by a single-pole transfer function. In subsequent sections we consider multipole transfer functions.

447

Single-pole Transfer Function The gain A of a single-pole amplifier is given by

$$A = \frac{A_o}{1 + j(f/f_H)} \tag{14-2}$$

where A_o (real and negative) is the midband gain without feedback, and f_H is the high 3-dB frequency. The gain with feedback is given by Eq. (14-1), or using Eq. (14-2),

$$A_f = \frac{A_o/[1 + j(f/f_H)]}{1 + \beta A_o/[1 + j(f/f_H)]} = \frac{A_o}{1 + \beta A_o + j(f/f_H)}$$

By dividing numerator and denominator by $1 + \beta A_o$, this equation may be put in the form

$$A_f = \frac{A_{of}}{1 + j(f/f_{Hf})} \tag{14-3}$$

where

$$A_{of} \equiv \frac{A_o}{1 + \beta A_o} \qquad \text{and} \qquad f_{Hf} \equiv f_H(1 + \beta A_o) \tag{14-4}$$

We see that the *midband amplification with feedback* A_{of} equals the midband amplification without feedback A_o divided by $1 + \beta A_o$. Also, the *high 3-dB frequency with feedback* f_{Hf} equals the corresponding 3-dB frequency without feedback f_H multiplied by the same factor, $1 + \beta A_o$. The gain-frequency product has not been changed by feedback because, from Eqs. (14-4),

$$A_{of}f_{Hf} = A_o f_H \tag{14-5}$$

By starting with Eq. (12-2) for the low-frequency gain of a single RC-coupled stage and proceeding as above, we can show that the *low 3-dB frequency with feedback* f_{Lf} is decreased by the same factor as is the gain, or

$$f_{Lf} = \frac{f_L}{1 + \beta A_o} \tag{14-6}$$

For an audio or video amplifier, $f_H \gg f_L$, and hence the bandwidth is $f_H - f_L \approx f_H$. Under these circumstances, Eq. (14-5) may be interpreted to mean that the gain-bandwidth product is the same with or without feedback. Figure 14-1a is a plot of A and A_f versus frequency, whereas in Fig. 14-1b we show the Bode plots of both A and A_f in decibels vs. log f. The intersection of the two curves in Fig. 14-1b takes place at the frequencies $f = f_{Lf}$ and $f = f_{Hf}$ on the low end and high end, respectively. To verify this we equate the magnitude of Eq. (14-4) to the magnitude of Eq. (14-2) and solve for the value of f. Thus, if $f \gg f_H$ at the intersection point, we have

$$20 \log \left| \frac{A_o}{1 + \beta A_o} \right| = 20 \log \frac{|A_o|}{\sqrt{1 + (f/f_H)^2}} \approx 20 \log \frac{|A_o|}{f/f_H}$$

or

$$f = f_H(1 + \beta A_o) = f_{Hf} \tag{14-7}$$

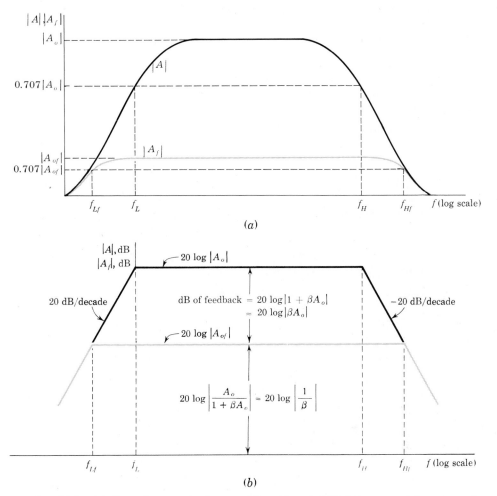

Fig. 14-1 (a) Transfer gain is decreased and bandwidth is increased for an amplifier using negative feedback. (b) Idealized Bode plot.

Similarly, we can show that the intersection at the low-frequency end occurs at $f = f_{Lf}$.

Bandwidth Improvement Equations (14-4) and (14-6) show how the upper and lower 3-dB frequencies are affected by negative feedback. We may obtain a physical feeling of the mechanism by which feedback extends bandwidth by considering the voltage-series feedback circuit of Fig. 14-2. The amplifier A_V has two input terminals 1 and 2, and the effective input voltage V_i is the difference in the voltages $V_1 = V_s$ and $V_2 = V_f$ applied to these two

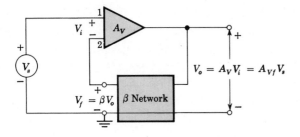

Fig. 14-2 Voltage-series feedback applied to an amplifier with voltage gain A_V.

terminals. Such a device is called a *difference amplifier* and is studied in detail in Chap. 15. Clearly, Fig. 14-2 has the voltage-series topology of Fig. 13-9a.

Let us assume that the midband voltage gain $A_{Vo} = -1,000$, $\beta = -0.1$, and $V_s = 0.1$ V. Under these conditions

$$A_{Vof} = \frac{A_{Vo}}{1 + A_{Vo}\beta} = \frac{-1,000}{1 + 100} = -9.90$$

$$V_o = A_{Vof}V_s = (-9.90)(0.1) = -0.99 \text{ V}$$

$$V_f = \beta V_o = (-0.1)(-0.99) = 0.099 \text{ V}$$

$$V_i = V_s - V_f = 0.1 - 0.099 = 0.001 \text{ V}$$

Note that almost the entire applied signal is canceled (*bucked out*) by the feedback signal, leaving a very small voltage V_i at the input terminals of the amplifier.

Now assume that at some higher frequency the gain of the amplifier (without feedback) has fallen to half its previous value, so that $A_V = -500$. Then, if V_s remains at 0.1 V,

$$A_{Vf} = \frac{A_V}{1 + A_V\beta} = \frac{-500}{1 + 50} = -9.80$$

$$V_o = A_{Vf}V_s = (-9.80)(0.1) = -0.98 \text{ V}$$

$$V_f = (-0.1)(-0.98) = 0.098 \text{ V}$$

$$V_i = 0.1 - 0.098 = 0.002 \text{ V}$$

Note that although the base amplifier gain has been halved, the amplification with feedback has changed by only 1 percent. In the second case, V_i has doubled to compensate for the drop in A_V. There exists a self-regulating action so that, if the open-loop voltage gain falls (as a function of frequency), the feedback voltage also falls. Therefore less of the input voltage is bucked out, permitting more voltage to be applied to the amplifier input, and V_o remains almost constant.

Step Response The transient behavior is that discussed in Sec. 12-5 for a single-pole transfer function. The output for a pulse input is given in Fig. 12-8, and for a square-wave input in Fig. 12-10.

14-2   DOUBLE–POLE TRANSFER FUNCTION WITH FEEDBACK

Let us consider a circuit where the basic amplifier gain (without feedback) has two poles on the negative real axis at $s_1 = -\omega_1$ and $s_2 = -\omega_2$ (ω_1 and ω_2 are positive), as shown in Fig. 14-3. In other words, $\omega_1/2\pi$ and $\omega_2/2\pi$ represent the *corner*, or *break*, frequencies of the linearized Bode plot (Sec. 12-4). If the midband gain is A_o, then the transfer gain is given by

$$A = \frac{A_o}{(1 - s/s_1)(1 - s/s_2)} = \frac{A_o}{(1 + s/\omega_1)(1 + s/\omega_2)} \tag{14-8}$$

If this expression for A is substituted into Eq. (14-1), we obtain for A_f, the transfer gain with feedback,

$$A_f = \frac{A_o\omega_1\omega_2}{s^2 + (\omega_1 + \omega_2)s + \omega_1\omega_2(1 + \beta A_o)} \tag{14-9}$$

or

$$A_f = \frac{A_{of}}{(s/\omega_o)^2 + (1/Q)(s/\omega_o) + 1} \tag{14-10}$$

where A_{of}, the midband gain with feedback, is given by Eq. (14-4) and ω_o and Q are defined by

$$\omega_o \equiv \sqrt{\omega_1\omega_2(1 + \beta A_o)} \qquad Q \equiv \frac{\omega_o}{\omega_1 + \omega_2} \tag{14-11}$$

The poles of A_f are

$$\frac{s}{\omega_o} = -\frac{1}{2Q} \pm \frac{1}{2}\sqrt{\frac{1}{Q^2} - 4} \tag{14-12}$$

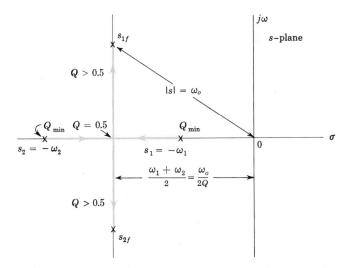

Fig. 14-3   Root locus of the two-pole transfer function in the $s = \sigma + j\omega$ plane. The value $Q_{\min}$ corresponds to $\beta A_o = 0$.

or

$$s = -\frac{\omega_1 + \omega_2}{2} \pm \frac{\omega_1 + \omega_2}{2}\sqrt{1 - 4Q^2} \tag{14-13}$$

where the value $\omega_o/Q = \omega_1 + \omega_2$ from Eq. (14-11) is used. For negative feedback we shall assume βA_o to be real and positive. Hence the minimum value of Q is obtained for $\beta A_o = 0$ or $Q_{min} = \sqrt{\omega_1\omega_2}/(\omega_1 + \omega_2)$. Substituting Q_{min} into Eq. (14-13) yields, after a little algebraic manipulation, the two values $s_{1f} = -\omega_1$ and $s_{2f} = -\omega_2$. Clearly, this result is correct, since if $\beta A_o = 0$, the poles with feedback (s_{1f} and s_{2f}) must coincide with the poles ($s_1 = -\omega_1$ and $s_2 = -\omega_2$) of the basic amplifier before feedback is added.

Root Locus[1,2] The movement of the poles in the s plane ($s = \sigma + j\omega$) as the feedback is increased is indicated in Fig. 14-3. The poles start at $-\omega_1$ and $-\omega_2$ at Q_{min} and move toward each other along the negative real axis as Q is increased until at $Q = 0.5$ the poles coincide. This behavior follows from Eq. (14-13), which shows that the values of s are real for $Q < 0.5$, and at $Q = 0.5$ the poles coincide at the value $-\frac{1}{2}(\omega_1 + \omega_2)$. For $Q > 0.5$ the two values of s become complex conjugates with the real part remaining at $-\frac{1}{2}(\omega_1 + \omega_2)$, as sketched in Fig. 14-3. Incidentally, the magnitude of a complex pole is $|s| = \omega_o$. This result is obtained by taking the magnitude of Eq. (14-12) for $Q \geq 0.5$. Thus

$$\left|\frac{s}{\omega_o}\right|^2 = \frac{1}{4Q^2} + \frac{1}{4}\left(4 - \frac{1}{Q^2}\right) = 1 \quad \text{or} \quad |s_{1f}| = |s_{2f}| = \omega_o$$

The shaded path in Fig. 14-3 is known as the *root locus* of the poles.

Note that for all positive values of βA_o the transfer function has poles which remain in the left-hand s plane (the poles have negative real parts). Therefore *the negative feedback amplifier is stable; independent of the amount of feedback*. This statement is true for a single-pole or double-pole transfer gain, but may not be true if three or more poles are present (Sec. 14-3).

Circuit Model We now demonstrate that the network in Fig. 14-4 is the analog of the two-pole feedback amplifier. The transfer function is found to be

$$\frac{V_o(s)}{V_i(s)} = \frac{1}{s^2LC + s(L/R) + 1} \tag{14-14}$$

Introducing

$$\omega_o \equiv \frac{1}{\sqrt{LC}} \qquad Q \equiv \frac{R}{\omega_o L} = R\sqrt{\frac{C}{L}} \tag{14-15}$$

leads to

$$\frac{V_o(s)}{V_i(s)} = \frac{1}{(s/\omega_o)^2 + (1/Q)(s/\omega_o) + 1} = \frac{A_f}{A_{of}} \tag{14-16}$$

Fig. 14-4 A circuit model for a two-pole
feedback amplifier.

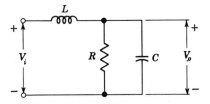

where the second equality follows from Eq. (14-10).   Clearly, Fig. 14-4 is a
circuit model of a two-pole feedback amplifier in the sense that both have the
same frequency, phase, and transient response; *the transfer gain of the amplifier
is A_{of} times the transfer function of the network.*   Physical meanings can now
be given to the symbols ω_o and Q, introduced in connection with the feedback
amplifier.   From Eqs. (14-15) it is evident that

ω_o = undamped ($R = \infty$) resonant angular frequency of oscillation
Q = quality factor (Q) at the resonant frequency

Frequency Response   If in Eq. (14-16) s is replaced by $j\omega$, then the
magnitude of this expression gives the frequency response of the two-pole
amplifier with feedback.   It is convenient to use the *damping factor* k in place
of Q.   These are related by

$$k \equiv \frac{1}{2Q} \tag{14-17}$$

Thus, from Eqs. (14-16) and (14-17), we obtain

$$\left| \frac{A_f}{A_{of}} \right| = \frac{1}{\sqrt{[1 - (\omega/\omega_o)^2]^2 + 4k^2(\omega/\omega_o)^2}} \tag{14-18}$$

The peaks of this function are obtained by setting the derivative of the quantity under the square-root sign equal to zero.   We find that a peak occurs at

$$\omega = \omega_o \sqrt{1 - 2k^2} \tag{14-19}$$

and the magnitude of the peak is given by

$$\left| \frac{A_f}{A_{of}} \right|_{\text{peak}} = \frac{1}{2k \sqrt{1 - k^2}} \tag{14-20}$$

Note that if $2k^2 > 1$ or $k > 0.707$ or $Q < 0.707$, the frequency response will not
exhibit a peak.   A plot of the normalized frequency response is given in Fig. 14-5.

Step Response   It has been proved in this section that regardless of how
much negative feedback is employed, a two-pole amplifier remains stable (its
poles are always in the left half s plane).   However, if the loop gain βA_o is too
large, the transient response of the amplifier may be entirely unsatisfactory.

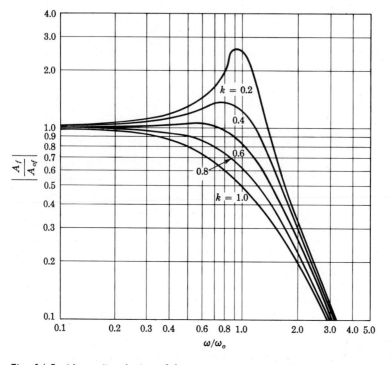

Fig. 14-5 Normalized plot of frequency response of a two-pole amplifier with feedback ($k = 1/2Q$).

For example, in Fig. 14-6 there is indicated one possible response to a voltage step. Note that the output overshoots its final value by 37 percent and oscillates before settling down to the steady-state voltage. For most applications such a violent response is not acceptable.

The important parameters of the waveform are indicated in Fig. 14-6 and are defined as follows:

Rise time = time for waveform to rise from 0.1 to 0.9 of its steady-state value

Delay time = time for waveform to rise from 0 to 0.5 of its steady-state value

Overshoot = peak excursion above the steady-state value

Damped period = time interval for one cycle of oscillation

Settling time = time for response to settle to within $\pm P$ percent of the steady-state value (P specified for a particular application, say $P = 0.1$)

Analytical expressions for the response of the amplifier to a step of amplitude V is obtained by setting $V_i(s) = V/s$ into Eq. (14-16) and solving for the inverse Laplace transform. Recalling from Eq. (14-17) that $Q = 1/2k$, the poles, given in Eq. (14-12), can be put into the form

$$s = -k\omega_o \pm \omega_o \sqrt{k^2 - 1} \qquad (14-21)$$

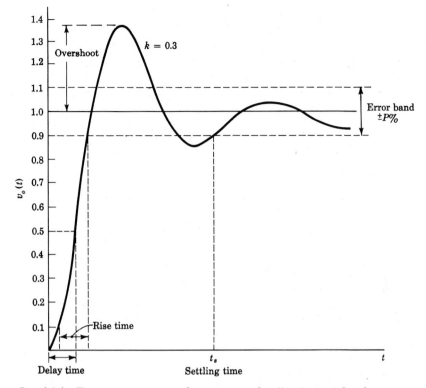

Fig. 14-6 The step response of a two-pole feedback amplifier for a damping factor $k = 0.3$.

If $k = 1$, the two poles coincide, corresponding to the *critically damped* case. If $k < 1$, the poles are complex conjugates, corresponding to an *underdamped* condition, where the response is a sinusoid whose amplitude decays with time. If $k > 1$, both poles are real and negative, corresponding to an *overdamped* circuit, where the response approaches its final value monotonically (without oscillation). For the underdamped case it is convenient to introduce the damped frequency

$$\omega_d \equiv \sqrt{1 - k^2}\, \omega_o \tag{14-22}$$

and the response $v_o(t)$ to a step of magnitude V into an amplifier of midband gain A_{of} is given by the following equations:

Critical damping, $k = 1$:

$$\frac{v_o(t)}{V A_{of}} = 1 - (1 + \omega_o t)\epsilon^{-\omega_o t} \tag{14-23}$$

Overdamped, $k > 1$:

$$\frac{v_o(t)}{V A_{of}} = 1 - \frac{1}{2\sqrt{k^2 - 1}} \left(\frac{1}{k_1} \epsilon^{-k_1 \omega_o t} - \frac{1}{k_2} \epsilon^{-k_2 \omega_o t} \right) \tag{14-24}$$

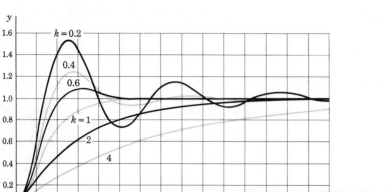

Fig. 14-7 The response of a two-pole feedback amplifier to an input step [$y \equiv v_o(t)/VA_{of}$ and $x \equiv \omega_o t/2\pi = t/T_o$].

where

$$k_1 \equiv k - \sqrt{k^2 - 1} \quad \text{and} \quad k_2 \equiv k + \sqrt{k^2 - 1}$$

If $4k^2 \gg 1$, the response may be approximated by

$$\frac{v_o(t)}{VA_{of}} \approx 1 - \epsilon^{-\omega_o t /2k} \tag{14-25}$$

Underdamped, $k < 1$:

$$\frac{v_o(t)}{VA_{of}} = 1 - \left(\frac{k\omega_o}{\omega_d} \sin \omega_d t + \cos \omega_d t\right) \epsilon^{-k\omega_o t} \tag{14-26}$$

These equations are plotted in Fig. 14-7 using the normalized coordinates $x \equiv t/T_o$ and $y \equiv v_o(t)/VA_{of}$, where $T_o \equiv 2\pi/\omega_o$ is the undamped period. If the derivative of Eq. (14-26) is set equal to zero, the positions $x = x_m$ and magnitudes $y = y_m$ of the maxima and minima are obtained. The results are

$$x_m = \frac{\omega_o t_m}{2\pi} = \frac{m}{2(1 - k^2)^{\frac{1}{2}}} \qquad y_m = \frac{v_o(t_m)}{VA_{of}} = 1 - (-1)^m \epsilon^{-2\pi k x_m} \tag{14-27}$$

where m is an integer. The maxima occur for odd values of m, and the minima are obtained for even values of m. By using Eq. (14-27) the waveshape of the underdamped output may be sketched very rapidly. From Eq. (14-27) it follows that the *overshoot* is given by $\exp[-\pi km/(1 - k^2)^{\frac{1}{2}}]$.

Note that for heavy damping (k large or Q small) the rise time t_r is very long. As k is decreased (Q or βA_o increased), t_r decreases. For the critically damped case we find from Fig. 14-7 that $t_r = 0.53T_o = 3.33/\omega_o$. If the feedback is increased so that $k < 1$, the rise time is decreased further, but this improvement is obtained at the expense of a ringing (oscillatory) response which may be unacceptable for some applications. Often $k \geq 0.707$ ($Q \leq 0.707$)

is specified as a satisfactory response (corresponding to an overshoot of 4.3 percent or less).

14-3   THREE–POLE TRANSFER FUNCTION WITH FEEDBACK

In the previous two cases of single-pole and double-pole transfer functions we verify that the feedback-amplifier transfer function always has poles which lie in the left-hand plane.   We now consider a three-pole transfer function and find that if the loop gain is sufficiently large, the poles of the feedback amplifier move into the right-hand plane, and thus the circuit becomes unstable.

Let us assume the open-loop gain to be given by

$$A(s) = \frac{A_o}{(1 + s/\omega_1)(1 + s/\omega_2)(1 + s/\omega_3)} \tag{14-28}$$

Using Eq. (14-1) for the gain with feedback, we find

$$A_f(s) = \frac{A_{of}}{(s/\omega_o)^3 + a_2(s/\omega_o)^2 + a_1(s/\omega_o) + 1} \tag{14-29}$$

where A_{of} is the midband gain with feedback and ω_o, a_2, and a_1 are given in Prob. 14-14.   The stability of the feedback amplifier is determined by the poles of its transfer function.   The techniques for the construction of the root locus of Eq. (14-29) are given in Ref. 1.   The general shape of the root locus is shown in Fig. 14-8.   It is clear that the poles start at $-\omega_1$, $-\omega_2$, and $-\omega_3$ when $\beta A_o = 0$.   As βA_o is increased one pole, s_{3f} increases in magnitude but always remains on the negative real axis, while the other two poles, s_{2f} and s_{1f},

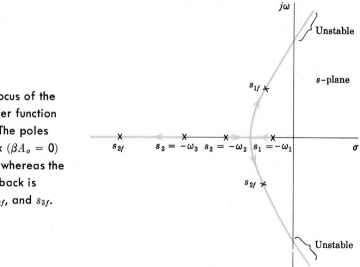

Fig. 14-8   Root locus of the three-pole transfer function in the s plane.   The poles without feedback ($\beta A_o = 0$) are s_1, s_2, and s_3, whereas the poles after feedback is added are s_{1f}, s_{2f}, and s_{3f}.

approach each other, and then become complex conjugates when they break away from the real axis. The real part of s_{2f} and s_{1f} is negative when the roots coincide. However, as βA_o increases beyond this critical value, the real part of s_2 and s_1 becomes less negative. It is shown in Ref. 2 that if $\omega_1 = 0.1 \times 10^8$, $\omega_2 = 0.7 \times 10^8$, and $\omega_3 = 1.8 \times 10^8$ rad/s, the complex conjugate poles s_{2f} and s_{1f} move into the right-hand plane when $\beta A_o = 31$. We conclude that a three-pole amplifier can become unstable if sufficient negative feedback is applied to it.

14-4   APPROXIMATE ANALYSIS OF A MULTIPOLE FEEDBACK AMPLIFIER

In the general case, the response of a feedback system with three or more poles is so complicated that a computer must be used to obtain a solution. However, if the open-loop poles are widely separated, a simple approximate method of analysis is possible. We describe this technique now.

Assume that we have obtained the poles $|s_1| < |s_2| < |s_3| \cdots < |s_n|$ by removing the feedback. This can be done by measuring in the laboratory the corner frequencies of the Bode plot, or the poles may be calculated. We wish to determine the closed-loop response. In the preceding section we note that, for a three-pole amplifier, the effect of adding feedback is to bring the poles s_1 and s_2 closer together along the real axis and to separate further the poles s_2 and s_3 (Fig. 14-8). Hence, if the second and third poles without feedback are at least two octaves apart ($s_3/s_2 \geq 4$), they will be separated even further apart after feedback is added ($s_{3f}/s_{2f} > 4$). *The response of an amplifier with three (or more) poles is determined approximately by the two lowest poles, s_1 and s_2, provided that* $|s_3/s_2| \geq 4$.

The above conclusion is consistent with the fact that *for a set of widely separated poles, the higher poles will remain almost fixed for moderate amounts of feedback.*[2] Hence we can use the theory in Sec. 14-2 for a double-pole transfer function with feedback to describe a multipole amplifier with widely separated poles. In other words, k is calculated for the first two poles, and then the frequency response is found from Fig. 14-5 and the transient response from Fig. 14-7.

It is convenient to introduce n as the ratio of the first two open-loop poles:

$$n = \frac{s_2}{s_1} = \frac{\omega_2}{\omega_1} > 1 \tag{14-30}$$

and then from Eq. (14-11) the Q is given in terms of n, and the desensitivity $1 + \beta A_o$ by

$$Q = \frac{\sqrt{n(1 + \beta A_o)}}{n + 1} = \frac{1}{2k} \tag{14-31}$$

Dominant Pole   If the first two open-loop poles are widely separated and if the desensitivity is not too great, the first two closed-loop poles may be more than two octaves apart.   Under these circumstances the response is simply that of a single (dominant) pole, and is discussed in Sec. 12-4.

We now determine the maximum value of Q for which a dominant pole exists.   From Eq. (14-13), with $n = \omega_2/\omega_1$, the two lowest poles with feedback are

$$s_{1f} = \frac{-\omega_1(n+1)}{2}(1 - \sqrt{1 - 4Q^2})$$

$$s_{2f} = \frac{-\omega_1(n+1)}{2}(1 + \sqrt{1 - 4Q^2}) \tag{14-32}$$

Setting $s_{2f}/s_{1f} \geq 4$, we find $Q \leq 0.4 = Q_{\max}$.   *For the poles of an amplifier with feedback to be separated by at least two octaves, Q must be no larger than 0.4* (or $Q^2 < 0.16$).   Incidentally, if Q increases to 0.5, the two poles coincide.

If $Q < 0.4$, the dominant pole is s_{1f} in Eq. (14-32).   If $4Q^2 \ll 1$, this equation yields (Prob. 14-20), for the upper 3-dB frequency,

$$f_{Hf} \approx f_1\left(\frac{n}{n+1}\right)(1 + \beta A_o)(1 + Q^2) \tag{14-33}$$

In the following sections we analyze a number of feedback amplifiers by the approximate method outlined above and compare the results with exact computer solutions.

14-5   VOLTAGE–SHUNT FEEDBACK AMPLIFIER— FREQUENCY RESPONSE

We see in the preceding sections that negative feedback in general increases the bandwidth of the transfer function stabilized by the specific type of feedback (topology) used in a circuit.   For example, voltage-series feedback increases the upper 3-dB frequency f_H of A_V to the value $f_{Hf} = (1 + \beta A_V)f_H$ after feedback is employed, provided A_V is a single-pole transfer function. We shall now examine specific feedback-amplifier circuits and obtain their frequency response.   We start with a *voltage-shunt feedback amplifier.*

Figure 13-20a shows a common-emitter stage with a resistance R' connected from collector to base.   Clearly, this is a case of voltage-shunt feedback, and we expect the bandwidth of the transresistance $R_{Mf} = V_o/I_s$ to be improved due to the feedback through R'.   The example in Sec. 13-12 deals with this amplifier at low frequencies.   Let us now examine the frequency response of the transresistance R_{Mf} and of the voltage gain $A_{Vf} = V_o/V_s$.   The circuit of Fig. 14-9a is the same as Fig. 13-20a, with the transistor replaced by the hybrid-Π equivalent circuit.   The loading effects of the β network on

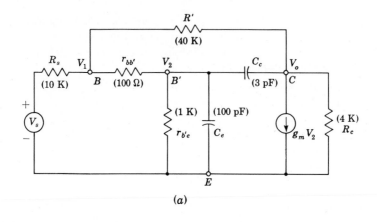

(a)

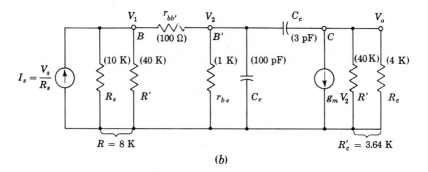

(b)

Fig. 14-9   (a) CE amplifier with voltage-shunt feedback; (b) the A network including the loading of R'.

the basic amplifier without feedback are included in Fig. 14-9b, which is
equivalent to Fig. 13-20b.   The voltage source is represented by its Norton's
equivalent current source $I_s = V_s/R_s$.   Since Fig. 14-9b is equivalent to
Fig. 11-10, then from Eq. (11-37) we find the transfer function R_M without
feedback:

$$R_M = \frac{V_o}{I_s} = \frac{-R'_c R G_1 (g_m - sC_c)}{s^2 C_e C_c R'_c + s[C_e + C_c + C_c R'_c (g_m + g_{b'e} + G_1)] + G_1 + g_{b'e}}$$

$$(14\text{-}34)$$

where the following abbreviations are introduced:

$$R'_c \equiv R_c \| R' = 4 \| 40 = 3.64 \text{ K}$$

$$R \equiv R_s \| R' = 10 \| 40 = 8.00 \text{ K}$$

$$R_1 \equiv R + r_{bb'} = 8.10 \text{ K} \equiv \frac{1}{G_1}$$

We assume that $C_e = 100$ pF, $C_c = 3$ pF, $r_{bb'} = 100\ \Omega$, and to be consistent with the low-frequency h parameters, we have from Eq. (11-16)

$$r_{b'e} = h_{ie} - r_{bb'} = 1{,}100 - 100 = 1{,}000\ \Omega$$

and

$$g_m = g_{b'e}h_{fe} = 50\ \text{mA/V}$$

With these numerical values, R_M is given by

$$R_M = \frac{V_o}{I_s} = 0.985 \times 10^{10} \frac{s - 16.6 \times 10^9}{(s + 600 \times 10^6)(s + 1.70 \times 10^6)} \qquad (14\text{-}35)$$

Dominant Pole   We see from Eq. (14-35) that we have a double-pole transfer function. However, the zero and one of the poles are much higher than $s_1 = -1.70 \times 10^6$ rad/s, and thus s_1 is a dominant open-loop pole. The question now arises whether the amplifier will have a dominant pole after feedback is applied to it. Thus we must check to see if $Q \leq Q_{\max} = 0.4$.

From Eq. (14-35), with $s = j2\pi f = 0$, we obtain the midband transresistance

$$R_{Mo} = -1.60 \times 10^5 \qquad (14\text{-}36)$$

and from Sec. 13-11, $\beta = -1/R' = -2.50 \times 10^{-5}$ and

$$1 + \beta R_{Mo} = 1 + 1.60 \times 2.50 = 5$$

Since

$$n = \frac{\omega_2}{\omega_1} = \frac{600}{1.70} = 353$$

then from Eq. (14-31)

$$Q^2 = \frac{n}{(n+1)^2}\,(1 + \beta R_{Mo}) = \frac{(353)(5)}{(354)^2} = 0.0141 \qquad (14\text{-}37)$$

Since $Q^2 < Q_{\max}{}^2 = 0.16$, a dominant pole does exist, and its value is given by Eq. (14-33) because $4Q^2 \ll 1$. Since

$$f_H = \frac{\omega_1}{2\pi} = \frac{1.70 \times 10^6}{2\pi}\ \text{Hz} = 0.271\ \text{MHz}$$

then the high 3-dB frequency with feedback is

$$f_{Hf} = f_H \frac{n}{n+1}\,(1 + \beta R_{Mo})(1 + Q^2) = (0.271)(\tfrac{353}{354})(5)(1 + 0.0141)$$
$$= 1.37\ \text{MHz}$$

Since

$$R_{Mof} = \frac{R_{Mo}}{1 + \beta R_{Mo}} = \frac{-1.60 \times 10^5}{5} = -3.20 \times 10^4$$

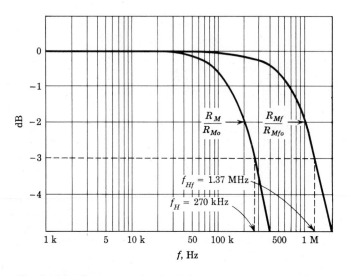

Fig. 14-10 Computer-obtained frequency response of normalized transfer functions R_M/R_{Mo} and R_{Mf}/R_{Mfo} of the amplifier shown in Fig. 14-9a. $R_{Mo} = 104$ dB and $R_{Mfo} = 90.1$ dB.

then, with f in megahertz,

$$R_{Mf} = \frac{-3.20 \times 10^4}{1 + j(f/1.37)} \qquad (14\text{-}38)$$

The voltage gain with feedback is given by

$$A_{Vf} = \frac{V_o}{V_s} = \frac{V_o}{I_s R_s} = \frac{R_{Mf}}{R_s} = \frac{-3.20}{1 + j(f/1.37)} \qquad (14\text{-}39)$$

From Eqs. (14-38) and (14-39) we see that R_{Mf} and A_{Vf} have the same upper 3-dB frequency.

In Fig. 14-9a we show the amplifier with feedback. Exact analysis of the circuit shown in Fig. 14-9b, using the CORNAP computer program,[3] yields the same values for the zero and two poles as in Eq. (14-35) for R_M. Computer analysis of the circuit shown in Fig. 14-9a yields for R_{Mf} two poles (in radians per second),

$$s_{1f} = -8.60 \times 10^6 \qquad s_{2f} = -5.95 \times 10^8$$

and one zero,

$$s_{3f} = 15.3 \times 10^9$$

Note that $s_{2f}/s_{1f} \gg 4$, verifying our conclusion of a dominant pole. The frequency response for R_M and R_{Mf} is plotted in Fig. 14-10. From these curves we find $f_H = 270$ kHz and $f_{Hf} = 1.37$ MHz, in excellent agreement with the values obtained with the dominant-pole approximate analysis.

14-6   CURRENT–SERIES FEEDBACK AMPLIFIER—
FREQUENCY RESPONSE

A common-emitter stage with an unbypassed emitter resistance represents a case of current-series feedback. The low-frequency operation of such a stage was studied in Sec. 13-10. In Fig. 14-11a the stage is shown with biasing and power supplies omitted. We shall obtain the frequency response of this amplifier using feedback concepts. Clearly, the transfer function which is stabilized is the transconductance $G_M = I_o/V_s$, and since $V_o = I_oR_c$, then $A_V = R_cG_M$ is also stabilized if R_c is a stable resistance. By referring to Fig. 13-16a and b, we construct the A network shown in Fig. 14-11b, which is used for the calculation of the open-loop gain, taking the loading of the β network into account. Note that R_e appears in both input and output loops and that the transistor is replaced by its hybrid-II model. From this figure we find for the transfer function $G_M = I_o/V_s$ without feedback

$$G_M = \frac{-G'_s(g_m - sC_c)}{s^2C_eC_cR_L + s[C_e + C_c + C_cR_L(g_m + g_{b'e} + G'_s)] + G'_s + g_{b'e}}$$

$$(14\text{-}40)$$

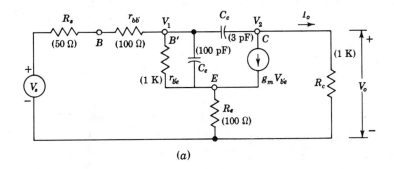

(a)

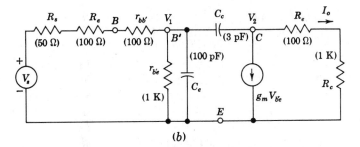

(b)

Fig. 14-11 (a) Current-series feedback amplifier at high frequencies. (b) The basic amplifier without feedback but taking the loading of the β network (R_e) into account.

where we make use of Eq. (11-37) and introduce

$$R_L \equiv R_e + R_c = 1.1 \text{ K}$$

and

$$G_s' \equiv \frac{1}{R_s + r_{bb'} + R_e} = \frac{1}{250} = 4 \times 10^{-3} \text{ A/V}$$

With $g_m = 50 \times 10^{-3}$ A/V, we obtain

$$G_M = -3.67 \times 10^4 \frac{167 \times 10^8 - s}{(s + 0.182 \times 10^8)(s + 8.45 \times 10^8)} \tag{14-41}$$

Dominant Pole   We now check to see if $s = -0.182 \times 10^8$ rad/s results in a dominant pole after the loop is closed.   The transconductance G_{Mo} at low frequencies is found by substituting $s = 0$ in Eq. (14-41) or by using the value given in Eq. (13-53).   In either case we find

$$G_{Mo} = \frac{-G_s' g_m}{G_s' + g_{b'e}} = -\frac{4 \times 10^{-3} \times 50 \times 10^{-3}}{4 \times 10^{-3} + 10^{-3}} = -4 \times 10^{-2} \text{ A/V} \tag{14-42}$$

Since from Sec. 13-10, $\beta = -R_e = -100$ Ω, then $1 + \beta G_{Mo} = 1 + 4 = 5$. With

$$n = \frac{\omega_2}{\omega_1} = \frac{8.45}{0.182} = 46.8$$

then from Eq. (14-31)

$$Q^2 = \frac{n}{(n+1)^2}(1 + \beta G_{Mo}) = \frac{46.8 \times 5}{(47.8)^2} = 0.102$$

Since $Q^2 < Q_{\max}^2 = 0.16$, a closed-loop dominant pole does exist.   The open-loop dominant-pole high 3-dB frequency is

$$f_H = \frac{0.182 \times 10^8}{2\pi} \text{ Hz} = 2.90 \text{ MHz}$$

and the closed-loop dominant frequency is given by Eq. (14-32):

$$f_{Hf} = \frac{f_H}{2}(n+1)(1 - \sqrt{1 - 4Q^2})$$

$$= \frac{2.90 \times 47.8}{2}(1 - \sqrt{1 - 0.408}) = 16.0 \text{ MHz}$$

Since

$$G_{Mof} = \frac{G_{Mo}}{1 + \beta G_{Mo}} = \frac{-4 \times 10^{-2}}{5} = -8 \times 10^{-3} \text{ A/V} \tag{14-43}$$

then, with f in megahertz,

$$G_{Mf} = \frac{-8 \times 10^{-3}}{1 + j(f/16.0)} \tag{14-44}$$

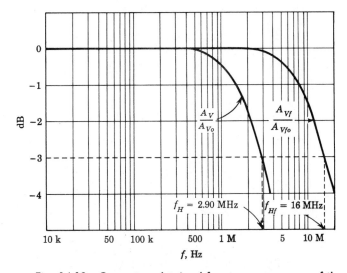

Fig. 14-12 Computer-obtained frequency response of the amplifier shown in Fig. 14-11a.

The voltage gain $A_{Vf} = R_c G_{Mf} = 1,000 G_{Mf}$ has the same value of upper 3-dB frequency as the transconductance gain. The upper 3-dB frequency is $f_{Hf} = 16.0$ MHz. An exact computer analysis[3] of the circuit shown in Fig. 14-11b yields the same value of the zero and the two poles as in Eq. (14-41) for G_M. For G_{Mf} the computer analysis yields two poles (in radians per second),

$$s_{1f} = -1.025 \times 10^8 \qquad s_{2f} = -7.65 \times 10^8$$

and two zeros,

$$s_{3f} = 10.2 \times 10^8 \qquad s_{4f} = -16.3 \times 10^8$$

The frequency response for A_{Vf} is plotted in Fig. 14-12, from which we read $f_{Hf} = 16.0$ MHz which is identical with that obtained using the dominant-pole approximation.

14-7   CURRENT–SHUNT FEEDBACK PAIR—FREQUENCY RESPONSE

In Sec. 13-11 the low-frequency response of a current-shunt feedback pair shown in Fig. 13-18 is analyzed. This amplifier stabilizes the current gain $A_{If} = I_o/I_s$, and since from Eq. (13-70) we have $A_{Vf} = A_{If} R_{c2}/R_s$, then the voltage gain is also stabilized if R_{c2} and R_s are stable resistors. In other words, A_{If} and A_{Vf} have the same dependence on frequency. Figure 13-18 is redrawn in Fig. 14-13a for high-frequency calculations using the hybrid-II equivalent circuit, with $g_m = 50$ mA/V, $C_e = 100$ pF, $C_c = 3$ pF, $r_{bb'} = 100\ \Omega$, and $r_{b'e} = 1$ K.

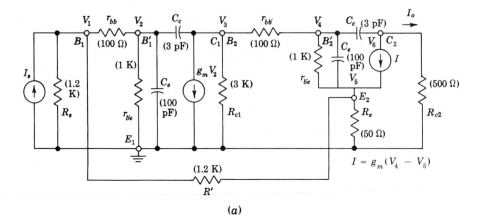

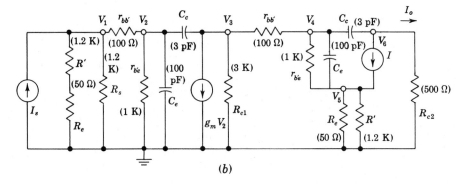

Fig. 14-13 (a) Current-shunt feedback pair at high frequencies. (b) The A network taking the loading of R' into account.

We desire to obtain the frequency response of A_{If} of the feedback amplifier shown in Fig. 14-13a, with $R_s = R' = 1.2$ K, $R_{c2} = 500$ Ω, $R_{c1} = 3$ K, and $R_e = 50$ Ω. These are the same parameters used in the example in Sec. 13-11.

The gain A_I without feedback, but with the loading effects of the feedback network included, can be obtained from the A network of Fig. 14-13b. Since this network has four independent capacitors, we expect the transfer function for current gain to have four poles. In addition, the presence of the C_e capacitor from node V_2 to ground will cause the gain to go to zero as $1/s$ as $s \to \infty$; thus we must also have three zeros in the transfer function. The complexity of the network shown in Fig. 14-13b is such that it becomes necessary to use the computer for the calculation of the poles and zeros. Using the CORNAP computer program,[3] we find

$$A_I = \frac{I_o}{I_s} = K' \frac{(s - s_5)(s - s_6)(s - s_7)}{(s - s_1)(s - s_2)(s - s_3)(s - s_4)} \tag{14-45}$$

where the poles in radians per second are

$$s_1 = -46.2 \times 10^5 \qquad s_2 = -45.9 \times 10^6$$

$$s_3 = -11.4 \times 10^8 \qquad s_4 = -30.4 \times 10^8$$

and the zeros are

$$s_5 = 16.65 \times 10^9 \qquad s_6 = 15.4 \times 10^8 \qquad s_7 = -22.55 \times 10^8$$

The exact transfer function of the amplifier with feedback shown in Fig. 14-13a is of the form

$$A_{If} = \frac{I_o}{I_s} = K \frac{(s - s_{5f})(s - s_{6f})(s - s_{7f})(s - s_{8f})}{(s - s_{1f})(s - s_{2f})(s - s_{3f})(s - s_{4f})} \tag{14-46}$$

Using the CORNAP program,[3] we find the poles (in radians per second)

$$s_{1f} = -29.2 \times 10^6 + j5.40 \times 10^7$$

$$s_{2f} = -29.2 \times 10^6 - j5.40 \times 10^7$$

$$s_{3f} = -11.5 \times 10^8$$

$$s_{4f} = -30.2 \times 10^8$$

and the zeros

$$s_{5f} = 18.35 \times 10^8 - j9.75 \times 10^8$$

$$s_{6f} = 18.35 \times 10^8 + j9.75 \times 10^8$$

$$s_{7f} = -21.5 \times 10^8$$

$$s_{8f} = -7.40 \times 10^9$$

Note that the two highest poles are modified hardly at all by feedback, as predicted in Sec. 14-4. Equations (14-45) and (14-46) are plotted in Fig. 14-14. We observe that the peak in the frequency response comes at 7.0 MHz and the overshoot is 1.5 dB.

EXAMPLE (a) Calculate the two lowest poles of the amplifier with feedback shown in Fig. 14-13a, using the fact that poles s_1 and s_2 in Eq. (14-45) are much lower than poles s_3 and s_4 ($|s_3/s_2| > 4$). (b) Compute the frequency at which the frequency response peaks, and find the overshoot in decibels.

Solution a. We approximate the transfer function of Eq. (14-45), using the lowest two poles, and refer to our discussion of the two-pole transfer function in Sec. 14-2. Thus

$$A_I = \frac{I_o}{I_s} \approx \frac{K''}{(s + 46.2 \times 10^5)(s + 45.9 \times 10^6)}$$

where $K'' = -K' s_5 s_6 s_7 / s_3 s_4$.

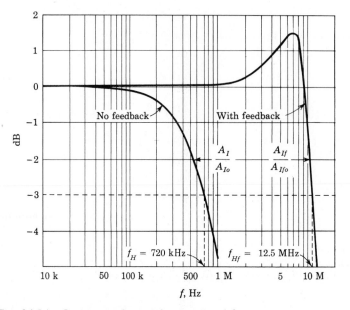

Fig. 14-14 Computer-obtained normalized frequency response of the amplifier shown in Fig. 14-13a and b. A_{Io} = **52.2 dB** and A_{Ifo} = **27.2 dB**.

The same example is considered at low frequencies in Sec. 13-11, where we find $1 + \beta A_{Io} = 17.4$. Since $n = 459/46.2 = 9.92$, we find from Eq. (14-31)

$$Q^2 = \frac{n}{(n + 1)^2}(1 + \beta A_{Io}) = \frac{9.92 \times 17.4}{(10.92)^2} = 1.44 \qquad \text{or} \qquad Q = 1.20$$

and from Eq. (14-31) $k = 1/2Q = 1/2.40 = 0.417$ and $k^2 = 0.174$.

The poles of the amplifier with feedback can be found using Eq. (14-13) or Eq. (14-32). Hence s_{1f} and s_{2f} are given by

$$-(46.2 \times 10^5)\left(\frac{10.92}{2}\right)(1 \mp j\sqrt{5.76 - 1}) = (-25.3 \pm j55.2) \times 10^6 \text{ rad/s}$$

as compared with the values $(-29.2 \pm j54.0) \times 10^6$ rad/s obtained using the computer and the exact analysis.

b. The frequency response peak occurs at the frequency $\omega = \omega_o \sqrt{1 - 2k^2}$, given by Eq. (14-19), where

$$\omega_o = Q(\omega_1 + \omega_2) = 1.20(45.9 + 4.62) \times 10^6 = 60.05 \times 10^6$$

as given by Eq. (14-11). Thus the response peaks at

$$f_{\text{peak}} = \frac{60.05}{6.28}\sqrt{1 - 2 \times 0.174} = 7.7 \text{ MHz}$$

The magnitude of the peak is given by Eq. (14-20), or at the peak we have

$$\left| \frac{A_f}{A_{of}} \right| = \frac{1}{2k \sqrt{1 - k^2}} = \frac{1}{2 \times 0.417 \sqrt{1 - 0.174}} = 1.32$$

or 20 log 1.32 = 2.4 dB.

The exact analysis required the use of a computer and gave $f_{\text{peak}} = 7.0$ MHz and overshoot of 1.5 dB.

14-8   VOLTAGE–SERIES FEEDBACK PAIR—FREQUENCY RESPONSE

In this section we examine the frequency response of the voltage-series feedback pair of Fig. 13-15. This amplifier is analyzed for low frequencies in Sec. 13-9. In Fig. 14-15a we show the amplifier prepared for high-frequency analysis, where the hybrid-II equivalent circuit is used to represent transistors $Q1$ and $Q2$. The same values of transistor parameters are assumed as in the preceding section. The 6.6-K load of Q_1 represents the parallel combination of $R_{c1} = 10$ K and the biasing resistors of 47 K and 33 K for $Q2$. In the analysis, we need, as before, the open-loop gain, with the loading effects of the feedback network taken into account. Figure 14-15b shows this A network. From Sec. 13-5 we know that this type of feedback stabilizes the voltage gain $A_V = V_o/V_s$. Thus we expect to increase the bandwidth of the voltage-gain transfer function. The A network of Fig. 14-15b is similar to the network of Fig. 14-13b, except for the interchange of the unbypassed emitter resistance from the second stage to the first stage. Hence the expression for the voltage gain A_V is similar to A_I in Eq. (14-45); that is, it has three zeros and four poles.

Using the CORNAP program,[3] we find from Fig. 14-15b that the poles (in radians per second) are

$$s_1 = -24.4 \times 10^5 \qquad s_2 = -26.8 \times 10^7$$
$$s_3 = -6.45 \times 10^8 \qquad s_4 = -26.3 \times 10^8$$

and the zeros are

$$s_5 = -16.4 \times 10^8 \qquad s_6 = 10.3 \times 10^8 \qquad s_7 = 16.6 \times 10^9$$

Approximate Solution   Since s_1 is much smaller in magnitude than all other poles or zeros, we test to see if $f_1 = -s_1/2\pi = 390$ kHz results in a dominant pole after feedback. We find (Prob. 14-23) that $4Q^2 = 0.67$, which slightly exceeds $4Q^2_{\text{max}} = 0.64$. This means that the second pole will not be quite two octaves away. If we ignore the presence of this second pole, the approximate dominant-pole solution of the closed-loop amplifier is

$$A_{Vf} = \frac{45.4}{1 + j(f/9.05)} \tag{14-47}$$

where f is in megahertz.

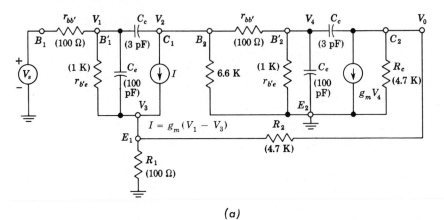

(a)

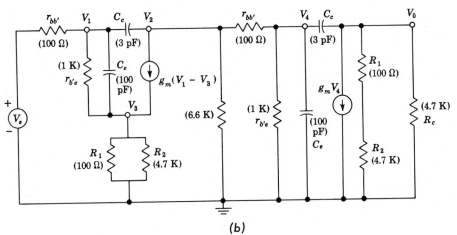

(b)

Fig. 14-15 (a) Voltage-series feedback pair at high-frequencies. (b) The basic amplifier taking the loading of the β network into account.

Exact Solution The exact transfer function A_{Vf} obtained from Fig. 14-15a, using the CORNAP program,[3] has three zeros and four poles. The poles (in radians per second) are

$$s_{1f} = -6.55 \times 10^7 \qquad s_{2f} = -17.4 \times 10^7$$
$$s_{3f} = -6.90 \times 10^8 \qquad s_{4f} = -26.2 \times 10^8$$

and the zeros are

$$s_{5f} = -17.0 \times 10^8 \qquad s_{6f} = 11.9 \times 10^8 \qquad s_{7f} = 8.10 \times 10^9$$

Note that the two highest poles are modified only slightly by feedback. The open-loop and closed-loop frequency responses obtained with the aid of the computer are plotted in Fig. 14-16, from which we find the upper 3-dB frequency without feedback to be 400 kHz, and the corresponding frequency

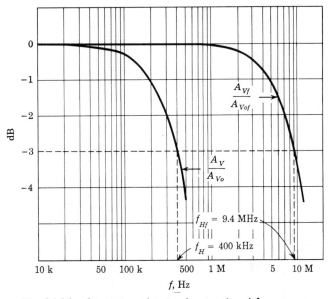

Fig. 14-16 Computer-obtained normalized frequency response of the amplifiers shown in Fig. 14-15a and b. $A_{Vo} = 58.4$ dB and $A_{Vof} = 33.1$ dB.

of the amplifier with feedback is 9.4 MHz, which should be compared with the approximate value of 390 kHz and 9.05 MHz.

14-9   STABILITY[4]

Negative feedback for which $|1 + A\beta| > 1$ has been considered in some detail in the foregoing sections. If $|1 + A\beta| < 1$, the feedback is termed *positive*, or *regenerative*. Under these circumstances, the resultant transfer gain A_f will be greater than A, the nominal gain without feedback, since $|A_f| = |A|/|1 + A\beta| > |A|$. Regeneration as an effective means of increasing the amplification of an amplifier was first suggested by Armstrong.[5] Because of the reduced stability of an amplifier with positive feedback, this method is seldom used.

To illustrate the instability in an amplifier with positive feedback, consider the following situation: No signal is applied, but because of some transient disturbance, a signal X_o appears at the output terminals. A portion of this signal, $-\beta X_o$, will be fed back to the input circuit, and will appear in the output as an increased signal $-A\beta X_o$. If this term just equals X_o, then the spurious output has regenerated itself. In other words, if $-A\beta X_o = X_o$ (that is, if $-A\beta = 1$), the amplifier will oscillate (Sec. 14-15). Hence, if an attempt is made to obtain large gain by making $|A\beta|$ almost equal to unity, there is the possibility that the amplifier may break out into spontaneous oscillation. This would occur if, because of variation in supply voltages, aging of transistors, etc., $-A\beta$ becomes equal to unity. There is little point in attempting to achieve amplification at the expense of stability. In fact, because of all the advantages enumerated in Sec. 13-2, feedback in amplifiers is almost always negative. However, combinations of positive and negative feedback are used.

The Condition for Stability If an amplifier is designed to have negative feedback in a particular frequency range but breaks out into oscillation at some high or low frequency, it is useless as an amplifier. Hence, in the design of a feedback amplifier, it must be ascertained that the circuit is stable at *all* frequencies, and not merely over the frequency range of interest. In the sense used here, the system is stable if a transient disturbance results in a response which dies out. A system is unstable if a transient disturbance persists indefinitely or increases until it is limited only by some nonlinearity in the circuit. In Sec. 14-3 it is shown that the question of stability involves a study of the poles of the transfer function since these determine the transient behavior of the network. If a pole exists with a positive real part, this will result in a disturbance increasing exponentially with time. Hence the condition which must be satisfied, if a system is to be stable, is that the poles of the transfer function must all lie in the left-hand half of the complex-frequency plane. If the system without feedback is stable, the poles of A do lie in the left-hand half plane. It follows from Eq. (14-1), therefore, that *the stability condition requires that the zeros of $1 + A\beta$ all lie in the left-hand half of the complex-frequency plane.*

The Nyquist Criterion Nyquist[4,6] has obtained an alternative but equivalent condition for stability which may be expressed in terms of the steady-state, or frequency-response, characteristic. It is given here without proof: Since the product $A\beta$ is a complex number, it may be represented as a point in the complex plane, the real component being plotted along the X axis, and the j component along the Y axis. Furthermore, $A\beta$ is a function of frequency. Consequently, points in the complex plane are obtained for the values of $A\beta$ corresponding to all values of f from $-\infty$ to $+\infty$. The locus of all these points forms a closed curve. The criterion of Nyquist is that *the amplifier is unstable if this curve encloses the point $-1 + j0$, and the amplifier is stable if the curve does not enclose this point.*

The criterion for positive or negative feedback may also be represented in the complex plane. From Fig. 14-17 we see that $|1 + A\beta| = 1$ represents a circle of unit radius, with its center at the point $-1 + j0$. If, for any frequency, $A\beta$ extends outside this circle, the feedback is negative, since then $|1 + A\beta| > 1$. If, however, $A\beta$ lies within this circle, then $|1 + A\beta| < 1$, and the feedback is positive. In the latter case the system will not oscillate unless Nyquist's criterion is satisfied.

Fig. 14-17   The locus of $|1 + A\beta| = 1$ is a
circle of unit radius, with its center at $-1 + j0$.
If the vector $A\beta$ ends in the shaded region,
the feedback is positive.

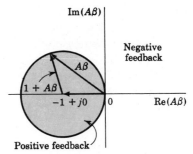

Illustration   Consider an amplifier with an open-loop dominant pole so
that the transfer gain is represented by Eq. (14-2).   Hence

$$\beta A = \frac{\beta A_o}{1 + j(f/f_H)} \tag{14-48}$$

$$\beta A + j\beta A \frac{f}{f_H} = \beta A_o \tag{14-49}$$

Since βA is a vector in the complex βA plane, then $j\beta A (f/f_H)$ is at right
angles to βA, and from Eq. (14-49) the sum of these two vectors is βA_o, which
is a constant for β independent of frequency.   Hence we see from Fig. 14-18
that βA_o is the diameter of a circle, and the bottom half of this circle is the
locus of βA as a function of positive values of frequency.   In other words, the
bottom semicircle (in the right half plane) is the polar plot of βA for positive

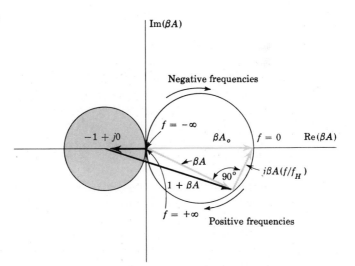

Fig. 14-18   For a dominant-pole amplifier, the locus (for all
values of frequency) of βA in the complex βA plane is a
circle in the right half plane.

frequencies. In a similar manner it can be shown that the upper semicircle is the locus of βA for negative frequencies.

Since the locus βA does not enclose the point $-1 + j0$, the amplifier is stable and the feedback is negative for all frequencies. Alternatively, it is noted from the diagram that $|1 + \beta A| > 1$ for all frequencies, which is the condition for negative feedback.

14-10   GAIN AND PHASE MARGINS

In the preceding section, we examine two criteria for determinine whether a feedback amplifier is stable or unstable. Often it is difficult to apply either of the above conditions for stability to a practical amplifier. It should be clear from the foregoing discussion that *no oscillations are possible if the magnitude of the loop gain $|A\beta|$ is less than unity when the phase angle of $A\beta$ is 180°.* This condition is sought for in practice to ensure that the amplifier will be stable.

Consider, for example, a three-pole transfer function given by Eq. (14-28). To be specific, we can assume that this gain represents three cascaded stages, each with a dominant pole due to shunt capacitance. To simplify the discussion, consider that the amplifier stages are noninteracting, and that the poles are equal; $\omega_1 = \omega_2 = \omega_3$. (The general case of nonidentical poles is considered in the next section.) There is a definite maximum value of the feedback fraction $|\beta|$ allowable for stable operation. To see this, note that at low frequencies there is a 180° phase shift in each stage and 540°, or equivalently 180°, for the three stages. In other words, the midband gain A_o in Eq. (14-28) is negative. Since we are considering negative feedback, then $1 + \beta A_o > 1$ and β must be negative (it is assumed to be real). At high frequencies there is a phase shift due to the shunting capacitances, and at the frequency for which the phase shift per stage is 60°, the total phase shift of A is zero, and of $A\beta$ is 180°. If the magnitude of the gain at this frequency is called A_{60}, then β must be chosen so that $A_{60}|\beta|$ is less than unity, if the possibility of oscillations is to be avoided. Similarly, because of the phase shift introduced by the blocking capacitors between stages, there is a low frequency for which the phase shift per stage is also 60°, and hence there is the possibility of oscillation at this low frequency also, unless the maximum value of β is restricted as outlined above.

It should now be apparent that instead of plotting the product $A\beta$ in the complex plane, it is more convenient to plot the magnitude, usually in decibels, and also the phase of $A\beta$ as a function of frequency. If we can show that $|A\beta|$ is less than unity when the phase angle of $A\beta$ is 180°, the closed-loop amplifier will be stable.

Gain Margin   The gain margin is defined as the value of $|A\beta|$ in decibels at the frequency at which the phase angle of $A\beta$ is 180°. If the gain margin is negative, this gives the decibel rise in open-loop gain, which is theoretically

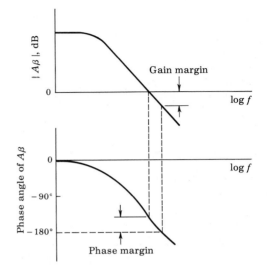

Fig. 14-19 Bode plots relating to the definitions of gain and phase margins.

permissible without oscillation. If the gain margin is positive, the amplifier is potentially unstable. This definition is illustrated in Fig. 14-19.

Phase Margin The phase margin is 180° minus the magnitude of the angle of $A\beta$ at the frequency at which $|A\beta|$ is unity (zero decibels). The magnitudes of these quantities give an indication of how stable an amplifier is. For example, a linear amplifier of good stability requires gain and phase margins of at least 10 dB and 50°, respectively. This definition is illustrated in Fig. 14-19.

14-11   COMPENSATION[7]

In the preceding section we establish that if $\beta A = 1\underline{/180°}$, the feedback amplifier will be unstable. For this condition to occur with resistive feedback, it is necessary that the A network, or forward transfer function, must have more than two poles. For a multistage amplifier, if the open-loop gain $|\beta A|$ is unity when the phase shift is 180°, the closed-loop amplifier will oscillate. Compensating techniques reduce the amplifier gain A at those frequencies for which phase shift is high.

Three-pole Amplifier To illustrate the stability problem in greater detail than in the preceding section, consider an amplifier with voltage gain

$$A_V = \frac{-10^3}{\left(1 + j\frac{f}{1}\right)\left(1 + j\frac{f}{10}\right)\left(1 + j\frac{f}{50}\right)} \qquad (14\text{-}50)$$

where f is in megahertz.   The three poles are at $f_1 = 1$, $f_2 = 10$, and $f_3 = 50$ MHz.   If we sample the output voltage V_o and return a fraction βV_o in series opposition to the input voltage, the feedback-amplifier voltage gain A_{Vf} will be given by Eq. (14-29).   For negative feedback, β must be a real negative number, and we assume it is independent of frequency.   *This amplifier will oscillate when* $-A_V|\beta| = 1\underline{/180°}$, *or when the magnitude of the gain* $|A_V|$ *is equal to* $|1/\beta|$ *and the phase of* $-A_V$ *is* 180°.

In Fig. 14-20 we show the ideal Bode plot of $|A_V|$ and the phase of $-A_V$ versus frequency.   The breaks in the magnitude plot occur at f_1, f_2, and f_3, and the slope of the lines increases by 20 dB per decade (6 dB per octave) after each break, as discussed in Sec. 12-4.   The phase curve is more complicated, and hence the contribution to the phase θ by each pole is indicated separately by the dashed lines.   For example, the curve marked θ_2 corresponds to the pole at f_2.   Consistent with the discussion in Sec. 12-4, $\theta_2 = 0$ for $f \leq 0.1f_2$ and $\theta_2 = -90°$ for $f \geq 10f_2$, and θ_2 decreases by 45° per decade for $0.1f_2 \leq f \leq 10f_2$. At $f = f_2 = 10$ MHz, $\theta_2 = -45°$.   The resultant phase at any frequency is the sum of the three dashed curves at that frequency, and is indicated by the shaded line plot.

We see from Fig. 14-20 that the phase of $-A_V$ is 180° at $f = 22$ MHz. Hence, if $|1/\beta| = |A_V| = 26$ dB, the amplifier will oscillate at $f = 22$ MHz. If we desire to maintain a phase margin of 45°, which occurs at $f = 8$ MHz, we must have

$$\left|\frac{1}{\beta}\right| = |A_V|(f = 8 \text{ MHz}) = 42 \text{ dB}$$

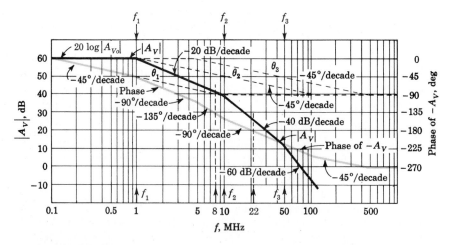

Fig. 14-20 The open-loop voltage gain $|A_V|$ (solid lines) and phase of $-A_V$ (shaded lines) of a three-pole amplifier.   The dashed lines are the phase contributions from each pole.

Since

$$20 \log |\beta A_{Vo}| = 20 \log |A_{Vo}| - 20 \log \left|\frac{1}{\beta}\right| \qquad (14\text{-}51)$$

then the maximum midband loop gain for a phase margin of 45° is

$$|\beta A_{Vo}|_{max} = 60 - 42 = 18 \text{ dB}$$

In some applications the desensitivity afforded by this loop gain may not be sufficient, and then compensation is used to increase the maximum loop gain while maintaining the same phase margin.

General Methods of Compensation   The essential idea of compensation is to reshape the magnitude and phase plots of βA so that $|\beta A| < 1$ when the angle of βA is 180°.   There are three general methods of accomplishing this goal:

   1. *Dominant-pole, or lag, compensation.*   This method inserts an extra pole into the transfer function at a lower frequency than the existing poles.   Such a circuit introduces a phase lag into the amplifier.
   2. *Lead compensation.*   The amplifier or the feedback network is modified so as to add a zero to the transfer function, thereby increasing the phase.
   3. *Pole-zero, or lag-lead, compensation.*   This technique adds both a pole (a lag) and a zero (a lead) to the transfer gain.   The zero is chosen so as to cancel the lowest pole.

These methods are discussed in detail in the following sections.

14-12   DOMINANT–POLE COMPENSATION

The amplifier is modified by adding a dominant pole, that is, a pole much smaller in magnitude than all other poles in the forward transfer function. Consequently, the loop gain drops to 0 dB with a slope of 6 dB per octave at a frequency where the poles of A_V contribute negligible phase shift.

   Suppose we modify A_V of Eq. (14-50) by adding a dominant pole so that the new forward transfer function becomes

$$A_V' = \frac{1}{1 + j(f/f_d)} A_V \qquad (14\text{-}52)$$

where $f_d \ll 1$ MHz.   This can be accomplished by a simple RC network placed in the forward amplifier, as shown in Fig. 14-21, or by connecting a capacitance C from a suitable high-resistance point to ground.   Let us assume $f_d = 1$ kHz, and the Bode plots are then given in Fig. 14-22.   We now see that a phase margin of 45° occurs at $f = 1$ MHz and that $|A_V| = |1/\beta|$ at this frequency is 0 dB.   Hence, from Eq. (14-51), the maximum loop gain $|\beta A_{Vo}|_{max}$ for a phase margin of 45° is 60 dB.

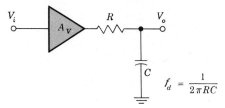

Fig. 14-21 Dominant-pole compensation.

$$f_d = \frac{1}{2\pi RC}$$

It is interesting to compare the bandwidth of the feedback amplifier with and without compensation. For an amount of feedback of $|\beta A_{Vo}| = 10$ dB, Fig. 14-20 yields $f_{Hf} = 3$ MHz, and from Fig. 14-22, we find for the compensated amplifier $f_{Hf} = 3$ kHz. This clearly demonstrates that the dominant-pole compensation technique wastes the available bandwidth, and other methods of compensation must be used if we desire maximum bandwidth.

14-13 POLE–ZERO (LAG–LEAD) COMPENSATION

In this type of compensation the forward transfer function A is altered by adding both a pole and a zero, with the zero at a higher frequency than the pole. Figure 14-23 shows the forward amplifier with the pole-zero network cascaded with it. The transfer function of the phase network is found to be

$$\frac{V_3}{V_2} = \frac{1 + j(f/f_z)}{1 + j(f/f_p)} \qquad (14\text{-}53)$$

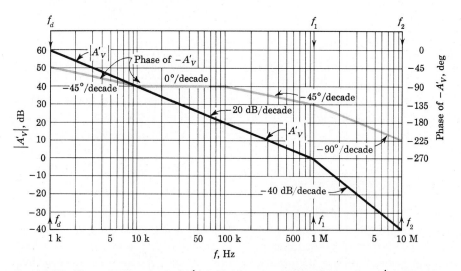

Fig. 14-22 The open-loop gain $|A'_V|$ (solid lines) and the phase of $-A'_V$ (shaded lines), where A'_V represents A_V of Fig. 14-20 augmented by a dominant pole at 1 kHz.

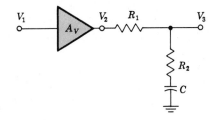

Fig. 14-23   Pole-zero compensation.

where

$$f_z = \frac{1}{2\pi R_2 C} \qquad f_p = \frac{1}{2\pi (R_1 + R_2)C} \tag{14-54}$$

If we assume that the compensation network does not load the amplifier, the modified forward transfer function becomes

$$A'_V = \frac{V_3}{V_1} = A_V \frac{1 + j(f/f_z)}{1 + j(f/f_p)} \tag{14-55}$$

Let us assume that A_V is given by Eq. (14-50); let us further select f_z to be equal to the smallest pole of A_V so as to effectively cancel it.

The forward transfer function now becomes (with f expressed in megahertz)

$$A'_V = \frac{-10^3}{\left(1 + j\dfrac{f}{f_p}\right)\left(1 + j\dfrac{f}{10}\right)\left(1 + j\dfrac{f}{50}\right)} \tag{14-56}$$

If we set the pole at $f_p = 0.2$ MHz, the Bode plots of the magnitude of A'_V and phase of $-A'_V$ are drawn in Fig. 14-24.   We see that the maximum

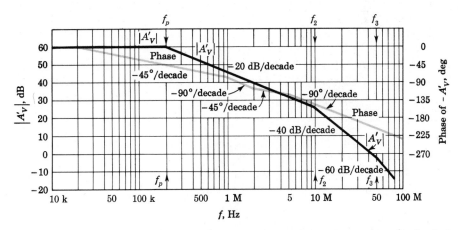

Fig. 14-24   The open-loop gain $|A'_V|$ (solid lines) and the phase of $-A'_V$ (shaded lines), where A'_V represents A_V of Fig. 14-20 augmented by a pole at 200 kHz and a zero at 1 MHz.

loop gain $|\beta A_{Vo}|_{\max}$ for a phase margin of 45° is $60 - 29 = 31$ dB. However, the bandwidth for a loop gain of 10 dB is seen to be 640 kHz, as compared with 3 kHz for the dominant-pole compensation.

A comparison of the dominant-pole and pole-zero compensation techniques is given in Fig. 14-25. The dominant pole is selected so that the compensated forward transfer function goes through 0 dB at the first pole f_1 of the uncompensated response. For the lead-lag compensation the zero is chosen equal to f_1, while the pole is selected so that the compensated forward transfer function goes through 0 dB at the second pole of the uncompensated transfer function. The bandwidth improvement is also shown in this figure.

14-14   COMPENSATION BY MODIFICATION OF THE β NETWORK[8]

Instead of shaping the response of the forward transfer function A, it is possible to alter the loop gain βA by adding reactive elements to the feedback network β and thus compensate the feedback amplifier. In Sec. 12-11 we examine the three-stage amplifier shown in Fig. 12-20a. This amplifier is redrawn in Fig. 14-26 as a voltage-shunt feedback amplifier.

We assume that the feedback network does not load the forward amplifier, so that we can use the open-circuit forward transfer function

$$R_M = V_o/I_s = (V_o/V_s)\, R_s$$

derived in Sec. 12-11.

The exact transfer function of Eq. (12-49) obtained with the aid of the computer yields six poles and three zeros. The loading of the reactive β network will introduce another pole much higher in frequency into the transfer function. For this reason we can use the three lowest poles obtained in Sec.

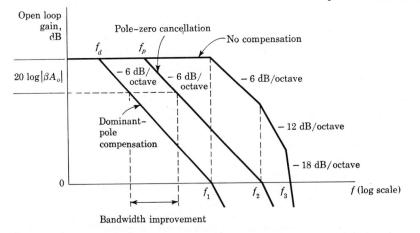

Fig. 14-25 Comparison of dominant-pole and pole-zero compensation techniques.

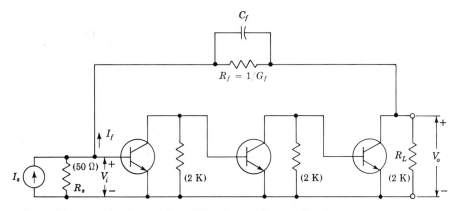

Fig. 14-26 Compensation by modification of the feedback network.

12-11 to approximate the transfer function R_M. Thus we have (with f in megahertz)

$$R_M = \frac{V_o}{I_s} = -\frac{45.25 \times 10^6}{\left(1 + j\,\dfrac{f}{0.438}\right)\left(1 + j\,\dfrac{f}{2.12}\right)\left(1 + j\,\dfrac{f}{12.3}\right)} \tag{14-57}$$

To find β, assume a high-gain amplifier so that $V_o \gg V_i$ and

$$I_f = (G_f + sC_f)(V_i - V_o) \approx -G_f\left(1 + s\,\frac{C_f}{G_f}\right)V_o = \beta V_o$$

Thus the negative of the loop gain becomes

$$\beta R_M = \frac{45.25 \times 10^6 G_f\left(1 + j\,\dfrac{f}{f_z}\right)}{\left(1 + j\,\dfrac{f}{0.438}\right)\left(1 + j\,\dfrac{f}{2.12}\right)\left(1 + j\,\dfrac{f}{12.3}\right)} \tag{14-58}$$

where $f_z = 1/2\pi C_f R_f$. The presence of C_f introduces a zero in the loop-gain expression, which can be used for *lead compensation* purposes. If $C_f = 0$, then $\beta = -G_f$, a real negative number, and the amplifier will oscillate when $-|\beta|R_M = 1\underline{/180°}$, or when $|R_M| = |1/\beta|$ and the phase of $-R_M$ is 180°.

The designer has the choice of where to place the zero of Eq. (14-58). It is found that if maximum bandwidth with no peaking in the frequency response is desired, there is an optimum location for the zero between the second-in-magnitude and third-in-magnitude poles, as we demonstrate in the following example.

EXAMPLE The amplifier of Fig. 14-26 is modified so that $R_L = 50\ \Omega$ and $C_f = 0$. Find (a) the poles and plot $A_V = V_o/V_s$, with $R_f = \infty$; (b) repeat

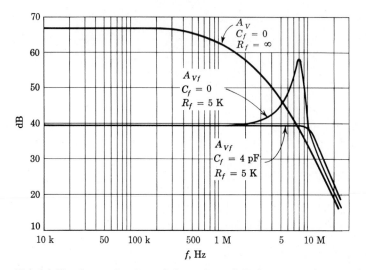

Fig. 14-27 Determination of the value of C_f for no peaking in the response.

part a, with $R_f = 5$ K; (c) find C_f so that the frequency response is flat with no peak (with $R_f = 5$ K).

Solution a. Clearly, the three-stage amplifier is too complicated for hand calculations. Using the CORNAP program and setting $R_f = \infty$, we find the poles (in radians per second):

$$s_{p1} = -5.05 \times 10^6 \qquad s_{p2} = -2.60 \times 10^7$$

$$s_{p3} = -7.95 \times 10^7 \qquad s_{p4} = -3.96 \times 10^9$$

$$s_{p5} = -4.41 \times 10^9 \qquad s_{p6} = -7.40 \times 10^9$$

The zeros are

$$s_{z1} = 16.65 \times 10^9$$

$$s_{z2} = 16.65 \times 10^9 + j11.15 \times 10^4$$

$$s_{z3} = 16.65 \times 10^9 - j11.15 \times 10^6$$

The low-frequency gain $A_V = V_o/V_s$ is found to be $-2{,}262$, or 67.09 dB. The transfer function $A_V(f)$ is plotted in Fig. 14-27.

b. Setting $R_f = 5$ K and using CORNAP, we find the poles with feedback (in radians per second):

$$s_{p1} = -8.85 \times 10^5 - j4.73 \times 10^7 \qquad s_{p2} = -8.85 \times 10^5 + j4.73 \times 10^7$$

$$s_{p3} = -10.9 \times 10^7 \qquad\qquad\qquad s_{p4} = -3.96 \times 10^9$$

$$s_{p5} = -4.41 \times 10^9 \qquad\qquad\qquad s_{p6} = -7.45 \times 10^9$$

and the low-frequency gain with feedback $A_{Vf} = -95.68$, or 39.61 dB. The complete response is shown in Fig. 14-27.

c. From the plot of part *b* we see that the voltage gain peaks at $f_o = 8$ MHz. If we place the zero of the β network at f_o, we find

$$C_f = \frac{1}{2\pi R_f f_o} \approx 4 \text{ pF}$$

The frequency response of A_{Vf} with $R_f = 5$ K and $C_f = 4$ pF is plotted in Fig. 14-27, from which we see that there is no peaking.

14-15   SINUSOIDAL OSCILLATORS

Many different circuit configurations deliver an essentially sinusoidal output waveform even without input-signal excitation. The basic principles governing all these oscillators are investigated. In addition to determining the conditions required for oscillation to take place, the frequency and amplitude stability are also studied.

Figure 14-28 shows an amplifier, a feedback network, and an input mixing circuit not yet connected to form a closed loop. The amplifier provides an output signal x_o as a consequence of the signal x_i applied directly to the amplifier input terminal. The output of the feedback network is $x_f = \beta x_o = A \beta x_i$, and the output of the mixing circuit (which is now simply an inverter) is

$$x_f' = -x_f = -A\beta x_i$$

From Fig. 14-28 the loop gain is

$$\text{Loop gain} = \frac{x_f'}{x_i} = \frac{-x_f}{x_i} = -\beta A \tag{14-59}$$

Suppose it should happen that matters are adjusted in such a way that the signal x_f' is *identically* equal to the externally applied input signal x_i. Since the amplifier has no means of distinguishing the source of the input signal applied to it, it would appear that, if the external source were removed and if terminal 2 were connected to terminal 1, the amplifier would continue to

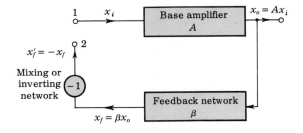

Fig. 14-28 An amplifier with transfer gain A and feedback network β not yet connected to form a closed loop. (Compare with Fig. 13-8.)

provide the same output signal x_o as before. Note, of course, that the statement $x_f' = x_i$ means that the instantaneous values of x_f' and x_i are exactly equal at all times. Note also that, since in the above discussion no restriction was made on the waveform, it need not be sinusoidal. The amplifier need not be linear, and the waveshape need not preserve its form as it is transmitted through the amplifier, provided only that the signal x_f' has the waveform and frequency of the input signal x_i. The condition $x_f' = x_i$ is equivalent to $-A\beta = 1$, or *the loop gain must equal unity.*

The Barkhausen Criterion We assume in this discussion of oscillators that the entire circuit operates linearly and that the amplifier or feedback network or both contain reactive elements. Under such circumstances, the only periodic waveform which will preserve its form is the sinusoid. For a sinusoidal waveform the condition $x_i = x_f'$ is equivalent to the condition that the *amplitude, phase,* and *frequency* of x_i and x_f' be identical. Since the phase shift introduced in a signal in being transmitted through a reactive network is invariably a function of the frequency, we have the following important principle:

The frequency at which a sinusoidal oscillator will operate is the frequency for which the total shift introduced, as a signal proceeds from the input terminals, through the amplifier and feedback network, and back again to the input, is precisely zero (or, of course, an integral multiple of 2π). Stated more simply, the frequency of a sinusoidal oscillator is determined by the condition that the loop-gain phase shift is zero.

Although other principles may be formulated which may serve equally to determine the frequency, these other principles may always be shown to be identical with that stated above. It might be noted parenthetically that it is not inconceivable that the above condition might be satisfied for more than a single frequency. In such a contingency there is the possibility of simultaneous oscillations at several frequencies or an oscillation at a single one of the allowed frequencies.

The condition given above determines the frequency, provided that the circuit will oscillate at all. Another condition which must clearly be met is that the magnitude of x_i and x_f' must be identical. This condition is then embodied in the following principle:

Oscillations will not be sustained if, at the oscillator frequency, the magnitude of the product of the transfer gain of the amplifier and the magnitude of the feedback factor of the feedback network (the magnitude of the loop gain) are less than unity.

The condition of *unity loop gain* $-A\beta = 1$ is called the *Barkhausen criterion.* This condition implies, of course, both that $|A\beta| = 1$ and that the phase of $-A\beta$ is zero. The above principles are consistent with the feedback formula $A_f = A/(1 + \beta A)$. For if $-\beta A = 1$, then $A_f \to \infty$, which may be interpreted to mean that there exists an output voltage even in the absence of an externally applied signal voltage.

Practical Considerations Referring to Fig. 14-28, it appears that if $|\beta A|$ at the oscillator frequency is precisely unity, then, with the feedback signal connected to the input terminals, the removal of the external generator will make no difference. If $|\beta A|$ is less than unity, the removal of the external generator will result in a cessation of oscillations. But now suppose that $|\beta A|$ is greater than unity. Then, for example, a 1-V signal appearing initially at the input terminals will, after a trip around the loop and back to the input terminals, appear there with an amplitude larger than 1 V. This larger voltage will then reappear as a still larger voltage, and so on. It seems, then, that if $|\beta A|$ is larger than unity, the amplitude of the oscillations will continue to increase without limit. But of course, such an increase in the amplitude can continue only as long as it is not limited by the onset of nonlinearity of operation in the active devices associated with the amplifier. Such a nonlinearity becomes more marked as the amplitude of oscillation increases. This onset of nonlinearity to limit the amplitude of oscillation is an essential feature of the operation of all practical oscillators, as the following considerations will show: The condition $|\beta A| = 1$ does not give a range of acceptable values of $|\beta A|$, but rather a single and precise value. Now suppose that initially it were even possible to satisfy this condition. Then, because circuit components and, more importantly, transistors change characteristics (drift) with age, temperature, voltage, etc., it is clear that if the entire oscillator is left to itself, in a very short time $|\beta A|$ will become either less or larger than unity. In the former case the oscillation simply stops, and in the latter case we are back to the point of requiring nonlinearity to limit the amplitude. An oscillator in which the loop gain is exactly unity is an abstraction completely unrealizable in practice. It is accordingly necessary, in the adjustment of a practical oscillator, always to arrange to have $|\beta A|$ somewhat larger (say 5 percent) than unity in order to ensure that, with incidental variations in transistor and circuit parameters, $|\beta A|$ shall not fall below unity. While the first two principles stated above must be satisfied on purely theoretical grounds, we may add a third general principle dictated by practical considerations, i.e.:

In every practical oscillator the loop gain is slightly larger than unity, and the amplitude of the oscillations is limited by the onset of nonlinearity.

The treatment of oscillators, taking into account the nonlinearity, is very difficult on account of the innate perverseness of nonlinearities generally. In many cases the extension into the range of nonlinear operation is small, and we simply neglect these nonlinearities altogether.

14-16   THE PHASE-SHIFT OSCILLATOR[9]

We select the so-called *phase-shift oscillator* (Fig. 14-29) as a first example because it exemplifies very simply the principles set forth above. Here an FET amplifier of conventional design is followed by three cascaded arrangements of a capacitor C and a resistor R, the output of the last RC combination

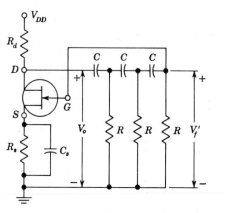

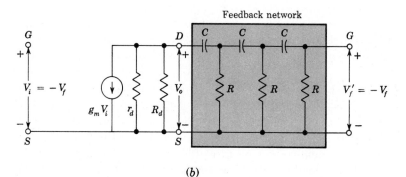

(b)

Fig. 14-29 (a) An FET phase-shift oscillator; (b) the small-signal equivalent circuit. The feedback network is indicated shaded, and $\beta = V_f'/V_o = -V_f/V_o$.

being returned to the gate. If the loading of the phase-shift network on the amplifier can be neglected, the amplifier shifts by 180° the phase of any voltage which appears on the gate, and the network of resistors and capacitors shifts the phase by an additional amount. At some frequency the phase shift introduced by the RC network will be precisely 180°, and at this frequency the total phase shift from the gate around the circuit and back to the gate will be exactly zero. This particular frequency will be the one at which the circuit will oscillate, provided that the magnitude of the amplification is sufficiently large.

From the mesh equations of the feedback network of Fig. 14-29b we find (Prob. 14-33) that the transfer function of the RC network, which is also the (negative of the) feedback factor, is

$$-\beta = \frac{V_f'}{V_o} = \frac{1}{1 - 5\alpha^2 - j(6\alpha - \alpha^3)} \qquad (14\text{-}60)$$

where $\alpha \equiv 1/\omega RC$. The phase shift of V_f'/V_o is 180° for $\alpha^2 = 6$, or

$$f = 1/(2\pi RC \sqrt{6})$$

At that frequency of oscillation, $\beta = +\frac{1}{29}$. In order that $|\beta A|$ shall not be less than unity, it is required that $|A|$ be at least 29. Hence an FET with $\mu < 29$ cannot be made to oscillate in such a circuit.

It should be pointed out that it is not always necessary to make use of an amplifier with transfer gain $|A| > 1$ to satisfy the Barkhausen criterion. It is only necessary that $|\beta A| > 1$. Passive network structures exist for which the transfer function $|\beta|$ is greater than unity at some particular frequency. In Prob. 14-41 we show an oscillator circuit consisting of a source follower and the RC circuit of Fig. 14-29 appropriately connected.

Transistor Phase-shift Oscillator If a transistor were used for the active element in Fig. 14-29, the output R of the feedback network would be shunted by the relatively low input resistance of the transistor. Hence, instead of employing voltage-series feedback as in Fig. 14-29, we use voltage-shunt feedback for a transistor phase-shift oscillator as indicated in Fig. 14-30a. For the circuit we assume that $h_{oe}R_c < 0.1$, so that we may use the approximate hybrid model to characterize the low-frequency small-signal behavior of the transistor, as in Fig. 14-30b. The resistor $R_3 = R - R_i$, where $R_i \approx h_{ie}$ is the input resistance of the transistor. This choice makes the three RC sections of the phase-shifting network alike and simplifies the calculations. We assume that the biasing resistors R_1, R_2, and R_e have no effect on the signal operation and neglect these in the following analysis. For a high-frequency oscillator the hybrid-Π equivalent circuit must be used in Fig. 14-30b.

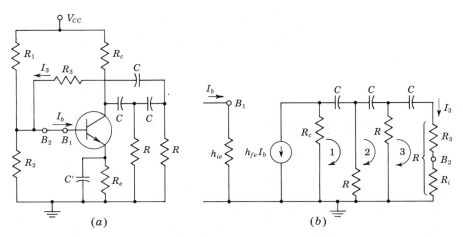

Fig. 14-30 (a) Transistor phase-shift oscillator. (b) The equivalent circuit from which to calculate the loop gain.

We imagine the loop broken at the base between B_1 and B_2, but in order not to change the loading on the feedback network, we place R_i from B_2 to ground. If we assume a current I_b to enter the base at B_1, then from Eq. (14-59) for the loop gain and Fig. 14-30 we have $x_f = -I_3$ and $x_i = I_b$. The loop current gain equals $-x_f/x_i = I_3/I_b$ and is found by writing Kirchhoff's voltage equation for the three meshes (Prob. 14-42). The Barkhausen condition that the phase of I_3/I_b must equal zero leads to the following expression for the frequency of oscillation:

$$f = \frac{1}{2\pi RC} \frac{1}{\sqrt{6 + 4k}} \tag{14-61}$$

where $k \equiv R_c/R$. The requirement that the magnitude of I_3/I_b must exceed unity in order for oscillations to start leads to the inequality

$$h_{fe} > 4k + 23 + \frac{29}{k} \tag{14-62}$$

The value of k which gives the minimum h_{fe} turns out to be 2.7, and for this optimum value of R_c/R, we find $h_{fe} = 44.5$. A transistor with a small-signal common-emitter short-circuit current gain less than 44.5 cannot be used in this phase-shift oscillator.

Variable-frequency Operation The phase-shift oscillator is particularly suited to the range of frequencies from several hertz to several hundred kilohertz, and so includes the range of audio frequencies. At frequencies in the megahertz range, it has no marked advantage over circuits (discussed in the following sections) employing tuned LC networks. The frequency of oscillation may be varied by changing any of the impedance elements in the phase-shifting network. For variations of frequency over a large range, the three capacitors are usually varied simultaneously. Such a variation keeps the input impedance to the phase-shifting network constant (Prob. 14-34) and also keeps constant the magnitude of β and $A\beta$. Hence the amplitude of oscillation will not be affected as the frequency is adjusted. The phase-shift oscillator is operated in class A in order to keep distortion to a minimum.

14-17   RESONANT–CIRCUIT OSCILLATORS

Figure 14-31a shows the *tuned-drain FET oscillator* in which a resonant circuit is used to determine the frequency. Other oscillators of this type are considered in Sec. 14-20. In Fig. 14-31a, r represents a resistance in series with the winding (of inductance L) in order to account for the losses in the transformer. If these losses are negligible, so that r can be neglected, then at the frequency $\omega = 1/\sqrt{LC}$, the impedance of the resonant circuit is arbitrarily large and purely resistive. In this case the voltage drop across the inductor from drain to ground is precisely 180° out of phase with the applied input voltage to the

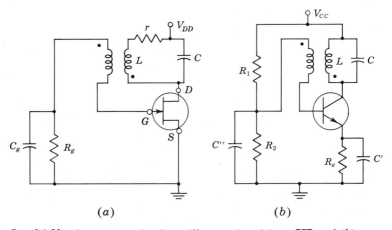

Fig. 14-31 A resonant-circuit oscillator using (a) an **FET** and (b) a bipolar transistor.

FET, independently of the magnitude of the FET drain resistance. If the direction of the winding of the secondary of the transformer (connected to the gate) is such as to introduce an additional phase shift of 180° (it is assumed that the secondary is not loaded), the total loop phase shift is exactly zero. At this frequency, then, the phase-shift condition for oscillation will have been satisfied. Again, since the transformer is considered to be unloaded, the ratio of the amplitude of the secondary to the primary voltage is M/L, where M is the mutual inductance. Since $A_V = -\mu$ for an amplifier with an infinite load impedance, the condition $-\beta A_V = 1$ is equivalent to $\mu = L/M$. More exactly, taking into account the finite magnitude of the resistance r, we find

$$\omega^2 = \frac{1}{LC}\left(1 + \frac{r}{r_d}\right) \tag{14-63}$$

as the frequency-determining condition and

$$g_m = \frac{\mu r C}{\mu M - L} \tag{14-64}$$

as the condition which is equivalent to $-\beta A = 1$.

Note that there is no a priori connection between the oscillation frequency and the steady-state "resonance" frequency. The frequency of oscillation is determined solely by the consideration that the loop phase shift is zero. In this sense, the suggestive near agreement of the frequency of the oscillator and the frequency of a natural oscillation or steady-state resonance is to be considered, superficially at least, as a pure coincidence. In the light of these last remarks it appears, too, that the designation of the oscillator of Sec. 14-16 as a "phase-shift oscillator," as opposed to the present designation, "resonant-circuit oscillator," is entirely artificial. All oscillators, those discussed above as well as those to be considered below, could be called phase-shift oscillators.

Self-bias and Amplitude Stabilization The bias for a resonant-circuit oscillator is obtained from an $R_g C_g$ parallel combination in series with the gate, as in Fig. 14-31a. The gate and source of the FET act as a rectifier, and if the $R_g C_g$ time constant is large compared with one period, the gate capacitor will charge up essentially to the peak gate swing. This voltage across C_g acts as the bias, and the gate is therefore driven positive (by the cut in voltage of the junction) only for a short interval at the peak of the swing. The voltage at the gate is a large sinusoid, and since its peak value is approximately at ground potential, we say that the gate is "clamped" to ground (Sec. 4-11). The operation is class C.

When the circuit is first energized, the gate bias is zero and the FET operates with a large g_m. The loop gain is therefore greater than unity, and the amplitude of oscillation starts to grow. As it does so, gate current is drawn, clamping takes place, and the bias automatically adjusts itself so that its magnitude equals the peak of the gate voltage. As the bias becomes more negative, the value of g_m decreases, and finally the amplitude stabilizes itself at that value for which the loop gain for the fundamental frequency is reduced to unity. Since the operation is class C, the use of the linear equivalent circuit is at best a rough approximation. In view of the foregoing discussion, the value of g_m in Eq. (14-64) may be considered to be the minimum value required at zero bias in order for oscillations to start. It may also be interpreted as the average value of transconductance which determines the amplitude of oscillation.

A Transistor Tuned-collector Oscillator The transistor circuit of Fig. 14-31b is analogous to the FET oscillator of Fig. 14-31a. The quiescent bias is determined by R_1, R_2, and R_e (Sec. 9-3). With R_1 in place, the transistor is biased in its active region, oscillations build up, and the self-bias is obtained from the $R_2 C''$ combination due to the flow of base current. As explained above, this action results in class C operation.

14-18   A GENERAL FORM OF OSCILLATOR CIRCUIT

Many oscillator circuits fall into the general form shown in Fig. 14-32a. The active device may be a bipolar transistor, an operational amplifier (discussed in Chap. 15), or an FET. In the analysis that follows we assume an active device with infinite input resistance such as an FET, or an operational amplifier. Figure 14-32b shows the linear equivalent circuit of Fig. 14-32a, using an amplifier with negative gain $-A_v$ and output resistance R_o. Clearly the topology of Fig. 14-32 is that of voltage-series feedback.

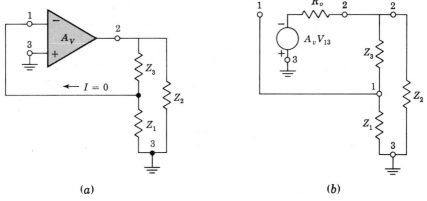

(a)   (b)

Fig. 14-32 (a) The basic configuration for many resonant-circuit oscillators. (b) The linear equivalent circuit using an operational amplifier.

The Loop Gain The value of $-A\beta$ will be obtained by considering the circuit of Fig. 14-32a to be a feedback amplifier with output taken from terminals 2 and 3 and with input terminals 1 and 3. The load impedance Z_L consists of Z_2 in parallel with the series combination of Z_1 and Z_3. The gain without feedback is $A = -A_v Z_L/(Z_L + R_o)$. The feedback factor is $\beta = -Z_1/(Z_1 + Z_3)$. The loop gain is found to be

$$-A\beta = \frac{-A_v Z_1 Z_2}{R_o(Z_1 + Z_2 + Z_3) + Z_2(Z_1 + Z_3)} \tag{14-65}$$

Reactive Elements Z_1, Z_2, and Z_3 If the impedances are pure reactances (either inductive or capacitive), then $Z_1 = jX_1$, $Z_2 = jX_2$, and $Z_3 = jX_3$. For an inductor, $X = \omega L$, and for a capacitor, $X = -1/\omega C$. Then

$$-A\beta = \frac{+A_v X_1 X_2}{jR_o(X_1 + X_2 + X_3) - X_2(X_1 + X_3)} \tag{14-66}$$

For the loop gain to be real (zero phase shift),

$$X_1 + X_2 + X_3 = 0 \tag{14-67}$$

and

$$-A\beta = \frac{A_v X_1 X_2}{-X_2(X_1 + X_3)} = \frac{-A_v X_1}{X_1 + X_3} \tag{14-68}$$

From Eq. (14-67) we see that the circuit will oscillate at the resonant frequency of the series combination of X_1, X_2, and X_3.

Using Eq. (14-67) in Eq. (14-68) yields

$$-A\beta = \frac{+A_v X_1}{X_2} \tag{14-69}$$

Since $-A\beta$ must be positive and at least unity in magnitude, then X_1 and X_2 must have the same sign (A_v is positive). In other words, they must be the

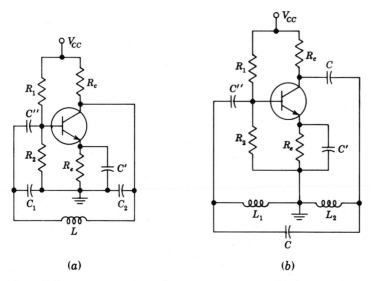

Fig. 14-33 (a) A transistor Colpitts oscillator. (b) A transistor Hartley oscillator.

same kind of reactance, either both inductive or both capacitive. Then, from Eq. (14-67), $X_3 = -(X_1 + X_2)$ must be inductive if X_1 and X_2 are capacitive, or vice versa.

If X_1 and X_2 are capacitors and X_3 is an inductor, the circuit is called a *Colpitts oscillator*. If X_1 and X_2 are inductors and X_3 is a capacitor, the circuit is called a *Hartley oscillator*. In this latter case, there may be mutual coupling between X_1 and X_2 (and the above equations will then not apply).

Transistor versions of the above types of LC oscillators are possible. As an example, a transistor Colpitts oscillator is indicated in Fig. 14-33a. Qualitatively, this circuit operates in the manner described above. However, the detailed analysis of a transistor oscillator circuit is more difficult, for two fundamental reasons. First, the low input impedance of the transistor shunts Z_1 in Fig. 14-32a, and hence complicates the expressions for the loop gain given above. Second, if the oscillation frequency is beyond the audio range, the simple low-frequency h-parameter model is no longer valid. Under these circumstances the more complicated high-frequency hybrid-II model of Fig. 11-5 must be used. A transistor Hartley oscillator is shown in Fig. 14-33b.

14-19   THE WIEN BRIDGE OSCILLATOR

An oscillator circuit in which a balanced bridge is used as the feedback network is the Wien bridge[10] oscillator shown in Fig. 14-34. The active element is an operational amplifier (Chap. 15) which has a very large positive voltage gain

$(V_o = A_V V_i)$, negligible output resistance, and very high (infinite) input resistance. We assume further that A_V is constant over the range of frequencies of operation of this circuit.

To find the loop gain $-\beta A$, we break the loop at the point marked P and apply an external voltage V_o' across terminals 3 and 4.

Since $V_o = A_V V_i$, the loop gain is given by

$$\text{Loop gain} = \frac{V_o}{V_o'} = \frac{V_i}{V_o'} A_V = -\beta A \tag{14-70}$$

Two auxiliary voltages V_1 and V_2 are indicated in Fig. 14-34 such that $V_i = V_2 - V_1$. From Eq. (14-70) $A = A_V$ and

$$-\beta = \frac{V_i}{V_o'} = \frac{V_2 - V_1}{V_o'} = \frac{Z_2}{Z_1 + Z_2} - \frac{R_2}{R_1 + R_2} \tag{14-71}$$

It is not difficult to show that Z_1 and Z_2 have the same phase angle at the frequency

$$f_o = \frac{1}{2\pi RC} \tag{14-72}$$

and that at this frequency $Z_1 = (1 - j)R$ and $Z_2 = (1 - j)R/2$. Hence $V_2 = \frac{1}{3}V_o'$, at $\omega = \omega_o$. If a null is desired, then R_1 and R_2 must be chosen so that $V_i = 0$. From Eq. (14-71) $R_2/(R_1 + R_2) = \frac{1}{3}$, or $R_1 = 2R_2$.

In the present case, where the bridge is used as the feedback network for an oscillator, the loop gain of Eq. (14-70) must equal unity and must have zero phase. Thus, since A_V is a positive number, the phase of $-\beta$ from Eq. (14-71) must be zero, but the magnitude must not be zero. This is accomplished by taking the ratio $R_2/(R_1 + R_2)$ smaller than $\frac{1}{3}$.

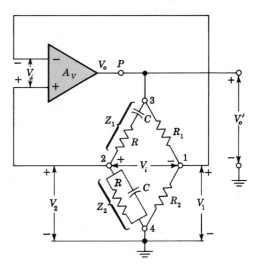

Fig. 14-34 Wien bridge oscillator using an operational amplifier as the active element.

If we let

$$\frac{V_1}{V_o'} = \frac{R_2}{R_1 + R_2} = \frac{1}{3} - \frac{1}{\delta}$$

where δ is a number greater than 3, then

$$-\beta = \frac{V_2 - V_1}{V_o'} = \frac{V_2}{V_o'} - \frac{1}{3} + \frac{1}{\delta} \qquad (14\text{-}73)$$

At $\omega = \omega_o$, $V_2/V_o' = \frac{1}{3}$ and $\beta = -1/\delta$. The condition $-\beta A = 1$ is now satisfied by making

$$\delta = A_V \qquad (14\text{-}74)$$

Note that the frequency of oscillation is precisely the null frequency of the balanced bridge, given by Eq. (14-72). At any other frequency V_2 is not in phase with V_o', and therefore $V_i = V_2 - V_1$ is not in phase with V_o', so that the condition $-\beta A = 1$ is satisfied only at the one frequency f_o.

Continuous variation of frequency is accomplished by varying simultaneously the two capacitors (ganged variable air capacitors). Changes in frequency range are accomplished by switching in different values for the two identical resistors R.

Amplitude Stabilization We consider modification of the circuit of Fig. 14-34, which serves to stabilize the amplitude against variations due to fluctuations occasioned by the aging of transistors, components, etc. The modification consists simply in replacing the resistor R_2 by a sensistor (Sec. 2-7) which has a positive thermal coefficient.

The amplitude of oscillation is determined by the extent to which the loop gain $-\beta A$ is greater than unity. If $-\beta$ is fixed, the amplitude is then determined by A, increasing as A increases, until further increase is limited by the amplifier nonlinearity. The regulation mechanism introduced by the sensistor operates by automatically changing β in such a direction as to keep βA more nearly constant if, as when the loading of the amplifier changes, the values of A should change. It will be recalled that the resistance of a sensistor increases with temperature, and the temperature is in turn determined by the root-mean-square value of the current which passes through it. If the root-mean-square value of the current changes, then, because of the thermal lag of the sensistor, the temperature will be determined by the average value over a large number of cycles of the root-mean-square value of the current.

Consider now that the amplitude has decreased because A has decreased. The value of R_2 will decrease, and as a consequence $-\beta = 1/\delta$ will increase, as indicated by Eq. (14-71). Or to put the matter another way, as A changes, the extent to which the Wien bridge is unbalanced will adjust itself in such a manner as to keep βA more nearly constant. An important fact to keep in mind about the mechanism just described is that, because of the thermal lag of the sensistor, the resistance of the sensistor during the course of a single cycle is very nearly absolutely constant. Therefore, at any fixed amplitude of oscillation, the sensistor behaves entirely like an ordinary linear resistor.

A thermistor which has a negative temperature coefficient can also be used, in place of R_1 rather than R_2.

14-20   CRYSTAL OSCILLATORS

If a piezoelectric crystal, usually quartz, has electrodes plated on opposite faces and if a potential is applied between these electrodes, forces will be exerted on the bound charges within the crystal. If this device is properly mounted, deformations take place within the crystal, and an electromechanical system is formed which will vibrate when properly excited. The resonant frequency and the Q depend upon the crystal dimensions, how the surfaces are oriented with respect to its axes, and how the device is mounted.[11] Frequencies ranging from a few kilohertz to a few megahertz, and Q's in the range from several thousand to several hundred thousand, are commercially available. These extraordinarily high values of Q and the fact that the characteristics of quartz are extremely stable with respect to time and temperature account for the exceptional frequency stability of oscillators incorporating crystals (Sec. 14-21).

The electrical equivalent circuit of a crystal is indicated in Fig. 14-35. The inductor L, capacitor C, and resistor R are the analogs of the mass, the compliance (the reciprocal of the spring constant), and the viscous-damping factor of the mechanical system. Typical values for a 90-kHz crystal are $L = 137$ H, $C = 0.0235$ pF, and $R = 15$ K, corresponding to $Q = 5,500$. The dimensions of such a crystal are 30 by 4 by 1.5 mm. Since C' represents the

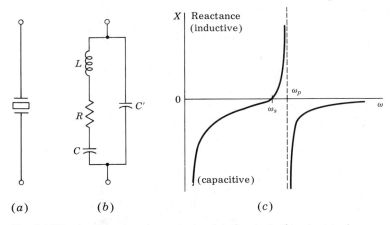

(a)   *(b)*   *(c)*

Fig. 14-35 A piezoelectric crystal. (a) Symbol; (b) electrical model; (c) the reactance function (if $R = 0$).

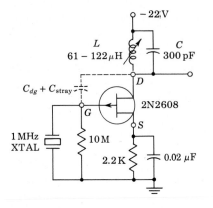

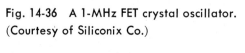

Fig. 14-36 A 1-MHz FET crystal oscillator.
(Courtesy of Siliconix Co.)

electrostatic capacitance between electrodes with the crystal as a dielectric, its magnitude ($\sim$3.5 pF) is very much larger than C.

If we neglect the resistance R, the impedance of the crystal is a reactance jX whose dependence upon frequency is given by

$$jX = -\frac{j}{\omega C'}\frac{\omega^2 - \omega_s^2}{\omega^2 - \omega_p^2} \tag{14-75}$$

where $\omega_s^2 = 1/LC$ is the series resonant frequency (the zero impedance frequency), and $\omega_p^2 = (1/L)(1/C + 1/C')$ is the parallel resonant frequency (the infinite impedance frequency). Since $C' \gg C$, then $\omega_p \approx \omega_s$. For the crystal whose parameters are specified above, the parallel frequency is only three-tenths of 1 percent higher than the series frequency. For $\omega_s < \omega < \omega_p$, the reactance is inductive, and outside this range it is capacitive, as indicated in Fig. 14-35.

A variety of crystal-oscillator circuits is possible. If in the basic configuration of Fig. 14-32a a crystal is used for Z_1, a tuned LC combination for Z_2, and the capacitance C_{dg} between drain and gate for Z_3, the resulting circuit is as indicated in Fig. 14-36. From the theory given in the preceding section, the crystal reactance, as well as that of the LC network, must be inductive. For the loop gain to be greater than unity, we see from Eq. (14-69) that X_1 cannot be too small. Hence the circuit will oscillate at a frequency which lies between ω_s and ω_p but close to the parallel-resonance value. Since $\omega_p \approx \omega_s$, the oscillator frequency is essentially determined by the crystal, and not by the rest of the circuit.

14-21   FREQUENCY STABILITY

An oscillator having initially been set at a particular frequency will invariably not maintain its initial frequency, but will instead drift and wander about in frequency, sometimes uniformly in one direction, sometimes quite erratically.

The *frequency stability* of an oscillator is a measure of its ability to maintain as nearly a fixed frequency as possible over as long a time interval as possible. These deviations of frequency arise because the values of the circuit features, on which the oscillator frequency depends, do not remain constant in time. (We use here the term "circuit features" to include circuit components, transistor parameters, supply voltages, stray capacitances, etc.) Accordingly, an obvious but clearly useless solution of the problem of making a frequency-stable oscillator is to keep constant all the circuit features. In the first place, the number of circuit features is very large, in general; second, some of the circuit features, such as transistor parameters, are inherently unstable and extremely difficult to keep constant; and third, it is hard enough to know where stray circuit elements and couplings are located and how to estimate their magnitudes without having to devise schemes to maintain them constant.

But we recognize also that in every oscillator circuit there are a relatively few circuit features on which the frequency is sensitively dependent, whereas the frequency dependence of the far larger number of remaining features is comparatively slight. For example, in the circuit of Fig. 14-29, the frequency is for the most part determined by R and C, and the other features of the circuit affect the frequency to a much smaller extent. We shall then have taken a long step in the direction toward frequency stability if we take pains to ensure the stability, at least, of these relatively few passive elements which influence the frequency markedly. The principal cause of drift in these is the variation of temperature. Measures for maintaining the temperature constant and for balancing the temperature-induced variation in one such element against that in another can be taken.[12]

Stability Criterion   *If in an oscillator there exists one set of elements which has the property that at the oscillation frequency these components introduce a large variation of phase θ with frequency, then dθ/dω serves as a measure of the independence of the frequency of all other features of the circuit. The frequency stability improves as dθ/dω increases. In the limit, as dθ/dω becomes infinite, the oscillator frequency depends only on this set of elements and becomes completely independent of all other features of the circuit.*

The proof of the foregoing principle is almost self-evident, and is readily arrived at from the following considerations: Suppose that a variation takes place in some one feature of the oscillator *other than one of the components of the set of elements described above.* Then, if initially the phase condition for oscillation was satisfied at the frequency of oscillation, it will, in general, no longer be satisfied after the alteration of the circuit feature. The frequency must accordingly shift in order once again to restore the loop phase shift to the exact value zero. If, however, there is a set of elements which, at the nominal oscillator frequency, produces a large phase shift for a small frequency change (that is, $d\theta/d\omega$ large), it is clear that the frequency shift required to restore the circuital phase shift to zero need be only very small.

In a parallel-resonant circuit the impedance changes from an inductive to a capacitive reactance as the frequency is increased through the resonant point. If the Q is infinite (an ideal inductor with zero series resistance), this change in phase is abrupt, $d\theta/d\omega \to \infty$, because the phase changes abruptly from -90 to $+90°$. Hence a tuned-circuit oscillator will have excellent frequency stability provided that Q is sufficiently high and that L and C are stable (independent of temperature, current, etc.).

These ideas about tuned-circuit oscillators can be carried over to account for the exceptional frequency stability of crystal oscillators. From Fig. 14-35c we see that, for a crystal with infinite Q, the phase changes discontinuously from -90 to $+90°$ as the frequency passes through ω_s and then abruptly back again from $+90$ to $-90°$ as ω passes through ω_p. Of course, infinite Q is unattainable, but since commercially available crystals have values of Q of tens or hundreds of thousands, very large values of $d\theta/d\omega$ are realizable. Hence, if a crystal is incorporated into a circuit (such as that of Fig. 14-36), an oscillator is obtained whose frequency depends essentially upon the crystal itself and nothing else. The crystal frequency does, however, still depend somewhat on the temperature, and regulated-temperature ovens must be employed where the highest stability is required.

To compare the frequency stability of two different types of oscillators, $d\theta/d\omega$ is evaluated for each at the operating frequency. The circuit giving the larger value of $d\theta/d\omega$ has the more stable oscillator frequency.

REFERENCES

1. Gray, P. E., and C. L. Searle: "Electronic Principles: Physics, Models, and Circuits," p. 677, John Wiley & Sons, Inc., New York, 1969.

2. Thornton, R. D., C. L. Searle, D. O. Peterson, R. B. Adler, and E. J. Angelo, Jr.: "Multistage Transistor Circuits," SEEC Committee Series, vol. 5, pp. 108–118, John Wiley & Sons, Inc., 1965.

3. Pottle, C.: A Textbook Computerized State Space Network Analysis Algorithm, *Cornell Univ. Elec. Eng. Res. Lab. Rep.*, September, 1968.

4. Bode, H. W.: "Network Analysis and Feedback Amplifier Design," chap. 7, D. Van Nostrand Company, Inc., New York, 1956.

5. Armstrong, E. H.: Some Recent Developments in the Audion Receiver, *Proc. IRE*, vol. 3, pp. 215–247, September, 1915.

6. Nyquist, H.: Regeneration Theory, *Bell System Tech. J.*, vol. 11, pp. 126–147, January, 1932.

7. Ref. 2, pp. 118, 120, 138.

8. Ref. 1, p. 699.

9. Sherr, S.: Generalized Equations for *R-C* Phase-shift Oscillators, *Proc. IRE*, vol. 42, pp. 1169–1172, July, 1954.

10. Clarke, K. K.: Wein-bridge Oscillator Design, *Proc. IRE*, vol. 41, pp. 246–249, February, 1953.

11. Fair, Z. E.: Piezoelectric Crystals in Oscillator Circuits, *Bell System Tech. J.*, vol. 24, pp. 161–216, April, 1945.

12. Chance, B., et al.: "Waveforms," Radiation Laboratory Series, McGraw-Hill Book Company, New York, 1949.

REVIEW QUESTIONS

14-1 Consider a feedback amplifier with a single-pole transfer function. (a) What is the relationship between the high 3-db frequency with and without feedback? (b) Repeat part a for the low 3-dB frequency. (c) Repeat part a for the gain-bandwidth product.

14-2 Consider a feedback amplifier with a double-pole transfer function. (a) Without proof sketch the locus of the poles in the s plane after feedback. (b) Why is this amplifier stable, independent of the amount of negative feedback?

14-3 (a) Indicate (without proof) a circuit having the same transfer function as the double-pole feedback amplifier. (b) Sketch the step response of this amplifier for both the underdamped and overdamped condition.

14-4 For an underdamped two-pole amplifier response define (a) *rise time*, (b) *delay time*, (c) *overshoot*, (d) *damped period*, (e) *settling time*.

14-5 (a) Sketch (without proof) the root locus of the poles of a three-pole amplifier after feedback is added. (b) Indicate where the amplifier becomes unstable.

14-6 Consider a multipole amplifier with $|s_1| < |s_2| < |s_3| \cdots < |s_n|$. Under what circumstances is the response with feedback determined by (a) s_1 and s_2 and (b) s_1 alone?

14-7 Explain in words (without equations) how to obtain the frequency response of a voltage-shunt *feedback amplifier* using feedback concepts. Consider a single stage with a resistor R' between collector and base.

14-8 Repeat Rev. 14-7 for a current-series feedback amplifier. Consider a CE stage with an emitter resistor R_e.

14-9 (a) Define *positive* feedback. (b) What is the relationship between A_f and A for positive feedback? (c) Illustrate positive feedback in the complex βA plane.

14-10 State the condition on $1 + A\beta$ which a feedback amplifier must satisfy in order to be stable.

14-11 State the *Nyquist criterion* for stability.

14-12 Consider an amplifier with an open-loop dominant pole and with β independent of frequency. Prove that the Nyquist plot is a circle.

14-13 State the stability condition in terms of Bode plots.

14-14 Consider an amplifier consisting of three identical noninteracting stages. Explain why a definite maximum value of $|\beta|$ exists before oscillations take place.

14-15 Define with the aid of graphs (a) gain margin and (b) phase margin.

14-16 (a) Describe *dominant-pole compensation*. (b) Show one circuit for accomplishing such compensation. (c) What is a disadvantage of this technique?

14-17 (a) Describe *pole-zero compensation*. (b) Show one circuit for accomplishing such compensation.

14-18 (*a*) Describe *lead compensation*. (*b*) Show one circuit for accomplishing such compensation.

14-19 Give the two Barkhausen conditions required in order for sinusoidal oscillations to be sustained.

14-20 Sketch the circuit of a phase-shift oscillator using (*a*) an FET and (*b*) a bipolar junction transistor.

14-21 Sketch the circuit of a tuned-collector oscillator.

14-22 (*a*) Sketch the topology for a generalized resonant-circuit oscillator, using impedances Z_1, Z_2, Z_3. (*b*) At what frequency will the circuit oscillate? (*c*) Under what conditions does the configuration reduce to a Colpitts oscillator? A Hartley oscillator?

14-23 (*a*) Sketch the circuit of a Wien bridge oscillator. (*b*) What determines the frequency of oscillation? (*c*) Will oscillations take place if the bridge is balanced? Explain.

14-24 (*a*) Draw the electrical model of a piezoelectric crystal. (*b*) Sketch the reactance vs. frequency function. (*c*) Over what portion of the reactance curve do we desire oscillations to take place when the crystal is used as part of a sinusoidal oscillator? Explain.

14-25 Sketch a circuit of a crystal-controlled oscillator.

14-26 State the frequency-stability criterion for a sinusoidal oscillator.

MPLIFIERS

d OP AMP) is a direct-coupled
s added to control its overall
rform a wide variety of linear
tions) and is often referred to
analog) *integrated circuit.* Its
lowing chapter.

r has gained wide acceptance
c system building block. It
integrated circuits: small size,
ure tracking, and low offset
efined carefully, later in this

tages in an OP AMP is made.
P AMP parameters are given.
compensation are discussed.

MPLIFIER

is shown in Fig. 15-1a, and the
number of operational ampli-
ges V_2 and V_1 applied to the
espectively. A single-ended
case where one of the input
AMPS have only one output

eal OP AMP has the following

1. Input resistance $R_i = \infty$
2. Output resistance $R_o = 0$
3. Voltage gain $A_v = -\infty$
4. Bandwidth $= \infty$
5. $V_o = 0$ when $V_1 = V_2$ independent of the magnitude of V_1
6. Characteristics do not drift with temperature.

In Fig. 15-2a we show the ideal operational amplifier with feedback impedances (Z and Z') and the $+$ terminal grounded. This is the basic inverting circuit. This topology represents voltage-shunt feedback, and is discussed in Sec. 13-12. The circuit of Fig. 15-2a is a generalization of Fig. 13-20a, with the single-transistor replaced by the multistage OP AMP and the resistors R_s and R' replaced by impedances Z and Z', respectively. From Eq. (13-79) the voltage gain A_{Vf} with feedback is given by

$$A_{Vf} = -\frac{Z'}{Z} \tag{15-1}$$

An instructive alternative proof of this equation is obtained as follows: Since $R_i \rightarrow \infty$, the current I through Z also passes through Z', as indicated in Fig. 15-2a. In addition, we note that $V_i = V_o/A_V \rightarrow 0$ as $|A_V| \rightarrow \infty$, so that the input is effectively shorted. Hence

$$A_{Vf} = \frac{V_o}{V_s} = \frac{-IZ'}{IZ} = -\frac{Z'}{Z}$$

in agreement with Eq. (15-1).

The operation of the circuit may now be described in the following terms: At the input to the amplifier proper there exists a *virtual ground*, or *short circuit*.

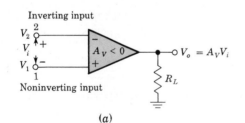

(a)

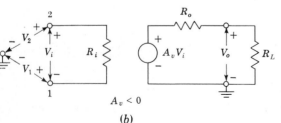

$A_v < 0$

(b)

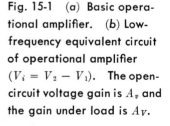

Fig. 15-1 (a) Basic operational amplifier. (b) Low-frequency equivalent circuit of operational amplifier ($V_i = V_2 - V_1$). The open-circuit voltage gain is A_v and the gain under load is A_V.

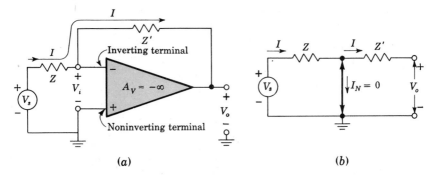

Fig. 15-2   (a) Inverting operational amplifier with added voltage-shunt feedback.   (b) Virtual ground in the operational amplifier.

The term "virtual" is used to imply that, although the feedback from output to input through Z' serves to keep the voltage V_i at zero, no current actually flows into this short.   The situation is depicted in Fig. 15-2b, where the virtual ground is represented by the heavy double-headed arrow.   This figure does not represent a physical circuit, but it is a convenient mnemonic aid from which to calculate the output voltage for a given input signal.   This symbolism is very useful in connection with analog computations discussed in Sec. 16-1.

Practical Inverting Operational Amplifier   Equation (15-1) is valid only if the voltage gain is infinite.   It is sometimes important to consider a physical amplifier which does not satisfy these restrictions.   In Fig. 15-3 the amplifier in Fig. 15-2a is replaced by its small-signal model, with $|A_V| \neq \infty$, $R_i \neq \infty$, and $R_o \neq 0$.   The symbol A_v is the *open-circuit (unloaded) voltage gain*.   The impedances shown shaded indicate the effect of Z' on the input and output of the amplifier, where use is made of the Miller theorem (Sec. 8-11).   Using these Miller impedances in place of Z' in Fig. 15-3, the following expression for the *closed-loop gain* is obtained (Prob. 15-1).

$$A_{Vf} = \frac{-Y}{Y' - (1/A_V)(Y' + Y + Y_i)} \qquad (15\text{-}2)$$

where the Y's are the admittances corresponding to the Z's (for example, $Y' = 1/Z'$) and where the voltage gain $A_V \equiv V_o/V_i$, taking the loading of Z' into account, is given by

$$A_V = \frac{A_v + R_o Y'}{1 + R_o Y'} \qquad (15\text{-}3)$$

Note that if $R_o = 0$ or $Y' = 0$ $(Z' = \infty)$, the loading is effectively removed and $A_V = A_v$.   Also observe that as $|A_v| \to \infty$, then $|A_V| \to \infty$ and

$$A_{Vf} \to -\frac{Y}{Y'} = -\frac{Z'}{Z}$$

in agreement with Eq. (15-1).

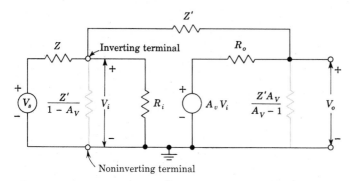

Fig. 15-3 Circuit model of the OP AMP of Fig. 15-1. (The shaded impedances are the Miller replacements for Z'.)

Noninverting Operational Amplifier Very often there is a need for an amplifier whose output is equal to, and in phase with, the input, and in addition $R_i = \infty$ and $R_o = 0$, so that the source and load are in effect isolated. An emitter follower approximates these specifications. More ideal characteristics can be obtained by using an operational amplifier having a noninverting terminal for signals and an inverting terminal for the feedback voltage, as shown in Fig. 15-4.

If we assume again that $R_i = \infty$, we have

$$V_2 = \frac{R}{R + R'} V_o$$

Since $V_o = A_V(V_2 - V_s)$, then for a finite V_o and $-A_V = \infty$ it follows that $V_s = V_2$ (there is a virtual short at the input terminals) and

$$A_{Vf} = \frac{V_o}{V_s} = \frac{V_o}{V_2} = \frac{R + R'}{R} \tag{15-4}$$

Hence the closed-loop gain is always greater than unity. If $R = \infty$ and/or $R' = 0$, then $A_{Vf} = +1$ and the amplifier acts as a *voltage follower*.

In the analysis of noninverting OP AMP circuits we shall use the facts that (1) *no current flows into either input* and (2) *the potentials of the two inputs are equal.*

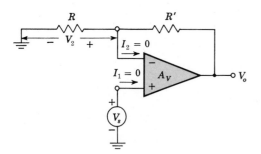

Fig. 15-4 Noninverting operational amplifier with resistive feedback. If $R \gg R'$, the output follows the input; $V_o \approx V_s$.

15-2   THE DIFFERENTIAL AMPLIFIER

The function of a differential amplifier[4] (abbreviated DIFF AMP) is, in general, to amplify the difference between two signals. The need for DIFF AMPS arises in many physical measurements where response from dc to many megahertz is required. It is also the basic stage of an integrated operational amplifier with differential input.

Figure 15-5 represents a linear active device with two input signals v_1, v_2 and one output signal v_o, each measured with respect to ground. In an ideal DIFF AMP the output signal v_o should be given by

$$v_o = A_d(v_1 - v_2) \tag{15-5}$$

where A_d is the gain of the differential amplifier. Thus it is seen that any signal which is common to both inputs will have no effect on the output voltage. However, a practical DIFF AMP cannot be described by Eq. (15-5), because, in general, the output depends not only upon the *difference signal* v_d of the two signals, but also upon the average level, called the *common-mode signal* v_c, where

$$v_d \equiv v_1 - v_2 \quad \text{and} \quad v_c \equiv \tfrac{1}{2}(v_1 + v_2) \tag{15-6}$$

For example, if one signal is $+50\ \mu V$ and the second is $-50\ \mu V$, the output will not be exactly the same as if $v_1 = 1{,}050\ \mu V$ and $v_2 = 950\ \mu V$, even though the difference $v_d = 100\ \mu V$ is the same in the two cases.

The Common-mode Rejection Ratio The foregoing statements are now clarified, and a figure of merit for a difference amplifier is introduced. The output of Fig. 15-5 can be expressed as a linear combination of the two input voltages

$$v_o = A_1 v_1 + A_2 v_2 \tag{15-7}$$

where $A_1(A_2)$ is the voltage amplification from input 1(2) to the output under the condition that input 2(1) is grounded. From Eqs. (15-6)

$$v_1 = v_c + \tfrac{1}{2}v_d \quad \text{and} \quad v_2 = v_c - \tfrac{1}{2}v_d \tag{15-8}$$

If these equations are substituted in Eq. (15-7), we obtain

$$v_o = A_d v_d + A_c v_c \tag{15-9}$$

where

$$A_d \equiv \tfrac{1}{2}(A_1 - A_2) \quad \text{and} \quad A_c \equiv A_1 + A_2 \tag{15-10}$$

Fig. 15-5 The output is a linear function of v_1 and v_2 for an ideal differential amplifier; $v_o = A_d(v_1 - v_2)$.

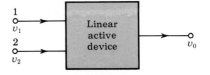

The voltage gain for the difference signal is A_d, and that for the common-mode signal is A_c. We can measure A_d directly by setting $v_1 = -v_2 = 0.5$ V, so that $v_d = 1$ V and $v_c = 0$. Under these conditions the measured output voltage v_o gives the gain A_d for the difference signal [Eq. (15-9)]. Similarly, if we set $v_1 = v_2 = 1$ V, then $v_d = 0$, $v_c = 1$ V, and $v_o = A_c$. The output voltage now is a direct measurement of the common-mode gain A_c.

Clearly, we should like to have A_d large, whereas ideally, A_c should equal zero. A quantity called the *common-mode rejection ratio*, which serves as a figure of merit for a DIFF AMP, is defined by

$$\text{CMRR} \equiv \rho \equiv \left| \frac{A_d}{A_c} \right| \qquad (15\text{-}11)$$

From Eqs. (15-9) and (15-11) we obtain an expression for the output in the following form:

$$v_o = A_d v_d \left(1 + \frac{1}{\rho} \frac{v_c}{v_d} \right) \qquad (15\text{-}12)$$

From this equation we see that the amplifier should be designed so that ρ is large compared with the ratio of the common-mode signal to the difference signal. For example, if $\rho = 1,000$, $v_c = 1$ mV, and $v_d = 1$ μV, the second term in Eq. (15-12) is equal to the first term. Hence, for an amplifier with a common-mode rejection ratio of 1,000, a 1-μV difference of potential between the two inputs gives the same output as a 1-mV signal applied with the same polarity to both inputs.

EXAMPLE (a) Consider the situation referred to above, where the first set of signals is $v_1 = +50$ μV and $v_2 = -50$ μV and the second set is $v_1 = 1,050$ μV and $v_2 = 950$ μV. If the common-mode rejection ratio is 100, calculate the percentage difference in output voltage obtained for the two sets of input signals. (b) Repeat part a if $\rho = 10,000$.

Solution a. In the first case, $v_d = 100$ μV and $v_c = 0$, so that, from Eq. (15-12), $v_o = 100A_d$ μV.

In the second case, $v_d = 100$ μV, the same value as in part a, but now $v_c = \frac{1}{2}(1,050 + 950) = 1,000$ μV, so that, from Eq. (15-12),

$$v_o = 100A_d \left(1 + \frac{10}{\rho} \right) = 100A_d \left(1 + \frac{10}{100} \right) \qquad \mu\text{V}$$

These two measurements differ by 10 percent.

b. For $\rho = 10,000$, the second set of signals results in an output

$$v_o = 100A_d(1 + 10 \times 10^{-4}) \qquad \mu\text{V}$$

whereas the first set of signals gives an output $v_o = 100A_d$ μV. Hence the two measurements now differ by only 0.1 percent.

15-3   THE EMITTER–COUPLED DIFFERENTIAL AMPLIFIER

The circuit of Fig. 15-6 is an excellent DIFF AMP if the emitter resistance R_e is large. This statement can be justified as follows: If $V_{s1} = V_{s2} = V_s$, then from Eqs. (15-6) and (15-9) we have $V_d = V_{s1} - V_{s2} = 0$ and $V_o = A_c V_s$. However, if $R_e = \infty$, then, because of the symmetry of Fig. 15-6, we obtain $I_{e1} = I_{e2} = 0$. If $I_{b2} \ll I_{c2}$, then $I_{c2} \approx I_{e2}$, and it follows that $V_o = 0$. Hence the common-mode gain A_c becomes very small, and the common-mode rejection ratio is very large for a very large value of R_e and a symmetrical circuit.

We now analyze the emitter-coupled circuit for a finite value of R_e. A_c can be evaluated by setting $V_{s1} = V_{s2} = V_s$ and making use of the symmetry of Fig. 15-6. This circuit can be bisected as in Fig. 15-7a. An analysis of this circuit (Prob. 15-10), using Eqs. (8-67) to (8-69) and neglecting the term in h_{re} in Eq. (8-68), yields

$$A_c = \frac{V_o}{V_s} = \frac{(2h_{oe}R_e - h_{fe})R_c}{2R_e(1 + h_{fe}) + (R_s + h_{ie})(2h_{oe}R_e + 1)} \tag{15-13}$$

provided that $h_{oe}R_c \ll 1$. Similarly, the difference mode gain A_d can be obtained by setting $V_{s1} = -V_{s2} = V_s/2$. From the symmetry of Fig. 15-6, we see that, if $V_{s1} = -V_{s2}$, then $I_{e1} = -I_{e2}$, the drop across R_e is zero, and E is grounded for small-signal operation. Under these conditions the circuit of Fig. 15-7b can be used to obtain A_d. Hence

$$A_d = \frac{V_o}{V_s} = \frac{1}{2} \frac{h_{fe}R_c}{R_s + h_{ie}} \tag{15-14}$$

provided $h_{oe}R_c \ll 1$. The common-mode rejection ratio can now be obtained using Eqs. (15-11), (15-13), and (15-14).

From Eq. (15-13) it is seen that the common-mode rejection ratio increases with R_e, as predicted above. There are, however, practical limitations on the magnitude of R_e because of the quiescent dc voltage drop across it; the emitter

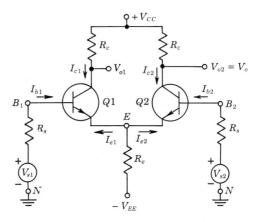

Fig. 15-6 Symmetrical emitter-coupled difference amplifier.

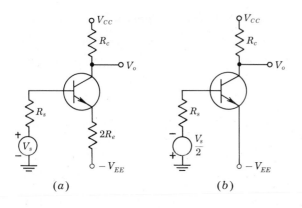

Fig. 15-7 Equivalent circuit for a symmetrical differential amplifier used to determine (a) the common-mode gain A_c and (b) the difference gain A_d.

(a)   (b)

supply V_{EE} must become larger as R_e is increased, in order to maintain the quiescent current at its proper value. If the operating currents of the transistors are allowed to decrease, this will lead to higher h_{ie} values and lower values of h_{fe}. Both of these effects will tend to decrease the common-mode rejection ratio.

Differential Amplifier Supplied with a Constant Current Frequently, in practice, R_e is replaced by a transistor circuit, as in Fig. 15-8, in which R_1, R_2, and R_3 can be adjusted to give the same quiescent conditions for $Q1$ and $Q2$ as the original circuit of Fig. 15-6. This modified circuit of Fig. 15-8 presents a very high effective emitter resistance R_e for the two transistors $Q1$ and $Q2$. Since R_e is also the effective resistance looking into the collector of transistor $Q3$, it is given by Eq. (8-70). In Sec. 8-15 it is verified that R_e is hundreds of kilohms even if R_3 is as small as 1 K.

We now verify that transistor $Q3$ acts as an approximately constant cur-

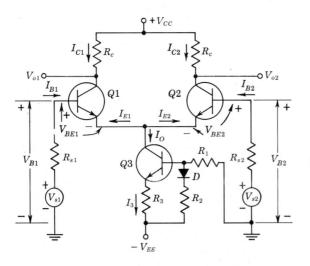

Fig. 15-8 Differential amplifier with constant-current stage in the emitter circuit. Nominally, $R_{s1} = R_{s2}$.

rent source, subject to the condition that the base current of $Q3$ is negligible. Applying KVL to the base circuit of $Q3$, we have

$$I_3R_3 + V_{BE3} = V_D + (V_{EE} - V_D)\frac{R_2}{R_1 + R_2} \tag{15-15}$$

where V_D is the diode voltage. Hence

$$I_O \approx I_3 = \frac{1}{R_3}\left(\frac{V_{EE}R_2}{R_1 + R_2} + \frac{V_D R_1}{R_1 + R_2} - V_{BE3}\right) \tag{15-16}$$

If the circuit parameters are chosen so that

$$\frac{V_D R_1}{R_1 + R_2} = V_{BE3} \tag{15-17}$$

then

$$I_O = \frac{V_{EE}R_2}{R_3(R_1 + R_2)} \tag{15-18}$$

Since this current is independent of the signal voltages V_{s1} and V_{s2}, then $Q3$ acts to supply the DIFF AMP consisting of $Q1$ and $Q2$ with the constant current I_O.

The above result for I_O has been rendered independent of temperature because of the added diode D. Without D the current would vary with temperature because V_{BE3} decreases approximately 2.5 mV/°C (Sec. 5-8). The diode has this same temperature dependence, and hence the two variations cancel each other and I_O does not vary appreciably with temperature. Since the cutin voltage V_D of a diode has approximately the same value as the base-to-emitter voltage V_{BE3} of a transistor, then Eq. (15-17) cannot be satisfied with a single diode. Hence two diodes in series are used for V_D (Fig. 15-11).

The above discussion assumes that the resistances do not vary with temperature T. Since these have a negative temperature coefficient of resistance then, in practice, Eq. (15-17) is not satisfied, but $R_1/(R_1 + R_2)$ is chosen experimentally so that I_O in Eq. (15-16) is almost independent of T.

Consider that $Q1$ and $Q2$ are identical and that $Q3$ is a true constant-current source. Under these circumstances we can demonstrate that the common-mode gain is zero. Assume that $V_{s1} = V_{s2} = V_s$, so that from the symmetry of the circuit, the collector current I_{c1} (the increase over the quiescent value for $V_s = 0$) in $Q1$ equals the current I_{c2} in $Q2$. However, since the total current increase $I_{c1} + I_{c2} = 0$ if $I_O = $ constant, then $I_{c1} = I_{c2} = 0$ and $A_c = V_{o2}/V_s = -I_{c2}R_c/V_s = 0$.

Practical Considerations[4] In some applications the choice of V_{s1} and V_{s2} as the input voltages is not realistic because the resistances R_{s1} and R_{s2} represent the output impedances of the voltage generators V_{s1} and V_{s2}. In such a case we use as input voltages the base-to-ground voltages V_{b1} and V_{b2} of $Q1$ and $Q2$, respectively. For the analysis of nonsymmetrical differential circuits the reader is referred to Ref. 4.

The differential amplifier is often used in dc applications. It is difficult to design dc amplifiers using transistors because of drift due to variations of h_{FE}, V_{BE}, and I_{CBO} with temperature. A shift in any of these quantities changes the output voltage and cannot be distinguished from a change in input-signal voltage. Using the techniques of integrated circuits (Chap. 7), it is possible to construct a DIFF AMP with $Q1$ and $Q2$ having almost identical characteristics. Under these conditions any parameter changes due to temperature will cancel and the output will not vary.

Differential amplifiers may be cascaded to obtain larger amplifications for the difference signal. Outputs V_{o1} and V_{o2} are taken from each collector (Fig. 15-8) and are coupled directly to the two bases, respectively, of the next stage (Fig. 15-11).

Finally, the differential amplifier may be used as an emitter-coupled phase inverter. For this application the signal is applied to one base, whereas the second base is not excited (but is, of course, properly biased). The output voltages taken from the collectors are equal in magnitude and 180° out of phase.

15-4   TRANSFER CHARACTERISTICS OF A DIFFERENTIAL AMPLIFIER

It is important to examine the transfer characteristic[5] (I_C versus $V_{B1} - V_{B2}$) of the DIFF AMP of Fig. 15-8 to understand its advantages and limitations. We first consider this circuit qualitatively. When V_{B1} is below the cutoff point of $Q1$, all the current I_O flows through $Q2$ (assume for this discussion that V_{B2} is constant). As V_{B1} carries $Q1$ above cutoff, the current in $Q1$ increases, while the current in $Q2$ decreases, and the sum of the currents in the two transistors remain constant and equal to I_O. The total range ΔV_O over which the output can follow the input is $R_C I_O$ and is therefore adjustable through an adjustment of I_O.

From Fig. 15-8 we have

$$I_{E1} + I_{E2} = -I_O \tag{15-19}$$

$$V_{B1} - V_{B2} = V_{BE1} - V_{BE2} \tag{15-20}$$

The emitter current I_E of each transistor is related to the voltage V_{BE} by the diode volt-ampere characteristic

$$I_E = I_S \epsilon^{V_{BE}/V_T} \tag{15-21}$$

where I_S is defined in terms of the Ebers-Moll parameters in Prob. 15-12.

If we assume that $Q1$ and $Q2$ are matched, it follows from Eqs. (15-19) to (15-21) that

$$I_{C1} \approx -I_{E1} = \frac{I_O}{1 + \exp\left[-(V_{B1} - V_{B2})/V_T\right]} \tag{15-22}$$

and I_{C2} is given by the same expression with V_{B1} and V_{B2} interchanged. The transfer characteristics described by Eq. (15-22) for the normalized collector currents I_{C1}/I_O (and I_{C2}/I_O) are shown in Fig. 15-9, where the abscissa is the normalized differential input $(V_{B1} - V_{B2})/V_T$.

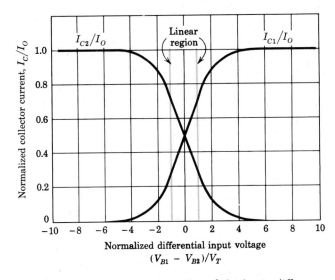

Fig. 15-9 Transfer characteristics of the basic differential-amplifier circuit.

If Eq. (15-22) is differentiated with respect to $V_{B1} - V_{B2}$, we have the transconductance g_{md} of the DIFF AMP with respect to the differential input voltage, or

$$\frac{dI_{C1}}{d(V_{B1} - V_{B2})} = g_{md} = \frac{I_O}{4V_T} \tag{15-23}$$

where g_{md} is evaluated at $V_{B1} = V_{B2}$. This equation indicates that, for the same value of I_O, the effective transconductance of the differential amplifier is one-fourth that of a single transistor [Eq. (11-4)]. An alternative proof of Eq. (15-23) is given in Prob. 15-11.

The following conclusions can be drawn from the transfer curves of Fig. 15-9:

1. The differential amplifier is a very good limiter, since when the input $(V_{B1} - V_{B2})$ exceeds $\pm 4V_T$ ($\approx \pm 100$ mV at room temperature), very little further increase in the output is possible.

2. The slope of these curves defines the transconductance, and it is clear that g_{md} starts from zero, reaches a maximum of $I_O/4V_T$ when $I_{C1} = I_{C2} = \frac{1}{2}I_O$, and again approaches zero.

3. The value of g_{md} is proportional to I_O [Eq. (15-23)]. Since the output voltage change V_{o2} is given by

$$V_{o2} = g_{md}R_c\Delta(V_{B1} - V_{B2}) = g_{md}R_c(V_{b1} - V_{b2}) \tag{15-24}$$

it is possible to change the differential gain by varying the value of the current I_O. This means that automatic gain control (AGC) is possible with the DIFF AMP.

4. The transfer characteristics are linear in a small region around the operating point where the input varies approximately $\pm V_T$ (± 26 mV at room temperature). In Prob. 15-14 we show that it is possible to increase the region of linearity by inserting two equal resistors R_e in series with the emitter leads of $Q1$ and $Q2$. This current-series feedback added to each transistor results in a smaller value of g_{md}. Reasonable values for R_e are $50 - 100$ Ω, since for large values, A_d is reduced too much. The insertion of R_e also increases the input impedance.

15-5   AN EXAMPLE OF AN IC OPERATIONAL AMPLIFIER

An integrated OP AMP usually consists of a cascade of four stages. As indicated in Fig. 15-10, the first stage is a DIFF AMP with a double-ended output, the second stage is a DIFF AMP with a single-ended output, the third stage is an emitter follower, and the last stage is a dc level translator and output driver.

In this section we examine in some detail an example[6] of an integrated operational amplifier, such as the Motorola MC1530 shown in Fig. 15-11. This amplifier is constructed to utilize the advantages of monolithic integrated circuits. It offers low offset voltage and current, small size, increased reliability, and excellent temperature tracking.

The purpose of this section is to analyze and evaluate the performance of this OP AMP.

Input Resistance   The first stage, A_{V1}, consists of $Q2$ and $Q3$, with $Q1$ used as a constant-current source to provide high common-mode rejection. The differential input resistance R_{id} to the total input signal V_1 is $2h_{ie}$, provided $R_s = 0$ and $h_{oe}R_c \leq 0.1$. This statement follows from the fact that, since $Q1$ acts as a constant current I_O, the emitters of $Q2$ and $Q3$ are floating. Hence the resistance between the two inputs 1 and 2 is $h_{ie2} + h_{ie3} = 2h_{ie}$. If input 2 is grounded, then input 1 is loaded by $2h_{ie}$.

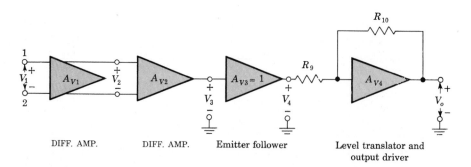

Fig. 15-10   The Motorola MC1530 as a four-stage cascaded amplifier.

If we neglect $r_{bb'}$ compared with $r_{b'e}$, then

$$h_{ie} \approx r_{b'e} = \frac{h_{fe}}{g_m} = \frac{h_{fe}V_T}{|I_C|} \tag{15-25}$$

where use is made of Eq. (11-11). We show further on in this section that $I_{C2} = I_{C3} = I_{C4} = I_{C5} \approx 0.5$ mA. Since $h_{fe} = 100$ for Q2 and Q3, the differential input resistance is

$$R_{id} = 2h_{ie} = \frac{2 \times 100 \times 26}{0.5}\,\Omega = 10.4 \text{ K}$$

If this resistance is too small for the applied signal source, it can be increased by modifying the input circuit. For example, some IC OP AMPS have a Darlington pair (Sec. 8-16) in place of Q2 and another in place of Q3. Another modification[7] is to add a matched discrete FET differential stage at the input, or preferably to fabricate an FET differential pair on the same chip with the rest of the OP AMP. Widlar[1] has designed OP AMPS (National Semiconductor Corp. LM108, for example) using supergain transistors (Sec. 7-6) in the input stage (current gains of 5,000 can be obtained at 1-μA collector current). For this transistor we find from Eq. (15-25)

$$h_{ie} = \frac{5,000 \times 26 \times 10^{-3}}{10^{-6}}\,\Omega = 130 \text{ M}$$

which is very high indeed for a bipolar transistor.

The differential input resistance of the second stage, consisting of the differential pair Q4 and Q5, is $2h_{ie}$. However, since double-ended signals are applied to Q4 and Q5, then the resistance looking into each base is half this value, or h_{ie}. This result follows from the equivalent circuit of Fig. 15-7b, which indicates that the emitter is effectively at ground potential. Since it is known that h_{fe} for transistor Q4 or Q5 is also 100, then $h_{ie} = 10.4/2 = 5.2$ K. This resistance is effectively connected from each collector of Q2 and Q3 to ground. Hence the equivalent collector-circuit load is

$$R_{L2} = R_{L3} = 7.75 \| 5.20 = 3.12 \text{ K}$$

Open-loop Voltage Gain The differential gain $A_d = A_{V1}$ is given by Eq. (15-14) multiplied by 2 (because the collector-to-collector output is twice the collector-to-ground output). Since $R_s = 0$, $h_{fe} = 100$, and

$$h_{ie} = 10.4/2 = 5.2 \text{ K}$$

for the first stage,

$$A_{V1} = \frac{V_2}{V_1} = \frac{h_{fe}R_{L2}}{h_{ie}} = \frac{100 \times 3.12}{5.20} = 60.0$$

For the second stage, $h_{fe} = 100$, $h_{ie} = 5.2$ K, and the load is $R_7 = 3$ K if we neglect the loading on Q5 of the emitter follower Q6 (whose input impedance is high compared with 3 K).   Since the second stage has a single-ended output, the differential gain is

$$A_{V2} = \frac{V_3}{V_2} = -\frac{1}{2}\frac{h_{fe}R_7}{h_{ie}} = -\frac{100 \times 3}{2 \times 5.2} = -28.9$$

For the emitter follower, $A_{V3} \approx 1$.   The output stage uses voltage-shunt feedback because of R_9 and R_{10}.   From Eq. (15-1)

$$A_{V4} \approx -\frac{R_{10}}{R_9} = -\frac{30}{6} = -5$$

Hence the overall OP AMP differential voltage gain is

$$A_V = (60.0)(-28.9)(-5) = +8{,}670$$

Note that node 1 is the noninverting input terminal.

DC Analysis   It is necessary to know the dc currents and voltages of the circuit to obtain the open-loop gain and the differential input resistance and to understand the operation of the level-translator output stage.

Let us start with the current source $Q1$: We assume that all base currents can be neglected and all diode forward voltages and base-to-emitter voltages are 0.7 V.   The dc voltage V_{BN1} of the base of $Q1$ with respect to ground N is (from Fig. 15-11)

$$V_{BN1} = \frac{[-V_{EE} + 2(0.7)]R_5}{R_4 + R_5} = \frac{(-6 + 1.4)(3.2)}{1.5 + 3.2} = -3.14 \text{ V}$$

and

$$I_O \approx I_1 = \frac{V_{EE} + (V_{BN1} - 0.7)}{R_1} = \frac{6 - 3.84}{2.2} = 0.99 \text{ mA}$$

If it is assumed that the integrated transistors $Q2$ and $Q3$ are identical, one-half of I_1 will flow through each:

$$I_{C2} = I_{C3} = 0.495 \text{ mA}$$

The dc voltage of the base of $Q4$ or $Q5$ with respect to ground is

$$V_{BN4} = V_{BN5} = V_{CC} - I_{C3}R_3 = 6 - 0.495 \times 7.75 \approx 2.18 \text{ V}$$

The dc voltage at the common emitter of $Q4$ and $Q5$ is

$$V_{EN4} = V_{BN4} - V_{BE4} = 2.18 - 0.7 = 1.48 \text{ V}$$

and the current in R_6 is

$$I_6 = \frac{V_{EN4}}{R_6} = \frac{1.48}{1.5} = 0.986 \text{ mA}$$

Since $I_6 = I_{C4} + I_{C5} = 2I_{C5}$, then $I_{C5} = 0.493$ and the base voltage of $Q6$, which equals the collector voltage V_3 of $Q5$, is

$$V_3 = V_{BN6} = V_{CN5} = V_{CC} - I_{C5}R_7 = 6 - (0.493)(3) = 4.52 \text{ V}$$

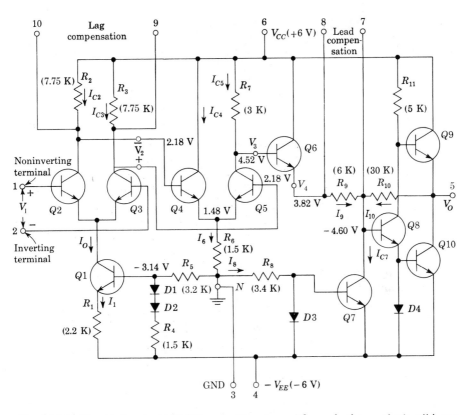

Fig. 15-11 The Motorola MC1530 operational amplifier. In the analysis all base currents are neglected.

The output V_4 of the emitter follower is

$$V_4 = V_{EN6} = V_{BN6} - V_{BE6} = 4.52 - 0.7 = 3.82 \text{ V}$$

The Output Stage The last stage provides level translation and a symmetrical output swing (at low impedance) with respect to ground. When the differential input voltage V_1 is zero, the output V_O should be zero. Of course, due to mismatch of V_{BE} and h_{FE}, there will be some nonzero output voltage, which we consider in Sec. 15-7.

The voltage $V_{EN6} = 3.82$ V must be reduced to zero at the amplifier output while dc coupling is maintained. We shall now demonstrate that this level translation can be accomplished with the circuit parameter values in Fig. 15-11. Note that $Q7$ is biased by $D3$ in the manner explained in Sec. 9-7. Hence, following our discussion in Sec. 9-7 with respect to Fig. 9-11a, we find

$$I_{C7} \approx I_8 = \frac{V_{EE} - V_{D3}}{R_8} = \frac{6.0 - 0.7}{3.4} = 1.56 \text{ mA}$$

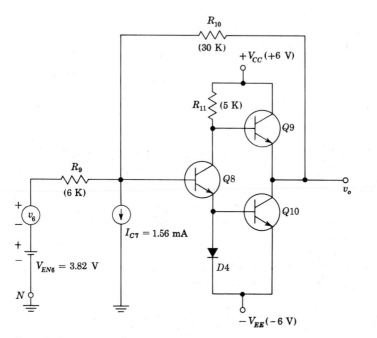

Fig. 15-12   The output stage of the MC1530.

The voltage from the base of $Q8$ to ground is

$$V_{BN8} = V_{BE8} + V_{D4} - V_{EE} = 0.7 + 0.7 - 6 = -4.60 \text{ V}$$

The currents in R_9 and R_{10} are

$$I_9 = \frac{V_{EN6} - V_{BN8}}{R_9} = \frac{3.82 + 4.60}{6} = 1.40 \text{ mA}$$

$$I_{10} = I_{C7} - I_9 = 1.56 - 1.40 = 0.16 \text{ mA}$$

Finally, the dc output voltage is

$$V_O = V_{BN8} + I_{10}R_{10} = -4.60 + (0.16)(30) = 0.20 \text{ V}$$

This calculated value for V_O is not to be taken too seriously, because we obtained I_{10} as the difference between two almost equal numbers. Such a subtraction can result in a large error in the small difference. Note that if I_{10} were 0.153 (instead of 0.16), then $V_O = 0$ (instead of 0.2 V). Also, I_{10} is greatly affected by small changes in the circuit parameter values. Hence a balancing technique (Sec. 15-6) is used to ensure that $V_O = 0$ for $V_1 = 0$.

To consider the output stage under conditions of an applied excitation, refer to Fig. 15-12, where v_6 represents the signal voltage at the emitter of $Q6$, V_{EN6} is the dc voltage at this emitter, and I_{C7} is the constant current supplied by $Q7$. The signal v_6 is amplified by $Q8$ and is transmitted to the *totem-pole* arrangement of $Q9$ and $Q10$. If v_6 is positive, then the current in $Q9$ is decreased and that in $Q10$ is increased, and current is taken from the load which is across the output and v_o decreases. Similarly, if v_6 is negative, the current in $Q9$ is increased, that in $Q10$ is decreased, current is delivered to the load, and v_o increases.

For a very large positive v_6, $Q9$ is cut off and $Q10$ is driven into saturation. Under these circumstances $v_o = -V_{EE} + V_{CE10,sat} = -6 + 0.2 = -5.8$ V. Similarly, for a very large negative v_6, $Q10$ is cut off and $Q9$ is driven into saturation, so that $v_o = V_{CC} - V_{CE9,sat} = 6 - 0.2 = +5.8$ V. The maximum peak-to-peak output swing is 11.6 V.

Note that the output stage is stabilized by means of the voltage-shunt feedback supplied by resistors R_9 and R_{10}.

Common-mode Voltage Swing   We are going to show now that V_{BN1} and V_{CN2} set a limit on the input *common-mode voltage swing* V_{iCM}. This parameter is defined as the maximum peak input voltage that can be applied to either input terminal without causing abnormal operation or damage. The positive limit of V_{iCM} depends on the collector voltage of the input stage; that is, $V_{CN2} = V_{BN4} \approx 2.2$ V. If V_{iCM} exceeds 2.2 V, then the collector-to-base junction of $Q2$ will become forward biased and $Q2$ may saturate. On the other hand, if V_{iCM} becomes more negative than

$$V_{BN1} + V_{BE2} = -3.14 + 0.7 = -2.44 \text{ V}$$

then the collector of $Q1$ will become forward biased, and this will result in abnormal operation. Therefore, when the power supplies are ± 6 V, the common-mode voltage swing for this amplifier should not exceed ± 2 V maximum.

15-6   OFFSET ERROR VOLTAGES AND CURRENTS

In Sec. 15-1 we observe that the ideal operational amplifier shown in Fig. 15-1a is perfectly balanced, that is, $V_o = 0$ when $V_1 = V_2$. A real operational amplifier exhibits an unbalance caused by a mismatch of the input transistors. This mismatch results in unequal bias currents flowing through the input terminals, and also requires that an input offset voltage be applied between the two input terminals to balance the amplifier output.

In this section we are concerned with the dc error voltages and currents that can be measured at the input and output terminals.

Input Bias Current   The input bias current is one-half the sum of the separate currents entering the two input terminals of a balanced amplifier, as shown in Fig. 15-13a. Since the input stage is that shown in Fig. 15-8, the input bias current is $I_B = (I_{B1} + I_{B2})/2$ when $V_o = 0$.

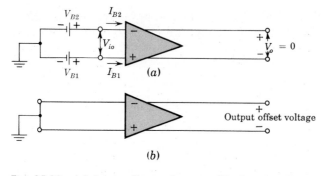

Fig. 15-13   (a) Input offset voltage.   (b) Output offset voltage.

Input Offset Current   The input offset current I_{io} is the difference between the separate currents entering the input terminals of a balanced amplifier.   As shown in Fig. 15-13a, we have $I_{io} = I_{B1} - I_{B2}$ when $V_o = 0$.

Input Offset Current Drift   The input offset current drift $\Delta I_{io}/\Delta T$ is the ratio of the change of input offset current to the change of temperature.

Input Offset Voltage   The input offset voltage V_{io} is that voltage which must be applied between the input terminals to balance the amplifier, as shown in Fig. 15-13a.

Input Offset Voltage Drift   The input offset voltage drift $\Delta V_{io}/\Delta T$ is the ratio of the change of input offset voltage to the change in temperature.

Output Offset Voltage   The output offset voltage is the difference between the dc voltages present at the two output terminals (or at the output terminal and ground for an amplifier with one output) when the two input terminals are grounded (Fig. 15-13b).

Power Supply Rejection Ratio   The power supply rejection ratio (PSRR) is the ratio of the change in input offset voltage to the corresponding change in one power supply voltage, with all remaining power supply voltages held constant.

Slew Rate   The slew rate is the time rate of change of the closed-loop amplifier output voltage under large-signal conditions.

The various parameters of a typical monolithic operational amplifier are given in Table 15-1.

TABLE 15-1   Typical parameters of monolithic
operational amplifier

Open-loop gain A_d....................... 50,000
Input offset voltage V_{io}................... 1 mV
Input offset current I_{io}.................... 10 nA
Input bias current I_B..................... 100 nA
Common-mode rejection ratio ρ........... 100 dB
PSRR................................ 20 μV/V
I_{io} drift............................... 0.1 nA/°C
V_{io} drift.............................. 1.0 μV/°C
Slew rate.............................. 1 V/μs

Universal Balancing Techniques   When we use an operational amplifier,
it is often necessary to balance the offset voltage.   This means that we must
apply a small dc voltage in the input so as to cause the dc output voltage to
become zero.   The techniques shown here allow offset-voltage balancing without regard to the internal circuitry of the amplifier.   The circuit shown in
Fig. 15-14a supplies a small voltage effectively in series with the noninverting

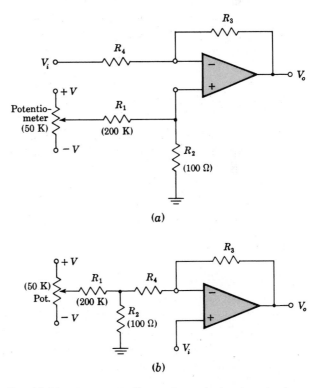

(a)

(b)

Fig. 15-14   Universal offset-voltage balancing circuits
for (a) inverting and (b) noninverting operational amplifiers.   (See Fig. 15-18 for balancing of a DIFF AMP.)

input terminal in the range $\pm V[R_2/(R_1 + R_2)] = \pm 7.5\,\mathrm{mV}$ if ± 15-V supplies are used and $R_1 = 200\,\mathrm{K}$, $R_2 = 100\,\Omega$. Thus this circuit is useful for balancing inverting amplifiers even when the feedback element R_3 is a capacitor or a nonlinear element. If the operational amplifier is used as a noninverting amplifier, the circuit of Fig. 15-14b is used for balancing the offset voltage.

15-7   TEMPERATURE DRIFT OF INPUT OFFSET VOLTAGE AND CURRENT

The most critical influence on the operation of the operational amplifier is exercised by the input differential stage because the equivalent error effects of the subsequent stages are reduced by the gain provided by the first stage. In this section we examine the input error signals and thermal drifts of the input differential stage, shown in Fig. 15-8.

From Fig. 15-8 we see that the input offset voltage is

$$V_{io} = V_{BE1} - V_{BE2} \tag{15-26}$$

where V_{BE1} and V_{BE2} correspond to $I_{C1} = I_{C2}$. It is possible to fabricate matched integrated transistor pairs, where the base-to-emitter voltage difference is approximately one millivolt.

The input offset voltage drift can be found using Eqs. (15-26) and (19-92):

$$\frac{dV_{io}}{dT} = \frac{dV_{BE1}}{dT} - \frac{dV_{BE2}}{dT} = \frac{V_{BE1} - V_{BE2}}{T}$$

or

$$\frac{dV_{io}}{dT} = \frac{V_{io}}{T} \tag{15-27}$$

From the differential equation (15-27) we find

$$V_{io} = CT \tag{15-28}$$

where C is a constant. Thus we see that the input offset voltage drift is independent of temperature. If we assume that $V_{io} = 1\,\mathrm{mV}$ at room temperature, then $dV_{io}/dT = 10^{-3}/300 = 3.3\,\mu\mathrm{V}/°\mathrm{C}$, as compared with

$$dV_{BE}/dT = -2.5\,\mathrm{mV}/°\mathrm{C}$$

for a single common-emitter transistor.

The input offset bias current has been defined as $I_{io} = I_{B1} - I_{B2}$, and if $I_{B1} \neq I_{B2}$, even with equal source resistances a differential input error voltage will be produced at the input of the first stage. Ideally, we would like to have $I_{B1} = I_{B2} = 0$, and it is for this reason primarily that very high input-resistance differential stages are used (Sec. 15-5).

Since

$$I_{io} = \frac{I_{C1}}{\beta_1} - \frac{I_{C2}}{\beta_2}$$

we find that the input current drift caused by mismatch in β_1 and β_2, if $I_{C1} = I_{C2}$, is given by

$$\frac{dI_{io}}{dT} = -\left(\frac{1}{\beta_1}\frac{d\beta_1}{dT}\right)I_{B1} + \left(\frac{1}{\beta_2}\frac{d\beta_2}{dT}\right)I_{B2} \tag{15-29}$$

For matched integrated transistors the beta temperature coefficients are almost equal and the beta variation is caused by changing minority-carrier lifetime in the base region.[8]   For typical silicon transistors it is found that

$$\frac{1}{\beta}\frac{d\beta}{dT} = \begin{cases} -0.005°\text{C}^{-1} & T > 25°\text{C} \\ -0.015°\text{C}^{-1} & T < 25°\text{C} \end{cases}$$

The drift expression now becomes

$$\frac{dI_{io}}{dT} \approx -\left(\frac{1}{\beta}\frac{d\beta}{dT}\right)I_{io} \tag{15-30}$$

It is possible to reduce the input bias current and the corresponding current drift by using Darlington pairs instead of $Q1$ and $Q2$ in Fig. 15-8.   However, because of the added pair of emitter junction voltages, there is an increase in the input offset voltage and voltage drift.

15-8   MEASUREMENT OF OPERATIONAL AMPLIFIER PARAMETERS

In this section we describe practical methods of measuring some of the important parameters of operational amplifiers.   Specifically, we examine (1) open-loop voltage gain A_V, (2) output resistance R_o without feedback, (3) differential input resistance R_i, (4) input offset voltage V_{io}, (5) input bias current I_B and input offset current I_{io}, (6) common-mode rejection ratio, and (7) slewing rate.

Open-loop Differential Voltage Gain $A_V = A_d$ The open-loop voltage gain is defined as the ratio of the output signal voltage to the input differential signal voltage V_i.   Figure 15-15 shows a technique of measuring this parameter.   It is essential that the effect of the input offset voltage be canceled as shown in the figure, since otherwise the high amplification of this voltage will result in output saturation.   The input excitation V_s is an ac signal; by varying its frequency we can obtain the frequency response A_V.   The input attenuator is essential so that V_i can be at a sufficiently low level for output swings no greater than about 30 percent of the output voltage rating (to ensure linear operation).

Output Resistance R_o The output resistance R_o of the operational amplifier can be obtained using the circuit of Fig. 15-15 and measuring the decrease in the low-frequency gain A_V caused by a load resistance R_L.   Then, from Fig. 15-1b

$$A_V = \frac{R_L}{R_L + R_o}A_v$$

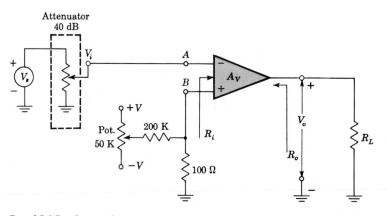

Fig. 15-15 Circuit for measuring A_V, R_o, and R_i.

and

$$R_o = \left(\frac{A_v}{A_V} - 1\right) R_L \qquad\qquad (15\text{-}31)$$

where A_v and A_V are the open-loop gains with $R_L = \infty$ and $R_L \neq \infty$, respectively.

Differential Input Resistance R_i The differential input resistance R_i can be measured by forming a voltage divider at the input of the amplifier in Fig. 15-15.   This is done by inserting two equal resistors R at points A and B in series with the inverting and noninverting terminals.   The new output is

$$V_o' = \frac{R_i}{R_i + 2R} V_o$$

where V_o is the value measured with $R = 0$.   Two resistors instead of one are used so that any stray coupling from the output will generate equal signals at the inverting and noninverting terminals.   These equal stray input signals will be prevented from reaching the output by the common-mode rejection. Often a capacitor C is placed across each of the resistors R to reduce high-frequency noise.
From the above equation we solve for R_i.

$$R_i = 2R \frac{V_o'}{V_o - V_o'} \qquad\qquad (15\text{-}32)$$

If the capacitors C are employed, the input signal frequency must be much less than $1/(2\pi RC)$.   If R_i is very high, as with FET input differential stages, then $V_o \approx V_o'$, and the measurement is not practical.

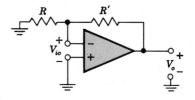

Fig. 15-16 Measurement of input offset voltage V_{io}.

Input Offset Voltage V_{io} The simple closed-loop circuit of Fig. 15-16 can be used for this measurement. We have from Eq. (15-4)

$$V_o = \frac{R + R'}{R} V_{io} \tag{15-33}$$

If $R = 100 \ \Omega$ and $R' = 100$ K, the small input offset voltage is multiplied at the output by a factor of 1,001, and thus it is easily measured.

Input Bias Current In Fig. 15-17 we allow the input bias currents I_{B1} and I_{B2} to flow through the two large resistors $R_B > 10$ M, while the amplifier is connected as a unity-gain noninverting amplifier. If resistors R_B are selected so that $I_B R_B > V_{io}$, then the voltages created by the bias currents are much larger than V_{io}.

If we connect terminals A and B, the measured output will be $V_o = I_{B2} R_B$; similarly, the measured output will be $V_o = -I_{B1} R_B$ if we connect terminals C and D. The large resistors R_B are bypassed with 0.01-μF capacitors to reduce high-frequency noise. The bias current I_B is defined in Sec. 15-6 as the average of I_{B1} and I_{B2}, while the input offset current I_{io} is the difference $I_{B1} - I_{B2}$ of the individual base currents.

Common-mode Rejection Ratio The common-mode rejection ratio ρ is defined in Eq. (15-11) by

$$\rho \equiv \left| \frac{A_d}{A_c} \right|$$

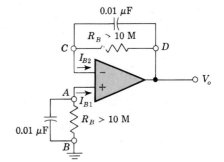

Fig. 15-17 Measurement of bias currents.

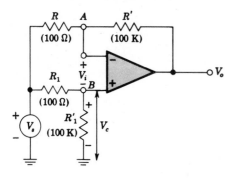

Fig. 15-18   Measurement of common-mode rejection ratio.   (The offset voltage is balanced with the network of Fig. 15-14; R_2 is inserted between the bottom of R_1' and ground.)

where A_d is the differential voltage gain, and A_c is the common-mode voltage gain.   For the circuit of Fig. 15-18, the signal at point A or B is essentially the common-mode signal V_c, where

$$V_c = \frac{R_1'}{R_1 + R_1'} V_s = \frac{R'}{R + R'} V_s \approx V_s$$

assuming that $R_1' \gg R_1$ and $R' \gg R$.   The voltage across R is $V_i - V_s R_1 / (R_1 + R_1')$ and that across R' is $V_o - V_i - V_s R_1'/(R_1 + R_1')$.   Equating the currents in R and R' and assuming $R = R_1$ and $R' = R_1'$ we obtain $V_i = [R/(R + R')]V_o$.   From Eq. (15-9)

$$V_o = A_d V_i + A_c V_c = \frac{A_d R V_o}{R + R'} + \frac{A_c R' V_s}{R + R'}$$

If $A_d R/(R + R') \gg 1$, then $A_d R V_o + A_c R' V_s \approx 0$, and

$$\rho = \left|\frac{A_d}{A_c}\right| = \frac{R'}{R}\left|\frac{V_s}{V_o}\right| \tag{15-34}$$

Note that for $A_d = 50,000$, $R' = 100K$, and $R = 100\Omega$, then $A_d R/(R + R') = 50$ which satisfies the inequality assumed above.   It is found that A_c is a nonlinear function of the magnitude of V_c.   For this reason it is important to make the above measurement at the rated common-mode voltage swing.

Slewing Rate[9]   The maximum rate of change of the output voltage when supplying the rated output is defined in Sec. 15-6 as the slewing rate.   This rate dV_o/dt can be measured using the noninverting circuit of Fig. 15-4, with $R = \infty$ and $R' = 0$, since this usually represents the worst case.   If the amplifier has a single-ended input, then the circuit of Fig. 15-2a is used, with $Z = R = 1$ K and $Z' = R' = 10$ K.   The input V_s is a high-frequency square wave, and the slopes with respect to time of the leading and trailing edges of the output signal are measured.   It is common to specify the slower of the two rates as the slewing rate of the device.

15-9   FREQUENCY RESPONSE OF OPERATIONAL AMPLIFIERS

The typical operational amplifier discussed in Sec. 15-5 consists of four stages, as shown in Fig. 15-11.   If we assume that the amplifier is driven by two signal sources which are equal and opposite in phase, there are only true difference signals present.   Thus we need consider only half of each differential pair for an analysis of the frequency response.   This response can be obtained by considering a cascade of two common-emitter stages, an emitter follower and the output stage.   The general method of obtaining the high-frequency response of such a chain of interacting stages was presented in Secs. 12-10 and 12-11.   The computational complexity is so great that computer-aided analysis is required.   The response may also be obtained by laboratory measurements.

The open-loop gain of the OP AMP has a transfer function with several poles and with zeros at much higher frequencies than the poles.   Experimentally, the poles can be found from the amplitude response curve (the plot of the magnitude of gain in decibels versus log f).   Tangent to this curve are drawn straight lines whose slopes are 0, -20 dB per decade, -40 dB per decade, . . . , as indicated in Figs. 12-4 and 12-6.   The pole frequencies f_1, f_2, . . . are then obtained from the *corner* frequencies, the values of f at which adjacent lines intersect.

The poles and zeros of $A(jf)$ are normally specified by the manufacturer in data sheets provided with commercial operational amplifiers.   In Fig. 15-19 we show the open-loop gain and phase response of a typical OP AMP (μA702), using the straight-line Bode approximation.   We see that the transfer function has three poles, one at 1 MHz, a second at 4 MHz, and a third at 40 MHz.   For the MC1530 the manufacturer gives the first three poles at 1, 6, and 22 MHz.

Stability of an OP AMP   For the inverting OP AMP of Fig. 15-2a, with $Z = R$, $Z' = R'$, and $R_i = \infty$, we obtain from Eq. (15-2)

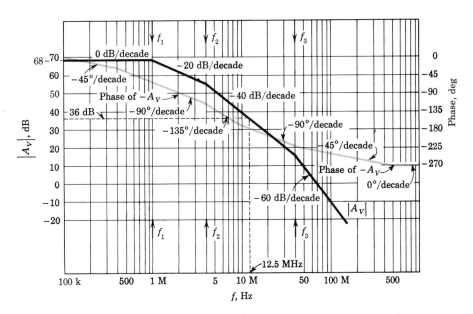

Fig. 15-19 Open-loop gain and phase shift characteristics of the μA702A.

$$A_{Vf} = \frac{V_o}{V_s} = \frac{R'}{R + R'} \frac{A_V}{1 - RA_V/(R + R')} \tag{15-35}$$

This equation may also be obtained by using the feedback concepts of Chap. 13, as we now demonstrate.

The topology corresponds to voltage-shunt feedback, and it is found (Prob. 15-24) that

$$\beta = -\frac{1}{R'} \qquad R_M = \frac{A_V RR'}{R + R'} \qquad A_{Vf} = \frac{R_{Mf}}{R} \tag{15-36}$$

Using the feedback formula [Eq. (13-4)]

$$R_{Mf} = \frac{R_M}{1 + \beta R_M}$$

and Eqs. (15-36), the expression for A_{Vf} in Eq. (15-35) is obtained.

The condition for oscillation is

$$\beta R_M = -\frac{RA_V}{R + R'} = 1\underline{/180°} \tag{15-37}$$

Similarly, for the noninverting amplifier of Fig. 15-4, we find

$$A_{Vf} = \frac{V_o}{V_s} = \frac{-A_V}{1 - [R/(R + R')]A_V} \tag{15-38}$$

We see from Eq. (15-38) that Eq. (15-37) represents the stability criterion for both the inverting and noninverting operational amplifiers.

It is important to point out that for negative feedback the gain A_V represents a negative real number at low frequencies. Hence we observe that if the product $[R/(R + R')]|A_V|$ becomes unity when the phase shift of $-A_V$ reaches 180°, the amplifier will oscillate. From the open-loop gain and phase shift of the μA702A shown in Fig. 15-19, we find that at the frequency of $f = 12.5$ MHz, where the phase shift of $-A_V$ equals 180°, the magnitude of A_V is 36 dB. Thus, from Eq. (15-37) we have

$$20 \log \frac{R}{R + R'} + 20 \log |A_V| = 20 \log 1 = 0 \tag{15-39}$$

$$20 \log \frac{R}{R + R'} = -36$$

$$\frac{R}{R + R'} = \frac{1}{63} \quad \text{and} \quad \frac{R}{R'} = \frac{1}{62}$$

For 45° phase margin we find, from Fig. 15-19, $20 \log |A_V| = 50$ dB, and from Eq. (15-39) we obtain $20 \log [R/(R + R')] = -50$ dB, or $R/R' \approx \frac{1}{316}$. Since the closed-loop voltage gain is approximately $A_{Vf} \approx -R'/R$, we see that the low-frequency gain of the inverting feedback amplifier cannot be less than 316 in magnitude for a phase margin of at least 45°.

15-10   DOMINANT–POLE COMPENSATION

In the preceding section we observed that the μA702A op amp will be unstable if sufficient feedback is used to obtain a low-frequency gain with feedback less than 62. By adding poles and zeros to the frequency response of the loop gain, we can compensate the phase shift introduced by β and/or A_V to ensure stability, limit any peaking in the closed-loop frequency response, and even reduce overshoot and ringing if a square wave is applied to the closed-loop operational amplifier. Our discussion in Sec. 14-11 on compensation of a feedback amplifier is equally applicable to an op amp. Thus we can cause the gain A_V to decrease as a function of frequency by inserting a single capacitor to shunt the signal path to ground. The capacitance is chosen such that it creates a dominant pole in A_V low enough in frequency so that the magnitude of the loop gain becomes less than unity at a frequency where the amplifier introduces negligible phase shift. Since the capacitor adds a phase shift smaller than 90°, the circuit will be stable. A possible point to which to connect such a compensating capacitor in Fig. 15-11 is from pin 5, the output point, to ground. This is done primarily to suppress internally generated broadband noise voltages. The more desirable location[6] for the compensation capacitor is between pins 9 and 10, since the slew rate decreases with increasing capacitance. A larger capacitance (for the same roll-off pole) is required at the output pin 5, due to the lower resistance seen by the capacitor at this node than between pins 9 and 10.

Figure 15-20a shows the open-loop voltage gain A_V versus frequency for three different values of roll-off capacitance between pins 9 and 10 for the MC1530 op amp. Figure 15-20b shows the decreasing slew rate with increasing compensating capacitance for the same amplifier. Figure 15-21a shows the closed-loop voltage gain with frequency for $A_{Vf_o} = 100$, 10, and 1 for the

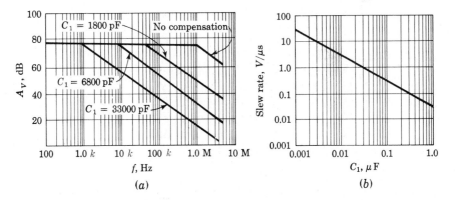

Fig. 15-20   (a) The MC1530 open-loop voltage gain A_V versus f for three different values of compensating capacitor C_1.   (b) Slew rate versus C_1. (Courtesy of Motorola, Inc.)

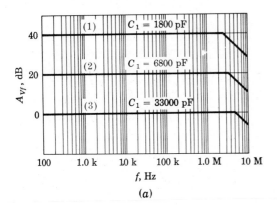

Fig. 15-21 (a) Closed-loop voltage gain A_{Vf} versus f for three different values of compensating capacitor. (b) Amplifier connection. Note:

Curve	R	R'	R_1
1	1	100	1 K
2	10	100	10 K
3	10	10	5 K

(Courtesy of Motorola, Inc.)

amplifier connected as shown in Fig. 15-21b. It is clear that the added capacitor modifies the open-loop gain A_V so that it becomes A'_V which has a much lower dominant pole f_d, determined from

$$f_d = \frac{1}{2\pi R_i C_1} \tag{15-40}$$

and R_i is the resistance seen looking into the terminals to which the capacitor is connected. The frequency f_d is found graphically by having A'_V pass through 0 dB with a slope of 20 dB per decade. Usually, f_d is selected so that A'_V passes through 0 dB at the first pole f_1 of the uncompensated A_V.

15-11   POLE–ZERO COMPENSATION

The simple dominant-pole compensation through a single capacitor reduces bandwidth drastically, as is seen from Fig. 15-20a. Clearly, this method is useful only for low-frequency (audio) amplifiers. Improvement in bandwidth may be obtained by introducing both a pole and a zero in the gain A'_V, where the zero is designed to correspond exactly with the first pole of A_V so that the zero can cancel the pole. This technique is discussed in Sec. 14-13. Three examples of pole-zero cancellation are given below.

Figure 15-22 shows an RC network connected to a node Y of the amplifier. Note that R_y is the resistance seen by the signal at the point Y. The modified open-loop voltage gain A_V' is given by Eq. (14-55), where f_z and f_p are defined in Eqs. (14-54), with $R_2 = R_c$, $C = C_c$, and $R_1 = R_y$. In designing this network, f_z is selected equal to the first pole of A_V, and the pole $f_p < f_z$ is placed on the Bode plot so that the desired phase margin is obtained when the loop gain becomes unity.

Modification of Open-loop Input Impedance It is found that the amplifier slewing rate (or the maximum time rate of output swing at high frequencies) increases if the RC network is connected at a point where the signal swing is small, and thus only a small current is required to charge the compensating capacitor. Since the input terminals of an OP AMP are virtually shorted together, it is advantageous to place the compensating R_cC_c network between terminals 1 and 2 instead of from point Y to ground in Fig. 15-22. Thus the network is connected in parallel with the operational amplifier input impedance Z_i. Let us assume that $R' \gg R_o$ and $|Z_c| \ll |Z_i|$, where $Z_c = R_c + 1/j\omega C_c$. Then we find for the compensated gain without feedback

$$A_V'(\omega) = A_V(\omega) \frac{1 + j\omega R_c C_c}{1 + j\omega C_c(R_c + R||R' + R_1)} \frac{R'}{R + R'} \qquad (15\text{-}41)$$

This equation is of the same form as Eq. (14-55), and thus this procedure allows compensation by means of pole-zero cancellation.

Miller Effect Compensation Adding a feedback capacitance around an intermediate stage of the operational amplifier is another method of providing phase compensation by means of pole-zero cancellation. Due to the Miller effect (Sec. 11-8), a response zero coincident with the pole of that stage is developed, as we now demonstrate. Consider a differential amplifier input

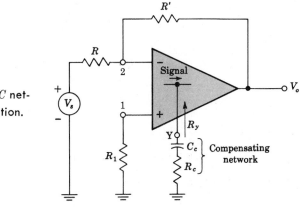

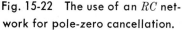

Fig. 15-22 The use of an RC network for pole-zero cancellation.

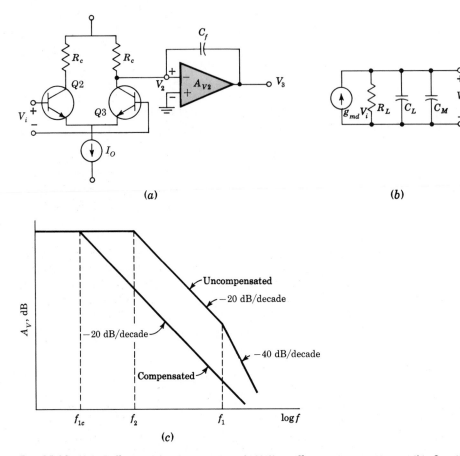

Fig. 15-23 (a) Differential input stage with Miller effect compensation. (b) Small-signal equivalent circuit of first stage. (c) Overall response of the two stages.

stage followed by a second stage across which is the compensating capacitance C_f, as indicated in Fig. 15-23a. *In the absence of C_f*, we assume that the input stage has a dominant pole f_1, and hence its voltage gain is approximated by

$$A_{V1} = \frac{V_2}{V_i} = \frac{A_{Vo1}}{1 + j(f/f_1)} \tag{15-42}$$

In Fig. 15-23b is shown the small-signal model of the input DIFF AMP which results in the transfer function of Eq. (15-42). In this circuit R_L is the effective load resistance and C_L the effective load capacitance from collector to ground of the input stage, g_{md} is the differential transconductance, and C_M is the Miller capacitance (which is zero for $C_f = 0$). Solving for V_2/V_i from this circuit, we obtain (with $C_M = 0$) Eq. (15-42), with

$$A_{Vo1} = g_{md}R_L \qquad f_1 = \frac{1}{2\pi R_L C_L} \tag{15-43}$$

We assume that the second stage also has a dominant pole f_2 (with C_f in place), and hence its voltage gain is approximated by

$$A_{V2} = \frac{V_3}{V_2} = \frac{A_{Vo2}}{1 + j(f/f_2)} \qquad (15\text{-}44)$$

For negative feedback we assume $A_{Vo2} < 0$. The Miller effect indicates that $C_M = (1 - A_{V2})C_f$, and using Eq. (15-44), we find that

$$C_L + C_M = \frac{C_o\left(1 + \dfrac{jf}{f_2}\dfrac{C_L + C_f}{C_o}\right)}{1 + jf/f_2} \approx \frac{-A_{Vo2}C_f}{1 + jf/f_2} \qquad (15\text{-}45)$$

where $C_o \equiv C_L + (1 - A_{Vo2})C_f \approx -A_{Vo2}C_f$, since $|A_{Vo2}C_f| \gg C_L + C_f$.

From Fig. 15-23b, with $C_M \neq 0$, we obtain the voltage gain A'_{V1} of the first stage after the compensating capacitor C_f has been added.

$$A'_{V1} = \frac{g_{md}R_L}{1 + j2\pi f R_L(C_L + C_M)} \qquad (15\text{-}46)$$

From Eqs. (15-45) and (15-46) is obtained

$$A'_{V1} = \frac{g_{md}R_L[1 + j(f/f_2)]}{1 + j(f/f_2) + j(f/f_{1C})} \approx \frac{g_{md}R_L[1 + j(f/f_2)]}{1 + j(f/f_{1C})} \qquad (15\text{-}47)$$

where the compensated pole of the first stage f_{1C} is given by

$$f_{1C} \equiv \frac{-1}{2\pi R_L A_{Vo2}C_f}$$

and C_f is chosen so that $f_{1C} \ll f_2$. Note that the effect of the compensating capacitor C_f is to change the pole of the first stage from f_1 to a much smaller value f_{1C} and *to add to the gain function of the first stage a zero which exactly equals the pole of the second stage.* Hence there is a pole-zero cancellation, and the overall gain A_V of the two stages is

$$A_V = A'_{V1}A_{V2} = \frac{g_{md}R_L A_{Vo2}}{1 + j(f/f_{1C})} \qquad (15\text{-}48)$$

The relative positions of f_{1C}, f_1, and f_2 are indicated in Fig. 15-23c, where the Bode plots of both stages A_V are shown, uncompensated and compensated. Since the compensated response has a slope of -20 dB per decade when it crosses the 0-dB line, the amplifier is unconditionally stable.

15-12   LEAD COMPENSATION

Lead compensation is generally provided by modifying the β network, specifically, by shunting resistor R' with a capacitance C', as shown in Fig. 15-24, so that the new loop gain will have an added positive phase shift in the frequency range near the unity-loop-gain crossover point. Equation (15-37)

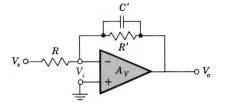

Fig. 15-24 Lead-compensated operational amplifier.

gives the loop gain for the uncompensated amplifier. If we substitute for R' an impedance Z', which is the parallel combination of R' and C', Eq. (15-37) becomes

$$\beta R_M = -\frac{R A_V}{R + Z'} = \frac{-R A A_V}{R + R'} \tag{15-49}$$

where we find that A is given by

$$A \equiv \frac{1 + j(f/f_z)}{1 + j(f/f_p)} \tag{15-50}$$

$$f_z \equiv \frac{1}{2\pi C' R'} \quad \text{and} \quad f_p \equiv \frac{R + R'}{R} f_z \tag{15-51}$$

Since $f_p \gg f_z$ then $A \approx 1 + j(f/f_z)$ which is lead compensation.

EXAMPLE Design the amplifier of Fig. 15-24 with the μA702A, using 32 dB of feedback at low frequencies. Find C' for a phase margin of 45°.

Solution Figure 15-19 shows the open-loop gain A_V of the μA702A, from which we see that with 32 dB of feedback, the phase margin is zero and the circuit will oscillate. By adding C', we introduce a phase lead due to the compensating zero, and thus we may shape the phase-shift curve so as to obtain the 45° phase margin desired. Optimum values for f_z and f_p must be found graphically from the Bode plot. Of course, f_p cannot be placed independently of f_z since they are related by Eq. (15-51).

To find the ratio f_p/f_z, we must calculate $(R + R')/R$. Since a desensitivity D of 32 dB at low frequencies is desired, then

$$20 \log D = 20 \log |1 + \beta R_{Mo}| \approx 20 \log |\beta R_{Mo}| = 32 \tag{15-52}$$

or using the values of β and R_{Mo} from Eq. (15-36) and noting that $20 \log |A_{Vo}| = 68$ dB, from Fig. 15-19, we have

$$20 \log \frac{R}{R + R'} + 20 \log |A_{Vo}| = 20 \log \frac{R}{R + R'} + 68 = 32 \tag{15-53}$$

from which we find

$$\frac{R + R'}{R} = 63 \quad \text{and} \quad f_p = 63 f_z \tag{15-54}$$

Note that the pole is located at a very much higher frequency than the zero.

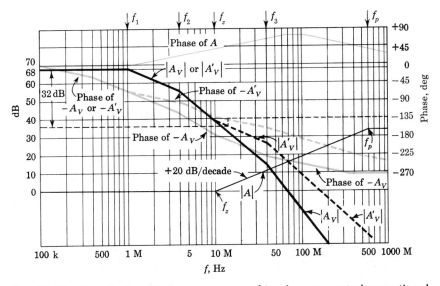

Fig. 15-25 Magnitude and phase response of lead-compensated operational amplifier ($A'_V = AA_V$).

In Fig. 15-25 is indicated the idealized Bode plot for the open-loop gain A_V of the µA702A OP AMP, the transfer function A of the compensating network, and the product of these two, namely, $A'_V \equiv AA_V$. The magnitude of A_V is indicated by solid heavy lines, of A by solid lightweight lines, and of A'_V by solid dashed lines. The phase lines are drawn in a corresponding manner, except that they are shaded. The magnitude and phase of A_V are identical with the corresponding Bode plots in Fig. 15-19.

Since $f_p \gg f_z$, then the pole has practically no influence on the resultant gain A'_V in the neighborhood of the phase of $-180°$. Hence we must locate f_z by trial and error so that the overall response A'_V satisfies the design requirement of 45° phase margin at 32 dB low-frequency feedback. It takes only a few trials to settle upon $f_z = 10.0$ MHz. The plots in Fig. 15-25 correspond to this value of the zero (and to $f_p = 63f_z = 630$ MHz), and we see that the desired specification is satisfied.

To determine the value of C' we may arbitrarily choose R, and from Eq. (15-54) $R' = 62R$, and then C' is given by Eq. (15-51). For example, for $R = 1$ K, then $R' = 62$ K, and we find

$$C' = \frac{1}{2\pi f_z R'} = \frac{1}{6.28 \times 10^7 \times 62 \times 10^3} \text{F} = 0.25 \text{ pF}$$

Since this value of C' is impractically small, we choose $R = 100 \ \Omega$, $R' = 6.2$ K, and then $C' = 2.5$ pF.

Note that since the input to the OP AMP terminals is a virtual ground, the impedance seen by the signal source V_s in Fig. 15-24 is only 100 Ω.

Lead compensation can also be provided by modifying the open-loop voltage gain of the basic amplifier, provided that appropriate leads are available from the IC chip. For example, a capacitor C' can be connected between terminals 7 and 8 of the IC OP AMP shown in Fig. 15-11. If resistor R_9 is paralleled by a capacitor C' then the $R_9 C'$ parallel combination has an impedance $Z_9 = R_9/(1 + sC'R_9)$. The voltage gain of the output driver becomes $V_o/V_4 = -R_{10}/Z_9 = -R_{10}(1 + sC'R_9)/R_9$ and, thus, a zero is introduced in the open-loop voltage gain transfer function of the basic amplifier.

15-13   STEP RESPONSE OF OPERATIONAL AMPLIFIERS

In Chap. 14 it is demonstrated that many feedback amplifiers have a single dominant pole or two dominant poles, with all other poles at least two octaves away. If an OP AMP is represented by a two-pole transfer function, the small-signal response is that discussed in Sec. 14-2 and plotted in Fig. 14-7.

Large-signal Step Response If the output voltage exceeds 1 V, the transient response is altered by nonlinear operation such as bias disturbances and the slewing-rate[9] limit determined by the circuit capacitances. If we connect, for example, a compensation capacitor C_1 from collector to collector of the first differential stage, this capacitor must charge during the large-signal transient response, and the maximum available current is $2I_C$, where I_C is the dc current of either transistor. Thus

$$\frac{dV_{C1}}{dt} = \frac{2I_C}{C_1} \tag{15-55}$$

and this limit distorts the large output signal. In general, the nonlinear operation and slewing-rate limit increase the settling time (Fig. 14-6).

In practice, the selection of the compensation capacitor is greatly aided by observing the amplifier step response with a square-wave input and varying the compensating capacitance at various signal levels to obtain satisfactory compensation.

REFERENCES

1. Widlar, R. J.: Design Techniques for Monolithic Operational Amplifiers, *IEEE J. Solid-state Circuits*, vol. SC-4, pp. 184–191, August, 1969.

2. Eimbinder, J.: "Designing with Linear Integrated Circuits," pp. 13–31, John Wiley & Sons, Inc., New York, 1969.

3. Tobey, M., et al. (eds.), Burr-Brown Research Corp.: "Operational Amplifiers: Design and Applications," McGraw-Hill Book Company, New York, 1971.

4. Giacoletto, L. J.: "Differential Amplifiers," Wiley-Interscience, a Division of John Wiley & Sons, Inc., New York, 1970.

5. "RCA Linear Integrated Circuits," pp. 28–43, Radio Corporation of America, Harrison, N.J., 1967.

6. Wissernan, L., and J. J. Robertson: High Performance Integrated Operational Amplifiers, Motorola Semiconductor Products, Inc., *Application Note* AN-204.

7. Wilson, G. R.: A Monolithic Junction FET—*n-p-n* Operational Amplifier, *IEEE J. Solid-state Circuits*, vol. SC-3, pp. 341–348, December, 1968.

8. Searle, C. L., A. R. Boothroyd, E. J. Angelo, Jr., P. E. Gray, and D. O. Pederson: "Elementary Circuit Properties of Transistors," pp. 142–143, John Wiley & Sons, Inc., New York, 1964.

9. Hearn, W. E.: Fast Slewing Monolithic Operational Amplifier, *IEEE J. Solid-state Circuits*, vol. SC-6, pp. 20–24, February, 1971.

REVIEW QUESTIONS

15-1 (*a*) Draw the schematic block diagram of the basic OP AMP with inverting and noninverting inputs. (*b*) Indicate its equivalent circuit.

15-2 List six characteristics of the ideal OP AMP.

15-3 (*a*) Draw the schematic diagram of an ideal inverting OP AMP with voltage-shunt feedback impedances Z and Z'. (*b*) Indicate the virtual-ground model for calculating the gain.

15-4 For the OP AMP of Rev. 15-3, assume finite A_v and R_i and nonzero R_o. Draw the equivalent circuit using Miller's theorem.

15-5 (*a*) Draw the schematic diagram of an ideal noninverting OP AMP with voltage series feedback. (*b*) Derive the expression for the voltage gain.

15-6 (*a*) Define an ideal DIFF AMP. (*b*) Define difference signal v_d and common-mode signal v_c.

15-7 (*a*) Draw the circuit of an emitter-coupled DIFF AMP. (*b*) Explain why the CMRR → ∞ for a symmetrical circuit with $R_e \to \infty$.

15-8 (*a*) Draw the equivalent circuit from which to calculate A_c for the emitter-coupled DIFF AMP. (*b*) Repeat for A_d.

15-9 (*a*) Why is R_e in an emitter-coupled DIFF AMP replaced by a constant-current source? (*b*) Draw such a circuit. (*c*) Explain why the network replacing R_e acts as an approximately constant current I_O. (*d*) Explain how I_O is made to be independent of temperature.

15-10 Explain why the CMRR is infinite if a true constant-current source is used in a symmetrical emitter-coupled DIFF AMP.

15-11 (*a*) Sketch the transfer characteristics of a DIFF AMP. (*b*) Over what differential voltage is the DIFF AMP a good limiter? (*c*) Over what differential voltage is the transfer characteristic quite linear? (*d*) How does the transconductance vary (qualitatively) with differential voltage? (*e*) Explain why AGC is possible with the DIFF AMP.

15-12 (*a*) Draw an IC OP AMP in block-diagram form. (*b*) Identify each stage by function.

15-13 Define (a) *input bias current,* (b) *input offset current,* (c) *input offset voltage,* (d) *output offset voltage,* (e) *power supply rejection ratio, and* (f) *slew rate* for an OP AMP.

15-14 Show the balancing arrangement for (a) an inverting and (b) a noninverting OP AMP.

15-15 Show the circuit and explain how to measure (a) A_V, (b) R_o, and (c) R_i of an OP AMP.

15-16 Repeat Rev. 15-15 for V_{io}.

15-17 Repeat Rev. 15-15 for I_{io}.

15-18 Repeat Rev. 15-15 for CMRR $= \rho$.

15-19 Repeat Rev. 15-15 for the slewing rate.

15-20 Explain how the poles of an OP AMP may be determined experimentally.

15-21 Discuss dominant-pole compensation of an OP AMP.

15-22 Indicate three methods of implementing pole-zero compensation of an OP AMP.

15-23 (a) Draw the circuit which applies lead compensation to an inverting OP AMP. (b) Verify that a phase lead is introduced by the circuit element added.

16 / INTEGRATED CIRCUITS AS ANALOG SYSTEM BUILDING BLOCKS

Many analog systems (both linear and nonlinear) are constructed with the OP AMP or DIFF AMP as the basic building block. These IC's augmented by a few external discrete components, either singly or in combination, are used in the following *linear* systems: analog computers, voltage-to-current and current-to-voltage converters, amplifiers of various types (for example, dc instrumentation, tuned, and video amplifiers), voltage followers, active filters, and delay equalizers.

Among the *nonlinear* analog system configurations discussed in this chapter are the following: amplitude modulators, logarithmic amplifiers and analog multipliers, sample-and-hold circuits, comparators, and square-wave and triangle-waveform generators.

I. LINEAR ANALOG SYSTEMS

16-1   BASIC OPERATIONAL AMPLIFIER APPLICATIONS[1]

An OP AMP may be used to perform many mathematical operations. This feature accounts for the name *operational amplifier*. Some of the basic applications are given in this section. Consider the ideal OP AMP of Fig. 15-2a, which is repeated for convenience in Fig. 16-1a. Recalling (Sec. 15-1) that the equivalent circuit of Fig. 16-1b has a virtual ground (which takes no current), it follows that the voltage gain is given by

$$A_{Vf} = \frac{V_o}{V_s} = - \frac{Z'}{Z} \tag{16-1}$$

Based upon this equation we can readily obtain an *analog inverter*, a *scale changer*, a *phase shifter*, and an *adder*.

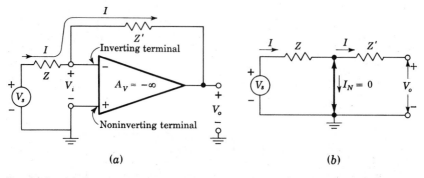

Fig. 16-1 (a) Inverting operational amplifier with voltage-shunt feedback. (b) Virtual ground in the OP AMP.

Sign Changer, or Inverter   If $Z = Z'$ in Fig. 16-1, then $A_{Vf} = -1$, and the sign of the input signal has been changed.   Hence such a circuit acts as a phase inverter.   If two such amplifiers are connected in cascade, the output from the second stage equals the signal input without change of sign. Hence the outputs from the two stages are equal in magnitude but opposite in phase, and such a system is an excellent *paraphase amplifier*.

Scale Changer   If the ratio $Z'/Z = k$, a real constant, then $A_{Vf} = -k$, and the scale has been multiplied by a factor $-k$.   Usually, in such a case of multiplication by a constant, -1 or $-k$, Z and Z' are selected as precision resistors.

Phase Shifter   Assume that Z and Z' are equal in magnitude but differ in angle.   Then the operational amplifier shifts the phase of a sinusoidal input voltage while at the same time preserving its amplitude.   Any phase shift from 0 to 360° (or ±180°) may be obtained.

Adder, or Summing Amplifier   The arrangement of Fig. 16-2 may be used to obtain an output which is a linear combination of a number of input signals. Since a virtual ground exists at the OP AMP input, then

$$i = \frac{v_1}{R_1} + \frac{v_2}{R_2} + \cdots + \frac{v_n}{R_n}$$

and

$$v_o = -R'i = -\left(\frac{R'}{R_1} v_1 + \frac{R'}{R_2} v_2 + \cdots + \frac{R'}{R_n} v_n\right) \tag{16-2a}$$

If $R_1 = R_2 = \cdots = R_n$, then

$$v_o = -\frac{R'}{R_1} (v_1 + v_2 + \cdots + v_n) \tag{16-2b}$$

and the output is proportional to the sum of the inputs.

Fig. 16-2 Operational adder, or summing amplifier.

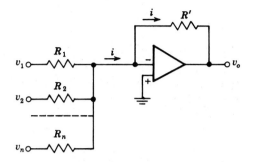

Many other methods may, of course, be used to combine signals. The present method has the advantage that it may be extended to a very large number of inputs requiring only one additional resistor for each additional input. The result depends, in the limiting case of large amplifier gain, only on the resistors involved, and because of the virtual ground, there is a minimum of interaction between input sources.

Voltage-to-current Converter Often it is desirable to convert a voltage signal to a proportional output current. This is required, for example, when we drive a deflection coil in a television tube. If the load impedance has neither side grounded (if it is floating), the simple circuit of Fig. 16-2 with R' replaced by the load impedance Z_L is an excellent *voltage-to-current converter*. For a single input $v_1 = v_s(t)$, the current in Z_L is

$$ i_L = \frac{v_s(t)}{R} \tag{16-3} $$

Note that i is independent of the load Z_L, because of the virtual ground of the operational amplifier input. Since the same current flows through the signal source and the load, it is important that the signal source be capable of providing this load current. On the other hand, the amplifier of Fig. 16-3a requires

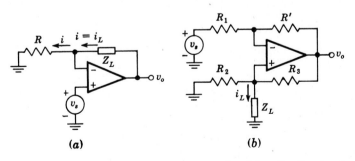

(a)   (b)

Fig. 16-3 Voltage-to-current converter for (a) a floating load
and (b) a grounded load Z_L.

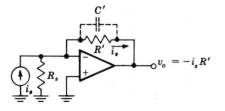

Fig. 16-4   Current-to-voltage converter.

very little current from the signal source due to the very large input resistance seen by the noninverting terminal.

If the load Z_L is grounded, the circuit of Fig. 16-3b can be used. In Prob. 16-7 we show that if $R_3/R_2 = R'/R_1$, then

$$i_L(t) = -\frac{v_s(t)}{R_2} \qquad (16\text{-}4)$$

Current-to-voltage Converter   Photocells and photomultiplier tubes give an output current which is independent of the load.   The circuit in Fig. 16-4 shows an operational amplifier used as a current-to-voltage converter.   Due to the virtual ground at the amplifier input, the current in R_s is zero and i_s flows through the feedback resistor R'.   Thus the output voltage v_o is $v_o = -i_s R'$.   It must be pointed out that the lower limit on current measurement with this circuit is set by the bias current of the inverting input.   It is common to parallel R' with a capacitance C' to reduce high-frequency noise.

DC Voltage Follower   The simple configuration of Fig. 16-5 approaches the ideal *voltage follower*.   Because the two inputs are tied together (virtually), then $V_o = V_s$ and *the output follows the input*.   The LM 102 (National Semiconductor Corporation) is specifically designed for voltage-follower usage and has very high input resistance (10,000 M), very low input current ($\sim$3 nA), and very low output resistance ($\sim$0 Ω).

16-2   DIFFERENTIAL DC AMPLIFIER[2]

The differential-input single-ended-output instrumentation amplifier is often used to amplify inputs from transducers which convert a physical parameter and its variations into an electric signal.   Such transducers are strain-gauge bridges, thermocouples, etc.   The circuit shown in Fig. 16-6 is very simple

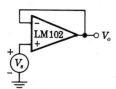

Fig. 16-5   A voltage follower, $V_o = V_s$.

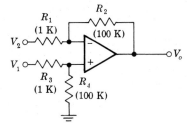

Fig. 16-6 Differential amplifier using one OP AMP. (The offset-voltage balancing arrangement is not indicated.)

and uses only one OP AMP. In Prob. 16-8 we show that if $R_2/R_1 = R_4/R_3$, then

$$V_o = \frac{R_2}{R_1} (V_1 - V_2) \tag{16-5}$$

If the signals V_1 and V_2 have source resistances R_{s1} and R_{s2}, then these resistances add to R_3 and R_1, respectively. Note that the signal source V_1 sees a resistance $R_3 + R_4 = 101$ K. If $V_1 = 0$, the inverting input is at ground potential and hence V_2 is loaded by $R_1 = 1$ K. If this is too heavy a load for the transducer, a voltage follower may be used as a buffer. If the configuration of Fig. 16-5 precedes each input in Fig. 16-6, the resulting topology of three IC OP AMPS represents a very high-input-resistance, high-performance, low-cost, dc amplifier system.

Bridge Amplifier A differential amplifier is often used to amplify the output from a transducer bridge, as shown in Fig. 16-7. Nominally, the four arms of the bridge have equal resistances R. However, one of the branches has a resistance which changes to $R + \Delta R$ with temperature or some other physical parameter. The goal of the measurement is to obtain the fractional change δ of the resistance value of the active arm, or $\delta = \Delta R/R$.

In Prob. 16-11 we find that for the circuit of Fig. 16-7, the output V_o is given by

$$V_o = -\frac{A_d V}{4} \frac{\delta}{1 + \delta/2} \tag{16-6}$$

For small changes in R ($\delta \ll 1$) Eq. (16-6) reduces to

$$V_o = -\frac{A_d V}{4} \delta \tag{16-7}$$

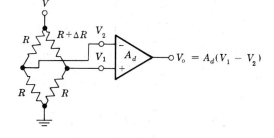

Fig. 16-7 Differential bridge amplifier.

16-3 STABLE AC–COUPLED AMPLIFIER

In some applications the need arises for the amplification of an ac signal, while any dc signal present must be blocked. A very simple and stable ac amplifier is shown in Fig. 16-8a, where capacitor C blocks the dc component of the input signal and together with the resistor R sets the low-frequency 3-dB response for the overall amplifier.

The output voltage V_o as a function of the complex variable s is found from the equivalent circuit of Fig. 16-8b (where the double-ended heavy arrow represents the virtual ground) to be

$$V_o = -IR' = -\frac{V_s}{R + 1/sC} R'$$

and

$$A_{Vf} = \frac{V_o}{V_s} = -\frac{R'}{R} \frac{s}{s + 1/RC} \tag{16-8}$$

From Eq. (16-8) we see that the low 3-dB frequency is

$$f_L = \frac{1}{2\pi RC} \tag{16-9}$$

The high-frequency response is determined by the frequency characteristics of the operational amplifier A_V and the amount of voltage-shunt feedback present (Sec. 14-5). The midband gain is, from Eq. (16-1), $A_{Vf} = -R'/R$.

AC Voltage Follower The ac voltage follower is used to provide impedance buffering, that is, to connect a signal source with high internal source resistance to a load of low impedance, which may even be capacitive. In Fig. 16-9 is shown a practical high-input impedance ac voltage follower using the LM 102 operational amplifier. We assume that C_1 and C_2 represent short circuits at all frequencies of operation of this circuit. Resistors R_1 and R_2 are used to provide RC coupling and allow a path for the dc input current into the noninverting terminal. In the absence of the bootstrapping capacitor C_2, the ac signal source would see an input resistance of only $R_1 + R_2 = 200$ K. Since

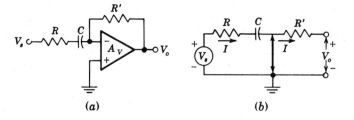

(a)   (b)

Fig. 16-8 (a) AC stable feedback amplifier. (b) Equivalent circuit when $|A_V| = \infty$.

Fig. 16-9 AC voltage follower. (Courtesy of National Semiconductor Corporation.)

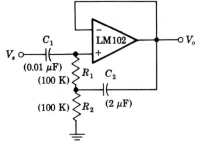

the LM 102 is connected as a voltage follower, the voltage gain A_V between the output terminal and the noninverting terminal is very close to unity. Thus, from our discussion in Sec. 8-16, the input resistance the source sees becomes, approximately, $R_1/(1 - A_V)$, which is measured to be 12 M at 100 Hz and increases to 100 M at 1 kHz.

16-4   ANALOG INTEGRATION AND DIFFERENTIATION[1]

The analog integrator is very useful in many applications which require the generation or processing of analog signals. If, in Fig. 16-1, $Z = R$ and a capacitor C is used for Z', as in Fig. 16-10, we can show that the circuit performs the mathematical operation of integration. The input need not be sinusoidal, and hence is represented by the lowercase symbol $v = v(t)$. (The subscript s is now omitted, for simplicity.) In Fig. 16-10b, the double-headed arrow represents a virtual ground. Hence $i = v/R$, and

$$v_o = -\frac{1}{C} \int i\, dt = -\frac{1}{RC} \int v\, dt \qquad (16\text{-}10)$$

The amplifier therefore provides an output voltage proportional to the integral of the input voltage.

    If the input voltage is a constant, $v = V$, then the output will be a ramp, $v_o = -Vt/RC$. Such an integrator makes an excellent sweep circuit for a cathode-ray-tube oscilloscope, and is called a *Miller integrator*, or *Miller sweep*.[3]

    DC Offset and Bias Current   The input stage of the operational amplifier used in Fig. 16-10 is usually a DIFF AMP. The dc input offset voltage V_{io} appears across the amplifier input, and this voltage will be integrated and will appear at the output as a linearly increasing voltage. Part of the input bias current will also flow through the feedback capacitor, charging it and producing an additional linearly increasing voltage at the output. These two ramp voltages continue to increase until the amplifier reaches its saturation point.

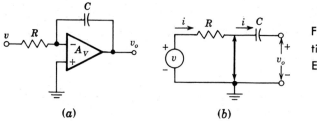

Fig. 16-10 (a) Opera-
tional integrator. (b)
Equivalent circuit.

(a)   (b)

We see then that a limit is set on the feasible integration time by the above error components. The effect of the bias current can be minimized by increasing the feedback capacitor C while simultaneously decreasing the value of R for a given value of the time constant RC.

Finite Gain and Bandwidth   The integrator supplies an output voltage proportional to the integral of the input voltage, provided the operational amplifier shown in Fig. 16-10a has infinite gain $|A_V| \to \infty$ and infinite bandwidth. The voltage gain as a function of the complex variable s is, from Eq. (16-1),

$$A_{Vf}(s) = \frac{V_o(s)}{V(s)} = -\frac{Z'}{Z} = -\frac{1}{RCs} \tag{16-11}$$

and it is clear that the ideal integrator has a pole at the origin.

Let us assume that in the absence of C the operational amplifier has a dominant pole at f_1, or $s_1 \equiv -2\pi f_1$. Hence its voltage gain A_v is approximated by

$$A_v = \frac{A_{vo}}{1 + j(f/f_1)} = \frac{A_{vo}}{1 - s/s_1} \tag{16-12}$$

If we further assume that $R_o = 0$ in Fig. 15-3, then $A_v = A_V$. Substituting Eq. (16-12) in Eq. (15-2) with $R_i = \infty$ and using $|A_{Vo}| \gg 1$, $|A_{Vo}|RC \gg 1/|s_1|$, we find

$$A_{Vf} = -\frac{s_1}{RC} A_{Vo} \frac{1}{(s + A_{Vo}s_1)(s - 1/RCA_{Vo})} \tag{16-13}$$

where A_{Vo} is a negative number and represents the low-frequency voltage gain of the operational amplifier.

The above transfer function has two poles on the negative real axis as compared with one pole at the origin for the ideal integrator. In Fig. 16-11 we show the Bode plots of Eqs. (16-11) to (16-13). We note that the response of the real integrator departs from the ideal at both low and high frequencies. At high frequencies the integrator performance is affected by the finite bandwidth $(-s_1/2\pi)$ of the operational amplifier, while at low frequencies the integration is limited by the finite gain of the OP AMP.

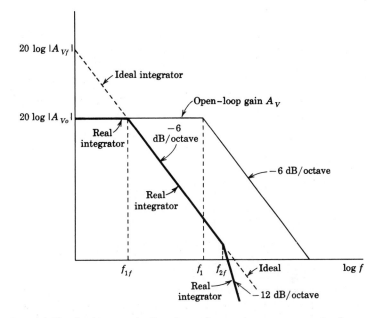

Fig. 16-11 Bode magnitude plots of open-loop OP AMP gain A_V, ideal integrator and real integrator. Note that $f_{1f} = 1/2\pi RC|A_{Vo}|$ and $f_{2f} = A_{Vo}s_1/2\pi$.

Practical Circuit A practical integrator must be provided with an external circuit to introduce initial conditions, as shown in Fig. 16-12. When switch S is in position 1, the input is zero and capacitor C is charged to the voltage V, setting an initial condition of $v_o = V$. When switch S is in position 2, the amplifier is connected as an integrator and its output will be V plus a constant times the time integral of the input voltage v. In using this circuit, care must be exercised to stabilize the amplifier and $I_{B2}R_2$ must be equal to $I_{B1}R_1$ to minimize the error due to bias current.

Fig. 16-12 Practical integrator circuit. For minimum offset error due to input bias current it is required that $I_{B1}R_1 = I_{B2}R_2$. (Courtesy of National Semiconductor Corporation.)

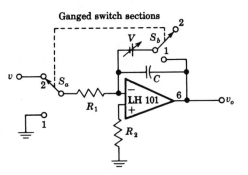

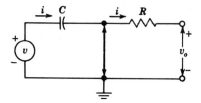

Fig. 16-13 Equivalent circuit of the operational differentiator.

Differentiator If Z is a capacitor C and if $Z' = R$, we see from the equivalent circuit of Fig. 16-13 that $i = C \, dv/dt$ and

$$v_o = -Ri = -RC \frac{dv}{dt} \qquad (16\text{-}14)$$

Hence the output is proportional to the time derivative of the input. If the input signal is $v = \sin \omega t$, then the output will be $v_o = -RC\omega \cos \omega t$. Thus the magnitude of the output increases linearly with increasing frequency, and the differentiator circuit has high gain at high frequencies. This results in amplification of the high-frequency components of amplifier noise, and the noise output may completely obscure the differentiated signal.

The General Case In the important cases considered above, Z and Z' have been simple elements such as a single R or C. In general, they may be any series or parallel combinations of R, L, or C. Using the methods of Laplace transform analysis, Z and Z' can be written in their operational form as $Z(s)$ and $Z'(s)$, where s is the complex-frequency variable. In this notation the reactance of an inductor is written formally as Ls and that of a capacitor as $1/sC$. The current $I(s)$ is then $V(s)/Z(s)$, and the output is

$$V_o(s) = -\frac{Z'(s)}{Z(s)} V(s) \qquad (16\text{-}15)$$

The amplifier thus solves this operational equation.

16-5   ELECTRONIC ANALOG COMPUTATION[1]

The OP AMP is the fundamental building block in an electronic analog computer. As an illustration, let us consider how to program the differential equation

$$\frac{d^2v}{dt^2} + K_1 \frac{dv}{dt} + K_2 v - v_1 = 0 \qquad (16\text{-}16)$$

where v_1 is a given function of time, and K_1 and K_2 are real positive constants.
We begin by assuming that d^2v/dt^2 is available in the form of a voltage. Then, by means of an integrator, a voltage proportional to dv/dt is obtained. A second integrator gives a voltage proportional to v. Then an adder (and

scale changer) gives $-K_1(dv/dt) - K_2v + v_1$. From the differential equation (16-16), this equals d^2v/dt^2, and hence the output of this summing amplifier is fed to the input terminal, where we had assumed that d^2v/dt^2 was available in the first place.

The procedure outlined above is carried out in Fig. 16-14. The voltage d^2v/dt^2 is assumed to be available at an input terminal. The integrator (1) has a time constant $RC = 1$ s, and hence its output at terminal 1 is $-dv/dt$. This voltage is fed to a similar integrator (2), and the voltage at terminal 2 is $+v$. The voltage at terminal 1 is fed to the inverter and scale changer (3), and its output at terminal 3 is $+K_1(dv/dt)$. This same operational amplifier (3) is used as an adder. Hence, if the given voltage $v_1(t)$ is also fed into it as shown, the output at terminal 3 also contains the term $-v_1$, or the net output is $+K_1(dv/dt) - v_1$. Scale changer–adder (4) is fed from terminals 2 and 3, and hence delivers a resultant voltage $-K_2v - K_1(dv/dt) + v_1$ at terminal 4. By Eq. (16-16) this must equal d^2v/dt^2, which is the voltage that was assumed to exist at the input terminal. Hence the computer is completed by connecting terminal 4 to the input terminal. (This last step is omitted from Fig. 16-14 for the sake of clarity of explanation.)

The specified initial conditions (the value of dv/dt and v at $t = 0$) must now be inserted into the computer. We note that the voltages at terminals 1 and 2 in Fig. 16-14 are proportional to dv/dt and v, respectively. Hence initial conditions are taken care of (as in Fig. 16-12) by applying the correct voltages V_1 and V_2 across the capacitors in integrators 1 and 2, respectively.

The solution is obtained by opening switches S_1 and S_2 and simultaneously

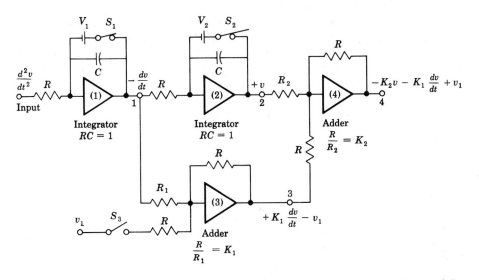

Fig. 16-14   A block diagram of an electronic analog computer. At $t = 0$, S_1 and S_2 are opened and S_3 is closed. Each OP AMP input is as in Fig. 16-12.

closing S_3 (by means of relays) at $t = 0$ and observing the waveform at terminal 2. If the derivative dv/dt is also desired, its waveform is available at terminal 1. The indicator may be a cathode-ray tube (with a triggered sweep) or a recorder or, for qualitative analysis with slowly varying quantities, a high-impedance voltmeter.

The solution of Eq. (16-16) can also be obtained with a computer which contains differentiators instead of integrators. However, integrators are almost invariably preferred over differentiators in analog-computer applications, for the following reasons: Since the gain of an integrator decreases with frequency whereas the gain of a differentiator increases nominally linearly with frequency, it is easier to stabilize the former than the latter with respect to spurious oscillations. As a result of its limited bandwidth, an integrator is less sensitive to noise voltages than a differentiator. Further, if the input waveform changes rapidly, the amplifier of a differentiator may overload. Finally, as a matter of practice, it is convenient to introduce initial conditions in an integrator.

16-6   ACTIVE FILTERS[4]

Consider the ideal low-pass-filter response shown in Fig. 16-15a. In this plot all signals within the band $0 \leq f \leq f_o$ are transmitted without loss, whereas inputs with frequencies $f > f_o$ give zero output. It is known[5] that such an ideal characteristic is unrealizable with physical elements, and thus it is

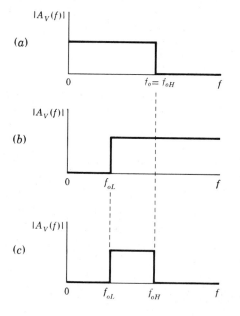

Fig. 16-15   Ideal filter characteristics. (a) Low-pass, (b) high-pass, and (c) bandpass.

necessary to approximate it.   An approximation for an ideal low-pass filter is of the form

$$A_V(s) = \frac{1}{P_n(s)} \tag{16-17}$$

where $P_n(s)$ is a polynomial in the variable s with zeros in the left-hand plane. Active filters permit the realization of arbitrary left-hand poles for $A_V(s)$, using the operational amplifier as the active element and only resistors and capacitors for the passive elements.

Since commercially available OP AMPS have unity gain-bandwidth products as high as 100 MHz, it is possible to design active filters up to frequencies of several MHz.   The limiting factor for full-power response at those high frequencies is the slewing rate (Sec. 15-6) of the operational amplifier. (Commercial integrated OP AMPS are available with slewing rates as high as 100 V/μs.)

Butterworth Filter[6]   A common approximation of Eq. (16-17) uses the Butterworth polynomials $B_n(s)$, where

$$A_V(s) = \frac{A_{Vo}}{B_n(s)} \tag{16-18}$$

and with $s = j\omega$,

$$|A_V(s)|^2 = A_V(s)A_V(-s) = \frac{A_{Vo}^2}{1 + (\omega/\omega_o)^{2n}} \tag{16-19}$$

From Eqs. (16-18) and (16-19) we note that the magnitude of $B_n(\omega)$ is given by

$$|B_n(\omega)| = \sqrt{1 + \left(\frac{\omega}{\omega_o}\right)^{2n}} \tag{16-20}$$

The Butterworth response [Eq. (16-19)] for various values of n is plotted in Fig. 16-16.   Note that the magnitude of A_V is down 3 dB at $\omega = \omega_o$ for all n. The larger the value of n, the more closely the curve approximates the ideal low-pass response of Fig. 16-15a.

If we normalize the frequency by assuming $\omega_o = 1$ rad/s, then Table 16-1 gives the Butterworth polynomials for n up to 8.   Note that for n even, the polynomials are the products of quadratic forms, and for n odd, there is present the additional factor $s + 1$.   The zeros of the normalized Butterworth polynomials are either -1 or complex conjugate and are found on the so-called *Butterworth circle* of unit radius shown in Fig. 16-17.   The *damping factor k* is defined as one-half the coefficient of s in each quadratic factor in Table 16-1.   For example, for $n = 4$, there are two damping factors, namely, $0.765/2 = 0.383$ and $1.848/2 = 0.924$.   It turns out (Prob. 16-20) that k is given by

$$k = \cos\theta \tag{16-21}$$

where θ is as defined in Fig. 16-17a for n even and Fig. 16-17b for n odd.

TABLE 16-1   Normalized Butterworth polynominals

n	Factors of polynomial $B_n(s)$
1	$(s + 1)$
2	$(s^2 + 1.414s + 1)$
3	$(s + 1)(s^2 + s + 1)$
4	$(s^2 + 0.765s + 1)(s^2 + 1.848s + 1)$
5	$(s + 1)(s^2 + 0.618s + 1)(s^2 + 1.618s + 1)$
6	$(s^2 + 0.518s + 1)(s^2 + 1.414s + 1)(s^2 + 1.932s + 1)$
7	$(s + 1)(s^2 + 0.445s + 1)(s^2 + 1.247s + 1)(s^2 + 1.802s + 1)$
8	$(s^2 + 0.390s + 1)(s^2 + 1.111s + 1)(s^2 + 1.663s + 1)(s^2 + 1.962s + 1)$

From the table and Eq. (16-18) we see that the typical second-order Butterworth filter transfer function is of the form

$$\frac{A_V(s)}{A_{Vo}} = \frac{1}{(s/\omega_o)^2 + 2k(s/\omega_o) + 1} \tag{16-22}$$

where $\omega_o = 2\pi f_o$ is the high-frequency 3-dB point. Similarly, the first-order filter is

$$\frac{A_V(s)}{A_{Vo}} = \frac{1}{s/\omega_o + 1} \tag{16-23}$$

Practical Realization Consider the circuit shown in Fig. 16-18a, where the active element is an operational amplifier whose stable midband gain

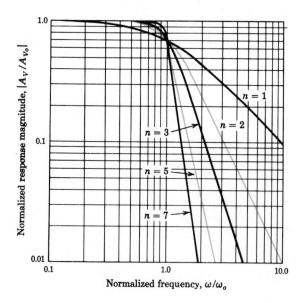

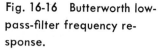

Fig. 16-16 Butterworth low-pass-filter frequency response.

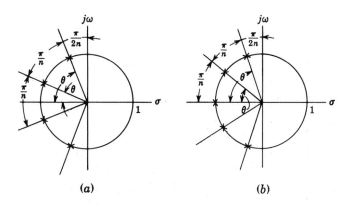

(a)   (b)

Fig. 16-17   The Butterworth circle for (a) n even and (b)
n odd.   Note that for n odd, one of the zeros is at $s = -1$.

$V_o/V_i = A_{Vo} = (R_1 + R_1')/R_1$ [Eq. (15-4)] is to be determined.   We assume
that the amplifier input current is zero, and we show in Prob. 16-25 that

$$A_V(s) = \frac{V_o}{V_s} = \frac{A_{Vo}Z_3Z_4}{Z_3(Z_1 + Z_2 + Z_4) + Z_1Z_2 + Z_1Z_4(1 - A_{Vo})} \quad (16\text{-}24)$$

If this network is to be a low-pass filter, then Z_1 and Z_2 are resistances and
Z_3 and Z_4 are capacitances.   Let us assume $Z_1 = Z_2 = R$ and $C_3 = C_4 = C$,
as shown in Fig. 16-18b.   The transfer function of this network takes the form

$$A_V(s) = A_{Vo}\frac{(1/RC)^2}{s^2 + \left(\dfrac{3 - A_{Vo}}{RC}\right)s + \left(\dfrac{1}{RC}\right)^2} \quad (16\text{-}25)$$

Comparing Eq. (16-25) with Eq. (16-22), we find

$$\omega_o = \frac{1}{RC} \quad (16\text{-}26)$$

and

$$2k = 3 - A_{Vo} \quad \text{or} \quad A_{Vo} = 3 - 2k \quad (16\text{-}27)$$

We are now in a position to synthesize even-order Butterworth filters
by cascading prototypes of the form shown in Fig. 16-18b, using identical
R's and C's and selecting the gain A_{Vo} of each operational amplifier to satisfy
Eq. (16-27) and the damping factors from Table 16-1.

To realize odd-order filters, it is necessary to cascade the first-order filter
of Eq. (16-23) with second-order sections such as indicated in Fig. 16-18b.
The first-order prototype of Fig. 16-18c has the transfer function of Eq.
(16-23) for arbitrary A_{Vo} provided that ω_o is given by Eq. (16-26).   For
example, a third-order Butterworth active filter consists of the circuit in
Fig. 16-18b in cascade with the circuit of Fig. 16-18c, with R and C chosen so
that $RC = 1/\omega_o$, with A_{Vo} in Fig. 16-18b selected to give $k = 0.5$ (Table 16-1,
$n = 3$), and A_{Vo} in Fig. 16-18c chosen arbitrarily.

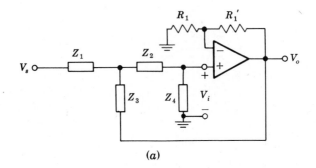

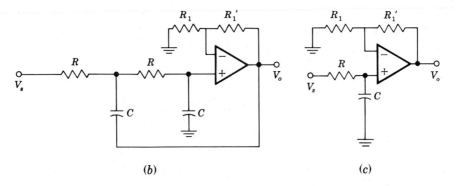

Fig. 16-18 (a) Generalized active-filter prototype. (b) Second-order low-pass section. (c) First-order low-pass section.

EXAMPLE Design a fourth-order Butterworth low-pass filter with a cutoff frequency of 1 kHz.

Solution We cascade two second-order prototypes as shown in Fig. 16-19. For $n = 4$ we have from Table 16-1 and Eq. (16-27)

$$A_{V1} = 3 - 2k_1 = 3 - 0.765 = 2.235$$

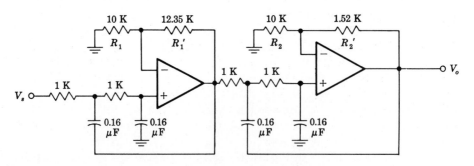

Fig. 16-19 Fourth-order Butterworth low-pass filter with $f_o = 1$ kHz.

and

$$A_{V2} = 3 - 2k_2 = 3 - 1.848 = 1.152$$

From Eq. (15-4), $A_{V1} = (R_1 + R_1')/R_1$. If we arbitrarily choose $R_1 = 10$ K, then for $A_{V1} = 2.235$, we find $R_1' = 12.35$ K, whereas for $A_{V2} = 1.152$, we find $R_2' = 1.520$ K and $R_2 = 10$ K. To satisfy the cutoff-frequency requirement, we have, from Eq. (16-26), $f_o = 1/2\pi RC$. We take $R = 1$ K and find $C = 0.16$ μF. Figure 16-19 shows the complete fourth-order low-pass Butterworth filter.

High-pass Prototype An idealized high-pass-filter characteristic is indicated in Fig. 16-15b. The high-pass second-order filter is obtained from the low-pass second-order prototype of Eq. (16-22) by applying the transformation

$$\left.\frac{s}{\omega_o}\right|_{\text{low-pass}} \longrightarrow \left.\frac{\omega_o}{s}\right|_{\text{high-pass}} \tag{16-28}$$

Thus, interchanging R's and C's in Fig. 16-18b results in a second-order high-pass active filter.

Bandpass Filter A second-order bandpass prototype is obtained by cascading a low-pass second-order section whose cutoff frequency is f_{oH} with a high-pass second-order section whose cutoff frequency is f_{oL}, provided $f_{oH} > f_{oL}$, as indicated in Fig. 16-15c.

Band-reject Filter Figure 16-20 shows that a band-reject filter is obtained by paralleling a high-pass section whose cutoff frequency is f_{oL} with a low-pass

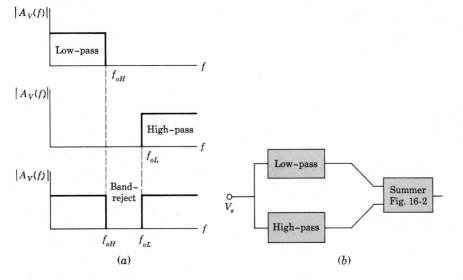

Fig. 16-20   (a) Ideal band-reject-filter frequency response.   (b) Parallel combination of low-pass and high-pass filters results in a band-reject filter.

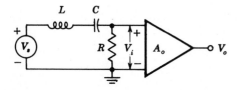

Fig. 16-21   A resonant circuit.

section whose cutoff frequency is f_{oH}.   Note that for band-reject characteristics it is required that $f_{oH} < f_{oL}$.

16-7   ACTIVE RESONANT BANDPASS FILTERS[7]

The idealized bandpass filter of Fig. 16-15c has a constant response for $f_{oL} < f < f_{oH}$ and zero gain outside this range.   An infinite number of Butterworth sections are required to obtain this filter response.   A very simple approximation to a narrowband characteristic is obtained using a single LC resonant circuit.   Such a bandpass filter has a response which peaks at some center frequency f_o and drops off with frequency on both sides of f_o.   A basic prototype for a resonant filter is the second-order section shown in Fig. 16-21, whose transfer function we now derive.

If we assume that the amplifier provides a gain $A_o = V_o/V_i$ which is positive and constant for all frequencies, we find

$$A_V(j\omega) = \frac{V_o}{V_s} = \frac{V_o V_i}{V_i V_s} = \frac{RA_o}{R + j(\omega L - 1/\omega C)} \tag{16-29}$$

The *center*, or *resonant*, frequency $f_o = \omega_o/2\pi$ is defined as that frequency at which the inductance resonates with the capacitance; in other words, the inductive and capacitive reactances are equal (in magnitude), or

$$\omega_o{}^2 = \frac{1}{LC} \tag{16-30}$$

It is convenient to define the *quality factor* Q of this circuit by

$$Q \equiv \frac{\omega_o L}{R} = \frac{1}{\omega_o C R} = \frac{1}{R}\sqrt{\frac{L}{C}} \tag{16-31}$$

Substituting Eq. (16-31) in Eq. (16-29), we obtain the magnitude and phase of the transfer function

$$|A_V(j\omega)| = \frac{A_o}{\left[1 + Q^2\left(\dfrac{\omega}{\omega_0} - \dfrac{\omega_o}{\omega}\right)^2\right]^{\frac{1}{2}}} \tag{16-32}$$

$$\theta(\omega) = -\arctan Q\left(\frac{\omega}{\omega_o} - \frac{\omega_o}{\omega}\right) \tag{16-33}$$

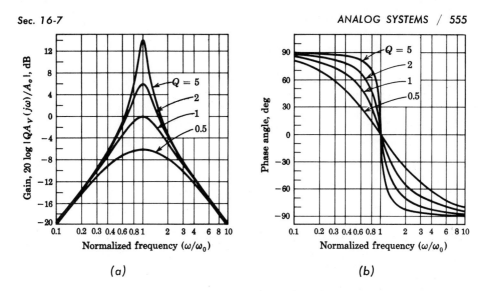

Fig. 16-22 The bandpass characteristics of a tuned circuit. (a) Amplitude and (b) phase response.

Normalized Eqs. (16-32) and (16-33) are plotted in Fig. 16-22 for different values of the parameter Q.

Geometric Symmetry In the $|A_V(j\omega)|$ curves of Fig. 16-22 it is seen that, for every frequency $\omega' < \omega_o$, there exists a frequency $\omega'' > \omega_o$ for which $|A_V(j\omega)|$ has the same value. We now show that these frequencies have ω_o as their geometric mean; that is, $\omega_o^2 = \omega'\omega''$.

Setting $|A_V(j\omega')| = |A_V(j\omega'')|$, we obtain

$$\frac{\omega'}{\omega_o} - \frac{\omega_o}{\omega'} = -\left(\frac{\omega''}{\omega_o} - \frac{\omega_o}{\omega''}\right) \tag{16-34}$$

where the minus sign is required outside the parentheses because $\omega' < \omega_o < \omega''$. From Eq. (16-34) we find

$$\omega_o^2 = \omega'\omega'' \tag{16-35}$$

Bandwidth Let $\omega_1 < \omega_o$ and $\omega_2 > \omega_o$ be the two frequencies on either side of ω_o for which the gain drops by 3 dB from its value A_o at ω_o. Then the bandwidth is defined by

$$B \equiv \frac{\omega_2 - \omega_1}{2\pi} = \frac{1}{2\pi}\left(\omega_2 - \frac{\omega_o^2}{\omega_2}\right) \tag{16-36}$$

where use is made of Eq. (16-35). The frequency ω_2 is found by setting

$$\left|\frac{A_V(j\omega)}{A_o}\right| = \frac{1}{\sqrt{2}} \tag{16-37}$$

From Eq. (16-32) it follows that

$$Q\left(\frac{\omega_2}{\omega_o} - \frac{\omega_o}{\omega_2}\right) = 1 = \frac{Q}{\omega_o}\left(\omega_2 - \frac{\omega_o^2}{\omega_2}\right) \qquad (16\text{-}38)$$

Comparing Eq. (16-36) with Eq. (16-38), we see that

$$B = \frac{1}{2\pi}\frac{\omega_o}{Q} = \frac{f_o}{Q} \qquad (16\text{-}39)$$

The bandwidth is given by the center frequency divided by Q.

Substituting Eq. (16-31) in Eq. (16-39), we find an alternative expression for B, namely,

$$B = \frac{1}{2\pi}\frac{\omega_o R}{\omega_o L} = \frac{1}{2\pi}\frac{R}{L} \qquad (16\text{-}40)$$

Active RC Bandpass Filter   The general form for the second-order bandpass filter is obtained if we let $s = j\omega$ in Eq. (16-29).

$$A_V(s) = \frac{RA_o}{R + sL + 1/sC} = \frac{(R/L)A_o s}{s^2 + s(R/L) + 1/LC} \qquad (16\text{-}41)$$

Substituting Eqs. (16-30) and (16-31) into (16-41) yields

$$A_V(s) = \frac{(\omega_o/Q)A_o s}{s^2 + (\omega_o/Q)s + \omega_o^2} \qquad (16\text{-}42)$$

The transfer function of Eq. (16-42) obtained from the *RLC* circuit shown in Fig. 16-21 can be implemented with the multiple-feedback circuit of Fig. 16-23, which uses two capacitors, three resistors, and one OP AMP, but *no inductors.* If we assume that the OP AMP voltage gain is infinite, we show in Prob. 16-29 that

$$\frac{V_o(s)}{V_s} = \frac{-s/R_1 C_1}{s^2 + \dfrac{C_1 + C_2}{R_3 C_1 C_2}s + \dfrac{1}{R' R_3 C_1 C_2}} \qquad (16\text{-}43)$$

where $R' = R_1 \| R_2$, or

$$R' \equiv \frac{R_1 R_2}{R_1 + R_2} \qquad (16\text{-}44)$$

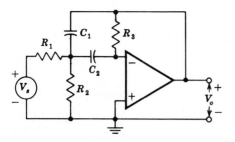

Fig. 16-23   An active resonant filter without an inductance.

Equating the corresponding coefficients in the three transfer functions of Eqs. (16-41), (16-42), and (16-43) yields

$$-R_1C_1 = \frac{L}{RA_o} = \frac{Q}{\omega_o A_o} \tag{16-45}$$

$$R_3 \frac{C_1 C_2}{C_1 + C_2} = \frac{L}{R} = \frac{Q}{\omega_o} \tag{16-46}$$

$$R'R_3 C_1 C_2 = LC = \frac{1}{\omega_o{}^2} \tag{16-47}$$

Any real positive values for R_1, R', R_3, C_1, and C_2 which satisfy Eqs. (16-45) to (16-47) are acceptable for the design of the active bandpass filter. Since we have only three equations for the five parameters, two of these (say, C_1 and C_2) may be chosen arbitrarily.

EXAMPLE Design a second-order bandpass filter with a midband voltage gain $-A_o = 50$ (34 dB), a center frequency $f_o = 160$ Hz, and a 3-dB bandwidth $B = 16$ Hz.

Solution From Eq. (16-39) we see that the required $Q = 160/16 = 10$. The center angular frequency is $\omega_o = 2\pi f_o = 2\pi \times 160 \approx 1,000$ rad/s. Assume $C_1 = C_2 = 0.1$ μF. From Eq. (16-45)

$$R_1 = \frac{Q}{-A_o \omega_o C_1} = \frac{10}{50 \times 10^3 \times 0.1 \times 10^{-6}} \; \Omega = 2 \text{ K}$$

From Eq. (16-46)

$$R_3 = \frac{Q}{\omega_o \left(\dfrac{C_1 C_2}{C_1 + C_2}\right)} = \frac{10}{1,000 \left(\dfrac{0.1 \times 0.1}{0.2}\right) \times 10^{-6}} \; \Omega = 200 \text{ K}$$

From Eq. (16-47)

$$R' = \frac{1}{\omega_o{}^2 R_3 C_1 C_2} = \frac{1}{10^6 \times 2 \times 10^5 \times 10^{-14}} = 500 \; \Omega$$

Finally, from Eq. (16-44)

$$R_2 = \frac{R_1 R'}{R_1 - R'} = \frac{2,000 \times 500}{2,000 - 500} = 667 \; \Omega$$

If the above specifications were to be met with the *RLC* circuit of Fig. 16-21, an unreasonably large value of inductance would be required (Prob. 16-32).

16-8   DELAY EQUALIZER

Signals such as digital data pulses transmitted over telephone wires suffer from delay distortion, discussed in Sec. 12-2. For the compensation of this distortion,

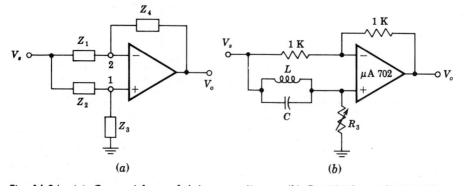

Fig. 16-24 (a) General form of delay equalizer. (b) Practical equalizer section using the μA702. (Courtesy of Fairchild Semiconductor, Inc.)

corrective networks known as *delay equalizers* are required. A delay equalizer is an all-pass network whose transfer function is of the form

$$A_V = \frac{V_o}{V_s} = \frac{R - jX}{R + jX} \tag{16-48}$$

We see from Eq. (16-48) that the amplitude of A_V is unity throughout the useful frequency range, and the delay D is given by the derivative of the phase of A_V with respect to frequency, or

$$D(\omega) = -2\frac{d}{d\omega}\left[\arctan\frac{X(\omega)}{R}\right] \tag{16-49}$$

A delay equalizer using an operational amplifier is shown in Fig. 16-24a. The transfer function for this configuration is found in Prob. 16-33 to be

$$A_V = \frac{Z_1 Z_3 - Z_2 Z_4}{Z_1 (Z_3 + Z_2)} \tag{16-50}$$

For $Z_1 = Z_4 = R = 1$ K, $Z_3 = R_3$, and $Z_2 = jX$, Eq. (16-50) becomes

$$A_V = \frac{R_3 - jX}{R_3 + jX} \tag{16-51}$$

which is the desired all-pass characteristic of the delay equalizer. A practical delay equalizer is shown in Fig. 16-24b using the Fairchild μA702 OP AMP. The low offset voltage of this amplifier allows a larger number of sections to be directly coupled. This advantage is particularly significant when we consider the fact that in many applications eight or more sections in cascade are required to compensate for the delay distortion.

16-9   INTEGRATED CIRCUIT TUNED AMPLIFIER

The differential amplifier stage in monolithic integrated form (Fig. 16-25) is an excellent basic building block for the design of a tuned amplifier (including

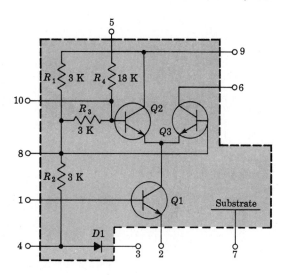

Fig. 16-25 The MC 1550 integrated circuit. (Courtesy of Motorola Semiconductor Inc.)

automatic gain control), an amplitude modulator, or a video amplifier. We now discuss these applications.

Operation of a Tuned Amplifier This circuit is designed to amplify a signal over a narrow band of frequencies centered at f_o. The *simplified* schematic diagram shown in Fig. 16-26 is used to explain the operation of this circuit. The external leads 1, 2, 3, . . . of the IC in this figure correspond to those in Fig. 16-25. The input signal is applied through the tuned trans-

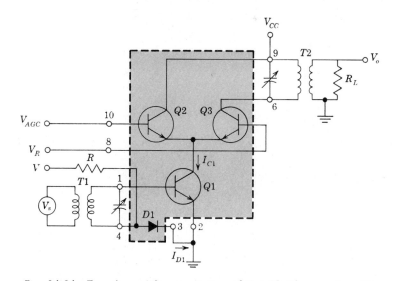

Fig. 16-26 Tuned amplifier consisting of the $Q1$-$Q3$ cascode, with the gain controlled by $Q2$.

former $T1$ to the base of $Q1$. The load R_L is applied across the tuned transformer $T2$ in the collector circuit of $Q3$. The amplification is performed by the transistors $Q1$ and $Q3$, whereas the magnitude of the gain is controlled by $Q2$. The combination of $Q1$-$Q3$ acts as a common-emitter common-base (CE-CB) pair, known as a *cascode* combination. In Prob. 8-39 we show that the input resistance and the current gain of a cascode circuit are essentially the same as those of a CE stage, the output resistance is the same as that of a CB stage, and the reverse-open-circuit voltage amplification is given by $h_r \approx h_{re}h_{rb} \approx 10^{-7}$. The extremely small value of h_r for the cascode transistor pair makes this circuit especially useful in tuned-amplifier design. The reduction in the "reverse internal feedback" of the compound device simplifies tuning, reduces the possibility of oscillation, and results in improved stability of the amplifier.

The voltage V_{AGC} applied to the base of $Q2$ is used to provide automatic gain control. From Fig. 15-9 we see that if V_{AGC} is at least 120 mV greater than V_R, $Q3$ is cut off and all the current of $Q1$ flows through $Q2$. Since $Q3$ is cut off, its transconductance is zero and the gain $A_V = V_o/V_s$ becomes zero. If V_{AGC} is less than V_R by more than 120 mV, $Q2$ is cut off and the collector current of $Q1$ flows through $Q3$, increasing the transconductance of $Q3$ and resulting in maximum voltage gain A_V.

An important advantage of this amplifier is its ability to vary the value of A_V by changing V_{AGC} without detuning the input circuit. This follows from the fact that variations in V_{AGC} cause changes in the division of the current between $Q2$ and $Q3$ without affecting significantly the collector current of $Q1$. Thus the input impedance of $Q1$ remains constant and the input circuit is not detuned.

Biasing of this integrated amplifier is obtained using a technique similar to that discussed in Sec. 9-7. The voltage V and resistor R establish the dc current I_{D1} through the diode $D1$. Since the diode and transistor $Q1$ are on the same silicon chip, very close to each other, and with $V_{D1} = V_{BE1}$, the collector current I_{C1} of $Q1$ is within ± 5 percent of I_{D1}.

y-parameters In the design of tuned amplifiers, it is convenient to characterize the amplifiers as a two-port network and measure the y-parameters at the frequency of operation. These y-parameters are defined by choosing the input and output voltages V_1 and V_2 as independent variables and expressing the currents I_1 and I_2 in Fig. 16-27a in terms of these two voltages. Thus

$$I_1 = y_{11}V_1 + y_{12}V_2 \tag{16-52}$$

$$I_2 = y_{21}V_1 + y_{22}V_2 \tag{16-53}$$

where the I's and V's represent rms values of the small-signal currents and voltages. The circuit model satisfying these equations is indicated in Fig. 16-27b.

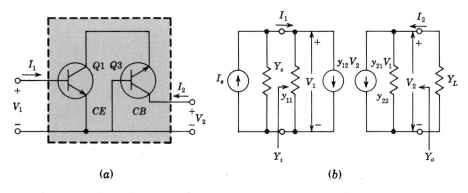

Fig. 16-27 (a) Cascode pair. (b) y-parameter two-port model.

The y-parameter in Eqs. (16-52) and (16-53) and Fig. 16-27b are complex-valued functions of frequency which are defined as follows:

$$y_{11} \equiv G_{11} + jB_{11} \equiv \frac{I_1}{V_1}\bigg|_{V_2=0} = \text{short-circuit input admittance}$$

$$y_{12} \equiv G_{12} + jB_{12} \equiv \frac{I_1}{V_2}\bigg|_{V_1=0} = \text{short-circuit reverse transfer admittance}$$

$$y_{21} \equiv G_{21} - jB_{21} \equiv \frac{I_2}{V_1}\bigg|_{V_2=0} = \text{short-circuit forward transfer admittance}$$

$$y_{22} \equiv G_{22} + jB_{22} \equiv \frac{I_2}{V_2}\bigg|_{V_1=0} = \text{short-circuit output admittance}$$

For a given device, single transistor or cascode pair, these parameters may be specified as explicit functions of frequency, or, as is more often the case, as graphs of the real and imaginary parts, the conductance G and the susceptance B, versus frequency. The data sheet of the MC 1550 gives the y-parameters measured on the General Radio 1607A immittance bridge. Typical measured values are shown in Fig. 16-28. The internal feedback factor y_{12} is not shown because it was found to be less than 0.001 mA/V (m℧) and is neglected.

Let us consider the two-port network of Fig. 16-27b terminated at the output by a load admittance Y_L and driven by a current source I_s with source admittance Y_s. The equivalent admittance seen by the current source is $Y_{eq} = Y_s + Y_i$. In Prob. 16-36 we show that

$$Y_i = \frac{I_1}{V_1} = y_{11} - \frac{y_{12}y_{21}}{y_{22} + Y_L} \tag{16-54}$$

and the output admittance is

$$Y_o = \frac{I_2}{V_2} = y_{22} - \frac{y_{12}y_{21}}{y_{11} + Y_s} \tag{16-55}$$

Since $y_{12} \approx 0$, then $Y_i \approx y_{11}$ and $Y_o \approx y_{22}$.

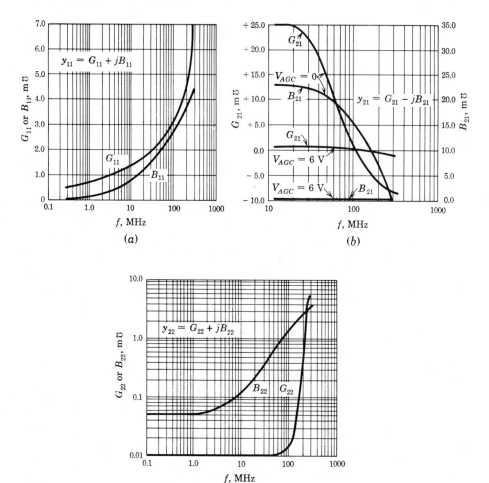

Fig. 16-28 The y-parameters of the MC 1550 for $V_{AGC} = 6.0$ V and $V_{AGC} = 0$ V as functions of frequency. (a) The parameter y_{11}, (b) the parameter y_{21}, (c) the parameter y_{22}. (Courtesy of Motorola, Inc.)

The average power P_{av} delivered by the current source to the two-port is the power dissipated in the conductive part of Y_i, or

$$P_{av} = |V_1|^2 \, \text{Re} \, [Y_i]\dagger \qquad (16\text{-}56)$$

If $\text{Re} \, [Y_i]$ becomes negative at some frequency ω_1, the network absorbs negative power; in other words, power is supplied to the source by the network. We note from Eq. (16-54) that if $y_{12} \approx 0$ and $\text{Re} \, [y_{11}] > 0$ and $\text{Re} \, [Y_s] > 0$, the circuit cannot oscillate.

† $\text{Re} \, [Y_i]$ means the real part of Y_i and V_1 is the rms value of the input voltage.

The current gain A_I and voltage gain A_V are found in Prob. 16-37 to be given by

$$A_I = -\frac{I_2}{I_1} = -\frac{y_{21}Y_L}{y_{11}y_{22} - y_{12}y_{21} + y_{11}Y_L} \tag{16-57}$$

and

$$A_V = \frac{V_2}{V_1} = -\frac{y_{21}}{y_{22} + Y_L} \tag{16-58}$$

A Practical Tuned Amplifier A hybrid monolithic circuit which embodies the principles discussed above is indicated in Fig. 16-29. (For the moment, assume that the audio generator V_a is not present; $V_a = 0$.) The shaded block is the MC 1550 IC chip of Fig. 16-25. All other components are discrete elements added externally. Resistors R_1 and R_2 bias the diode $D1$ (and hence determine the collector current of $Q1$). These resistors also establish the bias voltage for $Q3$. Resistors R_3 and R_4 serve to "widen" the AGC voltage range from 120 mV to approximately 850 mV, thus rendering the AGC terminal less susceptible to external noise pickup.

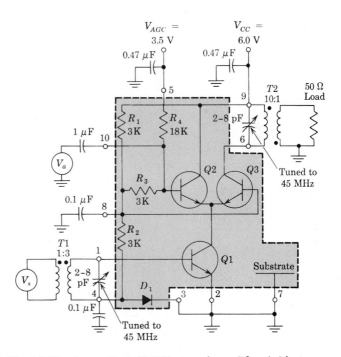

Fig. 16-29 A practical 45-MHz tuned amplifier (with $V_a = 0$), or an RF modulator if $V_a \neq 0$. (Courtesy of Motorola Semiconductor, Inc.)

The source V_s is a 45-MHz RF (radio-frequency) generator whose resistance is 50 Ω. The transformers are wound with No. 32 wire on T12-2 cores; T1 with 6:18 turns has a magnetizing inductance $L_M = 1.1$ μH, and T2 with 30:3 turns gives $L_M = 2.5$ μH. The variable capacitors in Fig. 16-29 are adjusted so as to resonate with these inductors at 45 MHz.

For maximum power gain through an amplifier the source admittance and load admittance must be selected to be the complex conjugates of the input admittance $Y_i \approx y_{11}$ and of the output admittance $Y_o \approx y_{22}$, respectively. In place of transformers, LC networks may be used to obtain this matching. In Fig. 16-30, the input network consisting of C_1, C_2, and L_1 transforms the 50-Ω resistance of the source into the complex conjugate of y_{11}. The values of C_1, C_2, and L_1 are calculated using techniques in Ref. 8. The output network consists of C_3, C_4, and L_2 and transforms the 50-Ω load resistance into the complex conjugate of y_{22}. The center frequency for this amplifier is 60 MHz, the bandwidth is 500 kHz, and the power gain is measured to be 30 dB.

Amplitude Modulator The RF carrier signal V_s may be varied in amplitude by changing the AGC voltage. Hence, if an audio signal V_a is applied to terminal 10 in Fig. 16-29 so as to modify the AGC voltage, the output will be the amplitude-modulated waveform indicated in Fig. 4-27.

The gain of the $Q1$-$Q3$ cascode is proportional to $|y_{21}|$, as indicated in Eq. (16-58). The parameter y_{21} depends upon the collector current of $Q3$, which can be varied by changing V_{AGC}. Figure 16-31 shows the variation of $|y_{21}|$ versus V_{AGC} at the frequency of 45 MHz. From the curve we see that between $V_{AGC} = 2.75$ V and $V_{AGC} = 4.25$ V, $|y_{21}|$ is linear with V_{AGC}. By biasing the AGC line to 3.5 V (point B on the curve) and impressing an audio sinusoidal signal V_a on the base of $Q2$ (as indicated in Fig. 16-29), $|y_{21}|$ will vary sinusoidally. From Eq. (16-58) the gain A_V will also vary sinusoidally with the audio signal. Thus the amplifier output will be an amplitude-modulated signal.

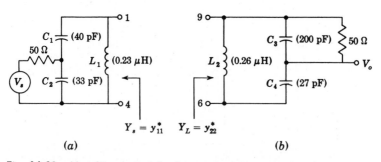

(a)   (b)

Fig. 16-30 Matching networks for maximum power transfer in a 60-MHz tuned amplifier (a) at the input and (b) at the output.

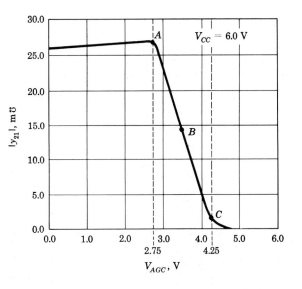

Fig. 16-31 Variation of $|y_{21}|$ versus V_{AGC} at 45 MHz for the MC 1550. (Courtesy of Motorola Semiconductor, Inc.)

16-10   A CASCODE VIDEO AMPLIFIER

A video amplifier, as opposed to a tuned amplifier, must amplify signals over a wide band of frequencies, say up to 20 MHz. The MC 1550 can be used as a cascode video amplifier (Fig. 16-32a) by avoiding tuned input and output circuits. Between pins 1 and 4, 50 Ω is inserted in order to properly terminate the coaxial cable carrying the video signal. This small resistance has negligible effect on the biasing of $Q1$. The load R_L is placed directly in the collector lead of $Q3$.

The small-signal analysis of this video amplifier can be made using the approximate equivalent circuit of Fig. 16-32c. If both transistors $Q2$ and $Q3$ are operating in their active region, the collector of $Q1$ sees the very small input resistance $(r_{e2}\|r_{e3})$ of two common-base stages in parallel. We can represent $Q1$ with its hybrid-Π model, and due to the very low load on $Q1$, we can neglect the effect of C_c. The video output is taken from the collector of $Q3$, which is operating as a common-base stage. We shall assume that $Q3$ can be represented by a current source $\alpha_3 I_3$, where I_3 is the emitter signal current of $Q3$ and $\alpha_3 \approx 1$ and is independent of frequency over the band of frequencies under consideration. C_s represents the capacitance from the collector of $Q1$ and $Q3$ to the substrate (ground).

From Fig. 16-32c we find (Prob. 16-42)

$$\frac{V_o}{V_i} = -\frac{\alpha_3 g_m}{r_{e3}r_{bb'}} \frac{1}{\left(\dfrac{1}{r_{bb'}}+\dfrac{1}{r_{b'e}}+sC_e\right)\left(\dfrac{1}{r_{e2}}+\dfrac{1}{r_{e3}}+sC_s\right)\left[\dfrac{1}{R_L}+s(C_s+C_L)\right]}$$

$$(16\text{-}59)$$

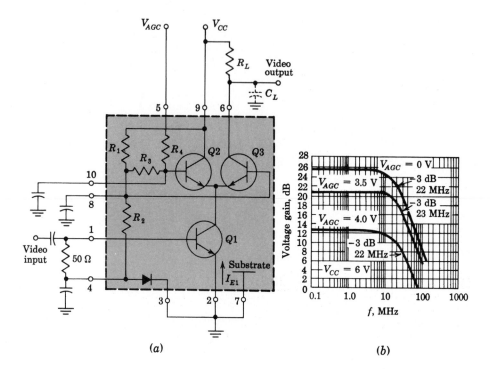

(a)

(b)

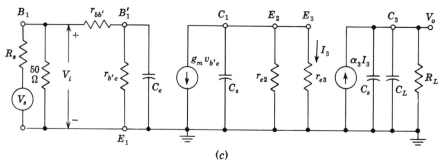

(c)

Fig. 16-32 (a) The MC 1550 used as a video amplifier; (b) frequency response for three different values of V_{AGC}; (c) approximate small-signal equivalent circuit.

From Eq. (3-14) the emitter-base-diode incremental resistance r_e (r_{e2} or r_{e3}) is given by $r_e = \eta V_T/I_E$, where I_E is the quiescent emitter current.

EXAMPLE Design a video amplifier, using the circuit of Fig. 16-32 and the MC 1550, to provide voltage gain $A_V = V_o/V_i = -25$ and bandwidth greater than 20 MHz when $V_{AGC} = 0$. Assume $V_{CC} = 6$ V, $h_{fe} = 50$, $r_{bb'} = 50$ Ω, $C_s = 5$ pF, $C_L = 5$ pF, $I_{E1} = 1$ mA, and $f_T = 900$ MHz.

Solution   The low-frequency voltage gain V_o/V_i is found from Eq. (16-59) by letting $s = 0$. When $V_{AGC} = 0$, transistor $Q2$ is cut off and all the collector current of $Q1$ flows through $Q3$. Since $I_{E2} = 0$, then $r_{e2} = \eta V_T/I_{E2} = \infty$ and

$$r_{e3} = \eta \frac{V_T}{I_{E3}} = \frac{52 \text{ mV}}{1 \text{ mA}} = 52 \ \Omega \text{ at } 25°\text{C}$$

From Sec. 11-2 we obtain

$$g_m = \frac{I_{E1}}{V_T} = \frac{1}{26} = 38.5 \times 10^{-3} \text{ A/V}$$

$$r_{b'e} = \frac{h_{fe}}{g_m} = \frac{50}{38.5 \times 10^{-3}} \ \Omega = 1.30 \text{ K}$$

and

$$C_e = \frac{g_m}{2\pi f_T} = \frac{38.5 \times 10^{-3}}{2 \times 3.14 \times 900 \times 10^{-6}} \text{ F} = 6.80 \text{ pF}$$

If we assume $\alpha_3 \approx 1$, we find from Eq. (16-59)

$$A_{Vo} = \left. \frac{V_o}{V_i} \right|_{s=0} = -\frac{38.5 \times 10^{-3}}{52 \times 50} \ \frac{1}{\left(\dfrac{1}{50} + \dfrac{1}{1,300}\right) \times \dfrac{1}{52} \times \dfrac{1}{R_L}}$$

$$= -37.2 \times 10^{-3} R_L$$

Thus

$$A_{Vo} = -25 = -37.2 \times 10^{-3} \times R_L$$

or

$$R_L = \frac{25}{37.2} \times 10^3 = 675 \ \Omega$$

The voltage transfer function of Eq. (16-59) has three poles, and the corresponding 3-dB frequencies are

$$f_1 = \frac{1}{2\pi R_L (C_s + C_L)} = \frac{1}{2 \times 3.14 \times 675 \times 10 \times 10^{-12}} \text{ Hz} = 23.6 \text{ MHz}$$

$$f_2 = \frac{1}{2\pi C_e (r_{b'e} \| r_{bb'})} = \frac{1}{2 \times 3.14 \times 6.80 \times 10^{-12} \times 48} \text{ Hz} = 490 \text{ MHz}$$

$$f_3 = \frac{1}{2\pi C_s (r_{e2} \| r_{e3})} = \frac{1}{2 \times 3.14 \times 5 \times 10^{-12} \times 52} \text{ Hz} = 610 \text{ MHz}$$

We conclude that f_1 is a dominant pole and $f_H \approx f_1 = 23.6$ MHz. Figure 16-32b shows measured data for three values of V_{AGC}. We see that, although we used a simplified model to analyze the circuit, we obtained excellent agreement with experiment.

II. NONLINEAR ANALOG SYSTEMS

16-11   COMPARATORS

With the exception of amplitude modulation and automatic gain control, all the systems discussed thus far in this chapter have operated linearly. The remainder of the chapter is concerned with nonlinear OP AMP functions.

The comparator, introduced in Sec. 4-6, is a circuit which compares an input signal $v_i(t)$ with a reference voltage V_R. When the input v_i exceeds V_R, the comparator output v_o takes on a value which is very different from the magnitude of v_o when v_i is smaller than V_R. The DIFF AMP input-output curve of Fig. 15-9 approximates this comparator characteristic. Note that the total input swing between the two extreme output voltages is $\sim 8V_T = 200$ mV. This range may be reduced considerably by cascading two DIFF AMPS as in Fig. 15-11. This MC 1530 OP AMP serves as a comparator if connected open-loop, as shown in Fig. 16-33a. The transfer characteristic is given in Fig. 16-33b, and it is now observed that the change in output state takes place with a variation in input of only 2 mV. Note that the input offset voltage contributes an error in the point of comparison between v_i and V_R of the order of 1 mV. The reference V_R may be any voltage, provided that it does not exceed the maximum common-mode range.

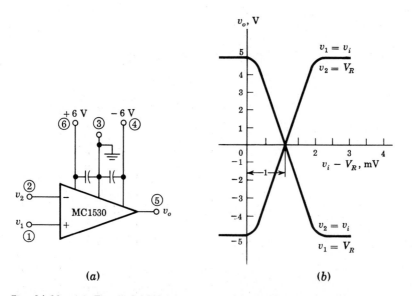

(a)   (b)

Fig. 16-33   (a) The MC 1530 operational amplifier as a comparator.
(b) The transfer characteristic.

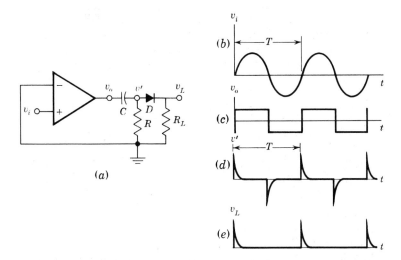

Fig. 16-34 A zero-crossing detector converts a sinusoid v_i into a square wave v_o. The pulse waveforms v' and v_L result if v_o is fed into a short time-constant RC circuit in cascade with a diode clipper.

Zero-crossing Detector If V_R is set equal to zero, the output will respond almost discontinuously every time the input passes through zero. Such an arrangement is called a *zero-crossing detector*.

Some of the most important systems using comparators will now be listed. Other applications are discussed in Secs. 16-15 and 17-20.

Square Waves from a Sine Wave If the input to an OP AMP comparator is a sine wave, the output is a square wave. If a zero-crossing detector is used, a symmetrical square wave results, as shown in Fig. 16-34c. This idealized waveform has vertical sides which, in reality, should extend over a range of a few millivolts of input voltage v_i.

Timing-markers Generator from a Sine Wave The square-wave output v_o of the preceding application is applied to the input of an RC series circuit. If the time constant RC is very small compared with the period T of the sine-wave input, the voltage v' across R is a series of positive and negative pulses, as indicated in Fig. 16-34d. If v' is applied to a clipper with an ideal diode (Fig. 16-34a), the load voltage v_L contains only positive pulses (Fig. 16-34e). Thus the sinusoid has been converted into a train of positive pulses whose spacing is T. These may be used for timing markers (on the sweep voltage of a cathode-ray tube, for example).

Phasemeter   The phase angle between two voltages can be measured by a method based on the circuit of Fig. 16-34.   Both voltages are converted into pulses, and the time interval between the pulse of one wave and that obtained from the second sine wave is measured.   This time interval is proportional to the phase difference.   Such a phasemeter can measure angles from 0 to 360°.

Amplitude-distribution Analyzer   A comparator is a basic building block in a system used to analyze the amplitude distribution of the noise generated in an active device or the voltage spectrum of the pulses developed by a nuclear-radiation detector, etc.   To be more specific, suppose that the output of the comparator is 10 V if $v_i > V_R$ and 0 V if $v_i < V_R$.   Let the input to the comparator be noise.   A dc meter is used to measure the average value of the output square wave.   For example, if V_R is set at zero, the meter will read 10 V, which is interpreted to mean that the probability that the amplitude is greater than zero is 100 percent.   If V_R is set at some value V'_R and the meter reads 7 V, this is interpreted to mean that the probability that the amplitude of the noise is greater than V'_R is 70 percent, etc.   In this way the cumulative amplitude probability distribution of the noise is obtained by recording meter readings as a function of V_R.

Pulse-time Modulation   If a periodic sweep waveform is applied to a comparator whose reference voltage V_R is not constant but rather is modulated by an audio signal, it is possible to obtain a succession of pulses whose relative spacing reflects the input information.   The result is a *time-modulation system* of communication.

16-12   SAMPLE–AND–HOLD CIRCUITS[9]

A typical data-acquisition system receives signals from a number of different sources and transmits these signals in suitable form to a computer or a communication channel.   A multiplexer (Sec. 17-5) selects each signal in sequence, and then the analog information is converted into a constant voltage over the gating-time interval by means of a *sample-and-hold circuit*.   The constant output of the sample-and-hold may then be converted to a digital signal by means of an analog-to-digital (A/D) converter (Sec. 17-20) for digital transmission.

A sample-and-hold circuit in its simplest form is a switch S in series with a capacitor, as in Fig. 16-35a.   The voltage across the capacitor tracks the input signal during the time T_g when a logic control gate closes S, and holds the instantaneous value attained at the end of the interval T_g when the control gate opens S.   The switch may be a relay (for very slow waveforms), a sampling diode-bridge gate (Sec. 4-7), a bipolar transistor switch,[10] or a MOSFET controlled by a gating signal.   The MOSFET makes an excellent

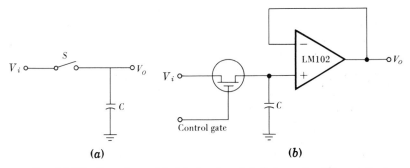

Fig. 16-35 Sample-and-hold circuit. (a) Schematic, (b) practical.

chopper because its *offset voltage* when ON ($\sim$5 μV) is much smaller than that of a bipolar junction transistor.

The circuit shown in Fig. 16-35b is one of the simplest practical sample-and-hold circuits. A negative pulse at the gate of the p-channel MOSFET will turn the switch ON, and the holding capacitor C will charge with a time constant $R_{ON}C$ to the instantaneous value of the input voltage. In the absence of a negative pulse, the switch is turned OFF and the capacitor is isolated from any load through the LM 102 OP AMP. Thus it will hold the voltage impressed upon it. It is recommended that a capacitor with polycarbonate, polyethylene, or Teflon dielectric be used. Most other capacitors do not retain the stored voltage, due to a polarization phenomenon[11] which causes the stored voltage to decrease with a time constant of several seconds. Even if the polarization phenomenon does not occur, the OFF current of the switch ($\sim$1 nA) and the bias current of the OP AMP will flow through C. Since the maximum input bias current for the LM 102 is 10 nA, it follows that with a 10-μF capacitance the drift rate during the HOLD period will be less than 1 mV/s.

Two additional factors influence the operation of the circuit: the reaction time, called *aperture time* (typically less than 100 ns), which is the delay between the time that the pulse is applied to the switch and the actual time the switch closes, and the *acquisition time*, which is the time it takes for the capacitor to change from one level of holding voltage to the new value of input voltage after the switch has closed.

When the hold capacitor is larger than 0.05 μF, an isolation resistor of approximately 10 K should be included between the capacitor and the + input of the OP AMP. This resistor is required to protect the amplifier in case the output is short-circuited or the power supplies are abruptly shut down while the capacitor is charged.

16-13   PRECISION AC/DC CONVERTERS[12]

If a sinusoid whose peak value is less than the threshold or cutin voltage V_γ ($\sim$0.6 V) is applied to the rectifier circuit of Fig. 4-6, we see that the output

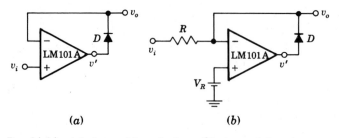

(a)   (b)

Fig. 16-36   (a) A precision diode.   (b) A precision clamp.

is zero for all times.   In order to be able to rectify millivolt signals, it is clearly necessary to reduce V_γ.   By placing the diode in the feedback loop of an OP AMP, the cutin voltage is divided by the open-loop gain A_V of the amplifier. Hence V_γ is virtually eliminated and the diode approaches the ideal rectifying component.   If in Fig. 16-36a the input v_i goes positive by at least V_γ/A_V, then v' exceeds V_γ and D conducts.   Because of the virtual connection between the noninverting and inverting inputs (due to the feedback with D ON), $v_o \approx v_i$. Therefore the circuit acts as a voltage follower for positive signals (in excess of approximately 0.1 mV).   When v_i swings negatively, D is OFF and no current is delivered to the external load except for the small bias current of the LM 101A and the diode reverse saturation current.

Precision Clamp   By modifying the circuit of Fig. 16-36a, as indicated in Fig. 16-36b, an almost ideal clamp (Sec. 4-5) is obtained.   If $v_i < V_R$, then v' is positive and D conducts.   As explained above, under these conditions the output equals the voltage at the noninverting terminal, or $v_o = V_R$.   If $v_i > V_R$, then v' is negative, D is OFF, and $v_o = v_i$.   In summary: The output follows the input for $v_i > V_R$ and v_o is clamped to V_R if v_i is less than V_R by about 0.1 mV.   When D is reverse-biased in Fig. 16-36a or b, a large differential voltage may appear between the inputs and the OP AMP must be able to withstand this voltage.   Also note that when $v_i > V_R$, the input stage saturates because the feedback through D is missing.

Fast Half-wave Rectifier   By adding R' and $D2$ to Fig. 16-36b and setting $V_R = 0$, we obtain the circuit of Fig. 16-37a.   If v_i goes negative, $D1$ is ON, $D2$ is OFF, and the circuit behaves as an inverting OP AMP, so that $v_o = -(R'/R)v_i$.   If v_i is positive, $D1$ is OFF and $D2$ is ON.   Because of the feedback through $D2$, a virtual ground exists at the input and $v_o = 0$.   If v_i is a sinusoid, the circuit performs half-wave rectification.   Because the amplifier does not saturate, it can provide rectification at frequencies up to 100 kHz.

An alternative configuration to that in Fig. 16-37a is to ground the left-hand side of R and to impress v_i at the noninverting terminal.   The output now has a value of $(R + R')/R$ times the input for positive voltages and $v_o = v_i$

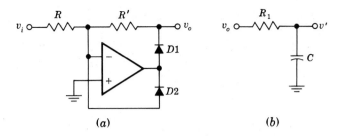

Fig. 16-37 (a) A half-wave rectifier. (b) A low-pass filter which can be cascaded with the circuit in (a) to obtain an average detector.

for negative inputs. Hence half-wave rectification is obtained if $R' \gg R$. A full-wave system is indicated in Prob. 16-43.

Active Average Detector Consider the circuit of Fig. 16-37a to be cascaded with the low-pass filter of Fig. 16-37b. If v_i is an amplitude-modulated carrier (Fig. 4-27), the R_1C filter removes the carrier and v' is proportional to the average value of the audio signal. In other words, this configuration represents an *average detector*.

Active Peak Detector If a capacitor is added at the output of the precision diode of Fig. 16-36a, a peak detector results. The capacitor in Fig. 16-38a will hold the output at $t = t'$ to the most positive value attained by the input v_i prior to t', as indicated in Fig. 16-38b. This operation follows from the fact that if $v_i > v_o$, the OP AMP output v' is positive, so that D conducts. The capacitor is then charged through D (by the output current of the amplifier) to the value of the input because the circuit is a voltage follower. When v_i falls below the capacitor voltage, the OP AMP output goes negative

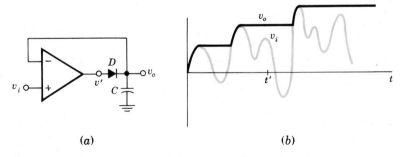

Fig. 16-38 (a) A positive peak detector. (b) An arbitrary input waveform v_i and the corresponding output v_o.

and the diode becomes reverse-biased. Thus the capacitor gets charged to the most positive value of the input.

This circuit is a special case of a sample-and-hold circuit, and the capacitor leakage current considerations given in Sec. 16-12 also apply to this configuration. If the output is loaded, a buffer voltage follower should be used to prevent the load from discharging C. To reset the circuit, a low-leakage switch such as a MOSFET gate must be placed across the capacitor.

16-14   LOGARITHMIC AMPLIFIERS[12]

In Fig. 16-39a there is indicated an OP AMP with the feedback resistor R' replaced by the diode $D1$. This amplifier is used when it is desired to have the output voltage proportional to the logarithm of the input voltage.

From Eq. (3-9) the volt-ampere diode characteristic is

$$I_f = I_o(\epsilon^{V_f/\eta V_T} - 1) \approx I_o \epsilon^{V_f/\eta V_T}$$

provided that $V_f/\eta V_T \gg 1$ or $I_f \gg I_o$. Hence

$$V_f = \eta V_T(\ln I_f - \ln I_o) \tag{16-60}$$

Since $I_f = I_s = V_s/R$ due to the virtual ground at the amplifier input, then

$$V_o = -V_f = -\eta V_T\left(\ln \frac{V_s}{R} - \ln I_o\right) \tag{16-61}$$

We note from Eq. (16-61) that the output voltage V_o is temperature-dependent due to the scale factor ηV_T and to the saturation current I_o. Both temperature effects can be reduced by using the circuit of Fig. 16-39b, where the diodes $D1$ and $D2$ are matched, R_T is temperature-dependent, and the constant source I is independent of T.

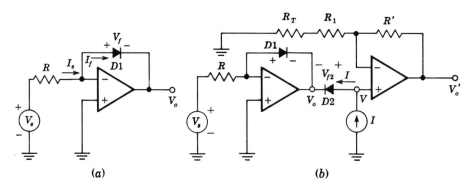

Fig. 16-39 (a) Logarithmic amplifier for positive input voltage V_s. (b) Temperature-compensated amplifier.

We have for this circuit, and using Eq. (16-61),

$$V = V_{f2} + V_o = \eta V_T \left(\ln I - \ln I_o - \ln \frac{V_s}{R} + \ln I_o \right) = -\eta V_T \ln \frac{V_s}{RI}$$

Thus the output voltage V_o' becomes

$$V_o' = -\frac{R_T + R_1 + R'}{R_1 + R_T} \eta V_T \ln \frac{V_s}{RI} \tag{16-62}$$

The temperature dependence of R_T is selected to compensate approximately for the factor ηV_T in Eq. (16-62).

Logarithmic Amplifier Using Matched Transistors Instead of two matched diodes, it is possible to use a matched pair of transistors connected as in Fig. 16-40a to remove from the expression for V_o' the factor η, whose value normally depends on the current flowing through the diode. In Fig. 16-40a, Q1 is used as the feedback element around the first op amp. If we neglect $V_{BE1} - V_{BE2}$ with respect to V_{CC}, and since $I_{B2} \ll I_{C2}$, then

$$I_{C2} \approx \frac{V_{CC}}{R_6} \quad \text{and} \quad I_{C1} = \frac{V_s}{R_1 + R_4} = \frac{V_s}{2R_1} \tag{16-63}$$

From Eq. (15-21) it follows that

$$V_{BE1} - V_{BE2} = V_T \ln I_{C1} - V_T \ln I_{C2} = V_T \ln \left(\frac{V_s}{2R_1} \frac{R_6}{V_{CC}} \right) \tag{16-64}$$

Since the base of Q1 is grounded, the negative of the above voltage appears at the noninverting terminal of the second operational amplifier, whose gain is determined by resistors R_7 and R_8. Hence

$$V_o = -V_T \frac{R_7 + R_8}{R_7} \ln \left(\frac{V_s}{2R_1} \frac{R_6}{V_{CC}} \right) \tag{16-65}$$

The above transfer function of the amplifier is plotted in Fig. 16-40b for various operating temperatures. It is seen that the dynamic range extends over 5 mV to 50 V of input voltage, or 80 dB. From Eq. (16-65),

$$\frac{dV_o}{d(\ln V_s)} = -V_T \left(\frac{R_7 + R_8}{R_7} \right) = -0.026 \times \frac{43.8}{0.52} = -2.20$$

which is in excellent agreement with the slope obtained from Fig. 16-40b.

Antilog Amplifier The amplifiers discussed above give an output V_o proportional to the natural logarithm of the input V_s, or

$$V_o = K_1 \ln K_2 V_s \tag{16-66}$$

Sometimes we desire an output proportional to the antilogarithm ($\ln^{-1}$) of the input; that is,

$$V_o = K_3 \ln^{-1} K_4 V_s \tag{16-67}$$

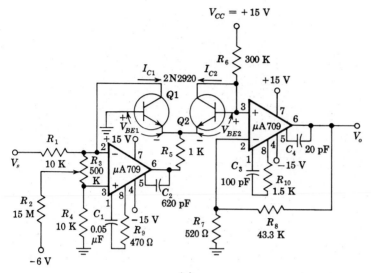

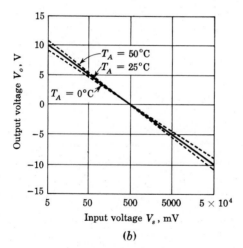

(b)

Fig. 16-40 (a) Logarithmic amplifier. (b) Transfer characteristic. (Courtesy of Fairchild Semiconductor, Inc.)

The circuit shown in Fig. 16-41 can be used as an antilog amplifier. If we assume infinite input resistance for A_1 and A_2 as well as zero differential input voltage for each operational amplifier, we obtain

$$V_2 = -V_f + V_1 = -\eta V_T(\ln I_f - \ln I_o) + \frac{R_1}{R_1 + R_2} V_s \qquad (16\text{-}68)$$

and since V_2 is the negative of the voltage across $D2$,

$$V_2 = -\eta V_T(\ln I_2 - \ln I_o) \qquad (16\text{-}69)$$

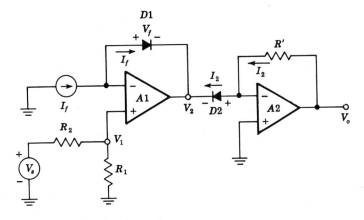

Fig. 16-41 Antilog amplifier.

Combining Eqs. (16-68) and (16-69) yields

$$V_s \frac{R_1}{R_1 + R_2} = \eta V_T \ln \frac{I_f}{I_2} = \eta V_T \ln \frac{I_f R'}{V_o} \tag{16-70}$$

because $V_o = I_2 R'$. Finally, from Eq. (16-70) it follows that

$$V_o = R' I_f \ln^{-1} \left[-V_s \left(\frac{R_1}{R_1 + R_2} \frac{1}{\eta V_T} \right) \right] \tag{16-71}$$

Equation (16-71) is of the form given in Eq. (16-67).

We show in Prob. 16-45 that it is possible to raise the input V_s to an arbitrary power by combining log and antilog amplifiers.

Logarithmic Multiplier The log and antilog amplifiers can be used for the multiplication or division of two analog signals V_{s1} and V_{s2}. In Fig. 16-42

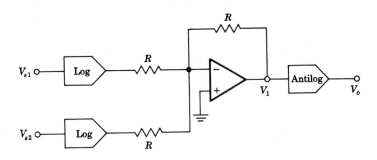

Fig. 16-42 Logarithmic multiplier of two analog signals ($V_o = K V_{s1} V_{s2}$).

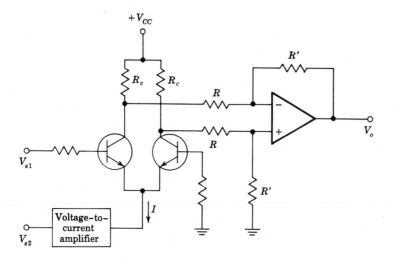

Fig. 16-43 Variable transconductance multiplier ($V_o = KV_{s1}V_{s2}$).

the logarithm of each input is taken, then the two logarithms are added, and finally the antilog of the sum yields the product of the two inputs. Thus

$$V_1 = K_1 \ln V_{s1} + K_1 \ln V_{s2} = K_1 \ln V_{s1}V_{s2} \qquad (16\text{-}72)$$

and

$$V_o = K_2 \ln^{-1} K_3 V_1 = K_2 \ln^{-1} (K_3 K_1 \ln V_{s1}V_{s2}) \qquad (16\text{-}73)$$

If $K_3 K_1 = 1$, then

$$V_o = K_2 V_{s1} V_{s2} \qquad (16\text{-}74)$$

The input signals can be divided if we subtract the logarithm of V_{s1} from that of V_{s2} and then take the antilog. We must point out that the logarithmic multiplier or divider is useful for unipolar inputs only. This is often called one-quadrant operation. Other techniques[13] are available for the accurate multiplication of two signals.

Differential Amplifier Multiplier From Eqs. (15-24) and (15-23) we observe that the output voltage of a differential amplifier depends on the current source I. If V_{s1} is applied to one input and V_{s2} is used to vary I, as in Fig. 16-43, the output will be proportional to the product of the two signals $V_{s1}V_{s2}$. The device AD 530 manufactured by Analog Devices, Inc., is a completely monolithic multiplier/divider with basic accuracy of 1 percent and bandwidth of 1 MHz. As a multiplier, the AD 530 has the transfer function $XY/10$ and as a divider $+10Z/X$. The X, Y, and Z input levels are ± 10 V for multiplication and the output is ± 10 V at 5 mA. As a divider, operation is restricted to two quadrants (where X is negative) only.

16-15   WAVEFORM GENERATORS[9]

The operational amplifier comparator, together with an integrator, can be used to generate a square wave, a pulse, or a triangle waveform, as we now demonstrate.

Square-wave Generator   In Fig. 16-44a, the output v_o is shunted to ground by two Zener diodes connected back to back and is limited to either $+V_{Z2}$ or $-V_{Z1}$, if $V_\gamma \ll V_Z$ (Fig. 4-11).   A fraction $\beta = R_3/(R_2 + R_3)$ of the output is fed back to the noninverting input terminal.   The differential input voltage v_i is given by

$$v_i = v_c - \beta v_o \tag{16-75}$$

From the transfer characteristic of the comparator given in Fig. 16-33 we see that if v_i is positive (by at least 1 mV), then $v_o = -V_{Z1}$, whereas if v_i is negative (by at least 1 mV), then $v_o = +V_{Z2}$.   Consider an instant of time when $v_i < 0$ or $v_c < \beta v_o = \beta V_{Z2}$.   The capacitor C now charges exponentially toward V_{Z2} through the integrating $R'C$ combination.   The output remains constant at V_{Z2} until v_c equals $+\beta V_{Z2}$, at which time the comparator output reverses to $-V_{Z1}$.   Now v_c charges exponentially toward $-V_{Z1}$.   The output voltage v_o and capacitor voltage v_c waveforms are shown in Fig. 16-44b for the special case $V_{Z1} = V_{Z2} = V_Z$.   If we let $t = 0$ when $v_c = -\beta V_Z$ for the first half cycle, we have

$$v_c(t) = V_Z[1 - (1 + \beta)\epsilon^{-t/R'C}] \tag{16-76}$$

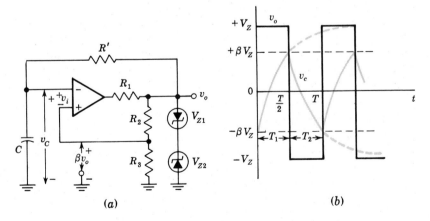

(a)	(b)

Fig. 16-44   (a) A square-wave generator.   (b) Output and capacitor voltage waveforms.

Since at $t = T/2$, $v_c(t) = +\beta V_Z$, we find T, solving Eq. (16-76), to be given by

$$T = 2R'C \ln \frac{1 + \beta}{1 - \beta} \qquad (16\text{-}77)$$

Note that T is independent of V_Z.

This square-wave generator is particularly useful in the frequency range 10 Hz to 10 kHz. At higher frequencies the delay time of the operational amplifier as it moves out of saturation, through its linear range, and back to saturation in the opposite direction, becomes significant. Also, the slew rate of the operational amplifier limits the slope of the output square wave. The frequency stability depends mainly upon the Zener-diode stability and the capacitor, whereas waveform symmetry depends on the matching of the two Zener diodes. If an unsymmetrical square wave is desired, then $V_{Z1} \neq V_{Z2}$.

The circuit will operate in essentially the same manner as described above if $R_1 = 0$ and the avalanche diodes are omitted. However, now the amplitude of the square wave depends upon the power supply voltage (± 5.8 V for the MC 1530, using ± 6 V supplies as in Sec. 15-5).

The circuit of Fig. 16-44 is called an *astable multivibrator* because it has two quasistable states. The output remains in one of these states for a time T_1 and then abruptly changes to the second state for a time T_2, and the cycle of period $T = T_1 + T_2$ repeats.

Pulse Generator   A *monostable multivibrator* has one stable state and one quasistable state. The circuit remains in its stable state until a triggering signal causes a transition to the quasistable state. Then, after a time T, the circuit returns to its stable state. Hence a single pulse has been generated, and the circuit is referred to as a *one-shot*.

The square-wave generator of Fig. 16-44 is modified in Fig. 16-45 to operate as a monostable multivibrator by adding a diode ($D1$) clamp across C and by introducing a narrow negative triggering pulse through $D2$ to the non-inverting terminal. To see how the circuit operates, assume that it is in its stable state with the output at $v_o = +V_Z$ and the capacitor clamped at

$$v_c = V_1 \approx 0.7 \text{ V}$$

(the ON voltage of $D1$ with $\beta V_Z > V_1$). If the trigger amplitude is greater than $\beta V_Z - V_1$, then it will cause the comparator to switch to an output $v_o = -V_Z$. The capacitor will now charge through R' toward $-V_Z$ because $D1$ becomes reverse biased. When v_c becomes more negative than $-\beta V_Z$, the comparator output swings back to $+V_Z$. The capacitor now starts charging toward $+V_Z$ through R' until v_c reaches V_1 and C becomes clamped again at $v_c = V_1$. In Prob. 16-48 we find that the pulse width T is given by

$$T = R'C \ln \frac{1 + (V_1/V_Z)}{1 - \beta} \qquad (16\text{-}78)$$

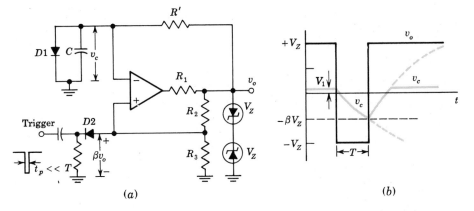

Fig. 16-45 (a) Monostable multivibrator. (b) Output and capacitor voltage waveforms.

If $V_Z \gg V_1$ and $R_2 = R_3$, so that $\beta = \frac{1}{2}$, then $T = 0.69 R'C$. For short pulse widths the switching times of the comparator become important and limit the operation of the circuit. If $R_1 = 0$ and the Zener diodes are omitted, Eq. (16-78) remains valid with $V_Z = V_{CC} - V_{CE,\text{sat}}$ (Sec. 15-5).

Triangle-wave Generator We observe from Fig. 16-44b that v_c has a triangular waveshape but that the sides of the triangles are exponentials rather than straight lines. To linearize the triangles, it is required that C be charged with a constant current rather than the exponential current supplied through R' in Fig. 16-44a. In Fig. 16-46 an OP AMP integrator is used to supply constant current to C so that the output is linear. Because of the inversion through the integrator, this voltage is fed back to the noninverting terminal of the comparator in this circuit rather than to the inverting terminal as in Fig. 16-44.

When the comparator has reached either the positive or negative saturation state, the matched Zener diodes will clamp the voltage V_A at either $+V_Z$ or $-V_Z$. Let us assume that $V_A = +V_Z$ at $t = t_o$. The current flowing into the integrator is

$$I^+ = \frac{V_Z}{R_3 + R_4} \tag{16-79}$$

and the integrator output becomes a negative-going ramp, or

$$v_o(t) = v_o(t_o) - \frac{1}{C} \int_{t_o}^{t} I^+ \, dt = v_o(t_o) - \frac{I^+}{C}(t - t_o) \tag{16-80}$$

The voltage at pin 3 of the threshold detector is, using superposition,

$$v_3(t) = \frac{R_5 V_Z}{R_1 + R_2 + R_5} + \frac{(R_1 + R_2)v_o(t)}{R_1 + R_2 + R_5} \tag{16-81}$$

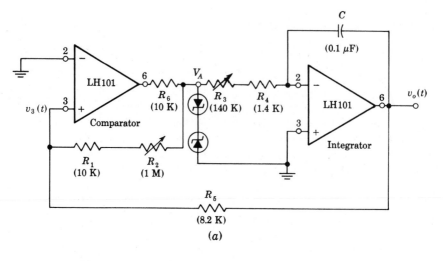

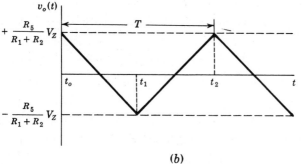

Fig. 16-46   (a) Practical triangle-wave generator.   (b) Output waveform.
(Courtesy of National Semiconductor Corp.)

When $v_3(t)$ goes through zero and becomes negative, the comparator output changes to the negative-output state and $V_A = -V_Z$. At this time, $t = t_1$, $v_3(t_1) = 0$, or from Eq. (16-81) we find

$$v_o(t_1) = -\frac{R_5}{R_1 + R_2} V_Z \qquad\qquad (16\text{-}82)$$

The current supplied to the integrator for $t_2 > t > t_1$ is

$$I^- = -\frac{V_Z}{R_3 + R_4} = -I^+$$

and the integrator output $v_o(t)$ becomes a positive-going ramp with the same slope as the negative-going ramp. At a time t_2, when

$$v_o(t_2) = +\frac{R_5}{R_1 + R_2} V_Z \qquad\qquad (16\text{-}83)$$

the comparator switches again to its positive output and the cycle repeats.

The frequency of the triangle wave is determined from Eq. (16-80) and Fig. (16-46) to be given by

$$f = \frac{R_1 + R_2}{4(R_3 + R_4)R_5C} \tag{16-84}$$

The amplitude can be controlled by the ratio $R_5 V_Z/(R_1 + R_2)$. The positive and negative peaks are equal if the Zener diodes are matched. It is possible to offset the triangle with respect to ground if we connect a dc voltage to the inverting terminal of the threshold detector or comparator.

The practical circuit shown in Fig. 16-46 makes use of the LH 101 OP AMP, which is internally compensated for unity-gain feed-back. This monolithic integrated OP AMP has maximum input offset voltage of 5 mV and maximum input bias current of 500 nA. For symmetry of operation the current into the integrator should be large with respect to I_{bias}, and the peak of the output triangle voltage should be large with respect to the input offset voltage.

The design of monostable and astable generators using discrete components is considered in detail in Ref. 3, Chap. 11. A one shot constructed from logic gates is indicated in Prob. 17-57.

16-16   REGENERATIVE COMPARATOR (SCHMITT TRIGGER)[14]

As indicated in Fig. 16-33, the transfer characteristic of the MC1530 DIFF AMP makes the change in output from -5 V to $+5$ V for a swing of 2 mV in input voltage. Hence the average slope of this curve or the large-signal voltage gain A_V is $A_V = 10/2 \times 10^{-3} = 5,000$. (The incremental gain at the center of the characteristic is calculated in Sec. 15-5 to be 8,670.) By employing positive (regenerative) voltage-series feedback, as is done in Figs. 16-44 and 16-45 for the astable and monostable multivibrators, the gain may be increased greatly. Consequently the total output excursion takes place in a time interval during which the input is changing by much less than 2 mV. Theoretically, if the loop gain $-\beta A_V$ is adjusted to be unity, then the gain with feedback A_{Vf} becomes infinite [Eq. (13-4)]. Such an idealized situation results in an abrupt (zero rise time) transition between the extreme values of output voltage. If a loop gain in excess of unity is chosen, the output waveform continues to be virtually discontinuous at the comparison voltage. However, the circuit now exhibits a phenomenon called *hysteresis*, or *backlash*, which is explained below.

The regenerative comparator of Fig. 16-47a is commonly referred to as a *Schmitt trigger* (after the inventor of a vacuum-tube version of this circuit). The input voltage is applied to the inverting terminal 2 and the feedback voltage to the noninverting terminal 1. The feedback factor is $\beta = R_2/(R_1 + R_2)$. For $R_2 = 100 \ \Omega$, $R_1 = 10$ K, and $A_V = -5,000$, the loop gain is

$$-\beta A_V = 0.1 \times \frac{5,000}{10.1} = 49.5 \gg 1$$

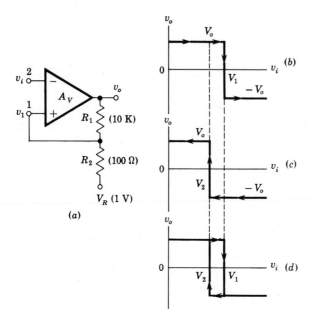

Fig. 16-47 (a) A Schmitt trigger. The transfer characteristics for (b) increasing v_i and (c) decreasing v_i. (d) The compositive input-output curve.

Assume that $v_i < v_1$, so that $v_o = +V_o$ (+5 V). Then, using superposition, we find from Fig. 16-47a that

$$v_1 = \frac{R_1 V_R}{R_1 + R_2} + \frac{R_2 V_o}{R_1 + R_2} \equiv V_1 \tag{16-85}$$

If v_i is now increased, then v_o remains constant at V_o, and $v_1 = V_1 = \text{constant}$ until $v_i = V_1$. At this *threshold, critical,* or *triggering voltage,* the output regeneratively switches to $v_o = -V_o$ and remains at this value as long as $v_i > V_1$. This transfer characteristic is indicated in Fig. 16-47b.

The voltage at the noninverting terminal for $v_i > V_1$ is

$$v_1 = \frac{R_1 V_R}{R_1 + R_2} - \frac{R_2 V_o}{R_1 + R_2} \equiv V_2 \tag{16-86}$$

For the parameter values given in Fig. 16-47 and with $V_o = 5$ V,

$$V_1 = 0.99 + 0.05 = 1.04 \text{ V}$$

$$V_2 = 0.99 - 0.05 = 0.94 \text{ V}$$

Note that $V_2 < V_1$, and the difference between these two values is called the *hysteresis* V_H.

$$V_H = V_1 - V_2 = \frac{2R_2 V_o}{R_1 + R_2} = 0.10 \text{ V} \tag{16-87}$$

If we now decrease v_i, then the output remains at $-V_o$ until v_i equals the voltage at terminal 1 or until $v_i = V_2$. At this voltage a regenerative transition takes place and, as indicated in Fig. 16-47c, the output returns to

$+V_o$ almost instantaneously. The complete transfer function is indicated in Fig. 16-47d, where the portions without arrows may be traversed in either direction, but the other segments can only be obtained if v_i varies as indicated by the arrows. Note that because of the hysteresis, the circuit triggers at a higher voltage for increasing than for decreasing signals.

We note above that transfer gain increases from 5,000 toward infinity as the loop gain increases from zero to unity, and that there is no hysteresis as long as $-\beta A_V \leq 1$. However, adjusting the gain precisely to unity is not feasible. The DIFF AMP parameters and, hence the gain A_V, are variable over the signal excursion. Hence an adjustment which ensures that the maximum loop gain is unity would result in voltage ranges where this amplification is less than unity, with a consequent loss in speed of response of the circuit. Furthermore, the circuit may not be stable enough to maintain a loop gain of precisely unity for a long period of time without frequent readjustment. In practice, therefore, a loop gain in excess of unity is chosen and a small amount of hysteresis is tolerated. In most cases a small value of V_H is not a matter of concern. In other applications a large backlash range will not allow the circuit to function properly. Thus if the peak-to-peak signal were smaller than V_H, then the Schmitt circuit, having responded at a threshold voltage by a transition in one direction, would never reset itself. In other words, once the output has jumped to, say, V_o, it would remain at this level and never return to $-V_o$.

The most important use made of the Schmitt trigger is to convert a very slowly varying input voltage into an output having an abrupt (almost discontinuous) waveform, occurring at a precise value of input voltage. This regenerative comparator may be used in all the applications listed in Sec. 16-11. For example, the use of the Schmitt trigger as a squaring circuit is illustrated in Fig. 16-48. The input signal is arbitrary except that it has a large enough excursion to carry the input beyond the limits of the hysteresis range V_H. The output is a square wave as shown, the amplitude of which is independent of the peak-to-peak value of the input waveform. The output has much faster leading and trailing edges than does the input.

The design of a Schmitt trigger from discrete components is explained in detail in Ref. 3, Secs. 10-11 and 10-13.

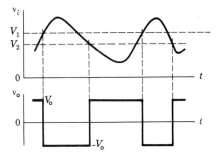

Fig. 16-48 Response of the Schmitt trigger to an arbitrary input signal.

16-17   EMITTER–COUPLED LOGIC (ECL)[15]

The transfer characteristic of the difference amplifier is discussed in Sec. 15-4. We find that the emitter current remains essentially constant and that this current is switched from one transistor to the other as the signal at the input transistor varies from about 0.1 V below to 0.1 V above the reference voltage V_{B2} at the base of the second transistor (Fig. 15-9). Except for a very narrow range of input voltage the output voltage takes on only one of two possible values and, hence, behaves as a binary circuit. Hence the DIFF AMF, which is considered in detail in this chapter on analog systems, is also important as a digital device. A logic family based upon this basic building block is called *emitter-coupled logic* (ECL) or *current-mode logic* (CML). Since in the DIFF AMP clipper or comparator neither transistor is allowed to go into saturation, the ECL is the fastest of all logic families (Table 6-5); a propagation delay time as low as 1 ns per gate is possible. The high speed (and high fan out) attainable with ECL is offset by the increased power dissipation per gate relative to that of the saturating logic families.

A 2-input OR (and also NOR) gate is drawn in Fig. 16-49a. This circuit is obtained from Fig. 15-6 by using two transistors in parallel at the input. Consider positive logic. If both A and B are low, then neither $Q1$ nor $Q2$ conducts whereas $Q3$ is in its active region. Under these circumstances Y is low and Y' is high. If either A or B is high, then the emitter current switches to the input transistor the base of which is high, and the collector current of $Q3$ drops approximately to zero. Hence Y goes high and Y' drops in voltage. Note that OR logic is performed at the output Y and NOR logic at Y', so that $Y' = \bar{Y}$. The logic symbol for such an OR gate with both true and false outputs is indicated in Fig. 16-49b. The availability of complementary out-

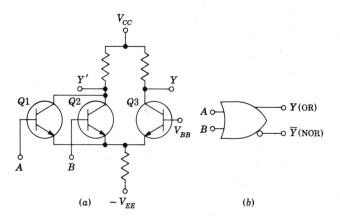

Fig. 16-49   (a) DIFF AMP converted into a 2-input emitter-coupled logic circuit.   (b) The symbol for a 2-input OR/NOR gate.

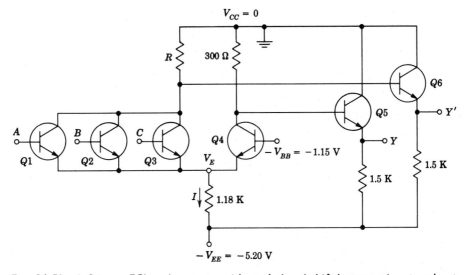

Fig. 16-50 A 3-input ECL OR/NOR gate, with no dc-level shift between input and output voltages.

puts is clearly an advantage to the logic design engineer since it avoids the necessity of adding gates simply as inverters.

One of the difficulties with ECL topology of Fig. 16-49a is that the $V(0)$ and $V(1)$ levels at the outputs differ from those at the inputs. Hence emitter followers $Q5$ and $Q6$ are used at the outputs to provide the proper dc-level shifts. The basic Motorola ECL 3-input gate is shown in Fig. 16-50. The reference voltage $-V_{BB}$ is obtained from a temperature-compensated network (not indicated). The quantitative operation of the gate is given in the following illustrative problem.

EXAMPLE (a) What are the logic levels at output Y of the ECL gate of Fig. 16-50? Assume a drop of 0.7 V between base and emitter of a conducting transistor. (b) Calculate the noise margins. (c) Verify that a conducting transistor is in its active region (*not* in saturation). (d) Calculate R so that the logic levels at Y' are the complements of those at Y. (e) Find the average power dissipated by the gate.

Solution (a) If all inputs are low, then assume transistors $Q1$, $Q2$, and $Q3$ are cut off and $Q4$ is conducting. The voltage at the common emitter is

$$V_E = -1.15 - 0.7 = -1.85 \text{ V}$$

The current I in the 1.18-K resistance is

$$I = \frac{-1.85 + 5.20}{1.18} = 2.84 \text{ mA}$$

Neglecting the base current compared with the emitter current, I is the current in the 300-Ω resistance and the output voltage at Y is

$$v_Y = -0.3I - V_{BE5} = -(0.3)(2.84) - 0.7 = -1.55 \text{ V} = V(0)$$

If all inputs are at $V(0) = -1.55$ V and $V_E = -1.85$ V, then the base-to-emitter voltage of an input transistor is

$$V_{BE} = -1.55 + 1.85 = 0.30 \text{ V}$$

Since the cutin voltage is $V_{BE,\text{cutin}} = 0.5$ V (Table 5-1), then the input transistors are nonconducting, as was assumed above.

If at least one input is high, then assume that the current in the 1.18-K resistance is switched to R, and $Q4$ is cut off. The drop in the 300-Ω resistance is then zero. Since the base and collector of $Q5$ are effectively tied together, $Q5$ now behaves as a diode. Assuming 0.7 V across $Q5$ as a first approximation, the diode current is $(5.20 - 0.7)/1.5 = 3.0$ mA. From Fig. 7-19a the diode voltage for 3.0 mA is 0.75 V. Hence

$$v_Y = -0.75 \text{ V} = V(1)$$

If one input is at -0.75 V, then $V_E = -0.75 - 0.7 = -1.45$ V, and

$$V_{BE4} = -1.15 + 1.45 = 0.30 \text{ V}$$

which verifies the assumption that $Q4$ is cutoff; since $V_{BE,\text{cutin}} = 0.5$ V.

Note that the total output swing between the two logic levels is only $1.55 - 0.75 = 0.80$ V (800 mV). This voltage is much smaller than the value (in excess of 4 V) obtained with a DTL or TTL gate.

(b) If all inputs are at $V(0)$, then the calculation in part (a) shows that an input transistor is within $0.50 - 0.30 = 0.20$ V of cutin. Hence a positive noise spike of 0.20 V will cause the gate to malfunction.

If one input is at $V(1)$, then we find in part (a) that $V_{BE4} = 0.30$ V. Hence a negative noise spike at the input of 0.20 V drops V_E by the same amount and brings V_{BE4} to 0.5 V, or to the edge of conduction. Note that the noise margins are quite small (± 200 mV).

(c) From part (a) we have that, when $Q4$ is conducting, its collector voltage with respect to ground is the drop in the 300-Ω resistance, or $V_{C4} = -(0.3)(2.84) = -0.85$ V. Hence the collector junction voltage is

$$V_{CB4} = V_{C4} - V_{B4} = -0.85 + 1.15 = +0.30 \text{ V}$$

For an n-p-n transistor this represents a reverse bias, and $Q4$ must be in its active region.

If any input, say A, is at $V(1) = -0.75$ V $= V_{B1}$, then $Q1$ is conducting and the output $Y' = \bar{Y} = V(0) = -1.55$ V. The collector of $Q1$ is more positive than $V(0)$ by V_{BE6}, or

$$V_{C1} = -1.55 + 0.7 = -0.85 \text{ V}$$

and

$$V_{CB1} = V_{C1} - V_{B1} = -0.85 + 0.75 = -0.10 \text{ V}$$

For an *n-p-n* transistor this represents a forward bias, but one whose magnitude is less than the cutin voltage of 0.5 V. Therefore $Q1$ is *not* in saturation; it is in its active region.

(*d*) If input A is at $V(1)$, then $Q1$ conducts and $Q4$ is OFF. Then

$$V_E = V(1) - V_{BE1} = -0.75 - 0.7 = -1.45 \text{ V}$$

$$I = \frac{V_E + V_{EE}}{1.18} = \frac{-1.45 + 5.20}{1.18} = 3.17 \text{ mA}$$

In part (*c*) we find that, if $Y' = \bar{Y}$, then $V_{C1} = -0.85$ V. This value represents the drop across R if we neglect the base current of $Q1$. Hence

$$R = \frac{0.85}{3.17} = 0.27 \text{ K} = 270 \text{ }\Omega$$

This value of R ensures that, if an input is $V(1)$, then $Y' = V(0)$. If all inputs are at $V(0) = -1.55$ V, then the current through R is zero and the output is -0.75 V $= V(1)$, independent of R.

Note that, if I had remained constant as the input changed state (true current-mode switching), then R would be identical to the collector resistance (300 Ω) of $Q4$. The above calculation shows that R is slightly smaller than this value.

(*e*) If the input is low, $I = 2.84$ mA (part *a*), whereas if the input is high, $I = 3.17$ mA (part *d*). The average I is $\frac{1}{2}(2.84 + 3.17) = 3.00$ mA. Since $V(0) = -0.75$ V and $V(1) = -1.55$ V, the currents in the two emitter followers are

$$\frac{5.20 - 0.75}{1.50} = 2.96 \text{ mA} \quad \text{and} \quad \frac{5.20 - 1.55}{1.50} = 2.40 \text{ mA}$$

The total power supply current drain is $3.00 + 2.96 + 2.40 = 8.36$ mA and the power dissipation is $(5.20)(8.36) = 43.5$ mW.

Note that the current drain from the power supply varies very little as the input switches from one state to the other. Hence power line spikes (of the type discussed in Sec. 6-12 for TTL gates) are virtually nonexistent.

The input resistance can be considered infinite if all inputs are low so that all input transistors are cut off. If an input is high, then $Q4$ is OFF, and the input resistance corresponds to a transistor with an emitter resistor $R_e = 1.18$ K, and from Eq. (8-55) a reasonable estimate is $R_i \approx 100$ K. The

Fig. 16-51 An implied-OR connection at the output of two ECL gates.

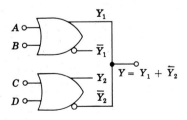

output resistance is that of an emitter-follower (or a diode) and a reasonable value is $R_o \approx 15 \ \Omega$.

If the outputs of two or more ECL gates are tied together as in Fig. 16-51, then wired-OR logic (Sec. 6-10) is obtained (Prob. 16-53). Open-emitter gates are available for use in this application.

Summary The principal characteristics of the ECL gate are summarized below:

Advantages

1. Since the transistors do not saturate, then the highest speed of any logic family is available.
2. Since the input resistance is very high and the output resistance is very low, a large fan out is possible.
3. Complementary outputs are available.
4. Current switching spikes are not present in the power supply leads.
5. Outputs can be tied together to give the implied-OR function.
6. There is little degradation of parameters with variations in temperature.
7. The number of functions available is high.
8. Easy data transmission over long distances by means of balanced twisted-pair 50-Ω lines is possible.[15]

Disadvantages

1. A small voltage difference (800 mV) exists between the two logic levels and the noise margin is only ± 200 mV.
2. The power dissipation is high relative to the other logic families.
3. Level shifters are required for interfacing with other families.
4. The gate is slowed down by heavy capacitive loading.

REFERENCES

1. Korn, G. A., and T. M. Korn: "Electronic Analog and Hybrid Computers," McGraw-Hill Book Company, New York, 1964.

2. Giacoletto, L. J.: "Differential Amplifiers," Wiley-Interscience, New York, 1970.

3. Millman, J., and H. Taub: "Pulse, Digital, and Switching Waveforms," pp. 536–548, McGraw-Hill Book Company, New York, 1965.

4. Huelsman, P. L.: "Active Filters," McGraw-Hill Book Company, New York, 1970.

5. Valley, Jr., C. E., and H. Wallman: "Vacuum Tube Amplifiers," appendix A, McGraw-Hill Book Company, New York, 1948.

6. Kuo, F. F.: "Network Analysis and Synthesis," John Wiley & Sons, Inc., New York, 1962.

7. Stremler, F. G.: "Design of Active Bandpass Filters," *Electronics*, vol. 44, no. 12, pp. 86–89, June 7, 1971.

8. Linvill, J. G., and J. F. Gibbons: "Transistors and Active Circuits," chap. 14, McGraw-Hill Book Company, 1961.

9. Graeme, J. G., and T. E. Tobey: "Operational Amplifiers. Design and Applications," McGraw-Hill Book Company, New York, 1971.

10. Ref. 3, pp. 649–658.

11. Dow, Jr., P. C.: An Analysis of Certain Errors in Electronic Differential Analyzers: Capacitor Dielectric Absorption, *IRE Trans. Electronic Computers*, March, 1958, pp. 17–22.

12. Dobkin, R. C.: "Linear Brief 8," National Semiconductor Corporation, August, 1969.

13. Gilbert, B.: A Precise Four-quadrant Multiplier with Subnanosecond Response, *IEEE J. Solid State Circuits*, December, 1968, p. 210.

14. Sifferlen T. P., and V. Vartanian: "Digital Electronics with Engineering Applications," Prentice-Hall Inc., Englewood Cliffs, N.J. 1970.

15. Garret, L. S.: "Integrated Circuit Digital Logic Families," *IEEE Spectrum*, vol. 7, no. 12, pp. 30–42, December 1970.

REVIEW QUESTIONS

16-1 Indicate an OP AMP connected as (*a*) an *inverter*, (*b*) a *scale changer*, (*c*) a *phase shifter*, and (*d*) an *adder*.

16-2 Draw the circuit of a *voltage-to-current converter* if the load is (*a*) floating and (*b*) grounded.

16-3 Draw the circuit of a *current-to-voltage converter*. Explain its operation.

16-4 Draw the circuit of a dc *voltage follower* and explain its operation.

16-5 Draw the circuit of a dc differential amplifier having (*a*) low input resistance and (*b*) high input resistance.

16-6 Draw the circuit of an ac *voltage follower* having very high input resistance. Explain its operation.

16-7 Draw the circuit of an OP AMP integrator and indicate how to apply the initial condition. Explain its operation.

16-8 Sketch the idealized characteristics for the following filter types: (*a*) low-pass, (*b*) high-pass, (*c*) bandpass, and (*d*) band-rejection.

16-9 Draw the prototype for a low-pass active-filter section of (*a*) first order, (*b*) second order, and (*c*) third order.

16-10 (*a*) Obtain the frequency response of an RLC circuit in terms of ω_o and Q. (*b*) Verify that the bandwidth is given by f_o/Q. (*c*) What is meant by an active resonant bandpass filter?

16-11 (*a*) A signal V_s is applied to the inverting terminal (2) of an OP AMP through Z_1 and to the noninverting terminal (1) through Z_2. From (1) to ground is an impedance Z_3, and between (2) and the output is Z_4. Derive the expression for the gain. (*b*) How should Z_1, Z_2, Z_3, and Z_4 be chosen so that the circuit behaves as a *delay equalizer?*

16-12 (*a*) Sketch the basic building block for an IC tuned amplifier. (*b*) Explain how automatic gain control (AGC) is obtained. (*c*) Why does AGC not cause detuning?

16-13 Define the *y*-parameters (*a*) by equations and (*b*) in words. (*c*) For a cascode circuit which *y*-parameter is negligible?

16-14 Draw the circuit of an *amplitude modulator* and explain its operation.

16-15 (*a*) Draw the circuit of an IC video amplifier with AGC. (*b*) Sketch the small-signal model.

16-16 (*a*) What does an IC comparator consist of? (*b*) Sketch the transfer characteristic and indicate typical voltage values.

16-17 Sketch the circuit for converting a sinusoid (*a*) into a square wave and (*b*) into a series of positive pulses, one per cycle.

16-18 Explain how to measure the phase difference between two sinusoids.

16-19 Sketch a *sample-and-hold* circuit and explain its operation.

16-20 Sketch the circuit of a precision (*a*) diode and (*b*) clamp and explain their operation.

16-21 (*a*) Sketch the circuit of a fast half-wave rectifier and explain its operation. (*b*) How is this circuit converted into an *average detector?*

16-22 Sketch the circuit of a *peak detector* and explain its operation.

16-23 (*a*) Sketch the circuit of a *logarithmic amplifier* using one OP AMP and explain its operation. (*b*) More complicated logarithmic amplifiers are given in Sec. 16-14. What purpose is served by these circuits?

16-24 In schematic form indicate how to multiply two analog voltages with log-antilog amplifiers.

16-25 Explain how to multiply two analog voltages using a DIFF AMP.

16-26 (*a*) Draw the circuit of a square-wave generator using an OP AMP. (*b*) Explain its operation by drawing the capacitor voltage waveform. (*c*) Derive the expression for the period of a symmetrical waveform.

16-27 (*a*) Draw the circuit of a pulse generator (a monostable multivibrator) using an OP AMP. (*b*) Explain its operation by referring to the capacitor waveform.

16-28 (*a*) Draw the circuit of a triangle generator using a comparator and an integrator. (*b*) Explain its operation by referring to the output waveform. (*c*) What is the peak amplitude?

16-29 (*a*) Sketch a regenerative comparator system and explain its operation. (*b*) What parameters determine the loop gain? (*c*) What parameters determine the hysteresis? (*d*) Sketch the transfer characteristic and indicate the hysteresis.

16-30 (*a*) Sketch a 2-input OR (and also NOR) ECL gate. (*b*) What parameters determine the noise margin? (*c*) Why are the two collector resistors unequal? (*d*) Explain why power line spikes are virtually nonexistent.

16-31 List and discuss at least four advantages and four disadvantages of the ECL gate.

17 / INTEGRATED CIRCUITS AS DIGITAL SYSTEM BUILDING BLOCKS

A digital system is constructed from very few types of basic network configurations, these elementary types being used over and over again in various topological combinations. As emphasized in Sec. 6-9, it is possible to perform all logic operations with a single type of circuit (for example, a NAND gate). A digital system must store binary numbers in addition to performing logic. To take care of this requirement, a memory cell, called a FLIP-FLOP, is introduced in this chapter.

Theoretically, any digital system can be constructed entirely from NAND gates and FLIP-FLOPS. Some functions (such as binary addition) are present in many systems, and hence the combination of gates and/or FLIP-FLOPS required to perform this function is available on a single chip. These integrated circuits form the practical (commercially available) basic building blocks for a digital system. The number of such different ICs is not large, and these blocks are discussed in this chapter. These chips perform the following functions: binary addition, decoding (demultiplexing), data selection (multiplexing), counting, storage of binary information (memories and registers), digital-to-analog (D/A) and analog-to-digital (A/D) conversion, and a few other related operations. These logic building blocks are treated as single components. Large-scale integration of MOSFETs for the storage and conversion of considerable information is also considered.

The topics in this chapter fall into four major divisions:

    I.   Combinational digital systems
    II. Sequential digital systems
  III. MOS/LSI digital systems
  IV. D/A and A/D systems

I. COMBINATIONAL DIGITAL SYSTEMS

17-1   STANDARD GATE ASSEMBLIES[1]

Since the fundamental gates are used in large numbers even in a relatively simple digital system, they are not packaged individually; rather, several gates are constructed within a single chip. The following list of standard digital IC gates is typical, but far from exhaustive:

Quad two-input NAND   Quad two-input NOR
Triple three-input NAND   Quad two-input AND
Dual four-input NAND   Hex inverter buffer
Single eight-input NAND   Dual two-wide, two-input AOI

These combinations are available in most logic families (TTL, DTL, etc.) listed in Sec. 6-15.   The limitation on the number of gates per chip is usually set by the number of pins available.   The most common packages are the *flat pack* and the *dual-in-line* (plastic package, type N, or ceramic package, type J), each of which has fourteen leads, seven brought out to each side of the IC.   The dimensions of the assembly, which is much larger than the chip size, are approximately 0.8 by 0.3 by 0.2 in.   A schematic of the triple three-input NAND is shown in Fig. 17-1a.   Note that there are $3 \times 3 = 9$ input leads, 3 output leads, a power-supply lead, and a ground lead; a total of 14 leads are used.

In Fig. 17-1b is indicated the dual two-wide, two-input AOI (Sec. 6-13). This combination needs 4 input leads and 1 output lead per AOI, or 10 for the dual array.   If 1 power-supply lead and 1 ground lead are added, we see that 12 of the 14 available pins are used.

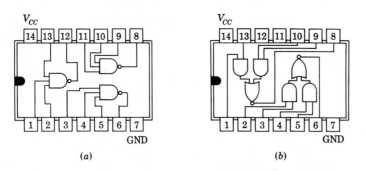

(a)   (b)

Fig. 17-1 The lead connections (top view) of (a) the TI 7410 triple three-input NAND and (b) the TI 7451 dual two-wide, two-input AOI gate.

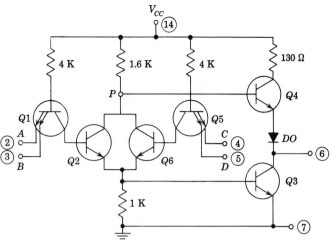

Fig. 17-2   The circuit configuration for a TTL AND-OR-INVERT
gate.

The circuit diagram for this AOI gate is given in Fig. 17-2, implemented
in TTL logic.   The operation of this network should be clear from the discussion in Chap. 6.   Thus $Q1$ and the input to $Q2$ (corresponding to the
similarly numbered transistors of Fig. 6-27) constitute an AND gate.   The
identical arrangement of $Q5$ and $Q6$ constitutes a second AND gate.   Since
the collectors of $Q2$ and $Q6$ are tied together at P, the output at this node
corresponds to either the inputs ② AND ③ OR ④ AND ⑤.   Also, because of the
inversion through a transistor, the NOT operation is performed at P.   The
result is AND-OR-INVERT (AOI) logic $(\overline{AB + CD})$.   Finally, note that $Q3$, DO,
and $Q4$ are the totem-pole output stage of Fig. 6-28.

An alternative way of analyzing the circuit of Fig. 17-2 is to consider $Q1$
and $Q2$ (with the output at P) to constitute a NAND circuit.   Similarly, $Q5$
and $Q6$ form a second NAND gate.   The outputs of these two NAND configurations are shorted together by the lead connecting the collectors of $Q2$ and $Q6$
to form a wired-AND (Sec. 6-13).   Hence the output at P is, using De Morgan's
law Eq. (6-27)

$$(\overline{AB})(\overline{CD}) = \overline{AB + CD}$$

which confirms that AOI logic is performed.

Some of the more complicated functions to be described in this chapter
require in excess of 14 pins, and these ICs are packaged with either 16 or
24 leads.   The latter has the dimensions 1.3 by 0.6 by 0.2 in.

The standard combinations considered in this section are examples of
small-scale integration (SSI).   Less than about 12 gates on a chip is considered SSI.   The FLIP-FLOPS discussed in Sec. 17-9 are also SSI packages.

Most other functions (using BJTs) discussed in this chapter are examples of *medium-scale integration* (MSI), defined to have more than 12, but less than 100, gates per chip. The BJT memories of Sec. 17-18 and the MOSFET arrays of Sec. 17-17 may contain in excess of 100 gates and are defined as *large-scale integration* (LSI) (Sec. 7-12).

17-2   BINARY ADDERS[2]

A digital computer must obviously contain circuits which will perform arithmetic operations, i.e., addition, subtraction, multiplication, and division. The basic operations are addition and subtraction, since multiplication is essentially repeated addition, and division is essentially repeated subtraction. It is entirely possible to build a computer in which an *adder* is the only arithmetic unit present. Multiplication, for example, may then be performed by *programming;* i.e., the computer may be given instructions telling it how to use the adder repeatedly to find the product of two numbers.

Suppose we wish to sum two numbers in decimal arithmetic and obtain, say, the hundreds digit. We must add together not only the hundreds digit of each number but also a carry from the tens digit (if one exists). Similarly, in binary arithmetic we must add not only the digit of like significance of the two numbers to be summed, but also the carry bit (should one be present) of the next lower significant digit. This operation may be carried out in two steps: first, add the two bits corresponding to the 2^n digit, and then add the resultant to the carry from the 2^{n-1} digit. A two-input adder is called a *half adder*, because to complete an addition requires two such half adders.

We shall first show how a *half adder* is constructed from the basic logic gates and then indicate how the *full, or complete, adder* is assembled. A half adder has two inputs—A and B—representing the bits to be added, and two outputs—D (for the digit of the same significance as A and B represent) and C (for the carry bit).

Half Adder   The symbol for a half adder is given in Fig. 17-3*a*, and the truth table in Fig. 17-3*b*. Note that the D column gives the sum of A and B as long as the sum can be represented by a single digit. When, however, the sum is larger than can be represented by a single digit, then D gives the digit in the result which is of the same significance as the individual digits being added. Thus, in the first three rows of the truth table, D gives the sum of A and B directly. Since the decimal equation "1 plus 1 equals 2" is written in binary form as "01 plus 01 equals 10," then in the last row $D = 0$. Because a 1 must now be carried to the place of next higher significance, $C = 1$.

From Fig. 17-3*b* we see that D obeys the EXCLUSIVE-OR function and C follows the logic of an AND gate. These functions are indicated in Fig. 17-3*c*, and may be implemented in many different ways with the circuitry discussed in Chap. 6. For example, the EXCLUSIVE-OR gate can be constructed with

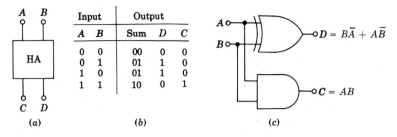

Fig. 17-3   (a) The symbol for a half adder; (b) the truth table for the digit D and the carry C; (c) the implementation for D with an EXCLU-SIVE-OR gate and for C with an AND gate.

any of the four topologies of Sec. 6-8 and in any of the logic families in Table 6-5.   The configuration in Fig. 6-15b $(Y = A\bar{B} + B\bar{A})$ is implemented in TTL logic with the AOI circuit of Fig. 17-2.   The inverter for B (or A) is a single-input NAND gate.   Since Y has an AND-OR (rather than an AND-OR-INVERT) topology, a transistor inverter is placed between node P and the base of $Q4$ of Fig. 17-2.

Parallel Operation   Two multidigit numbers may be added serially (one column at a time) or in parallel (all columns simultaneously).   Consider parallel operation first.   For an N-digit binary number there are (in addition to a common ground) N signal leads in the computer for each number.   The nth line for number A (or B) is excited by A_n (or B_n), the bit for the 2^n digit $(n = 0, 1, \ldots, N - 1)$.   A parallel binary adder is indicated in Fig. 17-4. Each digit except the least-significant one (2^0) requires a complete adder consisting of two half adders in cascade.   The sum digit for the 2^0 bit is $S_0 = D_0$

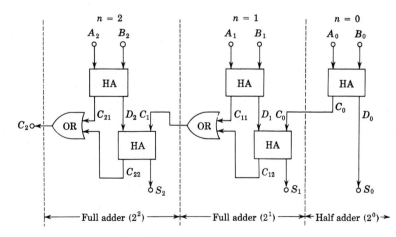

Fig. 17-4   A parallel binary adder consisting of half adders.

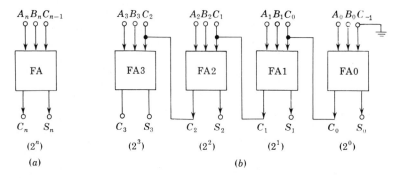

Fig. 17-5 (a) The symbol for a full adder. (b) A 4-bit parallel binary adder constructed from cascaded full adders.

of a half adder because there is no carry to be added to A_0 plus B_0. The sum S_n ($n \neq 0$) of A_n plus B_n is made in two steps. First the digit D_n is obtained from one half adder, and then D_n is summed with the carry C_{n-1} which may have resulted from the next lower place. As an example, consider $n = 2$ in Fig. 17-4. There the carry bit C_1 may be the result of the direct sum of A_1 plus B_1 if each of these is 1. This first carry is called C_{11} in Fig. 17-4. A second possibility is that $A_1 = 1$ and $B_1 = 0$ (or vice versa), so that $D_1 = 1$, but that there is a carry C_0 from the next lower significant bit. The sum of $D_1 = 1$ and $C_0 = 1$ gives rise to the carry bit designated C_{12}. It should be clear that C_{11} and C_{12} cannot both be 1, although they will both be 0 if $A_1 = 0$ and $B_1 = 0$. Since either C_{11} or C_{12} must be transmitted to the next stage, an OR gate must be interposed between stages, as indicated in Fig. 17-4.

Full Adder In integrated circuit implementation, addition is performed using a complete adder, which (for reasons of economy of components) is not constructed from two half adders. The symbol for the nth full adder (FA) is indicated in Fig. 17-5a. The circuit has three inputs: the addend A_n, the augend B_n, and the input carry C_{n-1} (from the next lower bit). The outputs are the sum S_n (sometimes designated Σ_n) and the output carry C_n. A parallel 4-bit adder is indicated in Fig. 17-5b. Since FA0 represents the least significant bit (LSB), it has no input carry; hence $C_{-1} = 0$.

The circuitry within the block FA may be determined from Fig. 17-6, which is the truth table for adding 3 binary bits. From this table we can verify that the Boolean expressions for S_n and C_n are given by

$$S_n = \bar{A}_n\bar{B}_nC_{n-1} + \bar{A}_nB_n\bar{C}_{n-1} + A_n\bar{B}_n\bar{C}_{n-1} + A_nB_nC_{n-1} \tag{17-1}$$

$$C_n = \bar{A}_nB_nC_{n-1} + A_n\bar{B}_nC_{n-1} + A_nB_n\bar{C}_{n-1} + A_nB_nC_{n-1} \tag{17-2}$$

Note that the first term of S_n corresponds to line 1 of the table, the second term to line 2, the third term to line 4, and the last term to line 7. (These

Line	Inputs			Outputs	
	A_n	B_n	C_{n-1}	S_n	C_n
0	0	0	0	0	0
1	0	0	1	1	0
2	0	1	0	1	0
3	0	1	1	0	1
4	1	0	0	1	0
5	1	0	1	0	1
6	1	1	0	0	1
7	1	1	1	1	1

Fig. 17-6 Truth table for a three-input adder.

are the only rows where $S_n = 1$.)   Similarly, the first term of C_n corresponds to the line 3 (where $C_n = 1$), the second term to the line 5, etc.

The AND operation ABC is sometimes called the *product* of A and B and C. Also, the OR operation $+$ is referred to as *summation*.   Hence expressions such as those in Eqs. (17-1) and (17-2) represent a *Boolean sum of products*.   Such an equation is said to be in a *standard*, or *canonical*, *form*, and each term in the equation is called a *minterm*.   A minterm contains the product of all Boolean variables, or their complements.

The expression for C_n can be simplified considerably as follows: Since $Y + Y + Y = Y$, then Eq. (17-2), with $Y = A_n B_n C_{n-1}$, becomes

$$C_n = (\bar{A}_n B_n C_{n-1} + A_n B_n C_{n-1}) + (A_n \bar{B}_n C_{n-1} + A_n B_n C_{n-1})$$
$$+ (A_n B_n \bar{C}_{n-1} + A_n B_n C_{n-1}) \quad (17\text{-}3)$$

Since $\bar{X} + X = 1$ where $X = A_n$ for the first parentheses, $X = B_n$ for the second parentheses, and $X = C_{n-1}$ for the third parentheses, then Eq. (17-3) reduces to

$$C_n = B_n C_{n-1} + C_{n-1} A_n + A_n B_n \quad (17\text{-}4)$$

This expression could have been written down directly from the truth table of Fig. 17-6 by noting that $C_n = 1$ if and only if at least two out of the three inputs is 1.

It is interesting to note that if all 1s are changed to 0s and all 0s to 1s, then lines 0 and 7 are interchanged, as are 1 and 6, 2 and 5, and also 3 and 4. Because this switching of 1s and 0s leaves the truth table unchanged, whatever logic is represented by Fig. 17-6 is equally valid if all inputs and outputs are complemented.   Therefore Eq. (17-3) is true if all variables are negated, or

$$\bar{C}_n = \bar{B}_n \bar{C}_{n-1} + \bar{C}_{n-1} \bar{A}_n + \bar{A}_n \bar{B}_n \quad (17\text{-}5)$$

This same result is obtained (Prob. 17-3) by Boolean manipulation of Eq. (17-4).

By evaluating $D_n \equiv (A_n + B_n + C_{n-1}) \bar{C}_n$ and comparing the result with Eq. (17-1), we find that $S_n \equiv D_n + A_n B_n C_{n-1}$, or

$$S_n = A_n \bar{C}_n + B_n \bar{C}_n + C_{n-1} \bar{C}_n + A_n B_n C_{n-1} \quad (17\text{-}6)$$

Equations (17-4) and (17-6) are implemented in Fig. 17-7 using AOI gates of the type shown in Fig. 17-2.

MSI Adders   There are commercially available 1-bit, 2-bit, and 4-bit full adders, each in one package.   In Fig. 17-8 is indicated the logic topology for 2-bit addition.   The inputs to the first stage are A_0 and A_1; the input marked C_{-1} is grounded.   The output is the sum S_0.   The carry C_0 is connected internally and is not brought to an output pin.   This 2^0 stage (LSB) is identical with that in Fig. 17-7 with $n = 0$.   The abbreviation LSB means least significant bit.

Since the carry from the first stage is $\bar{C}_0$, it should be negated before it is fed to the 2^1 stage.   However, the delay introduced by this inversion is undesirable, because the limitation upon the maximum speed of operation is the propagation delay (Sec. 6-15) of the carry through all the bits in the adder. The NOT-gate delay is eliminated completely in the carry by connecting $\bar{C}_0$ directly to the following stage and by complementing the inputs A_1 and B_1 before feeding these to this stage.   This latter method is used in Fig. 17-8. Note that now the outputs S_1 and C_1 are obtained directly without requiring inverters.   The logic followed by this second stage for the carry is given by Eq. (17-5), and for the sum by the modified form of Eq. (17-6), where each symbol is replaced by its complement.

In a 4-bit adder C_1 is not brought out but is internally connected to the third stage, which is identical with the first stage.   Similarly, the fourth and

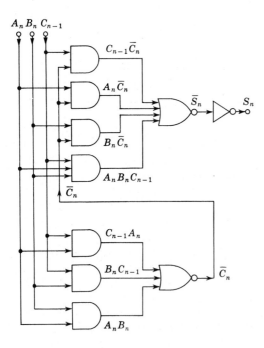

Fig. 17-7   Block-diagram implementation of the nth stage of a full adder.

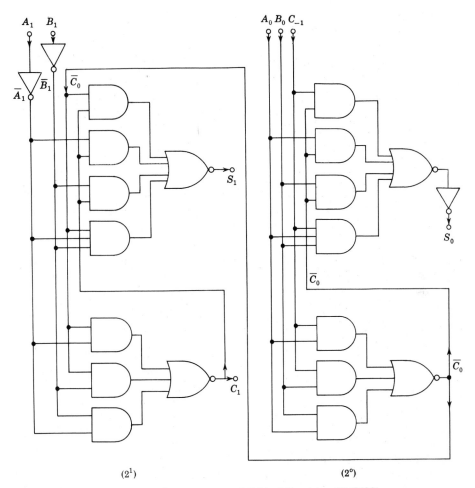

Fig. 17-8   Logic diagram of an integrated 2-bit full adder (TI 7482).

second stages have identical logic topologies.   A 4-bit adder requires a 16-pin package: 8 inputs, 4 sum outputs, a carry output, a carry input, the power-supply input, and ground.   The carry input is needed only if two arithmetic units are cascaded; for example, cascading a 2-bit with a 4-bit adder gives the sum of two 6-bit numbers.   If the 2-bit unit is used for the 2^4 and 2^5 digits, then 4 must be added to all the subscripts in Fig. 17-8.   For example, C_{-1} is now called C_3 and is obtained from the output carry of the 4-bit adder.

The MSI chip (TI 74LS83†) for a 4-bit binary full adder contains somewhat over 200 components (resistors, diodes, or transistors).   For high-speed,

† The specific designations given in this chapter refer to Texas Instrument units.[1] However, equivalent units are available from other vendors, such as Fairchild Semiconductor, Motorola, Inc., National Semiconductor, RCA, Signetics, etc.

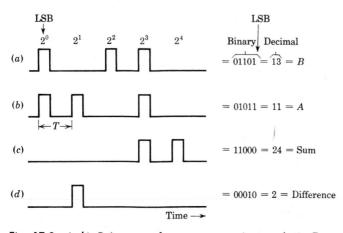

Fig. 17-9   (a,b) Pulse waveforms representing numbers B and A; (c,d) waveforms representing sum and difference. (LSB = least significant bit.)

low-power operation, Schottky transistors and diodes (Sec. 7-13) are used in each AOI block of Fig. 17-2, and each gate output contains a Darlington pair (Fig. 6-28c). The NOT circuit for S_0 is simply a transistor inverter placed between node P and the base of $Q4$ of Fig. 17-2. The NOT circuit for inverting A_1 (or B_1) is a single-input NAND gate. The propagation delay time of the carry is typically 50 ns, and the power dissipation is 75 mW.

Serial Operation   In a serial adder the inputs A and B are synchronous pulse trains on two lines in the computer. Figure 17-9a and b shows typical pulse trains representing, respectively, the decimal numbers 13 and 11. Pulse trains representing the sum (24) and difference (2) are shown in Fig. 17-9c and d, respectively. A serial *adder* is a device which will take as inputs the two waveforms of Fig. 17-9a and b and deliver the output waveform in Fig. 17-9c. Similarly, a *subtractor* (Sec. 17-3) will yield the output shown in Fig. 17-9d.

We have already emphasized that the sum of two multidigit numbers may be formed by adding to the sum of the digits of like significance the carry (if any) which may have resulted from the next lower place. With respect to the pulse trains of Fig. 17-9, the above statement is equivalent to saying that, at any instant of time, we must add (in binary form) to the pulses A and B the carry pulse (if any) which comes from the resultant formed one period T earlier. The logic outlined above is performed by the full-adder circuit of Fig. 17-10. This circuit differs from the configuration in the parallel adder of Fig. 17-5 by the inclusion of a time delay TD which is equal to the time T between pulses. Hence the carry pulse is delayed a time T and added to the digit pulses in A and B, exactly as it should be.

A comparison of Figs. 17-5 and 17-10 indicates that parallel addition is

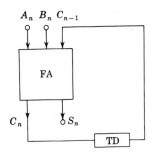

Fig. 17-10 A serial binary full adder.

faster than serial because all digits are added simultaneously in the former, but in sequence in the latter. However, whereas only one full adder is needed for serial arithmetic, we must use a full adder for each bit in parallel addition. Hence parallel addition is much more expensive than serial operation.

The time delay unit TD is a type D FLIP-FLOP, and the serial numbers A_n, B_n, and S_n are stored in *shift registers* (Secs. 17-10 and 17-11).

17-3   ARITHMETIC FUNCTIONS

In this section other arithmetic units besides the adder are discussed. These include the *subtractor*, the *digital comparator*, the *parity checker*, etc.

True/Complement, Zero/One Element It is sometimes required to select either a bit A or its complement $\bar{A}$ by means of control signals. It may also be desired to obtain an output which is either 0 or 1, independently of the value of A. To take care of these four possibilities requires a 2-bit signal, or *code*. In Fig. 17-11a the control code is applied to lines L and M. It is not difficult to verify (Prob. 17-5) that this logic block diagram satisfies the truth table in Fig. 17-11b. An MSI chip with four such networks is available in a single package (TI 74H87).

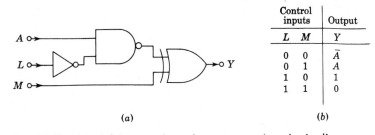

	Control inputs		Output
	L	M	Y
	0	0	$\bar{A}$
	0	1	A
	1	0	1
	1	1	0

(a)   (b)

Fig. 17-11   (a) A 1-bit true/complement, zero/one logic diagram.
(b) The truth table.

Binary Subtraction[2] The process of subtraction (B minus A) is equivalent to addition if the complement $\bar{A}$ of the subtrahend is used. To justify this statement consider the following argument (applied specifically to a 4-bit number). The NOT function changes a 1 to a 0, and vice versa. Therefore†

$$A \text{ plus } \bar{A} = 1111$$

and

$$A \text{ plus } \bar{A} \text{ plus } 1 = 1111 \text{ plus } 0001 = 10000$$

so that

$$A = 10000 \text{ minus } \bar{A} \text{ minus } 1$$

Finally,

$$B \text{ minus } A = (B \text{ plus } \bar{A} \text{ plus } 1) \text{ minus } 10000 \qquad (17\text{-}7)$$

This equation indicates that to subtract a 4-bit number A from a 4-bit number B it is only required to add B, $\bar{A}$, and 1 (a 2^0 bit). The operation B minus A must yield a 4-bit answer. The term "minus 10000" in Eq. (17-7) infers that the addition (B plus $\bar{A}$ plus 1) results in a fifth bit, which must be ignored.

EXAMPLE   Verify Eq. (17-7) for $B = 1100$ and $A = 1001$ (decimal 12 and 9).

Solution

$$B \text{ plus } \bar{A} \text{ plus } 1 = 1100 \text{ plus } 0110 \text{ plus } 0001 = 10011$$

The four (less significant) bits 0011 represent decimal 3 and the fifth bit 1 is a generated carry. Since, in decimal notation, B minus $A = 12 - 9 = 3$, then the correct answer is obtained by evaluating the sum in the parenthesis of Eq. (17-7), provided that the carry is ignored.

In Eq. (17-7) the 1 in 10000 is the output carry $C_3 = 1$ from the 4-bit adder, and may be used to supply the 1 which must be added to $\bar{A}$. This bit is called the *end-around carry* (EAC) because this carry out is fed back to the carry input C_{-1} (Fig. 17-8) of the least significant bit of A. This process of subtraction by means of an adder is indicated schematically in Fig. 17-12a. A more detailed diagram of an adder-subtractor is given in Fig. 17-12b. The *true/complement* unit determines whether addition or subtraction is performed; if $M = 0$, the system subtracts, because the input A is complemented (Fig. 17-11b), whereas if $M = 1$, the system adds. If addition is to be carried out, the EAC loop must be opened, and this logic is accomplished with the AND circuit whose inputs are C_3 and $\bar{M}$.

The 1s complement method of subtraction just described is valid only if B is greater than A, so that a positive difference results and a carry is generated

---

† To avoid confusion with the OR operation, the word *plus* (*minus*) is used in place of + (−) in the following equations.

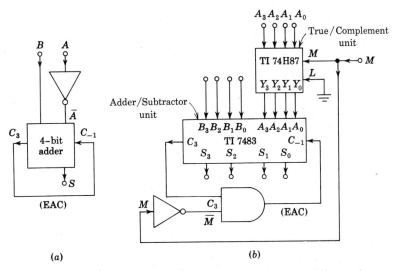

Fig. 17-12 (a) A simplified sketch of a 4-bit adder used as a subtractor. (b) More detailed block diagram of a system which adds if $M = 1$ and subtracts if $M = 0$.

from $(B$ plus $\bar{A}$ plus 1). If B is less than A, then the most significant bit (MSB) of B (which differs from the corresponding bit of A) is 0 and that of A is 1. Since $\bar{A} = 0$, the MSB of $(B$ plus $\bar{A})$ is 0. Hence no carry results from the sum $(B$ plus $\bar{A}$ plus 0001), and the method indicated in Fig. 17-12 must be modified. We now demonstrate that if no carry results in the system of Fig. 17-12, the correct answer for B minus A is negative, and is obtained by complementing the sum digits S_o, S_1, S_2, and S_3. From Eq. (17-7)

$$
\begin{aligned}
B \text{ minus } A &= (B \text{ plus } \bar{A}) \text{ minus } 1111 \\
&= \text{minus } [1111 \text{ minus } (B \text{ plus } \bar{A})] \\
&= \text{minus } (\overline{B \text{ plus } \bar{A}})
\end{aligned}
\tag{17-8}
$$

because 1111 minus a 4-bit binary number is the complement of the number. *In summary:* The circuit of Fig. 17-12b with $M = 0$ can be used for both positive and negative differences. If a carry results, then $(B$ minus $A)$ is positive and is given by the S bits. However, if no carry is obtained, then $(B$ minus $A)$ is negative and is given by the complement of the S bits.

Other systems of addition and subtraction (2s complement, binary-coded-decimal, etc.) are described in Ref. 1.

Digital Comparator   It is sometimes necessary to know whether a binary number A is greater than, equal to, or less than another number B. The system for making this determination is called a *magnitude digital* (or *binary*) *comparator.* Consider single bit numbers first. As mentioned in Sec. 6-7, the EXCLUSIVE-NOR gate is an *equality detector* because

$$E = \overline{A\bar{B} + \bar{A}B} = \begin{cases} 1 & A = B \\ 0 & A \neq B \end{cases} \tag{17-9}$$

The condition $A > B$ is given by

$$C = A\bar{B} = 1 \tag{17-10}$$

because if $A > B$, then $A = 1$ and $B = 0$, so that $C = 1$. On the other hand, if $A = B$ or $A < B$ ($A = 0$, $B = 1$), then $C = 0$.

Similarly, the restriction $A < B$ is determined from

$$D = \bar{A}B = 1 \tag{17-11}$$

The logic block diagram for the nth bit drawn in Fig. 17-13 has all three desired outputs C_n, D_n, and E_n. It consists of two inverters, two AND gates, and the AOI circuit of Fig. 17-2. Alternatively, Fig. 17-13 may be considered to consist of an EXCLUSIVE-NOR and two AND gates. (Note that the outputs of the AND gates in the AOI block of Fig. 17-2 are not available, and hence additional AND gates must be fabricated to give C_n and D_n.)

Consider now a 4-bit comparator. $A = B$ requires that

$$A_3 = B_3 \quad \text{and} \quad A_2 = B_2 \quad \text{and} \quad A_1 = B_1 \quad \text{and} \quad A_0 = B_0$$

Hence the AND gate E in Fig. 17-14 described by

$$E = E_3 E_2 E_1 E_0 \tag{17-12}$$

implies $A = B$ if $E = 1$ and $A \neq B$ if $E = 0$. (Assume that the input E' is held high; $E' = 1$.)

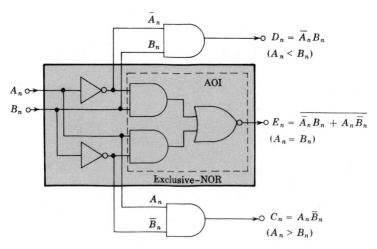

Fig. 17-13 A 1-bit digital comparator.

The inequality $A > B$ requires that

$$A_3 > B_3 \qquad \text{(MSB)}$$

or

$$A_3 = B_3 \qquad \text{and} \qquad A_2 > B_2$$

or

$$A_3 = B_3 \qquad \text{and} \qquad A_2 = B_2 \qquad \text{and} \qquad A_1 > B_1$$

or

$$A_3 = B_3 \qquad \text{and} \qquad A_2 = B_2 \qquad \text{and} \qquad A_1 = B_1 \qquad \text{and} \qquad A_0 > B_0$$

The above conditions are satisfied by the Boolean expression

$$C = A_3\bar{B}_3 + E_3 A_2 \bar{B}_2 + E_3 E_2 A_1 \bar{B}_1 + E_3 E_2 E_1 A_0 \bar{B}_0 \qquad (17\text{-}13)$$

if and only if $C = 1$. The AND-OR gate for C is indicated in Fig. 17-14. (Assume that $C' = 0$.)

The condition that $A < B$ is obtained from Eq. (17-13) by interchanging A and B. Thus

$$D = \bar{A}_3 B_3 + E_3 \bar{A}_2 B_2 + E_3 E_2 \bar{A}_1 B_1 + E_3 E_2 E_1 \bar{A}_0 B_0 \qquad (17\text{-}14)$$

implies that $A < B$ if and only if $D = 1$. This portion of the system is obtained from Fig. 17-14 by changing A to B, B to A, and C to D. Alternatively, D may be obtained from $D = \bar{E}\bar{C}$ because, if $A \neq B$ ($E = 0$) and if $A \not> B$ ($C = 0$), then $A < B$ ($D = 1$). However, this implementation for D introduces the additional propagation delay of an inverter and an AND gate. Hence the logic indicated in Eq. (17-14) for D is fabricated on the same chip as that for C in Eq. (17-13) and E in Eq. (17-12).

The TI 54L85 is an MSI package which performs 4-bit-magnitude comparison. If numbers of greater length are to be compared, several such units can be cascaded. Consider an 8-bit comparator. Designate the $A = B$ output terminal of the stage handling the less significant bits by E_L, the $A > B$ output terminal of this stage by C_L, and the $A < B$ output by D_L. Then the connections $E' = E_L$, $C' = C_L$, and $D' = D_L$ (Fig. 17-14) must be made to the stage with the more significant bits (Prob. 17-7). For the stage handling the less significant bits, the outputs C' and D' are grounded ($C' = 0$ and $D' = 0$) and the input E' is tied to the supply voltage ($E' = 1$). Why?

Parity Checker/Generator Another arithmetic operation that is often invoked in a digital system is that of determining whether the sum of the binary bits in a word is odd (called *odd parity*) or even (designated *even parity*). The output of an EXCLUSIVE-OR gate is 1 if and only if one input is 1 and the other is 0. Alternatively stated, the output is 1 if the sum of the digits is 1. An extension of this concept to the EXCLUSIVE-OR tree of Fig. 17-15 leads to the conclusion that $Z = 1$ (or $Y = 0$) if the sum of the input bits A, B, C, and D is odd. Hence, if the input P' is grounded ($P' = 0$), then $P = 0$ for odd parity and $P = 1$ for even parity.

The system of Fig. 17-15 is not only a parity checker, but it may also be

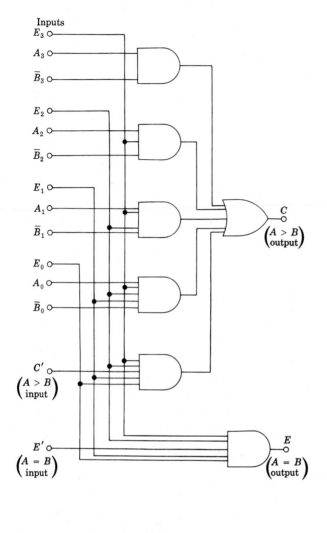

Fig. 17-14 A 4-bit magnitude comparator. (Assume that $C' = 0$ and $E' = 1$.) If $E = 1$, then $A = B$, and if $C = 1$, then $A > B$. If $D = 1$, then $A < B$, where D has the same logic topology as C but with A and B interchanged. The inputs $\bar{A}_n$, B_n, and D' $(A < B)$ are not indicated. The inputs E_n are obtained from Fig. 17-13.

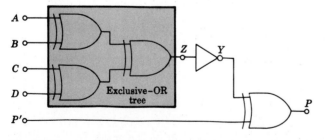

Fig. 17-15 An odd-parity checker, or parity-bit generator system, for a 4-bit input word. Assume $P' = 0$ and then $P = 0(1)$ represents odd (even) parity.

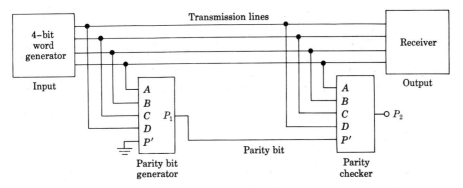

Fig. 17-16 Binary data transmission is tested by generating a parity bit at the input to a line and checking the parity of the transmitted bits plus the generated bit at the receiving end of the system.

used to generate a parity bit P. Independently of the parity of the 4-bit input word, the parity of the 5-bit code A, B, C, D, and P is odd. This statement follows from the fact that if the sum of A, B, C, and D is odd (even), then P is 0 (1), and therefore the sum of A, B, C, D, and P is always odd.

The use of a parity code is an effective way of increasing the reliability of transmission of binary information. As indicated in Fig. 17-16, a parity bit P_1 is generated and transmitted along with the N-bit input word. At the receiver the parity of the augmented $(N + 1)$-bit signal is tested, and if the output P_2 of the checker is 0, it is assumed that no error has been made in transmitting the message, whereas $P_2 = 1$ is an indication that (say, due to noise) the received word is in error. Note that only errors in an odd number of digits can be detected with a single parity check.

An MSI 8-bit parity generator/checker is available (TI 74180) with control inputs so that it may be used in either odd- or even-parity applications (Prob. 17-10). For words of length greater than 8 bits, several such units may be cascaded (Prob. 17-12).

The MSI unit TI SN7486 contains four two-input EXCLUSIVE-OR gates.

17-4   DECODER/DEMULTIPLEXER[3]

In a digital system, instructions as well as numbers are conveyed by means of binary levels or pulse trains. If, say, 4 bits of a character are set aside to convey instructions, then 16 different instructions are possible. This information is *coded* in binary form. Frequently a need arises for a multiposition switch which may be operated in accordance with this code. In other words, for each of the 16 codes, one and only one line is to be excited. This process of identifying a particular code is called *decoding*.

Binary-coded-decimal (BCD) System   This code translates decimal numbers by replacing each decimal digit with a combination of 4 binary digits. Since there are 16 distinct ways in which the 4 binary digits can be arranged in a row, any 10 combinations can be used to represent the decimal digits from 0 to 9.   Thus we have a wide choice of BCD codes.   One of these, called the "natural binary-coded-decimal," is the 8421 code illustrated by the first 10 entries in Table 6-2.   This is a weighted code because the decimal digit in the 8421 code is equal to the sum of the products of the bits in the coded word times the successive powers of two starting from the right (LSB). We need N 4-bit sets to represent in BCD notation an N-digit decimal number.   The first 4-bit set on the right represents units, the second represents tens, the third hundreds, and so on.   For example, the decimal number 264 requires three 4-bit sets, as shown in Table 17-1.   Note that this three-decade

TABLE 17-1   BCD representation for the decimal number 264

Weighting factor..........	800	400	200	100	80	40	20	10	8	4	2	1
BCD code...............	0	0	1	0	0	1	1	0	0	1	0	0
Decimal digits...........			2				6				4	

BCD code can represent any number between 0 and 999; hence it has a resolution of 1 part in 1,000, or 0.1 percent.   It requires 12 bits, which in a straight binary code can resolve one part in $2^{12} = 4,096$, or 0.025 percent.

BCD-to-decimal Decoder   Suppose we wish to decode a BCD instruction representing one decimal digit, say 5.   This operation may be carried out with a four-input AND gate excited by the 4 BCD bits.   For example, the output of the AND gate in Fig. 17-17 is 1 if and only if the BCD inputs are $A = 1$ (LSB), $B = 0$, $C = 1$, and $D = 0$.   Since this code represents the decimal number 5, the output is labeled "line 5."

A BCD-to-decimal decoder is indicated in Fig. 17-18.   This MSI unit has four inputs, A, B, C, and D, and 10 output lines.   (Ignore the dashed lines, for the moment.)   In addition, there must be a ground and a power-supply connection, and hence a 16-pin package is required.   The complementary inputs $\bar{A}$, $\bar{B}$, $\bar{C}$, and $\bar{D}$ are obtained from inverters on the chip.   Since NAND gates are used, an output is 0 (low) for the correct BCD code and is 1 (high) for any other (invalid) code.   The system in Fig. 17-18 is also referred to as a "4-to-10-line decoder" designating that a 4-bit input code selects 1 of 10 output lines.   In other words, the decoder acts as a 10-position switch which responds to a BCD input instruction.

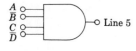

Fig. 17-17   The output is 1 if the BCD input is 0101 and is 0 for any other input instruction.

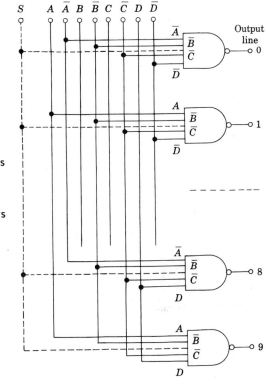

Fig. 17-18 A BCD-to-decimal decoder; Assume that $S = 1$. (Lines
2 to 7 are not indicated.)   The
dashed lines convert the system
into a demultiplexer if S represents
the input signal.

It is sometimes desired to decode only during certain intervals of time. In such applications an additional input, called a *strobe*, is added to each NAND gate. All strobe inputs are tied together and are excited by a binary signal S, as indicated by the dashed lines in Fig. 17-18. If $S = 1$, a gate is *enabled* and decoding takes place, whereas if $S = 0$, no coincidence is possible and decoding is inhibited. The strobe input can be used with a decoder having any number of inputs or outputs.

Demultiplexer A *demultiplexer* is a system for transmitting a binary signal (serial data) on one of N lines, the particular line being selected by means of an address. A decoder is converted into a demultiplexer by means of the dashed connections in Fig. 17-18. If the data signal is applied at S, then the output will be the complement of this signal (because the output is 0 if all inputs are 1) and will appear only on the addressed line.

An enabling signal may be applied to a demultiplexer by cascading the system of Fig. 17-18 with that indicated in Fig. 17-19. If the *enable* input is 0, then S is the complement of the data. Hence, the data will appear (without inversion) on the line with the desired code. If the enable input is 1, $S = 0$, the data are inhibited from appearing on any line and all inputs remain at 1.

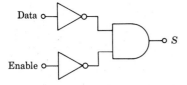

Data o—

Enable o—

—o S

Fig. 17-19 A decoder is converted into a demulti-plexer (with an enabling input) if the S terminal in Fig. 17-18 is obtained from the above AND gate output.

4-to-16-line Decoder/Demultiplexer If an address corresponding to a decimal number in excess of 9 is applied to the inputs in Fig. 17-18, this instruction is rejected; that is, all 10 outputs remain at 1. If it is desired to select 1 of 16 output lines, the system is expanded by adding 6 more NAND gates and using all 16 codes possible with 4 binary bits.

The TI 74154 is a 4-to-16-line decoder/demultiplexer. It has 4 address lines, 16 output lines, an enable input, a data input, a ground pin, and a power-supply lead, so that a 24-pin package is required.

A 2-to-4-line and a 3-to-8-line decoder/demultiplexer are also available in individual IC packages.

Decoder/Lamp Driver Some decoders are equipped with special output stages so that they can drive lamps such as the Burroughs Nixie tube. A Nixie indicator is a cold-cathode gas-discharge tube with a single anode and 10 cathodes, which are wires shaped in the form of numerals 0 to 9. These cathodes are connected to output lines 0 to 9, respectively, and the anode is tied to a fixed supply voltage. The decoder/lamp driver/Nixie indicator combination makes visible the decimal number corresponding to the BCD number applied. Thus, if the input is 0101, the numeral 5 will glow in the lamp.

A decoder for seven-segment numerals made visible by using light-emitting diodes is discussed in Sec. 17-7.

17-5   DATA SELECTOR/MULTIPLEXER

The function performed by a *multiplexer* is to select 1 out of N input data sources and to transmit the selected data to a single information channel. Since in a demultiplexer there is only one input line and these data are caused to appear on 1 out of N output lines, a multiplexer performs the inverse process of a demultiplexer.

The demultiplexer of Fig. 17-18 is converted into a multiplexer by making the following two changes: (1) Add a NAND gate whose inputs include all N outputs of Fig. 17-18 and (2) augment each NAND gate with an individual data input $X_0, X_1, \ldots, X_N$. The logic system for a 4-to-1-line data-selector multiplexer is drawn in Fig. 17-20. This AND-OR logic is equivalent to the NAND-NAND logic as described in the above steps 1 and 2. (See Fig. 6-22.) Note that the same decoding configuration is used in both the multiplexer and demultiplexer. If the select code is 01, then X_1 appears at the output Y, if the address is 11, then $Y = X_3$, etc., provided that the system is enabled ($S = 0$). Multiplexers are also available for selecting 1 of 8 or 1 of 16 data sources. The latter (TI 74150) is a 24-pin IC with 16 data inputs, a 4-bit select code, a strobe input, one output, a power-supply lead, and a ground terminal. For a 16 line-to-1 line multiplexer, Fig. 17-20 is extended from four 4-input AND gates to sixteen 6-input AND circuits. Two 16-data-input multiplexers may be interconnected to select 1 out of 32 information sources (Prob. 17-18).

Parallel-to-Serial Conversion   Consider a 16-bit word available in parallel form so that X_0 represents the 2^0 bit, X_1 the 2^1 bit, etc. By means of a counter (Sec. 17-13), it is possible to change the select code so that it is 0000 for the first T s, 0001 for the second T s, 0010 for the third interval T, etc. With such excitation of the address, the output of the multiplexer will be X_0 for the first T s, X_1 for the next interval T, X_2 for the third period, etc. The output Y is a waveform which represents serially the binary data applied in parallel at the input. In other words, a parallel-to-serial conversion is accomplished of one 16-bit word. This process takes 16 T s.

Sequential Data Selection   By changing the address with a counter in the manner indicated in the preceding paragraph, the operation of an electromechanical stepping switch is simulated. If the data inputs are pulse trains, this information will appear sequentially on the output channel: in other words, pulse train X_0 will appear for T s, followed by X_1 for the next T s, etc. If the number of data sources is M, then X_0 is again selected during the interval $MT < t < (M + 1)T$.

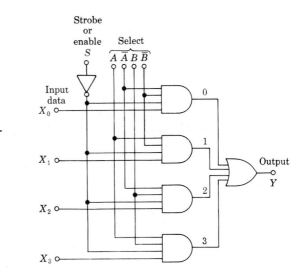

Fig. 17-20   A 4-to-1-line multiplexer.

17-6   ENCODER

A decoder is a system which accepts an M-bit word and establishes the state 1 on one (and only one) of 2^M output lines (Sec. 17-4). In other words, a decoder identifies (recognizes) a particular code. The inverse process is called *encoding*. An encoder has a number of inputs, only one of which is in the 1 state, and an N-bit code is *generated*, depending upon which of the inputs is excited.

Consider, for example, that it is required that a binary code be transmitted with every stroke of an alphanumeric keyboard (a typewriter or teletype). There are 26 lowercase and 26 capital letters, 10 numerals, and about 22 special characters on such a keyboard so that the total number of codes necessary is approximately 84. This condition can be satisfied with a minimum of 7 bits ($2^7 = 128$, but $2^6 = 64$). Let us modify the keyboard so that, if a key is depressed, a switch is closed, thereby connecting a 5-V supply (corresponding to the 1 state) to an input line. A block diagram of such an encoder is indicated in Fig. 17-21. Inside the shaded block there is a rectangular array (or matrix) of wires, and we must determine how to interconnect these wires so as to generate the desired codes.

To illustrate the design procedure for constructing an encoder, let us simplify the above example by limiting the keyboard to only 10 keys, the numerals 0, 1, . . . , 9. A 4-bit output code is sufficient in this case, and let us choose BCD words for the output codes. The truth table defining this encoding is given in Table 17-2. Input W_n ($n = 0, 1, . . . , 9$) represents the nth key. When $W_n = 1$, key n is depressed. Since it is assumed that no more than one key is activated simultaneously, then in any row every input except one is a 0. From this truth table we conclude that $Y_0 = 1$ if

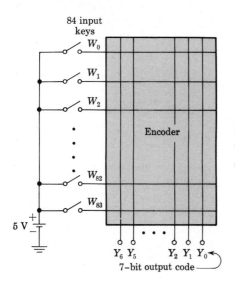

Fig. 17-21   A block diagram of an encoder for generating an output code (word) for every character on a keyboard.

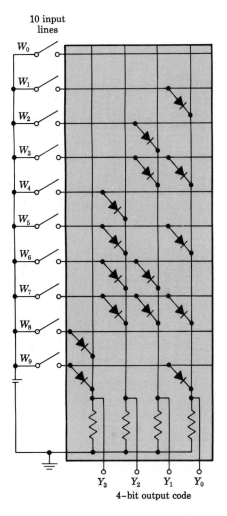

Fig. 17-22 An encoding matrix to transform a decimal number into a binary code (BCD).

10 input lines

4-bit output code

$W_1 = 1$ or if $W_3 = 1$ or if $W_5 = 1$ or if $W_7 = 1$ or if $W_9 = 1$. Hence, in Boolean notation,

$$Y_0 = W_1 + W_3 + W_5 + W_7 + W_9 \qquad (17\text{-}15)$$

Similarly,

$$
\begin{aligned}
Y_1 &= W_2 + W_3 + W_6 + W_7 \\
Y_2 &= W_4 + W_5 + W_6 + W_7 \qquad (17\text{-}16) \\
Y_3 &= W_8 + W_9
\end{aligned}
$$

The OR gates in Eqs. (17-15) and (17-16) are implemented with diodes in Fig. 17-22. (Compare with Fig. 6-3, but with the diodes reversed, because

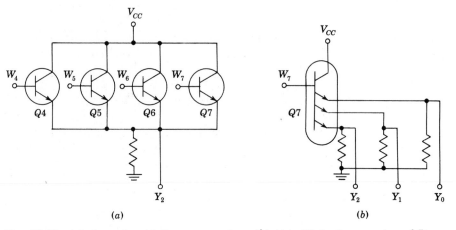

Fig. 17-23   (a) An emitter-follower OR gate.   (b) Line W_7 in the encoder of Fig. 17-22 is connected to the base of a three-emitter transistor.

we are now considering positive logic.)   An encoder array such as that in Fig. 17-22 is called a *rectangular diode matrix*.

TABLE 17-2   The truth table for encoding the decimal numbers 0 to 9

Inputs										Outputs			
W_9	W_8	W_7	W_6	W_5	W_4	W_3	W_2	W_1	W_0	Y_3	Y_2	Y_1	Y_0
0	0	0	0	0	0	0	0	0	1	0	0	0	0
0	0	0	0	0	0	0	0	1	0	0	0	0	1
0	0	0	0	0	0	0	1	0	0	0	0	1	0
0	0	0	0	0	0	1	0	0	0	0	0	1	1
0	0	0	0	0	1	0	0	0	0	0	1	0	0
0	0	0	0	1	0	0	0	0	0	0	1	0	1
0	0	0	1	0	0	0	0	0	0	0	1	1	0
0	0	1	0	0	0	0	0	0	0	0	1	1	1
0	1	0	0	0	0	0	0	0	0	1	0	0	0
1	0	0	0	0	0	0	0	0	0	1	0	0	1

Incidentally, a decoder can also be constructed as a rectangular diode matrix (Prob. 17-21).   This statement follows from the fact that a decoder consists of AND gates (Fig. 17-17), and it is possible to implement AND gates with diodes (Fig. 6-5b).

Each diode of the encoder of Fig. 17-22 may be replaced by the base-emitter diode of a transistor.   If the collector is tied to the supply voltage V_{CC}, then an emitter-follower OR gate results.   Such a configuration is indicated in Fig. 17-23a for the output Y_2.   Note that if either W_4 or W_5 or W_6

or W_7 is high, then the emitter-follower output is high, thus verifying that $Y_2 = W_4 + W_5 + W_6 + W_7$, as required by Eq. (17-16).

Only one transistor (with multiple emitters) is required for each encoder input. The base is tied to the input line, and each emitter is connected to a different output line, as dictated by the encoder logic. For example, since in Fig. 17-22 line W_7 is tied to three diodes whose cathodes go to Y_0, Y_1, and Y_2, then this combination may be replaced by the three-emitter transistor $Q7$ connected as in Fig. 17-23b. The maximum number of emitters that may be required equals the number of bits in the output code. For the particular encoder sketched in Fig. 17-22, Q1, Q2, Q4, and Q8 each have one emitter, Q3, Q5, Q6, and Q9 have two emitters each, and Q7 has three emitters.

17-7   READ–ONLY MEMORY (ROM)[4]

Consider the problem of converting one binary code into another. Such a code-conversion system (designated ROM and sketched in Fig. 17-24a) has M inputs (X_0, X_1, . . . , X_{M-1}) and N outputs (Y_0, Y_1, . . . , Y_{N-1}), where N may be greater than, equal to, or less than M. A definite M-bit code is to result in a specific output code of N bits. This code translation is achieved, as indicated in Fig. 17-24b, by first decoding the M inputs onto $2^M \equiv \mu$ word lines (W_0, W_1, . . . , $W_{\mu-1}$) and then encoding each line into the desired output word. If the inputs assume all possible combinations of 1s and 0s, then μ N-bit words are "read" at the output (not all these 2^M words need be unique, since it may be desirable to have the same output code for several different input words).

The functional relationship between output and input words is built into hardware in the encoder block of Fig. 17-24. Since this information is thus stored permanently, we say that the system has "memory." The *memory elements* are the diodes in Fig. 17-22 or the emitters of transistors in Fig. 17-23. The output word for any input code may be read as often as desired. However, since the stored relationship between output and input codes cannot be modified without adding or subtracting memory elements (hardware), this system is called a *read-only memory*, abbreviated ROM.

The largest bipolar ROM available in 1971 (MM 6280, available from Monolithic Memories) has $M = 10$ and $N = 8$, resulting in $2^M = 2^{10} = 1,024$ words of 8 bits each. This size is referred to as a $8 \times 1,024 = 8,192$-bit memory, and is an example of large-scale integration (LSI). Read-only memories using MOSFETs as memory elements are discussed in Sec. 17-17.

Code Converters   The truth table for translating from a binary to a Gray code[8] is given in Table 17-3. In the progression from one line to the next of the Gray code, 1 and only 1 bit is changed from 0 to 1, or vice versa. (This property does not uniquely define a code, and hence a number of Gray codes may be constructed.) The input bits are decoded in an ROM into the word

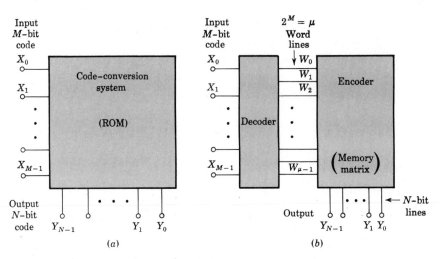

Fig. 17-24 (a) A block diagram of a system for converting one code into another; a read-only memory (ROM). (b) An ROM may be considered to be a decoder for the input code followed by an encoder for the output code.

lines $W_0, W_1, \ldots, W_{15}$, as indicated in Fig. 17-24b, and then are encoded into the desired Gray code $Y_3 Y_2 Y_1 Y_0$. The W's are the minterm outputs of the decoder. For example,

$$W_0 = \bar{X}_3 \bar{X}_2 \bar{X}_1 \bar{X}_0 \qquad W_5 = \bar{X}_3 X_2 \bar{X}_1 X_0 \qquad W_9 = X_3 \bar{X}_2 \bar{X}_1 X_0 \quad (17\text{-}17)$$

From the truth table 17-3, we conclude that

$$Y_0 = W_1 + W_2 + W_5 + W_6 + W_9 + W_{10} + W_{13} + W_{14} \qquad (17\text{-}18)$$

This equation is implemented by connecting eight diodes with their cathodes all tied to Y_0 and their anodes connected to the decoder lines $W_1, W_2, W_5,$ $W_6, W_9, W_{10}, W_{13},$ and W_{14}, respectively (or the base-emitter diodes of transistors may be used in an analogous manner to form an emitter-follower OR gate, as in Fig. 17-23a). Similarly, from Table 17-3, we may write the Boolean expressions for the other output bits. For example,

$$Y_3 = W_8 + W_9 + W_{10} + W_{11} + W_{12} + W_{13} + W_{14} + W_{15} \quad (17\text{-}19)$$

Consider the inverse code translation, from Gray to binary. The Gray code inputs are arranged in the order $W_0, W_1, \ldots, W_{15}$ (corresponding to decimal numbers 0 to 15). The binary code corresponding to a given input word W_n is listed as the output code for that line. For example, from Table 17-3, we find that the Gray code 1001 corresponds to the binary code 1110, and this relationship is maintained in Table 17-4 on line W_9. From this table we obtain the relationship between output and input bits. For example,

$$Y_0 = W_1 + W_2 + W_4 + W_7 + W_8 + W_{11} + W_{13} + W_{14} \qquad (17\text{-}20)$$

TABLE 17-3   Conversion from a binary to a Gray code

Binary code inputs				Decoded word	Gray code outputs			
X_3	X_2	X_1	X_0	W_n	Y_3	Y_2	Y_1	Y_0
0	0	0	0	W_0	0	0	0	0
0	0	0	1	W_1	0	0	0	1
0	0	1	0	W_2	0	0	1	1
0	0	1	1	W_3	0	0	1	0
0	1	0	0	W_4	0	1	1	0
0	1	0	1	W_5	0	1	1	1
0	1	1	0	W_6	0	1	0	1
0	1	1	1	W_7	0	1	0	0
1	0	0	0	W_8	1	1	0	0
1	0	0	1	W_9	1	1	0	1
1	0	1	0	W_{10}	1	1	1	1
1	0	1	1	W_{11}	1	1	1	0
1	1	0	0	W_{12}	1	0	1	0
1	1	0	1	W_{13}	1	0	1	1
1	1	1	0	W_{14}	1	0	0	1
1	1	1	1	W_{15}	1	0	0	0

TABLE 17-4   Conversion from a Gray to a binary code

Gray code inputs				Decoded word	Binary code outputs			
X_3	X_2	X_1	X_0	W_n	Y_3	Y_2	Y_1	Y_0
0	0	0	0	W_0	0	0	0	0
0	0	0	1	W_1	0	0	0	1
0	0	1	0	W_2	0	0	1	1
0	0	1	1	W_3	0	0	1	0
0	1	0	0	W_4	0	1	1	1
0	1	0	1	W_5	0	1	1	0
0	1	1	0	W_6	0	1	0	0
0	1	1	1	W_7	0	1	0	1
1	0	0	0	W_8	1	1	1	1
1	0	0	1	W_9	1	1	1	0
1	0	1	0	W_{10}	1	1	0	0
1	0	1	1	W_{11}	1	1	0	1
1	1	0	0	W_{12}	1	0	0	0
1	1	0	1	W_{13}	1	0	0	1
1	1	1	0	W_{14}	1	0	1	1
1	1	1	1	W_{15}	1	0	1	0

This equation defines how the memory elements are to be arranged in the encoder. Note that the ROM for Table 17-4 uses the same decoding arrangement as that for Table 17-3, but the encoders are completely different. In other words, the IC chips for these two ROMs are quite distinct since individual masks must be used for the encoder matrix of memory elements.

Programming the ROM Consider a 256-bit read-only memory (TI 7488A) arranged in 32 words of 8 bits each. The decoder input is a 5-bit binary select code, and its outputs are the 32 word lines. The encoder consists of 32 transistors (each base is tied to a different line) and with 8 emitters in each transistor. The customer fills out the truth table he wishes the ROM to satisfy, and then the vendor makes a mask for the metallization so as to connect one emitter of each transistor to the proper output line, or alternatively, to leave it floating. For example, for the Gray-to-binary-code conversion, Eq. (17-20) indicates that one emitter from each of transistors $Q1$, $Q2$, $Q4$, $Q7$, $Q8$, $Q11$, $Q13$, and $Q14$ is connected to line Y_0, whereas the corresponding emitter on each of the other transistors $Q0$, $Q3$, $Q5$, $Q6$, . . . is left unconnected. The process just described is called *custom programming*, or *mask programming*, of an ROM. Note that *hardware* (not *software*) programming is under consideration. If the sales demand for a particular code is sufficient, this ROM becomes available as an "off-the-shelf" item.

For small quantities of an ROM, the mask cost may be prohibitive, and also the delivery time may be too long. Hence some manufacturers† supply *field-programmable* ROMs,[4,5] abbreviated pROM, or ROMP (read-only memory, programmable). Such an IC chip has an encoder matrix made with all connections which may possibly be required. For example, the 256-bit memory discussed above is constructed as a pROM with 32 transistors, each having eight emitters (labeled E_0, E_1, . . . , E_7) and with E_0 from each transistor tied to output Y_0, E_1 to Y_1, etc. In series with each emitter there is incorporated a narrow aluminum or nichrome strip which acts as fuse and opens up when a current in excess of a maximum value is passed through this memory element. The user can easily fuse, or "zap," in the field those memory-element links which must be opened in order that the ROM perform the desired functional relationship between output and input.

Diode matrices are also available with fusible links, and these can be used for the encoder portion of a pROM, or also as a decoder (Prob. 17-22).

17-8   ROM APPLICATIONS

As emphasized in the preceding section, an ROM is a code-conversion unit. However, many different practical systems represent a translation from one

---

† For example, Harris Semiconductor, Intel, Intersil, Monolithic Memories, Motorola, Signetics, and Texas Instruments.

code to another. The most important of these ROM applications are discussed below.

Look-up Tables Routine calculations such as trigonometric functions, logarithms, exponentials, square roots, etc., are sometimes required of a computer. If these are repeated often enough, it is more economical to include an ROM as a *look-up table*, rather than to use a subroutine or a software program to perform the calculation. A look-up table for $Y = \sin X$ is a code-conversion system between the input code representing the argument X in binary notation (to whatever accuracy is desired) and the output code giving the corresponding values of the sine function. Clearly, any calculation for which a truth table can be written may be implemented with an ROM—a different ROM for each truth table.

Sequence Generators In a digital system (such as a computer, a data communications system, etc.) several pulse trains are often required for control (gating) purposes. The ROM may be used to supply these binary sequences if the address is changed by means of a counter. As mentioned in Sec. 17-5, the input to the encoder changes from W_0 to W_1 to W_2, etc. every T s. Under this excitation the output Y_1 of the ROM represented by Table 17-4 is

$$Y_1 = 1100001100111100 \qquad \text{(LSB)} \qquad\qquad (17\text{-}21)$$

This equation is obtained by reading the digits in the Y_1 column from top to bottom. It indicates that for the first $2T$ s, Y_1 remains low; for the following $4T$ s, Y_1 is high; for the next $2T$ s, Y_1 is low; for the next $2T$ s, Y_1 is high; for the following $4T$ s, Y_1 is low; for the last $2T$ s, Y_1 is high; and after these $16T$ s, this sequence is repeated (as long as pulses are fed to the counter).

Simultaneously with Y_1, three other synchronous pulse trains, Y_0, Y_2, and Y_3, are created. In general, the number of sequences obtained equals the number of outputs from the ROM. Any desired serial binary waveforms are generated if the truth table is properly specified, i.e., if the ROM is correctly programmed.

Seven-segment Visible Display It is common practice to make visible the reading of a digital instrument (a frequency meter, digital voltmeter, etc.) by means of the seven-segment numeric indicator sketched in Fig. 17-25a. A wide variety[7] of readouts are commercially available. A solid-state indicator in which the segments obtain their luminosity from light-emitting gallium arsenide or phosphide diodes (Sec. 3-15) is operated at low voltage and low power and hence may be driven directly from IC logic gates.

The first 10 displays in Fig. 17-25b are the numerals 0 to 9, which, in the digital instrument, are represented in BCD form. Such a 4-bit code has

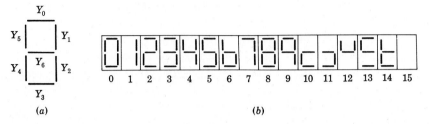

(a)

(b)

Fig. 17-25 (a) Identification of the segments in a seven-segment LED visible indicator. (b) The display which results from each of the sixteen 4-bit input codes.

16 possible states, and the displays 10 to 15 of Fig. 17-25b are unique symbols used to identify a nonvalid BCD condition.

The problem of converting from a BCD input to the seven-segment outputs of Fig. 17-25 is easily solved using an ROM. If an excited (luminous) segment is identified as state 0 and a dark segment as the 1 state, then truth table 17-5 is obtained. This table is verified as follows: For word W_0 (cor-

TABLE 17-5   Conversion from a BCD to a seven-segment-indicator code

Binary-coded-decimal inputs				Decoded word	Seven-segment-indicator code outputs						
$X_3 = D$	$X_2 = C$	$X_1 = B$	$X_0 = A$	W_n	Y_6	Y_5	Y_4	Y_3	Y_2	Y_1	Y_0
0	0	0	0	W_0	1	0	0	0	0	0	0
0	0	0	1	W_1	1	1	1	1	0	0	1
0	0	1	0	W_2	0	1	0	0	1	0	0
0	0	1	1	W_3	0	1	1	0	0	0	0
0	1	0	0	W_4	0	0	1	1	0	0	1
0	1	0	1	W_5	0	0	1	0	0	1	0
0	1	1	0	W_6	0	0	0	0	0	1	1
0	1	1	1	W_7	1	1	1	1	0	0	0
1	0	0	0	W_8	0	0	0	0	0	0	0
1	0	0	1	W_9	0	0	1	1	0	0	0
1	0	1	0	W_{10}	0	1	0	0	1	1	1
1	0	1	1	W_{11}	0	1	1	0	0	1	1
1	1	0	0	W_{12}	0	0	1	1	1	0	1
1	1	0	1	W_{13}	0	0	1	0	1	1	0
1	1	1	0	W_{14}	0	0	0	0	1	1	1
1	1	1	1	W_{15}	1	1	1	1	1	1	1

responding to numeral 0) we see from Fig. 17-25 that $Y_6 = 1$ and all other Y's are 0. For word W_4 (corresponding to the numeral 4) $Y_0 = Y_3 = Y_4 = 1$

and $Y_1 = Y_2 = Y_5 = Y_6 = 0$, and so forth. The ROM is programmed as explained in Sec. 17-7 to satisfy this truth table. For example,

$$Y_0 = W_1 + W_4 + W_6 + W_{10} + W_{11} + W_{12} + W_{14} + W_{15} \quad (17\text{-}22)$$

It should be pointed out that an ROM may not use the smallest number of gates to carry out a particular code conversion. Consider Eq. (17-22) written as a sum of products. Replacing the minterm W_1 by $\bar{X}_3\bar{X}_2\bar{X}_1X_0 \equiv \bar{D}\bar{C}\bar{B}A$ and using analogous expressions for the outputs of the other decoders, Eq. (17-22) becomes

$$Y_0 = \bar{D}\bar{C}\bar{B}A + \bar{D}C\bar{B}\bar{A} + \bar{D}CB\bar{A} + D\bar{C}B\bar{A} + D\bar{C}BA$$
$$+ DC\bar{B}\bar{A} + DCB\bar{A} + DCBA \quad (17\text{-}23)$$

There are a number of algebraic and graphical techniques[8,9] and computer programs for minimizing such Boolean expressions. Note, for example, that the second and third minterms can be simplified to

$$\bar{D}C\bar{B}\bar{A} + \bar{D}CB\bar{A} = \bar{D}C\bar{A}$$

because $\bar{B} + B = 1$. Proceeding in this manner (Prob. 17-24), the following minimized form of Y_0 is obtained:

$$Y_0 = \bar{D}\bar{C}\bar{B}A + C\bar{A} + DB \quad (17\text{-}24)$$

Using the minimized expressions for $Y_0, Y_1, \ldots, Y_6$ results in some saving (about 20 percent) of components over those required in the ROM. A chip fabricated in this manner (for example, TI 7446A) is designated a "BCD–to–seven-segment decoder/driver."

Minimization of Boolean equations (particularly if the number of variables in each product exceeds five) is tedious and time-consuming. The engineering man-hours cost for minimization and for designing the special IC chip to realize the savings in components must be compared with that of simply programming an existing ROM. Unless a tremendous number of units are to be manufactured (and particularly if the matrix size is large), the ROM is the more economical procedure.

Combinational Logic If N logic equations of M variables are given in the sum-of-products canonical form, these equations may be implemented with an M-input, N-output ROM. As explained above, this is an economical solution if M and N are large (particularly if M is large). However, in the logic design of one stage of a full adder, where $M = 3$ and $N = 2$ (small numbers), and where this unit is sold in considerable quantities, using distinct gate combinations as in Fig. 17-7 is more economical than using an ROM.

Character Generator Alphanumeric characters may be "written" on the face of a cathode-ray tube (a television-type display) with the aid of an ROM. This very important application is discussed in Sec. 17-21.

II. SEQUENTIAL DIGITAL SYSTEMS

17-9   A 1-BIT MEMORY[10]

All the systems discussed in the preceding sections of this chapter are based upon combinational logic; the outputs at a given instant of time depend only upon the values of the inputs at the same moment. Such a system is said to have no memory. Note that an ROM is a combinational circuit and, according to the above definition, it has no memory. *The memory of an ROM refers to the fact that it "memorizes" the functional relationship between the output variables and the input variables.* It does *not* store bits of information.

A Sequential System Many digital systems are pulsed or clocked; i.e., they operate in synchronism with a pulse train of period T, called the system *clock* (abbreviated Ck), such as that indicated in Fig. 17-26. The pulse width t_p is assumed small compared with T. The binary values at each node in the system are assumed to remain constant in each interval between pulses. A transition from one state of the system to another may take place only with the application of a clock pulse. Let Q_n be the output (0 or 1) at a given node in the nth interval (bit time n) preceding the nth clock pulse (Fig. 17-26). Then Q_{n+1} is the corresponding output in the interval immediately after the nth pulse. Such a system where the values $Q_1, Q_2, Q_3, \ldots ,$ of Q_n are obtained in time sequence at intervals T is called a *sequential* (to distinguish it from a *combinational*) logic system. The value of Q_{n+1} may depend upon the nodal values during the previous (nth) bit time. Under these circumstances a sequential circuit possesses memory.

A 1-bit Storage Cell The basic digital memory circuit is obtained by cross-coupling two NOT circuits $N1$ and $N2$ (single-input NAND gates) in the manner shown in Fig. 17-27a. The output of each gate is connected to the input of the other, and this feedback combination is called a FLIP-FLOP. The most important property of the FLIP-FLOP is that it can exist in one of

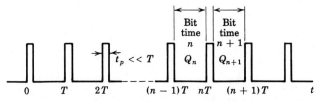

Fig. 17-26   The output of a master oscillator used as a clock pulse train to synchronize a digital sequential system.

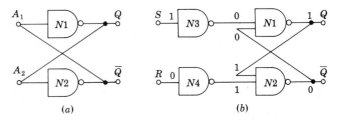

Fig. 17-27   (a) A 1-bit memory or latch.   (b) The FLIP-FLOP
provided with means for presetting or for clearing the state
of the cell.

two stable states, either $Q = 1$ ($\bar{Q} = 0$), called the 1 state, or $Q = 0$ ($\bar{Q} = 1$),
referred to as the 0 state.   The existence of these stable states is consistent
with the interconnections shown in Fig. 17-27a.   For example, if the output
of $N1$ is $Q = 1$, so also is A_2, the input to $N2$.   This inverter then has the
state 0 at its output $\bar{Q}$.   Since $\bar{Q}$ is tied to A_1, then the input to $N1$ is 0, and
the corresponding output is $Q = 1$.   This result confirms our original assumption that $Q = 1$.   A similar argument leads to the conclusion that $Q = 0$;
$\bar{Q} = 1$ is also a possible state.   It is readily verified that the situation in which
both outputs are in the same state (both 1 or both 0) is not consistent with
the interconnection.

Since the FLIP-FLOP has two stable states, it is also called a *binary*, or
bistable MULTI.   Since it may store one bit of information (either $Q = 1$ or
$Q = 0$), it is a *1-bit memory unit*, or a *1-bit storage cell*.   Since this information
is locked, or latched, in place, this FLIP-FLOP is also known as a *latch*.

Suppose it is desired to store a specific state, say $Q = 1$, in the latch.
Or conversely, we may wish to remember the state $Q = 0$.   We may "write"
a 1 or 0 into the memory cell by changing the NOT gates of Fig. 17-27a to
two-input NAND gates, $N1$ and $N2$, and by feeding this latch through two
NOT gates, $N3$ and $N4$, whose inputs are S and R, as in Fig. 17-27b.   If we
assume that $S = 1$ and $R = 0$, then the state of each gate input and output
is indicated on the logic diagram.   Since $Q = 1$, we have thus verified that
to set the FLIP-FLOP to the 1 state requires inputs $S = 1$ and $R = 0$.   The
input S is called the *set*, or *preset*, input.   In a similar manner it can be demonstrated that to enter a 0 into the memory, it is necessary to choose $S = 0$ and
$R = 1$.   Hence R is referred to as the *reset*, or *clear*, input.   The input combination $S = R = 0$ leads to an undetermined state (Q could be either 1 or 0).
Also $S = R = 1$ is not allowed (Prob. 17-27).

The Clocked S-R FLIP-FLOP   In a sequential system it is required to set or
reset a FLIP-FLOP in synchronism with clock pulses.   This is accomplished by
changing $N3$ and $N4$ in Fig. 17-27b to two-input NAND gates and by applying
the clock pulse train Ck simultaneously to $N3$ and $N4$.   Such a triggered set-

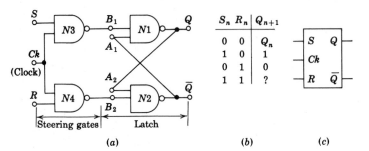

S_n	R_n	Q_{n+1}
0	0	Q_n
1	0	1
0	1	0
1	1	?

(a)   *(b)*   *(c)*

Fig. 17-28 (a) An S-R clocked FLIP-FLOP; (b) the truth table (the question mark in the last row indicates that this state cannot be predicted); (c) the logic symbol.

reset (abbreviated S-R, or R-S) FLIP-FLOP is indicated in Fig. 17-28a. The gates $N1$ and $N2$ form a latch, whereas $N3$ and $N4$ are the *control*, or *steering*, gates which program the state of the FLIP-FLOP after the pulse appears.

Note that between clock pulses ($Ck = 0$), the outputs of $N3$ and $N4$ are 1 independently of the values of R or S. Hence the circuit is equivalent to the latch of Fig. 17-27a. If $Q = 1$, it remains 1, whereas if $Q = 0$, it remains 0. In other words, *the* FLIP-FLOP *does not change state between clock pulses;* it is invariant within a bit time.

Now consider the time $t = nT(+)$ when a clock pulse is present ($Ck = 1$). If $S = 0$ and $R = 0$, then the outputs of $N3$ and $N4$ are 1. By the argument given in the preceding paragraph, the state Q_n of the FLIP-FLOP does not change. Hence, after the pulse passes (in the bit time $n + 1$), the state Q_{n+1} is identical with Q_n. If we denote the values of R and S in the interval just before $t = nT$ by R_n and S_n, then $Q_{n+1} = Q_n$ if $S_n = 0$ and $R_n = 0$. This relationship is indicated in the first row of the truth table of Fig. 17-28b.

If $Ck = 1$, $S_n = 1$, and $R_n = 0$, then the situation is that pictured in Fig. 17-27b and the output state is 1. Hence, after the clock pulse passes (at bit time $n + 1$), we find $Q_{n+1} = 1$, confirming the second row of the truth table. If R and S are interchanged and if simultaneously Q is interchanged with $\bar{Q}$, then the logic diagram of Fig. 17-28a is unaltered. Hence the third row of Fig. 17-28b follows from the second row.

If $Ck = 1$, $S_n = 1$, and $R_n = 1$, then the outputs of the NAND gates $N3$ and $N4$ are both 0. Hence the input B_1 of $N1$ as well as B_2 of $N2$ is 0, so that the outputs of *both* $N1$ and $N2$ must be 1. This condition is logically inconsistent with our labeling the two outputs Q and $\bar{Q}$. We must conclude that the output transistor ($Q3$ of Fig. 6-28) of each gate $N1$ and $N2$ is cut off, resulting in both outputs being high (1). At the end of the pulse the inputs at B_1 and B_2 rise from 0 toward 1. Depending upon which input increases faster and upon circuit parameter asymmetries, either the stable state $Q = 1$ ($\bar{Q} = 0$) or $Q = 0$ ($\bar{Q} = 1$) will result. Therefore we have indicated a question mark for Q_{n+1} in the fourth row of the truth table of Fig. 17-28b. This state is said to be *indeterminate, ambiguous,* or *undefined,* and the condition $S_n = 1$ and $R_n = 1$ is forbidden; it must be prevented from taking place.

17-10   FLIP-FLOPS[10]

In addition to the S-R FLIP-FLOP, three other variations of this basic 1-bit memory are commercially available: the J-K, T, and D types. The J-K FLIP-FLOP removes the ambiguity in the truth table of Fig. 17-28b. The T FLIP-FLOP acts as a toggle switch and changes the output state with each clock pulse; $Q_{n+1} = \bar{Q}_n$. The D type acts as a delay unit which causes the output Q to follow the input D, but delayed by 1 bit time; $Q_{n+1} = D_n$. We now discuss each of these three FLIP-FLOP types.

The J-K FLIP-FLOP This building block is obtained by augmenting the S-R FLIP-FLOP with two AND gates $A1$ and $A2$ (Fig. 17-29a). Data input J and the output $\bar{Q}$ are applied to $A1$. Since its output feeds S, then $S = J\bar{Q}$. Similarly, data input K and the output Q are applied to $A2$, and hence $R = KQ$. The logic followed by this system is given in the truth table of Fig. 17-29b. This logic can be verified by referring to Table 17-6. There are four possible combinations for the two data inputs J and K. For each of these there are two possible states for Q, and hence Table 17-6 has eight rows. From the J_n, K_n, Q_n, and $\bar{Q}_n$ bits in each row, $S_n = J_n\bar{Q}_n$ and $R_n = K_nQ_n$ are calculated and are entered into the fifth and sixth columns of the table. Using these values of S_n and R_n and referring to the S-R FLIP-FLOP truth table of Fig. 17-28b, the seventh column is obtained. Finally, column 8 follows from column 7 because $Q_n = 1$ in row 4, $Q_n = 0$ in row 5, $\bar{Q}_n = 1$ in row 7, and $\bar{Q}_n = 0$ in row 8.

Columns 1, 2, and 8 of Table 17-6 form the J-K FLIP-FLOP truth table of Fig. 17-29b. Note that *the first three rows of a J-K are identical with the corresponding rows for an S-R truth table* (Fig. 17-28b). However, the ambiguity of the state $S_n = 1 = R_n$ is now replaced by $Q_{n+1} = \bar{Q}_n$ for $J_n = 1 = K_n$. *If the two data inputs in the J-K FLIP-FLOP are high, the output will be complemented by the clock pulse.*

It is really not necessary to use the AND gates $A1$ and $A2$ of Fig. 17-29a, since the same function can be performed by adding an extra input terminal

Fig. 17-29   (a) An S-R FLIP-FLOP is converted into a J-K FLIP-FLOP; (b) the truth table.

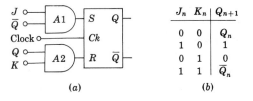

J_n	K_n	Q_{n+1}
0	0	Q_n
1	0	1
0	1	0
1	1	$\bar{Q}_n$

(a)   (b)

TABLE 17-6   Truth table for Fig. 17-29a

Column	1	2	3	4	5	6	7	8
Row	J_n	K_n	Q_n	$\bar{Q}_n$	S_n	R_n	Q_{n+1}	
1	0	0	0	1	0	0	Q_n	Q_n
2	0	0	1	0	0	0	Q_n	
3	1	0	0	1	1	0	1	1
4	1	0	1	0	0	0	Q_n	
5	0	1	0	1	0	0	Q_n	0
6	0	1	1	0	0	1	0	
7	1	1	0	1	1	0	1	$\bar{Q}_n$
8	1	1	1	0	0	1	0	

to each NAND gate $N3$ and $N4$ of Fig. 17-28a. This simplification is indicated in Fig. 17-30. (Ignore the dashed inputs; i.e., assume that they are both 1.) Note that Q and $\bar{Q}$ at the inputs are obtained by the feedback connections (drawn heavy) from the outputs.

Preset and Clear The truth table of Fig. 17-29b tells us what happens to the output with the application of a clock pulse, as a function of the data inputs J and K. However, the value of the output before the pulse is applied is arbitrary. The addition of the dashed inputs in Fig. 17-30 allows the initial state of the FLIP-FLOP to be assigned. For example, it may be required to *clear* the latch, i.e., to specify that $Q = 0$ when $Ck = 0$.

The clear operation may be accomplished by programming the clear input to 0 and the *preset* input to 1; $Cr = 0$, $Pr = 1$, $Ck = 0$. Since $Cr = 0$, the output of $N2$ (Fig. 17-30) is $\bar{Q} = 1$. Since $Ck = 0$, the output of $N3$ is 1, and hence all inputs to $N1$ are 1 and $Q = 0$, as desired. Similarly, if it is

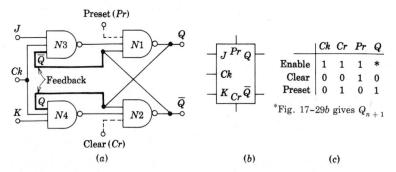

Fig. 17-30   (a) A J-K FLIP-FLOP; (b) the logic symbol; (c) the necessary conditions for synchronous operation (row 1) or for asynchronous clearing (row 2) or presetting (row 3).

required to preset the latch into the 1 state, it is necessary to choose $Pr = 0$, $Cr = 1$, $Ck = 0$. The preset and clear data are called *direct*, or *asynchronous*, inputs; i.e., they are not in synchronism with the clock, but may be applied at any time in between clock pulses. Once the state of the FLIP-FLOP is established asynchronously, the direct inputs must be maintained at $Pr = 1$, $Cr = 1$, before the next pulse arrives in order to *enable* the FLIP-FLOP. The data $Pr = 0$, $Cr = 0$, must not be used since they lead to an ambiguous state. Why?

The logic symbol for the J-K FLIP-FLOP is indicated in Fig. 17-30b, and the inputs for proper operation are given in Fig. 17-30c.

The Race-around Condition   There is a possible physical difficulty with the J-K FLIP-FLOP constructed as in Fig. 17-30. Truth table 17-6 is based upon combinational logic, which assumes that the inputs are independent of the outputs. However, because of the feedback connection $Q(\bar{Q})$ at the input to $K(J)$, the input will change during the clock pulse ($Ck = 1$) if the output changes state. Consider, for example, that the inputs to Fig. 17-30 are $J = 1$, $K = 1$, and $Q = 0$. When the pulse is applied, the output becomes $Q = 1$ (according to row 7 of Table 17-6), this change taking place after a time interval Δt equal to the propagation delay (Sec. 6-15) through two NAND gates in series in Fig. 17-30. Now $J = 1$, $K = 1$, and $Q = 1$, and from row 8 of Table 17-6, we find that the input changes back to $Q = 0$. Hence we must conclude that for the duration t_p (Fig. 17-26) of the pulse (while $Ck = 1$), the output will oscillate back and forth between 0 and 1. At the end of the pulse ($Ck = 0$), the value of Q is ambiguous.

The situation just described is called a *race-around condition*. It can be avoided if $t_p < \Delta t < T$. However, with IC components the propagation delay is very small, usually much less than the pulse width t_p. Hence the above inequality is *not* satisfied, and the output is indeterminate. Lumped delay lines can be used in series with the feedback connections of Fig. 17-30 in order to increase the loop delay beyond t_p, and hence to prevent the race-around difficulty. A more practical IC solution is now to be described.

The Master-Slave J-K FLIP-FLOP   In Fig. 17-31 is shown a cascade of two S-R FLIP-FLOPs with feedback from the output of the second (called the *slave*) to the input of the first (called the *master*). Positive clock pulses are applied to the master, and these are inverted before being used to excite the slave. For $Pr = 1$, $Cr = 1$, and $Ck = 1$, the master is enabled and its operation follows the J-K truth table of Fig. 17-29b. Furthermore, since $\overline{Ck} = 0$, the slave S-R FLIP-FLOP is inhibited (cannot change state), so that Q_n is invariant for the pulse duration t_p. Clearly, the race-around difficulty is circumvented with the master-slave topology. After the pulse passes, $Ck = 0$, so that the master is inhibited and $\overline{Ck} = 1$, which causes the slave to be enabled. The slave is an S-R FLIP-FLOP, which follows the logic in Fig. 17-28b. If

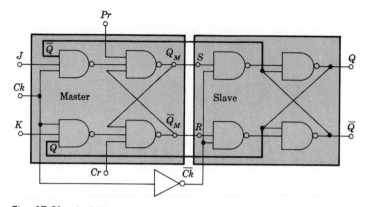

Fig. 17-31   A J-K master-slave FLIP-FLOP.

$S = Q_M = 1$ and $R = \bar{Q}_M = 0$, then $Q = 1$ and $\bar{Q} = 0$. Similarly, if $S = Q_M = 0$ and $R = \bar{Q}_M = 1$, then $Q = 0$ and $\bar{Q} = 1$. In other words, in the interval between clock pulses, the value of Q_M is transferred to the output Q. In summary, during a clock pulse the output Q does not change but Q_M follows J-K logic; at the end of the pulse, the value of Q_M is transferred to Q.

It should be emphasized that the data J and K must remain constant for the pulse duration or an erroneous output may result (Prob. 17-33). Also note that some commercially available FLIP-FLOPS have internal AND gates at the inputs to provide multiple J and K inputs, thereby avoiding the necessity of external gates in applications where these may be required.

The D-type FLIP-FLOP If a J-K FLIP-FLOP is modified by the addition of an inverter as in Fig. 17-32a, so that K is the complement of J, the unit is called a D (delay) FLIP-FLOP. From the J-K truth table of Fig. 17-29b, $Q_{n+1} = 1$ for $D_n = J_n = \bar{K}_n = 1$ and $Q_{n+1} = 0$ for $D_n = J_n = \bar{K}_n = 0$. Hence $Q_{n+1} = D_n$. The output Q_{n+1} after the pulse (bit time $n + 1$) equals the input D_n before the pulse (bit time n), as indicated in the truth table of Fig.

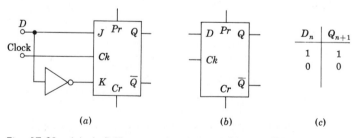

Fig. 17-32   (a) A J-K FLIP-FLOP is converted into a D-type latch; (b) the logic symbol; and (c) the truth table.

17-32c. If the FLIP-FLOP in Fig. 17-32a is of the S-R type, the unit also functions as a D-type latch. There is no ambiguous state because $J = K = 1$ is not possible.

The D-type FLIP-FLOP is a binary used to provide delay. The bit on the D line is transferred to the output at the next clock pulse, and hence this unit functions as a 1-bit delay device. There is available an MSI package (TI 74100) with two independent quadruple D-type latches, so that a total of 8 bits can be stored and transferred.

The T-type FLIP-FLOP This unit changes state with each clock pulse, and hence it acts as a toggle switch. If $J = K = 1$, then $Q_{n+1} = \bar{Q}_n$, so that the J-K FLIP-FLOP is converted into a T-type FLIP-FLOP. In Fig. 17-33a such a system is indicated with a data input T. The logic symbol is shown in Fig. 17-33b, and the truth table in Fig. 17-33c. The S-R- and the D-type latches can also be converted into toggle, or complementing, FLIP-FLOPS (Prob. 17-35).

Summary Four FLIP-FLOP configurations S-R, J-K, D, and T are important. The logic satisfied by each type is repeated for easy reference in Table 17-7. An IC FLIP-FLOP is driven synchronously by a clock, and in addition it may (or may not) have direct inputs for asynchronous operation, preset (Pr) and clear (Cr). A direct input can be 0 only in the interval between clock pulses when $Ck = 0$. When $Ck = 1$, both asynchronous inputs must be high; $Pr = 1$ and $Cr = 1$. The inputs must remain constant during a pulse width, $Ck = 1$. For a master-slave FLIP-FLOP the output Q remains constant for the pulse duration and changes only after Ck changes from 1 to 0, at the negative-going edge of the pulse.

The toggle, or complementing, FLIP-FLOP is not available commercially because a J-K can be used as a T type by connecting the J and K inputs together (Fig. 17-33).

The FLIP-FLOP is available in all the IC digital families, and the maximum frequencies of operation are given in Table 6-5.

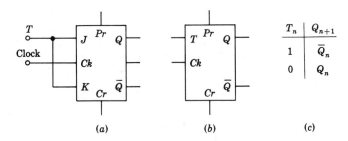

(a)   (b)   (c)

Fig. 17-33 A J-K FLIP-FLOP is converted into a T-type FLIP-FLOP, with a data input T; (b) the logic symbol; and (c) the truth table.

TABLE 17-7   FLIP-FLOP truth tables

S-R			J-K			D		T		Direct inputs			
S_n	R_n	Q_{n+1}	J_n	K_n	Q_{n+1}	D_n	Q_{n+1}	T_n	Q_{n+1}	Ck	Cr	Pr	Q
0	0	Q_n	0	0	Q_n	1	1	1	$\bar{Q}_n$	0	1	0	1
1	0	1	1	0	1	0	0	0	Q_n	0	0	1	0
0	1	0	0	1	0					1	1	1	†
1	1	?	1	1	$\bar{Q}_n$								
Fig. 17-28			Fig. 17-31			Fig. 17-32		Fig. 17-33					

† Refer to truth table S-R, J-K, D, or T for Q_{n+1} as a function of the inputs.

17-11   SHIFT REGISTERS

Since a binary is a 1-bit memory, then n FLIP-FLOPS can store an n-bit word. This combination is referred to as a *register*.   To allow the data in the word to be read into the register serially, the output of one FLIP-FLOP is connected to the input of the following binary. Such a configuration, called a *shift register*, is indicated in Fig. 17-34. Each FLIP-FLOP is of the S-R (or J-K) master-slave type. Note that the stage which is to store the most significant bit (MSB) is converted into a D-type latch (Fig. 17-32) by connecting S and R through an inverter. The 5-bit shift register indicated in Fig. 17-34 is available on a single chip in a 16-pin package (medium-scale integration). We shall now explain the operation of this system by assuming that the serial data 01011 (LSB) is to be registered. (The least significant bit is the right-most digit, which in this case is a 1.)

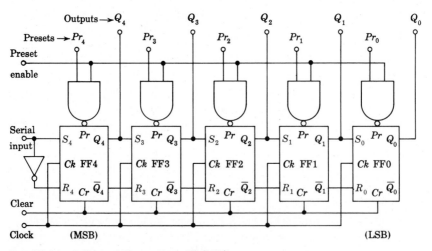

Fig. 17-34   A 5-bit shift register (TI 7496).

Serial-to-Parallel Converter   The FLIP-FLOPS are cleared by applying a 0 to the *clear* input so that every output Q_0, Q_1, . . . , Q_4 is 0.   Then Cr is set to 1 and Pr is held constant at 1 (by keeping the preset enable at 0).   The serial data train and the synchronous clock are now applied.   The least significant bit (LSB) is entered into FF4 when Ck changes from a 0 to a 1 by the action of a D-type FLIP-FLOP.   After the clock pulse, $Q_4 = 1$, while all other outputs remain at 0.

At the second clock pulse the state of Q_4 is transferred to the master latch of FF3 by the action of an S-R FLIP-FLOP.   Simultaneously, the next bit (a 1 in the 01011 word) enters the master of FF4.   After the second clock pulse the bit in each master transfers to its slave and $Q_4 = 1$, $Q_3 = 1$, and the other outputs remain 0.   The readings of the register *after* each pulse are given in Table 17-8.   For example, after the third pulse, Q_3 has shifted to Q_2, Q_4 to Q_3,

TABLE 17-8   Reading of shift register after each clock pulse

Clock pulse	Word bit	Q_4	Q_3	Q_2	Q_1	Q_0
1	1 → 1	0	0	0	0
2	1 → 1	1	0	0	0
3	0 → 0	1	1	0	0
4	1 → 1	0	1	1	0
5	0 → 0	1	0	1	1

and the third input bit (0) has entered FF4, so that $Q_4 = 0$.   We may easily follow this procedure and see that by registering each bit in the MSB FLIP-FLOP and then shifting to the right to make room for the next digit, the input word becomes installed in the register after the nth clock pulse (for an n-bit code). Of course, the clock pulses must stop at the moment the word is registered. Each output is available on a separate line, and they may be read simultaneously.   Since the data entered the system serially and came out in parallel, this shift register is a *serial-to-parallel* converter.   It is also referred to as a *series-in, parallel-out register*.   A *temporal code* (a time arrangement of bits) has been changed to a *spacial code* (information stored in a static memory).

Master-slave FLIP-FLOPS are required because of the race problem between stages (Sec. 17-10).   If all FLIP-FLOPS were to change states simultaneously, there would be an ambiguity as to what data would transfer from the preceding stage.   For example, at the third clock pulse, Q_4 changes from 1 to 0, and it would be questionable as to whether Q_3 would become a 1 or a 0.   Hence it is necessary that Q_4 remain a 1 until this bit is entered into FF3, and only then may it change to 0.   The master-slave configuration provides just this action. If in Fig. 17-31, the $J(K)$ input is called $S(R)$ and if the (heavy) feedback connections are omitted an S-R master-slave FLIP-FLOP results.

Series-in, Series-out Register We may take the output at Q_0 and read the register serially if we apply n clock pulses, for an n-bit word. After the nth pulse each FLIP-FLOP reads 0. Note that the shift-out clock rate may be greater or smaller than the original pulse frequency. Hence here is a method for changing the spacing in time of a binary code, a process referred to as *buffering*.

Parallel-to-Serial Converter Consider the situation where the word bits are available in parallel, e.g., at the outputs from an ROM (Sec. 17-7). It is desired to present this code, say 01011, in serial form.

The LSB is applied to Pr_o, the 2^1 bit to Pr_1, . . . , so that $Pr_0 = 1$, $Pr_1 = 1$, $Pr_2 = 0$, $Pr_3 = 1$, and $Pr_4 = 0$. The register is first cleared by $Cr = 0$, and then $Cr = 1$ is maintained. A 1 at the *preset enable* input activates all kth input NAND gates for which $Pr_k = 1$. The preset of the corresponding kth FLIP-FLOP is $Pr = 0$, and this stage is therefore preset to 1 (Table 17-7). In the present illustration FF0, FF1, and FF3 are preset and the input word 01011 is written into the register, all bits in parallel, by the preset enable pulse.

As explained above, the stored word may be read serially at Q_0 by applying five clock pulses. This is a *parallel-to-serial*, or a *spacial-to-temporal*, *converter*.

Parallel In, Parallel Out The data are entered as explained above by applying a 1 at the preset enable, or *write*, terminal. It is then available in parallel form at the outputs Q_0, Q_1, ... If it is desired to *read* the register during a selected time, each output Q_k is applied to one input of a two-input AND gate N_k, and the second input of each AND is excited by a read pulse. The output of N_k is 0 except for the pulse duration, when it reads 1 if $Q_k = 1$. (The gates N_k are not shown in Fig. 17-34.)

Note that in this application the system is not operating as a shift register since there is no clock required (and no serial input). Each FLIP-FLOP is simply used as an isolated 1-bit read/write memory.

Right-shift, Left-shift Register Some commercial shift registers are equipped with gates which allow shifting the data from right to left as well as in the reverse direction. One application for such a system is to perform multiplication or division by multiples of 2, as will now be explained. Consider first a right-shift register as in Fig. 17-34 and that the serial input is held low.

Assume that a binary number is stored in a shift register, with the least-significant bit stored in FF0. Now apply one clock pulse. Each bit then moves to the next lower significant place, and hence is divided by 2. The number now held in the register is half the original number, provided that FF0 was originally 0. Since the 2^0 bit is lost in the shift to the right, then if FF0 was originally in the 1 state, corresponding to the decimal number 1, after the shift the register is in error by the decimal number 0.5. The next clock pulse causes another division by 2, etc.

Consider now that the system is wired so that each clock pulse causes a shift to the left. Each bit now moves to the next higher significant digit, and the number stored is multiplied by 2.

Digital Delay Line A shift register may be used to introduce a time delay Δ into a system, where Δ is an integral multiple of the clock period T. Thus an input pulse train appears at the output of an n-stage register delayed by a time $(n - 1)T = \Delta$.

Sequence Generator An important application of a shift register is to generate a binary sequence. This system is also called a *word, code,* or *character, generator.* The shift register FLIP-FLOPS are preset to give the desired code. Then the clock applies shift pulses, and the output of the shift register gives the temporal pattern corresponding to the specified sequence. Clearly, we have just described a parallel-in, series-out register. For test purposes it is often necessary that the code be repeated continuously. This mode of operation is easily obtained by feeding the output Q_0 of the register back into the serial input to form a "reentrant shift register." Such a configuration is called a *dynamic,* or *circulating, memory,* or a *shift-register read-only memory.*

Sequence generators without presetting are possible[11] by feeding back to the input not simply Q_0, but rather the output from some combinational logic circuit whose inputs are obtained from $Q_0, Q_1, Q_2, \ldots$. Also, the output may be delivered from another combinational logic circuit whose inputs are $Q_0, Q_1, Q_2, \ldots$.

Shift-register Ring Counter[12] Consider the 5-bit shift register (Fig. 17-34) with Q_0 connected to the serial input. Such a circulating memory forms a *ring counter.* Assume that all FLIP-FLOPS are cleared and then that FF0 is preset so that $Q_0 = 1$ and $Q_4 = Q_3 = Q_2 = Q_1 = 0$. The first clock pulse transfers the state of FF0 to FF4, so that after the pulse $Q_4 = 1$ and

$$Q_3 = Q_2 = Q_1 = Q_0 = 0$$

Succeeding pulses will transfer the state 1 progressively around the ring. The count is read by noting which FLIP-FLOP is in state 1; no decoding is necessary.

Consider a ring counter with N stages. If the interval between triggers is T, then the output from any binary stage is a pulse train of period NT, with each pulse of duration T. The output pulse of one stage is delayed by a time T from a pulse in the preceding stage. These pulses may be used where a set of sequential gating waveforms is required. Thus a ring counter is analogous to a stepping switch, where each triggering pulse causes an advance of the switch by one step.

Since there is one output pulse for each N clock pulses, the counter is also a *divide-by-N* unit, or an $N{:}1$ *scaler.* Typically, TTL shift-register counters operate at frequencies as high as 25 MHz.

Twisted-ring Counter[12]   The topology where $\bar{Q}_0$ (rather than Q_0) is fed back to the input of the shift register is called a *twisted-ring*, or *Johnson, counter*. This system is a $2N{:}1$ scaler.   To verify this statement consider that initially all stages in Fig. 17-34 are in the 0 state.   Since $S_4 = \bar{Q}_0 = 1$, the first pulse puts FF4 into the 1 state; $Q_4 = 1$, and all other FLIP-FLOPS remain in the 0 state.   Since now $S_3 = Q_4 = 1$ and S_4 remains in the 1 state, then after the next pulse there results $Q_4 = 1$, $Q_3 = 1$, $Q_2 = 0$, $Q_1 = 0$, and $Q_0 = 0$.   In other words, pulse 1 causes only Q_4 to change state, and pulse 2 causes only Q_3 to change from 0 to 1.   Continuing the analysis, we see that pulses 3, 4, and 5 cause Q_2, Q_1, and Q_0, in turn, to switch from the 0 to the 1 state.   At the end of five pulses all FLIP-FLOPS are in the 1 state.

After pulse 5, $S_4 = \bar{Q}_0$ changes from 1 to 0.   Hence the sixth pulse causes Q_4 to change to 0.   The seventh pulse resets Q_3 to 0, and so on, until, at the tenth pulse, all stages have been returned to the 0 state, and the counting cycle is complete.   We have demonstrated that this five-stage twisted-ring configuration is a $10{:}1$ counter.   To read the count requires a 5-to-10-line decoder, but because of the unique waveforms generated, only two-input AND gates are required (Prob. 17-37).

MOS shift registers are considered in Sec. 17-16.

17-12   RIPPLE (ASYNCHRONOUS) COUNTERS[13]

The ring counters discussed in the preceding section do not make efficient use of the FLIP-FLOPS.   A $5{:}1$ counter (or $10{:}1$ with the Johnson ring) is obtained with five stages, whereas five FLIP-FLOPS define $2^5 = 32$ states.   By modifying the interconnections between stages (*not* using the shift-register topology), we now demonstrate that n binaries can function as a $2^n{:}1$ counter.

Ripple Counter   Consider a chain of four J-K master-slave FLIP-FLOPS with the output Q of each stage connected to the clock input of the following binary, as in Fig. 17-35.   The pulses to be counted are applied to the clock

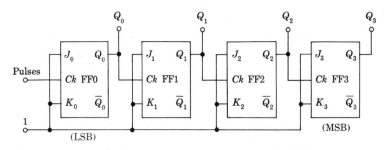

Fig. 17-35   A chain of FLIP-FLOPS connected as a ripple counter.

input of FF0.   For all stages J and K are tied to the supply voltage, so that $J = K = 1$.   This connection converts each stage to a T-type FLIP-FLOP (Fig. 17-33), with $T = 1$.

It should be recalled that, for a T-type binary with $T = 1$, the master changes state every time the waveform at its clock input changes from 0 to 1 and that the new state of the master is transferred to the slave when the clock falls from 1 to 0.   This operation requires that

1. Q_0 changes state at the *falling* edge of each pulse.
2. All other Q's make a transition when and only when the output of the preceding FLIP-FLOP changes from 1 to 0.   This negative transition "ripples" through the counter from the LSB to the MSB.

Following these two rules, the waveforms in Fig. 17-36 are obtained. Table 17-9 lists the states of all the binaries of the chain as a function of the

TABLE 17-9   States of the FLIP-FLOPS
in Fig. 17-35

Number of input pulses	FLIP-FLOP outputs			
	Q_3	Q_2	Q_1	Q_0
0	0	0	0	0
1	0	0	0	1
2	0	0	1	0
3	0	0	1	1
4	0	1	0	0
5	0	1	0	1
6	0	1	1	0
7	0	1	1	1
8	1	0	0	0
9	1	0	0	1
10	1	0	1	0
11	1	0	1	1
12	1	1	0	0
13	1	1	0	1
14	1	1	1	0
15	1	1	1	1
16	0	0	0	0

number of externally applied pulses.   This table may be verified directly by comparison with the waveform chart of Fig. 17-36.   Note that in Table 17-9 the FLIP-FLOPs have been ordered in the reverse direction from their order in Fig. 17-35.   We observe that the ordered array of states 0 and 1 in any row in

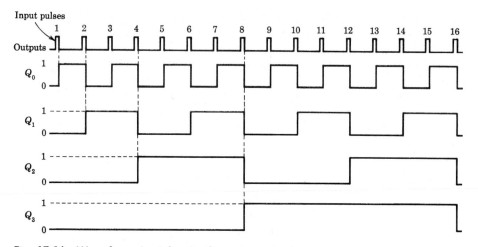

Fig. 17-36 Waveform chart for the four-stage ripple counter.

Table 17-9 is precisely the binary representation of the number of input pulses as given in Table 6-2. Thus the chain of FLIP-FLOPS *counts* in the binary system.

A chain of n binaries will count up to the number 2^n before it resets itself into its original state. Such a chain is referred to as a counter *modulo* 2^n. To read the counter, the 4-bit words (numbers) in Table 17-9 are recognized with a decoder, which in turn drives visible numerical indicators (Sec. 17-4). Spikes are possible in any counter unless all FLIP-FLOPS change state simultaneously. To eliminate the spikes at the decoder output, a strobe pulse is used (S in Fig. 17-20) so that the counter is read only after the spikes have decayed and a steady state is reached.

Up-Down Counter A counter which can be made to count in either the forward or reverse direction is called an *up-down*, a *reversible*, or a *forward-backward*, counter. Forward counting is accomplished, as we have seen, when the trigger input of a succeeding binary is coupled to the Q output of a preceding binary. The count will proceed in the reverse direction if the coupling is made instead to the $\bar{Q}$ output, as we shall now verify.

If a binary makes a transition from state 0 to 1, the output $\bar{Q}$ will make a transition from state 1 to 0. This negative-going transition in $\bar{Q}$ will induce a change in state in the succeeding binary. Hence, for the reversing connection, the following rules apply:

1. FLIP-FLOP FF0 makes a transition at each externally applied pulse.
2. Each of the other binaries makes a transition when and only when the preceding FLIP-FLOP goes from state 0 to state 1.

If these rules are applied to any of the numbers in Table 17-9, the next smaller number in the table results. For example, consider the number 12, which is 1100 in binary form. At the next pulse, the rightmost 0 (corresponding to Q_0) becomes 1. This change of state from 0 to 1 causes Q_1 to change state from 0 to 1, which in turn causes Q_2 to change state from 1 to 0. This last transition is in the direction not to affect the following binary, and hence Q_3 remains in state 1. The net result is that the counter reads 1011, which represents the number 11. Since we started with 12 and ended with 11, a reverse count has taken place.

The logic block diagram for an up-down counter is indicated in Fig. 17-37. For simplicity in drawing, no connections to J and K are indicated. For a ripple counter it is always to be understood that $J = K = 1$ as in Fig. 17-35. The two-level AND-OR gates CG1 and CG2 between stages control the direction of the counter. Note that this logic combination is equivalent to a NAND-NAND configuration (Fig. 6-22). If the input X is a 1 (0), then Q ($\bar{Q}$) is effectively connected to the following FLIP-FLOP and pulses are added (subtracted). In other words, $X = 1$ converts the system to an *up* counter and $X = 0$ to a *down* counter. The control X may not be changed from 1 to 0 (or 0 to 1) between input pulses, because a spurious count may be introduced by this transition. (The synchronous counter of Fig. 17-40 does not have this difficulty.)

Divide-by-N Counter   It may be desired to count to a base N which is not a power of 2. We may prefer, for example, to count to the base 10, since the decimal system is the one with which we are most familiar. To construct such a counter, start with a ripple chain of n FLIP-FLOPS such that n is the smallest number for which $2^n > N$. Add a feedback gate so that at count N all binaries are reset to zero. This feedback circuit is simply a NAND gate whose output feeds all *clear* inputs in parallel. Each input to the NAND gate is a FLIP-FLOP output Q which becomes 1 at the count N.

Let us illustrate the above procedure for a decade counter. Since the smallest value of n for which $2^n > 10$ is $n = 4$, then four FLIP-FLOPS are required. The decimal number 10 is the binary number 1010 (LSB), and hence $Q_0 = 0$, $Q_1 = 1$, $Q_2 = 0$, and $Q_3 = 1$. The inputs to the feedback NAND gate are therefore Q_1 and Q_3, and the complete circuit is shown in

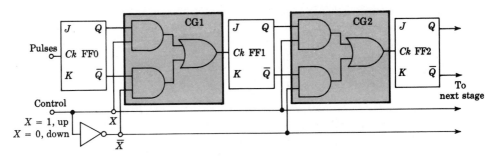

Fig. 17-37   An up-down ripple counter.   (It is understood that $J = K = 1$.)

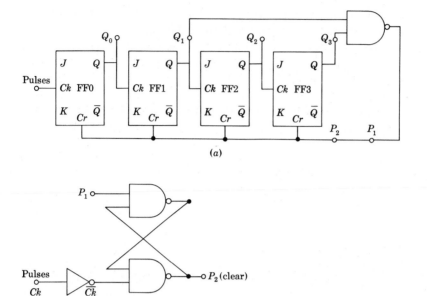

(a)

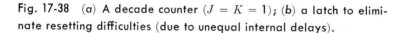

(b)

Fig. 17-38 (a) A decade counter ($J = K = 1$); (b) a latch to eliminate resetting difficulties (due to unequal internal delays).

Fig. 17-38a. Note that after the tenth pulse Q_1 and Q_3 both go to 1, the
output of the NAND gate becomes 0, and all FLIP-FLOPS are cleared (reset to 0).
(Note that Q_1 and Q_3 first become 1 and then return to 0 after pulse 10,
thereby generating a narrow spike.)

If the propagation delay from the clear input to the FLIP-FLOP output
varies from stage to stage, the clear operation may not be reliable. In the
above example, if FF3 takes appreciably longer time to reset than FF1, then
when Q_1 returns to 0, the output of the NAND gate goes to 1, so that $Cr = 1$
and Q_3 will not reset. Wide variations in reset propagation time may occur
if the counter outputs are unevenly loaded. A method of eliminating the
difficulty with resetting is to use a latch to memorize the output of the NAND
gate at the Nth pulse. The lead in Fig. 17-38a between the NAND output P_1
and the clear input P_2 is opened, and the circuit drawn in Fig. 17-38b is inserted
between these two points. The operation of the latch is considered in detail
in Prob. 17-41.

A divide-by-6 counter is obtained using a 3-bit ripple counter, and since
for $N = 6$, $Q_1 = 1 = Q_2$, then Q_1 and Q_2 are the inputs to the feedback
NAND gate. Similarly, a divide-by-7 counter requires a three-input NAND gate
with inputs Q_0, Q_1, and Q_2.

In some applications it is important to be able to program the count (the
value of N) of a divide-by-N counter, either by means of switches or through

control data inputs at the preset terminals. Such a *programmable* or *pre-settable* counter is indicated in the figure of Prob. 17-42.

Consider that it is required to count up to 10,000 and to indicate the count visually in the decimal system. Since $10,000 = 10^4$, then four decade-counter units, such as in Fig. 17-38, are cascaded. A BCD–to–decimal decoder/lamp driver (Sec. 17-4) or a BCD–to–seven-segment display decoder (Sec. 17-8) is used with each unit to make visible the four decimal digits giving the count.

17-13   SYNCHRONOUS COUNTERS[13]

The *carry propagation delay* is the time required for a counter to complete its response to an input pulse. The carry time of a ripple counter is longest when each stage is in the 1 state. For in this situation, the next pulse must cause all previous FLIP-FLOPS to change state. Any particular binary will not respond until the preceding stage has nominally completed its transition. The clock pulse effectively "ripples" through the chain. Hence the carry time will be of the order of magnitude of the sum of the propagation delay times (Sec. 6-15) of all the binaries. If the chain is long, the carry time may well be longer than the interval between input pulses. In such a case, it will not be possible to read the counter between pulses.

If the asynchronous operation of a counter is changed so that all FLIP-FLOPS are clocked simultaneously (synchronously) by the input pulses, the propagation delay time may be reduced considerably. Repetition rate is limited by the delay of any one FLIP-FLOP plus the propagation times of any control gates required. Typically, the maximum frequency of operation of a 4-bit synchronous counter using TTL logic is 32 MHz, which is about twice that of a ripple counter. Another advantage of the synchronous counter is that no decoding spikes appear at the output since all FLIP-FLOPS change state at the same time. Hence no strobe pulse is required when decoding a synchronous counter.

Series Carry A 5-bit synchronous counter is indicated in Fig. 17-39. Each FLIP-FLOP is a T type, obtained by tying the J terminal to the K terminal of a J-K FLIP-FLOP (Fig. 17-33). If $T = 0$, there is no change of state when the binary is clocked, and if $T = 1$, the FLIP-FLOP output is complemented with each pulse.

The connections to be made to the T inputs are deduced from the waveform chart of Fig. 17-36.

Q_0 toggles with each pulse: $\qquad\qquad T_0 = 1$

Q_1 complements only if $Q_0 = 1$: $\qquad T_1 = Q_0$

Q_2 becomes $\bar{Q}_2$ only if $Q_0 = Q_1 = 1$: $\quad T_2 = Q_0 Q_1$

Q_3 toggles only if $Q_0 = Q_1 = Q_2 = 1$: $\quad T_3 = Q_0 Q_1 Q_2$

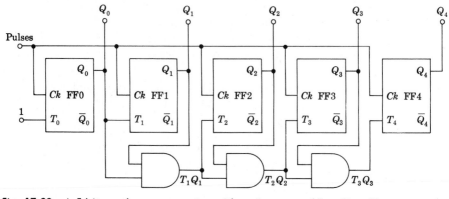

Fig. 17-39 A 5-bit synchronous counter with series carry ($J = K = T$).

Extending this logic to Q_4, we conclude that $T_4 = Q_0Q_1Q_2Q_3$, etc. Therefore the T logic is given by

$$T_0 = 1 \qquad T_1 = Q_0 \qquad T_2 = T_1Q_1 \qquad T_3 = T_2Q_2 \qquad T_4 = T_3Q_3$$
$$(17\text{-}25)$$

Clearly, the two-input AND gates of Fig. 17-39 perform this logic.

The minimum time $T_{\min}$ between pulses is the interval required for each J and K node to reach its steady-state value and is given by

$$T_{\min} = T_F + (n - 2)T_G \qquad\qquad (17\text{-}26)$$

where T_F is the propagation delay of one FLIP-FLOP, and T_G is the propagation delay of one AND gate (actually, a NAND gate plus an INVERTER). The maximum pulse frequency for series carry is the reciprocal of $T_{\min}$.

Parallel Carry Since the carry passes through all the control gates in series in Fig. 17-39, this is a synchronous counter with *series*, or *ripple, carry*. The maximum frequency of operation can be improved by using parallel, or *look-ahead, carry*, where the toggle input to each binary comes from a multi-input AND gate excited by the outputs from every preceding FLIP-FLOP. From Eq. (17-25) it follows that

$$T_1 = Q_0 \qquad T_2 = Q_0Q_1 \qquad T_3 = Q_0Q_1Q_2 \qquad T_4 = Q_0Q_1Q_2Q_3 \quad (17\text{-}27)$$

Hence T_4 is obtained from a four-input AND gate fed by Q_0, Q_1, Q_2, and Q_3. Clearly, for parallel carry,

$$T_{\min} = T_F + T_G \qquad\qquad (17\text{-}28)$$

which may be considerably smaller than the corresponding time for series carry given by Eq. (17-26), particularly if n is large (high division ratios).

The disadvantages of a parallel-carry counter are: (1) The large fan-in of the gates; the gate feeding T_k requires k inputs. (2) The heavy loading of the

FLIP-FLOPS at the beginning of the chain; the fan-out of Q_0 is $n - 1$, since it must feed the carry gates of every succeeding stage.

Up-Down Synchronous Counter with Parallel Carry   As explained in the preceding section, a counter is reversed if $\bar{Q}$ is used in place of Q in the coupling from stage to stage.   Hence a synchronous up-down counter is obtained if the control gates CG of Fig. 17-37 are interposed between the FLIP-FLOPS of Fig. 17-39.   This modification to an UP synchronous counter is made in Fig. 17-40, where CG is now indicated as a NAND-NAND gate (equivalent to the AND-OR logic of Fig. 17-37).   Note that CG1 is identical in Figs. 17-37 and 17-40.   All control gates in the ripple counter are two-input gates, whereas in the synchronous counter the fan-in for CG2 is 3, for CG3 is 4, etc.   The extra input leads to the gates, as required by Eq. (17-27), are used for the parallel carry. In other words, the CG blocks in Fig. 17-40 perform both the up-down and the parallel-carry logic.

Synchronous Decade Counter   Design of a system which is to divide by a number that is not a multiple of 2 is much more difficult for a synchronous than for a ripple counter.   Control matrices (Karnaugh maps) are used[13,14] to simplify the procedure.

With a great deal of patience and intuition, the design may be carried out from direct observation of the waveform chart.   Consider, for example, the synthesis of a synchronous decade counter with parallel carry.   The waveform chart is that given in Fig. 17-36 except that *after the tenth pulse all waveforms return to 0*.   Since $Q_0 = 0$ and $Q_2 = 0$ after the tenth pulse, FF0 and FF2 are excited as in the 16:1 synchronous counter.   Hence, from Eq. (17-25),

$$T_0 = J_0 = K_0 = 1 \qquad T_2 = J_2 = K_2 = Q_0 Q_1 \qquad (17\text{-}29)$$

We note from Fig. 17-36 that FF1 toggles if $Q_0 = 1$.   However, to prevent Q_1 from going to 1 after the tenth pulse, it is inhibited by Q_3.   These statements are equivalent to the statement

$$T_1 = J_1 = K_1 = Q_0 \bar{Q}_3 \qquad (17\text{-}30)$$

Finally, we wish FF3 to change state from 0 to 1 after the eighth pulse and to return to 0 after the tenth pulse.   If

$$J_3 = Q_0 Q_1 Q_2 \qquad K_3 = Q_0 \qquad (17\text{-}31)$$

then the desired logic is followed because $Q_0 = Q_1 = Q_2 = 1$, so that $J_3 = 1$, $K_3 = 1$, before pulse 8, whereas $Q_0 = 1$, $Q_1 = 0$, and $Q_2 = 0$, so that $J_3 = 0$, $K_3 = 1$, before pulse 10.   The implementation of Eqs. (17-29) to (17-31) is given in the logic block diagrams of Fig. 17-41.

Synchronous up-down decade counters are available commercially (for example, TI 74192) on a single chip.   Such a counter has the complexity of 55 equivalent gates, and hence is an example of MSI.   The FLIP-FLOPS are pro-

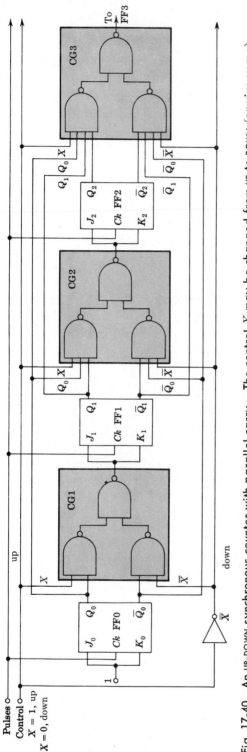

Fig. 17-40 An UP-DOWN synchronous counter with parallel carry. The control X may be changed from UP to DOWN (or vice versa) between input pulses without introducing spurious counts, because the counter responds only upon the application of a clock pulse.

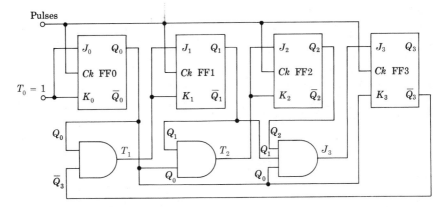

Fig. 17-41   A synchronous decade counter with parallel carry.

vided with *preset* and *clear* inputs not indicated in Fig. 17-41.   Division by a number other than 10 or a multiple of 2 is usually not commercially available and must be programmed as explained above.

17-14   APPLICATIONS OF COUNTERS

Many systems, including digital computers, data handling, and industrial control systems, use counters.   We describe briefly some of the fundamental applications.

Direct Counting   Direct counting finds application in many industrial processes.   Counters will operate with reliability where human counters fail because of fatigue or limitations of speed.   It is required, of course, that the event which is to be counted first be converted into an electrical signal, but this requirement usually imposes no important limitation.   For example, objects may be counted by passing them single-file on a conveyor belt between a photoelectric cell and a light source.

The *preset* input allows control of industrial processes.   The counter may be preset so that it will deliver an output pulse when the count reaches a predetermined number.   Such a counter may be used, for example, to count the number of pills dropped into a bottle.   When the preset count is attained, the output pulse is used to divert the pills to the next container and at the same time to reset the counter for counting the next batch.

Divide-by-N   There are many applications where it is desired to change the frequency of a square wave from f to f/N, where N is some multiple of 2. From the waveforms of Fig. 17-36 it is seen that a counter performs this function.

If instead of square waves it is required to use narrow pulses or spikes for system synchronization, these may be obtained from the waveforms of Fig. 17-36. A small *RC* coupling combination at the counter output, as in Fig. 16-34a, causes a positive pulse to appear at each transition from 0 to 1 and a negative pulse at each transition from 1 to 0. If now we count only the positive pulses (the negative pulses may be eliminated, say, by using a diode), it appears (Fig. 16-34e) that each binary divides by 2 the number of positive pulses applied to it. The four binaries together accomplish a division by a factor $2^4 = 16$. A single negative pulse will appear at the output for each 16 pulses applied at the input. A chain of n binaries used for this purpose of dividing or scaling down the number of pulses is referred to as a *scaler*. Thus a chain of four FLIP-FLOPS constitutes a scale-of-16 circuit, etc.

Measurement of Frequency   The basic principle by which counters are used for the precise determination of frequency is illustrated in Fig. 17-42. The input signal whose frequency is to be measured is converted into pulses by means of the zero crossing detector and applied through an AND gate to a counter. To determine the frequency, it is now only required to keep the gate open for transmission for a known time interval. If, say, the gating time is 1 s, the counter will yield the frequency directly in cycles per second (hertz). The *clock* for timing the gate interval is an accurate crystal oscillator whose frequency is, say, 1 MHz. The crystal oscillator drives a scale-of-10^6 circuit which divides the crystal frequency by a factor of 1 million. The divider output consists of a 1-Hz signal whose period is as accurately maintained as the crystal frequency. This divider output signal controls the gating time by setting a toggle FLIP-FLOP to the 1 state for 1 s. The system is susceptible to only slight errors. One source of error results from the fact that a variation of ± 1 count may be obtained, depending on the instant when the first and last pulses occur in relation to the sampling time. Beyond these, of course, the accuracy depends on the accuracy of the crystal oscillator.

Measurement of Time   The time interval between two pulses may also be measured with the circuit of Fig. 17-42. The FLIP-FLOP is now converted into set-reset type, with the first input pulse applied to the S terminal, the second pulse to the R terminal, and no connection made to Ck. With this configuration the first pulse opens the AND gate for transmission and the second pulse closes it. The crystal-oscillator signal (or some lower frequency from the divider chain) is converted into pulses, and these are passed through the gate into the counter. The number of counts recorded is proportional to the length of time the gate is open and hence gives the desired time interval.

Measurement of Distance   In radar or sonar systems a pulse is transmitted and a reflected pulse is received delayed by a time T. Since the velocity of light (or sound) is known, a measurement of the interval T, as outlined above, gives the distance from the transmitter to the object from which the reflection was received.

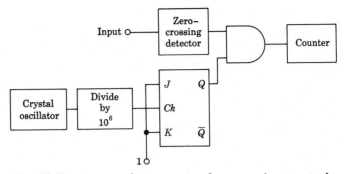

Fig. 17-42 A system for measuring frequency by means of a counter.

Measurement of Speed A speed determination may be converted into a time measurement. For example, if two photocell-light-source combinations are set a fixed distance apart, the average speed of an object passing between these points is inversely proportional to the time interval between the generated pulses. Projectile velocities have been measured in this manner.

Digital Computer In a digital computer a problem is solved by subjecting data to a sequence of operations in accordance with the program of instructions introduced into the computer. Counters may be used to count the operations as they are performed and to call forth the next operation from the memory when the preceding one has been completed.

Waveform Generation The waveforms which occur at the collectors or bases of binary counters may be combined either directly or in connection with logic gates to generate desirable pulse-type waveforms. Such waveforms are used for sequential data selection and parallel-to-serial conversion, as described in Sec. 17-5.

Conversion between Analog and Digital Information These systems are considered in Secs. 17-19 and 17-20.

III. MOS/LSI DIGITAL SYSTEMS

17-15   DYNAMIC MOS CIRCUITS

The first commercial MOSFET (abbreviated MOS) circuits appeared in 1964, and since then they have continued to increase in popularity because of their high packing density, small power consumption, and low cost. As many as 5,000 MOS devices are fabricated on a chip 150 by 150 mils square. Such LSI construction makes possible very large MOS shift registers and memories, described in Secs. 17-16 to 17-18. Many of these systems are operated synchronously. Hence a discussion of such clocked (called *dynamic* MOS) circuits follows. The operation of nonclocked (dc stable) digital MOS gates is described in Sec. 10-6 and should be reread before proceeding further. Note that *p*-channel devices are used with negative logic so that the 0 state is 0 V and the 1 state is -10 V.

Dynamic MOS circuits make use of the parasitic capacitance between gate and substrate to provide temporary storage. This storage can be made permanent using refreshing operations by means of clock waveforms. Since the leakage of the gate circuit is extremely low, the time constants are of the order of milliseconds, and to maintain the stored data, the refreshing or cycling must not be allowed to fall below some minimum rate, typically 1 kHz.

Dynamic MOS Inverter[15,16] The circuit of Fig. 17-43 shows a dynamic MOS inverter which requires a train of clock pulses ϕ for proper operation. Compare this circuit with the static MOS inverter of Fig. 10-15. When the clock ϕ is at 0 V, transistors $Q2$ and $Q3$ are OFF and the power supply is disconnected from the circuit and delivers essentially no power. When the clock pulse is at -10 V, both $Q2$ and $Q3$ are ON and inversion of the input V_i takes place. Thus, if $V_i = -10$ V, $Q1$ is ON and the output is $V_o \approx 0$, whereas if $V_i = 0$ V, then $Q1$ is OFF and the output becomes $V_o \approx -V_{DD}$ (say -10 V). Note that $Q3$ is a bidirectional switch; ① acts as the source when C charges to -10 V, whereas ② acts as the source when C discharges to 0 V. During the time the clock is at 0 V, the output capacitor C retains its charge. This parasitic capacitance of $Q3$ between source and ground has a typical value of 0.5 pF.

The inverter discussed above has been called a *ratio inverter*. The name derives from the fact that when the input is low and the clock is low, transistors $Q1$ and $Q2$ form a voltage divider between $-V_{DD}$ and ground. Therefore the output voltage V_o depends on the ratio of the ON resistance of $Q1$ and the effective load resistance of $Q2$ (typically, <1:5). This ratio is related to the physical size of $Q1$ and $Q2$ and is often referred to as the *aspect ratio*.

Figure 17-44 shows a dynamic *p*-channel MOS NAND gate, corresponding to the static NAND of Fig. 10-17. A dynamic MOS NOR is constructed by

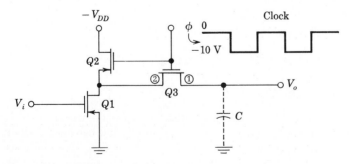

Fig. 17-43 Dynamic MOS inverter. (V_o is read only when $\phi = -10$ V.)

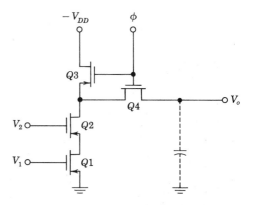

Fig. 17-44   Dynamic MOS NAND gate.

modifying the circuit of Fig. 10-18 in a similar manner.   The reader should verify that the dynamic circuits dissipate less power than the corresponding static circuits.

Two-phase MOS   The inverter shown in Fig. 17-45a consists of three p-channel enhancement devices and dissipates almost no power due to the fact that two clock trains are used in the phase sequence shown in Fig. 17-45b. During the clock pulse ϕ_1 (precharge clock), the parasitic capacitor C is charged to $-V_{DD}$.   Clock pulse ϕ_2, which comes after ϕ_1, performs the inversion.   If V_i is at -10 V, $Q3$ is turned ON, and since $Q2$ is also ON, the capacitor discharges to ground and the output becomes 0 V.   If $V_i = 0$ V, then $Q3$ is OFF, $Q2$ is ON, and there is no path to ground for C to discharge.   Thus V_o remains at $-V_{DD}$. We note that, except during switching, all of the transistors $Q1$, $Q2$, and $Q3$ are always OFF and quiescent power dissipation is of the order of 10 nW for $-V_{DD} = -10$ V.   The circuit has no dc current path regardless of the state of the clocks or the data stored on the parasitic capacitor C.   Such a circuit has the significant advantage in that its output does not depend on the ratio of the

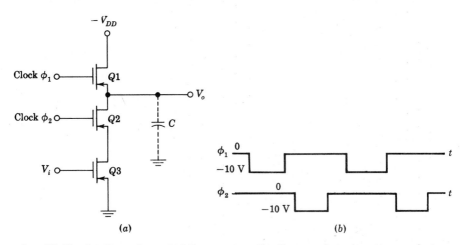

Fig. 17-45   (a) Two-phase MOS inverter.   (b) Timing of clock pulses ϕ_1 and ϕ_2.

resistances of any of its devices; therefore all devices can be of minimum geometry, reducing the chip size for a given number of gates. Circuits that use the above feature are called *ratioless powerless*.

17-16   MOS SHIFT REGISTERS

Bipolar and MOS shift registers can be used for the same purpose. However, if a large number of bits are involved, the MOS shift registers are preferred over bipolar circuits because they are more economical (in power and cost) and are smaller in size. Typical applications for MOS shift registers are calculators, display systems, refresh memories, scratch-pad memories, buffer memories, communication equipment, computer peripherals, and delay lines. MOS shift registers are available (1971) with from several hundred up to 1,024 bits of storage and shifting rates which run up to 5 MHz for p-channel devices and up to 15 MHz for n-channel MOS transistors. Various configurations can be supplied, such as serial-in, serial-out or serial-in, parallel-out or parallel-in, serial-out. For the sake of simplicity and because of their popularity, we shall treat only serial-in, serial-out registers in this section.

Dynamic MOS Shift Register[16,17]   There are two types of MOS shift registers, dynamic and static (more properly called *dc stable*). In a dynamic shift register each information bit is stored on the gate capacitance of a device and is transferred by pulsing the subsequent inverter. A typical MOS dynamic shift register stage is shown in Fig. 17-46. It uses the 2-phase clock pulses of Fig. 17-45b. Each stage of the register requires six MOSFETs. The input to the stage is the charge on the gate capacitance, such as C_1 of $Q1$, deposited there by the previous stage. When clock ϕ_1 goes negative (for p-channel devices), transistors $Q1$ and $Q2$ form an inverter. The common node of $Q1$, $Q2$, and $Q3$ approaches $+V_{SS}$ if the charge on C_1 is negative enough to turn $Q1$ ON, or it approaches $-V_{DD}$ if the charge is positive and $Q1$ is OFF. We should note that during the time ϕ_1 is negative, $Q3$ is ON and the gate capacitance C_2 of $Q4$ gets charged to either $+V_{SS}$ or $-V_{DD}$. When ϕ_1 goes high, C_2 retains the charge. Pulsing ϕ_2 negative then shifts and inverts the data, depositing the charge on C_3. The information at C_3 is identical with that on C_1 at A, but delayed by an amount predetermined by the clock period. The combination $Q1Q2Q3$ can be called the *master* inverter, and $Q4Q5Q6$ the *slave* section. To retain data stored in the register, the rate at which the data are clocked through the circuit must not fall below some minimum value.

A typical two-phase dynamic MOS shift register is the Texas Instruments TI 3401LC 512-bit register. The entire device is constructed using MOS p-channel, low-threshold devices. It operates at a minimum clock rate of 20 kHz and maximum rate of 5 MHz. Both input and output are directly compatible with TTL integrated circuits, and the power dissipation is 0.2 mW/bit at 1 MHz.

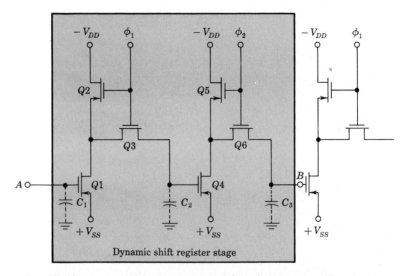

Dynamic shift register stage

Fig. 17-46 A two-phase dynamic MOS shift register. The clock pulses have the waveshapes indicated in Fig. 17-45b, but the binary levels are +5 V and −12 V.

Static MOS Shift Registers A "static" shift register is dc stable and can operate without a minimum clock rate. That is, it can store data indefinitely provided that power is supplied to the circuit. However, static shift-register cells are larger than the dynamic cells and consume more power. As shown in Fig. 17-47, the stage consists of a pair of static inverters with unclocked loads (Fig. 10-17) cross-connected through two transmission gates $Q3$ and $Q6$.

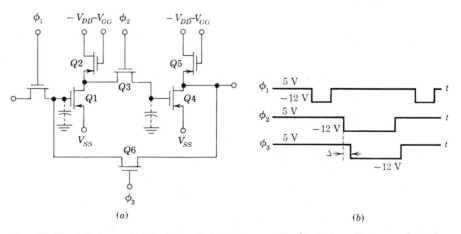

(a)   (b)

Fig. 17-47 (a) Basic MOS static shift-register cell; (b) timing diagram of static shift-register clocks.

When both clocks ϕ_2 and ϕ_3 are low, the feedback loop is closed and the two inverters form a FLIP-FLOP, or *latch* (Fig. 17-27a). In this condition the cell will hold information indefinitely.

Under normal conditions information is shifted by pulsing ϕ_1 and ϕ_2, very much as in the two-phase shift register of Fig. 17-46. As long as the clocking frequency is high, the static shift register operates in the same manner as the dynamic shift register, with the feedback loop open (clock ϕ_3 at a high value). When the frequency falls below a certain level, ϕ_3 is generated internally (on the chip). Clock ϕ_3 is identical with ϕ_2, except that it is delayed (slightly) by Δ with respect to ϕ_2. This signal ϕ_3 is used to close the feedback loop. An example of a static shift register is the TI 3101LC, which is a dual 100-bit unit. Each register has independent input and output terminals, common clocks, and power leads (nominal values $V_{SS} = +5$ V, $V_{DD} = 0$ V, and $-V_{GG} = -12$ V), and can operate from dc to 2.5 MHz. The MOS gate inputs are protected with Zener diodes and can be driven directly from DTL/TTL voltage levels, and the register outputs can drive DTL/TTL circuits without the addition of external components. Two external clocks ϕ_1 and ϕ_2 are required for operation. Data are transferred into the register when clock pulse ϕ_1 is at low level, and data are shifted when clock pulse ϕ_1 is returned to high level (typically, $+5$ V) and clock pulse ϕ_2 is pulsed to low level. For long-term storage, clock pulsed ϕ_1 must be held at a high level and clock pulses ϕ_2 and ϕ_3 at a low level.

Four-phase Shift Register[17]   Four-phase shift registers are used for very high density circuits operating at very high speeds. Figure 17-48 shows the basic cell of a four-phase dynamic shift register. When clock ϕ_1 is low, $Q1$ is ON and capacitor C_1 charges essentially to the low value of the clock. Clock pulse ϕ_2 turns ON transistor $Q2$, which is biased by the dc voltage on capacitor

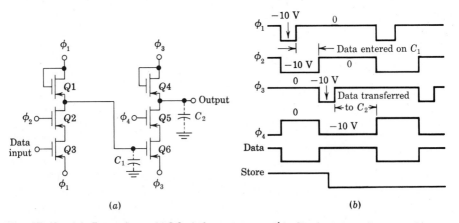

(a)   (b)

Fig. 17-48   (a) Four-phase MOS shift register.   (b) Clock timing diagram (the store input refers to Fig. 17-49).

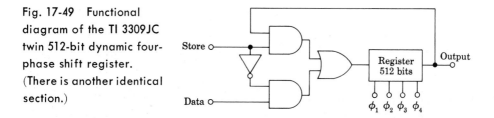

Fig. 17-49 Functional diagram of the TI 3309JC twin 512-bit dynamic four-phase shift register. (There is another identical section.)

C_1. At the presence of an input data signal the output on capacitor C_1 takes a voltage level which corresponds to the complement of the input. The above process is repeated by the slave section $Q4Q5Q6$, so that the information from capacitor C_1 is transferred to C_2 at the output, after clock pulse ϕ_3 and during clock pulse ϕ_4. The reader should verify that even when the clocks overlap, there is no dc current path in this circuit. The only power dissipated arises from the charge and discharge of the various parasitic capacitances. An example of a four-phase dynamic MOS shift register is the TI 3309JC shown in Fig. 17-49. This device consists of two 512-bit registers constructed on a monolithic chip using p-channel enhancement-mode transistors. The chip contains independent control logic for recirculating information for each register and separate clock lines. Operation at repetition rates from 10 kHz to 5 MHz is possible, and power dissipation is less than 90 μW/bit at 1 MHz. The timing diagram for the clocks, the data, and the store (see below) are given in Fig. 17-48. Input data are transferred into the register after the end of clock pulse ϕ_1 and before the end of clock pulse ϕ_2. Output data appear after the end of clock pulse ϕ_3 and before the termination of the ϕ_4 clock pulse. Data stored in the shift register can be recirculated during the time interval T_s if a store pulse of length T_s is applied at the *store* terminal (Fig. 17-49), provided the store pulse overlaps the trailing edge of clock pulse ϕ_3. The TI 3309JC is mounted in a 16-pin hermetically sealed dual-in-line package. It is interesting to note that this monolithic device contains $2 \times 512 \times 6 = 6,144$ MOSFETs, exclusive of the control circuitry.

17-17   MOS READ–ONLY MEMORY[17]

The ROM is discussed in Sec. 17-7, where it is seen (Fig. 17-24) to consist of a decoder, followed by an encoder (memory matrix). Consider, for example, a 10-bit input code, resulting in $2^{10} = 1,024$ word lines, and with 4 bits per output code. The memory matrix for this system consists of $1,024 \times 4$ intersections, as indicated schematically in Fig. 17-50. The code conversion to be performed by the ROM is permanently programmed during the fabrication process by using a custom-designed mask so as to construct or omit an MOS transistor at each matrix intersection. Such an encoder is indicated in Fig. 17-50, which shows how the memory FETs are connected between *word* and *bit* lines.

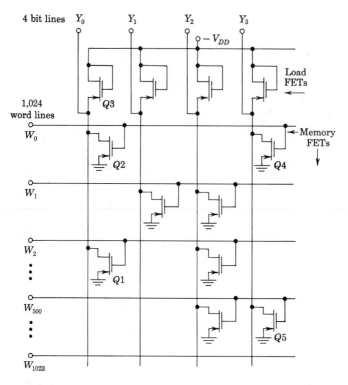

Fig. 17-50 MOS read-only-memory encoder. (Although there are a total of 1,024 word lines present, only 5 of these are indicated.)

In Sec. 17-7 it is demonstrated that the relationship between the output bits Y and the word lines W is satisfied by the logic OR function. Consider, for example, that it is required by the desired code conversion that

$$\bar{Y}_0 = W_0 + W_2 \qquad\qquad \bar{Y}_1 = W_1$$
$$\bar{Y}_2 = W_1 + W_2 + W_{500} \qquad \bar{Y}_3 = W_0 + W_{500} \tag{17-32}$$

These relationships are satisfied by the connections in Fig. 17-50. The NOR gate for Y_0 of Eq. (17-32) is precisely that drawn in Fig. 10-18, with $Q3$ as the load FET and with signals W_0 and W_2 applied to the gates of $Q2$ and $Q1$, respectively.

The presence or absence of a MOS memory cell at a matrix intersection is determined during fabrication in the oxide-gate mask step. If the MOSFET has a normal thin-oxide gate, its threshold voltage V_T is low; if the gate oxide is thick, then V_T is high. In response to a negative pulse on the word line, the low-threshold device will conduct and a logic 0 (because of inverter action) will be detected on the bit line. On the other hand, if a negative pulse is applied

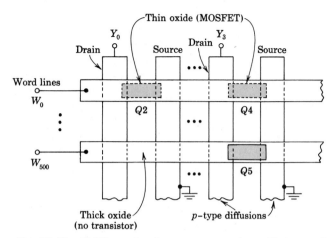

Fig. 17-51   MOS read-only-memory matrix. (Only word lines W_0 and W_{500} and bit lines Y_0 and Y_3 corresponding to Fig. 17-50 are indicated.)

to the thick-oxide gate (high-threshold device), it does not conduct; it is effectively missing from the circuit. In other words, growing a thick-oxide gate at a matrix location is equivalent to *not* constructing an MOSFET at this position, as shown in Fig. 17-51.

In a static ROM no clocks are needed. The time required for a valid output to appear on the bit lines from the time an input address is applied to the memory is defined as the access time ($\sim$300 ns to 5 μs). In a static ROM the output is available as long as the input address remains valid. An example of a static MOS ROM is the TI 2800JC (16-pin dual-in-line package) which consists of 256 words of 4 bits each for a total of 1,024 bits and with maximum access time of 900 ns.

The decoder in a static MOS ROM (Fig. 17-24) contains NAND gates which are static. Power dissipation as a result is relatively high. In the case of the TI 2800JC, power dissipation is typically 170 mW if all outputs are at their most positive voltage (logic 0). A dynamic ROM uses clocked or dynamic inverters in the decoder and requires a minimum clock rate, since otherwise the information is lost. However, its power dissipation is lower than for a static ROM. Most commercial ROMs are static because of the advantages of requiring no clocks and of giving an output which remains valid as long as the input address is applied.

17-18   RANDOM–ACCESS MEMORY (RAM) [17,18]

The random-access memory, abbreviated RAM, is an array of storage cells that memorize information in binary form. In such a memory, as contrasted

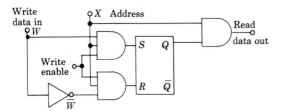

Fig. 17-52 A 1-bit read/write memory.

with an ROM, information can be randomly written into or read out of each storage element as required, and hence the name *random-access*, or *read/write, memory*. The basic monolithic storage cell is the latch, or FLIP-FLOP, discussed in Sec. 17-9.

Linear Selection To understand how the RAM operates we examine the simple 1-bit S-R FLIP-FLOP circuit shown in Fig. 17-52, with data input and output lines. From the figure we see that to read data out of or to write data into the cell, it is necessary to excite the *address line* ($X = 1$). To perform the write operation, the *write enable line* must also be excited. If the write input is a logic 1(0), then $S = 1(0)$ and $R = 0(1)$. Hence $Q = 1(0)$, and the data read out is 1(0), corresponding to that written in.

Suppose that we wish to read/write 16 words of 8 bits each. This system requires eight data inputs and eight data output lines. A total of $16 \times 8 = 128$ storage cells must be used. Of this number, 8 cells are arranged in a horizontal line, all excited by the same address line. There are 16 such lines, each excited by a different address. In other words, addressing is provided by exciting 1 of 16 lines. This type of addressing is called *linear selection* (Prob. 17-51).

Coincident Selection An RAM memory of sixteen 8-bit words has 16 lines with 8 storage cells per line, if linear addressing is used. A more commonly used topology is to arrange 16 memory elements in a rectangular 4×4 array, each cell now storing one bit of one word. Eight such matrix planes are required, one for each of the 8 bits in each word.

One plane of the above-described arrangement of cells is indicated in Fig. 17-53. Each bit (indicated as a shaded rectangle) is located by addressing an X address line and a Y address line; the intersection of the two lines locates a point in the two-dimensional matrix, thus identifying the storage cell under consideration. Such two-dimensional addressing is called X-Y, or *coincident, selection*.

Basic RAM Elements In the 1-bit memory of Fig. 17-52 separate read and write leads are required. For either the bipolar or the MOS RAM it is possible to construct a FLIP-FLOP (as we demonstrate in Figs. 17-56 and 17-57) which has a common terminal for both writing and reading, such as terminals 1 and 2 in Fig. 17-54. This configuration requires the use not only of the write

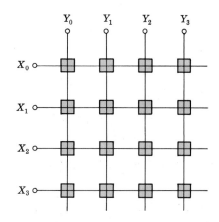

Fig. 17-53 Illustrating coincident addressing to locate 1 bit of a 16-word memory. The shaded squares represent schematically the storage cells, or memory units.

data W (write 1), but also of its complement $\bar{W}$ (write 0). At the cell terminal to which $W(\bar{W})$ is applied, there is obtained the read or sense data output $S(\bar{S})$. Such a memory unit is indicated schematically in Fig. 17-54, where we note that a total of four input/output leads to the storage cell are required, two for X-Y addressing and two for read/write data (true and complemented).

The basic elements of which an RAM is constructed are indicated in Fig. 17-55. These include the rectangular array of storage cells, the X and Y decoders, the write amplifiers to drive the memory, and the sense amplifiers to detect (read) the stored digital information. Some RAMs include a write enable input. For such a unit, the write amplifiers of Fig. 17-55 are two-input AND gates, as in Fig. 17-54. Each word is identified by the matrix number X-Y in the (shaded) memory cell. For M-bit words there will be M planes, like the one indicated in Fig. 17-55.

Magnetic cores have been used as storage elements for many years. Semiconductor memories are now becoming increasingly popular. Monolithic RAMs are constructed using integrated circuit technology and employing either bipolar or MOS transistors for the storage and supporting circuits. Some of the advantages of semiconductor over core memories are low cost, small size,

Fig. 17-54 A basic storage cell can be constructed with complementary inputs and outputs and with the write and sense amplifiers meeting at a common node (1) for true data and (2) for complementary data.

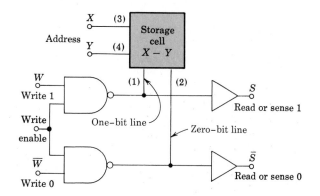

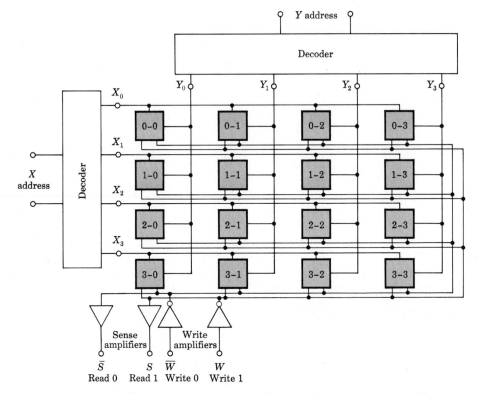

Fig. 17-55 Random-access memory (RAM) with coincident selection and 16 words by 1 bit. The shaded squares give the memory locations of one bit of each word.

and nondestructive reading of the array. On the other hand, the disadvantages include the *volatility of storage*, which means that all stored information is lost when the power supplies fail, and the power dissipation necessary in order to retain the information stored in a FLIP-FLOP.

Bipolar RAM In a bipolar RAM the FLIP-FLOPS are usually TTL high-speed circuits. Figure 17-56 shows the basic bipolar RAM storage cell consisting of two cross-coupled three-emitter transistors. One of the emitters of the right transistor serves to sense or write a logic 1 ($Q2$ conducting). Similarly, one emitter of the left transistor serves to read or write a logic 0 ($Q1$ ON). The two remaining emitters of each transistor are connected to the X and Y lines, respectively, as shown in Figs. 17-56 and 17-55. Address lines are normally held low (logic 0), and currents from all conducting transistors flow out of these address lines.

To address a specific FLIP-FLOP in a matrix array, the corresponding X and Y lines are taken to logic 1. All other FLIP-FLOPS except the one being addressed must have at least one address line at logic 0 and no change will occur in those

FLIP-FLOPS. In the addressed cell the current in the conducting transistor diverts from the address lines (which have just moved from logic 0 to logic 1) to the appropriate sense or read line and then to one of the read amplifiers. Thus, depending on whether a 1 or a 0 bit was stored, the read amplifier associated with read 1 or read 0 will be activated. When this occurs, the output of the activated amplifier indicates a logic 1 level.

To write a 1 or a 0 in a specific FLIP-FLOP it is necessary to address it and apply a logic 1 to the appropriate write amplifier input. The output of this amplifier will then drop to a logic 0 level. A logic 0 voltage on the output of a write amplifier will apply the same low voltage to all the FLIP-FLOP emitters connected to that amplifier. This low voltage will not affect the binaries which are not addressed, because at least one more emitter in those FLIP-FLOPS is held at a low voltage (logic 0) by the address lines. For the binary which is being addressed, there are two possibilities: The FLIP-FLOP may already be in the desired state, and in that case no change occurs. If the binary is not in the desired state and must be changed to the desired state, the low voltage applied to the emitter of the transistor which is OFF will turn it ON, thus causing the other transistor to turn OFF. *In summary:* If $W = 1(0)$, then the write $1(0)$ emitter of the addressed FLIP-FLOP is held at a low voltage, thus causing $Q2$ $(Q1)$ to conduct. This action stores a $1(0)$ in this selected FLIP-FLOP.

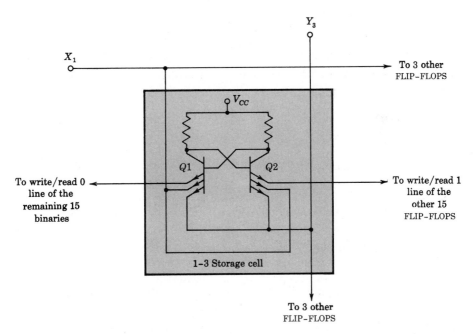

Fig. 17-56 Bipolar RAM storage cell 1–3 showing the X_1 and Y_3 address lines and also the read/write lines for the configuration of Fig. 17-55. The two cross-coupled NOT gates form a 1-bit latch (Fig. 17-27). A logic 1 is stored if $Q2$ conducts.

Since in Fig. 17-54 the output of the write amplifier is connected to the input of the read amplifier, then clearly the sense amplifiers must not be used to supply information on the state of a memory cell while a write amplifier is excited.

An example of a 16-bit bipolar RAM having the pattern of Fig. 17-55 is the TI 7481. Average power dissipation is 275 mW, and reading propagation delay is typically 20 ns. An example of a larger RAM is the IM 5503 (Intersil Memory Corporation) which has a $16 \times 16 = 256$-word by 1-bit organization. It has an access time of 75 ns.

Static MOS RAM[5]   The MOS FLIP-FLOP shown in Fig. 17-57 is used to store the binary information, and clocks are not needed. As long as power is maintained, the data will remain in storage. Devices $Q1$–$Q4$ form a bistable cross-coupled FLIP-FLOP circuit, whereas devices $Q5$–$Q8$ form the gating network through which the interior nodes N1 and N2 are connected to the ONE-bit line and the ZERO-bit line. Note that devices $Q5$ and $Q6$ or $Q7$ and $Q8$ form AND gates to which the X and Y address drives are applied for *coincident selection* of the storage cell. If the *linear-selection* scheme is used, devices $Q6$ and $Q8$ are omitted and the X address line represents the word line.

In the quiescent state both the X and the Y address lines are at ground potential, isolating the storage FLIP-FLOP from the bit lines. Assume that $Q2$ is ON and $Q1$ is OFF, so that node N1 is at $-V_{DD}$ and N2 is at 0 V. To read the cell, both address lines are pulsed (negative for p-channel MOS devices), turning on gating devices $Q5$-$Q8$. Current will flow into the ONE-bit line, which is kept at $-V_{DD}$ through devices $Q7$, $Q8$, and the ON device $Q2$. Very little or essentially no current will flow through the ZERO-bit line, which is also kept at

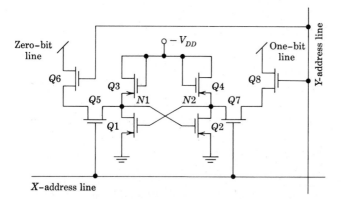

Fig. 17-57 Single MOS RAM cell including address gating. The inverter $Q1$–$Q3$ is cross-coupled to the inverter $Q2$–$Q4$ to form a 1-bit latch. A logic 1 is stored if $Q2$ conducts.

$-V_{DD}$ since device $Q1$ is OFF. The state of the storage cell is thus determined by detecting on which bit line the sense current flows.

To write 1 into the cell, the address lines are again pulsed and the ONE-bit line is grounded. Since the ONE-bit line is pulsed from $-V_{DD}$ to ground, this is interpreted to mean that we desire to write a 1 in the cell. However, the cell is already in the state to which it was to be written, and no change occurs because $Q2$ is already ON. If the ZERO-bit line is pulsed by grounding it, node N1 is pulled toward ground, turning device $Q2$ OFF and device $Q1$ ON through the regenerative process of the FLIP-FLOP. Thus the cell changes state, and we have written the logic 0 into the FLIP-FLOP. We observe that the reading process is nondestructive.

Typical cycle times for fully decoded bit-organized MOS chips generally lie in the range 500 ns to 1 μs. An example of a static MOS RAM is the MK 4002P (Mostek, Inc.). The unit is a 256-bit RAM, organized as 64 words of 4 bits each. It is TTL/DTL compatible because it uses low-threshhold p-channel MOS devices. Address decoding is performed on the chip from a binary 6-bit address which specifies each 4-bit word. The power supplies required are $+5$ and -12 V, and the device is available in a 24-pin dual-in-line package.

Dynamic MOS RAM[17] Instead of using an eight-device cell, it is possible to use a simpler three-device storage cell in which information is stored on the parasitic gate-to-substrate capacitance. Thus, at the expense of requiring a refresh operation to replenish the charge leaking off the storing capacitance, we achieve far greater density of storage cells on the same chip area.

An example of a dynamic MOS/LSI RAM is the 1103 manufactured by Intel Corp. This 1,024-bit memory is fully decoded and is organized as a 1,024-word by 1-bit array. Refreshing of all bits is required every 2 ms, and cycle time is 580 ns. The device is available in an 18-pin dual-in-line package. Power dissipation at room temperature is 400 mW.

IV. D/A AND A/D SYSTEMS

17-19   DIGITAL-TO-ANALOG CONVERTERS[19]

Many systems accept a digital word as an input signal and translate or convert it to an analog voltage or current. These systems are called *digital-to-analog*, or *D/A*, *converters*. The digital word is presented in a variety of codes, the most common being pure binary or binary-coded-decimal (BCD).

The output V_o of an N-bit D/A converter is given by the following equation:

$$V_o = (a_{N-1}2^{-1} + a_{N-2}2^{-2} + a_{N-3}2^{-3} + \cdots + a_0 2^{-N})V_R \qquad (17\text{-}33)$$

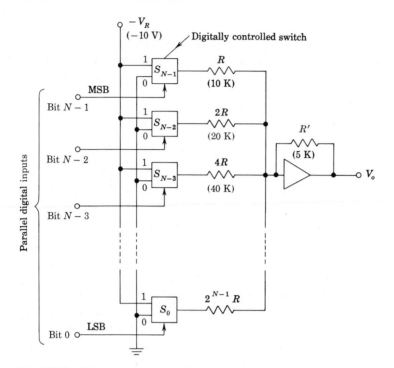

Fig. 17-58 D/A converter with binary weighted resistors.

where the coefficients a_n represent the binary word and $a_n = 1(0)$ if the nth bit is $1(0)$. The voltage V_R is a stable reference voltage used in the circuit. The most significant bit (MSB) is that corresponding to a_{N-1}, and its weight is $V_R/2$, while the least significant bit (LSB) corresponds to a_0, and its weight is $V_R/2^N$.

Consider, for example, a 5-bit word ($N = 5$) so that Eq. (17-33) becomes

$$V_o = (16a_4 + 8a_3 + 4a_2 + 2a_1 + a_0) \times \frac{V_R}{32} \tag{17-34}$$

For simplicity, assume $V_R = 32$ V. Then, if $a_0 = 1$ and all other a's are zero, we have $V_o = 1$. If $a_1 = 1$ and all other a's are zero, we obtain $V_o = 2$. If $a_0 = a_1 = 1$ and all other a's are zero, $V_o = 2 + 1 = 3$ V, etc. Clearly, V_o is an analog voltage proportional to the digital input.

A D/A converter is indicated schematically in Fig. 17-58. The blocks $S_0, S_1, S_2, \ldots, S_{N-1}$ in Fig. 17-58 are electronic switches which are digitally controlled. For example, when a 1 is present on the MSB line, switch S_{N-1} connects the 10-K resistor to the reference voltage $-V_R(-10$ V); conversely, when a 0 is present on the MSB line, the switch connects the resistor to the ground line. Thus the switch is a single-pole double-throw (SPDT) electronic switch. The operational amplifier acts as a current-to-voltage converter

(Sec. 16-1).   Using the numerical values shown in Fig. 17-58, we see that if the MSB is 1 and all other bits are 0, then the current through the 10-K resistor will be 1 mA and the output voltage will be $V_o = 5 = 16 \times \frac{5}{16}$ V.   Similarly, we see that the weight of the LSB (if $N = 5$) becomes $V_o = \frac{10}{160} \times 5 = 1 \times \frac{5}{16}$ V. If all five bits are 1, the output becomes

$$V_o = (1 + \tfrac{1}{2} + \tfrac{1}{4} + \tfrac{1}{8} + \tfrac{1}{16}) \times 5 = 31 \times \tfrac{5}{16}$$

Therefore the analog output V_o is proportional to the digital input; the proportionality factor is $\frac{5}{16}$ for the circuit of Fig. 17-58.

The implementation of the switching devices using p-channel MOS transistors is shown in Fig. 17-59.   The S-R FLIP-FLOP is also implemented with MOSFETs and holds the bit on the corresponding bit line.   Let us assume that logic 1 corresponds to -10 V and logic 0 corresponds to 0 V (negative logic).   A 1 on the bit line sets the FLIP-FLOP at $Q = 1$ and $\bar{Q} = 0$, and thus transistor $Q1$ is ON, connecting the resistor R_1 to the reference voltage $-V_R$, while transistor $Q2$ is kept OFF.   Similarly, a 0 at the input bit line will connect the resistor to the ground terminal.   The accuracy and stability of this D/A converter depend primarily on the absolute accuracy of the resistors and the tracking of each other with temperature.   Since all resistors are different and the largest is $2^{N-1}R$, where R is the smallest resistor, their values become excessively large, and it is very difficult and expensive to obtain stable, precise resistors of such values.

A Ladder-type D/A Converter   A circuit utilizing twice the number of resistors in Fig. 17-58 for the same number of bits (N) but of values R and $2R$

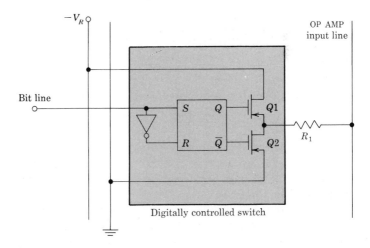

Fig. 17-59 An MOS FLIP-FLOP and a pair of totem-pole MOSFETs implement the single-pole double-throw switch of Fig. 17-58. The resistance R_1 depends upon the bit under consideration. For example for the $N - 3$ bit, $R_1 = 4R$ (Fig. 17-58).

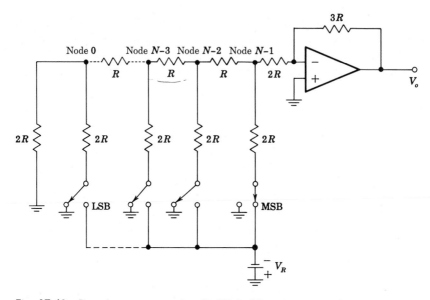

Fig. 17-60 D-to-A converter using R, $2R$ ladder.

only is shown in Fig. 17-60. The ladder used in this circuit is a current-splitting device, and thus the ratio of the resistors is more critical than their absolute value. We observe from the figure that at any of the ladder nodes the resistance is $2R$ looking to the left or the right or toward the switch. Hence the current will split equally toward the left and right, and this happens at every node. Considering node $N - 1$ and assuming the MSB turned on, the voltage at that node will be $-V_R/3$. Since the gain of the operational amplifier to node $N - 1$ is $-3R/2R$, the weight of the MSB becomes

$$V_o = (-V_R/3)(-3R/2R) = V_R/2$$

Similarly, we show in Prob. 17-54 that when the second MSB bit is on and all others are off, the output will be $V_o = +V_R/4$, the third MSB bit gives $+V_R/8$, and the LSB gives $+V_R/2^N$.

The circuits discussed so far use a negative reference voltage and give a positive analog output voltage. If negative binary numbers are to be converted, the sign-magnitude approach is used; an extra bit is added to the binary word to represent the sign, and this bit can be used to select the polarity of the reference voltage.

A typical 8-bit D/A converter by Zeltex Inc. is packaged in a module measuring 1.9 by 1.7 by 0.4 in. and includes the OP AMP, reference voltage, ladder network, and the switches. The 3750 D/A Converter (Fairchild Semiconductor) is a MOS/LSI 10-bit circuit using p-channel enhancement-mode transistors. The digital word can be entered serially or in parallel, and the output is available through 10 SPDT MOS switches. The user must provide the resistive ladder network which is connected to the poles of the 10 switches. The 3750 contains an input shift register in which the data are stored and a holding register which retains the state of the previous 10-bit input word and drives the output switches. The device is available in a 36-pin dual-in-line package.

Multiplying D/A Converter A D/A converter which uses a varying analog signal instead of a fixed reference voltage is called a *multiplying D/A converter*. From Eq. (17-34) we see that the output is the product of the digital word and the analog voltage V_R and its value depends on the binary word (which represents a number smaller than unity). This arrangement is often referred to as a *programmable attenuator* because the output V_o is a fraction of the input V_R and the attenuator setting can be controlled by a computer.

17-20   ANALOG-TO-DIGITAL CONVERTER[19]

It is often required that data taken in a physical system be converted into digital form. Such data would normally appear in electrical analog form. For example, a temperature difference would be represented by the output of a thermocouple, the strain of a mechanical member would be represented by the electrical unbalance of a strain-gauge bridge, etc. The need therefore arises for a device that converts analog information into digital form. A very large number of such devices have been invented. We shall consider below one such A/D converter.

In this system a continuous sequence of equally spaced pulses is passed through a gate. The gate is normally closed, and is opened at the instant of the beginning of a linear ramp. The gate remains open until the linear sweep voltage attains the reference voltage of a comparator, the level of which is set equal to the analog voltage to be converted. The number of pulses in the train that pass through the gate is therefore proportional to the analog voltage. If the analog voltage varies with time, it will of course not be possible to convert the analog data continuously, but it will be required that the analog data be sampled at intervals. The maximum value of the analog voltage will be represented by a number of pulses n. It is clear that n should be made as large as possible consistent with the requirement that the time interval between two successive pulses shall be larger than the timing error of the time modulator. The recurrence frequency of the pulses is equal, at a minimum, to the product of n and the sampling rate. Actually, the recurrence rate will be larger in order to allow time for the circuit to recover between samplings.

One form of digital voltmeter uses the above-described analog-to-digital converter. The number of pulses which pass through the gate is proportional to the voltage being measured. These pulses go to a counter whose reading is

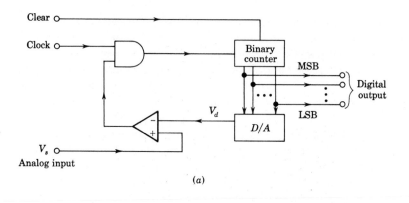

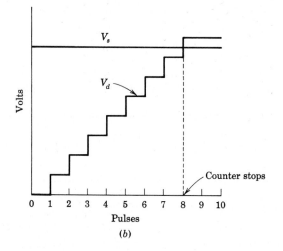

Fig. 17-61 (a) A/D converter using a counter; (b) counter ramp
waveform.

indicated visually by means of some form of luminous display (Secs. 17-8
and 17-21).

The principles discussed previously are used in the A/D converter shown
in Fig. 17-61a. The *clear* pulse resets the counter to the zero count. The
counter then records in binary form the number of pulses from the clock line.
The clock is a source of pulses equally spaced in time. Since the number of
pulses counted increases linearly with time, the binary word representing this
count is used as the input of a D/A converter whose output is shown in Fig.
17-61b. As long as the analog input V_s is greater than V_d, the comparator
output is high and the AND gate is open for the transmission of the clock pulses
to the counter. When V_d exceeds V_s, the comparator output changes to the
low value, and the AND gate is disabled. This stops the counting at the time
when $V_s \approx V_d$ and the counter can be read out as the digital word representing
the analog input voltage.

Successive-approximation A/D Converter[19]   The successive-approximation technique is another method to implement an A/D converter.   Instead
of a binary counter as shown in Fig. 17-61a, a programmer is used.   The
programmer sets the most significant bit (MSB) to 1, with all other bits to 0,
and the comparator compares the D/A output with the analog signal.   If the
D/A output is larger, the 1 is removed from the MSB, and it is tried in the next
most significant bit.   If the analog input is larger, the 1 remains in that bit.
Thus a 1 is tried in each bit of the D/A decoder until, at the end of the process,
the binary equivalent of the analog signal is obtained.   The 3751 MOS/LSI
circuit (Fairchild Semiconductor) is a 12-bit A/D converter monolithic circuit
which makes use of the successive-approximation technique.   The ladder
network must be supplied externally, and by choosing the appropriate coding
of the resistor values in the ladder, the output can be in either binary or BCD
digital form.   The device is available in a 36-pin dual-in-line package.

17-21   CHARACTER GENERATORS[20, 21]

This chapter concludes with a discussion of a two-dimensional alphanumeric
character generation and display system.   Many of the basic building blocks
introduced in the preceding sections are involved in this fairly complicated
system.   For example, included are an ROM, a decoder, a parallel-to-series
shift register, D/A converters, divide-by-N counters, and circulating shift-
register memories.
     The convenient display of information is a most important part of many
electronic systems.   Computers, calculators, business information systems, and
similar systems commonly display information using alphanumeric characters
on a cathode-ray tube (CRT), on seven-segment readouts, on indicator tubes,
on lamp arrays, or as "hard copy" from matrix-type printers, etc.   In many
of these display applications, characters are generated using ROMs and long
shift registers.   MOS read-only memories offer a versatile approach to character generation because the ease of programming allows the designer to choose
the characters and formats he desires.

Dot Matrix of a Character   There are many character formats that can
be designed into ROM character generators.   The 5×7 dot-matrix format is
easy to use and appears in many display systems.   Consider the letter E shown
in the 5×7 dot matrix of Fig. 17-62a.   If each of the dot positions corresponds
to a source of light or a lamp, the solid dots in the letter E are lamps which are
ON, and the open dots are OFF light sources.   The OFF lamps correspond to
logic 0 outputs from an ROM with a $5 \times 7 = 35$-bit output word, while the ON
lamps correspond to logic 1.   To store 64 alphanumeric characters and other
symbols, one ROM requires $64 \times 7 \times 5 = 2,240$ bits of storage.   Implemented with diodes or bipolar transistors, such a memory is bulky and expensive

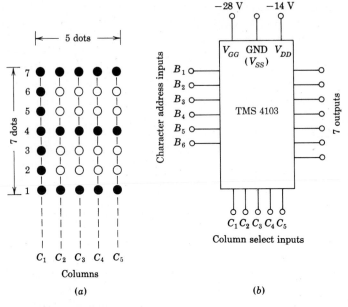

Fig. 17-62 (a) Letter E in 5×7 dot matrix. (b) Pin functions of the TMS 4103 character generator. (Courtesy of Texas Instruments Inc.)

to manufacture. MOS/LSI read-only memories offer significant advantages in cost, power consumption, size, and weight.

Figure 17-62b gives the pin functions of the TMS 4103 character generator ROM (Texas Instruments Inc.). This unit accepts what is known as USASCII input coding as a 6-bit binary word. It provides 64 standard alphanumeric characters in a 5×7 dot array (one column at a time). For example, for the letter E, the address is

$$B_1B_2B_3B_4B_5B_6 = 101000$$

Since there are only seven outputs to the TMS 4103, one column of the character is obtained at any particular time. If the third column is selected for display, the column select input becomes

$$C_1C_2C_3C_4C_5 = 00100$$

and the 7-bit parallel word which appears at the output is

$$\text{Output} = 1001001$$

Total power dissipation of the TMS 4103 is under 400 mW, and it can be accessed in less than 1.0 μs. To enter a logic 1 at a character address or column select input, the input voltage at that pin should be $V_i = -V_{DD} = -14$ V. Conversely, for a logic 0 the input should be at device ground.

Printed Characters The parallel 7-bit-output character generator can be used to obtain printed characters by connecting the seven outputs to a column of seven light-emitting diodes (LEDs, Sec. 3-15). The diodes are arranged over a strip of moving light-sensitive paper or film. The characters are formed by exposing the film to the radiation from the ON LEDs and spacing is obtained by uniformly moving the film or light-sensitive paper as the column select inputs are advanced from column C_1 to column C_5.

CRT Single-character Waveforms It is desired to display one character, say E, on a CRT. Since the screen fluoresces ("lights up") in the small region where the electron beam strikes it, only one dot in the character matrix can be generated at a given instant. Therefore the beam is kept at the position of one dot for a time T, and then stepped to the next dot quickly and held at the new position for another interval T, then moved to the third dot, etc. Hence the system is synchronized by a clock of period T.

At a given horizontal position ($X = 1$ in Fig. 17-63), the beam is swept vertically in synchronism with the clock in seven steps ($Y = 1, 2, \ldots, 7$ in Fig. 17-63). At each position, the intensity control (the grid of the CRT) turns the electron beam ON (unblanks it) if the dot in that position of the character is excited and turns the beam OFF (blanks it) if the spot should be dark. The intensity control, called the Z *axis*, obtains this information from the output of the ROM. Therefore the seven parallel outputs, corresponding to the seven vertical dots in a column, must be serialized. This serial waveform is used to intensity-modulate the Z axis. The bit stream for the first and second columns of the E in Fig. 17-62 are indicated as the Z waveform in

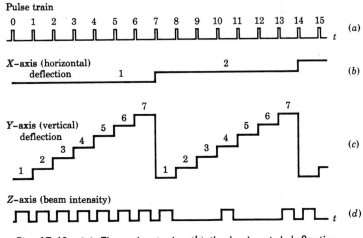

Fig. 17-63 (a) The pulse train; (b) the horizontal-deflection waveform; (c) the vertical-deflection waveform; (d) the unblanking waveform for the character generator system of Fig. 17-64.

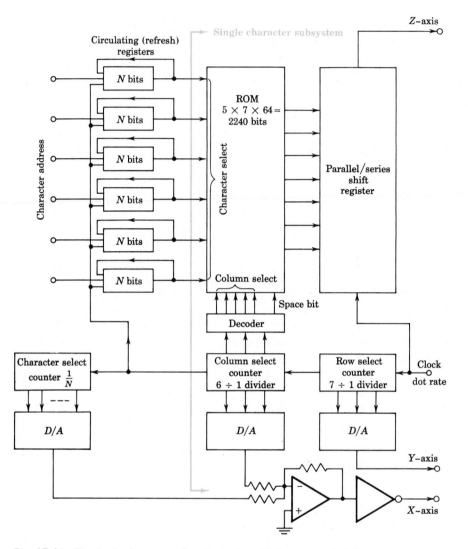

Fig. 17-64 The logic diagram of a single-line N character CRT display system.

Fig. 17-63. The Z-axis waveform is kept low when the beam is being stepped between dot positions by the blanking circuitry.

Display of One Character To understand how the X, Y, and Z wave-forms are generated, refer to the block diagram in Fig. 17-64. Concentrate first on that portion to the right of the shaded vertical line, since this portion of the figure is the subsystem which is required to display one character.

The clock frequency is that at which the output of the parallel-to-series register operates. This clock rate is also called the *dot rate* because the output bit stream corresponds to the dots of the display. Since there are seven dots in a column, a 3-bit counter is used to divide the *dot rate* by 7. The stair-step output waveform from the D/A converter connected to the row-select counter supplies the seven steps for the Y positions of the dots. By counting the rows of the character, the counter gives an output pulse every time the count reaches 7 and an entire column has been scanned. This pulse enters a second 3-bit counter which counts the five columns, and one space column between characters, for a total count of 6. The staircase output voltage from the D/A block connected to the column-select counter provides the six X-axis positions; one for each column. After every sixth X deflection, a new character must be generated.

Display of a Line of Characters Suppose we wish to display N characters in a horizontal line on the CRT. Consider now the blocks to the left of the shaded line in Fig. 17-64. The output pulse of the 6:1 divider is used to introduce the address into the ROM for the next character to be displayed; that is, to advance the input refresh shift registers in which we have stored the addresses of the N characters in the line. The contents of the six input shift registers are recirculated at a rate of 60 Hz so that the entire display can be refreshed at this rate. In the case of 30 characters/line, an entire scan of $7 \times 6 \times 30 = 1,260$ dot positions must be repeated rapidly enough so that flicker of the display is not apparent. A frequency of 60 Hz is considered adequate for all standard CRT phosphors and brightness levels. The output pulse of the column select counter signifies that an entire character (plus space) has been generated. This output pulse enters a third counter which divides by N, where N represents the number of characters per line.

The outputs of each counter feed a D/A converter. The outputs of the character D/A converter and the column D/A unit are summed and drive the horizontal-axis amplifier of the CRT. The D/A connected to the character counter produces N steps, each step having a height equal to six steps of the column select D/A. The D/A converter connected to the row select counter produces a number of steps equal to the number of rows in a character, or seven steps for a 5×7 dot format. Figure 17-63 shows the X-axis, Y-axis, and Z-axis waveforms corresponding to the first two columns of the letter E of Fig. 17-62.

The system described above can be extended so as to display an entire alphanumeric page on the face of a CRT (Prob. 17-56).

REFERENCES

1. "The Integrated Circuits Catalog for Design Engineers," Texas Instruments Inc., 1971.

2. Texas Instruments Staff: "Designing with TTL Integrated Circuits," chap. 9, McGraw-Hill Book Company, New York, 1971.

3. Ref. 2, chap. 8.

4. Rostky, G.: On Semiconductor Memories, *Electron. Design*, vol. 19, pp. 50–63, Sept. 16, 1971.

5. Vimari, D. C.: Field-programmable Read-only Memories and Applications, *Computer Design*, vol. 9, pp. 49–54, December, 1970.

6. Sifferlen, T. P., and V. Vartanian: "Digital Electronics with Engineering Applications," pp. 216–218, Prentice-Hall, Inc., Englewood Cliffs, N.J., 1970.

7. Baasch, T. L.: Selecting Alphanumeric Readouts, *Electron. Prod.*, Dec. 21, 1970, pp. 31–37.

8. McCluskey, E. J., Jr.: "Introduction to the Theory of Switching Circuits," McGraw-Hill Book Company, New York, 1965.
   Wickes, W. E.: "Logic Design with Integrated Circuits," chap. 3, John Wiley & Sons, Inc., New York, 1968.
   Ref. 6, chap. 1.

9. Peatman, J. B.: "The Design of Digital Systems," chap. 3, McGraw-Hill Book Company, New York, 1971.

10. Ref. 2, chap. 7.
    Ref. 6, chap. 3.
    Ref. 9, chap. 4.
    Millman, J., and H. Taub: "Pulse, Digital, and Switching Waveforms," chap. 10, McGraw-Hill Book Company, New York, 1965.

11. Ref. 6, chap. 8.

12. Ref. 2, sec. 11-2.

13. Ref. 2, chap. 10.

14. Ref. 9, chap. 5.
    Wickes, W. E.: Ref. 8, chap. 9.

15. Cobbold, R. S. C.: "Theory and Applications of Field Effect Transistors," Wiley-Interscience, New York, 1970.

16. Crawford, R. H.: "MOSFETs in Circuit Design," McGraw-Hill Book Company, New York, 1967.

17. Terman, L. M.: MOSFET Memory Circuits, *Proc. IEEE*, vol. 59, no. 7, pp. 1044–1057, July, 1971.

18. Talbert, C. D.: Simplify Random-access Memory Selection, *Electronics*, vol. 18, no. 17, pp. 70–74, Aug. 16, 1970.

19. Hoeschele, D. F., Jr.: "Analog-to-digital and Digital-to-analog Conversion Techniques," John Wiley & Sons, Inc., New York, 1968.

20. MOS Character Generators, *Texas Instruments Inc. Appl. Rept.* CA-145.

21. Carter, G., and D. Mrazek: The Systems Approach to Character Generators, *Natl. Semiconductor Appl. Note* AN-40, June, 1970.

REVIEW QUESTIONS

17-1 (*a*) How many input leads are needed for a chip containing quad two-input NOR gates? Explain. (*b*) Repeat part *a* for dual two-wide, two-input AOI gates.

17-2 Define SSI, MSI, and LSI.

17-3 Draw the circuit configuration for an IC TTL AOI gate. Explain its operation.

17-4 (*a*) Find the truth table for the *half adder*. (*b*) Show the implementation for the digit D and the carry C.

17-5 (*a*) Show the system of a three-bit *parallel binary full adder* consisting of half adders. (*b*) Explain its operation.

17-6 Show the system of a 4-bit parallel binary adder, constructed from single-bit full adders.

17-7 (*a*) Draw the truth table for a three-input adder. Explain clearly the meaning of the input and output symbols in the table. (*b*) Write the Boolean expressions for the sum and the carry. (Do not simplify these.)

17-8 (*a*) Show the system for a *serial* binary full adder. (*b*) Explain the operation.

17-9 (*a*) Consider two 4-bit numbers A and B with $B > A$. Verify that to subtract A from B it is only required to add B, $\bar{A}$, and 1. (*b*) Indicate in simple form a 4-bit subtractor obtained from a full adder.

17-10 Consider two 1-bit numbers A and B. What are the logic gates required to test for (*a*) $A = B$, (*b*) $A > B$, and (*c*) $A < B$?

17-11 (*a*) Consider two 4-bit numbers A and B. If $E = 1$ represents the equality $A = B$, write the Boolean expression for E. Explain. (*b*) If $C = 1$ represents the inequality $A > B$, write the Boolean expression for C. Explain.

17-12 Show the system for a 4-bit odd-parity checker.

17-13 (*a*) Show a system for increasing the reliability of transmission of binary information, using a parity check and generator. (*b*) Explain the operation of the system.

17-14 Write the decimal number 749 in the BCD system.

17-15 (*a*) Define a *decoder*. (*b*) Show how to decode the 4-bit code 1011 (LSB).

17-16 (*a*) Define a *demultiplexer*. (*b*) Show how to convert a decoder into a demultiplexer. (*c*) Indicate how to add a strobe to this system.

17-17 (*a*) Define a *multiplexer*. (*b*) Draw a logic block diagram of a 4-to-1-line multiplexer.

17-18 Show how a multiplexer may be used as (*a*) a *parallel-to-serial converter* and (*b*) a *sequential data selector*.

17-19 (*a*) Define an *encoder*. (*b*) Indicate a diode matrix encoder to transform a decimal number into a binary code.

17-20 (*a*) Indicate an encoder matrix using emitter followers. In particular, for an encoder to transform a decimal number into a binary code, show the connections (*b*) to the output Y_1 and (*c*) to the line W_5.

17-21   (*a*) Define a *read-only memory*.   (*b*) Show a block diagram of an ROM. (*c*) What is stored in the memory?   (*d*) What hardware constitutes the memory elements?

17-22   (*a*) Write the truth table for converting from a binary to a Gray code. (*b*) Write the first six lines of the truth table for converting a Gray into a binary code.

17-23   Explain what is meant by *mask-programming* an ROM.

17-24   (*a*) Explain what is meant by a pROM (or ROMP).   (*b*) How is the programming done in the field?

17-25   List three ROM applications and explain these very briefly.

17-26   (*a*) What is a *seven-segment visible display?*   (*b*) Show the following two lines in the conversion table from BCD to seven-segment-indicator code: 0011 and 1001.

17-27   (*a*) Define a *sequential* system.   (*b*) How does it differ from a *combinational* system?

17-28   (*a*) Define a *latch*.   (*b*) Show how to construct this unit from NOT gates. (*c*) Verify that the circuit of part *b* has two stable states.

17-29   (*a*) Sketch the logic system for a latch with set S (preset) and reset R (clear) inputs.   (*b*) Verify that if $S = 1$ and $R = 0$, the FLIP-FLOP is set to $Q = 1$.

17-30   (*a*) Sketch the logic system for a clocked S-R FLIP-FLOP.   (*b*) Verify that the state of the system does not change in between clock pulses.   (*c*) Give the truth table.   (*d*) Justify the entries in the truth table.

17-31   (*a*) Augment an S-R FLIP-FLOP with two AND gates to form a J-K FLIP-FLOP.   (*b*) Give the truth table.   (*c*) Verify part *b* by making a table of J_n, K_n, Q_n, $\bar{Q}_n$, S_n, R_n, and Q_{n+1}.

17-32   Explain what is meant by a *race-around* condition in connection with the J-K FLIP-FLOP of Rev. 17-31.

17-33   Draw a clocked J-K FLIP-FLOP system and include *preset* (Pr) and *clear* (Cr) inputs.   (*b*) Explain the clear operation.

17-34   (*a*) Draw a *master-slave* J-K FLIP-FLOP system.   (*b*) Explain its operation and show that the race-around condition is eliminated.

17-35   (*a*) Show how to convert a J-K FLIP-FLOP into a *delay* (D-type) unit. (*b*) Give the truth table.   (*c*) Verify this table.

17-36   Repeat Rev. 17-35 for a *toggle* (T-type) FLIP-FLOP.

17-37   Give the truth tables for each FLIP-FLOP type: (*a*) S-R, (*b*) J-K, (*c*) D, and (*d*) T.   What are the direct inputs Pr and Cr and the clock Ck for (*e*) presetting, (*f*) clearing, and (*g*) normal clocked operation?

17-38   (*a*) Define a *register*.   (*b*) Construct a shift register from S-R FLIP-FLOPS. (*c*) Explain its operation.

17-39   (*a*) Explain why there may be a race condition in a shift register.   (*b*) How is this difficulty bypassed?

17-40   Explain how a shift register is used as a converter from (*a*) *serial-to-parallel* data and (*b*) *parallel-to-serial* data.

17-41   Explain how a shift register is used as a *sequence generator*.

17-42   Explain how a shift register is used as a *read-only memory*.

17-43   (*a*) Explain how a shift register is used as a *ring counter*.   (*b*) Draw the output waveform from each FLIP-FLOP of a three-stage unit.

17-44   (*a*) Sketch the block diagram for a *Johnson (twisted ring) counter*.   (*b*) Draw the output waveform from each FLIP-FLOP of a three-stage unit.   (*c*) By what number N does this system divide?

17-45 (a) Draw the block diagram of a *ripple counter*. (b) Sketch the waveform at the output of each FLIP-FLOP for a three-stage counter. (c) Explain how this waveform chart is obtained. (d) By what number N does this system divide?

17-46 (a) Draw the block diagram for an *up-down counter*. (b) Explain its operation.

17-47 Explain how to modify a ripple counter so that it divides by N, where N is *not* a power of 2.

17-48 (a) Draw the block diagram of a decade ripple counter. (b) Explain its operation.

17-49 Repeat Rev. 17-48 for a divide-by-6 ripple counter.

17-50 What is the advantage of a *synchronous counter* over a ripple counter?

17-51 (a) Draw the block diagram of a four-stage synchronous counter with *series carry*. (b) Explain its operation. (c) What is the maximum frequency of operation? Define the symbols in your equation.

17-52 (a) Repeat Rev. 17-51 if the counter uses *parallel carry*. (b) What are the advantages and disadvantages of a parallel-carry counter?

17-53 Explain how to measure frequency by means of a counter.

17-54 List six applications of counters. Give no explanations.

17-55 (a) Draw the circuit of a *single-phase dynamic MOS inverter*. (b) Explain its operation.

17-56 Repeat Rev. 17-55 for a MOS NAND gate.

17-57 Repeat Rev. 17-55 for a *two-phase MOS inverter*. Draw the clocking waveforms.

17-58 (a) Draw the circuit of one stage of *a two-phase dynamic MOS shift register*. (b) Explain its operation. Draw the clocking waveforms.

17-59 (a) Draw the circuit of one stage of a *static MOS shift register*. (b) Draw the clocking waveforms. (c) Explain the operation of the circuit.

17-60 Repeat Rev. 17-59 for one stage of a four-phase MOS register. Identify the *master* and *slave* sections.

17-61 (a) Draw the circuit of the *encoder* of a MOS read-only memory. (b) Explain the operation of the circuit.

17-62 Explain how the MOS ROM is *programmed*.

17-63 (a) Draw the block diagram of a 1-bit *read/write memory*. (b) Explain its operation.

17-64 Explain *linear selection* in a *random-access memory* (RAM).

17-65 Repeat Rev. 17-64 for *coincident selection*.

17-66 (a) Draw in block-diagram form the basic elements of an RAM with coincident selection used to store four words of 1 bit each. (b) How is the system expanded to 3 bits/word? (c) How is the system expanded to 25 words of 3 bits/word?

17-67 List the advantages and disadvantages of a MOS RAM.

17-68 (a) Sketch the circuit for a *bipolar* RAM storage cell. (b) Explain its operation.

17-69 Repeat Rev. 17-68 for a MOS RAM memory cell.

17-70 (a) Draw a schematic diagram of a D/A converter. Use resistance values whose ratios are multiples of 2. (b) Explain the operation of the converter.

17-71 Repeat Rev. 17-70 for a ladder network whose resistances have one of two values, R or $2R$.

17-72 Indicate the circuit of the MOS switch in a D/A converter.

17-73   (a) Draw the block diagram for an A/D converter.   (b) Explain the operation of this system.

17-74   (a) Consider a 5 × 7 matrix to represent an alphanumeric character. Draw in block-diagram form a system for generating such characters and for writing them on a CRT.   (b) Indicate the horizontal (X), the vertical (Y), and the intensity (Z) waveforms for a single character.   (c) Explain the operation of the system.

17-75   (a) How is the system in Rev. 17-74 modified so as to write a line of N characters on the CRT?   (b) Calculate the clock frequency if each line of 40 characters is to be refreshed at a 60-Hz rate.

18 / POWER CIRCUITS AND SYSTEMS

This chapter considers power amplifiers, voltage regulators, and SCR power control circuits and systems.

An amplifying system usually consists of several stages in cascade. The input and intermediate stages operate in a small-signal class A mode. Their function is to amplify the small-input excitation to a value large enough to drive the final device. This output stage feeds a transducer such as a cathode-ray tube, a loudspeaker, a servomotor, etc., and hence must be capable of delivering a large voltage or current swing or an appreciable amount of power. In this chapter we study such large-signal amplifiers. Bias stabilization techniques and thermal runaway considerations are very important with power amplifiers. These topics are discussed in Chap. 9, and hence they are not considered here.

Almost all electronic circuits require a dc source of power. For portable low-power systems, batteries may be used. More frequently, however, electronic equipment is energized by a power supply, a circuit which converts the ac waveform of the power lines to direct voltage of constant amplitude. The process of ac-to-dc conversion is examined in Chap. 4. In this chapter we consider the regulator circuits used to control the amplitude of a dc supply voltage. These circuits can be considered as a special class of feedback amplifiers.

Finally, the use of solid-state switches such as p-n-p-n diodes, SCR's, and triacs in power control circuits is also examined.

18-1   CLASS A LARGE–SIGNAL AMPLIFIERS

A simple transistor amplifier that supplies power to a pure resistance load R_L is indicated in Fig. 18-1. Using the notation of Table 8-1, i_C

677

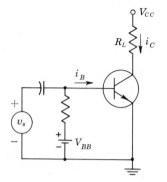

Fig. 18-1 The schematic wiring diagram of a simple series-fed transistor amplifier.

represents the total instantaneous collector current, i_c designates the instantaneous variation from the quiescent value I_C of the collector current. Similarly, i_B, i_b, and I_B represent corresponding base currents. The total instantaneous collector-to-emitter voltage is given by v_C, and the instantaneous variation from the quiescent value V_C is represented by v_c.

Let us assume that the static output characteristics are equidistant for equal increments of input base current i_b, as indicated in Fig. 18-2. Then, if the input signal i_b is a sinusoid, the output current and voltage are also sinusoidal, as shown. Under these circumstances the nonlinear distortion is negligible, and the power output may be found graphically as follows:

$$P = V_c I_c = I_c^2 R_L \tag{18-1}$$

where V_c and I_c are the rms values of the output voltage v_c and current i_c, respectively, and R_L is the load resistance. The numerical values of V_c and I_c can be determined graphically in terms of the maximum and mimimum voltage and current swings, as indicated in Fig. 18-2. If I_m (V_m) represents the peak sinusoidal current (voltage) swing, it is seen that

$$I_c = \frac{I_m}{\sqrt{2}} = \frac{I_{\max} - I_{\min}}{2\sqrt{2}} \tag{18-2}$$

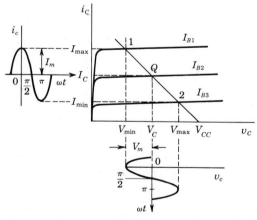

Fig. 18-2 The output characteristics and the current and voltage waveforms for a series-fed load for a transistor amplifier.

and

$$V_c = \frac{V_m}{\sqrt{2}} = \frac{V_{max} - V_{min}}{2\sqrt{2}} \tag{18-3}$$

so that the power becomes

$$P = \frac{V_m I_m}{2} = \frac{I_m{}^2 R_L}{2} = \frac{V_m{}^2}{2R_L} \tag{18-4}$$

which may also be written in the form

$$P = \frac{(V_{max} - V_{min})(I_{max} - I_{min})}{8} \tag{18-5}$$

This equation allows the output power to be calculated very simply. All that is necessary is to plot the load line on the volt-ampere characteristics of the device and to read off the values of V_{max}, V_{min}, I_{max}, and I_{min}.

18-2   SECOND–HARMONIC DISTORTION

In the preceding section the active device is idealized as a perfectly linear device. In general, however, the dynamic transfer characteristic (Sec. 4-1) is not a straight line. This nonlinearity arises because the static output characteristics are not equidistant straight lines for constant increments of input excitation. If the dynamic curve is nonlinear over the operating range, the waveform of the output voltage differs from that of the input signal. Distortion of this type is called *nonlinear*, or *amplitude*, *distortion*.

To investigate the magnitude of this distortion we assume that the dynamic curve with respect to the quiescent point Q can be represented by a parabola rather than a straight line. Thus, instead of relating the alternating output current i_c with the input excitation i_b by the equation $i_c = Gi_b$ resulting from a linear circuit, we assume that the relationship between i_c and i_b is given more accurately by the expression

$$i_c = G_1 i_b + G_2 i_b{}^2 \tag{18-6}$$

where the G's are constants. Actually, these two terms are the beginning of a power-series expansion of i_c as a function of i_b.

If the input waveform is sinusoidal and of the form

$$i_b = I_{bm} \cos \omega t \tag{18-7}$$

the substitution of this expression in Eq. (18-6) leads to

$$i_c = G_1 I_{bm} \cos \omega t + G_2 I_{bm}{}^2 \cos^2 \omega t$$

Since $\cos^2 \omega t = \frac{1}{2} + \frac{1}{2} \cos 2\omega t$, the expression for the instantaneous total current i_C reduces to the form

$$i_C = I_C + i_c = I_C + B_o + B_1 \cos \omega t + B_2 \cos 2\omega t \tag{18-8}$$

where the B's are constants which may be evaluated in terms of the G's. The physical meaning of this equation is evident. It shows that the application of a sinusoidal signal on a parabolic dynamic characteristic results in an output current which contains, in addition to a term of the same frequency as the input, a second-harmonic term, and also a constant current. This constant term B_o adds to the original dc value I_C to yield a total dc component of current $I_C + B_o$. *Parabolic nonlinear distortion introduces into the output a component whose frequency is twice that of the sinusoidal input excitation.* Also, *since a sinusoidal input signal changes the average value of the output current, rectification takes place.*

The amplitudes B_o, B_1, and B_2 for a given load resistor are readily determined from either the static or the dynamic characteristics. We observe from Fig. 18-2 that

$$\text{When } \omega t = 0: \quad i_C = I_{\max}$$

$$\text{When } \omega t = \frac{\pi}{2}: \quad i_C = I_C \tag{18-9}$$

$$\text{When } \omega t = \pi: \quad i_C = I_{\min}$$

By substituting these values in Eq. (18-8), there results

$$
\begin{aligned}
I_{\max} &= I_C + B_o + B_1 + B_2 \\
I_C &= I_C + B_o - B_2 \\
I_{\min} &= I_C + B_o - B_1 + B_2
\end{aligned}
\tag{18-10}
$$

This set of three equations determines the three unknowns B_o, B_1, and B_2. It follows from the second of this group that

$$B_o = B_2 \tag{18-11}$$

By subtracting the third equation from the first, there results

$$B_1 = \frac{I_{\max} - I_{\min}}{2} \tag{18-12}$$

With this value of B_1, the value for B_2 may be evaluated from either the first or the last of Eqs. (18-10) as

$$B_2 = B_o = \frac{I_{\max} + I_{\min} - 2I_C}{4} \tag{18-13}$$

The second-harmonic distortion D_2 is defined as

$$D_2 \equiv \frac{|B_2|}{|B_1|} \tag{18-14}$$

(To find the percent second-harmonic distortion, D_2 is multiplied by 100.) The quantities $I_{\max}$, $I_{\min}$, and I_C appearing in these equations are obtained directly from the characteristic curves of the transistor and from the load line.

If the dynamic characteristic is given by the parabolic form (18-6) and if the input contains two frequencies ω_1 and ω_2, then the output will consist of a dc term and sinusoidal components of frequencies ω_1, ω_2, $2\omega_1$, $2\omega_2$, $\omega_1 + \omega_2$, and $\omega_1 - \omega_2$ (Prob. 18-1). The sum and difference frequencies are called *intermodulation*, or *combination*, frequencies.

18-3 HIGHER-ORDER HARMONIC GENERATION

The analysis of the preceding section assumes a parabolic dynamic characteristic. This approximation is usually valid for amplifiers where the swing is small. For a power amplifier with a large input swing, however, it is necessary to express the dynamic transfer curve with respect to the Q point by a power series of the form

$$i_c = G_1 i_b + G_2 i_b{}^2 + G_3 i_b{}^3 + G_4 i_b{}^4 + \cdots \tag{18-15}$$

If we assume that the input wave is a simple cosine function of time, of the form in Eq. (18-7), the output current will be given by

$$i_C = I_C + B_o + B_1 \cos \omega t + B_2 \cos 2\omega t + B_3 \cos 3\omega t + \cdots \tag{18-16}$$

This equation results when Eq. (18-7) is inserted in Eq. (18-15) and the proper trigonometric transformations are made.

  That the output-current waveform must be expressible by a relationship of this form is made evident from an inspection of Fig. 18-2. It is observed from this figure that the output-current curve must possess *zero-axis symmetry*, or that the current is an *even* function of time. Expressed mathematically, $i(\omega t) = i(-\omega t)$. Physically, it means that the waveshape for every quarter cycle of the output-current curve as the operating point moves from point Q to point 1 is similar to the shape of the curve that is obtained as the operating point moves back from point 1 to point Q. Similarly, the waveshape of the current generated by the operating point as it moves from point Q to point 2 is symmetrical with that generated as it moves from point 2 back to point Q. These conditions are true regardless of the curvature of the characteristics. Since i_C is an even function of time, the Fourier series in Eq. (18-16), representing a periodic function possessing this symmetry, contains only cosine terms. (If any sine terms were present, they would destroy the symmetry since they are *odd*, and not *even*, functions of time.)

  If we assume, as is frequently done in the literature, that the excitation is a sine instead of a cosine function of time, the resulting output current is no longer expressed by a series of cosine terms only. Though a sine function differs from a cosine function in the shift of the time axis by an amount $\omega t = \pi/2$, nevertheless such a shift destroys the above-noted zero-axis symmetry. It is found in this case that the Fourier series representing the output current contains odd sine components and even cosine components.

Calculation of Fourier Components Any one of a number of methods[1] may be used to obtain the coefficients B_o, B_1, B_2, etc. The method due to Espley, which is simply an extension of the procedure of the last section, is described here. It is assumed in the foregoing section that only three terms, B_o, B_1, and B_2, of the Fourier series are different from zero. These three components are evaluated in terms of the three measured currents, I_{max}, I_{min}, and I_C. As the next approximation, it is assumed that only five terms, B_o, B_1, B_2, B_3, and B_4, exist in the resulting Fourier series. To evaluate these five coefficients, the values of the currents at five different values of i_b are needed. These are chosen at equal intervals in input swing. Thus I_{max}, $I_{\frac{1}{2}}$, I_C, $I_{-\frac{1}{2}}$, and I_{min} correspond, respectively, to the following values of i_b: the maximum positive value, one-half the maximum positive value, zero, one-half the maximum negative value, and the maximum negative value. These values are illustrated in Fig. 18-3, which is consistent with Fig. 18-5 for $R'_L = 10\ \Omega$.

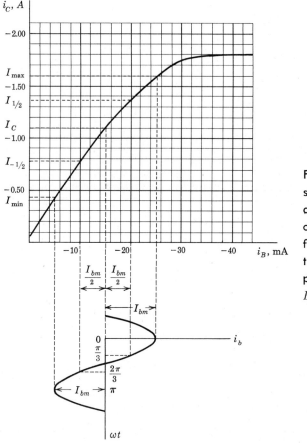

Fig. 18-3 The values of signal excitation and the corresponding values of output current used in the five-point schedule for determining the Fourier components B_0, B_1, B_2, B_3, and B_4 of the current.

Assuming an input signal of the form $i_b = I_{bm} \cos \omega t$ as illustrated, then

When $\omega t = 0$: $i_C = I_{\max}$

When $\omega t = \dfrac{\pi}{3}$: $i_C = I_{\frac{1}{2}}$

When $\omega t = \dfrac{\pi}{2}$: $i_C = I_C$ (18-17)

When $\omega t = \dfrac{2\pi}{3}$: $i_C = I_{-\frac{1}{2}}$

When $\omega t = \pi$: $i_C = I_{\min}$

By combining these conditions with Eq. (18-16), five equations containing five unknowns are obtained.   The solution of these equations yields

$$B_o = \tfrac{1}{6}(I_{\max} + 2I_{\frac{1}{2}} + 2I_{-\frac{1}{2}} + I_{\min}) - I_C$$
$$B_1 = \tfrac{1}{3}(I_{\max} + I_{\frac{1}{2}} - I_{-\frac{1}{2}} - I_{\min})$$
$$B_2 = \tfrac{1}{4}(I_{\max} - 2I_C + I_{\min}) \qquad\qquad (18\text{-}18)$$
$$B_3 = \tfrac{1}{6}(I_{\max} - 2I_{\frac{1}{2}} + 2I_{-\frac{1}{2}} - I_{\min})$$
$$B_4 = \tfrac{1}{12}(I_{\max} - 4I_{\frac{1}{2}} + 6I_C - 4I_{-\frac{1}{2}} + I_{\min})$$

The harmonic distortion is defined as

$$D_2 \equiv \frac{|B_2|}{|B_1|} \qquad D_3 \equiv \frac{|B_3|}{|B_1|} \qquad D_4 \equiv \frac{|B_4|}{|B_1|} \qquad (18\text{-}19)$$

where D_s ($s = 2, 3, 4, \ldots$) represents the distortion of the sth harmonic.

Power Output   If the distortion is not negligible, the power delivered at the fundamental frequency is

$$P_1 = \frac{B_1{}^2 R_L}{2} \qquad\qquad (18\text{-}20)$$

However, the total power output is

$$P = (B_1{}^2 + B_2{}^2 + B_3{}^2 + \cdots) \frac{R_L}{2} = (1 + D_2{}^2 + D_3{}^2 + \cdots)P_1$$

or

$$P = (1 + D^2)P_1 \qquad\qquad (18\text{-}21)$$

where *the total distortion*, or *distortion factor*, is defined as

$$D \equiv \sqrt{D_2{}^2 + D_3{}^2 + D_4{}^2 + \cdots} \qquad\qquad (18\text{-}22)$$

If the total distortion is 10 percent of the fundamental, then

$$P = [1 + (0.1)^2]P_1 = 1.01P_1$$

The total power output is only 1 percent higher than the fundamental power when the distortion is 10 percent. Hence little error is made in using only the fundamental term P_1 in calculating the power output. Considerable error may be made, however, if Eq. (18-5), rather than Eq. (18-20), is used to calculate the power. The former is based on the assumption that the fundamental component B_1 may be calculated from Eq. (18-12) rather than from the more accurate formula (18-18).

In passing, it should be noted that the total harmonic distortion is not necessarily indicative of the discomfort to someone listening to music. Usually, the same amount of distortion is more irritating, the higher the order of the harmonic frequency.

18-4   THE TRANSFORMER–COUPLED AUDIO POWER AMPLIFIER

If the load resistance is connected directly in the output circuit of the power stage, as shown in Fig. 18-1, the quiescent current passes through this resistance. This current represents a considerable waste of power, since it does not contribute to the ac (signal) component of power. Furthermore, it is generally inadvisable to pass the dc component of current through the output device, for example, the voice coil of a loudspeaker. For these reasons an arrangement using an output transformer is usually employed, as in Fig. 18-4. Although the input circuit also contains a transformer, it is possible to feed the excitation to the power stage through an RC coupling.

Impedance Matching   To transfer a significant amount of power to a load such as a loudspeaker with a voice-coil impedance of 5 to 15 Ω, it is necessary to use an output matching transformer. This follows from the fact that

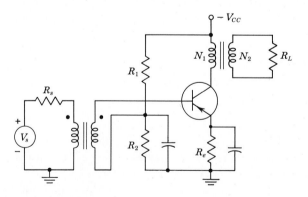

Fig. 18-4 A transformer-coupled transistor output stage.

the internal device resistance may be very much higher than that of the speaker, and so most of the power generated would be lost in the active device.

The impedance-matching properties of an ideal transformer follow from the simple transformer relations

$$V_1 = \frac{N_1}{N_2} V_2 \quad \text{and} \quad I_1 = \frac{N_2}{N_1} I_2 \tag{18-23}$$

where $V_1 (V_2)$ = primary (secondary) voltage
 $I_1 (I_2)$ = primary (secondary) current
 $N_1 (N_2)$ = number of primary (secondary) turns
When $N_2 < N_1$, these equations show that the transformer reduces the voltage in proportion to the turns ratio $n = N_2/N_1$ and steps the current up in the same ratio. The ratio of these equations yields

$$\frac{V_1}{I_1} = \frac{1}{n^2} \frac{V_2}{I_2}$$

Since, however, V_1/I_1 represents the effective input resistance R'_L, whereas V_2/I_2 is the output resistance R_L, then

$$R'_L = \frac{1}{n^2} R_L \tag{18-24}$$

Maximum Power Output A practical problem is to find the transformer turns ratio n (for a given value of R_L) in order that the power output be a maximum for a small allowable distortion. This problem is solved graphically as follows: First the quiescent operating point Q is located, taking into consideration the bounds discussed in Sec. 9-1 and indicated in Fig. 9-2. The quiescent current is $I_C = P_C/V_C$, where P_C is the value of collector dissipation specified by the manufacturer, and V_C is a value of quiescent collector voltage which locates Q somewhere near the center of the V_{CE} scale. The choice of V_C is somewhat arbitrary, but is subject to the restriction that V_{CE} must be less than $V_{C,\text{max}}$ even if the transistor is driven to cutoff. For the transistor whose characteristics are plotted in Fig. 18-5, the manufacturer specifies $P_C = 10$ W and $V_{CE,\text{max}} = 30$ V. A reasonable quiescent point Q is determined to be $V_C = -7.5$ V and $I_C = -1.1$ A. A static load line passing through this Q point with a slope corresponding to the small transformer dc primary resistance plus the small value of R_e is shown in Fig. 18-5a. The intersection of this line with the voltage axis gives the required power-supply voltage V_{CC}.

The base current at the Q point is seen to be -15 mA. If we were to drive the transistor too close to cutoff, an unacceptable amount of distortion would result. Hence the peak-to-peak voltage swing v_{be} is limited to 140 mV. We are here assuming that the input transformer in Fig. 18-4 represents voltage drive for the power transistor and that the source resistance R_s reflected into the secondary circuit of the input transformer is negligible. From the input

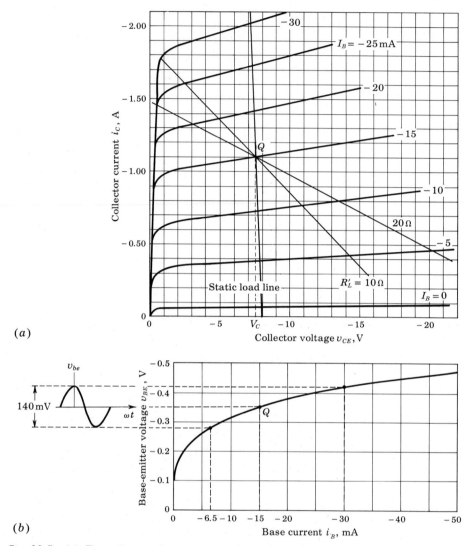

Fig. 18-5 (a) The collector characteristics for a power transistor. A static load line for a transformer-coupled load is indicated. Also shown are load lines for dynamic resistances of 10 and 20 Ω. (b) The input characteristic.

characteristic of Fig. 18-5b we see that the corresponding base current extremes are $I_{b,\max} = -30$ mA and $I_{b,\min} = -6.5$ mA. Note that the input current swing is not symmetric with respect to the quiescent point $I_B = -15$ mA. In Sec. 8-1 we show that the nonsymmetric base current swing compensates for the nonsymmetric collector voltage swing, and thus we have less distortion with voltage drive than with current drive. If the effect of R_s is not negligible, the input characteristic of Fig. 18-5b must be modified by constructing the dynamic

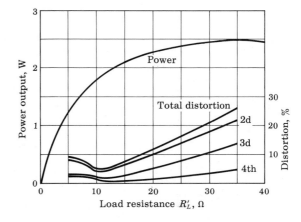

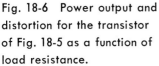

Fig. 18-6 Power output and distortion for the transistor of Fig. 18-5 as a function of load resistance.

input characteristic corresponding to the given R_s, as discussed in Sec. 4-1 and Fig. 4-2.

A series of load lines are drawn through Q for different values of R'_L. The two indicated in Fig. 18-5a correspond to $R'_L = 10$ and 20Ω. For each such load line the dynamic transfer characteristic of Fig. 18-3 is constructed using Fig. 18-5a and b, and the output power and distortion are calculated using the formulas in Sec. 18-3. For example, we see from Fig. 18-5b that when the input excitation voltage is at its maximum, the base current is $I_B = -30$ mA, and from Fig. 18-5a and the $R'_L = 20 \Omega$ load line, the maximum collector current is $I_{C,\max} = -1.45$ A. Similarly, we obtain the value $I_{C\frac{1}{2}} = -1.35$ A by noting from Fig. 18-5b that $I_b = -21$ mA when the input excitation voltage is at half its positive swing, or 35 mV above the Q point. The intersection of the load line $R'_L = 20 \Omega$ with the $I_b = -21$ mA base current line in Fig. 18-5a results in $I_{C\frac{1}{2}} = -1.35$ A. The results of such calculations are plotted in Fig. 18-6.

For R'_L very small, the voltage swing, and hence the power output P, approach zero. For R'_L very large, the current swing is small, and again P approaches zero. Therefore, in Fig. 18-6 the plot of P versus R'_L has a maximum. Note also that this maximum is quite broad. By choosing $R'_L = 15 \Omega$, a total distortion of less than 10 percent is obtained with a power output of 2.1 W, a value which is only 20 percent less than 2.5 W, the peak power possible.

18-5   EFFICIENCY

The various components of power in an amplifier circuit are now examined. Suppose that the stage is supplying power to a pure resistance load. The average power input from the dc supply is $V_{CC}I_C$. The power absorbed by the output circuit is $I_c{}^2 R_1 + I_c V_c$, where I_c and V_c are the rms output current and voltage, respectively, and where R_1 is the *static* load resistance. If P_D

denotes the average power dissipated by the active device, then, in accordance with the principle of the conservation of energy,

$$V_{CC}I_C = I_C^2 R_1 + I_c V_c + P_D \tag{18-25}$$

Since, however,

$$V_{CC} = V_C + I_C R_1$$

P_D may be written in the form

$$P_D = V_C I_C - V_c I_c \tag{18-26}$$

If the load is not a pure resistance, $V_c I_c$ must be replaced by $V_c I_c \cos \theta$, where $\cos \theta$ is the power factor of the load.

Equation (18-26) expresses the amount of power that must be dissipated by the active device. It represents the kinetic energy of the electrons which is converted into heat upon bombardment of the collector by these electrons. If the ac power output is zero, i.e., if no applied signal exists, then P_D has its maximum value of $V_C I_C$. Otherwise, the heating of the device is reduced by the amount of the ac power converted by the stage and supplied to the load. Hence a device is cooler when delivering power to a load than when there is no such ac power transfer. Obviously, then, the maximum dissipation is determined by the zero-excitation value.

Conversion Efficiency[2] A measure of the ability of an active device to convert the dc power of the supply into the ac (signal) power delivered to the load is called the *conversion efficiency*, or *theoretical efficiency*. This figure of merit, designated η, is also called the *collector-circuit efficiency* for a transistor amplifier. By definition, the percentage efficiency is

$$\eta \equiv \frac{\text{signal power delivered to load}}{\text{dc power supplied to output circuit}} \times 100\% \tag{18-27}$$

In general,

$$\eta = \frac{\frac{1}{2}B_1^2 R_L'}{V_{CC}(I_C + B_o)} \times 100\% \tag{18-28}$$

If the distortion components are negligible, then

$$\eta = \frac{\frac{1}{2}V_m I_m}{V_{CC}I_C} \times 100\% = 50 \frac{V_m I_m}{V_{CC}I_C} \% \tag{18-29}$$

The collector-circuit efficiency differs from the overall efficiency because the power taken by the base is not included in the denominator of Eq. (18-28).

Maximum Value of Efficiency It is possible to obtain an approximate expression for η if certain idealizations are made in the characteristic curves. These assumptions, of course, introduce errors in the analysis. However, the results permit a rapid estimate to be made of the numerical value of η and, in

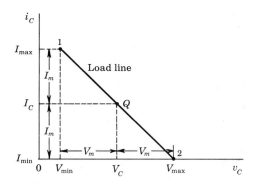

Fig. 18-7 Pertaining to the calculation of the conversion efficiency of
an ideal distortionless amplifier.

particular, furnish an upper limit for this figure of merit. It is assumed that
the static curves are equally spaced in the region of the load line for equal
increments in excitation (base current). Thus, in Fig. 18-7, the distance from
1 to Q is the same as that from Q to 2. It is also assumed that the excitation
is such as to give zero minimum current. The construction in Fig. 18-7 may
be used to analyze either a series-fed or a transformer-fed load. The only
difference between the two circuits is that the supply voltage V_{CC} equals V_{max}
in the series-fed case, whereas V_{CC} is equal to the quiescent voltage V_C (on the
assumption that the static dc drop is negligible) in the transformer-coupled
amplifier. The reader should compare Fig. 18-7 with Figs. 18-2 and 18-5.

Under the foregoing idealized conditions,

$$I_C = I_m \quad \text{and} \quad V_m = \frac{V_{max} - V_{min}}{2}$$

so that Eq. (18-29) becomes

$$\eta = \frac{25(V_{max} - V_{min})}{V_{CC}}\% \tag{18-30}$$

The type of coupling used must now be taken into account. For the series-fed
load, $V_{CC} = V_{max}$, and

$$\eta = \frac{25(V_{max} - V_{min})}{V_{max}}\% \tag{18-31}$$

This result indicates that the upper limit of the conversion efficiency is 25 percent, and even this low value is approached only if V_{min} is negligible compared
with V_{max}.

If the load is coupled to the stage through a transformer, then

$$V_{CC} = V_C = \frac{V_{max} + V_{min}}{2}$$

and Eq. (18-30) reduces to

$$\eta = 50 \frac{V_{max} - V_{min}}{V_{max} + V_{min}}\% \tag{18-32}$$

This result shows that the upper limit of the theoretical efficiency for a transformer-coupled power amplifier is 50 percent, or twice that of the series-fed circuit. For a transistor amplifier $V_{\min}$ occurs near the saturation region, and hence $V_{\min} \ll V_{\max}$, and the collector-circuit efficiency may approach the upper limit of 50 percent.

The numerical value of the conversion efficiency must be calculated from Eq. (18-28). The use of Eqs. (18-31) and (18-32) may lead to large errors in η since these equations are derived using the highly idealized conditions indicated in Fig. 18-7.

18-6   PUSH–PULL AMPLIFIERS[3]

A great deal of the distortion introduced by the nonlinearity of the dynamic transfer characteristic may be eliminated by the circuit shown in Fig. 18-8, known as a *push-pull configuration*. In the circuit the excitation is introduced through a center-tapped transformer. Thus, when the signal on transistor $Q1$ is positive, the signal on $Q2$ is negative by an equal amount. Any other circuit that provides two equal voltages which differ in phase by 180° may be used in place of the input transformer.

Consider an input signal (base current) of the form $i_{b1} = I_{bm} \cos \omega t$ applied to $Q1$. The output current of this transistor is given by Eq. (18-16) and is repeated here for convenience:

$$i_1 = I_C + B_o + B_1 \cos \omega t + B_2 \cos 2\omega t + B_3 \cos 3\omega t + \cdots \quad (18\text{-}33)$$

The corresponding input signal to $Q2$ is

$$i_{b2} = -i_{b1} = I_{bm} \cos (\omega t + \pi)$$

The output current of this transistor is obtained by replacing ωt by $\omega t + \pi$ in the expression for i_1. That is,

$$i_2(\omega t) = i_1(\omega t + \pi) \quad (18\text{-}34)$$

whence

$$i_2 = I_C + B_o + B_1 \cos (\omega t + \pi) + B_2 \cos 2(\omega t + \pi) + \cdots$$

which is

$$i_2 = I_C + B_o - B_1 \cos \omega t + B_2 \cos 2\omega t - B_3 \cos 3\omega t + \cdots \quad (18\text{-}35)$$

As illustrated in Fig. 18-8, the current i_1 and i_2 are in opposite directions through the output-transformer primary windings. The total output current is then proportional to the difference between the collector currents in the two transistors. That is,

$$i = k(i_1 - i_2) = 2k(B_1 \cos \omega t + B_3 \cos 3\omega t + \cdots) \quad (18\text{-}36)$$

This expression shows that a push-pull circuit will balance out all even harmonics in the output and will leave the third-harmonic term as the principal

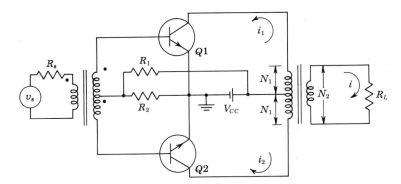

Fig. 18-8 Two transistors in a push-pull arrangement.

source of distortion.   This conclusion was reached on the assumption that the two transistors are identical.   If their characteristics differ appreciably, the appearance of even harmonics must be expected.

The fact that the output current contains no even-harmonic terms means that the push-pull system possesses "half-wave," or "mirror," symmetry, in addition to the zero-axis symmetry.   Half-wave symmetry requires that the bottom loop of the wave, when shifted 180° along the axis, will be the mirror image of the top loop.   The condition of mirror symmetry is represented mathematically by the relation

$$i(\omega t) = -i(\omega t + \pi) \tag{18-37}$$

If $\omega t + \pi$ is substituted for ωt in Eq. (18-36), it will be seen that Eq. (18-37) is satisfied.

Advantages of a Push-Pull System   Because no even harmonics are present in the output of a push-pull amplifier, such a circuit will give more output per active device for a given amount of distortion.   For the same reason, a push-pull arrangement may be used to obtain less distortion for a given power output per transistor.

Another feature of the push-pull system is evident from an inspection of Fig. 18-8.   It is noticed that the dc components of the collector current oppose each other magnetically in the transformer core.   This eliminates any tendency toward core saturation and consequent nonlinear distortion that might arise from the curvature of the transformer magnetization curve. Another advantage of this system is that the effects of ripple voltages that may be contained in the power supply because of inadequate filtering will be balanced out.   This cancellation results because the currents produced by this ripple voltage are in opposite directions in the transformer winding, and so will not appear in the load.   Of course, the power-supply hum will also act on the voltage-amplifier stages, and so will be part of the input to the power stage.   This hum will not be eliminated by the push-pull circuit.

18-7   CLASS B AMPLIFIERS[3]

The circuit for the class B push-pull system is the same as that for the class A system except that the devices are biased approximately at cutoff. The transistor circuit of Fig. 18-8 operates class B if $R_2 = 0$ because a silicon transistor is essentially at cutoff if the base is shorted to the emitter (Sec. 5-9). The advantages of class B as compared with class A operation are the following: It is possible to obtain greater power output, the efficiency is higher, and there is negligible power loss at no signal. For these reasons, in systems where the power supply is limited, such as those operating from solar cells or a battery, the output power is usually delivered through a push-pull class B transistor circuit. The disadvantages are that the harmonic distortion is higher, self-bias cannot be used, and the supply voltages must have good regulation.

Power Considerations To investigate the conversion efficiency of the system, it is assumed, as in Sec. 18-5, that the output characteristics are equally spaced for equal intervals of excitation, so that the dynamic transfer curve is a straight line. It is also assumed that the minimum current is zero. The graphical construction from which to determine the output current and voltage waveshapes for a single transistor operating as a class B stage is indicated in Fig. 18-9. Note that for a sinusoidal excitation the output is sinusoidal during one-half of each period and is zero during the second half cycle. The effective

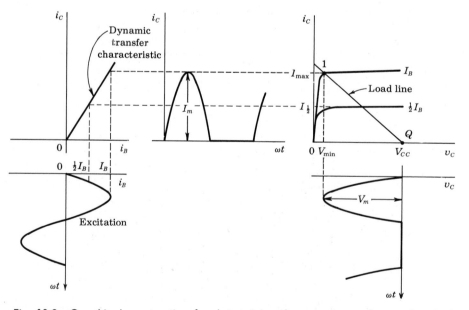

Fig. 18-9 Graphical construction for determining the output waveforms of a single class B transistor stage.

load resistance is $R'_L = (N_1/N_2)^2 R_L$. This expression for R'_L is the same as that in Eq. (18-24), where now N_1 represents the number of primary turns to the center tap (Fig. 18-8).

The waveforms illustrated in Fig. 18-9 represent one transistor $Q1$ only. The output of $Q2$ is, of course, a series of sine loop pulses that are 180° out of phase with those of $Q1$. The load current, which is proportional to the difference between the two collector currents, is therefore a perfect sine wave for the ideal conditions assumed. The power output is

$$P = \frac{I_m V_m}{2} = \frac{I_m}{2} (V_{CC} - V_{min}) \tag{18-38}$$

The corresponding direct collector current in each transistor under load is the average value of the half sine loop of Fig. 18-9. Since $I_{dc} = I_m/\pi$ for this waveform, the dc input power from the supply is

$$P_i = 2 \frac{I_m V_{CC}}{\pi} \tag{18-39}$$

The factor 2 in this expression arises because two transistors are used in the push-pull system.

Taking the ratio of Eqs. (18-38) and (18-39), we obtain for the collector-circuit efficiency

$$\eta \equiv \frac{P}{P_i} \times 100 = \frac{\pi}{4} \frac{V_m}{V_{CC}} = \frac{\pi}{4} \left(1 - \frac{V_{min}}{V_{CC}} \right) \times 100\% \tag{18-40}$$

This expression shows that the maximum possible conversion efficiency is $25\pi = 78.5$ percent for a class B system compared with 50 percent for class A operation. For a transistor circuit where $V_{min} \ll V_{CC}$, it is possible to approach this upper limit of efficiency. This large value of η results from the fact that there is no current in a class B system if there is no excitation, whereas there is a drain from the power supply in a class A system even at zero signal. We also note that in a class B amplifier the dissipation at the collectors is zero in the quiescent state and increases with excitation, whereas the heating of the collectors of a class A system is a maximum at zero input and decreases as the signal increases. Since the direct current increases with signal in a class B amplifier, the power supply must have good regulation.

The collector dissipation P_C (in both transistors) is the difference between the power input to the collector circuit and the power delivered to the load. Since $I_m = V_m/R'_L$,

$$P_C = P_i - P = \frac{2}{\pi} \frac{V_{CC} V_m}{R'_L} - \frac{V_m^2}{2R'_L} \tag{18-41}$$

This equation shows that the collector dissipation is zero at no signal ($V_m = 0$), rises as V_m increases, and passes through a maximum at $V_m = 2V_{CC}/\pi$ (Prob.

18-15). The peak dissipation is found to be

$$P_{C,\text{max}} = \frac{2V_{CC}{}^2}{\pi^2 R_L'} \tag{18-42}$$

The maximum power which can be delivered is obtained for $V_m = V_{CC}$ (if $V_{\text{min}} = 0$), or

$$P_{\text{max}} = \frac{V_{CC}{}^2}{2R_L'} \tag{18-43}$$

Hence

$$P_{C,\text{max}} = \frac{4}{\pi^2} P_{\text{max}} \approx 0.4 P_{\text{max}} \tag{18-44}$$

If, for example, we wish to deliver 10 W from a class B push-pull amplifier, then $P_{C,\text{max}} = 4$ W, or we must select transistors which have collector dissipations of approximately 2 W each. In other words, we can obtain a push-pull output of five times the specified power dissipation of a single transistor. On the other hand, if we paralleled two transistors and operated them class A to obtain 10 W out, the collector dissipation of each transistor would have to be at least 10 W (assuming 50 percent efficiency). This statement follows from the fact that $P_i = P_o/\eta = 10/0.5 = 20$ W. This input power must all be dissipated in the two collectors at no signal, or $P_C = 10$ W per transistor. Hence at no excitation there would be a steady loss of 10 W in each transistor, whereas in class B the standby (no-signal) dissipation is zero. This example clearly indicates the superiority of the push-pull over the parallel configuration.

Distortion The output of a push-pull system always possesses mirror symmetry (Sec. 18-6), so that $I_C = 0$, $I_{\text{max}} = -I_{\text{min}}$, and $I_{\frac{1}{2}} = -I_{-\frac{1}{2}}$. Under these circumstances, Eqs. (18-18) reduce to

$$B_o = B_2 = B_4 = 0 \qquad B_1 = \tfrac{2}{3}(I_{\text{max}} + I_{\frac{1}{2}}) \qquad B_3 = \tfrac{1}{3}(I_{\text{max}} - 2I_{\frac{1}{2}})$$
$$\tag{18-45}$$

Note that there is no even-harmonic distortion. The principal contribution to distortion is the third harmonic, given by $D_3 = |B_3|/|B_1|$. The values I_{max} and $I_{\frac{1}{2}}$ are found as follows: A load line corresponding to $R_L' = (N_1/N_2)^2 R_L$ is drawn on the collector characteristics through the point $I_C = 0$ and $V_{CE} = V_{CC}$. If the peak base current is I_B, then the intersection of the load line with the I_B curve is I_{max} and with the $I_B/2$ characteristic is $I_{\frac{1}{2}}$, as indicated in Fig. 18-9.

The power output, taking distortion into account, is

$$P = (1 + D_3{}^2)\frac{B_1{}^2 R_L'}{2} \tag{18-46}$$

Special Circuits[4] A class B configuration which dispenses with the output transformer is shown in Fig. 18-10. This arrangement requires a power supply whose center tap is grounded, a condition which is not difficult to obtain.

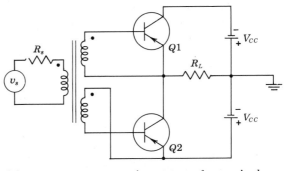

Fig. 18-10 A class B push-
pull circuit which does not
use an output transformer.

A circuit that requires neither an output nor an input transformer is shown
in Fig. 18-11.   This arrangement uses transistors having complementary symmetry (one *n-p-n* and one *p-n-p* type).   The difficulty with the circuit is that
of obtaining matched complementary transistors.   If there is an unbalance in
the characteristics of the two transistors in Fig. 18-11 (or also in Figs. 18-8
and 18-10), considerable distortion will be introduced; even harmonics will no
longer be canceled.   Very often negative feedback is used in power amplifiers
to reduce nonlinear distortion.

18-8   CLASS AB OPERATION

In addition to the distortion introduced by not using matched transistors and
that due to the nonlinearity of the collector characteristics, there is one more
source of distortion, that caused by nonlinearity of the input characteristic.
As pointed out in Sec. 5-9 and Fig. 5-16, no appreciable base current flows
until the emitter junction is forward-biased by the cutin voltage V_γ, which is
0.1 V for germanium and 0.5 V for silicon (Table 5-1).   Under these circumstances a sinusoidal base-voltage excitation will not result in a sinusoidal output
current.

The distortion caused by the nonlinear transistor input characteristic is

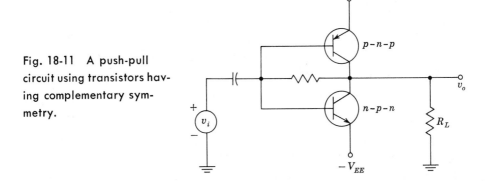

Fig. 18-11 A push-pull
circuit using transistors having complementary sym-
metry.

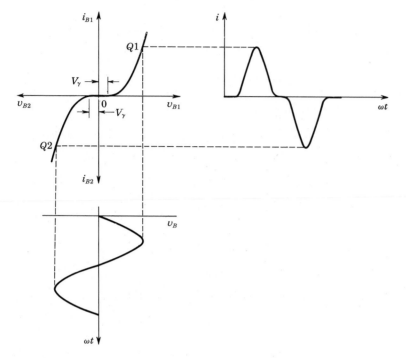

Fig. 18-12 Crossover distortion.

indicated in Fig. 18-12. The i_B-v_B curve for each transistor is drawn, and the construction used to obtain the output current (assumed proportional to the base current) is shown. In the region of small currents (for $v_B < V_\gamma$) the output is much smaller than it would be if the response were linear. This effect is called *crossover distortion*. Such distortion would not occur if the driver were a true current generator, in other words, if the base current (rather than the base voltage) were sinusoidal.

To minimize crossover distortion, the transistors must operate in a class AB mode, where a small standby current flows at zero excitation. In the circuit of Fig. 18-8, the voltage drop across R_2 is adjusted to be approximately equal to V_γ. Class AB operation results in less distortion than class B, but the price which must be paid for this improvement is a loss in efficiency and waste of standby power. The calculations of the distortion components in a class AB or class A push-pull amplifier due to the nonlinearity of the collector characteristics is somewhat involved since it requires the construction of composite output curves for the pair of transistors.[5]

A practical class AB power amplifier is shown in Fig. 18-13. In this circuit the input transformer is replaced by the CA 3007 integrated circuit audio driver. The IC driver is a balanced differential circuit with either single-ended or differential input and two push-pull emitter-follower outputs (pins 8 and 10). The CA 3007 is ac-coupled, and it does not affect the biasing

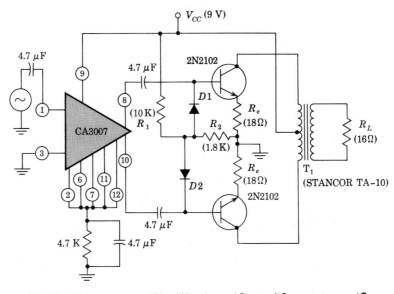

Fig. 18-13 A low-power (30-mW) class AB amplifier system.   (Courtesy of RCA, Electronic Components and Devices.)

of the output stage, which is set by the diodes $D1$, $D2$, and resistors R_1, R_2, and R_e. This power amplifier provides 30 mW output power for an audio input of 6.5 mV of signal voltage.

Integrated Circuit Power Amplifier The amplifier shown in Fig. 18-14 is a monolithic integrated circuit (IC) power amplifier. The MC 1554 is

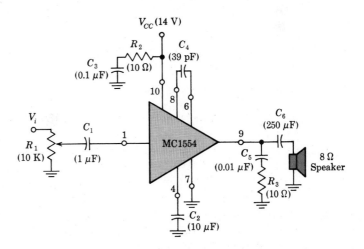

Fig. 18-14 A 1-W monolithic power amplifier system.
(Courtesy of Motorola Semiconductor Products, Inc.)

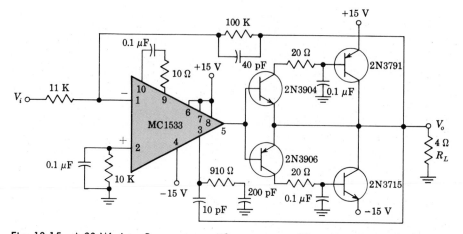

Fig. 18-15 A 20-W class B power amplifier system. (Courtesy Motorola Semiconductor Products, Inc.)

designed to amplify signals as high as 300 kHz and to deliver 1 W of power to a directly or capacitively coupled load. The component values shown result in total harmonic distortion at 1 kHz of less than 1 percent at 1-W output power into an 8-Ω load. When this device is used, care must be exercised to lay out the circuit properly and to avoid stray coupling or feedback from the output to the input, which may result in oscillations. To avoid oscillations, the input cable must be shielded and the compensating network R_3, C_5 must be connected from the output pin to ground. The R_2, C_3 network is a decoupling network used to cancel the effects of the inductance of the power-supply leads. The load can be d-c coupled provided a split power supply is used.

The circuit diagram of a 20-W amplifier[7] is shown in Fig. 18-15. The MC 1533 operational amplifier drives the complementary output transistors, and in the absence of an input signal, the output at pin 5 is zero volts, thus biasing the output transistors class B. The crossover distortion is eliminated by the use of negative feedback from the output to the input. The rated output power into a 4-Ω load is 20 W with less than 0.7 percent harmonic distortion. The frequency response is down 1.5 dB at 20 kHz, and the sensitivity for rated output power is 1 V rms input. The efficiency at rated power is 54 percent.

18-9   REGULATED POWER SUPPLIES

An ideal *regulated power supply* is an electronic circuit designed to provide a predetermined dc voltage V_o which is independent of the current I_L drawn from V_o, of the temperature, and also of any variations in the ac line voltage.

An unregulated power supply consists of a transformer, a rectifier, and a filter, as shown in Figs. 4-19 and 4-23.

There are three reasons why an unregulated power supply is not good enough for many applications. The first is its poor regulation; the output voltage is not constant as the load varies. The second is that the dc output voltage varies with the ac input. In some locations the line voltage (of nominal value 115 V) may vary over as wide a range as 90 to 130 V, and yet it is necessary that the dc voltage remain essentially constant. The third reason is that the dc output voltage varies with the temperature, particularly because semiconductor devices are used. The feedback circuit shown in Fig. 18-16 is used to overcome the above three shortcomings, and also to reduce the ripple voltage. Such a system is called a *regulated power supply*.[8] From Fig. 18-16 we see that the regulated power supply represents a cause of voltage-series feedback. If we assume that the voltage gain of the emitter follower $Q1$ ($Q1$ is also called the *pass element*) is approximately unity, then $V'_o \approx V_o$ and

$$V'_o = A_V V_i = A_V(V_R - \beta V_o) \approx V_o \qquad (18\text{-}47)$$

where the feedback factor is

$$\beta \equiv \frac{R_2}{R_1 + R_2} \qquad (18\text{-}48)$$

From Eq. (18-47) it follows that

$$V_o = V_R \frac{A_V}{1 + \beta A_V} \qquad (18\text{-}49)$$

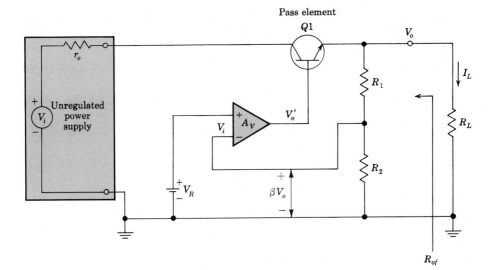

Fig. 18-16   Regulated power supply system.

The output voltage V_o can be changed by varying the feedback factor β. The emitter follower $Q1$ is used to provide current gain, because the current delivered by the amplifier A_V usually is not sufficient. The dc collector voltage required by the error amplifier A_V is obtained from the unregulated voltage.

Stabilization Since the output dc voltage V_o depends on the input unregulated dc voltage V_i, load current I_L, and temperature T, then the change ΔV_o in output voltage of a power supply can be expressed as follows:

$$\Delta V_o = \frac{\partial V_o}{\partial V_i} \Delta V_i + \frac{\partial V_o}{\partial I_L} \Delta I_L + \frac{\partial V_o}{\partial T} \Delta T$$

or

$$\Delta V_o = S_V \Delta V_i + R_o \Delta I_L + S_T \Delta T \tag{18-50}$$

where the three coefficients are defined as

Input regulation factor: $\quad S_V = \frac{\Delta V_o}{\Delta V_i} \Big|_{\substack{\Delta I_L = 0 \\ \Delta T = 0}}$ $\hspace{3cm}$ (18-51)

Output resistance: $\quad R_o = \frac{\Delta V_o}{\Delta I_L} \Big|_{\substack{\Delta V_i = 0 \\ \Delta T = 0}}$ $\hspace{3cm}$ (18-52)

Temperature coefficient: $\quad S_T = \frac{\Delta V_o}{\Delta T} \Big|_{\substack{\Delta V_i = 0 \\ \Delta I_L = 0}}$ $\hspace{3cm}$ (18-53)

The smaller the value of the three coefficients, the better the regulation of the power supply. The input-voltage change ΔV_i may be due to a change in ac line voltage or may be ripple because of inadequate filtering.

18-10   SERIES VOLTAGE REGULATOR

The physical reason for the improvement in voltage regulation with the circuit of Fig. 18-16 lies in the fact that a large fraction of the *increase* in input voltage appears across the pass element, so that the output voltage tries to remain constant. If the input increases, the output must also increase (but to a much smaller extent), because it is this increase in output that acts to bias the pass transistor toward less current. This stabilization is demonstrated with reference to Fig. 18-17 where $Q2$ is the comparison amplifier designated A_V in Fig. 18-16, and where the battery V_R is replaced by the breakdown diode D. Here a fraction of the output voltage βV_o is compared with the reference voltage V_R. The difference $\beta V_o - V_R$ is amplified by $Q2$. If the input voltage increases by ΔV_i (say, because the power-line voltage increases), then V_o need increase only slightly, and yet $Q2$ may cause a large current change in R_3. Thus it is possible for almost all of ΔV_i to appear across R_3 (and since the base-to-emitter voltage is small, also across $Q1$) and for V_o to remain essentially constant. These considerations are now made more quantitative.

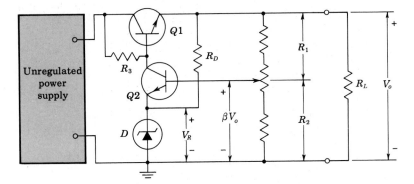

Fig. 18-17   A semiconductor-regulated power supply.   The series pass element or series regulator is $Q1$, the difference amplifier is $Q2$, and the reference avalanche diode is D.

Simplified Analysis   From Fig. 18-17 the output dc voltage V_o is given by

$$V_o = V_R + V_{BE2} + \frac{R_1}{R_1 + R_2} V_o$$

or using Eq. (18-48) for β

$$V_o = (V_R + V_{BE2})\left(1 + \frac{R_1}{R_2}\right) = (V_R + V_{BE2})/\beta \qquad (18\text{-}54)$$

Hence a convenient method for changing the output is to adjust the ratio R_1/R_2 by means of a resistance divider as indicated in Fig. 18-17.

An approximate expression for S_V (sufficiently accurate for most applications) is obtained as follows: The input-voltage change v_i is very much larger than the output change v_o.   Also, by the definition of Eq. (18-51), $\Delta I_L = 0$, and to a first approximation we can neglect the ac voltage drop across r_o.   Hence $\Delta V_i = v_i$ appears as shown in Fig. 18-18.   Neglecting the

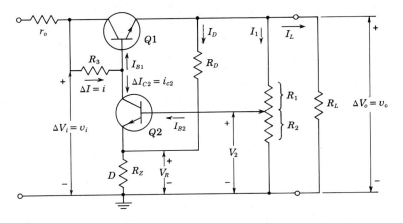

Fig. 18-18   Analysis of the series-regulated power supply.

small change in base-to-emitter voltage of $Q1$, the current change $\Delta I = i$ in R_3 is given by

$$i = \frac{v_i - v_o}{R_3} \approx \frac{v_i}{R_3} \tag{18-55}$$

Since R_L is fixed, constant output voltage requires that I_L, and hence I_{B1}, remain constant. Hence, for constant I_{B1},

$$i = \Delta I_{C2} = i_{c2} \tag{18-56}$$

In Prob. 18-23 we find, for small values of R_3, that $i_{c2} = G_m v_o$ where

$$G_m = h_{fe2} \frac{R_2}{R_1 + R_2} \frac{1}{(R_1\|R_2) + h_{ie2} + (1 + h_{fe2})R_Z} \tag{18-57}$$

where R_Z is the dynamic resistance of the Zener diode. Using Eqs. (18-55) to (18-57), we find, since $v_i \approx i_{c2}R_3$,

$$S_V = \frac{v_o}{v_i} = \frac{1}{G_m R_3} \tag{18-58}$$

In Prob. 18-24 the output resistance R_o of the circuit of Fig. 18-18 is found to be

$$R_o \approx \frac{r_o + (R_3 + h_{ie1})/(1 + h_{fe1})}{1 + G_m(R_3 + r_o)} \tag{18-59}$$

where $G_m \equiv i_{c2}/v_o$ is obtained from Eq. (18-57). A design procedure is indicated in the following illustrative example.

EXAMPLE (a) Design a series-regulated power supply to provide a nominal output voltage of 25 V and supply load current $I_L \leq 1$ A. The unregulated power supply has the following specifications: $V_i = 50 \pm 5$ V and $r_o = 10\ \Omega$. (b) Find the input regulation factor S_V. (c) Find the output resistance R_o. (d) Compute the change in output voltage ΔV_o due to input-voltage changes of ± 5 V and load current I_L variation from zero to 1 A.

Solution   a. Select a silicon reference diode with $V_R \approx V_o/2$. Two 1N755 diodes in series provide $V_R = 7.5 + 7.5 = 15$ V and $R_Z = 12\ \Omega$ at $I_Z = 20$ mA. Refer to Figs. 18-18 and 18-19. Choose $I_{C2} \approx I_{E2} = 10$ mA. The Texas Instruments 2N930 silicon transistor can provide the collector current of 10 mA. For this transistor the manufacturer specifies $I_{C,\max} = 30$ mA and $V_{CE,\max} = 45$ V.

At $I_{C2} = 10$ mA, the following parameters were measured:

$$h_{FE2} = 220 \qquad h_{fe2} = 200 \qquad h_{ie2} = 800\ \Omega$$

Choose $I_D = 10$ mA, so that $D1$, $D2$, operate at $I_z = 10 + 10 = 20$ mA. Then

$$R_D = \frac{V_o - V_R}{I_D} = \frac{25 - 15}{10} = 1\ \text{K}$$

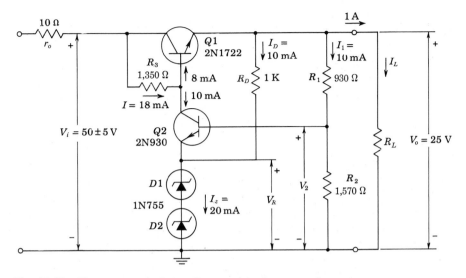

Fig. 18-19 The series regulator discussed in the example.

The ratio R_1/R_2 may be found from Eq. (18-54). Each resistor is determined as follows:

$$I_{B2} = \frac{I_{C2}}{h_{FE2}} = \frac{10 \text{ mA}}{220} = 45 \text{ } \mu A$$

Since we require $I_1 \gg I_{B2}$, we select $I_1 = 10$ mA; then, since $V_{BE} = 0.7$ V,

$$V_2 = V_{BE2} + V_R = 15.7 \text{ V}$$

$$R_1 = \frac{V_o - V_2}{I_1} = \frac{25 - 15.7}{10 \times 10^{-3}} = 930 \text{ } \Omega$$

$$R_2 \approx \frac{V_2}{I_1} = \frac{15.7}{10 \times 10^{-3}} = 1,570 \text{ } \Omega$$

If we select the Texas Instruments 2N1722 silicon power transistor for $Q1$, we measure at $I_{C1} = 1$ A the following parameters:

$$h_{FE1} = 125 \qquad h_{fe1} = 100 \qquad h_{ie1} = 20 \text{ } \Omega$$

We thus have

$$I_{B1} = \frac{I_L + I_1 + I_D}{h_{FE1}} = \frac{1,000 + 10 + 10}{125} \approx 8 \text{ mA}$$

The current I through resistor R_3 is $I = I_{B1} + I_{C2} = 8 + 10 = 18$ mA. The value for R_3 corresponding to $V_i = 45$ and to $I_L = 1$ A is given by

$$R_3 = \frac{V_i - (V_{BE1} + V_o)}{I} = \frac{50 - 25.7}{18 \times 10^{-3}} = 1,350 \text{ } \Omega$$

The complete circuit is shown in Fig. 18-19.

b. From Eq. (18-58) we find

$$S_V = \frac{2.50}{1.57} \times \frac{584 + 800 + (201)(12)}{(200)(1,350)} = 0.022$$

c. The output resistance is found from Eqs. (18-58) and (18-59). Since

$$G_m = \frac{1}{S_V R_3} = \frac{1}{0.022 \times 1,350} = 0.033$$

$$R_o = \frac{10 + (1,350 + 20)/101}{1 + (0.033)(1,350 + 10)} = 0.51 \; \Omega$$

d. The net change in output voltage, assuming constant temperature, is obtained using Eq. (18-50):

$$\Delta V_o = S_V \, \Delta V_i + R_o \, \Delta I_L = 0.022 \times 10 + 0.51 \times 1 = 0.22 + 0.51 = 0.73 \; V$$

The circuit designed in this example was built in the laboratory, and excellent agreement between measured and calculated values was obtained.

Very often it is necessary to design a power supply with much smaller value for S_V. From Eq. (18-58) we see that S_V can be improved if R_3 is increased. Since $R_3 \approx (V_i - V_o)/I$, we can increase R_3 by decreasing I. The current I can be decreased by using a Darlington pair (Fig. 8-29) for $Q1$. For even greater improvement in S_V, R_3 is replaced by a constant-current source (so that $R_3 \to \infty$), as shown in Fig. 18-20 (see also Sec. 15-3). For this circuit, which incorporates a Darlington pair, values of $S_V = 0.00014$ and $R_o = 0.1 \; \Omega$ have been obtained.[9] The constant-current source in Fig. 18-20 is often called a *transistor preregulator*. Other types of preregulators (Prob. 18-28) are possible.[9] The 0.01-μF capacitor in Fig. 18-20 is added to prevent high-frequency oscillation.

Practical Considerations The maximum dc load current of the power supply shown in Fig. 18-18 is restricted by the maximum allowable collector current of the series transistor. The difference between the output and input voltages of the regulator is applied across $Q1$, and thus the maximum allowable V_{CE} for a given $Q1$ and specified output voltage determines the maximum input voltage to the regulator. The product of the load current and V_{CE} is approximately equal to the power dissipated in the pass transistor. Consequently, the maximum allowable power dissipated in the series transistor further limits the combination of load current and input voltage of the regulator.

The reverse saturation current I_{CO1} of $Q1$ in Fig. 18-18 plays an important role in determining the minimum load of the regulator. If $I_{B1} = 0$, then

$$I_{C1} = -I_{E1} = I_{CO1}(1 + h_{FE1})$$

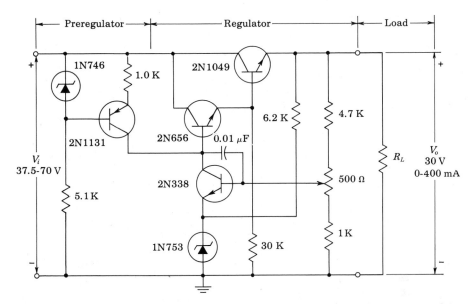

Fig. 18-20 Typical series regulator using preregulator and Darlington pair. (Courtesy of Texas Instruments, Inc.)

Hence, if the emitter current of $Q1$ ($I_L + I_D + I_1$) falls below $I_{CO1}(1 + h_{FE1})$, then V_{CE1} cannot be controlled by I_{B1}, and the regulator cannot function properly. We thus see that, at high temperatures, where I_{CO} and h_{FE} are high, the regulator may fail when the load current falls below a certain minimum level. Various techniques have been proposed[10] to reduce this minimum-load restriction due to I_{CO}. The 30-K resistor in Fig. 18-20 is added to allow operation at low load currents.

A power supply must be protected further from the possibility of damage through overload. In simple circuits protection is provided by using a fusible element in series with r_o. In more sophisticated equipment the series transistor is such that it can permit operation at any voltage from zero to the maximum output voltage. In case of an overload or short circuit, the circuit of Fig. 18-21 can provide protection. Here the diodes $D1$, $D2$ are nonconducting until the voltage drop across the sensing resistor R_S exceeds their forward threshold voltage V_γ. Thus, in the case of a short circuit, the current I_S would increase only up to a limiting point determined by

$$I_S = \frac{V_{\gamma 1} + V_{\gamma 2} - V_{BE1}}{R_S}$$

Under short-circuit conditions the load current would be, approximately,

$$I_L \approx \frac{V_i}{R_3} + \frac{V_{\gamma 1} + V_{\gamma 2} - V_{BE1}}{R_S} \tag{18-60}$$

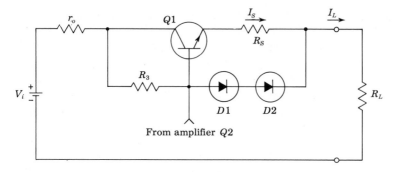

Fig. 18-21   Short-circuit overload-protection circuit.

Finally, an important practical consideration is the variation in output voltage with temperature. From Eq. (18-54) we see that, approximately,

$$\frac{\Delta V_o}{\Delta T} \approx \left(\frac{\Delta V_R}{\Delta T} + \frac{\Delta V_{BE2}}{\Delta T}\right)\left(1 + \frac{R_1}{R_2}\right) \qquad (18\text{-}61)$$

Thus cancellation of temperature coefficients between the reference diode $D1$ and the transistor $Q2$ can result in a very low $\Delta V_o/\Delta T$. The GE reference amplifiers RA-1, RA-2, and RA-3 have been designed for this purpose. They are integrated devices composed of a reference diode and n-p-n transistor in a single chip. Typical temperature coefficients for these units are better than ± 0.002 percent/°C.

18-11   MONOLITHIC REGULATORS[11]

The voltage regulator of Fig. 18-16 can be fabricated on a single silicon chip, thus offering all the benefits derived from integrated circuits: low cost, high performance, small size, and ease of use. Using monolithic regulators, it is possible to distribute unregulated voltage through electronic equipment and provide regulation locally, for example, on individual printed circuit (PC) boards. Among the advantages of this approach are greater flexibility in voltage levels and regulation for individual stages and improved isolation and decoupling of these stages. Monolithic regulators are available that operate in the voltage range from 0 to 1,000 V, and with external pass elements the current range extends to 60 A or more.

The circuit of a monolithic regulator is substantially more complex than the discrete configuration of Fig. 18-20. The added complexity is due to the fact that a multitransistor differential amplifier is used to provide the amplification A_V shown in Fig. 18-16, rather than a single transistor, as is the case in

Fig. 18-20. In addition, the Zener reference voltage is compensated with the use of proper circuitry so that it has zero temperature coefficient. Fortunately, it is relatively easy to add transistors and diodes on a monolithic chip without increasing significantly its cost.

An example of a 200-mA IC regulator is shown in Fig. 18-22. The National Semiconductor Corporation LM 105 contains the voltage reference and the error amplifier, as shown in Fig. 18-16. The external transistor 2N3740 increases the power-handling capacity of the IC to 200 mA, and the output voltage is set by R_1 and R_2. A fraction of the output voltage is compared by the error amplifier with an internal 1.8-V reference. Any error is amplified and used to drive the 2N3740 power transistor, which is the pass element. Resistor R_S provides current limiting because the voltage drop across it is applied to the emitter-to-base junction of a transistor in the IC. When that transistor is turned ON, it removes the drive from the pass element 2N3740, and the regulator output exhibits a constant-current characteristic; that is, the output voltage drops to zero when the load current exceeds a predetermined value. When used with external pass elements, the tantalum capacitor C_1 must be used to suppress any oscillations and provide low output impedance at high frequencies. External capacitor C_2 is used to compensate the internal error amplifier and thus avoid instability.

The output voltage of the LM 105 is adjustable from 4.5 to 40 V, and the output currents in excess of 10 A are possible by adding external series pass transistors. Ripple rejection is 0.01 percent, and full-load regulation with current limiting is 0.1 percent.

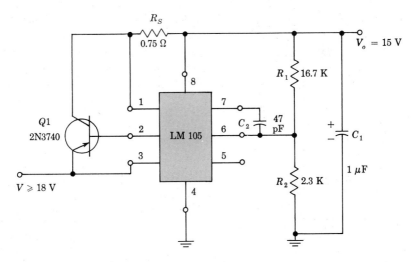

Fig. 18-22 A voltage regulator system using the LM 105 IC and with an external pass transistor $Q1$ for 200-mA operation.

18-12   THE FOUR–LAYER DIODE[12]

In the remaining sections of this chapter we examine the p-n-p-n diode, the silicon controlled switch, and the basic techniques used to control power with these solid-state devices.

The p-n-p-n diode when biased with the anode positive has two stable states. One is a very high resistance state, typically of the order of 100 M, and the other a very low resistance state, typically less than 10 Ω. When reverse-biased, this device acts like a typical p-n diode, having a very low leakage current. The device consists of four layers of silicon doped alternately with p- and n-type impurities, as shown in Fig. 18-23. Because of this structure it is called a p-n-p-n (often pronounced "pinpin") diode or switch. The terminal P region is the anode, or p emitter, and the terminal N region is the cathode, or n emitter. When an external voltage is applied to make the anode positive with respect to the cathode, junctions J_1 and J_3 are forward-biased and the center junction J_2 is reverse-biased. The externally impressed voltage appears principally across the reverse-biased junction, and the current which flows through the device is small. As the impressed voltage is increased, the current increases slowly until a voltage called the *firing*, or *breakover*, voltage V_{BO} is reached where the current increases abruptly and the voltage across the device decreases sharply. At this breakover point the p-n-p-n diode switches from its OFF (also called *blocking*) state to its ON state.

In Fig. 18-24a, the p-n-p-n switch has been split into two parts which have been displaced mechanically from one another but left electrically connected. This splitting is intended to illustrate that the device may be viewed as two transistors back to back. One transistor is a p-n-p type, whereas the second is an n-p-n type. The N region that is the base of one transistor is the collector of the other, and similarly for the adjoining P region. The junction J_2 is a common-collector junction for both transistors. In Fig. 18-24b the arrangement in Fig. 18-24a has been redrawn using transistor-circuit symbols, and a voltage source has been impressed through a resistor across the switch, giving

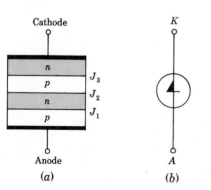

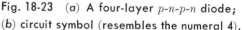

Fig. 18-23   (a) A four-layer p-n-p-n diode; (b) circuit symbol (resembles the numeral 4).

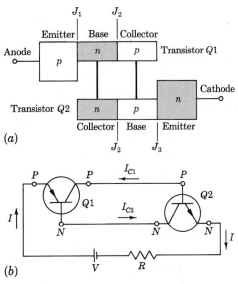
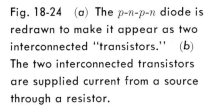

Fig. 18-24 (a) **The** p-n-p-n **diode is redrawn to make it appear as two interconnected "transistors."** (b) **The two interconnected transistors are supplied current from a source through a resistor.**

rise to a current I. Collector currents I_{C1} and I_{C2} for transistors $Q1$ and $Q2$ are indicated. In the active region the collector current is given by

$$I_C = -\alpha I_E + I_{CO} \tag{18-62}$$

with I_E the emitter current, I_{CO} the reverse saturation current, and α the short-circuit common-base forward current gain. We may apply Eq. (18-62), in turn, to $Q1$ and $Q2$. Since $I_{E1} = +I$ and $I_{E2} = -I$, we obtain

$$I_{C1} = -\alpha_1 I + I_{CO1} \tag{18-63}$$

$$I_{C2} = \alpha_2 I + I_{CO2} \tag{18-64}$$

For the p-n-p transistor I_{CO1} is negative, whereas for the n-p-n device I_{CO2} is positive. Hence we write $I_{CO2} = -I_{CO1} \equiv I_{CO}/2$. Setting equal to zero the sum of the currents into transistor $Q1$, we have

$$I + I_{C1} - I_{C2} = 0 \tag{18-65}$$

Combining Eqs. (18-63) through (18-65) we find

$$I = \frac{I_{CO2} - I_{CO1}}{1 - \alpha_1 - \alpha_2} = \frac{I_{CO}}{1 - \alpha_1 - \alpha_2} \tag{18-66}$$

We observe that as the sum $\alpha_1 + \alpha_2$ approaches unity, Eq. (18-66) indicates that the current I increases without limit; that is, the device breaks over. Such a development is not unexpected in view of the regenerative manner in which the two transistors are interconnected. The collector current of $Q1$ is furnished as the base current of $Q2$, and vice versa. When the p-n-p-n switch is operating in such a manner that the sum $\alpha_1 + \alpha_2$ is less than unity, the switch is in its OFF state and the current I is small. When the condition $\alpha_1 + \alpha_2 = 1$ is attained, the switch transfers to its ON state. The voltage across the switch drops to a low value and the current becomes large, being limited by the external resistance in series with the switch.

The reason why the device can exist in either of two states is that at very low currents α_1 and α_2 may be small enough so that $\alpha_1 + \alpha_2 < 1$, whereas at larger currents the α's increase, thereby making it possible to attain the condition $\alpha_1 + \alpha_2 = 1$. Thus, as the voltage across the switch is increased from zero, the current starts at a very small value and then increases because of avalanche multiplication (not avalanche breakdown) at the reverse-biased junction. This increase in current increases α_1 and α_2. *When the sum of the small-signal avalanche-enhanced alphas equals unity,* $\alpha_1 + \alpha_2 = 1$, *breakover occurs.* At this point, the current is large, and α_1 and α_2 might be expected individually to attain values in the neighborhood of unity. If such were the case, then Eq. (18-66) indicates that the current might be expected to reverse. What provides stability to the ON state of the switch is that in the ON state the center junction becomes forward-biased. Now all the transistors are in saturation and the current gain α is again small. Thus stability is attained by virtue of the fact that the transistors enter saturation to the extent necessary to maintain the condition $\alpha_1 + \alpha_2 = 1$.

In the ON state all junctions are forward-biased, and so the total voltage across the device is equal very nearly to the algebraic sum of these three saturation junction voltages. The voltage drop across the center junction J_2 is in a direction opposite to the voltages across the junctions J_1 and J_3. This feature serves additionally to keep quite low $(2V_{BE,\text{sat}} - V_{CB,\text{sat}} \approx 1.0 \text{ V})$ the total voltage drop across the switch in the ON state.

The operation of the p-n-p-n switch depends, as we have seen, on the fact that at low currents, the current gain α may be less than one-half, a condition which is necessary if the sum of two α's is to be less than unity. This characteristic of α is not encountered in germanium but is distinctive of silicon, where it results from the fact that at low currents an appreciable part of the current which crosses the emitter junction is caused by recombination of holes and electrons in the transition region rather than the injection of minority carriers across the junction from emitter to base. In germanium it is not practicable to establish $\alpha_1 + \alpha_2 < 1$. Therefore germanium structures incline to settle immediately in the ON state and have no stable OFF state. Accordingly, germanium p-n-p-n switches are not available. We shall see in the discussion below that the p-n-p-n structure and mechanism are basic to a large number of other switching devices.

18-13 p-n-p-n CHARACTERISTICS

The volt-ampere characteristic of a p-n-p-n diode, not drawn to scale, is shown in Fig. 18-25. When the voltage is applied in the reverse direction, the two

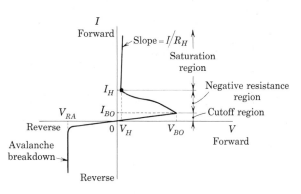

Fig. 18-25 Volt-ampere characteristic of the p-n-p-n diode.

outer junctions of the switch are reverse-biased. At an adequately large voltage, breakdown will occur at these junctions, as indicated, at the "reverse avalanche" voltage V_{RA}. However, no special interest is attached to operation in this reverse direction.

When a forward voltage is applied, only a small forward current will flow until the voltage attains the breakover voltage V_{BO}. The corresponding current is I_{BO}. If the voltage V, which is applied through a resistor as in Fig. 18-24, is increased beyond V_{BO}, the diode will switch from its OFF (blocked) state to its ON (saturation) state and will operate in the saturation region. The device is then said to *latch*. If the voltage is now reduced, the switch will remain ON until the current has decreased to I_H. This current and the corresponding voltage V_H are called the *holding*, or *latching*, current and voltage, respectively. The current I_H is the minimum current required to hold the switch in its ON state.

There are available p-n-p-n switches with voltages V_{BO} in the range from tens of volts to some hundreds of volts. The current I_{BO} is of the order, at most, of some hundreds of microamperes. In this OFF range up to breakover, the resistance of the switch is the range from some megohms to several hundred megohms.

The holding current varies, depending on the type, in the range from several milliamperes to several hundred milliamperes. The holding voltage is found to range from about 0.5 to about 20 V. The incremental resistance R_H in the saturation state is rarely in excess of 10 Ω and decreases with increasing current. At currents of the order of amperes (which can be sustained briefly under pulsed operation), the incremental resistance may drop to as low as some tenths of an ohm.

The switching parameters of the four-layer diode are somewhat temperature-dependent. A decrease in temperature from room temperature to $-60°C$ has negligible effect on V_{BO}, but a temperature increase to $+100°C$ will decrease V_{BO} by about 10 percent. I_H decreases substantially with increase in temperature and increases to a lesser extent with decrease in temperature.

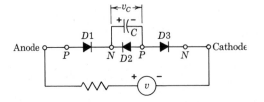

Fig. 18-26 p-n-p-n diode in OFF state to illustrate the origin of the rate effect.

Rate Effect We can see that the breakover voltage of a p-n-p-n switch depends on the rate[12] at which the applied voltage rises. In Fig. 18-26 we have represented the switch in the OFF state as a series combination of three diodes, two forward-biased and the center one reverse-biased. Across this latter diode we have placed a capacitance which represents the transition capacitance across this reverse-biased junction. When the applied voltage v increases slowly enough so that the current through C may be neglected, we must wait until the avalanche-increased current through $D2$ (which is also the current through $D1$ and $D3$) increases to the point where the current gains satisfy the condition $\alpha_1 + \alpha_2 = 1$. When, however, v changes rapidly, so that the capacitor voltage changes at the rate dv_C/dt, a current $C\,dv_C/dt$ passes through C and adds to the current in $D1$ and $D3$. The current through $D2$ need not be as large as before to attain breakover, and switching takes place at a lower voltage. The capacitance at the reverse-biased junction may lie in the range of some tens of picofarads to over 100 pF, and the reduction in switching voltage may well make itself felt for voltage rates of change dv_C/dt of the order of tens of volts per microsecond.

Bilateral Diode Switch[13] The p-n-p-n switch of Fig. 18-23 is limited to one direction of current flow when it is in the low-resistance or ON state. By arranging two p-n-p-n sections in parallel but in opposite order, as shown in Fig. 18-27a, it is possible to obtain bilateral current flow since there are two ON states. Thus, when the device is in the ON state in one direction, one section conducts current, whereas for the inverse ON state, the other section conducts and the current reverses. This symmetrical (bilateral) volt-ampere characteristic is indicated in Fig. 18-27c.

18-14   THE SILICON CONTROLLED RECTIFIER[14,15]

The structure of the silicon controlled rectifier (SCR) consists of four alternate p- and n-type layers, as in the four-layer diode. In the SCR (also called a *thyristor*) connections are made available to the inner layers which are not accessible in the diode. The circuit symbol for the SCR is shown in Fig. 18-28. The terminal connected to the P region nearest the cathode is called the *cathode gate*, or *p base*, and the terminal connected to the N region nearest the anode is called the *anode gate*, or *n base*. In very many switch types both gates are

Fig. 18-27 (a) Bilat-
eral diode switch; (b)
the circuit symbol; (c)
volt-ampere character-
istic is symmetrical
about the origin.

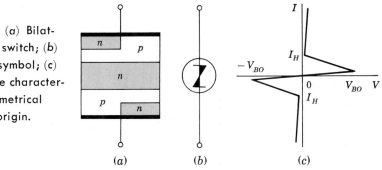

(a)   (b)   (c)

not brought out. Where only one gate terminal is available it is ordinarily
the cathode gate.

The usefulness of the gate terminals rests on the fact that currents introduced into one or both gate terminals may be used to control the anode-to-
cathode breakover voltage. Such behavior is to be expected on the basis
of the earlier discussion of the condition $\alpha_1 + \alpha_2 = 1$ which establishes the
firing point. If the current through one or both outer junctions is increased
as a result of currents introduced at the gate terminals, then α increases and
the breakover voltage will be decreased. In Fig. 18-29 the volt-ampere
characteristic of an SCR is shown for various cathode-gate currents. We
observe that the firing voltage is a function of the gate current, decreasing
with increasing gate current and increasing when the gate current is negative
and consequently in a direction to reverse-bias the cathode junction. The
current after breakdown may well be larger by a factor of 1,000 than the
current before breakdown. When the gate current is very large, breakover
may occur at so low a voltage that the characteristic has the appearance of a
simple p-n diode.

Suppose that a supply voltage is applied through a load resistor between
anode and cathode of a silicon controlled rectifier. Consider that the bias is

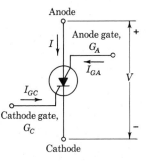

Fig. 18-28   Circuit symbol used for the SCR.

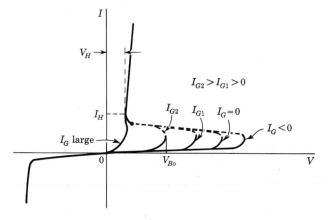

Fig. 18-29 Volt-ampere characteristics of a three-terminal SCR, illustrating that the forward breakover voltage is a function of the cathode-gate current. (Not drawn to scale.)

such that the applied voltage is less than breakover voltage. Then the rectifier will remain OFF and may be turned ON by the application to the gate of a triggering current or voltage adequate to lower the breakover voltage to less than the applied voltage. The rectifier having been turned on, it latches, and it is found to be impractical to stop the conduction by reverse-biasing the gate. For example, it may well be that the reverse gate current for turnoff is nearly equal to the anode current. Ordinarily, the most effective and commonly employed method for turnoff is, temporarily at least, to reduce the anode voltage below the holding voltage V_H, or equivalently to reduce the anode current below the holding current I_H. The gate will then again assume control of the breakover voltage of the switch.

Gate ON and OFF Times The process by which the SCR changes state occupies a finite time interval. When a triggering signal is applied to a gate to turn a switch ON, a time interval, the *turn-on* time, elapses before the transition is completed. This turn-on time decreases with increasing amplitude of trigger signal, increases with temperature, and increases also with increasing anode current. If the triggering signal is a pulse, then, to be effective, not only must the pulse amplitude be adequate, but the pulse duration must be at least as long as a critical value called *gate time to hold*. Otherwise, at the termination of the gating pulse, the SCR will fall back to its original state. A similar situation applies in driving the switch OFF by dropping the anode voltage. At a minimum, the anode voltage must drop below the holding voltage. If, however, the anode voltage is driven in the reverse direction, the *turnoff* time may thereby be reduced. The turnoff time increases with temperature and with increasing magnitude of anode current. Further, the anode voltage must be kept below the maintaining voltage for an interval at least as long as a critical value, called the *gate recovery time*, if the transition is to persist after the anode voltage rises.

In fast units, all the time intervals are in the range of tenths of microseconds, whereas in slower units, these times may be as long as several microseconds. In general, the time required to turn a switch OFF is longer than the time required to turn it ON.

Characteristics SCRs with currents in excess of 100 A and operating voltages up to about 1,000 V are available. The holding voltage is of the order of magnitude of 1 V. The ratio of the continuous allowable anode current to the forward gate current required to switch ON is rarely less than several thousand. For example, a gate current of less than 50 mA will turn on an anode current of 100 A.

A three-terminal *silicon controlled switch* (SCS) is a device which is similar to the SCR, except that it is mechanically smaller and is designed to operate at lower currents and voltages. The switches are intended for low-level applications. They have lower leakage and holding currents than SCRs, require small triggering signals, and have more uniform triggering characteristics from sample to sample of a given type.

Silicon controlled rectifiers suffer from the same rate effect as do four-layer diodes. The inclination to fire prematurely because of the rate effect may be suppressed by bypassing the gate to the cathode through a small capacitance. This component will shunt current past the cathode-gate junction in the presence of a rapidly varying applied voltage, but will have no effect on the dc operation of the switch.

The Triac, or Bilateral Triode Switch[16] The triac is a three-terminal silicon switch which can be triggered with either positive or negative gate pulses when the anode potentials are positive or negative, respectively. Thus the triac is an ac switch which can be made to conduct on both alterations (half cycles) of an ac voltage. The construction and operation of the triac is based on the SCS principle and relies on the strategic placement of junctions, as shown in Fig. 18-30a. Note that the triac is a five-layer n_1-p_1-n_2-p_2-n_3 device, which may be considered to consist of an n_1-p_1-n_2-p_2 section in parallel with a p_1-n_2-p_2-n_3 section, as indicated in Fig. 18-30a. An additional lateral n region serves as the control gate. The triac is therefore a double-ended SCR, or thyristor.

18-15   POWER CONTROL[17]

There are a number of applications which require a controlled amount of current. These include electric welding, lighting-control installations, motor-speed control, and a variety of other industrial control applications. It is possible to vary the amount of current supplied to the load either by controlling

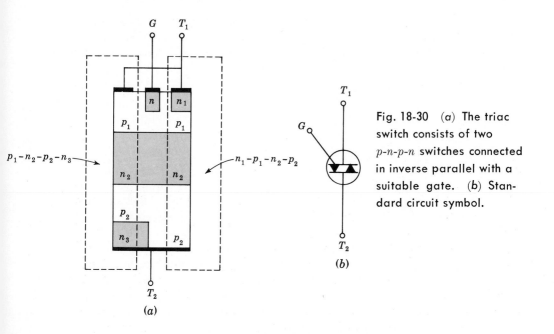

Fig. 18-30 (a) The triac switch consists of two p-n-p-n switches connected in inverse parallel with a suitable gate. (b) Standard circuit symbol.

the transformer secondary voltage or by inserting a controlling resistor in the output circuit. Neither of these methods is desirable. The first method may require expensive auxiliary equipment, and the second is characterized by poor efficiency. The development of SCRs and triacs has made control a relatively inexpensive process.

SCR Control If a sinusoidal potential is applied to the SCR anode, the device will be turned OFF once each alternate half cycle (when the voltage falls below the holding voltage) provided it is triggered ON regularly. The average rectified current can be varied over wide limits by controlling the point in each half cycle at which the SCR is turned ON.

To analyze the ON-OFF action of the SCR, we refer to Fig. 18-31. In this circuit the ac line voltage is used as the anode voltage for the SCR. A switching device, such as a p-n-p-n diode or a neon bulb, is connected in series with the control terminal. When the control voltage v_C exceeds the *trigger breakdown voltage* V_B of the switching device D, then D goes to its low-resistance state and the current through the SCR gate triggers it ON. In Fig. 18-31b the sine wave represents the input line voltage v_i as a function of time. The trigger breakdown curve is a straight line parallel to the time axis, indicating that V_B is independent of the anode potential.

Suppose that the circuit is so arranged that the control voltage v_C exceeds the trigger breakdown voltage at some angle, say φ, called the *delay angle*. Conduction will start at this point in the cycle. The voltage drop across the

SCR during conduction remains constant at a low value which is independent of current. This voltage drop ($\approx V_H$) is of the order of 1 V. The current through a pure resistance load R_L during the time the SCR is conducting is given by

$$i_L = \frac{V_m \sin \omega t - V_H}{R_L} \qquad (18\text{-}67)$$

where V_m is the maximum value of the applied potential.

The resulting form of the load current is illustrated in Fig. 18-31b. The current is seen to rise abruptly at the point corresponding to the angle φ and then follows the sine variation given in Eq. (18-67) until the supply voltage v_i falls below V_H at the phase $\pi - \varphi_o$. The current will remain zero until the phase φ is again reached in the next cycle.

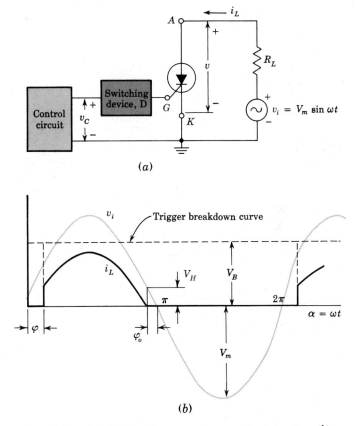

(a)

(b)

Fig. 18-31 (a) SCR half-wave power control circuit. (b) The waveshape of the load current i_L. Conduction starts at the angle φ and stops at $\pi - \varphi_o$ in each cycle.

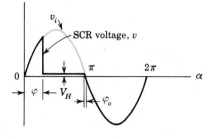

Fig. 18-32 The waveshapes of the SCR current and anode voltage. Conduction begins at an angle φ and ceases at $\pi - \varphi_o$ degrees. Since $V_m \gg V_o$, then $\varphi_o \approx 0$.

The average current [the value read on a dc ammeter, Eq. (4-11)]

$$I_{dc} = \frac{1}{2\pi} \int_{\varphi}^{\pi - \varphi_o} i_L \, d\alpha = \frac{V_m}{2\pi R_L} \int_{\varphi}^{\pi - \varphi_o} \left(\sin \alpha - \frac{V_H}{V_m} \right) d\alpha$$

which integrates to

$$I_{dc} = \frac{V_m}{2\pi R_L} \left[\cos \varphi + \cos \varphi_o - \frac{V_H}{V_m} (\pi - \varphi_o - \varphi) \right] \tag{18-68}$$

where $\alpha = \omega t$, and φ_o is the smallest angle defined by the relation

$$V_H = V_m \sin \varphi_o \tag{18-69}$$

If the ratio V_H/V_m is very small, then φ_o may be taken as zero, and Eq. (18-68) reduces to the form

$$I_{dc} = \frac{V_m}{2\pi R_L} (1 + \cos \varphi) \tag{18-70}$$

This analysis shows that the average rectified current can be controlled by varying the position at which the trigger voltage v_C exceeds the breakdown voltage V_B of the switching device D. The maximum current is obtained when the SCR is triggered ON at the beginning of each cycle; and the minimum current is obtained when no conduction occurs.

The voltage across the SCR is shown in Fig. 18-32. The applied voltage v_i appears across the SCR until conduction begins. After breakdown the SCR voltage drop is a constant equal to V_H. When the applied voltage falls below V_H, then the SCR voltage is again equal to the applied voltage.

The reading of a dc voltmeter placed across the SCR will be

$$V_{dc} = \frac{1}{2\pi} \int_{0}^{2\pi} v \, d\alpha$$

$$= \frac{1}{2\pi} \left(\int_{0}^{\varphi} V_m \sin \alpha \, d\alpha + \int_{\varphi}^{\pi - \varphi_o} V_H \, d\alpha + \int_{\pi - \varphi_o}^{2\pi} V_m \sin \alpha \, d\alpha \right)$$

This integrates to

$$V_{dc} = \frac{V_H}{2\pi} (\pi - \varphi_o - \varphi) - \frac{V_m}{2\pi} (\cos \varphi + \cos \varphi_o) \tag{18-71}$$

If $V_m \gg V_H$, this reduces to

$$V_{dc} \approx -\frac{V_m}{2\pi} (1 + \cos \varphi) \qquad\qquad (18\text{-}72)$$

The appearance of the negative sign means that the cathode is more positive than the anode for most of the cycle. It should be noted that the dc load voltage is the negative of the dc SCR voltage. This follows from the fact that the sum of the dc voltages around the circuit is zero.

EXAMPLE An SCR is connected according to Fig. 18-31 and supplies power to a 200-Ω load resistor from a 230-V source of supply. If the trigger voltage is adjusted so that conduction starts at 60° after the start of each cycle, calculate the readings of the following meters: (*a*) a true rms reading ammeter in series with the load, (*b*) a true rms reading voltmeter connected across the SCR, (*c*) a wattmeter inserted in the circuit so as to read the total power delivered by the ac supply. Neglect the SCR voltage drop V_H.

Solution Since $V_m \gg V_H$, no error is made by assuming that the conduction continues until the end of each positive half cycle. The instantaneous current through the SCR and the voltage across the SCR will have the forms shown in Figs. 18-31 and 18-32 with $\varphi = 60°$ or $\pi/3$ rad.

   a. In the interval between 60 and 180° the instantaneous current is given by

$$i_L = \frac{230 \sqrt{2} \sin \alpha}{200} = 1.625 \sin \alpha$$

An ac ammeter reads the rms value of the current [Eq. (4-15)]. For the wave sketched,

$$I_{rms} = \sqrt{\frac{1}{2\pi} \int_{\pi/3}^{\pi} (1.625 \sin \alpha)^2 \, d\alpha} = \sqrt{0.533} = 0.73 \text{ A}$$

The limits of integration are from 60 to 180°, the current being zero outside of this range.

   b. The ac voltmeter reads the rms value of the voltage wave sketched. It is noted that between 0 and $\pi/3$ the SCR voltage equals the line voltage; between $\pi/3$ and π, it is constant and $V_H \approx 0$; and between π and 2π, it again equals the line voltage. Thus

$$V_{rms} = \sqrt{\frac{1}{2\pi} \left[\int_0^{\pi/3} (230 \sqrt{2} \sin \alpha)^2 \, d\alpha + \int_{\pi}^{2\pi} (230 \sqrt{2} \sin \alpha)^2 \, d\alpha \right]}$$

$$= 178 \text{ V}$$

   c. The instantaneous power from the ac supply is the product of the instantaneous line current and the instantaneous line voltage. The wattmeter will read the average value of this product. Hence

$$P = \frac{1}{2\pi} \int_{\pi/3}^{\pi} (1.625 \sin \alpha)(230 \sqrt{2} \sin \alpha) \, d\alpha = 107 \text{ W}$$

The integration extends only from $\pi/3$ to π, for there can be no power when the current is zero.

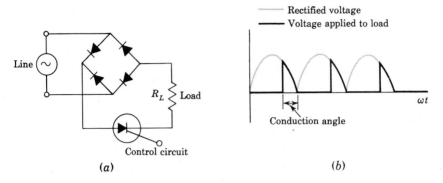

Fig. 18-33 (a) Full-wave-rectification power control. The small voltage drop V_H across the SCR is neglected. (b) Rectified voltage and voltage applied to the load.

Full-wave-rectified Operation In some applications it is desired to control large values of dc current. A diode bridge (Fig. 4-20) in series with one SCR, as shown in Fig. 18-33, can be used for this purpose by controlling the conduction angle for full-wave-rectified operation. The time during which the SCR is ON is called the *conduction angle*.

Control Circuit The most common control circuit is the relaxation oscillator used to provide phase control. This circuit is shown in Fig. 18-34a, where the switching device may be a neon bulb or a three-, four-, or five-layer trigger diode. The capacitor C is charged through the resistor R from voltage source V until the switching device reaches its breakover voltage V_B. At that time the switching device turns ON and the capacitor is discharged through the gate of the SCR to the holding voltage V_H *of the switching device,* as shown in Fig. 18-34b. The discharge current pulse turns the SCR ON at a time controlled by the RC time constant. The capacitor retains a voltage after discharge equal to the holding voltage V_H as long as the SCR is ON and provided that the current i_D exceeds the holding current I_H of the switching device. When the SCR anode voltage becomes negative, the gate-to-cathode junction becomes reverse-biased and the switching device D turns OFF. At this instant the capacitor starts charging again from the voltage V_H toward V with a time constant RC. If $t = 0$ is the time at which the charging of the capacitor starts, the voltage v_C across the capacitor is given by

$$v_C - V_H = (V - V_H)(1 - \epsilon^{-t/RC}) \tag{18-73}$$

This equation is consistent with the fact that at $t = 0$, v_C must equal the holding voltage V_H, and at $t = \infty$, v_C must equal V. In many circuits, for low-cost operation, instead of a dc supply V, it is common to use the ac line in

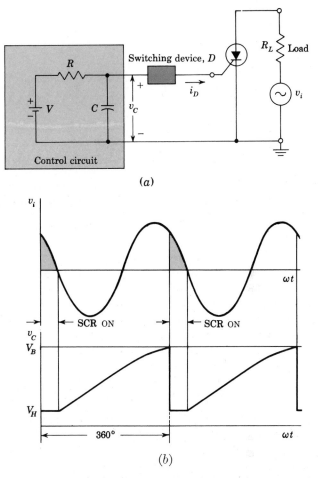

Fig. 18-34 (*a*) SCR relaxation oscillator phase control circuit; (*b*) anode voltage and control circuit waveform.

series with a rectifier to charge the RC network. In that case Eq. (18-73) cannot be used to find the capacitor voltage.

EXAMPLE A 60-Hz power source is connected to the anode of an SCR as shown in Fig. 18-34*a*. The SCR is the MCR 2304, and $C = 0.1\ \mu\text{F}$. Assume $V = 60$ V and $V_B = 32$ V is the required voltage to trigger the switching device for which $V_H = 10$ V and $I_H = 100\ \mu\text{A}$. Find R for a conduction angle of 45°.

Solution It is clear from Fig. 18-34 that the 45° conduction angle requires the capacitor to charge for $360 - 45 = 315°$, or

$$t_D = \frac{315}{360} \times \frac{1}{60}\ \text{s} = 14.6\ \text{ms}$$

If the anode voltage is positive, the SCR will fire when $v_C = 32$ V. Hence, using Eq. (18-73), we find

$$32 - 10 = (60 - 10)[1 - \epsilon^{-(14.6 \times 10^{-3})/RC}]$$

or

$$RC = 25.2 \times 10^{-3} \text{ s}$$

and

$$R = \frac{25.2 \times 10^{-3}}{0.1 \times 10^{-6}} = 252 \text{ K}$$

For the proper operation of the circuit in accordance with the waveforms shown in Fig. 18-34b, it is necessary that the current i_D exceed the holding current I_H of the switching device during the time interval that the SCR is ON. If i_D drops below I_H, the switching device D will turn OFF and the capacitor will start charging prematurely toward V.

Hence we require that

$$\frac{V - V_H}{R} > I_H$$

or

$$\frac{60 - 10}{252} \text{ mA} \approx 200 \text{ μA} > 100 \text{ μA}$$

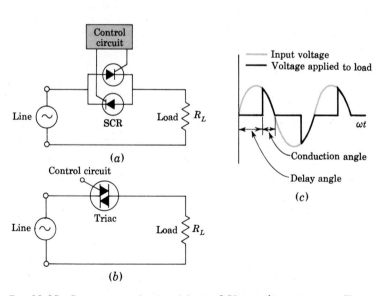

Fig. 18-35　Power control using (a) two SCRs or (b) one triac. The small (~1 V) voltage drop across the SCR or triac is neglected. (c) Input line voltage and voltage applied to the load, showing the delay and conduction angles.

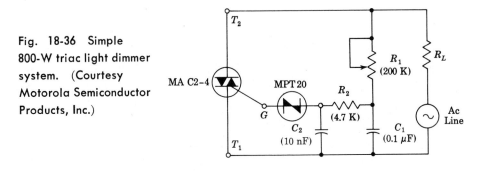

Fig. 18-36 Simple
800-W triac light dimmer
system. (Courtesy
Motorola Semiconductor
Products, Inc.)

AC Control In many instances it is desirable to control continuously the amount of power dissipated in a load. Two SCRs connected in inverse parallel or a single triac can be used to control both the negative and the positive half cycles of the ac power line. Figure 18-35a and b shows two simple power-control circuits. The phase-control circuit determines when the triac or one of the SCRs is triggered ON, connecting the line voltage to the load. Figure 18-36 shows a simple 800-W light dimmer. Output power, and thus light intensity, are varied by controlling the phase of conduction of the triac. The voltage across capacitor C_2 lags the ac line voltage by a number of degrees, determined by the value of the variable resistor R_1. The bilateral diode switch MPT 20 has a breakover voltage V_B equal to 20 V. When the voltage on capacitor C_2 reaches 20 V, the diode switches ON and the capacitor discharges into the gate of the triac and turns it ON (on both half cycles of the line voltage).

REFERENCES

1. Espley, D. C.: The Calculation of Harmonic Production in Thermionic Valves with Resistive Loads, *Proc. IRE*, vol. 21, pp. 1439–1446, October, 1933.

   Chaffee, E. L.: A Simplified Harmonic Analysis, *Rev. Sci. Instr.*, vol. 7, pp. 384–389, October, 1936.

   Block, A.: Distortion in Valves with Resistive Loads, *Wireless Eng.*, vol. 16, pp. 592–596, December, 1939.

2. Millman, J.: "Vacuum-tube and Semiconductor Electronics," p. 419, McGraw-Hill Book Company, New York, 1958.

3. Gordon, M.: Class B Audio Frequency Amplification, *Wireless Eng.*, vol. 16, pp. 457–459, September, 1939.

   Woll, H. J.: Low-frequency Amplifiers, in L. P. Hunter (ed.), "Handbook of Semiconductor Electronics," 2d ed., chap. 11, McGraw-Hill Book Company, New York, 1962.

   Lo, A. W., et al.: "Transistor Electronics," pp. 197–224, Prentice-Hall, Inc., Englewood Cliffs, N.J., 1956.

4. Lohman, R. D.: Complementary Symmetry Transistor Circuits, *Electronics*, vol. 26, pp. 140–143, September, 1953.

5. Ref. 2, pp. 424–430.

6. "Semiconductor Power Circuits Handbook," pp. 5-30 to 5-33, Motorola Semiconductor Products, Inc., Applications Engineering Department, Phoenix, Ariz., 1968.

7. Ehrsam, W.: Audio Power Generation Using IC Operational Amplifiers, *Motorola Semiconductor Products, Inc., Appl. Note* AN-275.

8. Wilson, E. C., and R. T. Windecker: DC Regulated Power Supply Design, *Solid-State J.*, November, 1961, pp. 37–46.

9. Texas Instruments, Inc.: "Transistor Circuit Design," chap. 9, McGraw-Hill Book Company, New York, 1963.

10. Moores, H. T.: Design Procedure for Power Transistors, Part 2, *Electron. Design*, September, 1955, pp. 43–45.

11. Kesner, D.: Monolithic Voltage Regulators, *IEEE Spectrum*, vol. 7, no. 4, pp. 24–32, April, 1970.

12. Moll, J. L., M. Tannenbaum, J. M. Goldey, and N. Holonyak: p-n-p-n Switches, *Proc. IRE*, vol. 44, pp. 1174–1182, 1956.

13. Gentry, F. E., et al.: "Semiconductor Controlled Rectifiers," pp. 139–141, Prentice-Hall, Inc., Englewood Cliffs, N.J., 1964.

14. "Transistor Manual," 7th ed., chap. 16, General Electric Company, Syracuse, N.Y., 1964.

15. "Semiconductor Power Circuits Handbook," Motorola Semiconductor Products, Inc., Phoenix, Ariz., 1968.

16. Ref. 13, pp. 142–148.

17. Gentry, F. E., et al.: "Semiconductor Controlled Rectifiers," chap. 8, Prentice-Hall, Inc., Englewood Cliffs, N.J., 1964.

REVIEW QUESTIONS

18-1 Derive an expression for the output power of a class A large-signal amplifier in terms of V_{max}, V_{min}, I_{max}, and I_{min}.

18-2 Discuss how rectification may take place in a power amplifier.

18-3 Define *intermodulation distortion*.

18-4 Describe the five-point method of computing harmonic distortion.

18-5 Draw the diagram of a transformer-coupled single-transistor output stage and explain the need for impedance matching.

18-6 Explain why the circuit of Rev. 18-5 exhibits a maximum in the power-output vs. load resistance curve.

18-7 (*a*) Define the *conversion efficiency* η of a power stage. (*b*) Derive a simple expression for η. (*c*) Compare the maximum efficiency of a series-fed and transformer-coupled class A single-transistor power stage.

18-8 (*a*) Explain why even harmonics are not present in a push-pull amplifier. (*b*) Give two additional advantages of this circuit over that of a single-transistor amplifier.

18-9 Derive a simple expression for the output power of an idealized class B push-pull power amplifier.

18-10 Show that the maximum conversion efficiency of the idealized class B push-pull circuit is 78.5 percent.

18-11 Using two complementary silicon transistors, draw a simple class B push-pull amplifier circuit which does not use an output transformer.

18-12 (*a*) Explain the origin of crossover distortion. (*b*) Describe a method to minimize this distortion.

18-13 Draw the circuit of a class AB power amplifier using an OP AMP with differential output and two discrete complementary power transistors.

18-14 (*a*) Draw a simplified circuit diagram of a regulated power supply. (*b*) What type of feedback is employed by this regulator?

18-15 List three reasons why an unregulated supply is not good enough for some applications.

18-16 Define *input regulation factor, output resistance,* and *temperature coefficient* for a voltage regulator.

18-17 Explain two methods of decreasing the value of S_V for a series-voltage regulator.

18-18 What is a transistor preregulator? Draw the circuit diagram.

18-19 Draw the short-circuit overload protection circuit, and explain its operation.

18-20 Give the voltage and current range for commercial IC voltage regulators.

18-21 How is short-circuit current protection provided for an IC regulator? Draw the circuit diagram.

18-22 Describe three advantages of IC voltage regulators.

18-23 (*a*) Explain the reason why the *p-n-p-n* silicon diode can exist in either of two states. (*b*) Why is it not possible to construct a germanium *p-n-p-n* switching diode?

18-24 Draw the volt-ampere characteristic of a four-layer diode.

18-25 Give the order of magnitude for the holding current and holding voltage of a four-layer diode.

18-26 Describe the rate effect in a *p-n-p-n* diode.

18-27 For the bilateral diode switch sketch (*a*) its cross section and (*b*) its volt-ampere characteristic.

18-28 Draw the volt-ampere characteristics of the silicon controlled rectifier (SCR) as a function of gate current.

18-29 (*a*) Describe a *triac*. (*b*) Draw a cross section of a triac.

18-30 (*a*) Draw a half-wave SCR circuit. (*b*) Indicate the current and voltage waveforms of the SCR.

18-31 (*a*) Draw the circuit for full-wave SCR power control. (*b*) Indicate the current and voltage waveform control.

18-32 (*a*) Draw the phase-shift SCR control circuit using an *RC* relaxation circuit. (*b*) Explain the operation of this circuit using appropriate waveforms.

18-33 (*a*) Give the circuit of a light dimmer, using a triac and employing phase control. (*b*) Explain how the lamp light intensity is controlled with this circuit.

19 / SEMICONDUCTOR-DEVICE PHYSICS

Fermi-Dirac statistics are applied to a metal and to a semiconductor. The band structure of a *p-n* junction diode is given, and the tunnel diode volt-ampere characteristic is explained. The equation of continuity is explored further. Quantitative analyses are made of the current components in a diode, and the Ebers-Moll equations for a transistor are derived.

19-1 ENERGY DISTRIBUTION OF ELECTRONS IN A METAL

It is important to know what energies are possessed by the mobile carriers in a solid. This relationship is called the *energy distribution function*. In this section the discussion is limited to free electrons in a metal, and in subsequent sections it is generalized to include mobile electrons and holes in a semiconductor. We first discuss the potential variation in a metal.

Simplified Potential-energy Picture of a Metal The region in which the free electrons find themselves is essentially a potential plateau, or equipotential region. It is only for distances close to an ion that there is any appreciable variation in potential. Since the regions of rapidly varying potential represent but a very small portion of the total volume of the metal, we henceforth assume that the field distribution within the metal is equipotential and the free electrons are subject to no forces whatsoever. The present viewpoint is therefore essentially that of classical electrostatics.

In Fig. 19-1 all potential-energy variations within the metal have been omitted and only the potential barrier at the surface is included. A conduction electron may move freely within the interior of the metal,

Fig. 19-1   For the free electrons, the interior
of a metal may be considered an equipotential volume, but there is a potential
barrier at the surface.

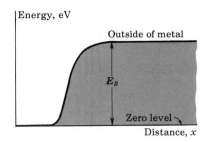

Energy, eV

but it cannot escape through the surface unless it has an energy at least equal
to E_B.   The existence of the surface barrier is readily understood.   If an
electron tries to escape from a metal, it will induce a positive charge on the
surface, because the metal was originally neutral.   There will be a force of
attraction between this induced charge and the electron.   Unless the electron
possesses sufficient energy to carry it out of the region of influence of this image
force, it will be forced back into the metal.   Clearly, a potential-energy barrier
exists at the surface.   We are interested in knowing if there are any electrons in
the interior with sufficient energy E_B to escape.   In other words, we need to
know the energy distribution of the free electrons within the metal.   We digress
briefly to make clear what is meant by a distribution function.

Age Density   Suppose that we were interested in the distribution in age
of the people in the United States.   A sensible way to indicate this relationship is shown in Fig. 19-2, where the abscissa is *age* and the ordinate is ρ_A, the
density of the population in age.   This density gives the number dn_A of people
whose ages lie in the range between A and $A + dA$, or

$$dn_A = \rho_A \, dA \tag{19-1}$$

The data for such a plot are obtained from census information.   We see, for
example, that the number of persons of ages between 10 and 12 years is represented by dn_A, with $\rho_A = 2.25$ million per year chosen as the mean ordinate

Fig. 19-2   The distribution
function in age of people in
the United States.

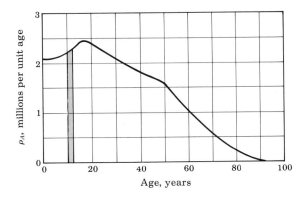

between 10 and 12 years, and dA is taken as $12 - 10 = 2$ years. Thus $dn_A = \rho_A\, dA = 4.50$ million. Geometrically, this is the shaded area of Fig. 19-2. Evidently, the total population n is given by

$$n = \int dn_A = \int \rho_A\, dA \qquad (19\text{-}2)$$

or simply the total area under the curve.

Energy Density We are now concerned with the distribution in energy of the free electrons in a metal. By analogy with Eq. (19-2), we may write

$$dn_E = \rho_E\, dE \qquad (19\text{-}3)$$

where dn_E represents the number of free electrons per cubic meter whose energies lie in the energy interval dE, and ρ_E gives the density of electrons in this interval. Since our interests are confined only to the free electrons, it is assumed that there are no potential variations within the metal. Hence there must be, a priori, the same number of electrons in each cubic meter of the metal. That is, the density in space (electrons per cubic meter) is a constant. However, within each unit volume of metal there will be electrons having all possible energies. It is this distribution in energy that is expressed by ρ_E (number of electrons per electron volt per cubic meter of metal).

The function ρ_E may be expressed as the product

$$\rho_E = f(E)N(E) \qquad (19\text{-}4)$$

where $N(E)$ is the density of states (number of states per electron volt per cubic meter) in the conduction band, and $f(E)$ is the probability that a quantum state with energy E is occupied by an electron.

The expression for $N(E)$ is derived in the following section and is given by

$$N(E) = \gamma E^{\frac{1}{2}} \qquad (19\text{-}5)$$

where γ is a constant defined by

$$\gamma \equiv \frac{4\pi}{h^3}(2m)^{\frac{3}{2}}(1.60 \times 10^{-19})^{\frac{3}{2}} = 6.82 \times 10^{27} \qquad (19\text{-}6)$$

The dimensions of γ are $(\text{m}^{-3})(\text{eV})^{-\frac{3}{2}}$; m is the mass of the electron in kilograms; and h is Planck's constant in joule-seconds.

19-2   THE FERMI–DIRAC FUNCTION

The equation for $f(E)$ is called *the Fermi-Dirac probability function*, and specifies the fraction of all states at energy E (electron volts) occupied under conditions of thermal equilibrium. From quantum statistics it is found[1,2] that

$$f(E) = \frac{1}{1 + \epsilon^{(E - E_F)/kT}} \qquad (19\text{-}7)$$

where k = Boltzmann constant, eV/°K

T = temperature, °K

E_F = Fermi level, or characteristic energy, for the crystal, eV

The Fermi level represents the energy state with 50 percent probability of being filled if no forbidden band exists. The reason for this last statement is that, if $E = E_F$, then $f(E) = \frac{1}{2}$ for any value of temperature. A plot of $f(E)$ versus $E - E_F$ is given in Fig. 19-3a and of $E - E_F$ versus $f(E)$ in Fig. 19-3b, both for $T = 0°$K and for larger values of temperatures. When $T = 0°$K, two possible conditions exist: (1) if $E > E_F$, the exponential term becomes infinite and $f(E) = 0$. Consequently, *there is no probability of finding an occupied quantum state of energy greater than E_F at absolute zero.* (2) If $E < E_F$, the exponential in Eq. (19-7) becomes zero and $f(E) = 1$. *All quantum levels with energies less than E_F will be occupied at $T = 0°$K.*

From Eqs. (19-4), (19-5), and (19-7), we obtain at *absolute zero temperature*

$$\rho_E = \begin{cases} \gamma E^{\frac{1}{2}} & \text{for } E < E_F \\ 0 & \text{for } E > E_F \end{cases} \qquad (19\text{-}8)$$

Clearly, there are no electrons at 0°K which have energies in excess of E_F. That is, the Fermi energy is the maximum energy that any electron may possess at absolute zero. The relationship represented by Eq. (19-8) is called the *completely degenerate energy distribution function.* Classically, all particles should have zero energy at 0°K. The fact that the electrons actually have energies extending from 0 to E_F at absolute zero is a consequence of the Pauli exclusion principle, which states that no two electrons may have the same set of quantum numbers (Sec. 1-6). Hence not all electrons can have the same energy even at 0°K.

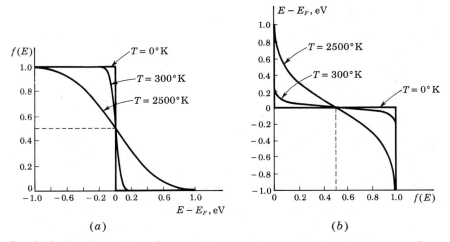

Fig. 19-3 The Fermi-Dirac function $f(E)$ gives the probability that a state of energy E is occupied.

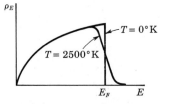

Fig. 19-4   Energy distribution in a metal at 0 and 2500°K.

A plot of the distribution in energy given by Eqs. (19-4) and (19-8) for a metal at $T = 0°$K and $T = 2500°$K is shown in Fig. 19-4. The area under each curve is simply the total number of free electrons per cubic meter of the metal; hence the two areas must be equal. Also, the curves for all temperatures must pass through the same ordinate, namely, $\rho_E = \gamma E_F^{\frac{1}{2}}/2$, at the point $E = E_F$, since, from Eq. (19-7), $f(E) = \frac{1}{2}$ for $E = E_F$.

A most important characteristic is to be noted, viz., the distribution function changes only very slightly with temperature, even though the temperature change is as great as 2500°K. The effect of the high temperature is merely to give those electrons having the high energies at absolute zero (those in the neighborhood of E_F) still higher energies, whereas those having lower energies have been left practically undisturbed. Since the curve for $T = 2500°$K approaches the energy axis asymptotically, a few electrons will have large values of energy.

The Fermi Level   An expression for E_F may be obtained on the basis of the completely degenerate function. The area under the curve of Fig. 19-4 represents the total number of free electrons (as always, per cubic meter of the metal). Thus

$$n = \int_0^{E_F} \gamma E^{\frac{1}{2}}\, dE = \tfrac{2}{3}\gamma E_F^{\frac{3}{2}}$$

or

$$E_F = \left(\frac{3n}{2\gamma}\right)^{\frac{2}{3}} \tag{19-9}$$

Inserting the numerical value (6.82×10^{27}) of the constant γ in this expression, there results

$$E_F = 3.64 \times 10^{-19} n^{\frac{2}{3}} \tag{19-10}$$

Since the density n varies from metal to metal, E_F will also vary among metals. Knowing the specific gravity, the atomic weight, and the number of free electrons per atom, it is a simple matter to calculate n, and so E_F. For most metals the numerical value of E_F is less than 10 eV.

19-3   THE DENSITY OF STATES

As a preliminary step in the derivation of the density function $N(E)$, we first show that the components of the momentum of an electron in a metal are quantized.   Consider a metal in the form of a cube, each side of which has a length L.   Assume that the interior of the metal is at a constant (zero) potential but that the potential-energy barrier (Fig. 19-1) at the surface is arbitrarily high, so that no electrons can escape.   Hence the wave functions [Sec. (1-5)] representing the electrons must be zero outside the metal and at the surface.   A one-dimensional model of the potential-energy diagram is given in Fig. 19-5a, and two possible wave functions are indicated in Fig. 19-5b and c.   Clearly, this situation is possible only if the dimension L is a half-integral multiple of the De Broglie wavelength λ, or

$$L = n_x \frac{\lambda}{2} \tag{19-11}$$

where n_x is a positive integer (not zero).   From the De Broglie relationship (1-15), $\lambda = h/p_x$ and the x component of momentum is

$$p_x = \frac{n_x h}{2L} \tag{19-12}$$

Hence the momentum is quantized since p_x can assume only values which are integral multiples of $h/2L$.

The energy W (in joules) of the electron in this one-dimensional problem is

$$W = \frac{p_x^2}{2m} = \frac{n_x^2 h^2}{8mL^2} \tag{19-13}$$

The wave nature of the electron has led to the conclusion that its energy must also be quantized.   Since $n_x = 1, 2, 3, \dots$, the lowest possible energy is $h^2/8mL^2$, the next energy level is $h^2/2mL^2$, etc.

Quantum States in a Metal   The above results may be generalized to three dimensions.   For an electron in a cube of metal, each component of momentum is quantized.   Thus

$$p_x = n_x \rho \qquad p_y = n_y \rho \qquad p_z = n_z \rho \tag{19-14}$$

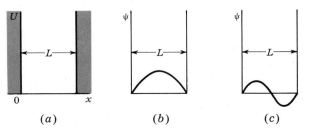

Fig. 19-5 (a) A one-dimensional problem in which the potential U is zero for a distance L but rises abruptly toward infinity at the boundaries $x = 0$ and $x = L$. (b,c) Two possible wave functions for an electron in the system described by (a).

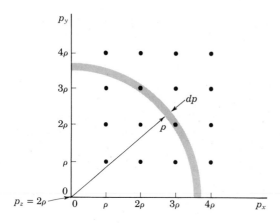

Fig. 19-6 The p_x-p_y plane in momentum space. Each dot represents three quantum numbers, n_x, n_y, and $n_z = 2$. There are two electrons per dot, corresponding to the two possible values of spin.

where $\rho \equiv h/2L$, and n_x, n_y, and n_z are positive integers. A convenient pictorial representation may be obtained by constructing three mutually perpendicular axes labeled p_x, p_y, and p_z. This "volume" is called *momentum space*. The only possible points which may be occupied by an electron in momentum space are those given by Eq. (19-14). These are indicated in Fig. 19-6, where for clarity we have indicated points only in a plane for a fixed value of p_z (say, $p_z = 2\rho$). By the Pauli exclusion principle (Sec. 1-6), no two electrons in a metal may have the same four quantum numbers, n_x, n_y, n_z, and the spin number s. Hence each dot in Fig. 19-6 represents two electrons, one for $s = \frac{1}{2}$ and the other for $s = -\frac{1}{2}$.

We now find the energy density function $N(E)$. Since in Fig. 19-6 there is one dot per volume ρ^3 of momentum space, the density of electrons in this space is $2/\rho^3$. The magnitude of the momentum is $p = (p_x^2 + p_y^2 + p_z^2)^{\frac{1}{2}}$. The electrons with momentum between p and $p + dp$ are those lying in the shaded spherical shell of Fig. 19-6. This number is

$$\frac{2}{\rho^3}(4\pi p^2\, dp)\left(\frac{1}{8}\right) = \frac{\pi p^2\, dp}{(h/2L)^3} = \frac{8\pi L^3 p^2\, dp}{h^3} \tag{19-15}$$

The factor $\frac{1}{8}$ introduced in this equation is due to the fact that only positive values of n_x, n_y, and n_z are permissible, and hence that only part of the shell in the first octant may be used.

If W is the energy (in joules), then $W = p^2/2m$. Hence

$$p = (2mW)^{\frac{1}{2}} \qquad p\, dp = m\, dW \qquad p^2\, dp = 2^{\frac{1}{2}} m^{\frac{3}{2}} W^{\frac{1}{2}}\, dW \tag{19-16}$$

If $N(W)$ is the density of states (per cubic meter), then, since the volume of the metal is L^3, it follows from Eq. (19-15) that

$$N(W)\, dW = \frac{8\pi p^2\, dp}{h^3} \tag{19-17}$$

gives the number of electrons with momenta between p and $p + dp$, corresponding to energies between W and $W + dW$. Substituting for $p^2\, dp$ from Eq. (19-16) in Eq. (19-17), we finally obtain

$$N(W)\, dW = \frac{4\pi}{h^3}\,(2m)^{\frac{3}{2}}W^{\frac{1}{2}}\, dW \qquad (19\text{-}18)$$

If we use electron volts E instead of joules W as the unit of energy, then, since $W = 1.60 \times 10^{-19}E$ (Sec. 1-3), the energy density $N(E)$ is given by Eq. (19-5), with γ defined in Eq. (19-6).

19-4   ELECTRON EMISSION FROM A METAL

In order for an electron to escape from the surface of a metal it must have an energy at least equal to the barrier height E_B in Fig. 19-1. The energy distribution curve of Fig. 19-4 indicates the energy of the electrons inside the metal. Hence these two diagrams may be combined so as to indicate which electrons may escape. If Fig. 19-4 is replotted so that the vertical axis represents energy and the horizontal axis energy density, it may be placed alongside of Fig. 19-1, whose vertical axis also signifies energy. This construction is indicated in Fig. 19-7.

Work Function At 0°K it is impossible for an electron to escape from the metal because this requires an amount of energy equal to E_B, and the maximum energy possessed by an electron is only E_F. It is necessary to supply an additional amount of energy equal to the difference between E_B and E_F in order to make this escape possible. This difference, written E_W, is known as the *work function* of the metal.

$$E_W \equiv E_B - E_F \qquad (19\text{-}19)$$

Thus the work function of a metal represents the minimum amount of energy that must be given to the fastest-moving electron at the absolute zero of temperature in order for this electron to be able to escape from the metal.

Thermionic Emission The curves of Fig. 19-7 show that the electrons in a metal at absolute zero are distributed among energies which range in value

Fig. 19-7 (a) The potential-energy barrier at the surface of a metal. (b) The energy distribution of the free electrons. Only those electrons in the tail of the curve, for $E > E_B$, can escape.

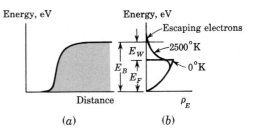

(a)

(b)

from zero to the maximum energy E_F. Since an electron must possess an amount of energy at least as great as E_B in order to be able to escape, no electrons can leave the metal. Suppose now that the metal, in the form of a filament, is heated by sending a current through it. Thermal energy is then supplied to the electrons from the lattice of the heated metal crystal. The energy distribution of the electrons changes, because of the increased temperature, as indicated in Fig. 19-7. Some of the electrons represented by the tail of the curve will have energies greater than E_B and so may be able to escape from the metal.

Using the analytical expression from the distribution function, it is possible to calculate the number of electrons which strike the surface of the metal per second with sufficient energy to be able to surmount the surface barrier and hence escape. Based upon such a calculation,[1,3] the thermionic current in amperes is given by

$$I_{th} = SA_o T^2 \epsilon^{-E_W/kT} \tag{19-20}$$

where S = area of filament, m²
A_o = a constant, whose dimensions are A/(m²)(°K²)
T = temperature, °K
k = Boltzmann constant, eV/°K
E_W = work function, eV

Equation (19-20) is called the *thermionic-emission, Dushman,* or *Richardson,* equation. The work function E_W is known also as the "latent heat of evaporation of electrons" from the metal, from the analogy of electron emission with the evaporation of molecules from a liquid.

The thermionic current is a very sensitive function of temperature. For tungsten ($E_W = 4.5$ eV) it is found that a 1 percent change in T results in a 24 percent change in I_{th} at the normal operating temperature of 2400°K. This statement may be verified by taking the derivative of the natural logarithm of Eq. (19-20), so as to obtain dI_{th}/I_{th} versus dT/T.

Energies of Emitted Electrons   Since the electrons inside a metal have a distribution of energies, those which escape from the metal will also have an energy distribution. It is easy to demonstrate this experimentally. Thus consider a plane emitter and a plane-parallel collector. The current is measured as a function of the retarding voltage V_r (the emitter positive with respect to the collector). If all the electrons left the cathode with the same energy, the current would remain constant until a definite voltage was reached, and then it would fall abruptly to zero. For example, if they all had 2-eV energy, then, when the retarding voltage was greater than 2 V, the electrons could not surmount the potential barrier between cathode and anode, and no particles would be collected. Experimentally, no such sudden falling off of current is found, but instead there is an exponential decrease of current I with voltage according to the equation

$$I = I_{th} \epsilon^{-V_r/V_T} \tag{19-21}$$

where V_T is the "volt equivalent of temperature," defined by

$$V_T \equiv \frac{\bar{k}T}{q} = \frac{T}{11,600} \qquad (19\text{-}22)$$

where $\bar{k}$ is the Boltzmann constant in joules per degree Kelvin. Note the distinction between $\bar{k}$ and k; the latter is the Boltzmann constant in electron volts per degree Kelvin. (Numerical values of $\bar{k}$ and k are given in Appendix A. From Sec. 1-3 it follows that $\bar{k} = 1.60 \times 10^{-19}k$.)

Equation (19-21) may be obtained theoretically as follows: Since I_{th} is the current for zero retarding voltage, the current obtained when the barrier height is increased by E_r is determined from the right-hand side of Eq. (19-20) by changing E_W to $E_W + E_r$. Hence

$$I = SA_oT^2\epsilon^{-(E_W+E_r)/kT} = I_{th}\epsilon^{-E_r/kT} \qquad (19\text{-}23)$$

where use was made of Eq. (19-20). From Eqs. (1-7) and (19-22)

$$\frac{E_r}{kT} = \frac{qV_r}{1.60 \times 10^{-19}kT} = \frac{V_r}{V_T} \qquad (19\text{-}24)$$

Hence Eq. (19-21) follows from Eq. (19-23).

From Eq. (19-21) it follows that only about 1.4 percent of the electrons from tungsten at 2700°K have energies in excess of 1 eV. If the emitter is an oxide-coated cathode operating at 1000°K, a similar calculation shows that only 0.001 percent of the electrons have an energy greater than 1 eV. These numerical values indicate that most of the electrons from a heated filament are emitted with extremely small initial speeds. A statistical analysis[2,3] shows that the average energy of an escaping electron is $2kT$, and hence, at operating temperatures of 2700 and 1000°K, the average energies of the emitted electrons are 0.47 and 0.17 eV, respectively.

19-5   CARRIER CONCENTRATIONS IN AN INTRINSIC SEMICONDUCTOR

To calculate the conductivity of a semiconductor from Eq. (2-17) it is necessary to know the concentration of free electrons n and the concentration of holes p. From Eqs. (19-3) and (19-4), with E in electron volts,

$$dn = N(E)f(E)\, dE \qquad (19\text{-}25)$$

where dn represents the number of conduction electrons per cubic meter whose energies lie between E and $E + dE$. The density of states $N(E)$ is derived in Sec. 19-3 on the assumption that the bottom of the conduction band is at zero potential. In a semiconductor the lowest energy in the conduction band is E_C, and hence Eq. (19-5) must be generalized as follows:

$$N(E) = \gamma(E - E_C)^{\frac{1}{2}} \qquad \text{for } E > E_C \qquad (19\text{-}26)$$

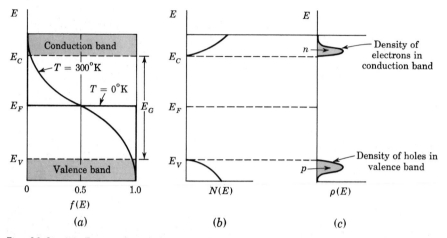

Fig. 19-8 (a) Energy-band diagram for an intrinsic semiconductor. The Fermi-Dirac probability function $f(E)$ is drawn at 0°K and at room temperature. (b) The density of states $N(E)$ in each band. (c) The density of carriers $\rho(E) = N(E)f(E)$ for electrons and $N(E)[1 - f(E)]$ for holes at room temperature. (Not drawn to scale.)

The Fermi function $f(E)$ is given by Eq. (19-3), namely,

$$f(E) = \frac{1}{1 + \epsilon^{(E-E_F)/kT}} \tag{19-27}$$

At room temperature $kT \approx 0.03$ eV, so that $f(E) = 0$ if $E - E_F \gg 0.03$ and $f(E) = 1$ if $E - E_F \ll 0.03$ (Fig. 19-3). We shall show that the Fermi level lies in the region of the energy gap midway between the valence and conduction bands, as indicated in Fig. 19-8a. This diagram shows the Fermi-Dirac distribution of Eq. (19-27) superimposed on the energy-band diagram of a semiconductor. At absolute zero ($T = 0°$K) the function is as shown in Fig. 19-8a. The probability of finding an electron in the conduction band is zero [$f(E) = 0$], and the probability of finding a hole in the valence band is zero [$1 - f(E) = 0$]. At room temperature some electrons are excited to higher energies and some states near the bottom of the conduction band E_C will be filled as indicated by the curve marked $T = 300°$K in Fig. 19-8a. Similarly, near the top of the valence band E_V, the probability of occupancy is decreased from unity since some electrons have escaped from their covalent bond and are now in the conduction band.

The density of states [Eqs. (19-25) and (19-26)] is plotted in Fig. 19-8b. The density of electrons is $N(E)f(E)$ and is indicated in Fig. 19-8c. The concentration of electrons in the conduction band is the area under this curve and is given by

$$n = \int_{E_C}^{\infty} N(E)f(E) \, dE \tag{19-28}$$

For $E \geq E_C$, $E - E_F \gg kT$ and Eq. (19-27) reduces to

$$f(E) = \epsilon^{-(E-E_F)/kT}$$

and

$$n = \int_{E_C}^{\infty} \gamma(E - E_C)^{\frac{1}{2}} \epsilon^{-(E-E_F)/kT} \, dE \tag{19-29}$$

This integral evaluates to

$$n = N_C \epsilon^{-(E_C-E_F)/kT} \tag{19-30}$$

where

$$N_C = 2 \left(\frac{2\pi m_n kT}{h^2} \right)^{\frac{3}{2}} (1.60 \times 10^{-19})^{\frac{3}{2}} = 2 \left(\frac{2\pi m_n \bar{k} T}{h^2} \right)^{\frac{3}{2}} \tag{19-31}$$

In deriving this equation the value of γ from Eq. (19-6) is used, k is given in electron volts per degree Kelvin, and $\bar{k}$ is expressed in joules per degree Kelvin. (The relationship between joules and electron volts is given in Sec. 1-3.) The mass m has been replaced by the symbol m_n, which represents the *effective mass* of the electron.

The Number of Holes in the Valence Band   Since the top of the valence band (the maximum energy) is E_V, the density of states [analogous to Eq. (19-26)] is given by

$$N(E) = \gamma(E_V - E)^{\frac{1}{2}} \qquad \text{for } E < E_V \tag{19-32}$$

Since a "hole" signifies an empty energy level, the Fermi function for a hole is $1 - f(E)$, where $f(E)$ is the probability that the level is occupied by an electron. For example, if the probability that a particular energy level is occupied by an electron is 0.2, the probability that it is empty (occupied by a hole) is 0.8. Using Eq. (19-27) for $f(E)$, we obtain

$$1 - f(E) = \frac{\epsilon^{(E-E_F)/kT}}{1 + \epsilon^{(E-E_F)/kT}} \approx \epsilon^{-(E_F-E)/kT} \tag{19-33}$$

where we have made use of the fact that $E_F - E \gg kT$ for $E \leq E_V$ (Fig. 19-8). Hence the number of holes per cubic meter in the valence band is

$$p = \int_{-\infty}^{E_V} \gamma(E_V - E)^{\frac{1}{2}} \epsilon^{-(E_F-E)/kT} \, dE \tag{19-34}$$

This integral, which represents the area under the bottom curve in Fig. 19-8c, evaluates to

$$p = N_V \epsilon^{-(E_F-E_V)/kT} \tag{19-35}$$

where N_V is given by Eq. (19-31), with m_n replaced by m_p, the effective mass of a hole.

The Fermi Level in an Intrinsic Semiconductor   It is important to note that Eqs. (19-30) and (19-35) apply to both intrinsic and extrinsic or impure semiconductors. In the case of intrinsic material the subscript i will be added to n and p. Since the crystal must be electrically neutral,

$$n_i = p_i \tag{19-36}$$

and we have from Eqs. (19-30) and (19-35)

$$N_C \epsilon^{-(E_C - E_F)/kT} = N_V \epsilon^{-(E_F - E_V)/kT}$$

Taking the logarithm of both sides, we obtain

$$\ln \frac{N_C}{N_V} = \frac{E_C + E_V - 2E_F}{kT}$$

Hence

$$E_F = \frac{E_C + E_V}{2} - \frac{kT}{2} \ln \frac{N_C}{N_V} \tag{19-37}$$

If the effective masses of a hole and a free electron are the same, $N_C = N_V$, and Eq. (19-37) yields

$$E_F = \frac{E_C + E_V}{2} \tag{19-38}$$

Hence the Fermi level lies in the center of the forbidden energy band, as shown in Fig. 19-8.

The Intrinsic Concentration Using Eqs. (19-30) and (19-35), we have for the product of electron-hole concentrations

$$np = N_C N_V \epsilon^{-(E_C - E_V)/kT} = N_C N_V \epsilon^{-E_G/kT} \tag{19-39}$$

Note that this product is independent of the Fermi level, but does depend upon the temperature and the energy gap $E_G \equiv E_C - E_V$. Equation (19-39) is valid for either an extrinsic or intrinsic material. Hence, writing $n = n_i$ and $p = p_i = n_i$, we have the important relationship (called the *mass-action law*)

$$np = n_i^2 \tag{19-40}$$

Note that, regardless of the donor or acceptor concentrations, or the individual magnitudes of n and p, the product is always a constant at a fixed temperature. Substituting numerical values for the physical constants in Eq. (19-31), we obtain

$$N_C = 4.82 \times 10^{21} \left(\frac{m_n}{m} \right)^{\frac{3}{2}} T^{\frac{3}{2}} \tag{19-41}$$

where N_C has the dimensions of a concentration (number per cubic meter). Note that N_V is given by right-hand side of Eq. (19-41) with m_n replaced by m_p. From Eqs. (19-39) to (19-41),

$$np = n_i^2 = (2.33 \times 10^{43}) \left(\frac{m_n m_p}{m^2} \right)^{\frac{3}{2}} T^3 \epsilon^{-E_G/kT} \tag{19-42}$$

As indicated in Eqs. (2-19) and (2-20), the energy gap decreases linearly with temperature, so that

$$E_G = E_{GO} - \beta T \tag{19-43}$$

where E_{GO} is the magnitude of the energy gap at $0°K$. Substituting this relationship into Eq. (19-42) gives an expression of the following form:

$$n_i{}^2 = A_o T^3 \epsilon^{-E_{GO}/kT} \tag{19-44}$$

This result has been verified experimentally.[4] The measured values of n_i and E_{GO} are given in Table 2-1.

19-6   FERMI LEVEL IN A SEMICONDUCTOR HAVING IMPURITIES

From Eqs. (2-16) and (2-17) it is seen that the electrical characteristics of a semiconductor material depend on the concentration of free electrons and holes. The expressions for n and p are given by Eqs. (19-30) and (19-35), respectively, and these are valid for both intrinsic semiconductors and semiconductors with impurities. The only parameter in Eqs. (19-30) and (19-35) which changes with impurities is the Fermi level E_F. In order to see how E_F depends on temperature and impurity concentration, we recall that, in the case of no impurities (an intrinsic semiconductor), E_F lies in the middle of the energy gap, indicating equal concentrations of free electrons and holes. If a donor-type impurity is added to the crystal, then, at a given temperature and assuming all donor atoms are ionized, the first N_D states in the conduction band will be filled. Hence it will be more difficult for the electrons from the valence band to bridge the energy gap by thermal agitation. Consequently, the number of electron-hole pairs thermally generated for that temperature will be reduced. Since the Fermi level is a measure of the probability of occupancy of the allowed energy states, it is clear that E_F must move closer to the conduction band to indicate that many of the energy states in that band are filled by the donor electrons, and fewer holes exist in the valence band. This situation is pictured in Fig. 19-9a for an n-type material. The same kind of argument leads to the conclusion that E_F must move from the center of the forbidden gap closer to the valence band for a p-type material, as indicated in Fig. 19-9b. If for a given concentration of impurities the temperature of, say, the n-type material increases, more electron-hole pairs will be formed, and since all donor atoms are ionized, it is possible that the concentration of thermally generated electrons in the conduction band may become much larger than the concentration of donor electrons. Under these conditions the concentrations of holes and electrons become almost equal and the crystal becomes essentially intrinsic. We can conclude that as the temperature of either n-type or p-type material increases, the Fermi level moves toward the center of the energy gap.

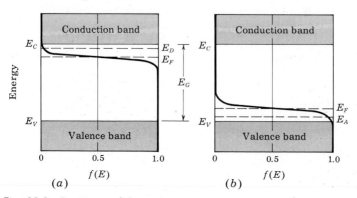

Fig. 19-9 Positions of Fermi level in (a) n-type and (b) p-type semiconductors.

A calculation of the exact position of the Fermi level in an n-type material can be made if we substitute $n = N_D$ from Eq. (2-12) into (19-30). We obtain

$$N_D = N_C \epsilon^{-(E_C - E_F)/kT} \tag{19-45}$$

or solving for E_F,

$$E_F = E_C - kT \ln \frac{N_C}{N_D} \tag{19-46}$$

Similarly, for p-type material, from Eqs. (2-15) and (19-35), we obtain

$$E_F = E_V + kT \ln \frac{N_V}{N_A} \tag{19-47}$$

Note that if $N_A = N_D$, Eqs. (19-46) and (19-47) added together (and divided by 2) yield Eq. (19-37).

19-7   BAND STRUCTURE OF AN OPEN–CIRCUIT p-n JUNCTION

We here consider that a p-n junction is formed by placing p- and n-type materials in intimate contact on an atomic scale. Under these conditions the Fermi level must be constant throughout the specimen at equilibrium. If this were not so, electrons on one side of the junction would have an average energy higher than those on the other side, and there would be a transfer of electrons and energy until the Fermi levels in the two sides did line up. In Sec. 19-6 it is verified that the Fermi level E_F is closer to the conduction band edge E_{Cn} in the n-type material and closer to the valence band edge E_{Vp} in the p side. Clearly, then, the conduction band edge E_{Cp} in the p material cannot be at the same level as E_{Cn}, nor can the valence band edge E_{Vn} in the n side line up with E_{Vp}. Hence the energy-band diagram for a p-n junction appear as shown in Fig. 19-10, where a shift in energy levels E_o is indicated. Note that

$$E_o = E_{Cp} - E_{Cn} = E_{Vp} - E_{Vn} = E_1 + E_2 \qquad (19\text{-}48)$$

This energy E_o represents the potential energy of the electrons at the junction, as is indicated in Fig. 3-1e.

The Contact Difference of Potential We now obtain an expression for E_o. From Fig. 19-10 we see that

$$E_F - E_{Vp} = \tfrac{1}{2}E_G - E_1 \qquad (19\text{-}49)$$

and

$$E_{Cn} - E_F = \tfrac{1}{2}E_G - E_2 \qquad (19\text{-}50)$$

Adding these two equations, we obtain

$$E_o = E_1 + E_2 = E_G - (E_{Cn} - E_F) - (E_F - E_{Vp}) \qquad (19\text{-}51)$$

From Eqs. (19-39) and (19-40),

$$E_G = kT \ln \frac{N_C N_V}{n_i^2} \qquad (19\text{-}52)$$

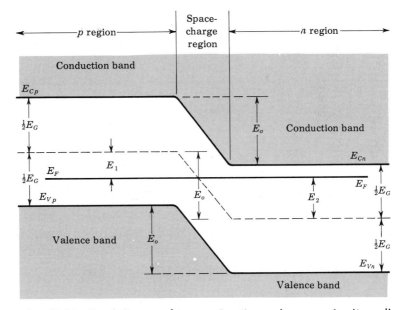

Fig. 19-10 Band diagram for a p-n junction under open-circuit conditions. This sketch corresponds to Fig. 3-1e and represents potential energy for electrons. The width of the forbidden gap is E_G in electron volts.

From Eqs. (19-30) and (2-15),

$$E_{Cn} - E_F = kT \ln \frac{N_C}{N_D} \tag{19-53}$$

From Eqs. (19-35) and (2-15),

$$E_F - E_{Vp} = kT \ln \frac{N_V}{N_A} \tag{19-54}$$

Substituting from Eqs. (19-52), (19-53), and (19-54) in Eq. (19-51) yields

$$E_o = kT \left(\ln \frac{N_C N_V}{n_i^2} - \ln \frac{N_C}{N_D} - \ln \frac{N_V}{N_A} \right)$$

$$= kT \ln \left(\frac{N_C N_V}{n_i^2} \frac{N_D}{N_C} \frac{N_A}{N_V} \right) = kT \ln \frac{N_D N_A}{n_i^2} \tag{19-55}$$

We emphasize that, in the above equations, the E's are expressed in electron volts and k has the dimensions of electron volts per degree Kelvin. The contact difference in potential V_o is expressed in volts and is *numerically* equal to E_o. Note that V_o depends only upon the equilibrium concentrations, and not at all upon the charge density in the transition region.

Other expressions for E_o are obtained by substituting Eqs. (2-13), (2-14), and (2-15) in Eq. (19-55). We find

$$E_o = kT \ln \frac{p_{po}}{p_{no}} = kT \ln \frac{n_{no}}{n_{po}} \tag{19-56}$$

where the subscripts o are added to the concentrations to indicate that these are obtained under conditions of thermal equilibrium.

The Einstein Relation   Since the net hole current is zero for an open-circuited junction, then from Eq. (2-39),

$$J_p = -qD_p \frac{dp}{dx} + q\mu_p p \mathcal{E} = 0$$

or

$$\frac{dp}{p} = \frac{\mu_p}{D_p} \mathcal{E} \, dx = -\frac{\mu_p}{D_p} \, dV \tag{19-57}$$

If this equation is integrated between limits which extend across the junction (Fig. 3-1) from the p side, where the equilibrium hole concentration is p_{po}, to the n side, where the hole density is p_{no}, the result is

$$\frac{D_p}{\mu_p} \ln \frac{p_{po}}{p_{no}} = V_o = \frac{1.60 \times 10^{-19} E_o}{q} \tag{19-58}$$

where Eq. (1-7) for the conversion from V_o (volts) to E_o (electron volts) has been used. Comparing Eqs. (19-56) and (19-58), it follows that

$$\frac{D_p}{\mu_p} = kT \frac{1.60 \times 10^{-19}}{q} = \frac{\bar{k}T}{q} = V_T \tag{19-59}$$

where $k(\text{eV}/°\text{T}) = \bar{k}(\text{J}/°\text{T})/(1.60 \times 10^{-19})$, and V_T is as defined in Eq. (19-22). Equation (19-59) is the Einstein relationship between the diffusion constant and mobility.

19-8   THE TUNNEL DIODE

When the concentration of impurity atoms in a p-n diode is very high (say, 1 part in 10^3), the depletion layer is reduced to about 100 Å. Classically, a carrier must have an energy at least equal to the potential-barrier height in order to cross the junction. However, quantum mechanics indicates that there is a nonzero probability that a particle may penetrate *through* a barrier as thin as that indicated above. This phenomenon is called *tunneling*, and hence these high-impurity-density p-n devices are called *tunnel diodes*, or *Esaki diodes*.[5] This same tunneling effect is responsible for radioactive emissions and high-field emission of electrons from a cold metal.

Energy-band Structure of a Highly Doped p-n Diode   The condition that the barrier be less than 100 Å thick is a necessary but not a sufficient condition for tunneling. It is also required that occupied energy states exist on the side from which the electron tunnels and that allowed empty states exist on the other side (into which the electron penetrates) at the same energy level. Hence we must now consider the energy-band picture when the impurity concentration is very high. In Fig. 19-10, drawn for the lightly doped p-n diode, the Fermi level E_F lies inside the forbidden energy gap. We shall now demonstrate that, for a diode which is doped heavily enough to make tunneling possible, E_F lies outside the forbidden band.

From Eq. (19-53),

$$E_F = E_C - kT \ln \frac{N_C}{N_D}$$

For a lightly doped semiconductor, $N_D < N_C$, so that $\ln (N_C/N_D)$ is a positive number. Hence $E_F < E_C$, and the Fermi level lies inside the forbidden band, as indicated in Fig. 19-10. Since $N_C \approx 10^{19}$ cm^{-3}, then, for donor concentrations in excess of this amount ($N_D > 10^{19}$ cm^{-3}, corresponding to a doping in excess of 1 part in 10^3), $\ln (N_C/N_D)$ is negative. Hence $E_F > E_C$, and the Fermi level in the n-type material lies in the conduction band. By similar reasoning we conclude that, for a heavily doped p region, $N_A > N_V$, and the Fermi level lies in the valence band [Eq. (19-54)]. A comparison of Eqs. (19-52) and (19-55) indicates that $E_o > E_G$, so that the contact difference of potential energy E_o now exceeds the forbidden-energy-gap voltage E_G. Hence, under open-circuit conditions, the band structure of a heavily doped p-n junction must be as pictured in Fig. 19-11a. The Fermi level E_F in the p side is at the same energy as the Fermi level E_F in the n side. Note that there are no filled states on one side of the junction which are at the same energy as empty

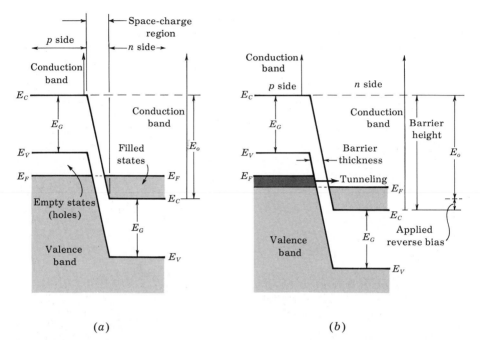

(a) (b)

Fig. 19-11 Energy bands in a heavily doped p-n diode (a) under open-circuited conditions and (b) with an applied reverse bias. (These diagrams are strictly valid only at 0°K, but are closely approximated at room temperature, as can be seen from Fig. 19-3.)

allowed states on the other side. Hence there can be no flow of charge in either direction across the junction, and the current is zero, an obviously correct conclusion for an open-circuited diode.

The Volt-Ampere Characteristic With the aid of the energy-band picture of Fig. 19-12 and the concept of quantum-mechanical tunneling, the tunnel-diode characteristic of Fig. 19-13 may be explained. Let us consider that the p material is grounded and that a voltage applied across the diode shifts the n side potential with respect to the p side. For example, if a reverse-bias voltage is applied, we know from Sec. 3-2 that the height of the barrier is increased above the open-circuit value E_o. Hence the n-side levels must shift downward with respect to the p-side levels, as indicated in Fig. 19-11b. We now observe that there are some energy states (the heavily shaded region) in the valence band of the p side which lie at the same level as allowed empty states in the conduction band of the n side. Hence these electrons will tunnel from the p to the n side, giving rise to a reverse diode current. As the magnitude of the reverse bias increases, the heavily shaded area grows in size, causing the reverse current to increase, as shown by section 1 of Fig. 19-13.

Consider now that a forward bias is applied to the diode so that the potential barrier is decreased below E_o.   Hence the n-side levels must shift upward with respect to those on the p side, and the energy-band picture for this situation is indicated in Fig. 19-12a.   It is now evident that there are occupied states in the conduction band of the n material (the heavily shaded levels) which are at the same energy as allowed empty states (holes) in the valence band of the p side.   Hence electrons will tunnel from the n to the p material, giving rise to the forward current of section 2 of Fig. 19-13.

As the forward base is increased further, the condition shown in Fig. 19-12b is reached.   Now the maximum number of electrons can leave occupied states on the right side of the junction, and tunnel through the barrier to empty states on the left side, giving rise to the peak current I_P in Fig. 19-13. If still more forward bias is applied, the situation in Fig. 19-12c is obtained,

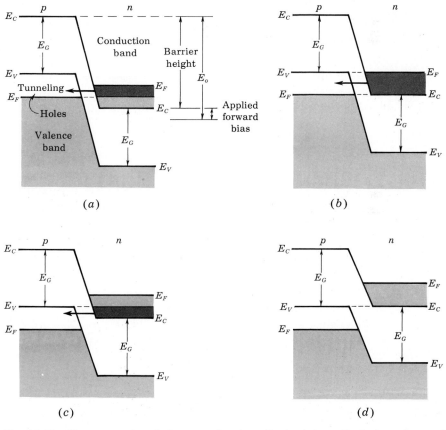

Fig. 19-12   The energy-band diagrams in a heavily doped p-n diode for a forward bias.   As the bias is increased, the band structure changes progressively from (a) to (d).

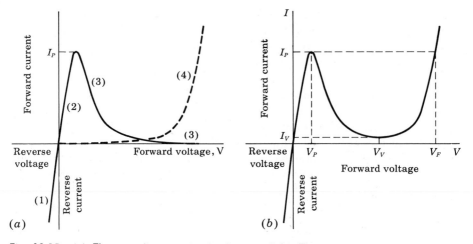

Fig. 19-13 (a) The tunneling current is shown solid. The injection current is the dashed curve. The sum of these two gives the tunnel-diode volt-ampere characteristic, which is shown in (b).

and the tunneling current decreases, giving rise to section 3 of Fig. 19-13. Finally, at an even larger forward bias, the band structure of Fig. 19-12d is valid. Since now there are no empty *allowed* states on one side of the junction at the same energy as occupied states on the other side, the tunneling current must drop to zero.

In addition to the quantum-mechanical current described above, the regular p-n junction injection current is also being collected. This current is given by Eq. (3-7) and is indicated by the dashed section 4 of Fig. 19-13. The curve in Fig. 19-13b is the sum of the solid and dashed curves of Fig. 19-13a, and this resultant is the tunnel-diode characteristic of Fig. 19-13b.

19-9   BASIC SEMICONDUCTOR EQUATIONS

The relationship which expresses the fact that, within any elemental volume in a semiconductor, charge can neither be created nor destroyed is derived in Sec. 2-10 and is given by Eq. (2-49). This *law of the conservation of charge,* or the *equation of continuity,* applies separately to electron and to hole concentrations. Thus

$$\frac{\partial p}{\partial t} = \frac{p_o - p}{\tau_p} - \frac{1}{q}\frac{\partial J_p}{\partial x} \tag{19-60}$$

$$\frac{\partial n}{\partial t} = \frac{n_o - n}{\tau_n} + \frac{1}{q}\frac{\partial J_n}{\partial x} \tag{19-61}$$

where $p(n)$ = hole (electron) concentration

$p_o(n_o)$ = thermal-equilibrium value of $p(n)$

$J_p(J_n)$ = hole (electron) current density

$\tau_p(\tau_n)$ = hole (electron) mean lifetime

q = magnitude of electronic charge

and for reference, in the following equations,

$\mu_p(\mu_n)$ = hole (electron) mobility

$D_p(D_n)$ = hole (electron) diffusion constant

$N_A(N_D)$ = acceptor (donor) concentration

$\mathcal{E}$ = electric field intensity

ϵ = permittivity of the semiconductor

If both drift and diffusion currents are to be taken into account, Eqs. (2-39) and (2-40) must be added to give

$$J_p = qp\mu_p\mathcal{E} - qD_p\frac{\partial p}{\partial x} \tag{19-62}$$

$$J_n = qn\mu_n\mathcal{E} + qD_n\frac{\partial n}{\partial x} \tag{19-63}$$

The electric field intensity $\mathcal{E}$ is related to the charge density ρ by Poisson's equation

$$\frac{\partial \mathcal{E}}{\partial x} = \frac{\rho}{\epsilon} = \frac{q}{\epsilon}(p + N_D - n - N_A) \tag{19-64}$$

where the net positive carrier concentration is $p + N_D - n - N_A$, assuming that all impurity atoms are ionized. Note that in any region where there is charge neutrality, $\rho = 0$. In the transition region of a p-n diode we do not have charge neutrality (Sec. 3-1).

Equations (19-60) through (19-64) are five relations for the five unknowns p, n, $\mathcal{E}$, J_p, and J_n. It has been assumed that a one-dimensional problem is under consideration. These equations are valid in three dimensions if derivatives in y and z are included corresponding to the derivative in x. The general solution to these nonlinear partial differential equations is quite complicated, and we shall be content with analyses of several important one-dimensional special cases.

For an n-type specimen we add the subscript n to p, p_o, and J_p. Substituting Eq. (19-62) into Eq. (19-60) yields

$$\frac{\partial p_n}{\partial t} = \frac{p_{no} - p_n}{\tau_p} - \mu_p\frac{\partial(p_n\mathcal{E})}{\partial x} + D_p\frac{\partial^2 p_n}{\partial x^2} \tag{19-65}$$

An analogous equation for minority carriers (electrons) in a p-type semiconductor is obtained by interchanging p and n and changing the sign of the term with μ_n. We now consider three special cases of the continuity equation.

Concentration Independent of x and with Zero Electric Field Since $\mathcal{E} = 0$ and $\partial p/\partial x = 0$, Eq. (19-65) reduces to

$$\frac{dp_n}{dt} = \frac{p_{no} - p_n}{\tau_p} \tag{19-66}$$

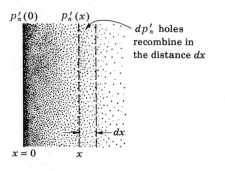

$p_n'(0)$ $p_n'(x)$

dp_n' holes recombine in the distance dx

dx

$x = 0$ x

Fig. 19-14 Relating to the injected hole concentration p_n' in n-type material.

This is identical with Eq. (2-32), and this situation is discussed in Sec. 2-8 and Fig. 2-13.

Concentration Independent of t and with Zero Electric Field Since $\mathcal{E} = 0$ and $\partial p/\partial t = 0$, Eq. (19-65) reduces to

$$\frac{d^2 p_n}{dx^2} = \frac{p_n - p_{no}}{L_p} \tag{19-67}$$

where $L_p \equiv \sqrt{D_p \tau_p}$ = the diffusion length for holes. Equation (19-67) is identical with Eq. (2-43) and is discussed in Sec. 2-11. Consider a very long semiconductor bar with one end at $x = 0$, where $p_n'(0)$ holes are injected (above the thermal-equilibrium value). The variation of injected concentration with distance is pictured in Figs. 2-16 and 19-14, and is given by Eq. (2-47), namely,

$$p_n'(x) \equiv p_n(x) - p_{no} = p_n'(0)\epsilon^{-x/L_p} \tag{19-68}$$

The diffusion length is the distance at which the injected carrier concentration drops to $1/\epsilon$ of its value at $x = 0$. We wish to demonstrate that L_p may also be interpreted as the average distance which an injected hole travels before recombining with an electron. This statement may be verified as follows: From Eq. (19-68)

$$|dp_n'| = \frac{p_n'(0)\epsilon^{-x/L_p}}{L_p}\, dx \tag{19-69}$$

$|dp_n'|$ gives the number of injected holes which recombine in the distance between x and $x + dx$. Since each hole has traveled a distance x, the total distance traveled by $|dp_n'|$ holes is $x\,|dp_n'|$. Hence the total distance covered by all the holes is $\int_0^\infty x\,|dp_n'|$. The average distance $\bar{x}$ equals this total distance divided by the total number $p_n'(0)$ of injected holes. Hence

$$\bar{x} \equiv \frac{\int_0^\infty x|dp_n'|}{p_n'(0)} = \frac{1}{L_p}\int_0^\infty x\epsilon^{-x/L_p}\, dx = L_p \tag{19-70}$$

thus confirming that the mean distance of travel of a hole before recombination is L_p.

Concentration Varies Sinusoidally with t and with Zero Electric Field
Let us retain the restriction $\mathcal{E} = 0$ but assume that the injected concentration varies sinusoidally with an angular frequency ω. Then, in phasor notation,

$$p'_n(x, t) = p'_n(x)\epsilon^{j\omega t} \tag{19-71}$$

where the space dependence of the injected concentration is given by $p'_n(x)$. If Eq. (19-71) is substituted into the continuity equation (19-65), the result is

$$j\omega p'_n(x) = -\frac{p'_n(x)}{\tau_p} + D_p\frac{d^2 p'_n(x)}{dx^2}$$

or

$$\frac{d^2 p'_n}{dx^2} = \frac{1 + j\omega\tau_p}{L_p{}^2}p'_n \tag{19-72}$$

where use has been made of Eq. (2-44). At zero frequency the equation for p'_n is given by Eq. (19-67). A comparison of Eq. (19-67) with Eq. (19-72) shows that the ac solution at frequency $\omega \neq 0$ can be obtained from the dc solution ($\omega = 0$) by replacing L_p by $L_p(1 + j\omega\tau_p)^{-\frac{1}{2}}$. For example, from Eq. (2-47), we conclude that for a long semiconductor sample

$$p'_n(x, t) = K\epsilon^{-(1+j\omega\tau_p)^{\frac{1}{2}}x/L_p}\epsilon^{j\omega t} \tag{19-73}$$

19-10   THE p-n DIODE VOLT–AMPERE EQUATION

Consider a p-n junction biased by an external voltage V in the forward direction as in Fig. 19-15. Under low-level injection conditions the minority diffusion current crossing the junction at $x = 0$ is, from Eq. (2-48),

$$I_{pn}(0) = -AqD_p\frac{dp_n}{dx} = \frac{AqD_p p'_n(0)}{L_p} \tag{19-74}$$

The injected carrier concentration $p'_n(0)$ at the junction depends upon the applied voltage V. We now find this relationship.

Fig. 19-15   A forward-biased diode. The right-hand edge of the transition region is defined as $x = 0$.

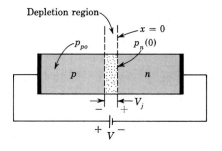

The Law of the Junction Across the junction the electric field is high, and hence so is the drift current. Also, the diffusion current is large because of the high-concentration gradient at the junction. In Eq. (19-62), namely

$$J_p = q p \mu_p \mathcal{E} - q D_p \frac{dp}{dx}$$

the magnitude of each term on the right-hand side is large compared with the low-level injection current J_p. If two large numbers are subtracted and we obtain a small difference, these two quantities are almost equal. Hence we may write approximately

$$q p \mu_p \mathcal{E} = q D_p \frac{dp}{dx} \tag{19-75}$$

Using the Einstein relationship, Eq. (19-75) becomes

$$\mathcal{E} = \frac{V_T}{p} \frac{dp}{dx} = -\frac{dV}{dx} \tag{19-76}$$

At the left-hand side of the p material the hole concentration equals the thermal-equilibrium value p_{po}. At the right-hand edge of the depletion region in Fig. 19-15 (in the n side) the hole concentration is $p_n(0)$. The junction voltage V_j is the barrier potential V_o decreased by the forward voltage V. Integrating Eq. (19-76) across the junction gives

$$\int_{p_{po}}^{p_n(0)} \frac{dp}{p} = -\int_0^{V_j} \frac{dV}{V_T} = -\int_0^{V_o - V} \frac{dV}{V_T} \tag{19-77}$$

Carrying out the integration yields

$$p_n(0) = p_{po} \epsilon^{-(V_o - V)/V_T} \tag{19-78}$$

From Eq. (2-59) with $p_1 = p_{po}$, $p_2 = p_{no}$, and $V_{21} = V_o$

$$p_{no} = p_{po} \epsilon^{-V_o/V_T} \tag{19-79}$$

Dividing Eq. (19-78) by Eq. (19-79) yields

$$p_n(0) = p_{no} \epsilon^{V/V_T} \tag{19-80}$$

This boundary condition is called the *law of the junction*. It indicates that for a forward bias ($V > 0$) and with $V \gg V_T$ (~ 26 mV at room temperature) the hole concentration $p_n(0)$ at the junction in the n side is greatly increased over the thermal-equilibrium value p_{no}. The argument given above is also valid for a reverse bias, and Eq. (19-80) indicates that for negative V with $|V| \gg V_T$, the concentration $p_n(0)$ is essentially zero.

The Current Components From Eq. (19-74), with $p'_n(0) = p_n(0) - p_{no}$

$$I_{pn}(0) = \frac{A q D_p p_{no}}{L_p} (\epsilon^{V/V_T} - 1) \tag{19-81}$$

The electron current $I_{np}(0)$ crossing the junction into the p side is obtained from Eq. (19-81) by interchanging n and p, or

$$I_{np}(0) = \frac{AqD_n n_{po}}{L_n} (\epsilon^{V/V_T} - 1) \tag{19-82}$$

Finally, from Eq. (3-6), the total diode current I is the sum of $I_{pn}(0)$ and $I_{np}(0)$, or

$$I = I_o(\epsilon^{V/V_T} - 1) \tag{19-83}$$

where

$$I_o \equiv \frac{AqD_p p_{no}}{L_p} + \frac{AqD_n n_{po}}{L_n} \tag{19-84}$$

If W_p and W_n are the widths of the p and n materials, respectively, the above derivation has implicity assumed that $W_p \gg L_n$ and $W_n \gg L_p$. If, as sometimes happens in a practical (short) diode, the widths are much smaller than the diffusion lengths, the expression for I_o remains valid provided that L_n and L_p are replaced by W_p and W_n, respectively.

The Reverse Saturation Current For a reverse bias whose magnitude is large compared with V_T, $I \rightarrow -I_o$. Hence I_o is called the *reverse saturation current*. Combining Eqs. (2-14), (2-15), and (19-84), we obtain

$$I_o = Aq \left(\frac{D_p}{L_p N_D} + \frac{D_n}{L_n N_A} \right) n_i^2 \tag{19-85}$$

where n_i^2 is as given by Eq. (19-44),

$$n_i^2 = A_o T^3 \epsilon^{-E_{Go}/kT} = A_o T^3 \epsilon^{-V_{Go}/V_T} \tag{19-86}$$

where V_{GO} is a voltage which is numerically equal to the forbidden-gap energy E_{GO} in electron volts, and V_T is the volt equivalent of temperature [Eq. (19-22)]. For germanium the diffusion constants D_p and D_n vary approximately[3] inversely proportional to T. Hence the temperature dependence of I_o is

$$I_o = K_1 T^2 \epsilon^{-V_{Go}/V_T} \tag{19-87}$$

where K_1 is a constant independent of temperature.

Throughout this section we have neglected carrier generation and recombination in the space-charge region. Such an assumption is valid for a germanium diode, but not for a silicon device. For the latter, the diffusion current is negligible compared with the transition-layer charge-generation[6,7] current, which is given approximately by

$$I = I_o(\epsilon^{V/\eta V_T} - 1) \tag{19-88}$$

where $\eta \approx 2$ for small (rated) currents and $\eta \approx 1$ for large currents. Also, I_o is now found to be proportional to n_i instead of n_i^2. Hence, if K_2 is a constant,

$$I_o = K_2 T^{1.5} \epsilon^{-V_{Go}/2V_T} \tag{19-89}$$

19-11   THE TEMPERATURE DEPENDENCE OF p-n CHARACTERISTICS

Let us inquire into the diode voltage variation with temperature at fixed current. This variation may be calculated from Eq. (19-88), where the temperature is contained implicitly in V_T and also in the reverse saturation current. The dependence of I_o on temperature T is, from Eqs. (19-87) and (19-89), given approximately by

$$I_o = KT^m \epsilon^{-V_{GO}/\eta V_T} \tag{19-90}$$

where K is a constant and qV_{GO} (q is the magnitude of the electronic charge) is the forbidden-gap energy in joules:

For Ge: $\eta = 1$ $m = 2$ $V_{GO} = 0.785$ V
For Si: $\eta = 2$ $m = 1.5$ $V_{GO} = 1.21$ V

Taking the derivative of the logarithm of Eq. (19-90), we find

$$\frac{1}{I_o}\frac{dI_o}{dT} = \frac{d(\ln I_o)}{dT} = \frac{m}{T} + \frac{V_{GO}}{\eta T V_T} \tag{19-91}$$

At room temperature, we deduce from Eq. (19-91) that $d(\ln I_o)/dT = 0.08°\text{C}^{-1}$ for Si and $0.11°\text{C}^{-1}$ for Ge. The performance of commercial diodes is only approximately consistent with these results. In Sec. 3-5 it is pointed out that a reasonable rule of thumb is that, for either silicon or germanium, the reverse saturation current approximately doubles for every 10°C rise in temperature.

From Eq. (19-88), dropping the unity in comparison with the exponential, we find, for constant I,

$$\frac{dV}{dT} = \frac{V}{T} - \eta V_T \left(\frac{1}{I_o}\frac{dI_o}{dT}\right) = \frac{V - (V_{GO} + m\eta V_T)}{T} \tag{19-92}$$

where use has been made of Eq. (19-91). Consider a diode operating at room temperature (300°K) and just beyond the threshold voltage V_γ (say, at 0.2 V for Ge and 0.6 for Si). Then we find, from Eq. (19-92),

$$\frac{dV}{dT} = \begin{cases} -2.1 \text{ mV/°C} & \text{for Ge} \\ -2.3 \text{ mV/°C} & \text{for Si} \end{cases} \tag{19-93}$$

Since these data are based on "average characteristics," it might be well for conservative design to assume a value of

$$\frac{dV}{dT} = -2.5 \text{ mV/°C} \tag{19-94}$$

for either Ge or Si at room temperature. Note from Eq. (19-92) that $|dV/dT|$ decreases with increasing T.

The temperature dependence of forward voltage is given in Eq. (19-92) as the difference between two terms. The positive term V/T on the right-hand side results from the temperature dependence of V_T. The negative term results from the temperature dependence of I_o, and does not depend on the voltage V across the diode.   The equation predicts that for increasing V, dV/dT should become less negative, reach zero at $V = V_{GO} + m\eta V_T$, and thereafter reverse sign and go positive.   This behavior is regularly exhibited by diodes.   Normally, however, the reversal takes place at a current which is higher than the maximum rated current.   The curves of Fig. 3-8 also suggest this behavior.   At high voltages the horizontal separation between curves of different temperatures is smaller than at low voltages.

19-12   THE DYNAMIC DIFFUSION CAPACITANCE

In Sec. 3-9 it is pointed out that if the excitation to a diode is sinusoidal, the diffusion capacitance C_D may be found from the reactive component of the current.   We now proceed to find C_D.

Assume for the sake of simplicity that the p side is doped much more heavily than the n region, so that we need only calculate the hole current. Consider that the external voltage is

$$V = V_1 + V_m \epsilon^{j\omega t} \tag{19-95}$$

where V_1 represents a bias voltage, and the second term is the sinusoidal voltage of peak value V_m.   If the diode has a dynamic conductance g and capacitance C_D, then the current must be of the form

$$I = I_1 + g V_m \epsilon^{j\omega t} + j\omega C_D V_m \epsilon^{j\omega t} \tag{19-96}$$

We now solve the equation of continuity to find $I = I_{pn}$, and then we shall be able to identify g and C_D.

The excess hole concentration in the n side consists of a dc term which, from Eqs. (2-47) and (19-80), is

$$p_n'(x) = [p_n(0) - p_{no}]\epsilon^{-x/L_p} = p_{no}(\epsilon^{V_1/V_T} - 1)\epsilon^{-x/L_p}$$

and to which must be added the ac term of Eq. (19-73), so that

$$p_n'(x) = p_{no}(\epsilon^{V_1/V_T} - 1)\epsilon^{-x/L_p} + K\epsilon^{-(1+j\omega\tau_p)^{\frac{1}{2}}x/L_p}\epsilon^{j\omega t} \tag{19-97}$$

To evaluate K, set $x = 0$ and also use the law of the junction with V given by Eq. (19-95).   Then Eq. (19-97) becomes

$$p_n(0) - p_{no} = p_{no}(\epsilon^{V_1/V_T} - 1) + K\epsilon^{j\omega t}$$

$$= p_{no}\epsilon^{(V_1+V_m\epsilon^{j\omega t})/V_T} - p_{no}$$

$$\approx p_{no}\epsilon^{V_1/V_T}\left(1 + \frac{V_m}{V_T}\epsilon^{j\omega t}\right) - p_{no} \tag{19-98}$$

where we have assumed that $V_m/V_T \ll 1$, so that $\epsilon^y \approx 1 + y$ for

$$y = \left(\frac{V_m}{V_T}\right)\epsilon^{j\omega t} \ll 1$$

From this equation, we find

$$K = p_{no} \frac{V_m}{V_T} \epsilon^{V_1/V_T} \tag{19-99}$$

The diffusion current crossing the junction,

$$I_{pn}(0) = -AqD_p \left(\frac{dp'_n}{dx}\right)\bigg|_{x=0}$$

is obtained from Eq. (19-74). Using the value of K in Eq. (19-99), we have

$$I_{pn}(0) = \frac{AqD_p p_{no}}{L_p} (\epsilon^{V_1/V_T} - 1) + \frac{AqD_p p_{no} V_m \epsilon^{V_1/V_T}}{V_T} \frac{(1 + j\omega\tau_p)^{\frac{1}{2}}}{L_p} \epsilon^{j\omega t} \tag{19-100}$$

We consider two special cases, very low and very high frequencies:

1. If $\omega\tau_p \ll 1$, then since $(1 + \psi)^{\frac{1}{2}} \approx 1 + \frac{1}{2}\psi$,

$$I_{pn}(0) = I_1 + \frac{AqD_p p_{no}}{L_p} \frac{V_m}{V_T} \epsilon^{V_1/V_T} \left(1 + j\frac{\omega\tau_p}{2}\right) \epsilon^{j\omega t} \tag{19-101}$$

where $I_1 \equiv (AqD_p p_{no})/L_p)(\epsilon^{V_1/V_T} - 1) =$ the bias current. Comparing Eq. (19-101) with Eq. (19-96), we have

$$g = \frac{AqD_p p_{no} \epsilon^{V_1/V_T}}{L_p V_T} = g_o = \text{low-frequency conductance} \tag{19-102}$$

and

$$C_D = \frac{g\tau_p}{2} \tag{19-103}$$

2. If $\omega\tau_p \gg 1$, then

$$(1 + j\omega\tau_p)^{\frac{1}{2}} \approx (j\omega\tau_p)^{\frac{1}{2}} = (\omega\tau_p)^{\frac{1}{2}}\epsilon^{j\pi/4} = \left(\frac{\omega\tau_p}{2}\right)^{\frac{1}{2}} (1 + j)$$

and from Eq. (19-100),

$$I_{pn}(0) = I_1 + g_o \left(\frac{\omega\tau_p}{2}\right)^{\frac{1}{2}} (1 + j) V_m \epsilon^{j\omega t} \tag{19-104}$$

Comparing this equation with Eq. (19-96) yields

$$g = g_o \left(\frac{\omega\tau_p}{2}\right)^{\frac{1}{2}} \tag{19-105}$$

and

$$C_D = g_o \left(\frac{\tau_p}{2\omega}\right)^{\frac{1}{2}} \tag{19-106}$$

Note that both the diffusion capacitance and the conductance are now functions of frequency.

19-13   THE CURRENTS IN A TRANSISTOR

This analysis follows in many respects that given in Sec. 19-10 for the current components in a junction diode.   From Eq. (3-6) we see that the net current crossing a junction equals the sum of the electron current I_{np} in the p side and the hole current I_{pn} in the n side, evaluated at the junction ($x = 0$).   For a p-n-p transistor (Fig. 5-1a) electrons are injected from the base region across the emitter junction in a p-region which is large compared with the diffusion length.   This is precisely the condition that exists in a junction diode, and hence the expression for I_{np} calculated previously is also valid for the transistor.   From Eq. (19-82) we find that at the junction,

$$I_{np}(0) = \frac{AqD_n n_{Eo}}{L_E} (\epsilon^{V_E/V_T} - 1) \tag{19-107}$$

where in Eq. (19-82) we have replaced V by V_E; we have changed n_{po} to n_{EO} because there are now two p regions and the emitter (E) is under consideration; we have changed L_n to L_E in order to refer to the diffusion length of the minority carriers in the emitter.   A summary of the new symbols used follows:

n_{Eo} (n_{Co}) = thermal-equilibrium electron concentration in the p-type material of the emitter (collector), m^{-3}

L_E (L_C) (L_B) = diffusion length for minority carriers in the emitter (collector) (base), m

V_E (V_C) = voltage drop across emitter (collector) junction; positive for a forward bias, i.e., for the p side positive with respect to the n side

The Hole Current in the n-type Base Region   The value of I_{pn} is not that found in Sec. 19-10 for a diode because, in the transistor, the hole current exists in a base region of small width, whereas in a diode, the n region extends over a distance large compared with L_n.   The hole concentration is given by Eq. (2-46):

$$p_n - p_{no} = K_1 \epsilon^{-x/L_B} + K_2 \epsilon^{+x/L_B} \tag{19-108}$$

where K_1 and K_2 are constants to be determined by the boundary conditions. The situation at each junction is exactly as for the diode junction, and the boundary condition is that given by Eq. (19-80), or

$$p_n = \begin{cases} p_{no}\epsilon^{V_E/V_T} & \text{at } x = 0 \\ p_{no}\epsilon^{V_C/V_T} & \text{at } x = W \end{cases} \tag{19-109}$$

The exact solution is not difficult to find (Prob. 19-32).   Usually, however, the base width W is small compared with L_B, and we can simplify the solution by introducing this inequality.   Since $0 \le x \le W$, we shall assume that $x/L_B \ll 1$, and then the exponentials in Eq. (19-108) can be expanded into a power series. If only the first two terms are retained, this equation has the form

$$p_n - p_{no} = K_3 + K_4 x \qquad (19\text{-}110)$$

where K_3 and K_4 are new (and, as yet, undetermined) constants. To this approximation, p_n is a linear function of distance in the base. Then, from Eqs. (2-36) and (19-110),

$$I_{pn} = -AqD_pK_4 = \text{constant} \qquad (19\text{-}111)$$

This result—that the minority-carrier current is a constant throughout the base region—is readily understood because we have assumed that $W \ll L_B$. Under these circumstances, little recombination can take place within the base, and hence the hole current entering the base at the emitter junction leaves the base at the collector junction unattenuated. Substituting the boundary conditions (19-109) in (19-110), we easily solve for K_4, and then find

$$I_{pn}(0) = -\frac{AqD_pp_{no}}{W}[(\epsilon^{V_C/V_T} - 1) - (\epsilon^{V_E/V_T} - 1)] \qquad (19\text{-}112)$$

The Ebers-Moll Equations From Fig. 5-4 we have for the emitter current

$$I_E = I_{pE} + I_{nE} = I_{pn}(0) + I_{np}(0) \qquad (19\text{-}113)$$

Using Eqs. (19-107), (19-112), and (19-113), we find

$$I_E = a_{11}(\epsilon^{V_E/V_T} - 1) + a_{12}(\epsilon^{V_C/V_T} - 1) \qquad (19\text{-}114)$$

where

$$a_{11} = Aq\left(\frac{D_pp_{no}}{W} + \frac{D_nn_{Eo}}{L_E}\right) \qquad a_{12} = -\frac{AqD_pp_{no}}{W} \qquad (19\text{-}115)$$

In a similar manner we can obtain

$$I_C = a_{21}(\epsilon^{V_E/V_T} - 1) + a_{22}(\epsilon^{V_C/V_T} - 1) \qquad (19\text{-}116)$$

where we can show (Prob. 19-31) that

$$a_{21} = -\frac{AqD_pp_{no}}{W} \qquad a_{22} = Aq\left(\frac{D_pp_{no}}{W} + \frac{D_nn_{Co}}{L_C}\right) \qquad (19\text{-}117)$$

We note that $a_{12} = a_{21}$. This result may be shown[8] to be valid for a transistor possessing any geometry. Equations (19-114) and (19-116) are valid for any positive or negative value of V_E or V_C, and they are known as the *Ebers-Moll equations*, relating the transistor currents to the junction voltages.

From the Ebers-Moll Eqs. (5-24) and (5-25) expressions for I_E and I_C in terms of V_E, V_C, and the parameters α_N, α_I, I_{CO}, and I_{EO} may be obtained. If these equations are put in the form of Eqs. (19-114) and (19-116) and a_{12} is equated to a_{21}, the result is $\alpha_N I_{EO} = \alpha_I I_{CO}$.

19-14   THE TRANSISTOR ALPHA

If V_E is eliminated from Eqs. (19-114) and (19-116), the result is

$$I_C = \frac{a_{21}}{a_{11}} I_E + \left(a_{22} - \frac{a_{21}a_{12}}{a_{11}}\right) (\epsilon^{V_C/V_T} - 1) \tag{19-118}$$

This equation has the same form as Eq. (5-6).   Hence we have, by comparison,

$$\alpha \equiv -\frac{a_{21}}{a_{11}} \tag{19-119}$$

$$I_{CO} \equiv \frac{a_{21}a_{12}}{a_{11}} - a_{22} \tag{19-120}$$

Using Eqs. (19-115) and (19-117), we obtain

$$\alpha = \frac{1}{1 + D_n n_{Eo} W / L_E D_p p_{no}} \tag{19-121}$$

Making use of Eq. (2-8) for the conductivity, Eq. (19-59) for the diffusion constant, and Eq. (2-10) for the concentration, Eq. (19-121) reduces to

$$\alpha = \frac{1}{1 + W\sigma_B / L_E \sigma_E} \tag{19-122}$$

where σ_B (σ_E) is the conductivity of the base (emitter).   We see that, in order to keep α close to unity, σ_E/σ_B should be large and W/L_E should be kept small.

The analysis of the preceding section is based upon the assumption that $W/L_B \ll 1$.   If this restriction is removed, the solution given in Prob. 19-31 is obtained.   We then find (Prob. 19-33) that $\alpha = \beta^*\gamma$, where

$$\gamma = \frac{1}{1 + (D_n L_B n_{Eo}/ D_p L_E p_{no}) \tanh (W/L_B)} \tag{19-123}$$

and

$$\beta^* = \operatorname{sech} \frac{W}{L_B} \tag{19-124}$$

If $W \ll L_B$, the hyperbolic secant and the hyperbolic tangent can be expanded in powers of W/L_B, and the first approximations are (Prob. 19-34)

$$\gamma \approx \frac{1}{1 + W\sigma_B/L_E\sigma_E} \approx 1 - \frac{W\sigma_B}{L_E\sigma_E} \tag{19-125}$$

$$\beta^* \approx 1 - \frac{1}{2}\left(\frac{W}{L_B}\right)^2 \tag{19-126}$$

and

$$\alpha = \beta^*\gamma \approx 1 - \frac{1}{2}\left(\frac{W}{L_B}\right)^2 - \frac{W\sigma_B}{L_E\sigma_E} \tag{19-127}$$

19-15 ANALYSIS OF TRANSISTOR CUTOFF AND SATURATION REGIONS

The currents and voltages in a transistor may be found from the Ebers-Moll equations of Sec. 5-12, which are repeated here for convenient reference.

$$I_C = -\alpha_N I_E - I_{CO}(\epsilon^{V_C/V_T} - 1) \tag{19-128}$$

$$I_E = -\alpha_I I_C - I_{EO}(\epsilon^{V_E/V_T} - 1) \tag{19-129}$$

$$\alpha_I I_{CO} = \alpha_N I_{EO} \tag{19-130}$$

$$V_E = V_T \ln\left(1 - \frac{I_E + \alpha_I I_C}{I_{EO}}\right) \tag{19-131}$$

$$V_C = V_T \ln\left(1 - \frac{I_C + \alpha_N I_E}{I_{CO}}\right) \tag{19-132}$$

The Cutoff Region   If we define *cutoff* as we did in Sec. 5-7 to mean zero emitter current and reverse saturation current in the collector, what emitter-junction voltage is required for cutoff? Equation (19-131) with $I_E = 0$ and $I_C = I_{CO}$ becomes

$$V_E = V_T \ln\left(1 - \frac{\alpha_I I_{CO}}{I_{EO}}\right) = V_T \ln(1 - \alpha_N) \tag{19-133}$$

where use was made of Eq. (19-130). At 25°C, $V_T = 26$ mV, and for $\alpha_N = 0.99$, $V_E = -120$ mV. Near cutoff we may expect that α_N may be smaller than the nominal value of 0.98. With $\alpha_N = 0.9$ for germanium, we find that $V_E = -60$ mV. For silicon near cutoff, $\alpha_N \approx 0$, and from Eq. (19-133), $V_E \approx V_T \ln 1 = 0$ V. The voltage V_E is the drop from the p to the n side of the emitter junction. To find the voltage which must be applied between base and emitter terminals, we must in principle take account of the drop across the base-spreading resistance $r_{bb'}$ in Fig. 5-20. If $r_{bb'} = 100$ Ω and $I_{CO} = 2$ μA, then $I_{CO}r_{bb'} = 0.2$ mV, which is negligible. Since the emitter current is zero, the potential V_E is called the *floating emitter potential*.

The foregoing analysis indicates that a reverse bias of approximately 0.1 V (0 V) will cut off a germanium (silicon) transistor. It is interesting to determine what currents will flow if a larger reverse input voltage is applied. Assuming that both V_E and V_C are negative and much larger than V_T, so that the exponentials may be neglected in comparison with unity, Eqs. (19-128) and (19-129) become

$$I_C = -\alpha_N I_E + I_{CO} \qquad I_E = -\alpha_I I_C + I_{EO} \tag{19-134}$$

Solving these equations and using Eq. (19-130), we obtain

$$I_C = \frac{I_{CO}(1 - \alpha_I)}{1 - \alpha_N \alpha_I} \qquad I_E = \frac{I_{EO}(1 - \alpha_N)}{1 - \alpha_N \alpha_I} \tag{19-135}$$

Since (for Ge) $\alpha_N \approx 1$, $I_C \approx I_{CO}$ and $I_E \approx 0$. Using $\alpha_N = 0.9$ and $\alpha_I = 0.5$, then $I_C = I_{CO}(0.50/0.55) = 0.91 I_{CO}$ and $I_E = I_{EO}(0.10/0.55) = 0.18 I_{EO}$ and represents a very small *reverse* current. Using $\alpha_I \approx 0$ and $\alpha_N \approx 0$ (for Si), we have that $I_C \approx I_{CO}$ and $I_E \approx I_{EO}$. Hence, increasing the magnitude of the reverse base-to-emitter bias beyond cutoff has very little effect (Fig. 5-16) on the very small transistor currents.

Short-circuited Base   Suppose that, instead of reverse-biasing the emitter junction, we simply short the base to the emitter terminal. The currents which now flow are found by setting $V_E = 0$ and by neglecting ϵ^{V_C/V_T} in the Ebers-Moll equations. The results are

$$I_C = \frac{I_{CO}}{1 - \alpha_N \alpha_I} \equiv I_{CES} \quad \text{and} \quad I_E = -\alpha_I I_{CES} \qquad (19\text{-}136)$$

where I_{CES} represents the collector current in the common-emitter configuration with a short-circuited base. If (for Ge) $\alpha_N = 0.9$ and $\alpha_I = 0.5$, then I_{CES} is about $1.8 I_{CO}$ and $I_E = -0.91 I_{CO}$. If (for Si) $\alpha_N \approx 0$ and $\alpha_I \approx 0$, then $I_{CES} \approx I_{CO}$ and $I_E \approx 0$. Hence, even with a short-circuited emitter junction, the transistor is virtually at cutoff (Fig. 5-16).

Open-circuited Base   If instead of a shorted base we allow the base to "float," so that $I_B = 0$, the cutoff condition is not reached. The collector current under this condition is called I_{CEO}, and is given by

$$I_{CEO} = \frac{I_{CO}}{1 - \alpha_N} \qquad (19\text{-}137)$$

It is interesting to find the emitter-junction voltage under this condition of a floating base. From Eq. (19-131), with $I_E = -I_C$, and using Eqs. (19-130) and (19-137),

$$V_E = V_T \ln \left[1 + \frac{\alpha_N (1 - \alpha_I)}{\alpha_I (1 - \alpha_N)} \right] \qquad (19\text{-}138)$$

For $\alpha_N = 0.9$ and $\alpha_I = 0.5$ (for Ge), we find $V_E = +60$ mV. For $\alpha_N \approx 10\alpha_I \approx 0$ (for Si), we have $V_E \approx V_T \ln 11 = +64$ mV. Hence an open-circuited base represents a slight *forward* bias (Fig 5-16).

The Cutin Voltage   We may estimate the cutin voltage V_γ in a typical case in the following manner: Assume that we are using a transistor as a switch, so that when the switch is ON it will carry a current of 20 mA. We may then consider that the cutin point has been reached when, say, the collector current equals 1 percent of the maximum current or a collector current $I_C = 0.2$ mA. Hence V_γ is the value of V_E given in Eq. (19-131), with

$$I_E = -(I_C + I_B) \approx -I_C = -0.2 \text{ mA}$$

Assume a germanium transistor with $\alpha_I = 0.5$ and $I_{EO} = 1\ \mu A$. Since at room temperature $V_T = 0.026$ V, we obtain from Eq. (19-131)

$$V_\gamma = (0.026)(2.30) \log \left[1 + \frac{0.2 \times 10^{-3}(1 - 0.5)}{10^{-6}} \right] = 0.12 \text{ V}$$

The Saturation Region Let us consider the 2N404 p-n-p germanium transistor operated with $I_C = -20$ mA, $I_B = -0.35$, and $I_E = +20.35$ mA. Assume the following reasonable values: $I_{CO} = -2.0\ \mu A$, $I_{EO} = -1.0\ \mu A$, and $\alpha_N = 0.99$. From Eq. (19-130), $\alpha_I = 0.50$. From Eqs. (19-131) and (19-132), we calculate that, at room temperature,

$$V_E = (0.026)(2.30) \log \left[1 - \frac{20.35 - (0.50)(20)}{-10^{-3}} \right] = 0.24 \text{ V}$$

and

$$V_C = (0.026)(2.30) \log \left[1 - \frac{-20 + 0.99(20.35)}{-(2)(10^{-3})} \right] = 0.11 \text{ V}$$

For a p-n-p transistor,

$$V_{CE} = V_C - V_E = 0.11 - 0.24 = -0.13 \text{ V}$$

Taking the voltage drop across $r_{bb'}$ ($\sim 100\ \Omega$) into account (Fig. 5-20),

$$V_{CB} = V_C - I_B r_{bb'} = 0.11 + 0.035 \approx 0.15 \text{ V}$$

and

$$V_{BE} = I_B r_{bb'} - V_E = -0.035 - 0.24 \approx -0.28 \text{ V}$$

Note that the base-spreading resistance does not enter into the calculation of the collector-to-emitter voltage. For a diffused-junction transistor the voltage drop resulting from the collector-spreading resistance may be significant for saturation currents. If so, this ohmic drop can no longer be neglected, as we have done above. For example, if the collector resistance is 5 Ω, then with a collector current of 20 mA, the ohmic drop is 0.10 V, and $|V_{CE}|$ increases from 0.13 to 0.23 V.

The values of I_{CO} and I_{EO} for silicon transistors are so very small that they are difficult to determine since they are masked by surface leakage currents. Hence no precise calculation of V_γ, $V_{CE,\text{sat}}$, and $V_{BE,\text{sat}}$ can be made. Experimentally determined values for silicon are given in Table 5-1, page 142.

REFERENCES

1. Fermi, E.: Zur Quantelung des idealen einatomigen Gases, *Z. Physik*, vol. 36, pp. 902–912, May, 1926.

Dirac, P. A. M.: On the Theory of Quantum Mechanics, *Proc. Roy. Soc. (London)*, vol. 112, pp. 661–677, October, 1926.

2. Sommerfeld, A., and H. Bethe: Elektronentheorie der Metalle, in "Handbuch der Physik," 2d ed., vol. 24, pt. 2, pp. 333–622, Springer Verlag OHG, Berlin, 1933.

Darrow, K. K.: Statistical Theories of Matter, Radiation and Electricity, *Bell System Tech. J.*, vol. 8, pp. 672–748, October, 1929.

3. Millman, J., and S. Seely: "Electronics," 2d ed., McGraw-Hill Book Company, New York, 1951.

4. Conwell, E. M.: Properties of Silicon and Germanium, *Proc. IRE*, vol. 46, pp. 1281–1300, June, 1958.

5. Esaki, L.: New Phenomenon in Narrow Ge *p-n* Junctions, *Phys. Rev.*, vol. 109, p. 603, 1958.

Nanavati, R. P.: "Introduction to Semiconductor Electronics," chap. 12, McGraw-Hill Book Company, New York, 1963.

6. Phillips, A. B.: "Transistor Engineering," pp. 129–133, McGraw-Hill Book Company, New York, 1962.

7. Moll, J.: "Physics of Semiconductors," pp. 117–121, McGraw-Hill Book Company, New York, 1964.

Sah, C. T., R. N. Noyce, and W. Shockley: Carrier-generation and Recombination in P-N Junctions and P-N Junction Characteristics, *Proc. IRE*, vol. 45, pp. 1228–1243, September, 1957.

8. Ebers, J. J., and J. L. Moll: Large-signal Behavior of Junction Transistors, *Proc. IRE*, vol. 42, pp. 1761–1772, December, 1954.

REVIEW QUESTIONS

19-1 (a) Draw the potential-energy picture of a metal. (b) Explain qualitatively the existence of the potential barrier at the surface.

19-2 Explain what is meant by a *distribution function*. Use as an example the distribution in age of people in the United States.

19-3 (a) Plot the *Fermi-Dirac probability function* $f(E)$ versus energy E at 0°K and, say, 2500°K. (b) What are the meanings of these plots?

19-4 (a) Plot the *energy distribution function* ρ_E in a metal at 0 and 2500°K. (b) Indicate the Fermi level E_F on your sketch and give its physical meaning.

19-5 The energy distribution function is given by the product of two factors. What is the interpretation to be given to each of these factors?

19-6 Define *work function*.

19-7 Indicate graphically which electrons in the energy distribution curve can escape from a metal at (a) 0°K and (b) 2500°K.

19-8 Give the equation which indicates that there is a distribution in energy of the electrons that escape from a metal. Explain the meaning of the equation.

19-9 (a) Define the *volt equivalent of temperature*. (b) What is its magnitude at room temperature?

19-10 How does the density of states $N(E)$ vary with E for (a) electrons in a metal, (b) electrons in a semiconductor, and (c) holes in a semiconductor?

19-11 Sketch curves of (a) E versus $f(E)$, (b) E versus $N(E)$, and (c) E versus $\rho(E)$ for electrons and holes in a semiconductor.

19-12 Sketch the energy-band picture for (a) an intrinsic, (b) an n-type, and (c) a p-type semiconductor. Indicate the positions of the Fermi, the donor, and the acceptor levels.

19-13 Sketch the energy-band picture for a p-n junction under open-circuit conditions. Indicate the Fermi level E_F and the contact difference of potential E_o.

19-14 State the *mass-action law*. Explain its meaning.

19-15 State Einstein's equation involving the diffusion constant.

19-16 Sketch the energy-band picture for a heavily doped p-n diode (a) under open-circuited conditions and (b) with an applied reverse bias.

19-17 Sketch the energy-band picture for a heavily doped p-n diode for (a) a small forward bias, (b) the forward bias which results in the peak tunneling current, and (c) the forward bias beyond which the tunneling current is zero.

19-18 For an Esaki diode, sketch (a) the tunneling current, (b) the injection current, and (c) the total current.

19-19 Write the *equation of continuity* for holes. Explain the physical meaning of each term in the equation.

19-20 Write the equation for the hole current density. Explain the physical meaning of each term.

19-21 What are the five equations from which to determine p, n, $\mathcal{E}$, J_p, and J_n?

19-22 (a) State the *law of the junction*. (b) Derive this equation by assuming that the diffusion current approximately equals the drift current for a forward-biased diode as well as for an open-circuited diode.

19-23 (a) From the expression for $p_n(x)$, evaluate $I_{pn}(0)$ as a function of $p_n'(0)$. (b) Using the law of the junction, evaluate $I_{pn}(0)$ as a function of applied voltage V. (c) Find the equation for the reverse saturation current I_o.

19-24 (a) Write the Ebers-Moll equation for I_C versus V_E and V_C. (b) What is the corresponding equation for inverse operation?

19-25 From the equations in Rev. 19-24, obtain the junction voltages in terms of the junction currents.

19-26 (a) Define *cutoff*. (b) Obtain an expression for the emitter-junction voltage at cutoff.

19-27 Obtain expressions for I_C and I_E if large reverse-biasing voltages appear across the junctions.

19-28 Obtain expressions for I_C and I_E for a short-circuited base.

19-29 Obtain expressions for I_C and V_E for an open-circuited base.

PROBABLE VALUES OF GENERAL PHYSICAL CONSTANTS[†]

Constant	Symbol	Value
Electronic charge	q	1.602×10^{-19} C
Electronic mass	m	9.109×10^{-31} kg
Ratio of charge to mass of an electron	q/m	1.759×10^{11} C/kg
Mass of atom of unit atomic weight (hypothetical)	......	1.660×10^{-27} kg
Mass of proton	m_p	1.673×10^{-27} kg
Ratio of proton to electron mass	m_p/m	1.837×10^3
Planck's constant	h	6.626×10^{-34} J-s
Boltzmann constant	$\bar{k}$	1.381×10^{-23} J/°K
	k	8.620×10^{-5} eV/°K
Stefan-Boltzmann constant	σ	5.670×10^{-8} W/(m²)(°K⁴)
Avogadro's number	N_A	6.023×10^{23} molecules/mole
Gas constant	R	8.314 J/(deg)(mole)
Velocity of light	c	2.998×10^8 m/s
Faraday's constant	F	9.649×10^3 C/mole
Volume per mole	V_o	2.241×10^{-2} m³
Acceleration of gravity	g	9.807 m/s²
Permeability of free space	μ_o	1.257×10^{-6} H/m
Permittivity of free space	ϵ_o	8.849×10^{-12} F/m

[†] E. A. Mechtly, "The International System of Units: Physical Constants and Conversion Factors," National Aeronautics and Space Administration, NASA SP-7012, Washington, D.C., 1964.

B / CONVERSION FACTORS AND PREFIXES

1 ampere (A)	$= 1$ C/s
1 angstrom unit (Å)	$= 10^{-10}$ m
	$= 10^{-4}$ μm
1 atmosphere pressure	$= 760$ mm Hg
1 coulomb (C)	$= 1$ A-s
1 electron volt (eV)	$= 1.60 \times 10^{-19}$ J
1 farad (F)	$= 1$ C/V
1 foot (ft)	$= 0.305$ m
1 gram-calorie	$= 4.185$ J
giga (G)	$= \times 10^9$
1 henry (H)	$= 1$ V-s/A
1 hertz (Hz)	$= 1$ cycle/s
1 inch (in.)	$= 2.54$ cm
1 joule (J)	$= 10^7$ ergs
	$= 1$ W-s
	$= 6.25 \times 10^{18}$ eV
	$= 1$ N-m
	$= 1$ C-V
kilo (k)	$= \times 10^3$
1 kilogram (kg)	$= 2.205$ lb
1 kilometer (km)	$= 0.622$ mile
1 lumen	$= 0.0016$ W
	(at 0.55 μm)

1 lumen per square foot	$= 1$ ft-candle (fc)
mega (M)	$= \times 10^6$
1 meter (m)	$= 39.37$ in.
micro (μ)	$= \times 10^{-6}$
1 micron	$= 10^{-6}$ m
	$= 1$ μm
1 mil	$= 10^{-3}$ in.
	$= 25$ μm
1 mile	$= 5,280$ ft
	$= 1.609$ km
milli (m)	$= \times 10^{-3}$
nano (n)	$= \times 10^{-9}$
1 newton (N)	$= 1$ kg-m/s^2
pico (p)	$= \times 10^{-12}$
1 pound (lb)	$= 453.6$ g
1 tesla (T)	$= 1$ Wb/m^2
1 ton	$= 2,000$ lb
1 volt (V)	$= 1$ W/A
1 watt (W)	$= 1$ J/s
1 weber (Wb)	$= 1$ V-s
1 weber per square meter (Wb/m^2)	$= 10^4$ gauss

APPENDIX

C / PROBLEMS

CHAPTER 1

1-1 (a) The distance between the plates of a plane-parallel capacitor is 1 cm. An electron starts at rest at the negative plate. If a direct voltage of 1,000 V is applied, how long will it take the electron to reach the positive plate?

(b) What is the magnitude of the force which is exerted on the electron at the beginning and at the end of its path?

(c) What is its final velocity?

(d) If a 60-Hz sinusoidal voltage of peak value 1,000 V is applied, how long will the time of transit be? Assume that the electron is released with zero velocity at the instant of time when the applied voltage is passing through zero. HINT: Expand the sine function into a power series. Thus $\sin \theta = \theta - \theta^3/3! + \theta^5/5! - \cdots$.

1-2 The plates of a parallel-plate capacitor are d m apart. At $t = 0$ an electron is released at the bottom plate with a velocity v_o (meters per second) normal to the plates. The potential of the top plate with respect to the bottom is $-V_m \sin \omega t$.

(a) Find the position of the electron at any time t.

(b) Find the value of the electric field intensity at the instant when the velocity of the electron is zero.

1-3 An electron is released with zero initial velocity from the lower of a pair of horizontal plates which are 3 cm apart. The accelerating potential between these plates increases from zero linearly with time at the rate of 10 V/μs. When the electron is 2.8 cm from the bottom plate, a reverse voltage of 50 V replaces the linearly rising voltage.

(a) What is the instantaneous potential between the plates at the time of the potential reversal?

(b) With which electrode does the electron collide?

(c) What is the time of flight?

(d) What is the impact velocity of the electron?

1-4 A 100-eV hydrogen ion is released in the center of the plates, as shown in the figure. The voltage between the plates varies linearly from 0 to 50 V in 10^{-7} s and then drops immediately to zero and remains at zero. The separation between the plates is 2 cm. If the ion enters the region between the plates at time $t = 0$, how far will it be displaced from the X axis upon emergence from between the plates?

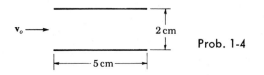

Prob. 1-4

1-5 Electrons are projected into the region of constant electric field intensity of magnitude 5×10^3 V/m that exists vertically. The electron-emitting device makes an angle of 30° with the horizontal. It ejects the electrons with an energy of 100 eV.

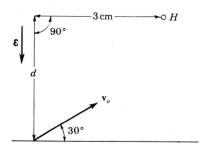

Prob. 1-5

(a) How long does it take an electron leaving the emitting device to pass through a hole H at a horizontal distance of 3 cm from the position of the emitting device? Refer to the figure. Assume that the field is downward.

(b) What must be the distance d in order that the particles emerge through the hole?

(c) Repeat parts a and b for the case where the field is upward.

1-6 (a) An electron is emitted from an electrode with a negligible initial velocity and is accelerated by a potential of 1,000 V. Calculate the final velocity of the particle.

(b) Repeat the problem for the case of a deuterium ion (heavy hydrogen ion—atomic weight 2.01) that has been introduced into the electric field with an initial velocity of 10^5 m/s.

1-7 An electron having an initial kinetic energy of 10^{-16} J at the surface of one of two parallel-plane electrodes and moving normal to the surface is slowed down by the retarding field caused by a 400-V potential applied between the electrodes.
(a) Will the electron reach the second electrode?
(b) What retarding potential would be required for the electron to reach the second electrode with zero velocity?

1-8 In a certain plane-parallel diode the potential V is given as a function of the distance x between electrodes by the equation

$$V = kx^{\frac{4}{3}}$$

where k is a constant.

(a) Find an expression for the time it will take an electron that leaves the electrode with the lower potential with zero initial velocity to reach the electrode with the higher potential, a distance d away.

(b) Find an expression for the velocity of this electron.

1-9 The essential features of the displaying tube of an oscilloscope are shown in the accompanying figure. The voltage difference between K and A is V_a and between P_1 and P_2 is V_p. Neither electric field affects the other one. The electrons are emitted from the electrode K with initial zero velocity, and they pass through a hole in the middle of electrode A. Because of the field between P_1 and P_2 they change direction while they pass through these plates and, after that, move with constant velocity toward the screen S. The distance between plates is d.

Prob. 1-9

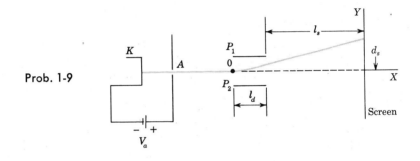

(a) Find the velocity v_x of the electrons as a function of V_a as they cross A.

(b) Find the Y-component of velocity v_y of the electrons as they come out of the field of plates P_1 and P_2 as a function of V_p, l_d, d, and v_x.

(c) Find the distance from the middle of the screen (d_s), when the electrons reach the screen, as a function of tube distances and applied voltages.

(d) For $V_a = 1.0$ kV, and $V_p = 10$ V, $l_d = 1.27$ cm, $d = 0.475$ cm, and $l_s = 19.4$ cm, find the numerical values of v_x, v_y, and d_s.

(e) If we want to have a deflection of $d_s = 10$ cm of the electron beam, what must be the value of V_a?

1-10 A diode consists of a plane emitter and a plane-parallel anode separated by a distance of 0.5 cm. The anode is maintained at a potential of 10 V negative with respect to the cathode.

(a) If an electron leaves the emitter with a speed of 10^6 m/s, and is directed toward the anode, at what distance from the cathode will it intersect the potential-energy barrier?

(b) With what speed must the electron leave the emitter in order to be able to reach the anode?

1-11 A particle when displaced from its equilibrium position is subject to a linear restoring force $f = -kx$, where x is the displacement measured from the equilibrium position. Show by the energy method that the particle will execute periodic vibrations with a maximum displacement which is proportional to the square root of the total energy of the particle.

1-12 A particle of mass m is projected vertically upward in the earth's gravitational field with a speed v_o.

(a) Show by the energy method that this particle will reverse its direction at the height of $v_o{}^2/2g$, where g is the acceleration of gravity.

(b) Show that the point of reversal corresponds to a "collision" with the potential-energy barrier.

1-13 (a) Prove Eq. (1-13).

(b) For the hydrogen atom show that the possible radii in meters are given by

$$r = \frac{h^2 \epsilon_o n^2}{\pi m q^2}$$

where n is any integer but not zero. For the ground state ($n = 1$) show that the radius is 0.53 Å.

1-14 Show that the time for one revolution of the electron in the hydrogen atom in a circular path about the nucleus is

$$T = \frac{m^{\frac{1}{2}} q^2}{4 \sqrt{2} \, \epsilon_o (-W)^{\frac{3}{2}}}$$

where the symbols are as defined in Sec. 1-4.

1-15 For the hydrogen atom show that the reciprocal of the wavelength (called the *wave number*) of the spectral lines is given, in waves per meter, by

$$\frac{1}{\lambda} = R \left(\frac{1}{n_2{}^2} - \frac{1}{n_1{}^2} \right)$$

where n_1 and n_2 are integers, with n_1 greater than n_2, and $R = mq^4/8\epsilon_o{}^2 h^3 c = 1.10 \times 10^7$ m^{-1} is called the *Rydberg constant*.

If $n_2 = 1$, this formula gives a series of lines in the ultraviolet, called the *Lyman series*. If $n_2 = 2$, the formula gives a series of lines in the visible, called the *Balmer series*. Similarly, the series for $n_2 = 3$ is called the *Paschen series*. These predicted lines are observed in the hydrogen spectrum.

1-16 Show that Eq. (1-14) is equivalent to Eq. (1-11).

1-17 (a) A photon of wavelength 1,026 Å is absorbed by hydrogen, and two other photons are emitted. If one of these is the 1,216 Å line, what is the wavelength of the second photon?

(b) If the result of bombardment of the hydrogen was the presence of the fluorescent lines 18,751 and 1,026 Å, what wavelength must have been present in the bombarding radiation?

1-18 The seven lowest energy levels of sodium vapor are 0, 2.10, 3.19, 3.60, 3.75, 4.10, and 4.26 eV. A photon of wavelength 3,300 Å is absorbed by an atom of the vapor.

(a) What are all the possible fluorescent lines that may appear?

(b) If three photons are emitted and one of these is the 11,380-Å line, what are the wavelengths of the other two photons?

(c) Between what energy states do the transitions take place in order to produce these lines?

1-19 (a) With what speed must an electron be traveling in a sodium-vapor lamp in order to excite the yellow line whose wavelength is 5,893 Å?
(b) What should be the frequency of a photon in order to excite the same yellow line?
(c) What would happen if the frequency of the photon was 530 or 490 THz (T = Tera = 10^{12})?
(d) What should be the minimum frequency of the photon in order to ionize an unexcited atom of sodium vapor?
(e) What should be the minimum speed of an electron in order to ionize an unexcited atom of sodium vapor?   Ionization of sodium vapor: 5.12 eV.

1-20 An x-ray tube is essentially a high-voltage diode. The electrons from the hot filament are accelerated by the plate supply voltage so that they fall upon the anode with considerable energy.   They are thus able to effect transitions among the tightly bound electrons of the atoms in the solid of which the target (the anode) is constructed.
(a) What is the minimum voltage that must be applied across the tube in order to produce x-rays having a wavelength of 0.5 Å?
(b) What is the minimum wavelength in the spectrum of an x-ray tube across which is maintained 60 kV?

1-21 Argon resonance radiation corresponding to an energy of 11.6 eV falls upon sodium vapor.   If a photon ionizes an unexcited sodium atom, with what speed is the electron ejected?   The ionization potential of sodium is 5.12 eV.

1-22 A radio transmitter radiates 1,000 W at a frequency of 10 MHz.
(a) What is the energy of each radiated quantum in electron volts?
(b) How many quanta are emitted per second?
(c) How many quanta are emitted in each period of oscillation of the electro-magnetic field?
(d) If each quantum acts as a particle, what is its momentum?

1-23 What is the wavelength of (a) a mass of 1 kg moving with a speed of 1 m/s, (b) an electron which has been accelerated from rest through a potential difference of 10 V?

1-24 Classical physics is valid as long as the physical dimensions of the system are much larger than the De Broglie wavelength.   Determine whether the particle is classical in each of the following cases:
(a) An electron accelerated through a potential of 300 V in a device whose dimensions are of the order of 1 cm.
(b) An electron in the electron beam of a cathode-ray tube (anode-cathode voltage = 25 kV).
(c) The electron in a hydrogen atom.

1-25 A photon of wavelength 1,216 Å excites a hydrogen atom which is at rest. Calculate
(a) The photon momentum imparted to the atom.
(b) The energy corresponding to this momentum and imparted to the hydrogen atom.
(c) The ratio of the energy found in part b to the energy of the photon.   HINT: Use conservation of momentum.

CHAPTER 2

2-1   Prove that the concentration n of free electrons per cubic meter of a metal is given by

$$ n = \frac{d\nu}{AM} = \frac{A_o d\nu \times 10^3}{A} $$

where d = density, kg/m³
    ν = valence, free electrons per atom
    A = atomic weight
    M = weight of atom of unit atomic weight, kg (Appendix A)
    A_o = Avogadro's number, molecules/mole

2-2   The specific density of tungsten is 18.8 g/cm³, and its atomic weight is 184.0. Assume that there are two free electrons per atom.   Calculate the concentration of free electrons.

2-3   (a) Compute the conductivity of copper for which $\mu = 34.8$ cm²/V-s and $d = 8.9$ g/cm³.   Use the result of Prob. 2-1.
(b) If an electric field is applied across such a copper bar with an intensity of 10 V/cm, find the average velocity of the free electrons.

2-4   Compute the mobility of the free electrons in aluminum for which the density is 2.70 g/cm³ and the resistivity is 3.44×10^{-6} Ω-cm.   Assume that aluminum has three valence electrons per atom.   Use the result of Prob. 2-1.

2-5   The resistance of No. 18 copper wire (diameter = 1.03 mm) is 6.51 Ω per 1,000 ft. The concentration of free electrons in copper is 8.4×10^{28} electrons/m³.   If the current is 2 A, find the (a) drift velocity, (b) mobility, (c) conductivity.

2-6   (a) Determine the concentration of free electrons and holes in a sample of germanium at 300°K which has a concentration of donor atoms equal to 2×10^{14} atoms/cm³ and a concentration of acceptor atoms equal to 3×10^{14} atoms/cm³. Is this p- or n-type germanium?   In other words, is the conductivity due primarily to holes or to electrons?
(b) Repeat part a for equal donor and acceptor concentrations of 10^{15} atoms/cm³. Is this p- or n-type germanium?
(c) Repeat part a for donor concentration of 10^{16} atoms/cm³ and acceptor concentration 10^{14} atoms/cm³.

2-7   (a) Find the concentration of holes and of electrons in p-type germanium at 300°K if the conductivity is 100 $(\Omega\text{-cm})^{-1}$.
(b) Repeat part a for n-type silicon if the conductivity is 0.1 $(\Omega\text{-cm})^{-1}$.

2-8   (a) Show that the resistivity of intrinsic germanium at 300°K is 45 Ω-cm.
(b) If a donor-type impurity is added to the extent of 1 atom per 10^8 germanium atoms, prove that the resistivity drops to 3.7 Ω-cm.

2-9   (a) Find the resistivity of intrinsic silicon at 300°K.
(b) If a donor-type impurity is added to the extent of 1 atom per 10^8 silicon atoms, find the resistivity.

2-10   Consider intrinsic germanium at room temperature (300°K).   By what percent does the conductivity increase per degree rise in temperature?

2-11   Repeat Prob. 2-10 for intrinsic silicon.

2-12 Repeat Prob. 2-6a for a temperature of 400°K, and show that the sample is essentially intrinsic.

2-13 A sample of germanium is doped to the extent of 10^{14} donor atoms/cm³ and 7×10^{13} acceptor atoms/cm³. At the temperature of the sample the resistivity of pure (intrinsic) germanium is 60 Ω-cm. If the applied electric field is 2 V/cm, find the total conduction current density.

2-14 (a) Find the magnitude of the Hall voltage V_H in an n-type germanium bar used in Fig. 2-10, having majority-carrier concentration $N_D = 10^{17}/\text{cm}^3$. Assume $B_z = 0.1$ Wb/m², $d = 3$ mm, and $\mathcal{E}_x = 5$ V/cm.
(b) What happens to V_H if an identical p-type germanium bar having $N_A = 10^{17}/\text{cm}^3$ is used in part a?

2-15 The Hall effect is used to determine the mobility of holes in a p-type silicon bar used in Fig. 2-10. Assume the bar resistivity is 200,000 Ω-cm, the magnetic field $B_z = 0.1$ Wb/m², and $d = w = 3$ mm. The measured values of the current and Hall voltage are 10 µA and 50 mV, respectively. Find μ_p.

2-16 A certain photosurface has a spectral sensitivity of 6 mA/W of incident radiation of wavelength 2,537 Å. How many electrons will be emitted photoelectrically by a pulse of radiation consisting of 10,000 photons of this wavelength?

2-17 (a) Consider the situation depicted in Fig. 2-13 with the light turned on. Show that the equation of conservation of charge is

$$\frac{dp}{dt} + \frac{p}{\tau} = \frac{\bar{p}}{\tau}$$

where the time axis in Fig. 2-13 is shifted to t'.
(b) Verify that the concentration is given by the equation

$$p = \bar{p} + (p_o - \bar{p})e^{-t/\tau}$$

2-18 The hole concentration in a semiconductor specimen is shown.
(a) Find an expression for and sketch the hole current density $J_p(x)$ for the case in which there is no externally applied electric field.

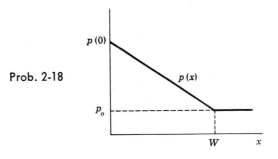

Prob. 2-18

(b) Find an expression for and sketch the built-in electric field that must exist if there is to be no net hole current associated with the distribution shown.
(c) Find the value of the potential between the points $x = 0$ and $x = W$ if $p(0)/p_o = 10^3$.

2-19 Given a 20 Ω-cm n-type germanium bar with material lifetime of 100 μs, cross section of 1 mm², and length of 1 cm. One side of the bar is illuminated with 10^{15} photons/s. Assume that each incident photon generates one electron-hole pair and that these are distributed uniformly throughout the bar. Find the bar resistance under continuous light excitation at room temperature.

2-20 (a) Consider an open-circuited graded semiconductor as in Fig. 2-17a. Verify the Boltzmann equation for electrons [Eq. (2-61)].

(b) For the step-graded semiconductor of Fig. 2-17b verify the expression for the contact potential V_o given in Eq. (2-63), starting with $J_n = 0$.

2-21 (a) Consider the step-graded germanium semiconductor of Fig. 2-17b with $N_D = 10^3 N_A$ and with N_A corresponding to 1 acceptor atom per 10^8 germanium atoms. Calculate the contact difference of potential V_o at room temperature.

(b) Repeat part a for a silicon p-n junction.

CHAPTER 3

3-1 (a) The resistivities of the two sides of a step-graded germanium diode are 2 Ω-cm (p side) and 1 Ω-cm (n side). Calculate the height E_o of the potential-energy barrier.

(b) Repeat part a for a silicon p-n junction.

3-2 (a) Sketch logarithmic and linear plots of carrier concentration vs. distance for an abrupt silicon junction if $N_D = 10^{15}$ atoms/cm³ and $N_A = 10^{16}$ atoms/cm³. Give numerical values for ordinates. Label the n, p, and depletion regions.

(b) Sketch the space-charge electric field and potential as a function of distance for this case (Fig. 3-1).

3-3 Repeat Prob. 3-2 for an abrupt germanium junction.

3-4 (a) Consider a p-n diode operating under low-level injection so that $p_n \ll n_n$. Assuming that the minority current is due entirely to diffusion, verify that the electric field in the n side is given by

$$\mathcal{E}(x) = \frac{I + (D_n/D_p - 1)I_{pn}(x)}{qn\mu_n A}$$

(b) Using this value of $\mathcal{E}$, find the next approximation to the drift hole current and show that it may indeed be neglected compared with the diffusion hole current.

(c) Sketch the following currents as a function of distance in the n side: (i) total diode current; (ii) minority-carrier current; (iii) majority diffusion current; (iv) majority drift current; (v) total majority-carrier current.

3-5 Starting with Eq. (3-5) for I_{pn} and the corresponding expression for I_{np}, prove that the ratio of hole to electron current crossing a p-n junction is given by

$$\frac{I_{pn}(0)}{I_{np}(0)} = \frac{\sigma_p L_n}{\sigma_n L_p}$$

where $\sigma_p(\sigma_n)$ = conductivity of $p(n)$ side. Note that this ratio depends upon the ratio of the conductivities. For example, if the p side is much more heavily doped than the n side, the hole current will be much larger than the electron current crossing the junction.

3-6 (a) Prove that the reverse saturation current in a p-n diode is given by

$$I_o = Aq \left(\frac{D_p}{L_p N_D} + \frac{D_n}{L_n N_A} \right) n_i^2$$

(b) Starting with the expression for I_o found in part a, verify that the reverse saturation current is given by

$$I_o = A V_T \frac{b \sigma_i^2}{(1 + b)^2} \left(\frac{1}{L_p \sigma_n} + \frac{1}{L_n \sigma_p} \right)$$

where $\sigma_n(\sigma_p)$ = conductivity of $n(p)$ side

σ_i = conductivity of intrinsic material

$b = \mu_n / \mu_p$

3-7 (a) Using the result of Prob. 3-6, find the reverse saturation current for a germanium p-n junction diode at room temperature, $300°K$. The cross-sectional area is 4.0 mm^2, and

$$\sigma_p = 1.0 \ (\Omega\text{-cm})^{-1} \qquad \sigma_n = 0.1 \ (\Omega\text{-cm})^{-1} \qquad L_n = L_p = 0.15 \text{ cm}$$

Other physical constants are given in Table 2-1.

(b) Repeat part a for a silicon p-n junction diode. Assume $L_n = L_p = 0.01$ cm and $\sigma_n = \sigma_p = 0.01 \ (\Omega\text{-cm})^{-1}$.

3-8 Find the ratio of the reverse saturation current in germanium to that in silicon, using the result of Prob. 3-6. Assume $L_n = L_p = 0.1$ cm and $\sigma_n = \sigma_p = 1.0$ $(\Omega\text{-cm})^{-1}$ for germanium, whereas the corresponding values are 0.01 cm and $0.01 \ (\Omega\text{-cm})^{-1}$ for silicon. See also Table 2-1.

3-9 (a) For what voltage will the reverse current in a p-n junction germanium diode reach 90 percent of its saturation value at room temperature?

(b) What is the ratio of the current for a forward bias of 0.05 V to the current for the same magnitude of reverse bias?

(c) If the reverse saturation current is $10 \ \mu$A, calculate the forward currents for voltages of 0.1, 0.2, and 0.3 V, respectively.

3-10 (a) Evaluate η in Eq. (3-9) from the slope of the plot in Fig. 3-8 for $T = 25°C$. Draw the best-fit line over the current range 0.01 to 10 mA.

(b) Repeat for $T = -55$ and $150°C$.

3-11 (a) Calculate the anticipated factor by which the reverse saturation current of a germanium diode is multiplied when the temperature is increased from 25 to $80°C$.

(b) Repeat part a for a silicon diode over the range 25 to $150°C$.

3-12 It is predicted that, for germanium, the reverse saturation current should increase by $0.11°C^{-1}$. It is found experimentally in a particular diode that at a reverse voltage of 10 V, the reverse current is $5 \ \mu$A and the temperature dependence is only $0.07°C^{-1}$. What is the leakage resistance shunting the diode?

3-13 A diode is mounted on a chassis in such a manner that, for each degree of temperature rise above ambient, 0.1 mW is thermally transferred from the diode to its surroundings. (The "thermal resistance" of the mechanical contact between the diode and its surroundings is 0.1 mW/°C.) The ambient temperature is 25°C. The diode temperature is not to be allowed to increase by more than 10°C above ambient. If the reverse saturation current is 5.0 μA at 25°C and increases at the rate 0.07°C^{-1}, what is the maximum reverse-bias voltage which may be maintained across the diode?

3-14 A silicon diode operates at a forward voltage of 0.4 V. Calculate the factor by which the current will be multiplied when the temperature is increased from 25 to 150°C. Compare the result with the plot of Fig. 3-8.

3-15 An ideal germanium p-n junction diode has at a temperature of 125°C a reverse saturation current of 30 μA. At a temperature of 125°C find the dynamic resistance for a 0.2 V bias in (a) the forward direction, (b) the reverse direction.

3-16 Prove that for an alloy p-n junction (with $N_A \ll N_D$), the width W of the depletion layer is given by

$$W = \left(\frac{2\epsilon\mu_p V_j}{\sigma_p} \right)^{\frac{1}{2}}$$

where V_j is the junction potential under the condition of an applied diode voltage V_d.

3-17 (a) Prove that for an alloy silicon p-n junction (with $N_A \ll N_D$), the depletion-layer capacitance in picofarads per square centimeter is given by

$$C_T = 2.9 \times 10^{-4} \left(\frac{N_A}{V_j} \right)^{\frac{1}{2}}$$

(b) If the resistivity of the p material is 3.5 Ω-cm, the barrier height V_o is 0.35 V, the applied reverse voltage is 5 V, and the cross-sectional area is circular of 40 mils diameter, find C_T.

3-18 (a) For the junction of Fig. 3-10, find the expression for the $\mathcal{E}$ and V as a function of x in the n-type side for the case where N_A and N_D are of comparable magnitude. HINT: Shift the origin of x so that $x = 0$ at the junction.
(b) Show that the total barrier voltage is given by Eq. (3-21) multiplied by $N_A/(N_A + N_D)$ and with $W = W_p + W_n$.
(c) Prove that $C_T = [qN_AN_D\epsilon/2(N_A + N_D)]^{\frac{1}{2}}V^{-\frac{1}{2}}$.
(d) Prove that $C_T = \epsilon A/(W_p + W_n)$.

3-19 Reverse-biased diodes are frequently employed as electrically controllable variable capacitors. The transition capacitance of an abrupt junction diode is 20 pF at 5 V. Compute the decrease in capacitance for a 1.0-V increase in bias.

3-20 Calculate the barrier capacitance of a germanium p-n junction whose area is 1 mm by 1 mm and whose space-charge thickness is 2×10^{-4} cm. The dielectric constant of germanium (relative to free space) is 16.

3-21 The zero-voltage barrier height at an alloy-germanium p-n junction is 0.2 V. The concentration N_A of acceptor atoms in the p side is much smaller than the concentration of donor atoms in the n material, and $N_A = 3 \times 10^{20}$ atoms/m^3. Calculate the width of the depletion layer for an applied reverse voltage of (a) 10 V and (b) 0.1 V and (c) for a forward bias of 0.1 V. (d) If the cross-sectional area of the diode is 1 mm², evaluate the space-charge capacitance corresponding to the values of applied voltage in (a) and (b).

3-22 (a) Consider a grown junction for which the uncovered charge density ρ varies linearly with distance. If $\rho = ax$, prove that the barrier voltage V_j is given by

$$V_j = \frac{aW^3}{12\epsilon}$$

(b) Verify that the barrier capacitance C_T is given by Eq. (3-23)

3-23 Given a forward-biased silicon diode with $I = 1$ mA. If the diffusion capacitance is $C_D = 1 \ \mu\text{F}$, what is the diffusion length L_p? Assume that the doping of the p side is much greater than that of the n side.

3-24 The derivation of Eq. (3-28) for the diffusion capacitance assumes that the p side is much more heavily doped than the n side, so that the current at the junction is entirely due to holes. Derive an expression for the total diffusion capacitance when this approximation is not made.

3-25 (a) Prove that the magnitude of the maximum electric field $\mathcal{E}_m$ at a step-graded junction with $N_A \gg N_D$ is given by

$$\mathcal{E}_m = \frac{2V_j}{W}$$

(b) It is found that Zener breakdown occurs when $\mathcal{E}_m = 2 \times 10^7$ V/m $\equiv \mathcal{E}_z$. Prove that Zener voltage V_z is given by

$$V_z = \frac{\epsilon\mathcal{E}_z{}^2}{2qN_D}$$

Note that the Zener breakdown voltage can be controlled by controlling the concentration of donor ions.

3-26 (a) Zener breakdown occurs in germanium at a field intensity of 2×10^7 V/m. Prove that the breakdown voltage is $V_Z = 51/\sigma_p$, where σ_p is the conductivity of the p material in $(\Omega\text{-cm})^{-1}$. Assume that $N_A \ll N_D$.

(b) If the p material is essentially intrinsic, calculate V_Z.

(c) For a doping of 1 part in 10^8 of p-type material, the resistivity drops to 3.7 Ω-cm. Calculate V_Z.

(d) For what resistivity of the p-type material will $V_Z = 1$ V?

3-27 (a) Two p-n germanium diodes are connected in series opposing. A 5-V battery is impressed upon this series arrangement. Find the voltage across each junction at room temperature. Assume that the magnitude of the Zener voltage is greater than 5 V.

Note that the result is independent of the reverse saturation current. Is it also independent of temperature?

HINT: Assume that reverse saturation current flows in the circuit, and then justify this assumption.

(b) If the magnitude of the Zener voltage is 4.9 V, what will be the current in the circuit? The reverse saturation current is 5 μA.

3-28 The Zener diode can be used to prevent overloading of sensitive meter movements without affecting meter linearity. The circuit shown represents a dc voltmeter which reads 20 V full scale. The meter resistance is 560 Ω, and $R_1 + R_2 = 99.5$ K. If the diode is a 16-V Zener, find R_1 and R_2 so that, when $V_i > 20$ V, the Zener diode conducts and the overload current is shunted away from the meter.

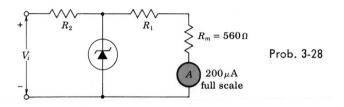

$R_m = 560\,\Omega$

Prob. 3-28

A $200\,\mu A$ full scale

3-29 A series combination of a 15-V avalanche diode and a forward-biased silicon diode is to be used to construct a zero-temperature-coefficient voltage reference. The temperature coefficient of the silicon diode is -1.7 mV/°C. Express in percent per degree centigrade the required temperature coefficient of the Zener diode.

3-30 The saturation currents of the two diodes are 1 and 2 μA. The breakdown voltages of the diodes are the same and are equal to 100 V.
(a) Calculate the current and voltage for each diode if $V = 90$ V and $V = 110$ V.
(b) Repeat part a if each diode is shunted by a 10-M resistor.

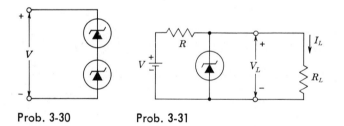

R

V_L

I_L

R_L

Prob. 3-30   **Prob. 3-31**

3-31 (a) The avalanche diode regulates at 50 V over a range of diode currents from 5 to 40 mA. The supply voltage $V = 200$ V. Calculate R to allow voltage regulation from a load current $I_L = 0$ up to I_{max}, the maximum possible value of I_L. What is I_{max}?
(b) If R is set as in part a and the load current is set at $I_L = 25$ mA, what are the limits between which V may vary without loss of regulation in the circuit?

3-32 (a) Consider a tunnel diode with $N_D = N_A$ and with the impurity concentration corresponding to 1 atom per 10^3 germanium atoms. At room temperature calculate (i) the height of the potential-energy barrier under open-circuit conditions (the contact potential energy), (ii) the width of the space-charge region.
(b) Repeat part a if the semiconductor is silicon instead of germanium.

3-33 The photocurrent I in a p-n junction photodiode as a function of the distance x of the light spot from the junction is given in Fig. 3-22. Prove that the slopes of ln I versus x are $-1/L_p$ and $-1/L_n$, respectively, on the n and p sides.   Note that L_p represents the diffusion length for holes in the n material.

3-34   (a) For the type LS 223 photovoltaic cell whose characteristics are given in Fig. 3-23, plot the power output vs. the load resistance R_L.
(b) What is the optimum value of R_L?

CHAPTER 4

4-1   (a) In the circuit of Prob. 3-27, the Zener breakdown voltage is 2.0 V.   The reverse saturation current is 5 μA.   If the silicon diode resistance could be neglected, what would be the current?
(b) If the ohmic resistance is 100 Ω, what is the current?
       NOTE: Answer part b by plotting Eq. (3-9) and drawing a load line.   Verify your answer analytically by a method of successive approximations.

4-2   A p-n germanium junction diode at room temperature has a reverse saturation current of 10 μA, negligible ohmic resistance, and a Zener breakdown voltage of 100 V.   A 1-K resistor is in series with this diode, and a 30-V battery is impressed across this combination.   Find the current (a) if the diode is forward-biased, (b) if the battery is inserted into the circuit with the reverse polarity. (c) Repeat parts a and b if the Zener breakdown voltage is 10 V.

4-3   Each diode is described by a linearized volt-ampere characteristic, with incremental resistance r and offset voltage V_γ.   Diode $D1$ is germanium with $V_\gamma =$

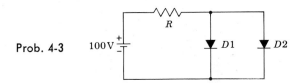

Prob. 4-3   100 V

0.2 V and $r = 20\ \Omega$, whereas $D2$ is silicon with $V_\gamma = 0.6$ V and $r = 15\ \Omega$.   Find the diode currents if (a) $R = 10$ K, (b) $R = 1$ K.

4-4   The photodiode whose characteristics are given in Fig. 3-21 is in series with a 30-V supply and a resistance R.   If the illumination is 3,000 fc, find the current for (a) $R = 0$, (b) $R = 50$ K, (c) $R = 100$ K.

4-5   (a) For the application in Sec. 4-3, plot the voltage across the diode for one cycle of the input voltage v_i.   Let $V_m = 2.4$ V, $V_\gamma = 0.6$ V, $R_f = 10\ \Omega$, and $R_L = 100\ \Omega$.
(b) By direct integration find the average value of the diode voltage and the load voltage.   Note that these two answers are numerically equal and explain why.

4-6   Calculate the break region over which the dynamic resistance of a diode is multiplied by a factor of 1,000.

4-7   For the diode clipping circuit of Fig. 4-9a assume that $V_R = 10$ V, $v_i = 20\sin \omega t$, and that the diode forward resistance is $R_f = 100\ \Omega$ while $R_r = \infty$ and $V_\gamma = 0$. Neglect all capacitances.   Draw to scale the input and output waveforms and label the maximum and minimum values if (a) $R = 100\ \Omega$, (b) $R = 1$ K, and (c) $R = 10$ K.

4-8 Repeat Prob. 4-7 for the case where the reverse resistance is $R_r = 10$ K.

4-9 In the diode clipping circuit of Fig. 4-9a and d, $v_i = 20 \sin \omega t$, $R = 1$ K, and $V_R = 10$ V.   The reference voltage is obtained from a tap on a 10-K divider connected to a 100-V source.   Neglect all capacitances.   The diode forward resistance is 50 Ω, $R_r = \infty$, and $V_\gamma = 0$.   In both cases draw the input and output waveforms to scale.   Which circuit is the better clipper?   HINT: Apply Thévenin's theorem to the reference-voltage divider network.

4-10 A symmetrical 5-kHz square wave whose output varies between $+10$ and -10 V is impressed upon the clipping circuit shown.   Assume $R_f = 0$, $R_r = 2$ M, and

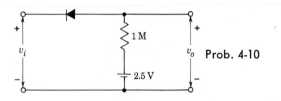

Prob. 4-10

$V_\gamma = 0$.   Sketch the steady-state output waveform, indicating numerical values of the maximum, minimum, and constant portions.

4-11 For the clipping circuits shown in Fig. 4-9b and d derive the transfer characteristic v_o versus v_i, taking into account R_f and V_γ and considering $R_r = \infty$.

4-12 The clipping circuit shown employs temperature compensation.   The dc voltage source V_γ represents the diode offset voltage; otherwise the diodes are assumed to be ideal with $R_f = 0$ and $R_r = \infty$.

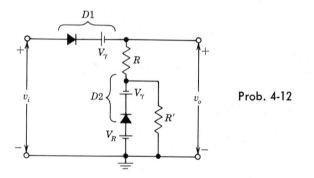

Prob. 4-12

(a) Sketch the transfer curve v_o versus v_i.

(b) Show that the maximum value of the input voltage v_i so that the current in D2 is always in the forward direction is

$$v_{i,\max} = V_R + \frac{R}{R'}(V_R - V_\gamma)$$

(c) What is the temperature dependence of the point on the input waveform at which clipping occurs?

4-13 (a) In the clipping circuit shown, D2 compensates for temperature variations. Assume that the diodes have infinite back resistance, a forward resistance of 50 Ω, and a break point at the origin ($V_\gamma = 0$). Calculate and plot the transfer characteristic v_o against v_i. Show that the circuit has an extended break point, that is, two break points close together.

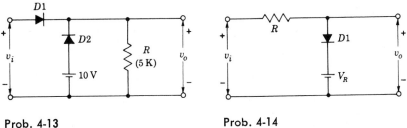

Prob. 4-13   Prob. 4-14

(b) Find the transfer characteristic that would result if D2 were removed and the resistor R were moved to replace D2.
(c) Show that the double break of part a would vanish and only the single break of part b would appear if the diode forward resistances were made vanishingly small in comparison with R.

4-14 (a) In the peak clipping circuit shown, add another diode D2 and a resistor R' in a manner that will compensate for drift with temperature.
(b) Show that the break point of the transmission curve occurs at V_R. Assume $R_r \gg R \gg R_f$.
(c) Show that if D2 is always to remain in conduction it is necessary that

$$v_i < v_{i,\text{max}} = V_R + \frac{R}{R'}(V_R - V_\gamma)$$

4-15 (a) The input voltage v_i to the two-level clipper shown in part a of the figure varies linearly from 0 to 150 V. Sketch the output voltage v_o to the same time scale as the input voltage. Assume ideal diodes.
(b) Repeat (a) for the circuit shown in part b of the figure.

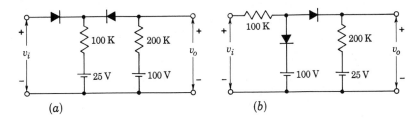

(a)   (b)

Prob. 4-15

4-16 The circuit of Fig. 4-10a is used to "square" a 10-kHz input sine wave whose peak value is 50 V. It is desired that the output voltage waveform be flat for 90 percent of the time. Diodes are used having a forward resistance of 100 Ω and a backward resistance of 100 K.

(a) Find the values of V_{R1} and V_{R2}.

(b) What is a reasonable value to use for R?

4-17 (a) The diodes are ideal. Write the transfer characteristic equations (v_o as a function of v_i).

(b) Plot v_o against v_i, indicating all intercepts, slopes, and voltage levels.

(c) Sketch v_o if $v_i = 40 \sin \omega t$. Indicate all voltage levels.

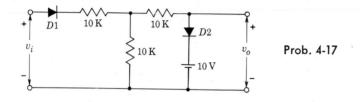

Prob. 4-17

4-18 (a) Repeat Prob. 4-17 for the circuit shown.

(b) Repeat for the case where the diodes have an offset voltage $V_\gamma = 1$ V.

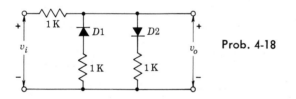

Prob. 4-18

4-19 Assume that the diodes are ideal. Make a plot of v_o against v_i for the range of v_i from 0 to 50 V. Indicate all slopes and voltage levels. Indicate, for each region, which diodes are conducting.

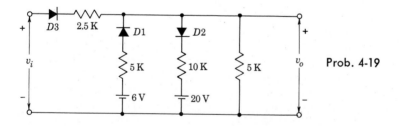

Prob. 4-19

4-20 The triangular waveform shown is to be converted into a sine wave by using clipping diodes. Consider the dashed waveform sketched as a first approximation to the sinusoid. The dashed waveform is coincident with the sinusoid at 0°,

30°, 60°, etc: Devise a circuit whose output is this broken-line waveform when the input is the triangular waveform. Assume ideal diodes and calculate the values of all supply voltages and resistances used. The peak value of the sinusoid is 50 V.

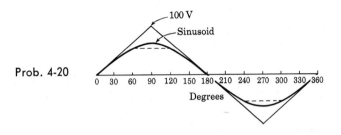

Prob. 4-20

4-21 Construct circuits which exhibit terminal characteristics as shown in parts *a* and *b* of the figure.

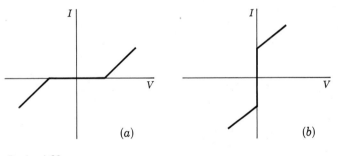

(a)   (b)

Prob. 4-21

4-22 The diode-resistor comparator of Fig. 4-13 is connected to a device which responds when the comparator output attains a level of 0.1 V. The input is a ramp which rises at the rate 10 V/μs. The germanium diode has a reverse saturation current of 1 μA. Initially, $R = 1$ K and the 0.1-V output level is attained at a time $t = t_1$. If we now set $R = 100$ K, what will be the corresponding change in t_1? $V_R = 0$.

4-23 For the four-diode sampling gate of Fig. 4-14 consider that v_s is at its most negative value, say, $v_s = -V_s$. Then verify that the expressions for $V_{n,\min}$ and $V_{c,\min}$ given in Eqs. (4-7) and (4-8) remain valid.

4-24 A balancing voltage divider is inserted between $D3$ and $D4$ in Fig. 4-14 so as to give zero output voltage for zero input. If the divider is assumed to be set at its midpoint, if its total resistance is R, and if R and R_f are both much less than R_c or R_L, show that

$$V_{c,\min} = V_s \left(2 + \frac{R_c}{R_L}\right)\left(1 + \frac{R}{4R_f}\right)$$

4-25 (a) Explain qualitatively the operation of the sampling gate shown. The supply voltage V is constant. The control voltage v_c is the square wave of Fig. 4-14b. Assume ideal diodes with $V_\gamma = 0$, $R_f = 0$, and $R_r = \infty$. HINT: When $v_c = V_c$, the diodes $D1$ and $D2$ conduct (if $V > V_{\min}$) and $D3$ and $D4$ are OFF. If $v_c = -V_n$, then $D3$ and $D4$ conduct and $D1$ and $D2$ are OFF.

Verify the following relationships:

(b) $V_{\min} = \dfrac{R_c}{R_2} \dfrac{R_1}{R_1 + 2R_L} V_s$

where $R_1 = R_c \| R_2$.

(c) $A = \dfrac{v_o}{v_s} = \dfrac{2R_L}{R_2} \dfrac{R_1}{R_1 + 2R_L}$

(d) $V_{n,\min} = V_s \dfrac{R_c}{R_c + R_2} - V \dfrac{R_2}{R_c + R_2}$

(e) $V_{c,\min} = A V_s$

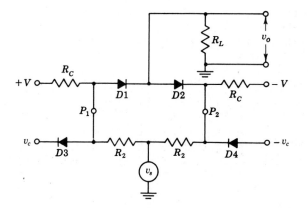

Prob. 4-25

4-26 A diode whose internal resistance is 20 Ω is to supply power to a 1,000-Ω load from a 110-V (rms) source of supply. Calculate (a) the peak load current, (b) the dc load current, (c) the ac load current, (d) the dc diode voltage, (e) the total input power to the circuit, (f) the percentage regulation from no load to the given load.

4-27 Show that the maximum dc output power $P_{dc} \equiv V_{dc}I_{dc}$ in a half-wave single-phase circuit occurs when the load resistance equals the diode resistance R_f.

4-28 The efficiency of rectification η_r is defined as the ratio of the dc output power $P_{dc} \equiv V_{dc}I_{dc}$ to the input power $P_i = (1/2\pi) \displaystyle\int_0^{2\pi} v_i i \, d\alpha$.

(a) Show that, for the half-wave-rectifier circuit,

$$\eta_r = \frac{40.6}{1 + R_f/R_L}\ \%$$

(b) Show that, for the full-wave rectifier, η_r has twice the value given in part a.

4-29 Prove that the regulation of both the half-wave and the full-wave rectifier is given by

$$\%\ \text{regulation} = \frac{R_f}{R_L} \times 100\%$$

4-30 (a) Prove Eqs. (4-21) and (4-22) for the dc voltage of a full-wave-rectifier circuit. (b) Find the dc voltage across a diode by direct integration.

4-31 A full-wave single-phase rectifier consists of a double-diode vacuum tube, the internal resistance of each element of which may be considered to be constant and equal to 500 Ω. These feed into a pure resistance load of 2,000 Ω. The secondary transformer voltage to center tap is 280 V. Calculate (a) the dc load current, (b) the direct current in each tube, (c) the ac voltage across each diode, (d) the dc output power, (e) the percentage regulation.

4-32 In the full-wave single-phase bridge, can the transformer and the load be interchanged? Explain carefully.

4-33 A 1-mA dc meter whose resistance is 10 Ω is calibrated to read rms volts when used in a bridge circuit with semiconductor diodes. The effective resistance of each element may be considered to be zero in the forward direction and infinite in the inverse direction. The sinusoidal input voltage is applied in series with a 5-K resistance. What is the full-scale reading of this meter?

4-34 The circuit shown is a half-wave voltage doubler. Analyze the operation of this circuit. Calculate (a) the maximum possible voltage across each capacitor, (b) the peak inverse voltage of each diode. Compare this circuit with the bridge voltage doubler of Fig. 4-22. In this circuit the output voltage is negative with respect to ground. Show that if the connections to the cathode and anode of each diode are interchanged, the output voltage will be positive with respect to ground.

Prob. 4-34

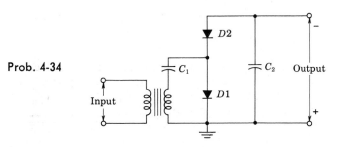

4-35 The circuit of Prob. 4-34 can be extended from a doubler to a quadrupler by adding two diodes and two capacitors as shown. In the figure, parts a and b are alternative ways of drawing the same circuit.

(a) Analyze the operation of this circuit.

(b) Answer the same questions as asked in Prob. 4-34.

(c) Generalize the circuit of this and of Prob. 4-34 so as to obtain n-fold multiplication when n is any even number. In particular, sketch the circuit for sixfold multiplication.

(d) Show that n-fold multiplication, with n odd, can also be obtained provided that the output is properly chosen.

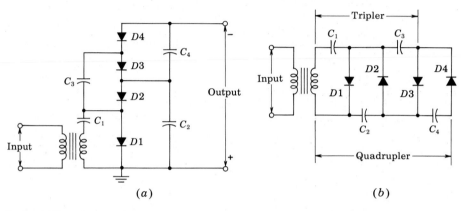

(a)   (b)

Prob. 4-35

4-36 A single-phase full-wave rectifier uses a semiconductor diode. The transformer voltage is 35 V rms to center tap. The load consists of a 40-μF capacitance in parallel with a 250-Ω resistor. The diode and the transformer resistances and leakage reactance may be neglected.

(a) Calculate the cutout angle.

(b) Plot to scale the output voltage and the diode current as in Fig. 4-25. Determine the cutin point graphically from this plot, and find the peak diode current corresponding to this point.

(c) Repeat parts a and b, using a 160-μF instead of a 40-μF capacitance.

CHAPTER 5

5-1 (a) Show that for an n-p-n silicon transistor of the alloy type in which the resistivity ρ_B of the base is much larger than that of the collector, the punch-through voltage V is given by $V = 6.1 \times 10^3 W_B^2 / \rho_B$, where V is in volts, ρ_B in ohm-centimeters, and W in mils. For punch-through, $W = W_B$ in Fig. 5-8a.

(b) Calculate the punch-through voltage if $W = 1$ μm and $\rho_B = 0.5$ Ω-cm.

5-2 The transistor of Fig. 5-3a has the characteristics given in Figs. 5-6 and 5-7. Let $V_{CC} = 6$ V, $R_L = 200$ Ω, and $I_E = 15$ mA.

(a) Find I_C and V_{CB}.

(b) Find V_{EB} and V_L.

(c) If I_E changes by $\Delta I_E = 10$ mA symmetrically around the point of part a and with constant V_{CC}, find the corresponding change in I_C.

5-3 The CB transistor used in the circuit of Fig. 5-3a has the characteristics given in Figs. 5-6 and 5-7. Let $I_C = -20$ mA, $V_{CB} = -4$ V, and $R_L = 200 \ \Omega$.

(a) Find V_{CC} and I_E.

(b) If the supply voltage V_{CC} decreases from its value in part a by 2 V while I_E retains its previous value, find the new values of I_C and V_{CB}.

5-4 The CE transistor used in the circuit shown has the characteristics given in Figs. 5-10 and 5-11.

(a) Find V_{BB} if $V_{CC} = 10$ V, $V_{CE} = -1$ V, and $R_L = 250 \ \Omega$.

(b) If $V_{CC} = 10$ V, find R_L so that $I_C = -20$ mA and $V_{CE} = -4$ V. Find V_{BB}.

Prob. 5-4

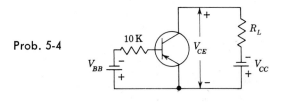

5-5 If $\alpha = 0.98$ and $V_{BE} = 0.7$ V, find R_1 in the circuit shown for an emitter current $I_E = -2$ mA. Neglect the reverse saturation current.

Prob. 5-5

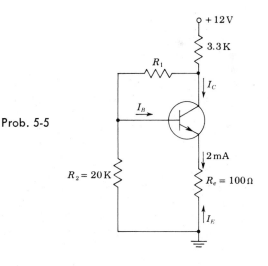

5-6 (a) Find R_e and R_b in the circuit of Fig. 5-12a if $V_{CC} = 10$ V and $V_{BB} = 5$ V, so that $I_C = 10$ mA and $V_{CE} = 5$ V. A silicon transistor with $\beta = 100$, $V_{BE} = 0.7$ V, and negligible reverse saturation current is under consideration.

(b) Repeat part a if a 100-Ω emitter resistor is added to the circuit.

5-7   In the circuit shown, $V_{CC} = 24$ V, $R_c = 10$ K, and $R_e = 270$ Ω. If a silicon
transistor is used with $\beta = 45$ and if $V_{CE} = 5$ V, find R. Neglect the reverse
saturation current.

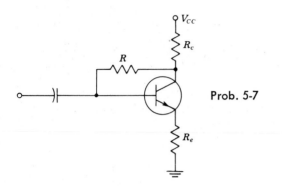

Prob. 5-7

5-8   For the circuit shown, transistors $Q1$ and $Q2$ operate in the active region with
$V_{BE1} = V_{BE2} = 0.7$ V, $\beta_1 = 100$, and $\beta_2 = 50$. The reverse saturation currents may be neglected.
(a) Find the currents I_{B2}, I_1, I_2, I_{C2}, I_{B1}, I_{C1}, and I_{E1}.
(b) Find the voltages V_{o1} and V_{o2}.

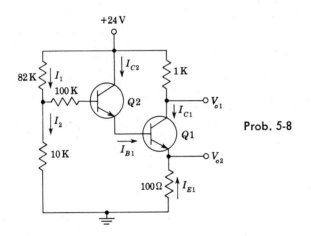

Prob. 5-8

5-9   (a) The reverse saturation current of the germanium transistor in Fig. 5-13 is
2 μA at room temperature (25°C) and increases by a factor of 2 for each temperature increase of 10°C. The bias $V_{BB} = 5$ V. Find the maximum allowable
value for R_B if the transistor is to remain cut off at a temperature of 75°C.
(b) If $V_{BB} = 1.0$ V and $R_B = 50$ K, how high may the temperature increase
before the transistor comes out of cutoff?

5-10   From the characteristic curves for the type 2N404 transistor given in Fig. 5-14,
find the voltages V_{BE}, V_{CE}, and V_{BC} for the circuit shown.

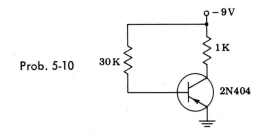

Prob. 5-10

5-11   A silicon transistor with $V_{BE,sat} = 0.8$ V, $\beta = h_{FE} = 100$, $V_{CE,sat} = 0.2$ V is used in the circuit shown.   Find the minimum value of R_c for which the transistor remains in saturation.

Prob. 5-11

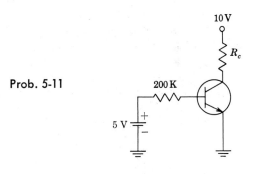

5-12   For the circuit shown, assume $\beta = h_{FE} = 100$.
(a) Find if the silicon transistor is in cutoff, saturation, or in the active region.
(b) Find V_o.
(c) Find the minimum value for the emitter resistor R_e for which the transistor operates in the active region.

Prob. 5-12

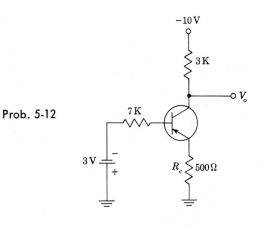

5-13 If the silicon transistor used in the circuit shown has a minimum value of $\beta = h_{FE}$ of 30 and if $I_{CBO} = 10$ nA at 25°C:

(a) Find V_o for $V_i = 12$ V and show that Q is in saturation.

(b) Find the minimum value of R_1 for which the transistor in part a is in the active region.

(c) If $R_1 = 15$ K and $V_i = 1$ V, find V_o and show that Q is at cutoff.

(d) Find the maximum temperature at which the transistor in part c remains at cutoff.

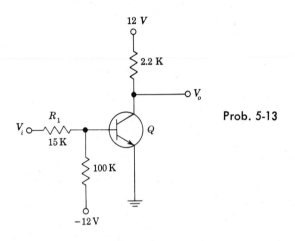

Prob. 5-13

5-14 Silicon transistors with $h_{FE} = 100$ are used in the circuit shown. Neglect the reverse saturation current.

(a) Find V_o when $V_i = 0$ V. Assume $Q1$ is OFF and justify the assumption.

(b) Find V_o when $V_i = 6$ V. Assume $Q2$ is OFF and justify this assumption.

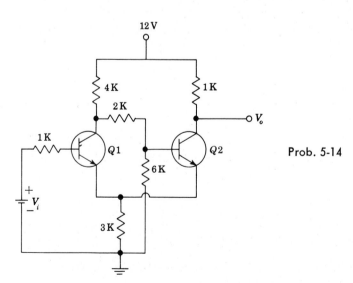

Prob. 5-14

5-15   For the circuit shown, $\alpha_1 = 0.98$, $\alpha_2 = 0.96$, $V_{CC} = 24$ V, $R_c = 120$ Ω, and $I_E = -100$ mA. Neglecting the reverse saturation currents, determine (a) the currents I_{C1}, I_{B1}, I_{E1}, I_{C2}, I_{B2}, and I_C; (b) V_{CE}; (c) I_C/I_B, I_C/I_E.

Prob. 5-15

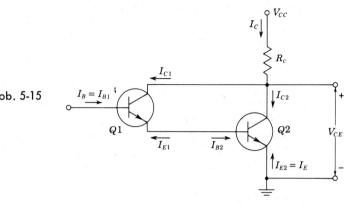

5-16   Derive from Eqs. (5-24) and (5-25) the explicit expressions for I_C and I_E in terms of V_C and V_E.

5-17   (a) Derive Eqs. (5-29) and (5-30).
(b) Derive Eq. (5-31).

5-18   Draw the Ebers-Moll model for an n-p-n transistor.

5-19   (a) Show that the exact expression for the CE output characteristics of a p-n-p transistor is

$$V_{CE} = V_T \ln \frac{\alpha_I}{\alpha_N} + V_T \ln \frac{I_{co} + \alpha_N I_B - I_C(1 - \alpha_N)}{I_{EO} + I_B + I_C(1 - \alpha_I)}$$

(b) Show that this reduces to Eq. (5-31) if $I_B \gg I_{EO}$ and $I_B \gg I_{co}/\alpha_N$.

5-20   (a) A transistor is operating in the cutoff region with both the emitter and collector junctions reverse-biased by at least a few tenths of a volt. Prove that the currents are given by

$$I_E = \frac{I_{EO}(1 - \alpha_N)}{1 - \alpha_N \alpha_I}$$

$$I_C = \frac{I_{co}(1 - \alpha_I)}{1 - \alpha_N \alpha_I}$$

(b) Prove that the emitter-junction voltage required just to produce cutoff ($I_E = 0$ and the collector back-biased) is

$$V_E = V_T \ln (1 - \alpha_N)$$

5-21   (a) Find the collector current for a transistor when both emitter and collector junctions are reverse-biased. Assume $I_{co} = 5$ μA, $I_{EO} = 3.57$ μA, and $\alpha_N = 0.98$.
(b) Find the emitter current I_E under the same conditions as in part a.

5-22 Show that the emitter volt-ampere characteristic of a transistor in the active region is given by

$$I_E \approx I_S \epsilon^{V_E/V_T}$$

where $I_S = -I_{EO}/(1 - \alpha_N\alpha_I)$. Note that this characteristic is that of a *p-n* junction diode.

5-23 (*a*) Given an *n-p-n* transistor for which (at room temperature) $\alpha_N = 0.98$, $I_{co} = 2$ μA, and $I_{EO} = 1.6$ μA. A common-emitter connection is used, and $V_{CC} = 12$ V and $R_L = 4.0$ K. What is the minimum base current required in order that the transistor enter its saturation region?
(*b*) Under the conditions in part *a*, find the voltages across each junction between each pair of terminals if the base-spreading resistance $r_{bb'}$ is neglected.
(*c*) Repeat part *b* if the base current is 200 μA.
(*d*) How are the above results modified if $r_{bb'} = 250$ Ω?

5-24 Plot the emitter current vs. emitter-to-base voltage for a transistor for which $\alpha_N = 0.98$, $I_{CO} = 2$ μA, and $I_{EO} = 1.6$ μA if (*a*) $V_C = 0$, (*b*) V_C is back-biased by more than a few tenths of a volt. Neglect the base-spreading resistance.

5-25 Plot carefully to scale the common-emitter characteristic I_C/I_B versus V_{CE} for a transistor with $\alpha_N = 0.90 = \alpha_I$.

5-26 Show that

$$I_{CES} = \frac{I_{co}}{1 - \alpha_N\alpha_I} \qquad I_{CEO} = \frac{I_{co}}{1 - \alpha_N}$$

5-27 A common method of calculating α_N and α_I is by measurement of I_{co}, I_{CEO}, and I_{CES}. Show that

$$(a) \ \alpha_N = \frac{I_{CEO} - I_{co}}{I_{CEO}} \qquad (b) \ \alpha_I = \frac{1 - I_{co}/I_{CES}}{1 - I_{co}/I_{CEO}}$$

5-28 The collector leakage current is measured as shown in the figure, with the emitter grounded and a resistor R connected between base and ground. If this current is designated as I_{CER}, show that

$$I_{CER} = \frac{I_{co}(1 + I_{EO}R/V_T)}{1 - \alpha_N\alpha_I + (I_{EO}R/V_T)(1 - \alpha_N)}$$

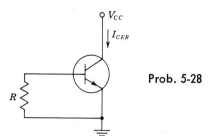

Prob. 5-28

5-29 For the circuit shown, verify that $V_o = V_{CC}$ when

$$I_B = \frac{V_{CC}}{R_e}\left(1 + \frac{\alpha_N}{\alpha_I}\frac{1 - \alpha_I}{1 - \alpha_N}\right) = \frac{V_{CC}}{R_e}\left(1 + \frac{\beta_N}{\beta_I}\right)$$

Under these conditions the base current exceeds the emitter current.

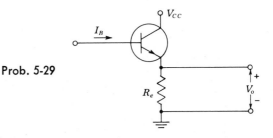

Prob. 5-29

5-30 For the circuit shown, prove that the floating emitter-to-base voltage is given by

$$V_{EBF} = V_T \ln (1 - \alpha_N)$$

Neglect $r_{bb'}$.

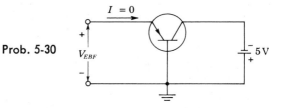

Prob. 5-30

5-31 For the "floating-base" connection shown, prove that

$$I_{CT} = \frac{2 - \alpha_N}{(1 - \alpha_N)^2} I_{CO}$$

Assume that the transistors are identical.

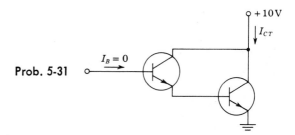

Prob. 5-31

5-32 (a) Show that if the collector junction is reverse-biased with $|V_{CB}| \gg V_T$, the voltage V_{BE} is related to the base current by

$$V_{BE} = I_B \left(r_{bb'} + \frac{R_E}{1 - \alpha_N} \right)$$
$$+ \left\{ \frac{I_{CO}R_E}{1 - \alpha_N} + V_T \ln \left[1 + \frac{I_B(1 - \alpha_N\alpha_I)}{I_{EO}(1 - \alpha_N)} + \frac{\alpha_N(1 - \alpha_I)}{\alpha_I(1 - \alpha_N)} \right] \right\}$$

where $r_{bb'}$ is the base-spreading resistance, and R_E is the emitter-body resistance. (b) Show that $V_{BE} = I_B(r_{bb'} + R_E) + V_T(1 + I_B/I_E)$ if the collector is open-circuited.

5-33 A transistor is operated at a forward emitter current of 2 mA and with the collector open-circuited. Find (a) the junction voltages V_C and V_E, (b) the collector-to-emitter voltage V_{CE}. Assume $I_{CO} = 2~\mu A$, $I_{EO} = 1.6~\mu A$, $\alpha_N = 0.98$. Is the transistor operating in saturation, at cutoff, or in the active region?

5-34 Photodiode 1N77 (Fig. 3-21) is used in the circuit shown. R_L represents the coil resistance of a relay for which the current required to close the relay is 6 mA. The transistor used is silicon with $V_{BE} = 0.7$ V and $h_{FE} = 100$.

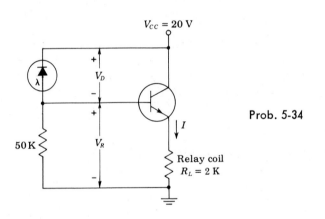

Prob. 5-34

(a) Find the voltage V_D at which switching of the relay occurs.
(b) Find the minimum illumination required to close the relay.
(c) If the relay coil is placed directly in series with the phototransistor of Fig. 5-25 across 20 V, find the illumination intensity required to close the relay.

CHAPTER 6

6-1 Convert the following decimal numbers to binary form: (a) 671, (b) 325, (c) 152.

6-2 The parameters in the diode OR circuit of Fig. 6-3 are $V(0) = +12$ V, $V(1) = -2$ V, $R_s = 600~\Omega$, $R = 10$ K, $R_f = 0$, $R_r = \infty$, and $V_\gamma = 0.6$ V. Calculate the output levels if one input is excited and if (a) $V_R = +12$ V, (b) $V_R = +10$ V, (c) $V_R = +14$ V, and (d) $V_R = 0$ V. For which of these cases is the OR function

satisfied (except possibly for a shift in level between input and output)? *(e)*
Repeat part *a* if three inputs are excited.

6-3 Consider a two-input positive-logic diode OR gate (Fig. 6-3 with the diodes
reversed) and with $V_R = 0$. The inputs are the square waves v_1 and v_2 indicated.
Sketch the output waveform if the ratio of the amplitude of v_2 to v_1 is *(a)* 2 and
(b) $\frac{1}{2}$. Assume ideal diodes ($R_f = 0$, $R_r = \infty$, and $V_\gamma = 0$) and $R_s = 0$.

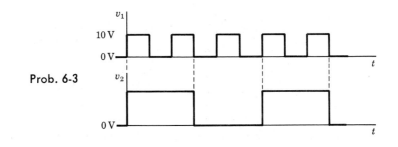

Prob. 6-3

6-4 Consider two signals, a 1-kHz sine wave and a 10-kHz square wave of zero average value, applied to the OR circuit of Fig. 6-3 with $V_R = 0$. Draw the output
waveform if the sine-wave amplitude *(a)* exceeds the square-wave amplitude,
(b) is less than the square-wave amplitude.

6-5 Consider a two-input positive-logic diode AND circuit (Fig. 6-5) with $V_R = 15$ V,
$R = 10$ K, $R_s = 1$ K. Assume ideal diodes and neglect all capacitances. A
square wave v_i, extending from -5 to $+5$ V with respect to ground, is applied
simultaneously to both inputs. *(a)* Sketch the output v_o and calculate the
maximum and minimum voltages with respect to ground. *(b)* If $v_1 = v_i$ and
$v_2 = -v_i$, calculate the voltage levels of v_o and plot.

6-6 *(a)* Using a clamping diode draw an AND diode gate whose output is V' when there
is no coincidence, and $V(1)$ when there is. Assume that $V_R \geq V(1)$ and that
positive logic is used.
(b) Find the minimum required value of V' and the maximum number of inputs
if the rated current of the catching diode is I_m.

6-7 *(a)* Indicate how to modify the circuit of Prob. 6-5 so that the minimum voltage
is zero.
(b) Repeat parts *a* and *b* of Prob. 6-5 assuming maximum rated current of the
catching diode 5 mA.
(c) Find the maximum number of inputs if rated current of catching diode is
50 mA.

6-8 Consider a two-input positive-logic diode AND circuit (Fig. 6-5*b*) with $V_R = 10$ V,
$R = 10$ K, and $R_s = 0$. Assume ideal diodes and neglect capacitances. The
input waveforms are v_1 and v_2 sketched in Prob. 6-3. Sketch the output wave-
form if the ratio of the amplitude of v_2 to v_1 is *(a)* 2, *(b)* 1, and *(c)* $\frac{1}{2}$. Repeat part
b if $R_s = 1$ K.

6-9 The binary input levels for the AND circuit shown are $V(0) = 0$ V and $V(1) =$
25 V. Assume ideal diodes. If $v_1 = V(0)$ and $v_2 = V(1)$, then v_o is to be at 5 V.
However, if $v_1 = v_2 = V(1)$, then v_o is to rise above 5 V.

(a) What is the maximum value of V_R which may be used?

(b) If $V_R = 20$ V, what is v_o at a coincidence [$v_1 = v_2 = V(1)$]? What are the diode currents?

(c) Repeat part b if $V_R = 40$ V.

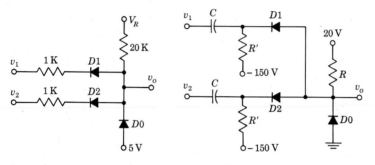

Prob. 6-9   Prob. 6-10

6-10 The two-input-diode AND circuit shown uses diodes with $R_f = 500\ \Omega$, $R_r = \infty$, and $V_\gamma = 0$. The quiescent current in $D0$ is 6 mA, and the currents in $D1$ and $D2$ are each 4 mA

(a) Calculate the quiescent output voltage v_o and the values of R and R'.

(b) Calculate the output voltage when one input diode is cut off. Calculate this result approximately by assuming that the currents through R and the remaining input diode do not change. Also, calculate the result exactly.

(c) Assume that diode $D0$ is omitted, that the currents in $D1$ and $D2$ remain 4 mA each, and that the output v_o is the same as that found in part a. Find R and R'.

(d) If the conditions are as indicated in part c but one of the diodes is cut off, find the output voltage v_o. Compare with the result in part b when $D0$ acts as a clamp.

6-11 Find v_o and v' if (a) there are no pulses at either A or B, (b) there is a 30-V positive pulse at A or B, and (c) there are positive pulses at both A and B. (d) What is the minimum pulse amplitude which must be applied in order that the circuit operate properly? Assume ideal diodes.

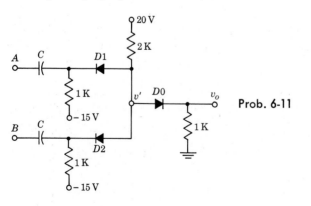

Prob. 6-11

6-12 (*a*) Verify that the circuit shown is an inverter by calculating the output levels corresponding to input levels of 0 and −6 V. What minimum value of h_{FE} is required? Neglect junction saturation voltages and assume an ideal diode.
(*b*) If the reverse collector saturation current at 25°C is 5 μA, what is the maximum temperature at which this inverter will operate properly?

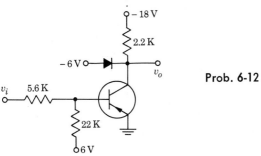

Prob. 6-12

6-13 For the circuit shown in Fig. 6-7, $V_{CC} = 8$ V, $V_{BB} = 8$ V, $V_{EE} = 0$, and $R_c = 2.2$ K. The inverter is to operate properly in the temperature range −25 to 125°C. The silicon transistor used has $(h_{FE})_{min} = 65$ at 25°C, 55 at −25°C, 85 at 125°C, and $I_{CBO} = 5$ nA at 25°C. The desired logic levels are $V(1) = 8 \pm 2$ V, $V(0) = 0.2 \pm 0.2$ V.
(*a*) Find the maximum value of R_1 if $R_2 = 100$ K.
(*b*) If the desired logic levels are $V(1) = 4 \pm 1$ V and $V(0) = 0.2 \pm 0.2$ V, what modification should you make to this circuit?

6-14 A half adder is a combination of OR, NOT and AND gates. It has two inputs and two outputs and the following truth table:

Input 1	Input 2	Output 1	Output 2
0	0	0	0
0	1	1	0
1	0	1	0
1	1	0	1

Draw the logic block diagram for a half adder.

6-15 The four inputs v_1, v_2, v_3, and v_4 are voltages from zero-impedance sources whose values are either $V(0) = 10$ V or $V(1) = 20$ V. The diodes are ideal. $V_R = 25$ V, $R_1 = 5$ K, and $R_2 = 10$ K.
(*a*) If $v_1 = v_2 = 10$ V and $v_3 = v_4 = 20$ V, find v_o and the currents in each diode.
(*b*) If $v_1 = v_3 = 10$ V and $v_2 = v_4 = 20$ V, find v_o and the currents in each diode.
(*c*) Sketch in block-diagram form the logic performed by this circuit.
(*d*) Verify that in order for the circuit to operate properly the following inequality must be satisfied:

$$R_2 > \frac{V_R - V(0)}{V(0)} R_1$$

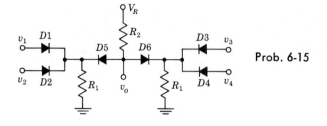

Prob. 6-15

6-16 (a) In block-diagram form indicate the logic performed by the diode system shown. The input levels are $V(0) = -8$ V and $V(1) = +2$ V. Neglect source resistance and assume that the diodes are ideal. Justify your answer by calculating the voltages v_A, v_B, and v_o (and indicating which diodes are conducting) under the following circumstances: (i) all inputs are at $V(0)$; (ii) some but not all inputs in A are at $V(1)$ and all inputs in B are at $V(0)$; (iii) all inputs in A are at $V(1)$ and some inputs in B are at $V(1)$; and (iv) all inputs are at $V(1)$. (b) If the 10-K resistance were increased, at what maximum value would the circuit no longer operate in the manner described above? (c) Indicate how to modify the circuit so that the output levels are -5 and 0 V, respectively.

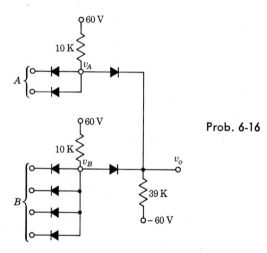

Prob. 6-16

6-17 (a) Verify De Morgan's law [Eq. (6-27)] in a manner analogous to that given in the text in connection with the proof of Eq. (6-25). (b) Prove Eq. (6-27) by constructing a truth table for each side and verifying that these two tables have the same outputs.

6-18 Verify the auxiliary Boolean identities in Table 6-4 (page 174).

6-19 Using Boolean algebra, verify
(a) $\overline{\bar{A} + B} + \overline{\bar{A} + \bar{B}} = A$
(b) $AB + AC + B\bar{C} = AC + B\bar{C}$
HINT: Multiply the first term on the left-hand side by $C + \bar{C} = 1$.
(c) $\overline{AB + BC + CA} = \bar{A}\bar{B} + \bar{B}\bar{C} + \bar{C}\bar{A}$

6-20 Using Boolean algebra, verify
  (a) $(A + B)(B + C)(C + A) = AB + BC + CA$
  (b) $(A + B)(\bar{A} + C) = AC + \bar{A}B$
  (c) $AB + \bar{B}\bar{C} + A\bar{C} = AB + \bar{B}\bar{C}$
  HINT: A term may be multiplied by $B + \bar{B} = 1$.

6-21 Given two N-bit characters which are available in parallel form. Indicate in block-diagram form a system whose output is 1 if and only if *all* corresponding bits are equal, that is, only if the two characters are equal.

6-22 A, B, and C represent the presence of pulses. The logic statement "A or B and C" can have two interpretations. Which are they? In block-diagram form draw the circuit to perform each of the two logic operations.

6-23 A circuit has three input and one output terminals. The output is 1 if any two of the three inputs are 1 and is 0 for any other combination of inputs. Draw a block diagram of this logic circuit.

6-24 In block-diagram form draw a circuit to perform the following logic: If pulses A_1, A_2, and A_3 occur simultaneously or if pulses B_1 and B_2 occur simultaneously, an output pulse is delivered, provided that pulse C does not occur at the same time. No output is to be obtained if A_1, A_2, A_3, B_1, and B_2 occur simultaneously.

6-25 A single-pole double-throw switch is to be simulated with AND, OR, and INHIBITOR circuits. Call the two signal inputs A and B. A third input C receives the switching instructions in the form of a code: 1 (a pulse is present) or 0 (no pulse exists). It is desired that $C = 1$ set the switch to A and $C = 0$ set the switch to B, as indicated schematically. In block-diagram form show the circuit for this switch.

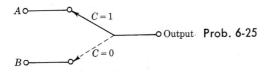
Prob. 6-25

6-26 In block-diagram form draw a circuit which satisfies simultaneously the conditions a, b, and c as follows:
  (a) The output is excited if any pair of inputs A_1, A_2, and A_3 is excited, provided that B is also excited.
  (b) The output is 1 if any one (and only one) of the inputs A_1, A_2, or A_3 is 1, provided that $B = 0$.
  (c) No output is excited if A_1, A_2, and A_3 are simultaneously excited.

6-27 (a) For the illustrative NAND gate of Fig. 6-19a calculate the minimum value of h_{FE} taking junction voltages into account.
  (b) What is the maximum noise voltage (superimposed upon the logic level) which will still permit the circuit to operate properly? Consider the following two cases: (i) a complete coincidence and (ii) all inputs but one in the 1 state.
  (c) What is the maximum value of the source resistance which will still permit proper circuit operation? Assume a 0.7-V drop across a conducting diode.

6-28 The circuit shown uses silicon diodes and a silicon transistor. The input A or B is obtained from the output Y of a similar gate.

(a) What are the logic levels? Take junction voltages into account.

(b) Verify that the circuit satisfies the NAND operation. Assume $h_{FE,min} = 15$.

(c) What is the maximum allowable value of I_{CBO}?

(d) Now neglect junction voltages and I_{CBO} and verify that the circuit satisfies the NOR operation.

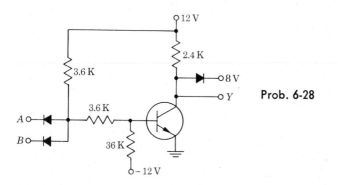

Prob. 6-28

6-29 Verify that the circuit shown is an EXCLUSIVE OR.

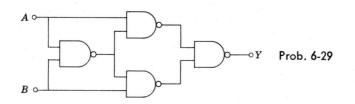

Prob. 6-29

6-30 Verify that the NOR-NOR topology is equivalent to an OR-AND system.

6-31 Verify that the logic operations OR, AND, and NOT may be implemented by using only NOR gates.

6-32 What logic operation is performed by the circuit shown, which consists of interconnected NOR gates?

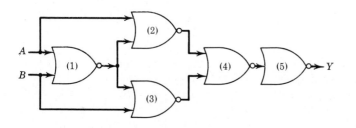

Prob. 6-32

6-33 (*a*) Implement the EXCLUSIVE OR gate using (i) NOR gates, (ii) NAND gates.

(*b*) Repeat part *a* for the half adder of Prob. 6-14.

6-34 (*a*) The discrete-components circuit of a DTL gate shown uses a silicon transistor with worst-case values of $V_{BE,\text{sat}} = 1.0$ V and $V_{CE,\text{sat}} = 0.5$ V. The voltage across any silicon diode (when conducting) is 0.7 V. Assume that $D1$ consists of two diodes in series. The circuit parameters are $V_{CC} = V_{BB} = 12$ V, $R = 15$ K, $R_2 = 100$ K, and $R_c = 2.2$ K. The inputs to this switch are obtained from the outputs of similar gates. Verify that the circuit functions as a positive NAND. In particular, for proper operation, calculate the minimum value of the clamping voltage V' and h_{FE}.

(*b*) Will the circuit operate properly if $D1$ is (i) a single diode or (ii) three diodes in series?

(*c*) Replace $D1$ by a 15-K resistance and repeat part *a*. Compare the binary levels in part *a* and *c*.

(*d*) What is the maximum allowed fan-in, assuming that the diodes are ideal? What is a practical limitation on fan-in?

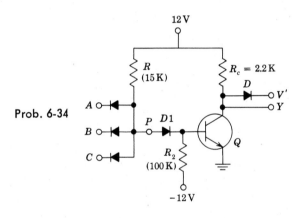

Prob. 6-34

6-35 The DTL shown uses silicon devices with $V_{BE,\text{sat}} = 0.8$ V, $V_{CE,\text{sat}} = 0.2$ V, $V_\gamma = 0.5$ V, and the drop across a conducting diode $= 0.7$ V. The inputs to this switch are obtained from the outputs of similar gates.

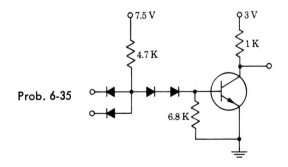

Prob. 6-35

(a) Verify that the circuit functions as a positive NAND and calculate $h_{FE,\min}$. Assume that the transistor is essentially cut off if the base-to-emitter voltage is at least 0.1 V smaller than the cutin voltage V_γ.

(b) Assume that the diode reverse saturation current is equal to the transistor reverse saturation collector current. Find $I_{CBO,\max}$.

(c) If all inputs are high, what is the magnitude of noise voltage at the input which will cause the gate to malfunction?

(d) Repeat part c if at least one input is low.

6-36 (a) Analyze the DTL circuit shown. Use the voltage drops given in Prob. 6-35.

(b) Find $h_{FE,\min}$ if two similar gates are to be driven by this circuit.

(c) Find the noise margins.

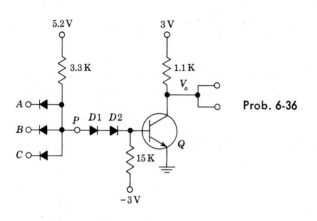

Prob. 6-36

6-37 (a) Analyze the DTL circuit shown. Use the voltage drops given in Prob. 6-35.

(b) If $h_{FE} = 25$, calculate the fan-out N.

(c) For a fan-out of 10 and assuming a diode reverse saturation current of 15 μA, what is $V(1)$?

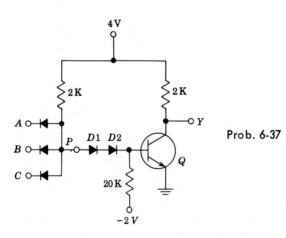

Prob. 6-37

6-38 The positive DTL NAND gate of Prob. 6-39 is to operate properly in the temperature range -50 to $160°C$. The silicon transistor has $h_{FE,min} = 50$ at $-50°C$, $h_{FE,min} = 65$ at $25°C$, and $h_{FE,min} = 100$ at $160°C$. The reverse saturation collector current of the transistor at $25°C$ is $I_{CBO} = 0.5$ nA, and it equal to the reverse saturation current of the silicon diode. The maximum current rating of the transistor is 50 mA. The gate will be used in a system with power supply voltage of 5 V, and the allowed variation in $V(1)$ is ± 0.5 V. The desired absolute value of the noise margin is 1.5 V and the desired fanout is 10. The transistor is considered OFF if $V_{BE} \leq 0.4$ V.
(a) Calculate the minimum required number of diodes between P and the base of the transistor.
(b) Calculate the maximum value of R_2.
(c) For the values found in (a) and (b) determine the range of values that R_c can take.
(d) Using the middle value of R_c found in part (c), specify the range of values that R_1 can take.

6-39 For the integrated positive DTL NAND gate shown, prove that
(a) The maximum number of diodes that can be used is given by $n_{max} = (V_{CC} - V_{BE,sat})/V_D$, where V_D is the voltage drop across a diode.
(b) The maximum fan-out is given by

$$N_{max} \approx h_{FE} - h_{FE}\left(n + \frac{R_1}{R_2}\right)\frac{V_D}{V_{CC}} - \frac{R_1}{R_c}\left(1 + \frac{V_D}{V_{CC}}\right)$$

Assume that $V_{BE,sat} \approx V_D$ and $V_{CC} - V_D \gg V_{CE,sat}$.

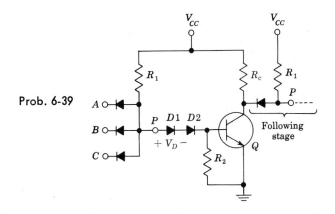

Prob. 6-39

6-40 For the modified integrated positive DTL NAND of Fig. 6-24 specify h_{FE} and the maximum current rating in order to have a fan-out of 50.

6-41 For the integrated positive DTL gate shown
(a) Verify its function as a NAND gate and specify the state of each transistor when at least one input is low and also at a coincidence.
(b) For $h_{FE,min} = 30$, calculate the fan-out of this gate. The inputs of this gate are obtained from the outputs of similar gates, and its output drives similar gates.

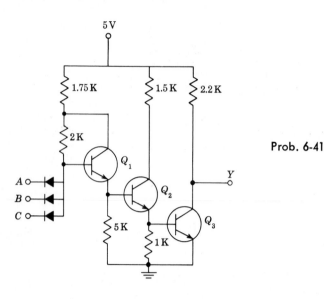

Prob. 6-41

6-42 In the integrated positive DTL NAND gate of Fig. 6-24 a Schottky diode is fabricated between base and collector to prevent the transistor Q2 from saturating. The anode of the Schottky diode is at the base and the drop across the diode when conducting is 0.4 V (Sec. 7-13).

(a) Explain why the transistor Q2 does not go into saturation.

(b) Verify the operation of the gate as a NAND gate and calculate noise margins.

(c) Find the logic levels and maximum fan-out if the inputs of this gate are obtained from similar gates and outputs drive similar gates and $h_{FE} = 30$.

6-43 Verify that this wired circuit performs the EXCLUSIVE OR function.

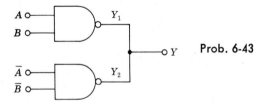

Prob. 6-43

6-44 (a) For the high-threshold logic NAND gate of Fig. 6-26, if V_2 of the diode is 6.9 V, verify that this circuit functions as a positive NAND and calculate $h_{FE,\min}$. The inputs of this gate are obtained from the output of similar gates.

(b) Calculate noise margins.

(c) Calculate the fan-out of this gate if $h_{FE,\min} = 40$.

6-45 If the output in Fig. 6-26 is capacitively loaded (by C), then the rise time as Y goes from its low to its high state will be long because of the high load resistance (15 K) of Q2. To reduce this time constant the active pull-up circuit indicated in the dashed block is added across the 15-K resistor.

(a) Explain how the circuit works.

(b) Why not simply replace 15 K by 1.5 K?

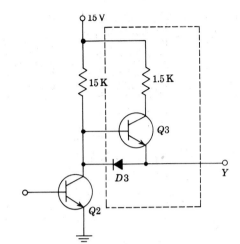

Prob. 6-45

6-46 (a) For the IC positive TTL NAND gate shown in Fig. 6-27, calculate $h_{FE,\,\min}$ for proper operation of the circuit.

(b) Calculate noise margins.

(c) Calculate the fan-out if $h_{FE,\,\min} = 30$.

6-47 For the IC positive NAND TTL gate shown, if the inputs are obtained from the outputs of similar gates and $h_{FE,\,\min}$ of the transistors is 30, verify its operation as a NAND gate when the fan-out is 10.

(a) At coincidence, find the state of each transistor and all currents and voltages of the circuit.

(b) Repeat part a if at least one input is low.

(c) Find the logic levels.

(d) Calculate the peak current drawn from the supply during the transient.

(e) Calculate maximum fan-out for proper operation of the gate.

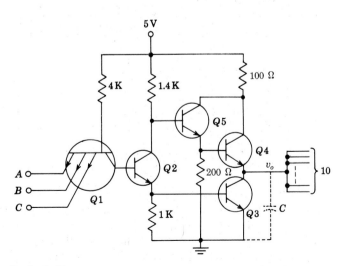

Prob. 6-47

6-48 For an RTL IC positive NOR gate prove that the maximum fan-out can be approximated by the formula

$$N_{max} = h_{FE.min} - h_{FE,min} \frac{0.6}{V_{CC}} - \frac{R_b}{R_c}$$

6-49 The inputs of the RTL IC positive NOR gate shown in Fig. 6-29 are obtained from the outputs of similar gates and the outputs drive similar gates. If the supply voltage of the system is 5 V and the temperature range for proper operation of the gate is −50 to 150°C, calculate the maximum permissible values of the resistances. Assume $h_{FE} = 30$ at −50°C, $I_{CBO} = 10$ nA at 25°C, and the desired fan-out is 10.

6-50 Verify that the DCTL circuit shown with the fan-in transistors in series satisfies the NAND operation. Assume that for the silicon transistors, $V_{CE,sat} = 0.2$ V and $V_{BE,sat} = 0.8$ V. Calculate the collector currents in each transistor when all inputs are high. The input to each base is taken from the output of a similar gate.

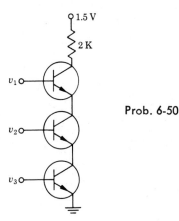

Prob. 6-50

CHAPTER 7

7-1 (*a*) Verify that Eq. (7-3) meets the stated boundary conditions.
(*b*) Verify that Eq. (7-5) satisfies the diffusion equation (7-2) and that it meets the stated boundary conditions.

7-2 A silicon wafer is uniformly doped with phosphorus to a concentration of 10^{15} cm^{-3}. Refer to Table 2-1 on page 29. At room temperature (300°K) find
(*a*) The percentage of phosphorus by weight in the wafer.
(*b*) The conductivity and resistivity.
(*c*) The concentration of boron, which, if added to the phosphorus-doped wafer, would halve the conductivity.

7-3 (*a*) Using the data of Fig. 7-8, calculate the percent maximum concentration of arsenic (atoms per cubic centimeter) that can be achieved in solid silicon. The concentration of pure silicon may be calculated from the data in Table 2-1 on page 29.

(b) Repeat part a for gold.

7-4 (a) How long would it take for a fixed amount of phosphorus distributed over one surface of a 25-μm-thick silicon wafer to become substantially uniformly distributed throughout the wafer at 1300°C? Consider that the concentration is sufficiently uniform if it does not differ by more than 10 percent from that at the surface.

(b) Repeat part a for gold, given that the diffusion coefficient of gold in silicon is 1.5×10^{-6} cm²/s at 1300°C.

7-5 Show that the junction depth x_j resulting from a Gaussian impurity diffusion into an oppositely doped material of background concentration N_{BC} is given by

$$x_j = \left(2Dt \ln \frac{Q^2}{N_{BC}^2 \pi Dt} \right)^{\frac{1}{2}}$$

7-6 A uniformly doped n-type silicon substrate of 0.1 Ω-cm resistivity is to be subjected to a boron diffusion with constant surface concentration of 4.8×10^{18} cm⁻³. The desired junction depth is 2.7 μm.

(a) Calculate the impurity concentration for the boron diffusion as a function of distance from the surface.

(b) How long will it take if the temperature at which this diffusion is conducted is 1100°C?

(c) An n-p-n transistor is to be completed by diffusing phosphorus at a surface concentration of 10^{21} cm⁻³. If the new junction is to be at a depth of 2 μm, calculate the concentration for the phosphorus diffusion as a function of distance from the surface.

(d) Plot the impurity concentrations (log scale) vs. distance (linear scale) for parts a and c, assuming that the boron stays put during the phosphorus diffusion. Indicate emitter, base, and collector on your plot.

(e) If the phosphorus diffusion takes 30 min, at what temperature is the apparatus operated?

7-7 List in order the steps required in fabricating a monolithic silicon integrated transistor by the epitaxial-diffused method. Sketch the cross section after each oxide growth. Label materials clearly. No buried layer is required.

7-8 Sketch to scale the cross section of a monolithic transistor fabricated on a 5-mil-thick silicon substrate. HINT: Refer to Sec. 7-1 and Figs. 7-12 and 7-13 for typical dimensions.

7-9 Sketch the five basic diode connections (in circuit form) for the monolithic integrated circuits. Which will have the lowest forward voltage drop? Highest breakdown voltage?

7-10 If the base sheet resistance can be held to within ± 10 percent and resistor line widths can be held to ± 0.1 mil, plot approximate tolerance of a diffused resistor as a function of line width w in mils over the range $0.5 \le w \le 5.0$. (Neglect contact-area and contact-placement errors.)

7-11 A 1-mil-thick silicon wafer has been doped uniformly with phosphorus to a concentration of 10^{16} cm⁻³, plus boron to a concentration of 2×10^{15} cm⁻³. Find its sheet resistance.

7-12 (a) Calculate the resistance of a diffused crossover 4 mils long, 1 mil wide, and 2 μm thick, given that its sheet resistance is 2.2 Ω/square.

(b) Repeat part a for an aluminum metalizing layer 0.5 μm thick of resistivity 2.8 $\times$ 10^{-6} Ω-cm. Note the advantage of avoiding diffused crossovers.

7-13 (a) What is the minimum number of isolation regions required to realize in monolithic form the logic gate shown?

(b) Draw a monolithic layout of the gate in the fashion of Fig. 7-25b.

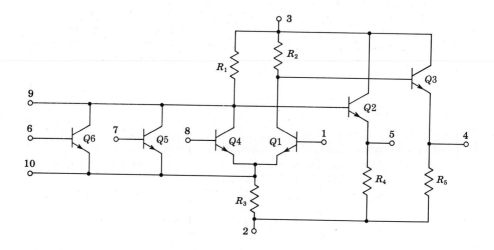

Prob. 7-13

7-14 Repeat Prob. 7-13 for the difference amplifier shown.

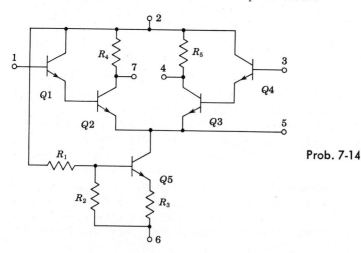

Prob. 7-14

7-15 For the circuit shown, find (a) the *minimum* number, (b) the *maximum* number, of isolation regions.

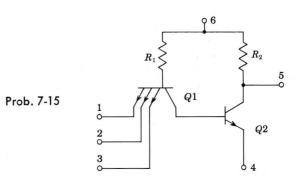

Prob. 7-15

7-16 For the circuit shown, (a) find the minimum number of isolation regions, and (b) draw a monolithic layout in the fashion of Fig. 7-26, given that (i) $Q1$, $Q2$, and $Q3$ should be single-base-stripe, 1- by 2-mil emitter, transistors, (ii) $R_1 = R_2 = R_3 = 400\ \Omega$, $R_4 = 600\ \Omega$. Use 1-mil-wide resistors.

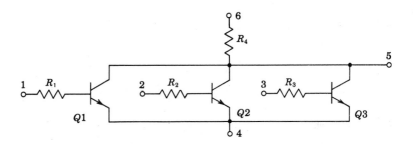

Prob. 7-16

7-17 An integrated junction capacitor has an area of 1,000 mils² and is operated at a reverse barrier potential of 1 V. The acceptor concentration of 10^{15} atoms/cm³ is much smaller than the donor concentration. Calculate the capacitance.

7-18 A thin-film capacitor has a capacitance of 0.4 pF/mil². The relative dielectric constant of silicon dioxide is 3.5. What is the thickness of the SiO_2 layer in angstroms?

7-19 The n-type epitaxial isolation region shown is 8 mils long, 6 mils wide, and 1 mil thick and has a resistivity of 0.1 Ω-cm. The resistivity of the p-type substrate is 10 Ω-cm. Find the parasitic capacitance between the isolation region and the substrate under 5-V reverse bias. Assume that the sidewalls contribute 0.1 pF/mil².

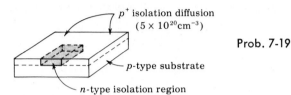

p^+ isolation diffusion
$(5 \times 10^{20}\mathrm{cm^{-3}})$

Prob. 7-19

p-type substrate

n-type isolation region

NOTE: *In the problems that follow, indicate your answer by giving the letter of the statement you consider correct.*

7-20 The typical number of diffusions used in making epitaxial-diffused silicon integrated circuits is (*a*) 1, (*b*) 2, (*c*) 3, (*d*) 4, (*e*) 5.

7-21 The "buried layer" in an integrated transistor is (*a*) p^+ doped, (*b*) located in the base region, (*c*) n^+ doped, (*d*) used to reduce the parasitic capacitance.

7-22 Epitaxial growth is used in integrated circuits (ICs)
(*a*) To grow selectively single-crystal p-doped silicon of one resistivity on a p-type substrate of a different resistivity.
(*b*) To grow single-crystal n-doped silicon on a single-crystal p-type substrate.
(*c*) Because it yields back-to-back isolating p-n junctions.
(*d*) Because it produces low parasitic capacitance.

7-23 Silicon dioxide (SiO_2) is used in ICs
(*a*) Because it facilitates the penetration of diffusants.
(*b*) Because of its high heat conduction.
(*c*) To control the location of diffusion and to protect and insulate the silicon surface.
(*d*) To control the concentration of diffusants.

7-24 The p-type substrate in a monolithic circuit should be connected to
(*a*) The most positive voltage available in the circuit.
(*b*) The most negative voltage available in the circuit.
(*c*) Any dc ground point.
(*d*) Nowhere, i.e., be left floating.

7-25 Monolithic integrated circuit systems offer greater reliability than discrete-component systems because
(*a*) There are fewer interconnections.
(*b*) High-temperature metalizing is used.
(*c*) Electric voltages are low.
(*d*) Electric elements are closely matched.

7-26 The collector-substrate junction in the epitaxial collector structure is, approximately,
(*a*) A step-graded junction.
(*b*) A linearly graded junction.
(*c*) An exponential junction.
(*d*) None of the above.

7-27 The sheet resistance of a semiconductor is
(*a*) An undesirable parasitic element.
(*b*) An important characteristic of a diffused region, especially when used to form diffused resistors.

(c) A characteristic whose value determines the required area for a given value of integrated capacitance.

(d) A parameter whose value is important in a thin-film resistance.

7-28 Isolation in ICs is required.

(a) To make it simpler to test circuits.

(b) To protect the components from mechanical damage.

(c) To protect the transistor from possible "thermal runaway."

(d) To minimize electrical interaction between circuit components.

7-29 Almost all resistors are made in a monolithic IC

(a) During the emitter diffusion.

(b) While growing the epitaxial layer.

(c) During the base diffusion.

(d) During the collector diffusion.

7-30 Increasing the yield of an integrated circuit

(a) Reduces individual circuit cost.

(b) Increases the cost of each good circuit.

(c) Results in a lower number of good chips per wafer.

(d) Means that more transistors can be fabricated on the same size wafer.

7-31 In a monolithic-type IC

(a) All isolation problems are eliminated.

(b) Resistors and capacitors of any value may be made.

(c) All components are fabricated into a single crystal of silicon.

(d) Each transistor is diffused into a separate isolation region.

7-32 The main purpose of the metalization process is

(a) To interconnect the various circuit elements

(b) To protect the chip from oxidation.

(c) To act as a heat sink.

(d) To supply a bonding surface for mounting the chip.

CHAPTER 8

NOTE: *Unless otherwise specified, all transistors in these problems are identical, and the numerical values of their h parameters are given in Table 8-2. Also assume that all capacitances are arbitrarily large.*

8-1 (a) Using Fig. 8-6b write the input and output equations.

(b) Draw the hybrid model for a CB transistor and write the input and output equations.

8-2 (a) Describe how to obtain h_{ie} from the CE input characteristics.

(b) Repeat part a for h_{re}. Explain why this procedure, although correct in principle, is inaccurate in practice.

8-3 The transistor whose input characteristics are shown in Fig. 8-2 is biased at $V_{CE} = -8$ V and $I_B = -300$ μA.

(a) Compute graphically h_{fe} and h_{oe} at the quiescent point specified above.

(b) Using the h parameters computed in part a, calculate h_{fb} and h_{ob}.

8-4 Justify the statement in the footnote to Table 8-3. HINT: Draw a CE transistor circuit with a signal voltage V_s between base and ground. Now interchange B and E and observe the resulting configuration.

8-5 (a) Find the CC h parameters in terms of the CE h parameters.

(b) Find the CE h parameters in terms of the CC parameters.

8-6 (a) Find the h_{rb} in terms of the CE h parameters.

(b) Find h_{ie} in terms of the CB h parameters.

8-7 (a) Show that the exact expression for h_{fe} in terms of the CB hybrid parameters is

$$h_{fe} = -\frac{h_{fb}(1 - h_{rb}) + h_{ib}h_{ob}}{(1 + h_{fb})(1 - h_{rb}) + h_{ob}h_{ib}}$$

(b) From this exact formula obtain the approximate expression for h_{fe}.

(c) Show that the exact expression for h_{fb} in terms of the CE hybrid parameters is

$$h_{fb} = -\frac{h_{fe}(1 - h_{re}) + h_{ie}h_{oe}}{(1 + h_{fe})(1 - h_{re}) + h_{oe}h_{ie}}$$

(d) From this exact formula obtain the approximate expression for h_{fb}.

8-8 For the circuit shown, verify that the modified h parameters (indicated by primes) are

(a) $h'_{ie} \approx h_{ie} + \dfrac{(1 + h_{fe})R_e}{1 + h_{oe}R_e}$ (b) $h'_{re} = \dfrac{h_{re} + h_{oe}R_e}{1 + h_{oe}R_e}$

(c) $h'_{fe} = \dfrac{h_{fe} - h_{oe}R_e}{1 + h_{oe}R_e}$ (d) $h'_{oe} = \dfrac{h_{oe}}{1 + h_{oe}R_e}$

(e) To what do these expressions reduce if $h_{oe}R_e \ll 1$?

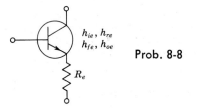

h_{ie}, h_{re}
h_{fe}, h_{oe}

R_e

Prob. 8-8

8-9 Show that the overall h parameters of the accompanying two-stage cascaded amplifier are

(a) $h_{11} = h'_{11} - \dfrac{h'_{12}h'_{21}}{1 + h'_{22}h''_{11}}h''_{11}$ (b) $h_{12} = \dfrac{h'_{12}h''_{12}}{1 + h'_{22}h''_{11}}$

(c) $h_{21} = -\dfrac{h'_{21}h''_{21}}{1 + h'_{22}h''_{11}}$ (d) $h_{22} = h''_{22} - \dfrac{h''_{12}h''_{21}}{1 + h'_{22}h''_{11}}h'_{22}$

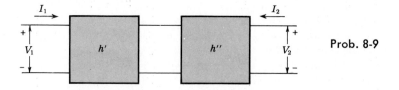

Prob. 8-9

8-10 Show that the overall h parameters for the composite transistor shown are

(a) $h_{ie} = h_{ie1} + \dfrac{(1 - h_{re1})(1 + h_{fe1})h_{ie2}}{1 + h_{oe1}h_{ie2}}$

(b) $h_{fe} = h_{fe1} + \dfrac{(h_{fe2} - h_{oe1}h_{ie2})(1 + h_{fe1})}{1 + h_{oe1}h_{ie2}}$

(c) $h_{oe} = h_{oe2} + \dfrac{(1 + h_{fe2})(1 - h_{re2})h_{oe1}}{1 + h_{oe1}h_{ie2}}$

(d) $h_{re} = h_{re2} + \dfrac{(h_{ie2}h_{oe1} + h_{re1})(1 - h_{re2})}{1 + h_{oe1}h_{ie2}}$

(e) Obtain numerical values for the h parameters of the composite transistor by assuming identical transistors $Q1$ and $Q2$ and using Table 8-2.

Prob. 8-10

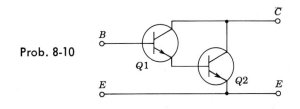

8-11 Given a single-stage transistor amplifier with the h parameters specified in Table 8-2, calculate A_I, A_V, A_{Vs}, R_i, and R_o for the CC transistor configuration, with $R_s = R_L = 10$ K.   Check your results with Fig. 8-16.

8-12 (a) Draw the equivalent circuit for the CE and CC configurations subject to the restriction that $R_L = 0$.   Show that the input impedances of the two circuits are identical.

(b) Draw the circuits for the CE and CC configurations subject to the restriction that the input is open-circuited.   Show that the output impedances of the two circuits are identical.

8-13 For any single-transistor amplifier prove that

$$R_i = \frac{h_i}{1 - h_r A_V}$$

8-14 Prove that

$$Y_o = h_o \left(\frac{R_s + R_{i\infty}}{R_s + R_{io}} \right)$$

where $R_{i\infty} \equiv R_i$ for $R_L = \infty$, and $R_{io} \equiv R_i$ for $R_L = 0$.

8-15 (a) For a CE configuration, what is the maximum value of R_L for which R_i differs by no more than 10 percent of its value at $R_L = 0$?   Use the transistor parameters given in Table 8-2.

(b) What is the maximum value of R_s for which R_o differs by no more than 10 percent of its value for $R_s = 0$?

(c) For the CB configuration, what is the maximum value of R_L for which R_i does not exceed 50 Ω?

8-16 Consider an emitter follower. Neglect h_{re} and show that as $R_e \rightarrow \infty$

(a) $R_i \rightarrow h_{ie} + \dfrac{1 + h_{fe}}{h_{oe}} \approx \dfrac{1}{h_{ob}}$

Explain the result physically.

(b) $1 - A_V \approx \dfrac{h_{ie}h_{oe}}{1 + h_{fe}}$

Evaluate A_V using the h-parameter values given in Table 8-2.

8-17 For the emitter follower with $R_s = 0.5$ K and $R_L = 5$ K, calculate A_I, R_i, A_V, A_{Vs}, R_o. Assume $h_{fe} = 50$, $h_{ie} = 1$ K, $h_{oe} = 25$ μA/V.

8-18 (a) Design an emitter follower having $R_i = 500$ K and $R_o = 20$ Ω. Assume $h_{fe} = 50$, $h_{ie} = 1$ K, $h_{oe} = 25$ μA/V.

(b) Find A_I and A_V for the emitter follower of part a.

(c) Find R_i and the necessary R_L so that $A_V = 0.999$.

8-19 For the transistor circuit in Fig. 8-12 show that

(a) $(A_{Is})_{\max} = -h_f$, if $R_L = 0$ and $R_s = \infty$.

(b) $R_i = h_i$, if $R_L = 0$.

(c) $R_i = \dfrac{h_i h_o - h_r h_f}{h_o}$, if $R_L = \infty$.

(d) $(A_{Vs})_{\max} = -\dfrac{h_f}{h_i h_o - h_r h_f}$, if $R_L = \infty$ and $R_s = 0$.

(e) $R_o = \dfrac{h_i}{h_i h_o - h_r h_f}$, if $R_s = 0$.

(f) $R_o = \dfrac{1}{h_o}$, if $R_s = \infty$.

8-20 Using the h-parameter values given in Table 8-2, calculate $(A_{Is})_{\max}$, R_i, $(A_{Vs})_{\max}$, and R_o derived in Prob. 8-19 (a) for a CE connection, (b) for a CB connection, (c) for a CC connection. Compare your answers with the values in Fig. 8-16.

8-21 Find the output impedance Z_o for the example in Sec. 8-6 by evaluating the current I_a drawn from an auxiliary voltage source V_a impressed across the output terminals (with zero input voltage and $R_L = \infty$). Then $Z_o = V_a / I_a$.

8-22 Find the voltage gain A_V for the example in Sec. 8-6 directly as the ratio V_o / V_i (without finding A_I or Z_i).

8-23 The transistor amplifier shown uses a transistor whose h parameters are given in Table 8-2. Calculate $A_I = I_o / I_i$, A_V, A_{Vs}, R_o, and R_i.

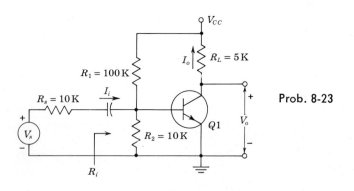

Prob. 8-23

8-24 (a) In the circuit shown, find the input impedance R_i in terms of the CE h parameters, R_L and R_e. HINT: Follow the rules given in Sec. 8-10.

(b) If $R_L = R_e = 1$ K and the h parameters are as given in Table 8-2, what is the value of R_i?

Prob. 8-24

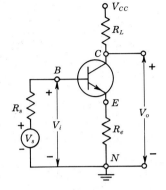

8-25 For the amplifier shown, using a transistor whose parameters are given in Table 8-2, compute $A_I = I_o/I_i$, A_V, A_{Vs}, and R_i. HINT: Follow the rules given in Sec. 8-10.

Prob. 8-25

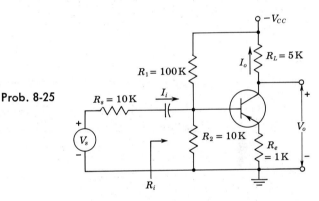

8-26 (a) Calculate R_i, A_V, and $A_I = -I_e/I_i$ for the circuit shown. Use the h-parameter values given in Table 8-2. HINT: Follow the rules given in Sec. 8-10.

(b) Repeat part a using the results in Prob. 8-10.

Prob. 8-26

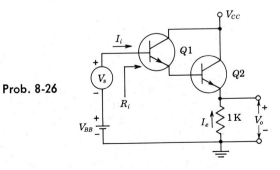

8-27   For the circuit shown, with the transistor parameters specified in Table 8-2, calculate $A_I = I_o/I_i$, A_V, A_{Vs}, and R_i.   HINT: Follow the rules given in Sec. 8-10.

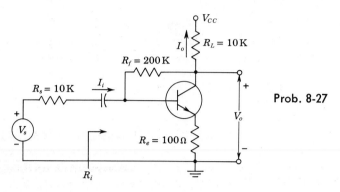

Prob. 8-27

8-28   Repeat Prob. 8-24 by applying the dual of Miller's theorem.

8-29   (a) For the two-transistor amplifier circuit shown (supply voltages are not indicated) calculate A_I, A_V, A_{Vs}, and R_i.   The transistors are identical, and their parameters are given in Table 8-2.   HINT: Follow rules given in Sec. 8-10.
(b) Repeat part a using the results given in Prob. 8-9.

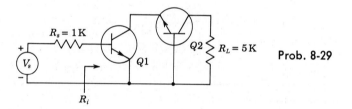

Prob. 8-29

8-30   (a) Find the voltage gain A_{Vs} of the amplifier shown.   Assume $h_{ie} = 1,000\ \Omega$, $h_{re} = 10^{-4}$, $h_{fe} = 50$, $h_{oe} = 10^{-4}\ \text{A/V}$.   (b) Find R_o'.

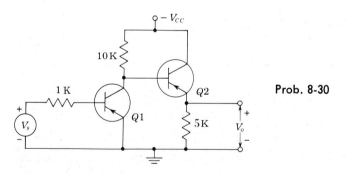

Prob. 8-30

8-31   The three-stage amplifier shown contains identical transistors.   Calculate the voltage gain of each stage and the overall voltage gain V_o/V_s.   See note on page 809.

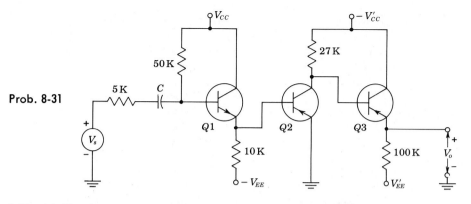

Prob. 8-31

8-32   (a) For the two-stage cascade shown, compute the input and output impedances and the individual and overall voltage and current gains, using the exact procedure of Sec. 8-12.   See note on page 809.

(b) Repeat part a using the approximate formulas in Table 8-7.

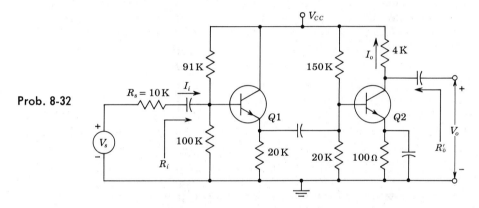

Prob. 8-32

8-33   (a) Compute A_I, A_V, A_{Vs}, R_i, and R_o' for the two-stage cascade shown, using the exact procedure of Sec. 8-12.   See note on page 809.

(b) Repeat part a using the approximate formulas in Table 8-7.

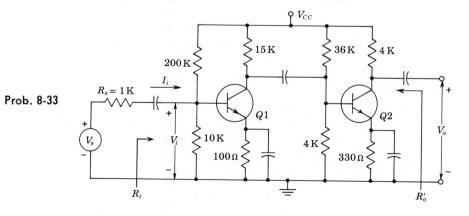

Prob. 8-33

8-34 For a CB connection derive the simplified expressions given in Table 8-7 and prove that they are in error by less than 10 percent from the exact formulas.

8-35 (a) Consider a CB connection with $R_s = 2$ K and $R_L = 4$ K. Find the exact and approximate values of A_I, A_V, A_{Vs}, R_i, and R'_c.
(b) Repeat part a for the CE connection.
(c) Repeat part a for the CC connection. See note on page 809.

8-36 For the circuit shown, compute A_I, A_V, A_{Vs}, R_i, and R'_o. See note on page 809.

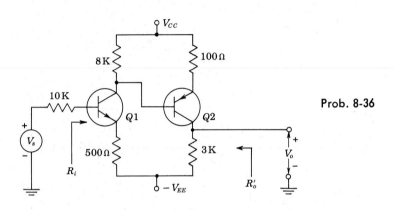

Prob. 8-36

8-37 For the two-stage amplifier shown calculate A_V, A_{Vs}, R_i, and R'_o. Neglect the effect of all capacitances. See note on page 809.

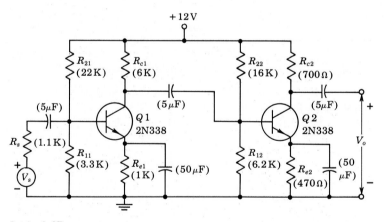

Prob. 8-37

8-38 In the circuit of Prob. 8-27 change R_L to 4 K. Find A_V and A_{Vs} by using Miller's theorem.

8-39 The cascode transistor configuration consists of a CE stage Q1 in series with a CB stage Q2 (the collector current of Q1 equals the emitter current of Q2). Verify that the cascode combination acts like a single CE transistor with negligible internal feedback and very small output conductance for an open-circuited input. In other words, verify that

$$h_{11} \approx h_{ie} \qquad h_{21} \approx h_{fe} \qquad h_{22} \approx h_{ob} \qquad h_{12} \approx h_{re} h_{rb}$$

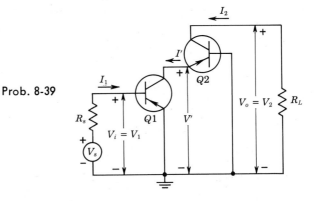

Prob. 8-39

8-40 Calculate $A_I = I_o/I_i$, A_V, A_{Vs}, R_i, and R_o' for the cascode circuit shown. See note on page 809. HINT: Use results of Prob. 8-39.

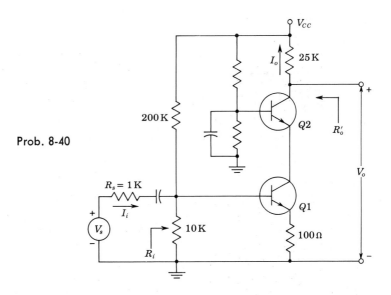

Prob. 8-40

8-41 The circuit shown is an amplifier using a p-n-p and an n-p-n transistor in parallel. The two transistors have identical characteristics. Find the expression for the voltage gain and the input resistance of the amplifier, using the simplified hybrid model.

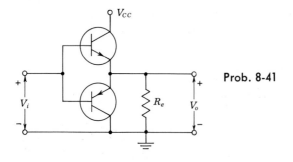

Prob. 8-41

8-42 For the two-stage cascade shown, find A_I, A_V, R_i, and R_o'. See note on page 809.

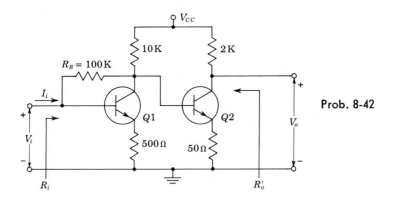

Prob. 8-42

8-43 Design a two-stage cascade using the configuration of Prob. 8-42, with $R_B = 100$ K, to meet the following specifications (see note on page 809):

$$125 \geq A_V \geq 100 \qquad 10\text{ K} \geq R_i \geq 5\text{ K} \qquad R_o' \leq 3\text{ K}$$

8-44 For the two-stage cascade shown, calculate A_I, A_V, R_i, and R_o'. See note on page 809.

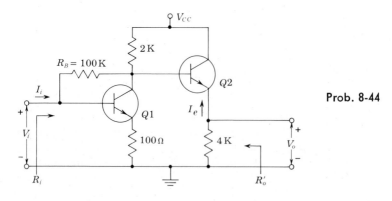

Prob. 8-44

8-45 Design a two-stage amplifier using the configuration of Prob. 8-44, with $R_B = 100$ K, to meet the following specifications (see note on page 809):

$$|A_V| \geq 15 \qquad R_i \geq 2 \text{ K} \qquad R_o' \leq 100 \; \Omega$$

8-46 For the circuit shown, find the voltage gain V_o/V_s and input impedance as a function of R_s, b, R_e, and R_L. Assume that $h_{oe}(R_e + R_L) \leq 0.1$.

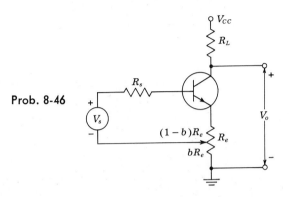

Prob. 8-46

8-47 Using the exact expressions of Eq. (8-67) for A_I and Eq. (8-68) for R_i, calculate the output resistance R_o in Fig. 8-28a as the ratio of open-circuit voltage V to short-circuit current I. Verify that R_o is given by Eq. (8-70). HINT: Note that $V = \lim_{R_L \to \infty} A_V V_s$.

8-48 The amplifier shown is made up of an n-p-n and a p-n-p transistor. The h parameters of the two transistors are identical, and are given as $h_{ie} = 1$ K, $h_{fe} = 100$, $h_{oe} = 0$, and $h_{re} = 0$.
(a) With the switch open, find $A_V = V_o/V_i$.
(b) With the switch closed, find (with the aid of Miller's theorem) A_V, A_{Vs}, R_i, and $A_I \equiv I_o/I_i$.

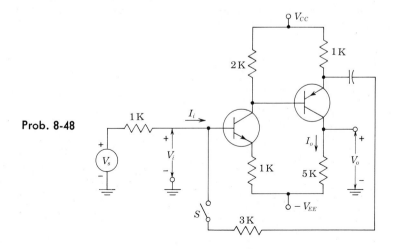

Prob. 8-48

8-49   The cascade configuration shown is known as the tandem emitter follower.   Find
the input resistance R_i if $h_{ie} = h_{re} = h_{oe} = 0$, and h_{fe} is the same for each of the
transistors $Q1$ to QN.

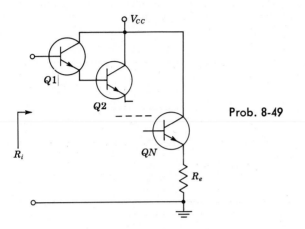

Prob. 8-49

8-50   (*a*) Verify Eq. (8-80) for the voltage gain of a Darlington emitter follower.
(*b*) Verify Eq. (8-81) for the output resistance.

8-51   Verify Eq. (8-84).

8-52   For the bootstrap circuit shown, calculate $A_I \equiv I_o/I_i$, R_i, and A_V.   The transistor parameters are $h_{ie} = 2$ K, $h_{fe} = 100$, $1/h_{oe} = 40$ K, and $h_{re} = 2.5 \times 10^{-4}$.

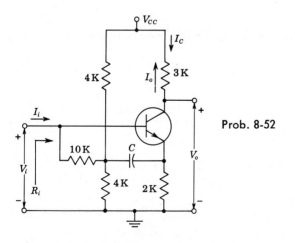

Prob. 8-52

8-53   The bootstrapped Darlington pair uses identical transistors with the following h
parameters; $h_{ie} = 1$ K, $h_{re} = 2.5 \times 10^{-4}$, $h_{oe} = 2.5 \times 10^{-5}$ A/V, and $h_{fe} = 100$.
Find I_{e1}/I_{b1}, V_{o2}/V_i, R_i, and V_{o1}/V_i.

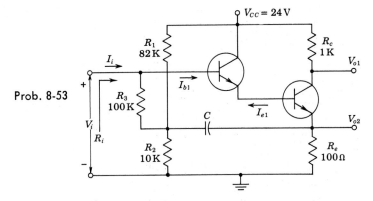

Prob. 8-53

8-54 Calculate A_I, A_V, R_i, and R'_o for the circuit shown.   See note on page 809.

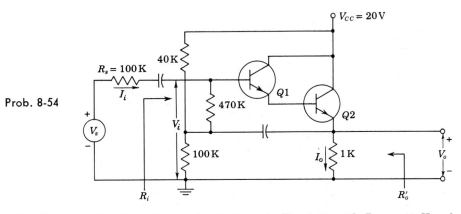

Prob. 8-54

8-55 Calculate R_i and A_V for the circuit shown in Fig. 8-32, with $R_{e1} = 100$ K and $R_{e2} = 1$ K.

8-56 For the circuit shown, find A_V, A_{Vs}, $A_I = I_o/I_i$, R_i, and R'_o.   See note on page 809.

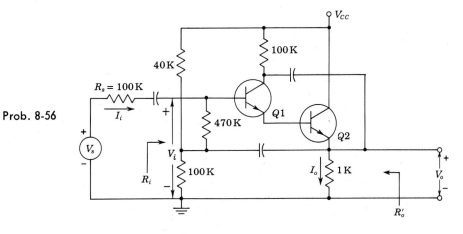

Prob. 8-56

CHAPTER 9

9-1 (a) Determine the quiescent currents and the collector-to-emitter voltage for a silicon transistor with $\beta = 50$ in the self-biasing arrangement of Fig. 9-5. The circuit component values are $V_{CC} = 20$ V, $R_c = 2$ K, $R_e = 0.1$ K, $R_1 = 100$ K, and $R_2 = 5$ K.

(b) Repeat (a) for a germanium transistor.

9-2 A p-n-p germanium transistor is used in the self-biasing arrangement of Fig. 9-5. The circuit component values are $V_{CC} = 4.5$ V, $R_c = 1.5$ K, $R_e = 0.27$ K, $R_2 = 2.7$ K, and $R_1 = 27$ K. If $\beta = 44$

(a) Find the quiescent point.

(b) Recalculate these values if the base-spreading resistance of 690 Ω is taken into account.

9-3 A p-n-p silicon transistor is used in a common-collector circuit (Fig. 9-5 with $R_c = 0$). The circuit component values are $V_{CC} = 3.0$ V, $R_e = 1$ K, $R_1 = R_2 = 5$ K. If $\beta = 44$

(a) Find the quiescent point.

(b) Recalculate these values, taking the base-spreading resistance of 690 Ω into account.

9-4 For the circuit shown

(a) Calculate I_B, I_C, and V_{CE} if a silicon transistor is used with $\beta = 50$.

(b) Specify a value for R_b so that $V_{CE} = 7$ V.

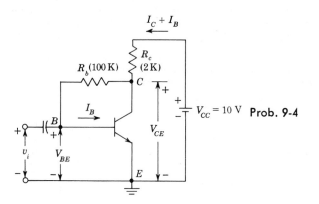

Prob. **9-4**

9-5 (a) Verify Eq. (9-13).

(b) Show that S may be put in the form

$$S = \frac{G_e + G_1 + G_2}{G_e/(1 + \beta) + G_1 + G_2}$$

where the G's are the conductances corresponding to the R's shown in Fig. 9-5a.

(c) Show that for the circuit of Prob. 9-4, S is given by

$$S = \frac{\beta + 1}{1 + \beta R_c/(R_c + R_b)}$$

9-6 (a) Find the stability factor S for the circuit of Prob. 9-1.
 (b) Repeat (a) for the circuit of Prob. 9-2.
 (c) Repeat (a) for the circuit of Prob. 9-3.

9-7 For the two-battery transistor circuit shown, prove that the stabilization factor S is given by

$$S = \frac{1 + \beta}{1 + \beta R_e / (R_e + R_b)}$$

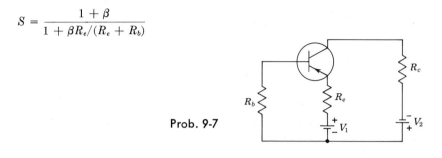

Prob. 9-7

9-8 Assume that a silicon transistor with $\beta = 50$, $V_{BE,\text{active}} = 0.7$, $V_{CC} = 22.5$ V, and $R_c = 5.6$ K is used in Fig. 9-5a. It is desired to establish a Q point at $V_{CE} = 12$ V, $I_C = 1.5$ mA, and stability factor $S \leq 3$. Find R_e, R_1, and R_2.

9-9 (a) A germanium transistor is used in the self-biasing arrangement of Fig. 9-5 with $V_{CC} = 16$ V and $R_c = 1.5$ K. The quiescent point is chosen to be $V_{CE} = 8$ V and $I_C = 4$ mA. A stability factor $S = 12$ is desired. If $\beta = 50$, find R_1, R_2, and R_e.
 (b) Repeat part a for $S = 3$.

9-10 Determine the stability factor S for the circuit shown.

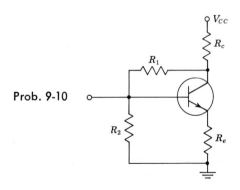

Prob. 9-10

9-11 In the circuit shown, $V_{CC} = 24$ V, $R_c = 10$ K, and $R_e = 270$ Ω. If a silicon transistor is used with $\beta = 45$ and if under quiescent conditions $V_{CE} = 5$ V, determine (a) R, (b) the stability factor S.

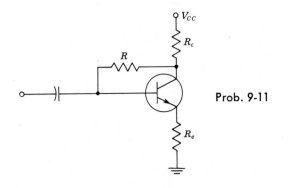

Prob. 9-11

9-12 In the transformer-coupled amplifier stage shown, $V_{BE} = 0.7$ V, $\beta = 50$, and the quiescent voltage is $V_{CE} = 4$ V. Determine (a) R_e, (b) the stability factor S.

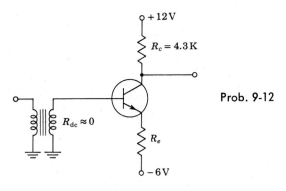

Prob. 9-12

9-13 In the two-stage circuit shown, assume $\beta = 100$ for each transistor.
(a) Determine R so that the quiescent conditions are $V_{CE1} = -4$ V and $V_{CE2} = -6$ V.
(b) Explain how quiescent-point stabilization is obtained. Assume $V_{BE} = 0.2$ V.

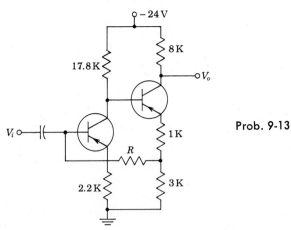

Prob. 9-13

9-14 In the Darlington stage shown, $V_{CC} = 24$ V, $\beta_1 = 24$, $\beta_2 = 39$, $V_{BE} = 0.7$, $R_c = 330\ \Omega$, and $R_e = 120\ \Omega$. If at the quiescent point $V_{CE2} = 6$ V, determine (a) R, (b) the stability factor defined as $S \equiv dI_C/dI_{CO1}$.

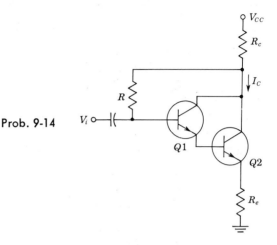

Prob. 9-14

9-15 (a) Prove that for the circuit of Fig. 9-5b the stability factor S' is given by

$$S' = \frac{-S}{R_b + R_e} \times \frac{\beta}{\beta + 1}$$

(b) Derive Eq. (9-22).

9-16 For the bias arrangement given in Prob. 9-4 prove that

$$S' = \frac{-\beta S}{1 + \beta} \times \frac{1}{R_e + R_b}$$

$$S'' = \frac{I_C}{\beta} \times \frac{S}{\beta + 1}$$

where S is the stabilization factor of this circuit.

9-17 If in Eq. (9-11) we do not assume $\beta \gg 1$ so that V' is now a function of β, verify that Eq. (9-22) is given by

$$S'' = \frac{(I_C - I_{CO})S}{\beta(1 + \beta)}$$

9-18 If in Eq. (9-11) we do not assume $\beta \gg 1$, so that V' is now a function of β, verify that Eq. (9-27) is given by

$$S'' = \frac{(I_{C1} - I_{CO1})S_2}{(\beta_1)(1 + \beta_2)}$$

HINT: Write the expression for $(I_{C2} - I_{CO2})/(I_{C1} - I_{CO1})$ and then subtract unity from both sides of the equation.

9-19 In the circuit of Fig. 9-5, let $R_c = 5.6$ K, $R_e = 1$ K, $R_1 = 90$ K, $R_2 = 10$ K, $I_c = 1.5$ mA at 25°C. Using the transistor of Table 9-1, find I_c at +175 and −65°C.

9-20 Repeat Prob. 9-19 for the transistor of Table 9-2 at +75 and −65°C.

9-21 In the emitter-follower circuit shown, $R_e = 1$ K and V_{CC} and V_{EE} are adjusted to give $I_c = 1.5$ mA at 25°C. Using the transistor of Table 9-1, find I_c at +175 and −65°C. Compare the results with those of Prob. 9-19.

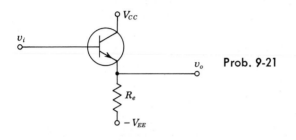

Prob. 9-21

9-22 Repeat Prob. 9-21 for the transistor of Table 9-2 at +75 and −65°C. Compare these results with those of Prob. 9-20.

9-23 For the self-bias circuit of Fig. 9-5a, $R_e = 1$ K and $R_b = R_1 \| R_2 = 7.75$ K. The collector supply voltage and R_c are adjusted to establish a collector current of 1.5 mA at 25°C. Determine the variation of I_c in the temperature range −65 to +175°C when the silicon transistor of Table 9-1 is used.

9-24 Repeat Prob. 9-23 for the range −65 to +75°C when the germanium transistor of Table 9-2 is used.

9-25 Design the emitter-follower circuit shown in Prob. 9-21 using the silicon transistor type 2N3565 to meet the specifications of the illustrative example on page 298.

9-26 Two identical silicon transistors with $\beta = 48$, $V_{BE} = 0.7$ V at $T = 25°C$, $V_{CC} = 20.7$ V, $R_1 = 10$ K, and $R_c = 5$ K are used in Fig. 9-11a.
(a) Find the currents I_{B1}, I_{B2}, I_{C1}, and I_{C2} at $T = 25°C$.
(b) Find I_{C2} at $T = 175°C$ when $\beta = 98$ and $V_{BE} = 0.3$ V. HINT: Assume $I_{B1} = I_{B2}$.

9-27 For the biasing arrangement of Fig. 9-10 and assuming that the reverse saturation currents of the diode and the transistor are equal, prove that

$$S = 1$$

$$S' = -\frac{\beta}{R_1}$$

$$S'' = \frac{\Delta(I_c - I_{co})}{\Delta\beta} = \frac{I_{C1} - I_{co1}}{\beta_1}$$

9-28 (a) For the self-bias circuit of Fig. 9-9, the operating point is $I_c = 1.5$ mA at 25°C, $R_e = 1$ K, $R_b = 7.75$ K, $V_{DD} = 15$ V, and $V_{D,ON} = 0.7$ V, find the maximum value of R_d for proper operation.

(b) For the circuit of part *a* determine the variation of I_C in the temperature range -65 for $175°C$ when the silicon transistor of Table 9-1 is used.

9-29 Prove Eq. (9-38).

9-30 (a) The circuit of Prob. 9-19 is modified by the addition of a thermistor as in Fig. 9-12. Find R_T, I_T, and V_{CE} for the modified circuit if $I_C = 1.5$ mA and $V_{CC} = 27.5$ V.
(b) It is desired that as the temperature changes from 25 to 175°, the variation of I_C be $+0.4$ mA. Calculate the temperature coefficient of the thermistor.

9-31 (a) Calculate the thermal resistance for the 2N338 transistor for which the manufacturer specifies $P_{C,\max} = 125$ mW at 25°C free-air temperature and maximum junction temperature $T_j = 150°C$.
(b) What is the junction temperature if the collector dissipation is 75 mW?

9-32 Show that the load line tangent to the constant-power-dissipation hyperbola of Fig. 9-14 is bisected by the tangency point, that is, $AC = BC$.

9-33 The transistor used in the circuit is at cutoff.
(a) Show that runaway will occur for values of I_{CO} in the range

$$\frac{V_{CC} - \sqrt{V_{CC}^2 - 8R_c/0.07\theta}}{4R_c} \le I_{CO} \le \frac{V_{CC} + \sqrt{V_{CC}^2 - 8R_c/0.07\theta}}{4R_c}$$

(b) Show that if runaway is not destructive, the collector current I_{CO} after runaway can never exceed $I_{CO} = V_{CC}/2R_c$.

Prob. 9-33

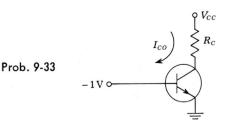

9-34 A germanium transistor with $\theta = 250°C/W$, $I_{CO} = 10$ μA at 25°C, $R_c = 1$ K, and $V_{CC} = 30$ V is used in the circuit of Prob. 9-33.
(a) Find I_{CO} at the point of runaway.
(b) Find the ambient temperature at which runaway will occur.

CHAPTER 10

10-1 The drain resistance R_d of an *n*-channel FET with the source grounded is 2 K. The FET is operating at a quiescent point $V_{DS} = 10$ V, and $I_{DS} = 3$ mA, and its characteristics are given in Fig. 10-3.
(a) To what value must the gate voltage be changed if the drain current is to change to 5 mA?

(b) To what value must the voltage V_{DD} be changed if the drain current is to be brought back to its previous value? The gate voltage is maintained constant at the value found in part a.

10-2 For a p-channel silicon FET with $a = 2 \times 10^{-4}$ cm and channel resistivity $\rho = 10$ Ω-cm
(a) Find the pinch-off voltage.
(b) Repeat (a) for a p-channel germanium FET with $\rho = 2$ Ω-cm.

10-3 (a) Plot the transfer characteristic curve of an FET as given by Eq. (10-8), with $I_{DSS} = 10$ mA and $V_P = -4$ V.
(b) The magnitude of the slope of this curve at $V_{GS} = 0$ is g_{mo} and is given by Eq. (10-17). If the slope is extended as a tangent, show that it intersects the V_{GS} axis at the point $V_{GS} = V_P/2$.

10-4 (a) Show that the transconductance g_m of a JFET is related to the drain current I_{DS} by

$$g_m = \frac{2}{|V_P|} \sqrt{I_{DSS} I_{DS}}$$

(b) If $V_P = -4$ V and $I_{DSS} = 4$ mA, plot g_m versus I_{DS}.

10-5 Show that for small values of V_{GS} compared with V_P, the drain current is given approximately by $I_D \approx I_{DSS} + g_{mo} V_{GS}$.

10-6 (a) For the FET whose characteristics are plotted in Fig. 10-3, determine r_d and g_m graphically at the quiescent point $V_{DS} = 10$ V and $V_{GS} = -1.5$ V. Also evaluate μ.
(b) Determine $r_{d,\text{ON}}$ for $V_{GS} = 0$.

10-7 (a) Verify Eq. (10-15).
(b) Starting with the definitions of g_m and r_d, show that if two identical FETs are connected in parallel, g_m is doubled and r_d is halved. Since $\mu = r_d g_m$, then μ remains unchanged.
(c) If the two FETs are not identical, show that

$$\frac{1}{r_d} = \frac{1}{r_{d1}} + \frac{1}{r_{d2}}$$

and that

$$\mu = \frac{\mu_1 r_{d2} + \mu_2 r_{d1}}{r_{d1} + r_{d2}}$$

10-8 Given the transfer characteristic of an FET, explain clearly how to determine g_m at a specified quiescent point.

10-9 (a) Using Eq. (10-18), prove that the load conductance of a MOSFET is given by

$$g_L = g_{do} \left(1 - \frac{V_{DS}}{V_T} \right) \qquad \text{where } g_{do} = \frac{2 I_{DSS}}{|V_T|}$$

(b) Prove that $g_L = g_m$ for a JFET. HINT: Use Eq. (10-8).

10-10 Draw the circuit of a MOSFET negative AND gate and explain its operation.

10-11 Consider the FLIP-FLOP circuit shown. Assume $V_T = V_{ON} = 0$ and $|V_{DD}| \gg V_T$.

(a) Assume $v_{i1} = v_{i2} = 0$. Verify that the circuit has two possible stable states; either $v_{o1} = 0$ and $v_{o2} = -V_{DD}$ or $v_{o1} = -V_{DD}$ and $v_{o2} = 0$.

(b) Show that the state of the FLIP-FLOP may be changed by momentarily allowing one of the inputs to go to $-V_{DD}$; in other words by applying a negative input pulse.

Prob. 10-11

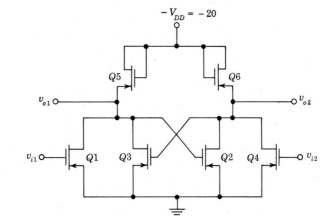

10-12 Draw a CMOS inverter using positive logic.

10-13 (a) The complementary MOS negative NAND gate is indicated. Explain its operation.

(b) Draw the corresponding positive NAND gate.

Prob. 10-13

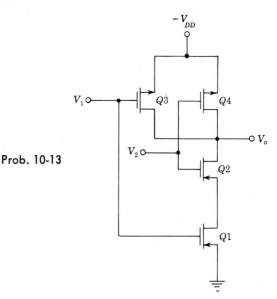

10-14 The circuit of a CMOS positive NOR gate is indicated. Explain its operation.

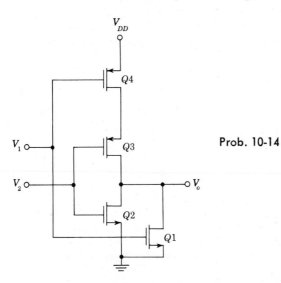

Prob. 10-14

10-15 Draw a MOSFET circuit satisfying the logic equation. $Y = \overline{A + BC}$, where Y is the output corresponding to the three inputs A, B, and C.

10-16 (a) Calculate the voltage gain $A_V = V_o/V_i$ at 1 kHz for the circuit shown. The FET parameters are $g_m = 2$ mA/V and $r_d = 10$ K. Neglect capacitances. (b) Repeat part a if the capacitance 0.003 μF is taken under consideration.

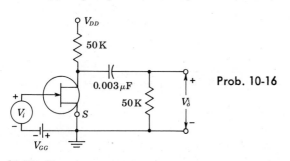

Prob. 10-16

10-17 If an input signal V_i is impressed between gate and ground, find the amplification $A_V = V_o/V_i$. Apply Miller's theorem to the 50-K resistor. The FET parameters are $\mu = 30$ and $r_d = 5$ K. Neglect capacitances.

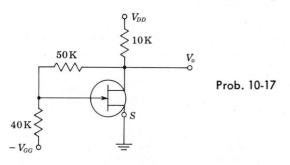

Prob. 10-17

10-18 If in Prob. 10-17 the signal V_i is impressed in series with the 40-K resistor (instead of from gate to ground), find $A_V = V_o/V_i$.

10-19 The circuit shown is called common-gate amplifier. For this circuit find (a) the voltage gain, (b) the input impedance, (c) the output impedance. Power supplies are omitted for simplicity. Neglect capacitances.

Prob. 10-19

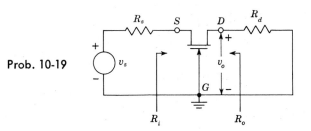

10-20 Find an expression for the signal voltage across R_L. The two FETs are identical, with parameters μ, r_d, and g_m. HINT: Use the equivalent circuits in Fig. 10-22 at S_2 and D_1.

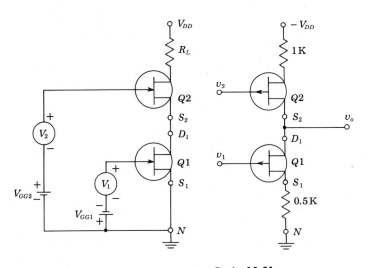

Prob. 10-20   Prob. 10-21

10-21 Each FET shown has the parameters $r_d = 10$ K and $g_m = 2$ mA/V. Using the equivalent circuits in Fig. 10-22 at S_2 and D_1, find the gain (a) v_o/v_1 if $v_2 = 0$, (b) v_o/v_2 if $v_1 = 0$.

10-22 (a) Prove that the magnitude of the signal current is the same in both FETs provided that

$$r = \frac{1}{g_m} + \frac{2R_L}{\mu}$$

Neglect the reactance of the capacitors.

(b) If r is chosen as in part a, prove that the voltage gain is given by

$$A = \frac{-\mu^2}{\mu+1} \frac{R_L}{R_L + r_d/2}$$

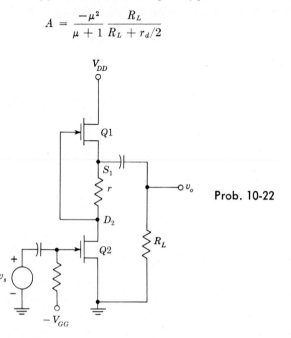

Prob. 10-22

10-23 (a) If $R_1 = R_2 = R$ and the two FETs have identical parameters, verify that the voltage amplification is $V_o/V_s = -\mu/2$ and the output impedance is $\frac{1}{2}[r_d + (\mu+1)R]$.

(b) Given $r_d = 62$ K, $\mu = 10$, $R_1 = 2$ K, and $R_2 = 1$ K. Find the voltage gain and the output impedance.

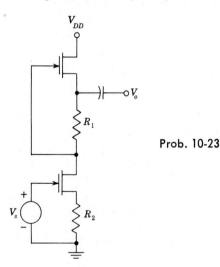

Prob. 10-23

10-24 (a) If in the amplifier stage shown the positive supply voltage V_{DD} changes by $\Delta V_{DD} = v_a$, how much does the drain-to-ground voltage change?
(b) How much does the source-to-ground voltage change under the conditions in part a?
(c) Repeat parts a and b if V_{DD} is constant but V_{SS} changes by $\Delta V_{SS} = v_s$.

Prob. 10-24

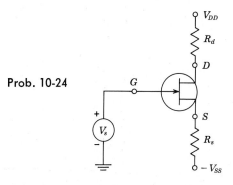

10-25 If in the circuit shown $V_2 = 0$, then this circuit becomes a source-coupled phase inverter, since $V_{o1} = -V_{o2}$. Solve for the current I_2 by drawing the equivalent circuit, looking into the source of $Q1$ (Fig. 10-22). Then replace $Q2$ by the equivalent circuit, looking into its drain. The source resistance R_s may be taken as arbitrarily large.

Prob. 10-25

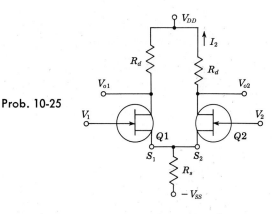

10-26 In the circuit of Prob. 10-25, assume that $V_2 = 0$, $R_d = r_d = 10$ K, $R_s = 1$ K, and $\mu = 19$. If the output is taken from the drain of $Q2$, find (a) the voltage gain, (b) the output impedance. HINT: Use the equivalent circuits in Fig. 10-22.

10-27 In the circuit of Prob. 10-25, $V_2 \neq V_1$, $R_d = 30$ K, $R_s = 2$ K, $\mu = 19$, and $r_d = 10$ K. Find (a) the voltage gains A_1 and A_2 defined by $V_{o2} = A_1 V_1 + A_2 V_2$. HINT: Use the equivalent circuits in Fig. 10-22. (b) If R_s is arbitrarily large, show that $A_2 = -A_1$. Note that the circuit now behaves as a difference amplifier.

10-28 The CS amplifier stage shown in Fig. 10-23 has the following parameters: $R_d = 12$ K, $R_g = 1$ M, $R_s = 470$ Ω, $V_{DD} = 30$ V, C_s is arbitrarily large, $I_{DSS} = 3$ mA, $V_P = -2.4$ V, and $r_d \gg R_d$. Determine (a) the gate-to-source bias voltage V_{GS}, (b) the drain current I_D, (c) the quiescent voltage V_{DS}, (d) the small-signal voltage gain A_V.

10-29 The amplifier stage shown uses an n-channel FET having $I_{DSS} = 1$ mA, $V_P = -1$ V. If the quiescent drain-to-ground voltage is 10 V, find R_1.

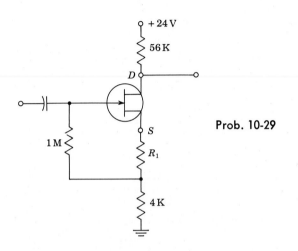

Prob. 10-29

10-30 The FET shown has the following parameters: $I_{DSS} = 5.6$ mA and $V_P = -4$ V.
(a) If $v_i = 0$, find v_o.
(b) If $v_i = 10$ V, find v_o.
(c) If $v_o = 0$, find v_i.
NOTE: v_i and v_o are constant voltages (and not small-signal voltages).

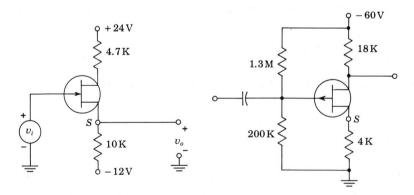

Prob. 10-30   Prob. 10-31

10-31 If $|I_{DSS}| = 4$ mA, $V_P = 4$ V, calculate the quiescent values of I_D, V_{GS}, and V_{DS}.

10-32 In the figure shown, two extreme transfer characteristics are indicated. The values of $V_{P,\text{max}}$ and $V_{P,\text{min}}$ are difficult to determine accurately. Hence these values are calculated from the experimental values of $I_{DSS,\text{max}}$, $I_{DSS,\text{min}}$, $g_{m,\text{max}}$, and $g_{m,\text{min}}$. Note that g_m is the slope of the transfer curve and that both $g_{m,\text{max}}$ and $g_{m,\text{min}}$ are measured at a drain current corresponding to $I_{DSS,\text{min}}$. Verify that

(a) $V_{P,\text{max}} = -\dfrac{2}{g_{m,\text{min}}} (I_{DSS,\text{max}} I_{DSS,\text{min}})^{\frac{1}{2}}$

(b) $V_{P,\text{min}} = -\dfrac{2 I_{DSS,\text{min}}}{g_{m,\text{max}}}$

(c) If for a given FET, $I_{DSS,\text{min}} = 2$ mA, $I_{DSS,\text{max}} = 6$ mA, $g_{m,\text{min}} = 1.5$ mA/V, and $g_{m,\text{max}} = 3$ mA/V, evaluate $V_{P,\text{max}}$ and $V_{P,\text{min}}$.

Prob. 10-32

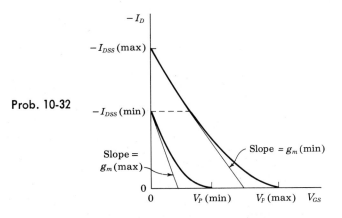

10-33 The drain current in milliamperes of the enhancement-type MOSFET shown is given by

$$I_D = 0.2(V_{GS} - V_P)^2$$

in the region $V_{DS} \geq V_{GS} - V_P$. If $V_P = +3$ V, calculate the quiescent values I_D, V_{GS}, and V_{DS}.

Prob. 10-33

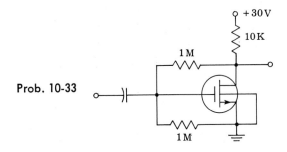

10-34 Show that if $R_L \ll 1/h_{ob2}$, the voltage gain of the hybrid cascode amplifier stage shown is given to a very good approximation by

$$A_V = g_m h_{fb} R_L$$

where g_m is the FET transconductance.

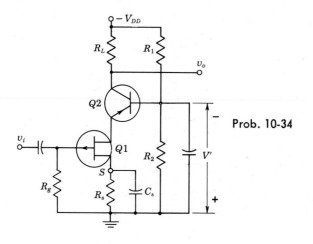

Prob. 10-34

10-35 If $h_{ie} \ll R_d$, $h_{ie} \ll r_d$, $h_{fe} \gg 1$, and $\mu \gg 1$ for the circuit, show that

(a) $A_{V1} = \dfrac{v_{o1}}{v_i} \approx \dfrac{g_m h_{fe} R_s}{1 + g_m h_{fe} R_s}$ (b) $A_{V2} = \dfrac{v_{o2}}{v_i} \approx \dfrac{g_m h_{fe}(R_s + R_c)}{1 + g_m h_{fe} R_s}$

where g_m is the FET transconductance.

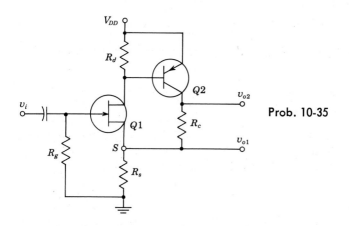

Prob. 10-35

10-36 If $r_d \gg R_1$, $R_2 \gg h_{ib3}$, $1/h_{oe2} \gg h_{ib3}$, $R' \gg R_3$, and $1/h_{ob3} \gg R_3$, show that the voltage gain at low frequencies is given by

$$A_o = \frac{v_o}{v_i} = g_m(1 + h_{fe2})h_{fb3} \frac{R_1 R_3}{R_1 + h_{ie2} + h_{ib3}(1 + h_{fe2})}$$

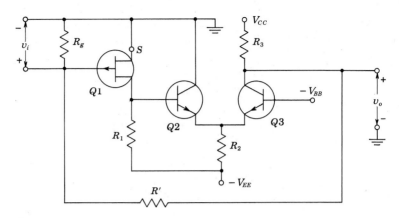

Prob. 10-36

10-37 In the circuit shown, the FET is used as an adjustable impedance element by varying the dc bias, and thereby the g_m of the FET.

(a) Assume that there is a generator V between the terminals A and B. Draw the equivalent circuit. Neglect interelectrode capacitances.

Prob. 10-37

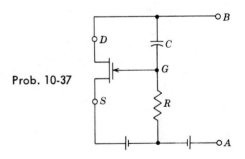

(b) Show that the input admittance between A and B is

$$Y_i = Y_d + (1 + g_m R) Y_{CR}$$

where Y_d is the admittance corresponding to r_d, and Y_{CR} is the admittance corresponding to R and C in series.

(c) If $g_m R \gg 1$, show that the effective input capacitance is

$$C_i = \frac{g_m \alpha}{\omega(1 + \alpha^2)}$$

and the effective input resistance is

$$R_i = \frac{(1 + \alpha^2) r_d}{1 + \alpha^2(1 + \mu)}$$

where $\alpha \equiv \omega CR$.

(d) At a given frequency, show that the maximum value of C_i (as either C or R is varied) is obtained when $\alpha = 1$, and

$$(C_i)_{max} = \frac{g_m}{2\omega}$$

Also show that the value of R_i corresponding to this C_i is

$$(R_i)_{max} = \frac{2r_d}{2 + \mu}$$

which, for $\mu \gg 2$, reduced to $(R_i)_{max} = 2/g_m$.

10-38 Solve Prob. 10-37 if the capacitance C is replaced by an inductance L.

10-39 (a) A MOSFET connected in the CS configuration works into a 100-K resistive load. Calculate the complex voltage gain and the input admittance of the system for frequencies of 100 and 100,000 Hz. Take the interelectrode capacitances into consideration. The MOSFET parameters are $\mu = 100$, $r_d = 40$ K, $g_m = 2.5$ mA/V, $C_{gs} = 4.0$ pF, $C_{ds} = 0.6$ pF, and $C_{gd} = 2.4$ pF. Compare these results with those obtained when the interelectrode capacitances are neglected.
(b) Calculate the input resistance and capacitance.

10-40 Calculate the input admittance of an FET at 10^3 and 10^6 Hz when the total drain circuit impedance is (a) a resistance of 50 K, (b) a capacitive reactance of 50 K at each frequency. Take the interelectrode capacitances into consideration. The FET parameters are $\mu = 20$, $r_d = 10$ K, $g_m = 2.0$ mA/V, $C_{gs} = 3.0$ pF, $C_{ds} = 1.0$ pF, and $C_{gd} = 2.0$ pF. Express the results in terms of the input resistance and capacitance.

10-41 (a) Starting with the circuit model of Fig. 10-31, verify Eq. (10-37) for the voltage gain of the source follower, taking interelectrode capacitances into account.
(b) Verify Eq. (10-39) for the input admittance.
(c) Verify Eq. (10-40) for the output admittance.
HINT: For part c, set $V_i = 0$ and impress an external voltage V_o from S to N; the current drawn from V_o divided by V_o is Y_o.

10-42 Starting with the circuit model of Fig. 10-8, show that, for the CG amplifier stage with $R_s = 0$ and $C_{ds} = 0$,

(a) $A_V = \dfrac{(g_m + g_d)R_d}{1 + R_d(g_d + j\omega C_{gd})}$ (b) $Y_i = g_m + g_d(1 - A_V) + j\omega C_{sg}$

(c) Repeat (a), taking the source resistance R_s into account.
(d) Repeat (b), taking the source resistance R_s into account.

10-43 (a) For the source follower with $g_m = 2$ mA/V, $R_s = 100$ K, $r_d = 50$ K, and with each internode capacitance 3 pF, find the frequency at which the reactive component of the output admittance equals the resistive component.
(b) At the frequency found in part a calculate the gain and compare it with the low-frequency value.

CHAPTER 11

11-1 Show that at low frequencies the hybrid-II model with $r_{b'c}$ and r_{ce} taken as infinite reduces to the approximate CE h-parameter model.

11-2 (a) Consider the hybrid-II circuit at low-frequencies, so that C_e and C_c may be neglected. Omit none of the other elements in the circuit. If the load resistance is $R_L = 1/g_L$, prove that

$$K \equiv \frac{V_{ce}}{V_{b'e}} = \frac{-g_m + g_{b'c}}{g_{b'c} + g_{ce} + g_L}$$

HINT: Use the theorem that the voltage between C and E equals the short-circuit current times the impedance seen between C and E, with the input voltage $V_{b'e}$ shorted [Eq. (8-36)].

(b) Using Miller's theorem, draw the equivalent circuit between C and E. Applying KCL to this network, show that the above value of K is obtained.

(c) Using Miller's theorem, draw the equivalent circuit between B and E. Prove that the current gain under load is

$$A_I = \frac{g_L}{(g_{b'c} + g_{b'e})/K - g_{b'c}}$$

(d) Using the results of parts a and c and the relationships between the hybrid-II and the h parameters, prove that

$$A_I = \frac{-h_{fe}}{1 + h_{oe}R_L}$$

which is the result [Eq. (8-18)] obtained directly from the low-frequency h-parameter model. HINT: Neglect $g_{b'c}$ compared with g_m or $g_{b'e}$ in A_I and in K. Justify these approximations.

11-3 The following low-frequency parameters are known for a given transistor at $I_C = 10$ mA, $V_{CE} = 10$ V, and at room temperature.

$$h_{ie} = 500 \ \Omega \qquad h_{oe} = 4 \times 10^{-5} \ \text{A/V}$$

$$h_{fe} = 100 \qquad h_{re} = 10^{-4}$$

At the same operating point, $f_T = 50$ MHz and $C_{ob} = 3$ pF, compute the values of all the hybrid-II parameters.

11-4 Given the following transistor measurements made at $I_C = 5$ mA, $V_{CE} = 10$ V, and at room temperature:

$$h_{fe} = 100 \qquad\qquad h_{ie} = 600 \ \Omega$$

$$[A_{ie}] = 10 \text{ at } 10 \text{ MHz} \qquad C_c = 3 \text{ pF}$$

Find f_β, f_T, C_e, $r_{b'e}$, and $r_{bb'}$.

11-5 A silicon p-n-p transistor has an $f_T = 400$ MHz. What is the base thickness?

11-6 Given a germanium p-n-p transistor whose base width is 10^{-4} cm. At room temperature and for a dc emitter current of 2 mA, find (a) the emitter diffusion capacitance, (b) f_T.

11-7 (a) At low frequencies the CE current gain β is related to the CB current gain α by

$$\alpha = \frac{\beta}{1 + \beta}$$

Assuming that this relationship remains valid at high frequencies and using

$$\beta = -A_i = \frac{\beta_o}{1 + j(f/f_\beta)}$$

show that α is given by

$$\alpha = \frac{\alpha_o}{1 + j(f/f_\alpha)}$$

where

$$\alpha_o = \frac{h_{fe}}{1 + h_{fe}} \qquad \text{and} \qquad f_\alpha = \frac{f_\beta}{1 - \alpha_o}$$

(b) Using the results of part a, verify that, for $\alpha_o \approx 1$, $f_\alpha \approx f_\beta h_{fe}$.

(c) Verify that

$$A_i = \frac{-\alpha_o}{1 - \alpha_o + jf/f_\alpha}$$

(d) To account for "excess phase" replace α_o by $\alpha_o \epsilon^{-jmf/f_\alpha}$. Prove that f_T, the frequency at which $|A_i| = 1$, is given implicitly by

$$1 + x^2 = 2\alpha_o(\cos mx - x \sin mx)$$

where $x = f_T/f_\alpha$.

(e) If $mx \ll 1$, expand the trigonometric functions and prove that

$$f_T \approx \frac{\alpha_o f_\alpha}{[1 + 2\alpha_o(m + m^2/2)]^{\frac{1}{2}}}$$

(f) If $\alpha_o = 1$ and $m = 0.2$, show that $f_T = f_\alpha/1.2$.

11-8 (a) Redraw the CE hybrid-Π equivalent circuit with the base as the common terminal and the output terminals, collector and base, short-circuited. Taking account of typical values of the transistor parameters, show that C_c, $r_{b'c}$, and r_{ce} may be neglected.

(b) Using the circuit in part a, prove that the CB short-circuit current gain is

$$A_{ib} = \frac{g_m}{g_{b'e} + g_m + j\omega C_e} = \frac{\alpha_o}{1 + jf/f_\alpha}$$

where

$$\alpha_o = \frac{h_{fe}}{1 + h_{fe}} \qquad \text{and} \qquad f_\alpha = \frac{g_m}{2\pi C_e \alpha_o} \approx \frac{f_\beta}{1 - \alpha_o}$$

11-9 The hybrid-Π parameters of the transistor used in the circuit shown are given in Sec. 11-1. Using Miller's theorem and the approximate analysis, compute
(a) The upper 3-dB frequency of the current gain $A_I = I_L/I_i$.
(b) The magnitude of the voltage gain $A_{Vs} = V_o/V_s$ at the frequency of part a.

Prob. 11-9

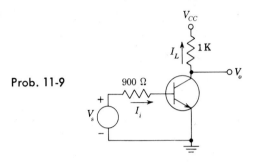

11-10 Consider a single-stage CE transistor amplifier with the load resistor R_L shunted by a capacitance C_L.
(a) Prove that the internal voltage gain $K = V_{ce}/V_{b'e}$ is

$$K \approx \frac{-g_m R_L}{1 + j\omega(C_c + C_L)R_L}$$

(b) Prove that the 3-dB frequency is given by

$$f_H \approx \frac{1}{2\pi(C_c + C_L)R_L}$$

provided that the following condition is satisfied:

$$g_{b'e}R_L(C_c + C_L) \gg C_e + C_c(1 + g_m R_L)$$

11-11 For a single-stage CE transistor amplifier whose hybrid-Π parameters have the average values given in Sec. 11-1, what value of source resistance R_s will give a 3-dB frequency f_H which is (a) half the value for $R_s = 0$, (b) twice the value for $R_s = \infty$? Do these values of R_s depend upon the magnitude of the load R_L? Use Miller's theorem and the approximate analysis.

11-12 A single-stage CE amplifier is measured to have a voltage-gain bandwidth f_H of 5 MHz with $R_L = 500\ \Omega$. Assume $h_{fe} = 100$, $g_m = 100\ \text{mA/V}$, $r_{bb'} = 100\ \Omega$, $C_c = 1\ \text{pF}$, and $f_T = 400\ \text{MHz}$.
(a) Find the value of the source resistance that will give the required bandwidth.
(b) With the value of R_s found in part a, find the midband voltage gain V_o/V_s.
HINT: Use the approximate analysis.

11-13 The hybrid-Π parameters of the transistor used in the circuit shown are given in Sec. 11-1. The input to the amplifier is an abrupt current step 0.2 mA in magnitude. Find the output voltage as a function of time (a) if $C_L = 0$. Neglect the output time constant. (b) If $C_L = 0.1\ \mu\text{F}$. Neglect the input time constant.

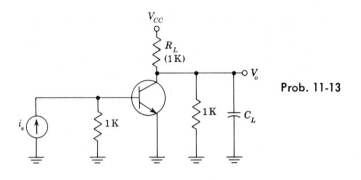

Prob. 11-13

11-14 (a) Verify the nodal equations for the single-stage CE amplifier of Sec. 11-8.
(b) Obtain Eq. (11-37) for the voltage gain V_o/V_s.

11-15 (a) Verify the values of K_1, s_o, s_1, and s_2 given in Sec. 11-8 for the CE stage of Fig. 11-10.
(b) Evaluate the gain at zero frequency.
(c) Evaluate the magnitude of the gain at 2 MHz and check with Fig. 11-11.
(d) Evaluate the phase of the gain at 2 MHz and check with Fig. 11-11.

11-16 (a) From the circuit of Fig. 11-12 (and not assuming that $|K| \gg 1$), prove that

$$K = \frac{-g_m R_L + j\omega C_c R_L}{1 + j\omega C_c R_L}$$

Why may the term $j\omega C_c R_L$ be neglected in the numerator but not in the denominator?
(b) The Miller admittance in the output circuit is given by

$$Y_o = j\omega C_c \left(1 - \frac{1}{K}\right)$$

Prove that this represents a capacitance C_o in parallel with a resistance R_o given by

$$C_o = C_c \frac{1 + g_m R_L}{g_m R_L} \qquad R_o = \frac{-g_m}{\omega^2 C_c^2}$$

Note that R_o is negative.
(c) Evaluate C_o and R_o at the 3-dB frequency of 3.0 MHz and verify that the effective output time constant remains $R_L C_c$ (approximately). Assume $g_m = 50$ mA/V, $R_L = 2$ K, $C_e = 100$ pF, and $C_c = 3$ pF.

11-17 Verify Eq. (11-54).

11-18 (a) Verify the nodal equations in Sec. 11-10 for the emitter follower.
(b) Find the gain V_e/V_s as a function of s.

11-19 Delete all capacitors from the emitter-follower equivalent circuit of Fig. 11-14b. Find (a) the input impedance and (b) the output impedance and (c) show that these results are consistent with the low-frequency equivalent circuits of Fig. 8-25.

11-20 (a) For the emitter follower of Fig. 11-14 at high frequencies, obtain $K = V_e/V_i'$ and (with $g \equiv g_m + g_{b'e}$) verify that

$$K = \frac{gR_L}{1 + gR_L} \frac{1 + j\omega(C_e/g)}{1 + j\omega \left(\dfrac{C_L + C_e}{1 + gR_L}\right) R_L}$$

(b) If $gR_L \gg 1$ and $C_L \gg C_e$, show that

$$K \approx \frac{1}{1 + jf/f_H}$$

where

$$f_H = \frac{1}{2\pi} \frac{g}{C_L + C_e} = \frac{1}{2\pi} \frac{g_m + g_{b'e}}{C_L + C_e}$$

CHAPTER 12

12-1 (a) To show the effect of phase shift on the image seen on a cathode-ray screen, consider the following example: The sinusoidal voltages applied to both sets of plates should be equal in phase and magnitude so that the maximum displacement in either direction on the screen is 2 in. Because of frequency distortion in the horizontal amplifier, the phase of the horizontal voltage is shifted 5° but the magnitude is changed inappreciably. Plot to scale the image that actually appears on the screen, and compare with the image that would be seen if there were no phase shift.
(b) If the phase shift in both amplifiers were the same, what would be seen on the cathode-ray screen?

12-2 The input to an amplifier consists of a voltage made up of a fundamental signal and a second-harmonic signal of half the magnitude and in phase with the fundamental. Plot the resultant.
    The output consists of the same magnitude of each component, but with the second harmonic shifted 90° (on the fundamental scale). This corresponds to perfect frequency response but bad phase-shift response. Plot the output and compare it with the input waveshape.

12-3 The bandwidth of an amplifier extends from 20 Hz to 20 kHz. Find the frequency range over which the voltage gain is down less than 1 dB from its midband value. Assume that the low- and high-frequency response is given by Eqs. (12-1) and (12-5) multiplied by a constant A_{Vo}.

12-4 Prove that over the range of frequencies from $10f_L$ to $0.1f_H$ the voltage amplification is constant to within 0.5 percent and the phase shift to within ± 0.1 rad. Make the same assumption as in Prob. 12-3.

12-5 (a) Show that the Bode magnitude plot for a two-pole transfer function is equal to the sum of the magnitude plots of each pole considered separately.
(b) Repeat part a for the Bode phase plot.

12-6 Sketch the idealized Bode amplitude and phase plots for a transfer function with one zero f_z and one pole f_p if (a) $f_p < f_z$ and (b) $f_p > f_z$.

12-7 Consider a transfer characteristic with two poles such that $f_{p2} = 4f_{p1}$.
(a) Plot the idealized and true Bode magnitude curves. Obtain the actual 3-dB frequency graphically.
(b) Plot the idealized and true Bode phase curves.

12-8 Repeat Prob. 12-7 for poles at $f_{p2} = 2f_{p1}$.

12-9 Consider the transfer function given in Eq. (12-1) which has one pole and a zero at $f = 0$. Draw the piecewise linear Bode plots for the pole, the zero, and the resultant for (a) amplitude, (b) phase.

12-10 An ideal 1-μs pulse is fed into an amplifier. Plot the output if the bandpass is (a) 10 MHz, (b) 1.0 MHz, (c) 0.1 MHz. Assume $f_L = 0$ and a single-pole amplifier.

12-11 (a) Prove that the response of a two-stage (identical and noninteracting) low-pass amplifier to a unit step is

$$v_o = A_o{}^2[1 - (1 + x)\epsilon^{-x}]$$

where A_o is the midband voltage gain and $x \equiv t/RC$.
(b) For $t \ll RC$, show that the output varies quadratically with time.

12-12 In Prob. 12-11, let the upper 3-dB frequency of a single stage be f_H and the rise time of the two stages in cascade be t_r. Show that $f_H t_r = 0.53$.

12-13 (a) For the transistor CE stage shown with $1/h_{oe} \approx \infty$, calculate the percentage tilt in the output if the input current I is a 100-Hz square wave.
(b) What is the lowest-frequency square wave which will suffer less than 1 percent tilt?

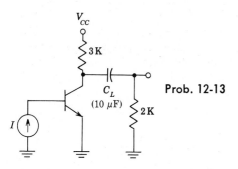

Prob. 12-13

12-14 Consider a transfer function with n poles and k zeros. Assume that all the zeros occur at much higher frequencies than the poles. Verify that the 3-dB frequency is given by Eq. (12-25).

12-15 The transfer function V_o/V_s of an amplifier has n poles $s_1, s_2, \ldots, s_n$ and k zeros $s_{z1}, s_{z2}, \ldots, s_{zk}$, as follows:

$$\frac{V_o}{V_s} = \frac{K(s - s_{z1})(s - s_{z2}) \cdots (s - s_{zk})}{(s - s_1)(s - s_2) \cdots (s - s_n)}$$

If the zeros are of much higher frequencies than the poles, show that
(a) An approximate expression for the high 3-dB frequency f_H^* is given by

$$\frac{1}{f_H^*} \approx \sqrt{\frac{1}{f_1^2} + \frac{1}{f_2^2} + \cdots + \frac{1}{f_n^2}}$$

(b) An expression which gives a more accurate result is

$$\frac{1}{f_H^*} \approx 1.1 \sqrt{\frac{1}{f_1^2} + \frac{1}{f_2^2} + \cdots + \frac{1}{f_n^2}}$$

Verify this, using Eq. (12-26) for the case of
   (i) Two identical poles $f_1 = f_2$.
   (ii) Three identical poles $f_1 = f_2 = f_3$.
Show that the error is within 10 percent.

12-16 Consider a transfer function with poles at 1 MHz and 2 MHz. Assume all other poles and zeros are much larger than 2 MHz. Calculate the high 3-dB frequency. Compare your result with the approximate value obtained from Eq. (12-31).

12-17 If two cascaded single-pole stages have very unequal bandpasses, show that the combined bandwidth is essentially that of the smaller. Assume noninteracting stages.

12-18 Three identical cascaded stages have an overall upper 3-dB frequency of 20 kHz and a lower 3-dB frequency of 20 Hz. What are f_L and f_H of each stage? Assume noninteracting stages.

12-19 It is desired that the voltage gain of an RC-coupled amplifier at 60 Hz should not decrease by more than 10 percent from its midband value. Show that the coupling capacitance C must be at least equal to $5.5/R'$, where $R' = R_o' + R_i'$ is expressed in kilohms, and C in microfarads.

12-20 The parameters of the transistors in the circuit shown are $h_{fe} = 50$, $h_{ie} = 1.1$ K, $h_{re} = h_{oe} = 0$. Find (a) the midband gain, (b) the value of C_b necessary to give a lower 3-dB frequency of 20 Hz. Assume that C_z represents a short-circuit at this frequency. (c) Find the value of C_b necessary to ensure less than 10 percent tilt for a 100-Hz square-wave input.

Prob. 12-20

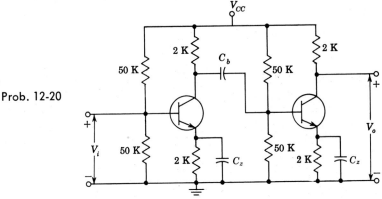

12-21 A two-stage FET RC-coupled amplifier has the following parameters: $g_m = 10$ mA/V, $r_d = 5.5$ K, $R_d = 10$ K, and $R_g = 0.5$ M for each stage. Assume C_s in Fig. 12-11b to be arbitrarily large.

(a) What must be the value of C_b in order that the frequency characteristic of each stage be flat within 1 dB down to 10 Hz?

(b) Repeat part a if the overall gain of both stages is to be down 1 dB at 10 Hz.

(c) What is the overall midband voltage gain?

12-22 A three-stage RC-coupled amplifier uses field-effect transistors (Fig. 12-11b), with the following parameters: $g_m = 2.6$ mA/V, $r_d = 7.7$ K, $R_d = 10$ K, $R_g = 0.1$ M, $C_b = 0.005\ \mu$F, and $C_s = \infty$. Evaluate (a) the overall midband voltage gain in decibels, (b) f_L of each individual stage, (c) the overall lower 3-dB frequency.

12-23 Plot the idealized Bode *phase* characteristic corresponding to the amplitude response of Fig. 12-15 [Eq. (12-39)].

12-24 (a) Show that the relative voltage gain of an amplifier with an emitter resistor R_e bypassed by a capacitor C_z may be expressed in the form

$$\frac{A_V}{A_o} = \frac{1 + j\omega R_e C_z}{B + j\omega R_e C_z}$$

where $B = 1 + R'/R$, $R' = R_e(1 + h_{fe})$, and $R = R_s + h_{ie}$.

(b) Prove that the lower 3-dB frequency is

$$f_L = \frac{\sqrt{B^2 - 2}}{2\pi R_e C_z}$$

What is the physical meaning of the condition $B < \sqrt{2}$?

(c) If $B^2 \gg 2$, show that $f_L \approx f_p$, the pole frequency as defined in Eq. (12-40).

12-25 In the circuit of Fig. 12-14a, let $R_s = 500\ \Omega$; $R_1 = R_2 = 50$ K; $R_c = R_e = 2$ K; $h_{ie} = 1.1$ K; $h_{fe} = 50$; $h_{re} = h_{oe} = 0$; $C_b = 5\ \mu$F.

(a) Neglecting the effects of C_z, find f_L for the transistor stage.

(b) Neglecting the effects of C_b, find expressions for f_p and f_o due to C_z alone.

(c) Find a value of C_z for which f_L is virtually unaffected by the presence of the emitter bypass capacitor.

12-26 Find the percentage tilt in the output of a transistor stage caused by a capacitor C_z bypassing an emitter resistor R_e. Use the following method: If V is the magnitude of the input step, then from Fig. 12-14 (and using lowercase letters for instantaneous values),

$$v_o = -h_{fe}i_bR_c = -h_{fe}R_c\frac{V - v_{en}}{R}$$

where $R \equiv R_s + h_{ie}$. Take as a first approximation $v_{en} = 0$. Calculate the corresponding current, and assuming that all the emitter current passes through C_z, calculate v_{en}, and then show that

$$v_o = -\frac{h_{fe}R_cV}{R}\left[1 - \frac{(1 + h_{fe})t}{RC_z}\right]$$

From this result verify Eq. (12-43).

12-27 Show that the low-frequency voltage gain of the FET stage shown with $r_d \gg R_L + R_s$ is given by

(a) $\dfrac{A_V}{A_o} = \dfrac{1}{1 + g_m R_s} \dfrac{1 + jf/f_o}{1 + jf/f_p}$

where

$A_o \equiv -g_m R_L$ $f_o \equiv \dfrac{1}{2\pi C_s R_s}$ $f_p = \dfrac{1 + g_m R_s}{2\pi C_s R_s}$

(b) If $g_m R_s \gg 1$ and $g_m = 5$ mA/V, find C_s so that a 50-Hz square-wave input will suffer no more than 10 percent tilt.

Prob. 12-27

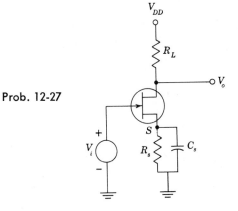

12-28 Verify that the transfer function of the two-stage interacting amplifier of Fig. 12-19 is given by Eq. (12-48).

12-29 Verify Eq. (12-50).

12-30 Justify Eq. (12-52).

12-31 (a) Find the noise bandwidth B_n for an amplifier for which $A_{Vo} = 1, f_L = 0$ Hz, and

$|A_V(f)| = \dfrac{1}{\sqrt{1 + (f/f_H)^2}}$

(b) Compute B_n if $f_H = 10$ kHz.

12-32 (a) Find the mean-square value V_o^2 of the output noise voltage for the circuit shown. The circle represents a generator supplying Johnson noise to the RC combination.

(b) Prove that

$\tfrac{1}{2}CV_o^2 = \tfrac{1}{2}kT$

This result is known as the *equipartition theorem.*

Prob. 12-32

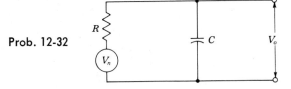

CHAPTER 13

13-1 For the circuit shown, with $R_c = 4$ K, $R_L = 4$ K, $R_b = 20$ K, $R_s = 1$ K, and the transistor parameters given in Table 8-2, find

(a) The current gain $I_L/I_s = A_I$.

(b) The voltage gain V_o/V_s, where $V_s \equiv I_s R_s$.

(c) The transconductance $I_L/V_s = G_M$.

(d) The transresistance $V_o/I_s = R_M$.

(e) The input resistance seen by the source.

(f) The output resistance seen by the load.

Make reasonable approximations. Neglect all capacitive effects.

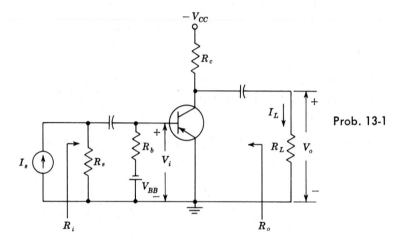

Prob. 13-1

13-2 Repeat Prob. 13-1 for the circuit shown, with $g_m = 5$ mA/V and $r_d = 100$ K. Note that $V_s \equiv I_s R_s$.

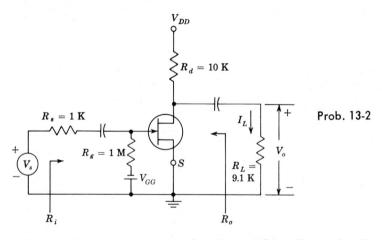

Prob. 13-2

13-3 (a) For the circuit shown, find the ac voltage V_i as a function of V_s and V_f. Assume that the inverting-amplifier input resistance is infinite, that $A = A_V =$

$-1,000$, $\beta = V_f/V_o = \frac{1}{100}$, $R_s = R_e = R_c = 1$ K, $h_{ie} = 1$ K, $h_{re} = h_{oe} = 0$, and $h_{fe} = 100$.   (b) Find $A_{Vf} = V_o/V_s = AV_i/V_s$.

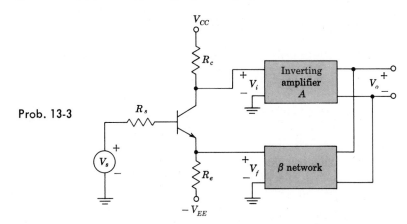

Prob. 13-3

13-4 An amplifier consists of three identical stages connected in cascade. The output voltage is sampled and returned to the input in series opposing. If it is specified that the relative change dA_f/A_f in the closed-loop voltage gain A_f must not exceed Ψ_f, show that the minimum value of the open-loop gain A of the amplifier is given by

$$A = 3A_f \frac{|\Psi_1|}{|\Psi_f|}$$

where $\Psi_1 \equiv dA_1/A_1$ is the relative change in the voltage gain of each stage of the amplifier.

13-5 An amplifier with open-loop voltage gain $A_V = 1,000 \pm 100$ is available. It is necessary to have an amplifier whose voltage gain varies by no more than ± 0.1 percent.
(a) Find the reverse transmission factor β of the feedback network used.
(b) Find the gain with feedback.

13-6 The figure shows the transfer characteristic of a nonlinear amplifier. Negative feedback is applied to this amplifier as shown. Find the new transfer characteristic x_o versus x_s if (a) $\beta = 0.1$, (b) $\beta = 0.05$. Plot the two transfer characteristics on the same figure.

Prob. 13-6

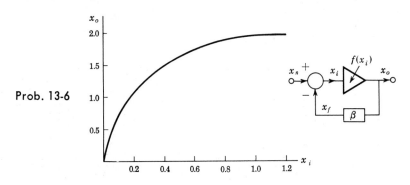

13-7 An amplifier without feedback gives a fundamental output of 36 V with 7 percent second-harmonic distortion when the input is 0.028 V.

(a) If 1.2 percent of the output is fed back into the input in a negative voltage-series feedback circuit, what is the output voltage?

(b) If the fundamental output is maintained at 36 V but the second-harmonic distortion is reduced to 1 percent, what is the input voltage?

13-8 An amplifier with an open-loop voltage gain of 1,000 delivers 10 W of output power at 10 percent second-harmonic distortion when the input signal is 10 mV. If 40-dB negative voltage-series feedback is applied and the output power is to remain at 10 W, determine (a) the required input signal, (b) the percent harmonic distortion.

13-9 (a) Verify Eq. (13-16) for the input impedance of the current-series feedback amplifier.

(b) Repeat part (a) for Eq. (13-25) for the voltage-shunt amplifier.

(c) Verify Eq. (13-32) for the output impedance of the voltage-shunt feedback amplifier.

(d) Repeat part (c) for Eq. (13-38) for the current-series feedback amplifier.

13-10 The output impedance may be calculated as the ratio of the open-circuit voltage to the short-circuit current. Using this method, evaluate R_{of} and R'_{of} for (a) voltage-series feedback, (b) current-series feedback, (c) current-shunt feedback, and (d) voltage-shunt feedback.

13-11 The h-parameter model of a transistor can be considered to represent a feedback amplifier due to the presence of the h_{re} source. Using feedback formulas, find (a) R_{if} and (b) $Y_{of} = 1/R_{of}$, representing the input and output resistances of a transistor stage taking h_{re}, h_{oe}, and a source resistance R_s into account.

13-12 Assume that the parameters of the circuit are $r_d = 10$ K, $R_g = 1$ M, $R_1 = 40$ Ω, $R_d = 50$ K, and $g_m = 6$ mA/V. Neglect the reactances of all capacitors. Find the voltage gain and output impedance of the circuit at the terminals (a) AN, (b) BN.

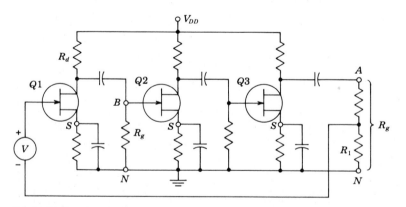

Prob. 13-12

13-13 Prove that for voltage-series feedback, with $R_s = 0$, $A_{If} = A_I$. HINT: $A_V = A_I R_L/R_i$.

13-14 The transistors in the feedback amplifier shown are identical, and their h parameters are as given in Table 8-2. Make reasonable approximations whenever appropriate, and neglect the reactance of the capacitors. Calculate $R_{if} = V_s/I_i$, $A_{If} = -I/I_i$, $A'_{Vf} = V_o/V_i$, $A_{Vf} = V_o/V_s$, and R'_{of}.

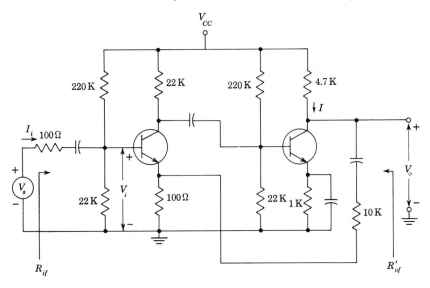

Prob. 13-14

13-15 A modified second-collector to first-emitter feedback pair is shown with dc biasing omitted for simplicity. All transistors are identical. Neglecting h_{re}, h_{rb}, h_{oe}, h_{ob} and assuming that $h_{fe} \gg 1$, $h_{fe}R_1 \gg R_s + h_{ie}$, and $R_2 \gg h_{ib3}$, show that
(a) The voltage gain $A_{Vf} = V_o/V_s \approx R_2/R_1$ (if $h_{fe}R_C \gg R_C + R_2$).
(b) The output resistance $R'_{of} \approx R_c \| (R_2/h_{fe})$.

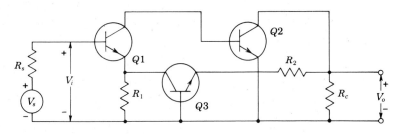

Prob. 13-15

13-16 Consider the transistor stage of Fig. 13-16a.
(a) Neglecting h_{re} and h_{oe} and assuming that $h_{fe} \gg 1$, show that the voltage gain is

$$A_{Vf} = \frac{V_o}{V_s} \approx \frac{-h_{fe}R_L}{R_s + h_{ie} + h_{fe}R_e}$$

(b) If the relative change dA_f/A_f of the voltage gain A_f must not exceed a specified value Ψ_f due to variations of h_{fe}, show that the minimum required value of the emitter resistor R_e is given by

$$R_e = \frac{R_s + h_{ie}}{h_{fe}} \left(\frac{dh_{fe}/h_{fe}}{\Psi_f} - 1 \right)$$

13-17 Solve the example in Sec. 13-11 on current-shunt feedback without using the feedback equations. Instead, apply Miller's theorem to the resistor R'. HINT: Assume the gain A'_V from V_{i1} to V_{e2} to be very large.

13-18 In the two-stage feedback amplifier shown, the transistors are identical and have the following parameters: $h_{fe} = 50$, $h_{ie} = 2$ K, $h_{re} = 0$, and $h_{oe} = 0$. Calculate

(a) $A_{If} = \dfrac{I_o}{I_s}$ (b) $R_{if} = \dfrac{V_i}{I_s}$ (c) $A'_{If} = \dfrac{I_o}{I'_i}$

(d) $A_{Vf} = \dfrac{V_o}{V_s}$ where $V_s = I_s R_s$

(e) Evaluate A_{Vf} from Eq. (13-70) and compare with the result obtained in part d.

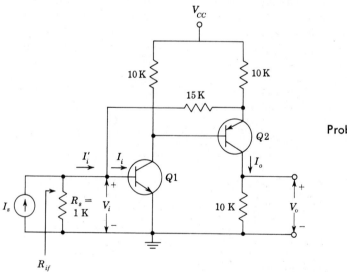

Prob. 13-18

13-19 For the circuit shown (and with the h-parameter values given in Prob. 13-18) find

(a) $A_{If} \equiv \dfrac{I_o}{I_s}$ (b) R_{if}

(c) $A_{Vf} \equiv \dfrac{V_o}{V_s}$ where $I_s \equiv \dfrac{V_s}{R_s}$

(d) $A'_{Vf} \equiv \dfrac{V_o}{V_i}$ (e) R'_{of}

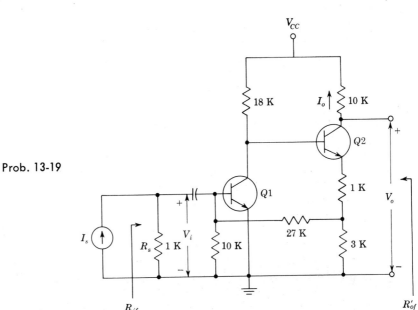

Prob. 13-19

13-20 The transistors in the feedback amplifier shown are identical, and their h parameters are given in Table 8-2. Make reasonable approximations where appropriate, and neglect the reactances of the capacitors. Calculate

(a) $A_{If} \equiv \dfrac{I_o}{I_s}$ (b) $A_{Vf} \equiv \dfrac{V_o}{V_s}$,   where $V_s \equiv I_s R_s$

(c) R_{if} (d) R_{of}

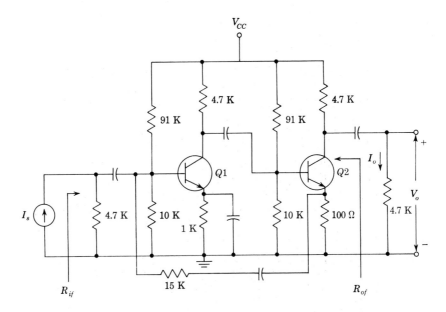

Prob. 13-20

13-21 Let h_{fe} of $Q1$ and $Q2$ of Prob. 13-20 increase to 100. If all other parameters remain constant, repeat Prob. 13-20.

13-22 For the transistor feedback-amplifier stage shown, $h_{fe} = 100$, $h_{ie} = 1$ K, while h_{re} and h_{oe} are negligible. Determine with $R_e = 0$

(a) $R_{Mf} = \dfrac{V_o}{I_s}$ where $I_s = \dfrac{V_s}{R_s}$

(b) $A_{Vf} = \dfrac{V_o}{V_s}$ (c) R_{if} (d) R'_{of}

(e) Repeat the four preceding calculations if $R_e = 1$ K.

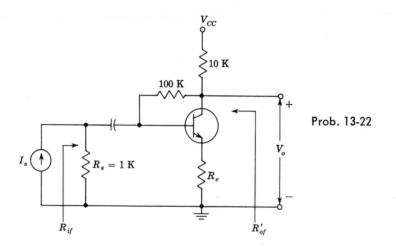

Prob. 13-22

13-23 Consider the illustrative example in Sec. 13-11 (current-shunt feedback) but with the output taken from the emitter of $Q2$. This configuration now represents voltage-shunt feedback and *not* current-shunt feedback. Analyze the circuit for (a) β; (b) R_M; (c) R_{Mf}; (d) A_{Vf}; (e) R_{if}; (f) R'_{of}.

13-24 For the circuit shown, prove that

$$A_{Vf} = \frac{V_o}{V_s} = -\frac{R'}{R} \cdot \frac{1}{1 + \dfrac{R'}{R_m}\left(\dfrac{R_i + R'}{R'} + \dfrac{R_i}{R}\right)}$$

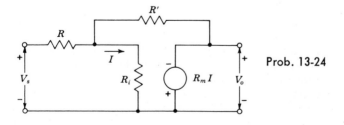

Prob. 13-24

13-25 For the voltage-shunt feedback circuit in the example in Sec. (13-12), replace the transistor by the low-frequency approximate model ($h_{oe} = h_{re} = 0$). Do *not* use feedback-analysis methods. Solve for A_{Vf}, R_{if}, and R_{of} exactly. Compare with results obtained in Sec. 13-12.

CHAPTER 14

14-1 A single-stage RC-coupled amplifier with a midband voltage gain of 1,000 is made into a feedback amplifier by feeding 10 percent of its output voltage in series with the input opposing. Assume that the amplifier gain without feedback may be approximated at low frequencies by Eq. (12-2) and at high frequencies by Eq. (14-2).

(a) As the frequency is varied, to what value does the voltage gain of the amplifier without feedback fall before gain of the amplifier with feedback falls 3 dB?

(b) What is the ratio of the half-power frequencies with feedback to those without feedback?

(c) If $f_L = 20$ Hz and $f_H = 50$ kHz for the amplifier without feedback, what are the corresponding values after feedback has been added?

14-2 (a) Verify Eqs. (14-9) and (14-10) for A_f for the two-pole transfer gain.

(b) Verify that for $Q = Q_{min}$, the roots are ω_1 and ω_2.

14-3 Verify Eqs. (14-14) and (14-16) for the transfer function of the circuit model of Fig. 14-4.

14-4 (a) Show that the two-pole closed-loop magnitude of the gain A_f is given by Eq. (14-18) as a function of frequency.

(b) Verify that the peak on the frequency response occurs at $\omega/\omega_o = \sqrt{1 - 2k^2}$ and has a value given by Eq. (14-20).

14-5 Plot the phase response (versus ω/ω_o) of a double-pole transfer function for $Q = 0.5, 1, 2, 5$.

14-6 Derive Eqs. (14-23), (14-24), (14-25), and (14-26), for the step response of the two-pole feedback amplifier. HINT: For the overdamped case, assume $k^2 \gg 1$ and expand $(1 - 1/k^2)^{\frac{1}{2}}$ in Taylor series.

14-7 Verify Eq. (14-27) for the positions x_m and magnitude y_m of the oscillatory-response maxima and minima.

14-8 Define the normalized settling time x_s to be the time at which the first peak (or dip) in Fig. 14-6 that is within the error band of $\pm P$ percent occurs. Show that the value of m corresponding to x_s is given by the smallest value of m that satisfies

$$100\epsilon^{-\pi km/(1-k^2)^{\frac{1}{2}}} \leq P$$

14-9 (a) Given a two-pole amplifier with corner frequencies at $\omega_1 = 1$ Mrad/s and $\omega_2 = 0.2$ Mrad/s. What are the maximum decibels of feedback which will give the fastest rise time without overshoot?

(b) What is the rise-time improvement for the condition in part a? In other words, find the ratio of the rise time for $k = 1$ to the rise time for zero feedback.

14-10 (a) For the amplifier in Prob. 14-9, find the maximum value of the loop gain for which the step-response overshoot will be 10 percent.
(b) At what time will the peak occur?
(c) Calculate the magnitude of the first minimum of the step response and the time at which it occurs.
(d) Verify that for $k = 0.707$ the maximum overshoot is 4.3 percent.

14-11 An amplifier has two poles on the negative real axis: $s_1 = -5 \ \mu s^{-1}$, $s_2 = -15 \ \mu s^{-1}$.
(a) Plot the root locus of the amplifier with negative feedback.
(b) Find the value of βA_o for which the maximum overshoot of the amplifier step response with feedback is 4.3 percent.

14-12 (a) If $k = 0.5$ ($Q = 1$), calculate the percent maximum overshoot in the step response for a two-pole feedback amplifier.
(b) If there is a 10 percent overshoot in the frequency response, what is the percent overshoot in the step response?

14-13 The roots of a closed-loop two-pole amplifier are $s_1 = -\sigma + j\omega$, $s_2 = -\sigma - j\omega$. Find the relationship between Q and $|\omega/\sigma|$.

14-14 For the three-pole feedback amplifier, verify Eq. (14-29) and show that

$$\omega_o{}^3 = \omega_1\omega_2\omega_3(1 + \beta A_o)$$

$$a_2 = \frac{\omega_1 + \omega_2 + \omega_3}{\omega_o}$$

$$a_1 = \frac{\omega_1\omega_2 + \omega_2\omega_3 + \omega_1\omega_3}{\omega_o{}^2}$$

14-15 (a) Consider a three-pole open-loop transfer function with all three poles at $s = -\omega_1$. Find an expression for the closed-loop gain.
(b) Show that as negative feedback is added, one pole s_{3f} moves along the negative real axis while the other two poles become complex conjugates and move toward the right-hand complex plane, as indicated in Fig. 14-8.
(c) Verify that the system is unstable for $|\beta A_o| > 8$; and that for $|\beta A_o| = 8$, the poles s_{1f} and s_{2f} are $\pm j\omega_1 \sqrt{3}$ and $s_{3f} = -3\omega_1$.

14-16 Consider the amplifier with a transfer function $A(s) = A_1/[s(s + 3)^2]$.
(a) Find the value of βA_1 corresponding to the breakaway point (the point in the complex plane where the real poles become complex).
(b) Find the value of βA_1 for which the amplifier with negative feedback becomes unstable.
(c) Plot approximately the root locus.

14-17 Consider a transfer function with three poles s_1, s_2, s_3. Find s_1, s_2, s_3 knowing that

$$|s_1| = |s_2| = |s_3| = 1$$

and the Q of the complex pole pair (s_2, s_3) is 1.

14-18 An amplifier has the following transfer function:

$$A(s) = \frac{A_o \times 7.2 \times 10^{-4}}{(s + 0.02)(s + 0.09)(s + 0.4)}$$

Feedback is applied to this amplifier. Find the poles and βA_o if the Q of the complex pole pair is 1.

14-19 (a) Verify that a two-pole feedback amplifier *cannot* have a closed-loop dominant pole if $Q > 0.4$.
(b) Find an expression for the maximum value of βA_o for which a closed-loop dominant pole exists.
(c) Calculate βA_o for $n = 4$. What is the physical interpretation of this result?

14-20 Verify Eq. (14-33), using Eqs. (14-32) and (14-31).

14-21 Verify the expression for R_M in Eq. (14-34).

14-22 Verify the expression for G_M in Eq. (14-40).

14-23 For the voltage-series feedback pair in Sec. 14-8 verify that a dominant pole exists and that A_{Vf} can be approximated by Eq. (14-47).

14-24 Show that for the dominant-pole-amplifier polar plot in Fig. 14-18
(a) The upper semicircle of the plot is the locus of βA for negative frequencies.
(b) The lower semicircle corresponds to positive frequencies.
(c) The points corresponding to $f = \pm f_H$ are at the midpoints of the two semicircles.

14-25 Consider a feedback amplifier for which the gain at low frequencies without feedback is given by Eq. (12-2).
(a) Show that the polar plot of the loop gain is a circle in the right half of the complex βA plane, as in Fig. 14-18.
(b) Show that the upper semicircle corresponds to values of $f > 0$, and that the lower semicircle corresponds to values of $f < 0$.

14-26 Consider a two-pole feedback amplifier for which the gain without feedback is given by Eq. (14-8). Sketch the polar plot of the loop gain βA (β is a real constant and $f_2 = 10f_1$) for this amplifier, indicating
(a) The section of the plot corresponding to $f > 0$ and that corresponding to $f < 0$.
(b) The points on the plot corresponding to $f = 0$; $f = \pm \infty$; $f = f_1 = \omega_1/2\pi$; $f = f_2 = \omega_2/2\pi$.

14-27 Sketch the polar plot of the loop gain βA for a three-pole feedback amplifier with a dc gain (without feedback). $A_o = -1,000$, and open-loop poles at $f_1 = 0.5$ MHz, $f_2 = 1$ MHz, and $f_3 = 2$ MHz, under the following conditions:
(a) $\beta = -0.005$ (b) $\beta = -0.02$
In each case indicate whether or not the closed-loop amplifier is stable.
(c) What is the maximum value of β for which the amplifier is stable?

14-28 A three-pole feedback amplifier has a dc gain without feedback of -10^4. All three open-loop poles are at $f = 2$ MHz.
(a) What is the maximum value of β for which the amplifier is stable?
(b) Assume that one of the poles is shifted to $f_1 = 100$ kHz.
Using the value of β found in part a, what is the gain margin of the modified circuit?

14-29 Verify Eq. (14-53) for the transfer function of the pole-zero-compensation network.

14-30 Pole-zero compensation is used, but the zero f_z of the compensating network does not exactly equal the lowest pole f_1 of the uncompensated amplifier.
(a) Choose $f_z = 1.1f_1$ and sketch on log-log paper the function

$$A = \frac{1 + j(f/f_z)}{1 + j(f/f_1)}$$

(b) Repeat part a if $f_z = 0.9f_1$.

14-31 A three-pole amplifier without feedback has a dc gain of -10^3 and poles located at $f_1 = 1$ MHz, $f_2 = 10$ MHz, and $f_3 = 30$ MHz. Dominant-pole compensation is applied to this amplifier.
(a) Find the location of the dominant pole so that the open-loop gain is first constant and then falls to 0 dB at a rate of -20 dB per decade for frequencies $f \leq 1$ MHz.
(b) What is the maximum value of β for which this compensated amplifier is stable?

14-32 Pole-zero compensation is used with an amplifier which has -10^3 dc gain and three poles at $f_1 = 1$ MHz, $f_2 = 10$ MHz, and $f_3 = 200$ MHz. The zero of the pole-zero network is selected to cancel the 1-MHz pole of the uncompensated amplifier.
(a) Find the pole of the compensating network so that the amplifier is stable with a 45° phase margin when $\beta = -0.1$.
HINT: Let $-|\beta|A'_V = 1/\underline{-135°}$ at $f = 10$ MHz.
(b) What is the bandwidth of the compensated amplifier with feedback?

14-33 Verify Eq. (14-60) for the feedback factor of the phase-shift network of Fig. 14-29, assuming that this network does not load the amplifier. Prove that the phase shift of V'_f/V_o is 180° for $\alpha^2 = 6$ and that at this frequency $\beta = \frac{1}{29}$.

14-34 (a) For the network of Prob. 14-33, show that the input impedance is given by

$$Z_i = R\frac{1 - 5\alpha^2 - j(6\alpha - \alpha^3)}{3 - \alpha^2 - j4\alpha}$$

(b) Show that the input impedance at the frequency of the oscillator, $\alpha = \sqrt{6}$, is $(0.83 - j2.70)R$.
Note that if the frequency is varied by varying C, the input impedance remains constant. However, if the frequency is varied by varying R, the impedance is varied in proportion to R.

14-35 Design a phase-shift oscillator to operate at a frequency of 5 kHz. Use a MOSFET with $\mu = 55$ and $r_d = 5.5$ K. The phase-shift network is not to load down the amplifier.
(a) Find the minimum value of the drain-circuit resistance R_d for which the circuit will oscillate.
(b) Find the product RC.
(c) Choose a reasonable value for R, and find C.

14-36 (a) A two-stage FET oscillator uses the phase-shifting network shown. Prove that

$$\frac{V'_f}{V} = \frac{1}{3 + j(\omega RC - 1/\omega RC)}$$

(b) Show that the frequency of oscillation is $f = 1/2\pi RC$ and that the gain must exceed 3.

Prob. 14-36

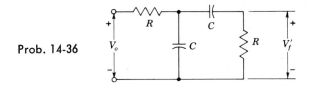

14-37 (a) Find V_f'/V_o for the network shown.
(b) Sketch the circuit of a phase-shift FET oscillator, using this feedback network.
(c) Find the expression for the frequency of oscillation, assuming that the network does not load down the amplifier.
(d) Find the minimum gain required for oscillation.

Prob. 14-37

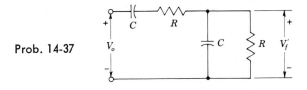

14-38 Consider the two-section RC network shown. Find the V_f'/V_o function, and verify that it is not possible to obtain 180° phase shift with a finite attenuation.

Prob. 14-38

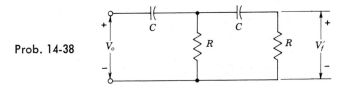

14-39 For the feedback network shown find (a) the transfer function, (b) the input impedance. (c) If this network is used in a phase-shift oscillator, find the frequency of oscillation and the minimum amplifier voltage gain. Assume that the network does not load down the amplifier.

Prob. 14-39

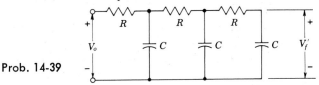

14-40 Take into account the loading of the RC network in the phase-shift oscillator of Fig. 14-29a. If R_o is the output impedance of the amplifier (assume that C_s is arbitrarily large), prove that the frequency of oscillation f and the minimum gain A are given by

$$ f = \frac{1}{2\pi RC} \frac{1}{\sqrt{6 + 4(R_o/R)}} \qquad A = 29 + 23\frac{R_o}{R} + 4\left(\frac{R_o}{R}\right)^2 $$

14-41 For the FET oscillator shown, find (a) V_f'/V_o, (b) the frequency of oscillations, (c) the minimum gain of the source follower required for oscillations.

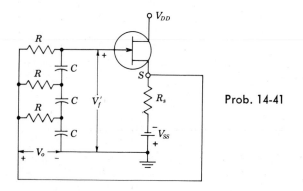

Prob. 14-41

14-42 Verify Eqs. (14-61) and (14-62) for the transistor phase-shift oscillator of Fig. 14-30.

14-43 Apply the Barkhausen criterion to the tuned-drain oscillator, and verify Eqs. (14-63) and (14-64).

14-44 (a) At what frequency will the circuit shown oscillate, if at all? (b) Find the minimum value of R needed to sustain oscillations. The FETs are identical with $g_m = 1.6$ mA/V and $r_d = 44$ K.

HINT: Assume a voltage V from gate G_1 of $Q1$ to ground but with the point G' not connected to the gate G_1. Calculate the loop gain from the equivalent circuit, obtained by looking into each source.

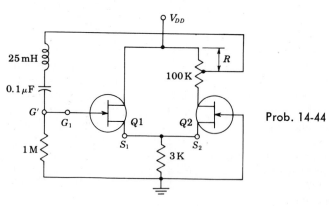

Prob. 14-44

14-45 (a) Consider a Colpitts oscillator, using the circuit of Fig. 14-32 and taking into account the resistance r_3 in series with the inductor L_3. Show that the frequency of oscillation is given by

$$\omega^2 = \frac{1}{L_3}\left[\frac{1}{C_1} + \frac{1}{C_2}\left(1 + \frac{r_3}{R_o}\right)\right]$$

(b) If $r_3/R_o \ll 1$, show that the minimum amplifier gain required for oscillations is

$$A_v = \frac{C_1}{C_2} + \frac{C_2 + C_1}{L_3}R_o r_3$$

14-46 (a) Consider the Hartley oscillator circuit shown (with bias and power supplies omitted for simplicity). If the resistances of the inductors are r_1 and r_2, respectively, find the frequency of oscillation.
(b) Find the value of R_s for which the value of the loop gain will just equal unity.

Prob. 14-46

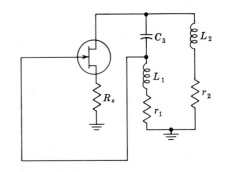

14-47 In the Wien bridge circuit of Fig. 14-34, add an inductor in series with R and C between points 2 and 3. Also, replace the parallel combination of R and C by a resistor R_3. Find the frequency of oscillation and the minimum gain of the amplifier if
(a) R_1 is infinite.
(b) R_1 is finite.

14-48 (a) Verify Eq. (14-75) for the reactance of a crystal.
(b) Prove that the ratio of the parallel- to series-resonant frequencies is given approximately by $1 + \frac{1}{2}C/C'$.
(c) If $C = 0.04$ pF and $C' = 2.0$ pF, by what percent is the parallel-resonant frequency greater than the series-resonant frequency?

14-49 A crystal has the following parameters: $L = 0.33$ H, $C = 0.065$ pF, $C' = 1.0$ pF, and $R = 5.5$ K.
(a) Find the series-resonant frequency.
(b) By what percent does the parallel-resonant frequency exceed the series-resonant frequency?
(c) Find the Q of the crystal.

CHAPTER 15

15-1 Verify Eqs. (15-2) and (15-3).

15-2 Find an expression for A_{Vf} in Fig. 15-3 by using feedback concepts. Show that this expression agrees with Eq. (15-2) if $A_v \gg R_o Y'$.

15-3 The amplifier shown uses an OP AMP with input resistance R_i, voltage gain $A_v < 0$, and zero output resistance. Assume also that the OP AMP is unilateral from input to output.

(a) Show that the amplifier satisfies the three fundamental assumptions of Sec. 13-3.

(b) Show that the transresistance of the amplifier without feedback is

$$\frac{A_v R_i R R'}{RR' + (R_i + R_1)(R + R')}$$

(c) Show that

$$A_{Vf} = \frac{V_o}{V_s} = \frac{A_v R_i R'}{RR' + (R_i + R_1)(R + R') - A_v R_i R}$$

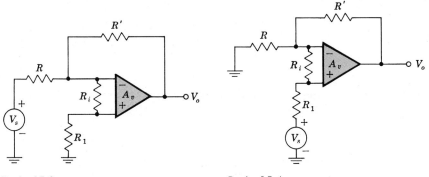

Prob. 15-3   Prob. 15-4

15-4 (a) Repeat Prob. 15-3 for the noninverting amplifier shown.

(b) Show that the voltage gain of the amplifier without feedback is

$$A_V = \frac{-A_v R_i (R + R')}{(R + R')(R_1 + R_i) + RR'}$$

(c) Show that

$$A_{Vf} = \frac{-A_v R_i (R + R')}{RR' + (R_i + R_1)(R + R') - A_v R R_i} = \frac{V_o}{V_s}$$

15-5 For the circuit of this problem with $R_i = \infty$, show that $Y_{of} = 1/R_{of}$ is given by

$$Y_{of} = \frac{1}{R_o}\left(1 - A_v \frac{R}{R + R'}\right) + \frac{1}{R + R'}$$

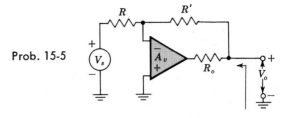

Prob. 15-5

15-6 The circuit shown is a differential amplifier using an ideal OP AMP.
  (a) Find the output voltage v_o.
  (b) Show that the output corresponding to the common-mode voltage $v_c = \frac{1}{2}(v_1 + v_2)$ is equal to zero if $R'/R = R_1/R_2$. Find v_o in this case.
  (c) Find the common-mode rejection ratio of the amplifier if $R'/R \neq R_1/R_2$.

Prob. 15-6

15-7 The circuit shown represents a dc feedback amplifier consisting of a differential input pair $Q1$-$Q2$ followed by two stages, $Q3$ and $Q4$.

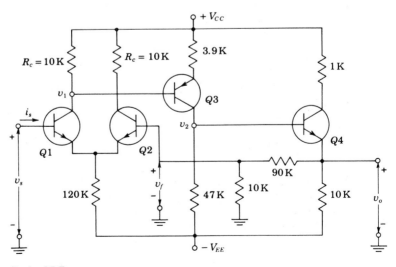

Prob. 15-7

All transistors are identical, and their parameters are

$$h_{ie} = 1 \text{ K} \qquad h_{oe} = 10 \text{ μ℧} \qquad h_{re} = 2.5 \times 10^{-4} \qquad h_{fe} = 100$$

Make reasonable approximations resulting in errors of no more than 10 percent. Compute the following quantities at low frequencies.

(a) The difference gain A_d and common-mode gain A_c for the differential amplifier defined by the equation

$$v_1 = A_d(v_f - v_s) + A_c \frac{v_f + v_s}{2}$$

Make use of the symmetry of the circuit.

(b) v_2/v_1, v_o/v_2, and $A = v_o/v_1$. Assume that $Q2$ does not load the 10-K resistance.

(c) $A_V = v_o/v_s$. Compare this result with that obtained using the feedback factor β.

15-8 For the circuit of Fig. 15-6 assume that $R_s = 0$, $h_{oe}(R_c + 2R_e) \ll 1$, $h_{fe} \gg 1$, and $h_{ie} \ll 2R_e h_{fe}$.

(a) Verify that the common-mode rejection ratio is given by

$$\rho = \frac{h_{fe}R_e}{h_{ie}}$$

(b) If $r_{bb'} \ll r_{b'e}$ verify that $\rho = g_m R_e \approx V/2V_T$, where V is the quiescent voltage across R_e.

15-9 Draw the h-parameter model for the common-mode gain in a DIFF AMP. Without solving for A_c show from the circuit that A_c must be zero if $h_{fe}/h_{oe} = 2R_e$ [in agreement with Eq. (15-13)].

15-10 Verify Eqs. (15-13) and (15-14) for the difference amplifier.

15-11 Starting with Eq. (15-14) for A_d and assuming $R_s \ll h_{ie}$ and $r_{bb'} \ll r_{b'e}$, verify that

$$A_d = \frac{1}{2} g_m R_c \qquad \text{and} \qquad g_{md} = \frac{I_o}{4V_T}$$

15-12 (a) Show that the emitter volt-ampere characteristic of a transistor in the active region is given by

$$I_E \approx I_S \epsilon^{V_E/V_T}$$

where $I_S = -I_{EO}/(1 - \alpha_N \alpha_I)$.

(b) Verify Eq. (15-22) for the transfer characteristic of the DIFF AMP.

(c) Verify Eq. (15-23) for g_{md}.

15-13 (a) From Eq. (15-22), for the transfer characteristic of the DIFF AMP find the range $\Delta V = \Delta(V_{B1} - V_{B2})$ over which each collector current increases from 0.1 to 0.9 its peak value.

(b) Repeat part a for a collector-current variation from 50 to 99 percent of I_o.

(c) Compare your result with Fig. 15-9.

15-14 The differential amplifier of Fig. 15-8 is modified by putting two resistors R_e in series with the emitter lead of $Q1$ and $Q2$.

(a) Express $V_{B1} - V_{B2}$ as a function of $V_{BE1} - V_{BE2}$ and I_{C1}.

(b) Find the transfer characteristic I_{C1}/I_O versus $(V_{B1} - V_{B2})/V_T$ if $R_e = 50 \ \Omega$ and $I_O = 2$ mA. Solve *graphically* by using Fig. 15-9 and part a.

(c) Find the transconductance

$$g'_{md} = \frac{dI_{C1}}{d(V_{B1} - V_{B2})}$$

evaluated at $V_{B1} = V_{B2}$.

(d) Express g'_{md} in terms of g_{md} given in Eq. (15-23).

15-15 (a) Calculate the common-mode rejection ratio ρ_1 for the first stage of the OP AMP in Fig. 15-11. Assume $1/h_{oe} = 100$ K and $h_{re} = 2.5 \times 10^{-4}$.

(b) Calculate ρ_2 for the second stage.

(c) What is the overall ρ (in decibels)?

15-16 The figure shows an inverting OP AMP with input resistance R_i and offset voltage V_{io}. Show that V_o is given by

$$V_o = \frac{-A_v R_i (R + R') V_{io}}{RR' + (R_i + R_1)(R + R') - RR_i A_v}$$

Prob. 15-16

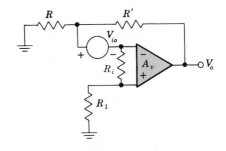

15-17 For the amplifier shown, V_1 and V_2 represent undesirable voltages. Show that, if $R_i = \infty$, $R_o = 0$, and $A_{v1} < 0$ and $A_{v2} < 0$,

$$V_o = A_{v2}[A_{v1}(V' - V_1) - V_2] \qquad \text{where } V' = V_o \frac{R}{R + R'}$$

Show also that, if $A_{v2}A_{v1}R/(R + R') \gg 1$,

$$V_o = -\left(1 + \frac{R'}{R}\right)\left(V_1 + \frac{V_2}{A_{v1}}\right)$$

Prob. 15-17

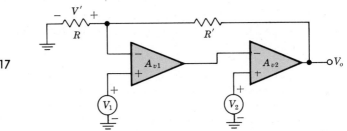

15-18 Consider the circuit shown. An attempt is made to find a value of R_1 which minimizes the offset-current effects.

(a) First show that $V_o = 0$ if $I_1[RR'/(R + R')] = I_2R_1$. Assume $|A_v| \gg 1$ and $R' \ll -RA_v$.

(b) Conclude that $R_1 = RR'/(R + R')$ is the optimum value of R_1 if $I_1/I_2 \approx 1$. Find V_o in this case. Assume $R_o = 0$.

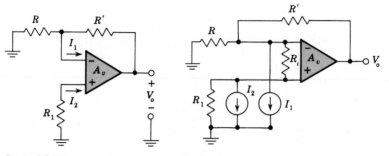

Prob. 15-18   Prob. 15-19

15-19 (a) For the amplifier shown (with $R_o = 0$) prove that

(i) The output voltage V_{o1} due to the bias current I_1 is

$$V_{o1} = \frac{-R'RR_iA_v}{(R_i + R_1)(R' + R) + RR' - A_vRR_i} I_1$$

(ii) The output voltage V_{o2} due to the bias current I_2 is

$$V_{o2} = \frac{R_iR_1(R + R')A_v}{(R + R')(R_1 + R_i) - A_vRR_i + RR'} I_2$$

(b) Show that if $I_1/I_2 \approx 1$, then $V_{o1} + V_{o2}$ is minimized by taking $R_1 = RR'/(R + R')$.

15-20 (a) Show that the gain A_{VfI} of the inverting OP AMP and the gain A_{VfNI} of the noninverting amplifier satisfy (assume $R_o = 0$)

$$\frac{A_{VfI}}{A_{VfNI}} = -\frac{R'}{R + R'}$$

HINT: Use the results of Probs. 15-3 and 15-4.

(b) Using (a), show that the input biasing currents I_1 and I_2 and the input offset voltage V_{io} produce an effective input error voltage V_{Ei} on the inverting terminal equal to

$$V_{EI} = -I_1R + \frac{R + R'}{R'} R_1I_2 - V_{io} \frac{R + R'}{R'}$$

(c) Plot the curve $V_{EI} = f(R')$ for $R_1 = RR'/(R + R')$, $R = 1$ K, and typical values of V_{io} and $I_2 - I_1$ (Table 15-1).

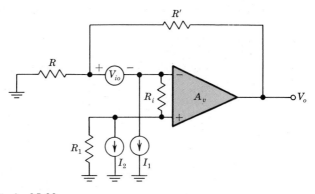

Prob. 15-20

15-21 (*a*) Repeat part *b* of Prob. 15-20 and show that the input offset signals produce an effective input error voltage V_{ENI} on the noninverting terminal equal to

$$V_{ENI} = V_{io} - R_1 I_2 + I_1 \frac{RR'}{R + R'}$$

(*b*) Repeat part *c* of Prob. 15-20.

15-22 Consider the inverting OP AMP of Prob. 15-20 with infinite open-loop gain and with an output-voltage swing of ±4.5 V.
(*a*) Plot the output voltage due to V_{io} as a function of R'/R.
(*b*) Plot the output voltage due to the biasing currents I_1 and I_2 as a function of R' (i) if $R_1 = RR'/(R + R')$, (ii) if $R_1 = 0$. Use typical values for the offset signals.   HINT: Use the results of Prob. 15-20.

15-23 The input offset voltage V_{io} of an OP AMP is equal to 1 mV at 25°C. The input offset-voltage drift of this amplifier is equal to 5 μV/°C. Assume that the open-loop voltage gain is infinite. Using Prob. 15-16, find the output offset voltage at temperature $T = 100$°C if (i) $R'/R = 1$, (ii) $R'/R = 100$. At which temperature is the output offset voltage equal to 0.2 V if $R'/R = 200$?

15-24 (*a*) For the inverting OP AMP with $R_i = \infty$, draw the model of the amplifier without feedback but taking the loading of R' into account (Sec. 13-7). Refer to Fig. 15-2 with $Z = R$ and $Z' = R'$.
(*b*) Verify that $\beta = -1/R'$ and $R_M = A_V R_{11}$, where $R_{11} = R \| R'$.
(*c*) Verify that

$$A_{Vf} = \frac{R_{Mf}}{R} = \frac{R_M/R}{1 + \beta R_M}$$

and show that this expression reduces to Eq. (15-35).

15-25 (*a*) For the noninverting OP AMP of Fig. 15-4 with $R_i = \infty$, draw the model of the amplifier without feedback but taking the loading of R' into account.
(*b*) Verify that $\beta = R/(R + R')$.
(*c*) Find A_{Vf}.

15-26 Without using feedback-amplifier concepts, verify that A_{Vf} for the inverting OP AMP of Prob. 15-24 is given by Eq. (15-35).

15-27 Without using feedback concepts, verify that A_{Vf} for the noninverting OP AMP of Prob. 15-25 is given by Eq. (15-38).

15-28 (a) The transfer function of an OP AMP has its first pole at 1 MHz and a low-frequency gain of 44 dB. Dominant-pole compensation is used for this OP AMP, and the gain of the compensated amplifier is zero dB at a frequency 1 MHz. Find the value f_d of the dominant pole.

(b) Repeat (a) if the low-frequency gain is 68 dB.

15-29 (a) In Fig. 15-11 find the resistance R_i seen between pins 9 and 10. Assume $1/h_{oe} = 100$ K and $h_{re} = 0$.

(b) A capacitor C_1 is connected between pins 9 and 10. For which value of C_1 is the first pole of the compensated amplifier equal to (i) $f_d = 200$ Hz, (ii) $f_d = 1$ kHz?

15-30 (a) Verify Eq. (15-41).

(b) Draw on the same figure the following Bode plots:

(i) Open-circuit voltage gain of the amplifier without compensation.

(ii) Open-circuit voltage gain of the amplifier if pole-zero cancellation is achieved, using Eq. (15-41).

In (i) assume that the amplifier has three poles.

15-31 In Prob. 15-30 the transfer function of the amplifier without compensation has three poles at 1, 4, and 40 MHz and a low-frequency open-loop gain of 72 dB.

(a) Find R_c and C_c as a function of R_1, R, R' if the gain of the compensated amplifier is zero dB at a frequency 4 MHz.

(b) Find R_c and C_c if $R_1 = RR'/(R + R')$, $R = 1$ K, and $R' \gg R$. Find also the bandwidth of the compensated amplifier without feedback.

15-32 (a) Show that pole-zero cancellation can be achieved by using the input circuit of the figure. Use the result of Prob. 15-3(b) and assume $R_i = \infty$.

(b) Draw the Bode plots of the open-loop gain for the compensated and the uncompensated amplifier. Assume that the OP AMP has three poles and that $R_2 \ll R$.

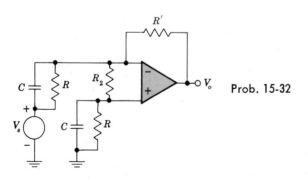

Prob. 15-32

15-33 Verify all the equations in Sec. 15-11 which relate to pole-zero cancellation using the Miller effect technique.

15-34 Verify Eq. (15-50) for the factor A introduced due to C' in Fig. 15-24.

15-35 The slew rate of an OP AMP is 6 V/μs when the closed-loop gain is unity. The amplified output signal is observed to be a pure sinusoid $v_o = V_m \cos \omega t$ provided the frequency of this signal does not exceed a certain limit.

Find the value of this limiting frequency before the output signal is distorted by the slew-rate limit if (a) $V_m = 1$ V, (b) $V_m = 10$ V.

CHAPTER 16

16-1 Design the circuit of Fig. 16-1 so that the output V_o (for a sinusoidal signal) is equal in magnitude to the input V_s and leads the input by 45°.

16-2 Consider the circuit of Fig. 16-1 with $A_V = -100$. If $Z = R$ and $Z' = -jX_C$ with $R = X_C$ at some specific frequency f, calculate the gain V_o/V_s as a complex number.

16-3 Given the operational amplifier circuit of Fig. 16-1 consisting of R and L in series for Z, and C for Z'. If the input voltage is a constant $v_s = V$, find the output v_o as a function of time. Assume an infinite open-loop gain.

16-4 For the given circuit, show that the output voltage is

$$-v_o = \frac{R_2}{R_1}v + \left(R_2C + \frac{L}{R_1}\right)\frac{dv}{dt} + LC\frac{d^2v}{dt^2}$$

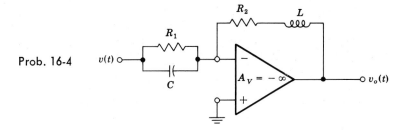

Prob. 16-4

16-5 Consider the operational amplifier circuit of Fig. 16-1 with Z consisting of a resistor R in parallel with a capacitor C, and Z' consisting of a resistor R'. The input is a sweep voltage $v = \alpha t$. Show that the output voltage v_o is a sweep voltage that starts with an initial step. Thus prove that

$$v_o = -\alpha R'C - \alpha \frac{R'}{R}t$$

Assume infinite open-loop gain.

16-6 Consider the operational amplifier circuit of Fig. 16-1 with Z consisting of a 100-K resistor and a series combination of a 50-K resistance with a 0.001-μF capacitance for Z'. If the capacitor is initially uncharged, and if at $t = 0$ the input voltage $v_s = 10\epsilon^{-t/\tau}$ with $\tau = 5 \times 10^{-4}$ s is applied, find $v_o(t)$.

16-7 In Fig. 16-3b show that i_L is equal to $-v_s/R_2$ if $R_3/R_2 = R'/R_1$.

16-8 The differential input operational amplifier shown consists of a base amplifier of infinite gain. Show that $V_o = (R_2/R_1)(V_2 - V_1)$.

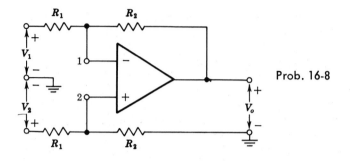

Prob. 16-8

16-9 Repeat Prob. 16-8 for the amplifier shown.

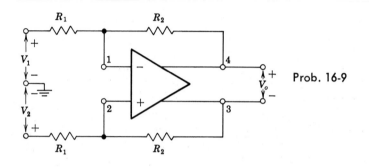

Prob. 16-9

16-10 For the base differential-input amplifier shown, assume infinite input resistance, zero output resistance, and finite differential gain $A_V = V_o/(V_1 - V_2)$.
(a) Obtain an expression for the gain $A_{Vf} = V_o/V_s$.
(b) Show that $\lim A_{Vf} = n + 1$, $A_V \to \infty$.

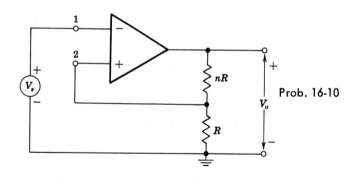

Prob. 16-10

16-11 Verify Eq. (16-6) for the bridge amplifier.
16-12 The circuit shown represents a low-pass dc-coupled amplifier. Assuming an

ideal operational amplifier determine (a) the high-frequency 3-dB point f_H; (b) the low-frequency gain $A_V = V_o/V_s$.

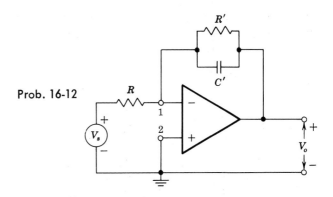

Prob. 16-12

16-13 (a) The input to the operational integrator of Fig. 16-10 is a step voltage of magnitude V. Show that the output is

$$v_o = A_V V(1 - \epsilon^{-t/RC(1-A_V)})$$

(b) Compare this result with the output obtained if the step voltage is impressed upon a simple RC integrating network (without the use of an operational amplifier). Show that for large values of RC, both solutions represent a voltage which varies approximately linearly with time. Verify that if $-A_V \gg 1$, the slope of the ramp output is approximately the same for both circuits. Also prove that the deviation from linearity for the amplifier circuit is $1/(1 - A_V)$ times that of the simple RC circuit.

16-14 Derive Eq. (16-13).

16-15 (a) The input to an operational differentiator whose open-loop gain $A_V \equiv A$ is infinite is a ramp voltage $v = \alpha t$. Show that the output is

$$v_o = \frac{A}{1 - A} \alpha RC(1 - \epsilon^{-t(1-A)/RC})$$

(b) Compare this result with that obtained if the same input is impressed upon a simple RC differentiating network (without the use of an amplifier). Show that, approximately, the same final constant output $RC\, dv/dt$ is obtained. Also show that the operational-amplifier output reaches this correct value of the differentiated input much more quickly than does the simple RC circuit.

16-16 Given an operational amplifier with Z consisting of R in series with C, and Z' consisting of R' in parallel with C'. The input is a step voltage of magnitude V. (a) Show by qualitative argument that the output voltage must start at zero, reach a maximum, and then again fall to zero.

(b) Show that if $R'C' \neq RC$, the output is given by

$$v_o = \frac{R'CV}{R'C' - RC} (\epsilon^{-t/RC} - \epsilon^{-t/R'C'})$$

16-17 Sketch an operational amplifier circuit having an input v and an output which is approximately $-5v - 3dv/dt$. Assume an ideal operational amplifier.

16-18 Sketch in block-diagram form a computer, using operational amplifiers, to solve the differential equation

$$\frac{dv}{dt} + 0.5v + 0.1 \sin \omega t = 0$$

An oscillator is available which will provide a signal $\sin \omega t$. Use only resistors and capacitors.

16-19 Set up a computer in block-diagram form, using operational amplifiers, to solve the following differential equation:

$$\frac{d^3y}{dt^3} + 2\frac{d^2y}{dt^2} - 4\frac{dy}{dt} + 2y = x(t)$$

where

$$y(0) = 0 \qquad \frac{dy}{dt}\bigg|_{t=0} = -2 \qquad \text{and} \qquad \frac{d^2y}{dt^2}\bigg|_{t=0} = 3$$

Assume that a generator is available which will provide the signal $x(t)$.

16-20 (a) Verify that the damping factor of each pair of complex poles of a Butterworth low-pass filter is given by $k = \cos \theta$, where θ is defined in Fig. 16-17.
(b) Define a damping factor k for the single pole at $s = -1$ which is consistent with Eq. (16-23).

16-21 Verify the entries in Table 16-1 for $n = 3$ by using Fig. 16-17b.

16-22 Using Eq. (16-22), show that the transfer function of a second-order Butterworth low-pass filter satisfies Eq. (16-19).

16-23 Use the value of $B_2(s)$ from Table 16-1 and verify that

$$B_2(s)B_2(-s)\bigg|_{s=j\omega} = 1 + \omega^4$$

16-24 Use the values of $B_3(s)$ from Table 16-1 and verify that

$$B_3(+s)B_3(-s)\bigg|_{s=j\omega} = 1 + \omega^6$$

16-25 Show that the voltage gain $A_V(s) = V_o/V_s$ in Fig. 16-18a is given by Eq. (16-24).

16-26 Design an active sixth-order Butterworth low-pass filter with a cutoff frequency (or upper 3-dB frequency) of 1 kHz.

16-27 The circuit shown uses an ideal OP AMP.
(a) Find the voltage gain $A_V = V_o/V_s$, the damping factor k, and the cutoff frequency ω_o.
(b) Using this circuit, design a second-order Butterworth low-pass filter with $f_o = 1$ kHz and low-frequency voltage gain equal to -1.

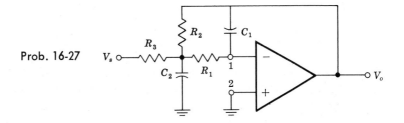

Prob. 16-27

16-28 Define the z parameters of a two-port network by the following relations:

$$V_1 = z_{11}I_1 + z_{12}I_2$$
$$V_2 = z_{21}I_1 + z_{22}I_2$$

For the circuit shown prove that the voltage gain $A_V = V_o/V_s$ is given by

$$\frac{V_o}{V_s} = -\frac{R_2 z_{21}}{(R_1 + R_2)(z_{11} - z_{21}) + R_1 R_2}$$

where z_{11} and z_{21} are the z parameters of the RC network.

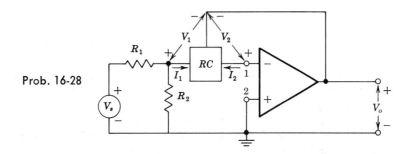

Prob. 16-28

16-29 The network shown is the RC network of Prob. 16-28.
(a) Find the parameters z_{11} and z_{21} of this network.
(b) Find the voltage gain V_o/V_s of the amplifier in Prob. 16-28 if this RC network is used.

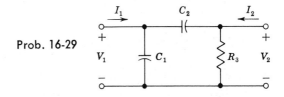

Prob. 16-29

16-30 Design a second-order active RC bandpass filter that has a midband voltage gain of 40 dB, center frequency of 100 Hz, and no specified bandwidth. However, the circuit must provide at least 20-dB rejection one decade from the center frequency and hold phase shift to $\pm 10°$ maximum for 10 percent change from the center frequency.   HINT: Use Fig. 16-22 to find Q and let $R_2 = \infty$, $R_1 = 1$ K.

16-31   Design a bandpass RC active filter with midband voltage gain of 30, center frequency of 200 Hz, and $Q = 5$.   HINT: Choose $C_1 = C_2 = 0.1$ μF.

16-32   Design the resonant RLC bandpass filter of Fig. 16-21 with $f_o = 160$ Hz, 3-dB bandwidth $B = 16$ Hz, and minimum input resistance seen by the voltage source V_s of 1,000 Ω.   Is this a practical circuit?

16-33   Verify Eq. (16-50) for the transfer function of the delay equalizer of Fig. 16-24a.

16-34   (a) Show that the circuit of the accompanying figure can simulate a grounded inductor if $R_1 > R_2$.   In other words, show that the reactive part of the input impedance of this circuit is positive if $R_1 > R_2$.

(b) Find the frequency range in which the $Q = \omega L/R$ of the inductor is greater than unity.

Assume that the unity gain amplifier has infinite input resistance and zero output resistance.

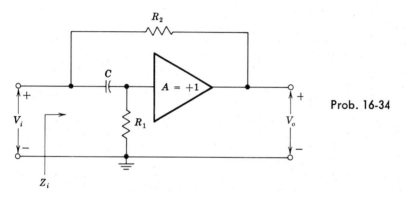

Prob. 16-34

16-35   (a) Show that the circuit of the given figure can simulate a grounded inductor if $A > 1$.   In other words, show that the reactive part of Z_i is positive.

(b) Show that the real part of Z_i becomes zero ($Q = \infty$) at the frequency

$$\omega = \frac{1}{R_2 C} \sqrt{\frac{R_1 + R_2}{R_1(A - 1)}}$$

Assume that the input resistance of the amplifier of gain A is infinite.

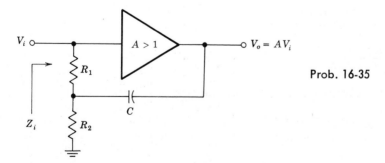

Prob. 16-35

16-36 Using Fig. 16-27b, derive Eqs. (16-54) and (16-55).

16-37 Using Fig. 16-27b, derive Eqs. (16-57) and (16-58).

16-38 The figure shows a circuit using an ideal OP AMP and an RC two-port network. The RC two-port is defined in terms of its y parameters (Sec. 16-9). Show that the voltage gain $A_V = V_o/V_s$ is given by

$$A_V = \frac{V_o}{V_s} = -\frac{y_{21}(1+k) + ky_{22}}{y_{22}} \qquad \text{where } k = \frac{R'}{R}$$

Prob. 16-38

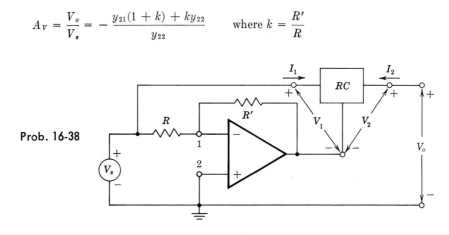

16-39 Repeat Prob. 16-38 for the circuit shown. Show that the expression for $A_V = V_o/V_s$ is the same as in Prob. 16-38.

Prob. 16-39

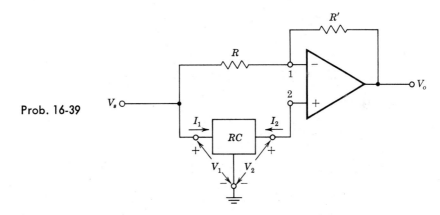

16-40 In Probs. 16-38 and 16-39 the RC two-port shown is used. (a) Find the parameters y_{21} and y_{22} of this two-port RC network. (b) Show that if

$$\frac{1}{k} = 2\left(\frac{R_2}{R_1} + \frac{C_1}{C_2}\right) + 1$$

then the two circuits are delay equalizers with transfer function

$$A_V = \frac{V_o}{V_s} = A_{V_o}\frac{(s - s_1)(s - s_2)}{(s + s_1)(s + s_2)}$$

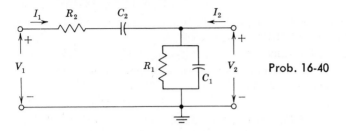

Prob. 16-40

16-41 Using the curve of Fig. 16-31 and assuming $V_{AGC} = 3.5$ V, with an audio modulating signal of 1.5 V peak to peak, calculate the modulation factor

$$k = \frac{V_{o,\text{max}} - V_{o,\text{min}}}{V_{o,\text{max}}}$$

16-42 Derive Eq. (16-59) for the gain of the cascode video amplifier.

16-43 (a) Verify that the circuit shown gives full-wave rectification provided that $R_2 = 2R_1$.

(b) What is the peak value of the rectified output?

(c) Draw carefully the waveforms $v_i = 10 \sin \omega t$, v_p, and v_o if $R_3 = 2R_1$.

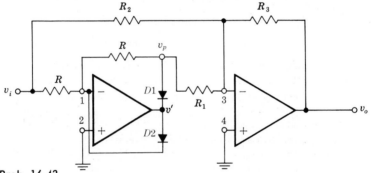

Prob. 16-43

16-44 If a waveform has a positive peak of magnitude V_1 and a negative peak of magnitude V_2, draw a circuit using two peak detectors whose output is equal to the peak-to-peak value $V_1 - V_2$.

16-45 Show that the given circuit can be used to raise the input V_s to an arbitrary power. Assume $V_1 = K_1 \ln K_2 V_s$, $V_o = K_3 \ln^{-1} K_4 V_2$, $V_2 = \alpha V_1$.

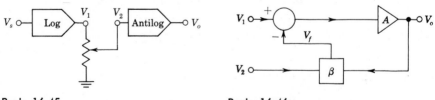

Prob. 16-45   Prob. 16-46

16-46 For the feedback circuit shown, the nonlinear feedback network β gives an output proportional to the product of the two inputs to this network, or $V_f = \beta V_2 V_o$. Prove that if $A = \infty$, then $V_o = KV_1/V_2$, where K is a constant.

16-47   (a) With the results of Prob. 16-46, draw the block diagram of a system used to obtain the square root of the voltage V_s.

(b) What should be the value of β if it is required that $V_o = \sqrt{V_s}$?

16-48   (a) Verify Eq. (16-78) for the pulse width of a monostable multivibrator.

(b) If $V_z \gg V_1$ and $\beta = 1/2$, what is T?

16-49   Verify Eq. (16-84) for the frequency of the triangle waveform.

16-50   The Schmitt trigger of Fig. 16-47 is modified to include two clamping Zener diodes across the output as in Fig. 16-45a. If $V_z = 4$ V and $A_v = 5,000$ and if the threshold levels desired are 6 ±0.5 V, find (a) R_2/R_1, (b) the loop gain, and (c) V_R.   (d) Is it possible to set the threshold voltage at a negative value? (e) In part (a) the ratio of R_2 to R_1 is obtained.   What physical conditions determine the choice of the individual resistances?

16-51   The input v_i to a Schmitt trigger is the set of pulses shown.   Plot v_o versus time.   Assume $V_1 = 3.2$ V, $V_2 = 2.8$ V, and $v_o = +5$ V at $t = 0$.

Prob. 16-51

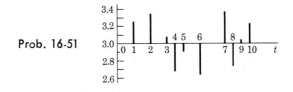

16-52   (a) Calculate the logic levels at output Y of the ECL Texas Instruments gate shown.   Assume that $V_{BE,\text{active}} = 0.7$ V.   To find the drop across an emitter follower when it behaves as a diode assume a piecewise-linear diode model with $V_\gamma = 0.6$ V and $R_f = 20\ \Omega$.

(b) Find the noise margin when the output Y is at $V(0)$ and also at $V(1)$.

(c) Verify that none of the transistors goes into saturation.

(d) Calculate R so that $Y' = \bar{Y}$.

(e) Find the average power taken from the power source.

Prob. 16-52

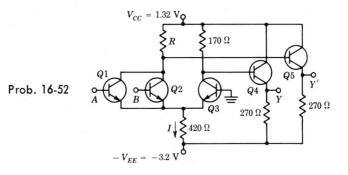

16-53   Verify that, if the outputs of two (or more) ECL gates are tied together as in Fig. 16-51, the OR function is satisfied.

16-54   (a) For the system in Fig. 16-51 obtain an expression for Y which contains three terms.

(b) If in Fig. 16-51 $\bar{Y}_1$ and $\bar{Y}_2$ are tied together, verify that the output is $Y = \bar{A}\bar{B} + \bar{C}\bar{D}$.

(c) If in Fig. 16-51 Y_1 and Y_2 are tied together and if the input to the lower ECL gate is $\bar{C}$ and $\bar{D}$ (instead of C and D), what is Y?

CHAPTER 17

17-1   Indicate how to implement S_n of Eq. (17-1) with AND, OR, and NOT gates.

17-2   Verify that the sum S_n in Eq. (17-1) for a full adder can be put in the form

$$S_n = A_n \oplus B_n \oplus C_{n-1}$$

17-3   (a) For convenience, let $A_n = A$, $B_n = B$, $C_{n-1} = C$, and $C_n = C^1$. Using Eq. (17-4) for C^1, verify Eq. (17-5) with the aid of the Boolean identities in Table 6-4; in other words, prove that

$$\bar{C}^1 = \bar{B}\bar{C} + \bar{C}\bar{A} + \bar{A}\bar{B}$$

(b) Evaluate $D \equiv (A + B + C)\bar{C}^1$ and prove that S_n in Eq. (17-1) is given by

$$S_n = D + ABC$$

17-4   (a) Verify that an EXCLUSIVE-OR gate is a true/complement unit.
(b) One input is A, the other (control) input is C, and the output is Y. Is $Y = A$ for $C = 1$ or $C = 0$?

17-5   For the system shown in Fig. 17-11a, verify the truth table in Fig. 17-11b.

17-6   (a) Make a truth table for a binary half subtractor A minus B (corresponding to the half adder of Fig. 17-3). Instead of a carry C, introduce a *borrow* P.
(b) Verify that the digit D is satisfied by an EXCLUSIVE-OR gate and that P follows the logic "B but not A."

17-7   Consider an 8-bit comparator. Justify the connections $C' = C_L$, $D' = D_L$, and $E' = E_L$ for the chip handling the more significant bits. HINT: Add 4 to each subscript in Fig. 17-14. Extend Eq. (17-12) for E and Eq. (17-13) for C to take all 8 bits into account.

17-8   (a) By means of a truth table verify the Boolean identity

$$Y = (A \oplus B) \oplus C = A \oplus (B \oplus C)$$

(b) Verify that $Y = 1(0)$ if an odd (even) number of variables equals 1. This result is *not* limited to three inputs, but is true for any number of inputs. It is used in Sec. 17-3 to construct a parity checker.

17-9   Construct the truth table for the EXCLUSIVE-OR tree of Fig. 17-15 for all possible inputs A, B, C, and D. Include $A \oplus B$ and $C \oplus D$ as well as the output Z. Verify that $Z = 1(0)$ for odd (even) parity.

17-10   (a) Draw the logic circuit diagram for an 8-bit parity check/generator system.
(b) Verify that the output is 0(1) for odd (even) parity.

17-11   (a) Verify that if $P' = 1$ in Fig. 17-15, this system is an even-parity check. In other words, demonstrate that with $P' = 1$, the output is $P = 0(1)$ for even (odd) parity of the inputs A, B, C, and D.
(b) Also verify that P generates the correct even-parity bit.

17-12   (a) Indicate an 8-bit parity checker as a block having 8 input bits (collectively designated A_1), an output P_1, and an input control P_1'. Consider a second 8-bit unit with inputs A_2, output P_2, and control P_2'. Show how to cascade the two packages in order to check for odd parity of a 16-bit word. Verify that the system operates properly if $P_1' = 1$. Consider the four possible parity combinations of A_1 and A_2.

(b) Show how to cascade three units to obtain the parity of a 24-bit word. Should $P_1' = 0$ or 1 for odd parity?

(c) Show how to cascade units to obtain the parity of a 10-bit word.

17-13 Draw a logic diagram of a 4-to-10-line decoder using OR gates instead of AND gates.

17-14 Draw a logic diagram for a 3-to-8-line decoder.

17-15 Explain how to convert a 4-to-10-line decoder unit into a 3-to-8-line decoder.

17-16 Draw a logic diagram for an 8-to-1-line multiplexer.

17-17 Write the Boolean expression for the output Y of a 4-to-1-line multiplexer with an enable input (Fig. 17-20).

17-18 The block diagram shows two data selectors being used to select 1 out of 32 data inputs. Explain the operation of the system.

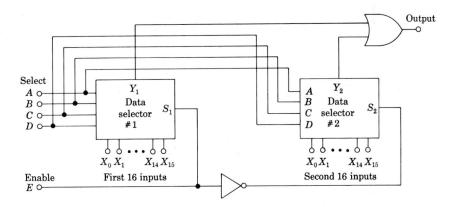

Prob. 17-18

17-19 Design an encoder satisfying the following truth table, using a diode matrix.

Inputs				Outputs			
W_3	W_2	W_1	W_0	Y_3	Y_2	Y_1	Y_0
0	0	0	1	0	1	1	1
0	0	1	0	1	1	0	0
0	1	0	0	1	1	0	1
1	0	0	0	0	0	1	0

17-20 (a) Design an encoder, using multiple-emitter transistors, to satisfy the following truth table.   (b) How many transistors are needed and how many emitters are there in each transistor?

Inputs			Outputs				
W_2	W_1	W_0	Y_4	Y_3	Y_2	Y_1	Y_0
0	0	1	1	1	0	1	0
0	1	0	1	0	0	0	1
1	0	0	0	1	1	1	1

17-21 A block diagram of a three-input (A, B, and C) and eight-output (Y_0 to Y_7) decoder matrix is indicated.   The bit Y_6 is to be 1 (5 V) if the input code is 110 corresponding to decimal 6.   (a) Indicate how diodes are to be connected to line Y_6.   (b) Repeat for Y_0, Y_1, and Y_7.

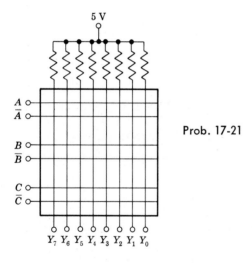

Prob. 17-21

17-22 (a) Write the expressions for Y_1 and Y_3 in the binary-to-Gray-code converter.
(b) Indicate how to implement the relationship for Y_1 with diodes.

17-23 (a) Give the relationships between the output and input bits for the Gray-to-binary-code translator for Y_1 and Y_2.
(b) Indicate how to implement the equation for Y_1 with transistors.

17-24 Minimize the number of terms in Eq. (17-23) and obtain Eq. (17-24).

17-25 (a) Write the sum-of-products canonical form for Y_4 of Table 17-5 for the seven-segment indicator code.
(b) Verify that this expression can be minimized to $Y_4 = A + C\bar{B}$.

17-26 How many AND, OR, and NOT gates are required if a three-input adder is implemented with an ROM? Compare these numbers with those used in a full-adder chip.

17-27 (a) Verify that it is not possible for both outputs in Fig. 17-27a to be in the same state.
(b) Verify that if $S = 0$ and $R = 1$ in Fig. 17-27b, the latch is reset to $Q = 0$.
(c) If $S = R = 0$, verify that the state of the latch is undetermined (it could be either $Q = 1$ or $Q = 0$).
(d) If $S = R = 1$, verify that both outputs would go to 1. Is this a valid situation?

17-28 Draw the logic diagram for an S-R FLIP-FLOP using AOI gates instead of NAND gates.

17-29 The excitation table for a J-K FLIP-FLOP is shown. An X in the table is to be interpreted to mean that it does not matter whether this entry is a 1 or a 0. It is referred to as a "don't care" condition. Thus the second row indicates that if the output is to change from 0 to 1, the J input must be 1, whereas K can be either 1 or 0. Verify this excitation table by referring to the truth table of Fig. 17-29b.

Q_n	Q_{n+1}	J_n	K_n
0	0	0	X
0	1	1	X
1	0	X	1
1	1	X	0

17-30 Verify that the J-K FLIP-FLOP truth table is satisfied by the difference equation

$$Q_{n+1} = J_n \bar{Q}_n + \bar{K}_n Q_n$$

17-31 (a) For the J-K FLIP-FLOP of Fig. 17-30, verify that for $Cr = 1$, $Pr = 0$, and $Ck = 0$, the 1 state is preset independent of the values of J_n and K_n.
(b) Repeat part (a) for $Ck = 1$, provided that $J_n = K_n = 0$.
(c) Verify that $Cr = Pr = Ck = 0$ leads to an indeterminate state; i.e., it may be 0 or 1.

17-32 (a) Verify that there is no race-around difficulty in the J-K circuit of Fig. 17-30 for any data input combination except $J = K = 1$.
(b) Explain why the race-around condition does not exist (even for $J = K = 1$) provided that $t_p < \Delta t < T$.

17-33 (a) For the master-slave J-K FLIP-FLOP of Fig. 17-31 assume $Q = 1$, $\bar{Q} = 0$, $Ck = 1$, $K = 0$, and J arbitrary. What is Q_M?
(b) If K changes to 1, what is Q_M?
(c) If K returns to 0, what is Q_M? Note that Q_M does *not* return to its initial value. Hence K (and J) must not vary during the pulse.

17-34 The indicated waveforms J, K, and Ck are applied to a J-K FLIP-FLOP. Plot the output waveform for Q and $\bar{Q}$ lined up with respect to the clock pulses.

NOTE: Assume that the output $Q = 0$ when the first clock pulse is applied and that $Pr = C_r = 1$.

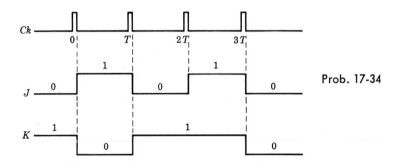

Prob. 17-34

17-35 (a) Verify that an S-R FLIP-FLOP is converted to a T type if S is connected to $\bar{Q}$ and R to Q.

(b) Verify that a D-type FLIP-FLOP becomes a T type if D is tied to $\bar{Q}$.

17-36 Augment the shift register of Fig. 17-34 with a four-input NOR gate whose output is connected to the *serial input* terminal. The NOR-gate inputs are Q_4, Q_3, Q_2, and Q_1.

(a) Verify that regardless of the initial state of each FLIP-FLOP, when power is applied, the register will assume correct operation as a ring counter after P clock pulses, where $P \leq 4$.

(b) If initially $Q_4 = 1$, $Q_3 = 1$, $Q_2 = 0$, $Q_1 = 0$, and $Q_0 = 1$, sketch the waveform at Q_0 for the first 16 pulses.

(c) Repeat part b if $Q_4 = 0$, $Q_3 = 1$, $Q_2 = 0$, $Q_1 = 0$, and $Q_0 = 0$.

17-37 (a) Draw a waveform chart for the twisted-ring counter; i.e., indicate the waveforms Q_4, Q_3, Q_2, Q_1, and Q_0 for, say, 12 pulses. Assume that initially $Q_0 = Q_1 = Q_2 = Q_3 = Q_4 = 0$.

(b) Write the truth table after each pulse.

(c) By inspection of the table show that two-input AND gates can be used for decoding. For example, pulse 1 is decoded by $Q_4\bar{Q}_3$. Why?

17-38 (a) For the modified ring counter shown, assume that initially $Q_0 = 1$, $Q_1 = 0$, and $Q_2 = 0$. Make a table of the readings Q_0, Q_1, Q_2, J_2, and K_2 after each clock pulse. How many pulses are required before the system begins to operate as a divide-by-N counter? What is N?

(b) Repeat part (a) if initially $Q_0 = 1$, $Q_1 = 0$, and $Q_2 = 1$.

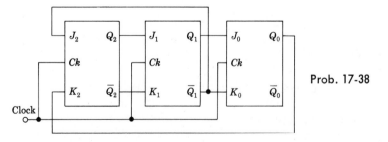

Prob. 17-38

17-39 A 50:1 ripple counter is desired. (*a*) How many FLIP-FLOPS are required?
(*b*) If 4-bit FLIP-FLOPS are available on a chip, how many chips are needed?
How are these interconnected?
(*c*) Indicate the feedback connections to the clear terminals.

17-40 (*a*) Indicate a divide-by-14 ripple-counter block diagram. Include a latch in
the clear input.
(*b*) What are the inputs to the feedback NAND gate for a 153:1 ripple counter?

17-41 Consider the operation of the latch in Fig. 17-38. Make a table of the quantities Ck, Q_1, Q_3, P_1, $\overline{Ck}$, and $P_2 = Cr$ for the following conditions:
(*a*) Immediately after the tenth pulse.
(*b*) After the tenth pulse and assuming Q_1 has reset before Q_3.
(*c*) During the eleventh pulse.
(*d*) After the eleventh pulse.
This table should demonstrate that
(*a*) The tenth pulse sets the latch to clear the counter.
(*b*) The latch remains set until all FLIP-FLOPS are cleared.
(*c*) The positive edge of the eleventh pulse resets the latch so that $Cr = 1$.
(*d*) The negative edge of the eleventh pulse initiates the new counting cycle.

17-42 (*a*) The circuit shown is a *programmable* ripple counter. It is understood that
$J = K = Cr = 1$ and that the latch in Fig. 17-38*b* exists between P_1 and P_2.
If $Pr_0 = Pr_1 = 1$ and $Pr_2 = Pr_3 = 0$, what is the count N?   Explain the operation of the system carefully.
(*b*) Why is the latch required?
(*c*) Generalize the result of part (*a*) as follows.   The counter has n stages
and is to divide by N, where $2^n > N > 2^{n-1}$.   How must the preset inputs be
programmed?

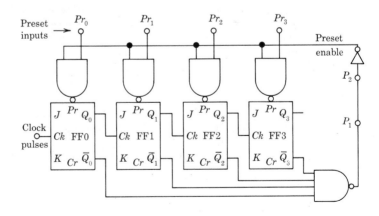

Prob. 17-42

17-43 Draw the logic diagram of a 5-bit UP-DOWN synchronous counter with series
carry.

17-44 For the logic diagram of the synchronous counter shown, write the truth table
of Q_0, Q_1, and Q_2 after each pulse and verify that this is a 5:1 counter.

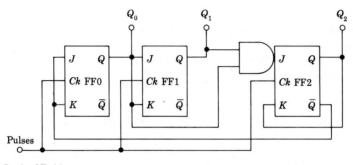

Prob. 17-44

17-45 Consider a two-stage synchronous counter (both stages receive the pulses at the Ck input). In each counter $K = 1$. If $J_0 = \bar{Q}_1$ and $J_1 = Q_0$, draw the circuit. From a truth table of Q_0 and Q_1 after each pulse, demonstrate that this is a 3:1 counter.

17-46 Draw the waveform chart for a 6:1 divider from Fig. 17-36 and deduce the connections for a synchronous counter. Draw the logic block diagram.

17-47 Solve Prob. 17-46 for a 5:1 divider.

17-48 (a) Verify that the circuit of Fig. 17-44 performs the function of a NAND gate. Let the voltage levels of V_1 and V_2 be 0 V or $-V_{DD}$.
(b) Verify that this circuit dissipates less power than the corresponding circuit of Fig. 10-17.
(c) Draw the circuit of a dynamic NOR gate corresponding to Fig. 10-18. Repeat parts a and b for this circuit.

17-49 Show that in the four-phase shift-register stage of Fig. 17-48 there is no dc current path to ground even when clocks ϕ_1 and ϕ_2 overlap or when clocks ϕ_3 and ϕ_4 overlap. Assume that the input terminal is maintained at zero volts.

17-50 Draw the logic diagrams of a recirculating or refresh memory to store 512 words each 4 bits long, using the TI 3309JC (shown in Fig. 17-49) as the basic building block. Input data are available in parallel form, and data output must be presented also in parallel form.

17-51 Draw a 16-word 4-bit RAM matrix using the basic 1-bit RAM of Fig. 17-52 and using linear selection.

17-52 The figure shows a 64-word 1-bit RAM with on-chip decoding. The memory accepts a 6-bit address word. Using the above memory unit as a building block, construct a 64-word by 4-bit memory.

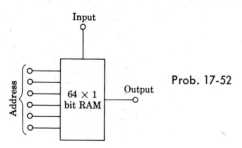

Prob. 17-52

17-53 The figure shows a 16-bit coincident memory matrix. A specific bit is selected by applying a logic 1 to the coincident X and Y address lines.
(a) Draw the diagram of a 16-word by N-bit memory (each word N bits long) using the above RAM as the basic building block.
(b) What determines the maximum value of N in this configuration?

Prob. 17-53

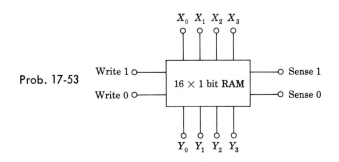

17-54 (a) For the D/A converter of Fig. 17-60 show that when the second most significant bit is 1 and all other bits are zero, the output is $V_o = V_R/4$.
(b) Find V_o if only the third MSB is 1.
(c) Find V_o if only the LSB is 1.

17-55 The figure shows a binary weighted resistor D/A converter.
(a) Show that the output resistance is independent of the digital word and that

$$R_o = \frac{2^{N-1}}{2^N - 1} R$$

(b) Show that the analog output voltage for the most significant bit is

$$V_o = \frac{2^{N-1}}{2^N - 1} V_R$$

(c) Show that the analog output voltage for the least significant bit is

$$V_o = \frac{1}{2^N - 1} V_R$$

Prob. 17-55

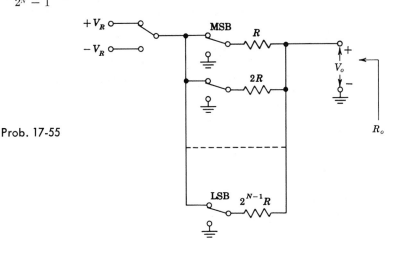

17-56 Modify the block diagram of Fig. 17-64 to display M lines of N characters each on the face of a CRT.

17-57 The circuit shown consists of two cross-coupled NAND gates. The coupling from the output of N_1 to the input of N_2 is direct (dc), whereas resistance-capacitance (ac) coupling is used from the output of N_2 to the input of N_1. Positive TTL logic is used and the levels are 0 and V_{CC}. Assume that a NAND gate changes

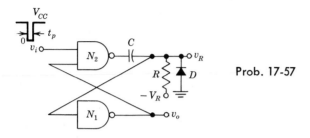

Prob. 17-57

state when its input voltage falls below V (≈ 1.6 V for a TTL gate). Neglect the drop across the clamping diode D. The input v_i is at V_{CC} and at $t = 0$ drops to 0 V for a short time t_p; that is, a negative narrow pulse is applied.
(a) Verify that the circuit behaves as a monostable multivibrator by drawing the waveforms v_R and v_o.
(b) Find the duration T of the output pulse, assuming $T > t_p$.

CHAPTER 18

18-1 (a) Nonlinear distortion results in the generation of frequencies in the output that are not present in the input. If the dynamic curve can be represented by Eq. (18-6), and if the input signal is given by

$$i_b = I_1 \cos \omega_1 t + I_2 \cos \omega_2 t$$

show that the output will contain a dc term and sinusoidal terms of frequency ω_1, ω_2, $2\omega_1$, $2\omega_2$, $\omega_1 + \omega_2$, and $\omega_1 - \omega_2$.
(b) Generalize the results of part a by showing that if the dynamic curve must be represented by higher-order terms in i_b, the output will contain intermodulation frequencies, given by the sum and difference of integral multiples of ω_1 and ω_2, for example, $2\omega_1 \pm 2\omega_2$, $2\omega_1 \pm \omega_2$, $3\omega_1 \pm \omega_2$, etc.

18-2 A transistor supplies 0.85 W to a 4-K load. The zero-signal dc collector current is 31 mA, and the dc collector current with signal is 34 mA. Determine the percent second-harmonic distortion.

18-3 The input excitation of an amplifier is $i_b = I_{bm} \sin \omega t$. Prove that the output current can be represented by a Fourier series which contains only odd sine components and even cosine components.

18-4 Supply the missing steps in the derivation of Eqs. (18-18).

18-5 Obtain a five-point schedule for determining B_0, B_1, B_2, B_3, and B_4 in terms of I_{max}, $I_{0.707}$, I_C, $I_{-0.707}$, and I_{min}.

18-6 The p-n-p transistor whose input and output characteristics are given in Fig. 18-5 is used in the circuit of Fig. 18-4, with $R_s = 0$ and $R'_L = (N_1/N_2)^2 R_L = 10\ \Omega$. The quiescent point is $I_C = -1.1$ A and $V_{CE} = -7.5$ V. The peak-to-peak 2,000-Hz sinusoidal base-to-emitter voltage is 140 mV.
(a) What is the fundamental current output?
(b) What is the percent second-, third-, and fourth-harmonic distortion?
(c) What is the output power?
(d) What is the rectification component B_o of the collector current?
Neglect any changes in the operating point.

18-7 Verify the data plotted in Fig. 18-6 for $R'_L = 20\ \Omega$.

18-8 For the operating conditions indicated in Fig. 18-5, calculate the fundamental power P_1 for (a) $R'_L = 5\ \Omega$, (b) $R'_L = 30\ \Omega$.

18-9 Repeat Prob. 18-6, but now assume a current drive (large R_s) so that the base current is sinusoidal, with a peak-to-peak value of 30 mA.

18-10 A power transistor operating class A in the circuit of Fig. 18-4 is to deliver a maximum of 5 W to a 4-Ω load ($R_L = 4\ \Omega$). The quiescent point is adjusted for symmetrical clipping, and the collector supply voltage is $V_{CC} = 20$ V. Assume ideal characteristics, as in Fig. 18-7, with $V_{min} = 0$.
(a) What is the transformer turns ratio $n = N_2/N_1$?
(b) What is the peak collector current I_m?
(c) What is the quiescent operating point I_C, V_{CE}?
(d) What is the collector-circuit efficiency?

18-11 Draw three transistor collector characteristics to correspond to base currents $I_B + I_{bm}$, I_B, $I_B - I_{bm}$. Draw the load line through the point $i_C = 0$, $v_{CE} = V_{CC}$, and the quiescent point $i_B = I_B$, $i_C = I_C$, and $v_{CE} = V_C$. This corresponds to a series-fed resistance load.
(a) Assuming that the input signal is zero, indicate on the i_C-v_{CE} plane the areas that represent the total input power to the collector circuit, the collector dissipation, and the power loss in the load resistance.
(b) Repeat part a if the input signal is sinusoidal, with a peak value equal to I_{bm}. Also, indicate the area that represents the output power.
(c) The ratio of what two areas gives the collector-circuit efficiency?
(d) Repeat parts a to c for a shunt-fed load. Assume that the static resistance is small but not zero.

18-12 In a push-pull system the input (base current) to transistor $Q1$ is $x_1 = X_m \cos \omega t$, and the input to transistor $Q2$ is $x_2 = -X_m \cos \omega t$. The collector current in each transistor may be expressed in terms of the input excitation by a series of the form

$$i_C = I_C + a_1 x + a_2 x^2 + a_3 x^3 + \cdots$$

(a) With the aid of this series, show that the output current contains only odd cosine terms.
(b) Show that the collector supply current contains only even harmonics, in addition to a dc term.

18-13 Prove, without recourse to a Fourier series, that mirror symmetry [Eq. (18-37)] exists in a push-pull amplifier. Start with $i = k(i_1 - i_2)$ and make use of Eq. (18-34).

18-14 A single transistor is operating as an ideal class B amplifier with a 1-K load. A dc meter in the collector circuit reads 10 mA. How much signal power is delivered to the load?

18-15 Given an ideal class B transistor amplifier whose characteristics are as in Fig. 18-9. The collector supply voltage V_{CC} and the effective load resistance $R'_L = (N_1/N_2)^2 R_L$ are fixed as the base-current excitation is varied. Show that the collector dissipation P_C is zero at no signal ($V_m = 0$), rises as V_m increases, and passes through a maximum given by Eq. (18-42)] at $V_m = 2V_{CC}/\pi$.

18-16 The idealized push-pull class B power amplifier shown in Fig. 18-8 has $R_2 = 0$, $V_{CC} = 20$ V, $N_2 = 2 N_1$, and $R_L = 20$ Ω, and the transistors have $h_{FE} = 20$. The input is a sinusoid. For the maximum output signal at $V_m = V_{CC}$, determine (a) the output signal power, (b) the collector dissipation in each transistor.

18-17 The power transistor whose characteristics are shown in Fig. 18-5 is used in the class B push-pull circuit of Fig. 18-8, with $R_2 = 0$ and $-V_{CC} = -20$ V. If the base current is sinusoidal, with a peak value of 20 mA and $R'_L = (N_1/N_2)^2 R_L = 15$ Ω, calculate (a) the third-harmonic distortion, (b) the power output, (c) the collector-circuit efficiency.

18-18 Repeat Prob. 18-17, using $-V_{CC} = -15$ V, $R'_L = 7.5$ Ω, and a peak base current of 30 mA.

18-19 The power transistor whose characteristics are shown in Fig. 18-5 is used in the class B push-pull circuit of Fig. 18-8, with $R_2 = 0$ and $-V_{CC} = -20$ V and $R'_L = 15$ Ω. If the base voltage is sinusoidal, with a peak value of 0.4 V, plot the output collector current. Note the crossover distortion.

18-20 Sketch the circuit of a push-pull class B transistor amplifier in the common-collector configuration (a) with an output transformer, (b) without an output transformer.

18-21 Discuss the push-pull complementary circuit of Fig. 18-11. In particular, show that no even harmonics are present.

18-22 The circuit shown represents a transformerless class B single-ended complementary-symmetry push-pull power amplifier. Transistors Q2 and Q3 are matched silicon devices, with $h_{FE} \approx h_{fe} = 100$ and $h_{ie} = 50$ Ω. Q1 is a silicon transistor whose small-signal h parameters are given in Table 8-2, and $h_{FE} = 50$.
(a) Explain the operation of this circuit. Note especially the role of the capacitor C_2. Neglect the reverse saturation currents.
(b) Calculate the quiescent currents in all the resistors, and determine the value of R_3 so that

$$|V_{CE3}| = |V_{CE2}|$$

(c) Find the output resistance R_o, assuming ideal class B operation.
(d) Calculate the maximum power that can be delivered to the 8-Ω speaker. Take the output resistance R_o into account, and assume $V_{CE,\text{sat}} \approx 0$.
    HINT: In parts c and d, assume that for class B operation $R_4 = 0$.

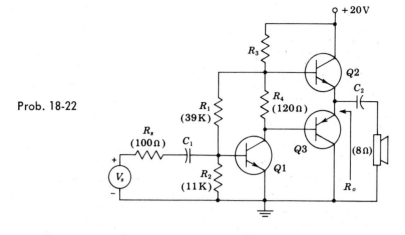

Prob. 18-22

18-23 Verify Eqs. (18-57) and (18-58).

18-24 Find the output resistance of the series-regulated power supply as given by Eq. (18-59). HINT: Short-circuit the input, $V_i = 0$, and derive the expression for the output current, using an auxiliary voltage source.

18-25 Design a regulated power supply as shown in Fig. 18-17 with the following specifications:

Nominal unregulated input voltage $V_i = 30$ V and $r_o = 8$ Ω
Nominal regulated output voltage $V_o = 12$ V
Maximum load current $I_{L,\max} = 200$ mA
Control transistor $Q1$ (silicon): $h_{FE} = h_{fe} = 100$, $h_{ie} = 200$ Ω
Amplifier transistor $Q2$ (silicon): $h_{FE} = h_{fe} = 200$, $h_{ie} = 1$ K
Reference avalanche diode D: $V_R = 6$ V, $R_z = 10$ Ω at $I_z = 20$ mA

(a) Sketch the complete circuit and obtain reasonable values for R_1, R_2 and R_3.
(b) Calculate the voltage stabilization factor S_V.
(c) Calculate the output resistance R_o.

18-26 In the circuit of Fig. 18-18, the control transistor $Q1$ is replaced by a Darlington pair $Q1$-$Q3$. The junction of R_3 and the collector of $Q2$ is connected to the base of $Q3$.
(a) Discuss the possible improvement in S_V over the value for the circuit of Fig. 18-18.
(b) Show that the output resistance is

$$R_o \approx \frac{r_o + \dfrac{R_3 + h_{fe3}h_{ie1}}{h_{fe1}h_{fe3}}}{1 + G_m(R_3 + r_o)}$$

where G_m is as given by Eq. (18-57).

18-27 Repeat Prob. 18-25 using the circuit of Prob. 18-26. Assume that $Q2$ and $Q3$ are identical.

18-28 The circuit shown employs a Zener diode preregulator.

(a) Explain carefully the operation of the circuit.

(b) Obtain an approximate expression for the input regulation factor S_V.
HINT: Assume $\Delta V_o \approx 0$ when $\Delta V_i \gg \Delta V_o$.

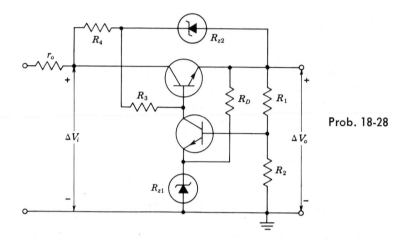

Prob. 18-28

18-29 Sketch the circuit of a regulated semiconductor power supply whose output is positive with respect to ground, using (a) p-n-p transistors, (b) complementary transistors.

18-30 Sketch the circuit of a regulated semiconductor power supply whose output is negative with respect to ground, using (a) p-n-p transistors, (b) n-p-n transistors, (c) complementary transistors.

18-31 If the V-I characteristic of the p-n-p-n diode is as shown, calculate and plot the output voltage v_o. Show all critical voltage and time values.

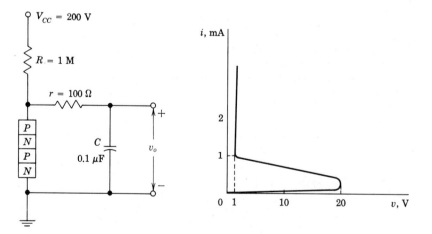

Prob. 18-31

18-32 The circuit of Fig. 18-31 is adjusted so that conduction commences 90° after the start of each positive half cycle of applied voltage. The SCR voltage drop is negligible. The applied voltage is 300 V sinusoidal rms, and the load is a 50-Ω resistor. Calculate

(a) The dc load current.

(b) The power dissipated by the load.

(c) The rms load current.

18-33 The circuit of Fig. 18-34 is adjusted so that the conduction angle is 60°. The rectifier and SCR voltage drops when conducting are negligible. The applied voltage is 300 V rms, and the load is a 10-Ω resistor. Calculate

(a) The reading of a true rms reading ammeter in series with the load.

(b) The reading of a dc ammeter in series with the load.

(c) The reading of a true rms reading voltmeter across the load.

(d) The reading of a dc voltmeter across the load.

(e) The dc load power.

(f) The total power dissipated by the load resistor.

18-34 The SCR is used to control the power delivered to the 50-Ω load by the sinusoidal source. If the gate supply V_{GG} is adjustable:

(a) Over what range may the conduction angle of the SCR be continuously varied?

(b) Over what range may the load dc current be continuously varied if the frequency is 60 Hz?

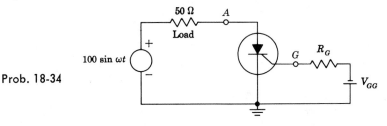

Prob. 18-34

18-35 The circuit shown is used to control the power dissipated in the 10-Ω load resistor. Assume that the bilateral switching diode has a breakdown voltage of ±2.8 V and that the holding voltages of the diode and the triac are negligible.

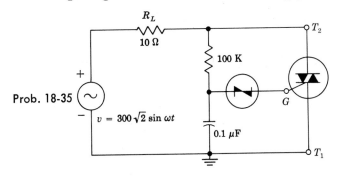

Prob. 18-35

The applied sinusoidal voltage is 300 V rms, at 60 Hz.

(a) Compute the conduction angle.

(b) Draw the waveform of the voltage applied to the load.

(c) Compute the total power dissipated by the load resistor.

CHAPTER 19

19-1 The specific gravity of tungsten is 18.8, and its atomic weight 184.0. Assume that there are two free electrons per atom. Calculate the numerical value of n and E_F.

19-2 How many electrons per cubic meter in metallic tungsten have energies between 8.5 and 8.6 eV (a) at 0°K, (b) at 2500°K?

19-3 (a) Calculate the maximum energy of the free electrons in metallic aluminum at absolute zero. Assume that there are three free electrons per atom. The specific gravity of aluminum is 2.7.

(b) Repeat part a for the electrons in metallic silver. The specific gravity of silver is 10.5. Assume that there is one free electron per atom.

19-4 (a) Show that the average energy E_{av} of the electrons in a metal is given by

$$E_{av} = \frac{\int E \, dn_E}{\int dn_E}$$

(b) Prove that the average energy at absolute zero is $3E_F/5$.

19-5 If the emission from a certain cathode is 10,000 times as great at 2000 as at 1500°K, what is the work function of this surface?

19-6 (a) If the temperature of a tungsten filament is raised from 2300 to 2320°K, by what percentage will the emission change?

(b) To what temperature must the filament be raised in order to double its emission at 2300°K?

19-7 Prove that the fractional change in thermionic current is given by

$$\frac{dI_{th}}{I_{th}} = \left(2 + \frac{E_W}{kT}\right)\frac{dT}{T}$$

19-8 If 10 percent of the thermionic-emission current is collected, what must be the retarding voltage at the surface of the metal? The filament temperature is 2000°K.

19-9 What fraction of the thermionic current will be obtained with zero applied voltage between the cathode and anode of a diode? The work function of the cathode is 4.50 V, and the work function of the anode is 4.75 V. The cathode temperature is 2000°K.

19-10 A plane cathode having a work function of 3.00 V is connected directly to a parallel-plane anode whose work function is 5.00 V. The distance between anode and cathode is 2.00 cm. If an electron leaves the cathode with a normal-to-surface velocity of 5.93×10^5 m/s, how close to the anode will it come?

19-11 A diode has an oxide-coated cathode operating at a temperature of 1000°K. With zero plate voltage the anode current is essentially zero, indicating that the contact potential is high enough to keep most of the electrons from reaching the plate. The applied voltage is increased so that a small current is drawn. Show that there is a tenfold increase in current for every 0.2-V increase in voltage.

19-12 A diode with plane-parallel electrodes is operated at a temperature of 1500°K. The filament is made of tungsten, the area being such that a thermionic current 10 μA is obtained. The contact difference of potential between cathode and anode is 0.5 V, with the cathode at the higher potential.
(a) What current is obtained with zero applied voltage?
(b) What applied voltage will yield a current of 1 μA?
(c) What fraction of the electrons emitted from this filament can move against an *applied retarding* field of 1 V?

19-13 Indicate by letter which of the following statements are true:
(a) The work function of a metal is always less than the potential barrier at the surface of a metal.
(b) The potential barrier at the surface of a metal is a solid hill made up of the material of the metal.
(c) The ionic structure of a metal shows that the inside of the metal is not an equipotential volume.
(d) At absolute zero the electrons in a metal all have zero energy.
(e) The ionic structure of a metal shows that the surface of a metal is not at a specific location.
(f) For an electron to escape from a metal, the potential barrier at the surface of the metal must first be broken down.
(g) The distribution function for the electrons in a metal shows how many electrons are close to a nucleus and how many are far away.

19-14 Indicate by letter which of the following statements are true:
(a) The potential energy as a function of distance along a row of ions *inside* a metal varies very rapidly in the immediate neighborhood of an ion but is almost constant everywhere else inside the metal.
(b) The potential-energy barrier at the surface of a metal *cannot* be explained on the basis of the modern crystal-structure picture of a metal, but it can be explained on the basis of classical electrostatics (image forces).
(c) To remove any one of the free electrons from a metal, it is necessary only to give this electron an amount of energy equal to the work function of the metal.
(d) The symbol E_F used in the energy distribution function represents the maximum number of free electrons per cubic meter of metal at absolute zero.
(e) The area under the energy distribution curve represents the total number of free electrons per cubic meter of metal at any temperature.
(f) The Dushman equation of thermionic emission gives the current that is obtained from a heated cathode as a function of applied plate voltage.

19-15 Evaluate n given by Eq. (19-29). HINT: Refer to a table of definite integrals.
19-16 Verify the expression for p in Eq. (19-35). HINT: Refer to a table of definite integrals.

19-17 If the effective mass of an electron is equal to twice the effective mass of a hole, find the distance (in electron volts) of the Fermi level in an intrinsic semiconductor from the center of the forbidden band at room temperature.

19-18 (a) Verify the numerical values in Eqs. (19-41) and (19-42).
(b) From Eq. (19-42) and the numerical values given in Table 2-1 evaluate $m_n m_p / m^2$.

19-19 (a) Prove that the fractional change in the conductivity of an intrinsic semiconductor is given by

$$\frac{d\sigma}{\sigma} = \frac{dn_i}{n_i} = \left(\frac{3}{2} + \frac{E_{Go}}{2kT} \right) \frac{dT}{T}$$

(b) Using the result of part (a), show that the conductivity of Ge (Si) at room temperature increases approximately 6 (8) percent per degree increase in temperature.

19-20 (a) In n-type germanium the donor concentration corresponds to 1 atom per 10^8 germanium atoms. Assume that the effective mass of the electron equals one-half the true mass. At room temperature, how far from the edge of the conduction band is the Fermi level? Is E_F above or below E_C?
(b) Repeat part (a) if impurities are added in the ratio of 1 donor atom per 10^3 germanium atoms.
(c) Under what circumstances will E_F coincide with E_C?

19-21 (a) In p-type silicon the acceptor concentration corresponds to 1 atom per 10^8 silicon atoms. Assume that $m_p = 0.6$ m. At room temperature, how far from the edge of the valence band is the Fermi level? Is E_F above or below E_V?
(b) Repeat part (a) if impurities are added in the ratio of 1 acceptor atom per 5×10^3 silicon atoms.
(c) Under what condition will E_F coincide with E_V?

19-22 In n-type silicon the donor concentration is 1 atom per 2×10^8 silicon atoms. Assume that the effective mass of the electron equals the true mass. At what temperature will the Fermi level coincide with the edge of the conduction band?

19-23 In p-type germanium at room temperature (300°K), for what doping concentration will the Fermi level coincide with the edge of the valence band? Assume $m_p = 0.4$ m.

19-24 (a) A germanium tunnel diode has impurity concentration at the p side of 3 parts in 10^3 atoms, and at the n side of 2 parts in 10^3 atoms. It $m_n = m_p = 0.4$ m, calculate E_G, E_o, and $E_F - E_{Cn}$ of this diode.
(b) Draw the energy bands for this diode using the results of part (a).

19-25 (a) A region of a semiconductor has a one-dimensional current flow in the x direction, with current density J A/m², due entirely to the hole-concentration gradient. J is constant with x and time; at $x = 0$, the hole concentration is $p(0)$. Find the hole concentration as a function of x. Recombination, field due to the stored charge, and conductivity modulation are all neglected. (This situation corresponds to the base region in a step-graded p-n-p transistor.)
(b) Now suppose that there is also an electric field of magnitude $\mathcal{E}$ V/m in the negative x direction. The same constant current J flows, but the hole concentration is now $p'(0)$ at $x = 0$. Noting that J is the diffusion current less

the conduction current, show that (if $\mathcal{E}$ is independent of x) the hole concentration is

$$p(x) = \epsilon^{-x/x_o} \left[p'(0) + \frac{J}{q\mu_p\mathcal{E}} \right] - \frac{J}{q\mu_p\mathcal{E}}$$

and find x_o.

(c) Show that for small x the formula in part b reduces to

$$p(0) = p'(0) - x \left[\frac{J}{qD_p} + p'(0) \frac{\mu_p\mathcal{E}}{D_p} \right]$$

where D_p is the diffusion constant for holes.

(d) Sketch the results of parts (a) and (b) on the same axes for $p'(0) > p(0)$. How do the slopes compare?

19-26 (a) Let Q be the excess minority charge stored within a volume of cross section A and length L. If there is no electric field within this volume, and if the current i flowing perpendicular to the section A is due exclusively to minority-carrier diffusion, show that the stored charge Q satisfies the equation

$$\frac{dQ}{dt} + \frac{Q}{\tau} = i$$

where τ is the mean lifetime of the minority carriers.

(b) Show that the steady-state current is

$$I_{ss} = \frac{Q}{\tau}$$

19-27 A semiconductor diode carries a dc current I_F in the forward direction. At $t = 0^+$ the current is changed abruptly to $-I_R$. Show that the time t_s required for the removal of the excess minority-carrier charge Q_o is

$$t_s = \tau \ln \left(1 + \frac{I_F}{I_R} \right)$$

where τ is the mean lifetime of holes and electrons. HINT: Use the results of Prob. 19-26 and note that at $t = 0$, $Q = Q_o = \tau I_F$.

19-28 (a) Verify Eq. (19-91).

(b) Calculate $[d(\ln I_o)]/dT$ for Ge and Si (for rated current).

19-29 (a) Consider a diode biased in the forward direction at a fixed voltage V. Prove that the fractional change in current with respect to temperature is

$$\frac{1}{I}\frac{dI}{dT} = \frac{V_{GO} - V}{\eta T V_T} + \frac{m}{T}$$

(b) Find the percentage change in current per degree centigrade for Ge at $V = 0.2$ V and for Si at $V = 0.6$ V.

19-30 Carry out in detail the derivation of the dynamic diffusion capacitance outlined in Sec. 19-12.

19-31 Verify Eq. (19-117).

19-32 (a) If it is not assumed that $W/L_B \ll 1$, prove that Eqs. (19-114) and (19-116) remain valid provided that

$$a_{11} = Aq\left(D_p \frac{p_{no}}{L_B} \coth \frac{W}{L_B} + \frac{D_n n_{EO}}{L_E}\right)$$

$$a_{12} = a_{21} = -AqD_p \frac{p_{no}}{L_B} \operatorname{csch} \frac{W}{L_B}$$

$$a_{22} = Aq\left(D_p \frac{p_{no}}{L_B} \coth \frac{W}{L_B} + \frac{D_n n_{CO}}{L_C}\right)$$

(b) Show that if $W/L_B \ll 1$, these expressions reduce to those given by Eqs. (19-115) and (19-117).

19-33 Using the results of Prob. 19-32a, verify that $\alpha = \beta^*\gamma$, where γ is given by Eq. (19-123) and β^* by Eq. (19-124).

19-34 If $W/L_B \ll 1$, verify that Eqs. (19-125) and (19-126) follow from Eqs. (19-123) and (19-124), respectively.

19-35 Starting with Eqs. (19-131) and (19-132) and assuming $I_B \gg I_{co}$, prove that

$$V_{CE} \approx \pm V_T \ln \frac{1 + \dfrac{I_C}{I_B}(1 - \alpha_I)}{\alpha_I\left(1 - \dfrac{I_C}{I_B}\dfrac{1 - \alpha_N}{\alpha_N}\right)} = \pm V_T \ln \frac{1 + h_{FEI} + \dfrac{I_C}{I_B}}{h_{FEI}\left(1 - \dfrac{I_C}{h_{FE}I_B}\right)}$$

where $h_{FE} = \alpha_N/(1 - \alpha_N)$ and $h_{FEI} = \alpha_I/(1 - \alpha_I)$. Compare with Eq. (5-31).

19-36 (a) The incremental resistance between collector and emitter for a grounded-emitter switch at constant base current may be computed as

$$r_{CE} = \left|\frac{dV_{CE}}{dI_C}\right|_{I_B} = \left|\frac{d(V_E - V_C)}{dI_C}\right|_{I_B}$$

where V_E and V_C are, respectively, the voltage drops across the emitter and collector junctions. Using Eqs. (19-131) and (19-132), show that

$$r_{CE} = V_T\left[\frac{1 - \alpha_N}{\alpha_N I_B - I_C(1 - \alpha_N) + I_{co}} + \frac{1 - \alpha_I}{I_B + I_C(1 - \alpha_I) + I_{EO}}\right]$$

(b) If $I_B \gg I_{co}$ and if $\dfrac{1 - \alpha_N}{\alpha_N}\dfrac{I_C}{I_B} \ll 1$, show that

$$r_{CE} \approx \frac{V_T}{I_B}\frac{1 - \alpha_N\alpha_I}{\alpha_N}$$

19-37 (a) Show that I_C is given approximately by

$$I_C = I_{CER} = \frac{\left[1 + \dfrac{I_{EO}(R + r_{bb'})}{V_T}\right]I_{co}}{1 - \alpha_N\alpha_I + (1 - \alpha_N)\dfrac{I_{EO}(R + r_{bb'})}{V_T}}$$

where $r_{bb'}$ is the base-spreading resistance.   HINT: Assume that the collector junction is reverse-biased and that the emitter junction is *slightly* forward-biased.   Take advantage of the approximations which are allowed because the forward bias is small.

(*b*) A germanium transistor operating at room temperature has $\alpha_N = 0.98$, $I_{CO} = 2 \ \mu\text{A}$, $I_{EO} = 1.6 \ \mu\text{A}$, and $r_{bb'} = 200 \ \Omega$.   Calculate I_C for $R = 0$ and $R = \infty$.

(*c*) What value of R will give a collector current midway between the currents corresponding to a shorted and open base?

Prob. 19-37

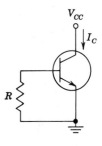

19-38 Use the Ebers-Moll equations to show that the transconductance of a transistor in the active region is given by

$$g_m = \frac{dI_C}{dV_E}\bigg|_{V_C=\text{const}} \approx \frac{1}{V_T}\left[I_C - \frac{(1-\alpha_I)I_{CO}}{1-\alpha_N\alpha_I}\right] \approx \frac{I_C}{V_T}$$

HINT: Assume $\epsilon^{V_C/V_T} \ll 1$.

19-39 A type 2N404 germanium transistor is operated at room temperature in the CE configuration.   The supply voltage is 6 V, the collector-circuit resistance is 200 Ω, and the base current is 20 percent higher than the minimum value required to drive the transistor into saturation.   Assume the following transistor parameters: $I_{CO} = -5 \ \mu\text{A}$, $I_{EO} = -2 \ \mu\text{A}$, $h_{FE} = 100$, and $r_{bb'} = 250 \ \Omega$.   Find $V_{BE,\text{sat}}$ and $V_{CE,\text{sat}}$.

19-40 The type 2N1708 double-diffused silicon transistor has parameters $h_{FE} = 30$, $h_{FEI} = 0.2$, $r_{bb'} = 30 \ \Omega$, and $I_{CO} = 13$ nA and has a collector body resistance of 6 Ω.   It operates with $I_B = 1$ mA and $I_C = 10$ mA.   Find V_{BE} and V_{CE} at room temperature.

INDEX

898